Theory of Groups and Symmetries

Representations of Groups and Lie Algebras, Applications

Other Related Titles from World Scientific

Theory of Groups and Symmetries: Finite Groups, Lie Groups, and Lie Algebras
by Alexey P Isaev and Valery A Rubakov
ISBN: 978-981-3236-85-1

Introduction to the Theory of the Early Universe: Hot Big Bang Theory Second Edition
by Valery A Rubakov and Dmitry S Gorbunov
ISBN: 978-981-3209-87-9
ISBN: 978-981-3209-88-6 (pbk)

Introduction to the Theory of the Early Universe: Cosmological Perturbations and Inflationary Theory Second Edition
by Valery A Rubakov and Dmitry S Gorbunov
ISBN: 978-981-3275-62-1
ISBN: 978-981-3276-69-7 (pbk)

Mathematical Foundations of Quantum Field Theory
by Albert Schwarz
ISBN: 978-981-3278-63-9
ISBN: 978-0-00-098924-6 (pbk)

Theory of Groups and Symmetries

Representations of Groups and Lie Algebras, Applications

Alexey P Isaev

N. N. Bogoliubov Laboratory of Theoretical Physics,
Joint Institute for Nuclear Research, Dubna, Russia
M. V. Lomonosov Moscow State University, Russia

Valery A Rubakov

Institute for Nuclear Research,
Russian Academy of Sciences, Moscow, Russia
M. V. Lomonosov Moscow State University, Russia

World Scientific

NEW JERSEY • LONDON • SINGAPORE • BEIJING • SHANGHAI • HONG KONG • TAIPEI • CHENNAI • TOKYO

Published by

World Scientific Publishing Co. Pte. Ltd.
5 Toh Tuck Link, Singapore 596224
USA office: 27 Warren Street, Suite 401-402, Hackensack, NJ 07601
UK office: 57 Shelton Street, Covent Garden, London WC2H 9HE

Library of Congress Cataloging-in-Publication Data
Names: Isaev, Alexey P., 1957– author. | Rubakov, V. A., author.
Title: Theory of groups and symmetries : representations of groups and lie algebras, applications / Alexey P. Isaev, N.N. Bogoliubov Laboratory of Theoretical Physics, Joint Institute for Nuclear Research, Dubna, Russia, M.V. Lomonosov Moscow State University, Russia, Valery A. Rubakov, Institute for Nuclear Research, Russian Academy of Sciences, Moscow, Russia, M.V. Lomonosov Moscow State University, Russia.
Description: New Jersey : World Scientific Publishing Co. Pte. Ltd., [2021] | Includes bibliographical references and index.
Identifiers: LCCN 2020019144 | ISBN 9789811217401 (hardcover) | ISBN 9789811217418 (ebook for institutions) | ISBN 9789811217425 (ebook for individuals)
Subjects: LCSH: Group theory. | Symmetry (Physics) | Group algebras. | Lie algebras.
Classification: LCC QC20.7.G76 I84 2021 | DDC 512/.2--dc23
LC record available at https://lccn.loc.gov/2020019144

British Library Cataloguing-in-Publication Data
A catalogue record for this book is available from the British Library.

For any available supplementary material, please visit
https://www.worldscientific.com/worldscibooks/10.1142/11749#t=suppl

Desk Editor: Ng Kah Fee

Typeset by Stallion Press
Email: enquiries@stallionpress.com

To Elvira and Helena

Contents

Preface

This book is a sequel to the book [1] by the same authors entitled *Theory of Groups and Symmetries: Finite Groups, Lie Groups, and Lie Algebras*, so this book can be viewed as the second part. References to sections and formulas from the first book are labeled by Roman numeral I, e.g., Section I-3.2.3, Eq. (I-7.2.37).

The second book is mainly devoted to the theory of representations of Lie groups and Lie algebras and also to applications. We begin with the presentation in Chapter 1 of the Dirac notation, which is illustrated by boson and fermion oscillator algebras and also Grassmann algebra. In Chapter 2, we give a detailed account of finite-dimensional representations of groups $SL(2,\mathbb{C})$ and $SU(2)$ and their Lie algebras. In Chapter 3, we consider the general theory of finite-dimensional irreducible representations of simple Lie algebras based on the construction of highest weight representations. We then give the classification of all finite-dimensional irreducible representations of the Lie algebras of the classical series $s\ell(n,\mathbb{C})$, $so(n,\mathbb{C})$ and $sp(2r,\mathbb{C})$.

In Chapter 4, finite-dimensional irreducible representations of linear groups $SL(N,\mathbb{C})$ and their compact real forms $SU(N)$ are constructed on the basis of the Schur–Weyl duality. In this approach, spaces of irreducible representations are viewed as invariant subspaces in the tensor products $\mathcal{V}^{\otimes r}$ of spaces $\mathcal{V}$ of defining representations. These invariant subspaces are singled out with the help of special elements (idempotents) of group algebra $\mathbb{C}[S_r]$ of the group of permutations (symmetric group) S_r. Here, special role is played by the theory of representations of the algebra $\mathbb{C}[S_r]$ (Schur–Frobenius theory, Okounkov–Vershik approach), based on combinatorics of Young diagrams and Young tableaux.

We then turn in Chapter 5 to pseudo-orthogonal groups $O(p,q)$ and $SO(p,q)$, including multidimensional Lorentz groups $O(1,N-1)$ and $SO(1,N-1)$ and their Lie algebras. We also consider symplectic groups $Sp(p,q)$, albeit in less detail. The presentation here is based on the appropriate modification of the techniques exposed in Chapter 4.

Finally, in Chapter 6 we study the covering groups $\mathsf{Spin}(p,q)$ for pseudo-orthogonal groups $SO^{\uparrow}(p,q)$. The groups $\mathsf{Spin}(p,q)$ are called spinor groups and are currently in active use in quantum field theory. The construction of the spinor groups $\mathsf{Spin}(p,q)$ requires the introduction of Clifford algebras in spaces $\mathbb{R}^{p,q}$ and the discussion of the representations of these algebras.

In this book, as done in the first one, we try to prove or at least give hints of proofs for the majority of facts. We emphasize, however, that, similar to the first book, the level of rigor in places is not to the standard of a mathematically oriented reader; our main goal is rather to make the material comprehensible. Text written in small font may be omitted at the first reading. When necessary, we give references to literature, where one can find more detailed analysis of one or another point.

This book started from lectures given at the Departments of Theoretical and Nuclear Physics of Dubna International University and at the Departments of Quantum Theory and High Energy Physics, Quantum Statistics and Field Theory, Particle Physics and Cosmology of Physics Faculty of M.V. Lomonosov Moscow State University. We are grateful to S.E. Derkachev, S.A. Fedoruk, E.A. Ivanov, S.O. Krivonos, S.K. Lando, S.A. Mironov, A.I. Molev, O.V. Ogievetsky, G.I. Olshansky, A.V. Silantiev, Y.M. Shnir, V.P. Spiridonov, A.O. Sutulin, V.O. Tarasov, S.V. Troitsky, N.A. Tyurin, M.A. Vasiliev and A.A. Vladimirov, as well as many other colleagues from the Theory Division of the Institute for Nuclear Research of the Russian Academy of Sciences, Moscow and N.N. Bogoliubov Laboratory for Theoretical Physics of the Joint Institute for Nuclear Research, Dubna, for numerous useful discussions of the material presented in this book.

Chapter 1

Dirac Notations

In this chapter, we introduce convenient notations which we use throughout the book for describing basis vectors and matrix elements of operators in the spaces of representations of Lie groups and Lie algebras. These notations were introduced by P. Dirac [30] and they are standard in quantum mechanics. We have used these notations in the first book [1]. Here, we give the general definitions and rules. We illustrate the convenience of these notations by considering examples of coherent states for quantum oscillator algebra and free fermion algebra.

1.1. Bra and Ket Vectors. Coordinate and Momentum Representations

Consider complex vector space $\mathcal{V}$ (finite dimensional or infinite dimensional). We denote vectors in $\mathcal{V}$ as $|\Psi\rangle$, where Ψ stands for a set of symbols, indices, numbers, etc., which characterize a given vector.[a] Consider also a complex vector space $\mathcal{V}^*$ dual to $\mathcal{V}$. The space $\mathcal{V}^*$ consists of linear functionals in $\mathcal{V}$. We denote vectors from $\mathcal{V}^*$ as $\langle\Phi|$. By definition, every $\langle\Phi| \in \mathcal{V}^*$ determines a function: $\mathcal{V} \to \mathbb{C}$, which we denote by

$$\langle\Phi|\Psi\rangle \in \mathbb{C}, \quad \forall|\Psi\rangle \in \mathcal{V}. \tag{1.1.1}$$

The expression $\langle\Phi|\Psi\rangle$ is called convolution of vectors $|\Psi\rangle$ and $\langle\Phi|$. The convolution (1.1.1) may be understood also from the dual viewpoint, namely, vectors $|\Psi\rangle \in \mathcal{V}$ define linear functionals in $\mathcal{V}^*$.

[a]In what follows, we disregard peculiarities occurring in the case of infinite-dimensional spaces, which are related, in particular, to the convergence of relevant integrals and sums; see, however, Definition 1.1.1 and discussion around it.

Let $\mathcal{V}$ have a basis $|\alpha\rangle$, where the index α that labels the basis vectors, is, in general, a set of indices (i.e., α is a multi-index). Indices entering the multi-index α can take both discrete and continuous values. The basis of $\mathcal{V}^*$ consisting of vectors $\langle\beta|$ is called dual to basis $|\alpha\rangle \in \mathcal{V}$ if β is a multi-index of the same type as α and the orthonormality condition holds:

$$\langle\beta|\alpha\rangle = \delta_{\alpha,\beta}. \tag{1.1.2}$$

Here $\delta_{\alpha,\beta}$ is a product of Kronecker symbols and delta-functions which involve indices entering α and β, depending on whether these indices take discrete of continuous values. In accordance with (1.1.2), vectors $|\Psi\rangle \in \mathcal{V}$ and $\langle\Phi| \in \mathcal{V}^*$ are expanded in basis vectors:

$$|\Psi\rangle = \sum_\alpha |\alpha\rangle\langle\alpha|\Psi\rangle, \quad \langle\Phi| = \sum_\alpha \langle\Phi|\alpha\rangle\langle\alpha|, \tag{1.1.3}$$

where $\sum_\alpha$ denotes a multiple integral and multiple sum over continuous and discrete indices entering the multi-index α.

For any linear operator T acting in $\mathcal{V}$, we define an operator T' which acts in $\mathcal{V}^*$ (and, conversely, given an operator acting in $\mathcal{V}^*$, we define an operator in $\mathcal{V}$) as follows:

$$\langle\Phi|(T\cdot|\Psi\rangle) = (\langle\Phi|\cdot T')|\Psi\rangle, \quad \forall|\Psi\rangle \in \mathcal{V}, \ \forall\langle\Phi| \in \mathcal{V}^*, \tag{1.1.4}$$

where it is convenient to think of the operator T as acting in $\mathcal{V}$ on the left, and T' acting in $\mathcal{V}^*$ on the right. In what follows, we use the same notation for T and T' to simplify formulas, and write $\langle\Phi|T|\Psi\rangle$ instead of (1.1.4). The relations (1.1.3) ensure that the unit operator I acting in spaces $\mathcal{V}$ and $\mathcal{V}^*$ is expanded in terms of the basis vectors in a simple way:

$$I = \sum_\alpha |\alpha\rangle\langle\alpha|, \tag{1.1.5}$$

while the convolution $\langle\Phi|\Psi\rangle$ takes the form

$$\langle\Phi|\Psi\rangle = \sum_\alpha \langle\Phi|\alpha\rangle\langle\alpha|\Psi\rangle. \tag{1.1.6}$$

Making use of the representation (1.1.5) of the unit operator, we write the action of linear operators T on the basis vectors $|\beta\rangle \in \mathcal{V}$ and on functions $\Psi(\alpha) \equiv \langle\alpha|\Psi\rangle$ as follows:

$$T|\beta\rangle = \sum_\alpha |\alpha\rangle\langle\alpha|T|\beta\rangle, \quad \langle\beta|T|\Psi\rangle = \sum_\alpha \langle\beta|T|\alpha\rangle\langle\alpha|\Psi\rangle. \tag{1.1.7}$$

This shows that the matrix of the operator T in the basis $\{|\beta\rangle\}$ has matrix elements $\langle\beta|T|\alpha\rangle$. In accordance with (1.1.3) and (1.1.7), function $\Psi(\alpha)$ can be understood as the set of coordinates of the vector $|\Psi\rangle \in \mathcal{V}$, where the index α "labels" these coordinates.

Let us establish one-to-one correspondence between vectors in spaces $\mathcal{V}$ and $\mathcal{V}^*$ by requiring that the convolution obeys

$$(\langle\Phi|\Psi\rangle)^* = \langle\Psi|\Phi\rangle, \quad \forall|\Psi\rangle \in \mathcal{V}, \ \forall\langle\Phi| \in \mathcal{V}^*, \tag{1.1.8}$$

where $*$ in the equality denotes complex conjugation. Then the coordinates of vectors $\langle\Phi|$ and $|\Phi\rangle$ are complex conjugate to each other:

$$(\langle\alpha|\Phi\rangle)^* = \langle\Phi|\alpha\rangle, \tag{1.1.9}$$

while the convolution (1.1.6) takes the form

$$\langle\Phi|\Psi\rangle = \sum_\alpha (\langle\alpha|\Phi\rangle)^* \langle\alpha|\Psi\rangle = \sum_\alpha \Phi(\alpha)^*\Psi(\alpha). \tag{1.1.10}$$

The relations (1.1.3) and (1.1.9) ensure that the vector $\langle\Phi| \in \mathcal{V}^*$ is Hermitian conjugate to the vector $|\Phi\rangle \in \mathcal{V}$, i.e., $(|\Phi\rangle)^\dagger = \langle\Phi|$, while the convolution

$$\langle\Phi|\Psi\rangle \equiv (|\Phi\rangle)^\dagger|\Psi\rangle \tag{1.1.11}$$

is interpreted as Hermitian scalar product in the complex space $\mathcal{V}$ (see Definition **I-1.2.9**). In order that the right-hand side of (1.1.10) makes sense, the functions $\Phi(\alpha) = \langle\alpha|\Phi\rangle$ must be square integrable and summable as functions of α. In the space $\mathcal{V}$ equipped with Hermitian scalar product (1.1.10), one naturally defines a norm of a vector, $||\Psi|| = \sqrt{\langle\Psi|\Psi\rangle}$.

Definition 1.1.1. If a complex vector space $\mathcal{V}$ with Hermitian scalar product (1.1.10) and norm $||\Psi|| = \sqrt{\langle\Psi|\Psi\rangle}$ is complete (i.e., any Cauchy sequence[b] in $\mathcal{V}$ converges, with respect to norm $||\Psi||$, to an element of $\mathcal{V}$), then the space $\mathcal{V}$ is called Hilbert space.

In general, square integrability and summability in (1.1.10) are understood in the sense of distributions, rather than conventional functions, and the space $\mathcal{V}$ does not have to be Hilbert space.

[b]Sequence of vectors $\{\Psi_k\}_{k=1}^\infty$ in space $\mathcal{V}$ equipped with norm is called Cauchy sequence (or fundamental sequence) if it obeys Cauchy property: for any arbitrarily small ϵ there exists a number N_ϵ such that $||\Psi_i - \Psi_k|| < \epsilon$ for all $i, k > N_\epsilon$.

Finally, making use of (1.1.4) and (1.1.11), we define operator $H^\dagger$, which is Hermitian conjugate to H, by the following equality:

$$\langle\Psi|H^\dagger|\Phi\rangle = (\langle\Phi|H|\Psi\rangle)^* \;\Rightarrow\; \langle\Phi|\cdot H^\dagger = (H\cdot|\Phi\rangle)^\dagger.$$

Operator H obeying

$$H^\dagger = H \Leftrightarrow \langle\alpha|H|\beta\rangle = (\langle\beta|H|\alpha\rangle)^*$$

is called Hermitian operator.

States in quantum mechanics are described by vectors from Hilbert space $\mathcal{V}$ and its Hermitian conjugate dual space $\mathcal{V}^*$, while observables like coordinate, momentum, energy, angular momentum, etc., are described by Hermitian operators acting in $\mathcal{V}$. When introducing notations $\langle\Phi|$ and $|\Psi\rangle$, Dirac [30] called them bra- and ket-vectors, which are two parts of the word "bracket".

It is often convenient in quantum mechanics to choose the basis in $\mathcal{V}$ in a particular way, namely, let us single out a complete set of commuting observables $H_i = H_i^\dagger$ $(i = 1, 2, \ldots)$, $[H_i, H_j] = 0$. Denote the eigenvalues of the operators H_i by α_i, and their eigenvectors by $|\alpha\rangle = |\alpha_1, \alpha_2, \ldots\rangle$:

$$H_i|\alpha\rangle = \alpha_i|\alpha\rangle. \tag{1.1.12}$$

Problem 1.1.1. Show that the eigenvalues α_i of Hermitian operators H_i are real, and that $\langle\alpha'|\alpha\rangle \propto \delta_{\alpha',\alpha}$.

The set of eigenvectors $|\alpha\rangle$ is precisely the particular basis used in quantum mechanics. It depends, of course, on the choice of the complete set of commuting operators H_i. One can choose the normalization of the vectors $|\alpha\rangle$ in such a way that they obey orthonormality and completeness conditions, Eqs. (1.1.2) and (1.1.5), respectively.

Example. Coordinate and momentum representations

States of a quantum mechanical particle in $\mathbb{R}^d$ are vectors $|\Psi\rangle$ in infinite-dimensional vector space $\mathcal{V}$ equipped with Hermitian scalar product (1.1.8). Coordinates of the particle are observables which are described by Hermitian operators $\hat{x}_k$ $(k = 1, \ldots, d)$, while the components of particle momenta are described by Hermitian operators $\hat{p}_k$ which obey commutation relations

$$[\hat{x}_k, \hat{x}_j] = 0 = [\hat{p}_k, \hat{p}_j], \quad [\hat{x}_k, \hat{p}_j] = iI\delta_{kj}, \tag{1.1.13}$$

where I is unit operator in $\mathcal{V}$. The algebra with $2d$ generators $\{\hat{x}_k, \hat{p}_k\}$ and defining relations (1.1.13) is called d-dimensional *Heisenberg algebra.* Eigenvectors of operators $\hat{x}_k$ and $\hat{p}_k$ in $\mathcal{V}$ are denoted by $|\vec{x}\rangle$ and $|\vec{p}\rangle$:

$$\hat{x}_k|\vec{x}\rangle = x_k|\vec{x}\rangle, \quad \hat{p}_k|\vec{p}\rangle = p_k|\vec{p}\rangle, \tag{1.1.14}$$

where $x_k, p_k \in \mathbb{R}$ are the eigenvalues. Particle states given by vectors $|\vec{x}\rangle$ and $|\vec{p}\rangle$ are states with definite coordinates and definite momenta, respectively. Let $|\vec{y}\rangle \in \mathcal{V}$ be an eigenvector of all operators $\hat{x}_k$ with eigenvalues y_k. Making use of the commutation relations (1.1.13), we obtain

$$\hat{x}_k \cdot e^{-ix_k\hat{p}_k}|\vec{y}\rangle = (x_k e^{-ix_k\hat{p}_k} + e^{-ix_k\hat{p}_k} \cdot \hat{x}_k)|\vec{y}\rangle = (x_k + y_k)e^{-ix_k\hat{p}_k}|\vec{y}\rangle.$$

Hence, vector $e^{-ix_k\hat{p}_k}|\vec{y}\rangle$ is the eigenvector of operators $\hat{x}_k$ with eigenvalues $(x_k + y_k)$:

$$|\vec{x} + \vec{y}\rangle = e^{-ix_k\hat{p}_k}|\vec{y}\rangle. \tag{1.1.15}$$

We differentiate the two sides of Eq. (1.1.15) over x_k and then set $\vec{y} = \vec{0}$ to obtain

$$i\partial_k|\vec{x}\rangle = i\partial_k e^{-ix_k\hat{p}_k}|\vec{0}\rangle = \hat{p}_k|\vec{x}\rangle \;\Rightarrow\; \langle\vec{x}|\hat{p}_k = -i\partial_k\langle\vec{x}|, \tag{1.1.16}$$

where $\partial_k = \partial/\partial x_k$, and the second relation is obtained from the first one by Hermitian conjugation.

Every state $|\Psi\rangle$ defines a complex-valued function $\langle\vec{x}|\Psi\rangle \equiv \Psi(\vec{x})$, which is coordinate representation of vector $|\Psi\rangle$ in basis $|\vec{x}\rangle$. It is called the wave function of the particle. This correspondence is one-to-one. The action of operators $\hat{x}_k$ and $\hat{p}_k$ on state $|\Psi\rangle$ is written in terms of the wave function $\langle\vec{x}|\Psi\rangle$ as follows (see (1.1.7)):

$$\begin{aligned}\langle\vec{x}|\hat{x}_k|\Psi\rangle &= \int d^d x' \langle\vec{x}|\hat{x}_k|\vec{x}'\rangle\langle\vec{x}'|\Psi\rangle = x_k\langle\vec{x}|\Psi\rangle,\\ \langle\vec{x}|\hat{p}_k|\Psi\rangle &= \int d^d x' \langle\vec{x}|\hat{p}_k|\vec{x}'\rangle\langle\vec{x}'|\Psi\rangle = -i\partial_k\langle\vec{x}|\Psi\rangle,\end{aligned} \tag{1.1.17}$$

where we use the completeness property (1.1.5):

$$\int d^d x|\vec{x}\rangle\langle\vec{x}| = I \tag{1.1.18}$$

and recall (1.1.14) and (1.1.16). From the viewpoint of the representation theory, Eqs. (1.1.17) define the (differential) coordinate representation ρ of the Heisenberg algebra (1.1.13):

$$\rho(\hat{x}_k) = x_k, \quad \rho(\hat{p}_k) = -i\partial_k, \tag{1.1.19}$$

which acts in the space of wave functions $\Psi(\vec{x}) = \langle\vec{x}|\Psi\rangle$.

> **Problem 1.1.2.** Construct (differential) momentum representation of the Heisenberg algebra (1.1.13), which acts on momentum space wave functions $\Psi(\vec{p}) = \langle\vec{p}|\Psi\rangle$.

The state $|\vec{p}\rangle$ of the particle with well-defined momentum $\vec{p} = (p_1, \ldots, p_d)$ is described by the wave function $\langle\vec{x}|\vec{p}\rangle$, which, according to (1.1.14) and (1.1.17), obeys the equations

$$-i\partial_k\langle\vec{x}|\vec{p}\rangle = p_k\langle\vec{x}|\vec{p}\rangle.$$

The solution to these equations is

$$\langle\vec{x}|\vec{p}\rangle = \frac{1}{(2\pi)^{d/2}} \exp(ip_k x_k), \tag{1.1.20}$$

where the normalization factor is determined by the orthonormality condition (1.1.2):

$$\langle\vec{p}'|\vec{p}\rangle = \int d^d x \langle\vec{p}'|\vec{x}\rangle\langle\vec{x}|\vec{p}\rangle = \delta^d(\vec{p} - \vec{p}'). \tag{1.1.21}$$

Here we again use the completeness property (1.1.18). Using (1.1.20), we obtain the wave function in momentum representation

$$\langle\vec{p}|\Psi\rangle = \int d^d x \langle\vec{p}|\vec{x}\rangle\langle\vec{x}|\Psi\rangle = \frac{1}{(2\pi)^{d/2}} \int d^d x \exp(-ip_k x_k)\langle\vec{x}|\Psi\rangle. \tag{1.1.22}$$

This relates the momentum and coordinate representations of the state vectors. The relation (1.1.22) is the well-known Fourier transformation from coordinate representation to momentum representation.

Another example of employing the Dirac notations is given in the following section.

1.2. Fock Representation. (Anti)holomorphic Representations

Besides coordinate and momentum representations, of particular importance in quantum mechanics are Fock representation and Bargmann–Fock representations (the latter dubbed holomorphic or antiholomorphic). In this section we present the standard material on these representations (see also books [15–18]). To define these representations, we introduce, instead of $\hat{x}_k$ and $\hat{p}_k$, new dynamical variables

$$a_k = \frac{1}{\sqrt{2}}(i\hat{p}_k + \hat{x}_k), \quad a_k^\dagger = \frac{1}{\sqrt{2}}(-i\hat{p}_k + \hat{x}_k), \tag{1.2.1}$$

which, according to (1.1.13), commute as follows:

$$[a_k, a_m] = 0, \quad [a_k^\dagger, a_m^\dagger] = 0, \quad [a_k, a_m^\dagger] = I\delta_{km}. \tag{1.2.2}$$

Further we denote the identity operator I by 1, and an element $c \cdot I$, where c is a constant, is written simply as c. Operators (1.2.1) are particularly convenient in studying *d-dimensional quantum oscillator* whose Hamiltonian reads

$$\hat{\mathcal{H}} = \frac{1}{2}\sum_{k=1}^{d}(\hat{p}_k^2 + \hat{x}_k^2) = \sum_{k=1}^{d}\left(a_k^\dagger a_k + \frac{1}{2}\right). \tag{1.2.3}$$

For this reason, the algebra with generators $a_k, a_m^\dagger$ and defining relations (1.2.2) is called *d-dimensional quantum oscillator algebra*.

The space $\mathcal{F}$ of states of the d-dimensional quantum oscillator, where the algebra with generators $\{a_k, a_m^\dagger\}$ acts, is called Fock space. This space is constructed in the following way. One introduces the normalized vacuum vector $|0\rangle \in \mathcal{F}$ such that

$$a_k|0\rangle = 0, \quad \forall k, \quad \langle 0\, 0\rangle = 1. \tag{1.2.4}$$

Accordingly, the conjugate vector $\langle 0| = (|0\rangle)^\dagger$ obeys $\langle 0|a_k^\dagger = 0$. Then one constructs normalized vectors

$$|n_1, n_2, \ldots, n_d\rangle = \frac{1}{\sqrt{n_1! \cdots n_d!}}(a_1^\dagger)^{n_1} \cdot (a_2^\dagger)^{n_2} \cdots (a_d^\dagger)^{n_d}|0\rangle, \tag{1.2.5}$$

which make basis in the Fock space $\mathcal{F}$, i.e., any vector in $\mathcal{F}$ can be written as

$$|\Psi\rangle = \sum_{n_1,\ldots,n_d=0}^{\infty} \psi_{n_1,\ldots,n_d}|n_1, n_2, \ldots, n_d\rangle \in \mathcal{F}, \quad \psi_{n_1,\ldots,n_d} \in \mathbb{C}.$$

The Fock space coincides with the space of all states of a quantum mechanical particle in $\mathbb{R}^d$: any state of that particle can be written in the above form. According to (1.2.2), (1.2.5) we have

$$\begin{aligned} a_k|n_1, \ldots, n_k, \ldots, n_d\rangle &= \sqrt{n_k}|n_1, \ldots, n_k - 1, \ldots, n_d\rangle, \\ a_k^\dagger|n_1, \ldots, n_k, \ldots, n_d\rangle &= \sqrt{n_k + 1}|n_1, \ldots, n_k + 1, \ldots, n_d\rangle, \end{aligned} \tag{1.2.6}$$

i.e., operator a_k decreases n_k by one, while $a_k^\dagger$ increases n_k by one. For this reason, the operators a_k and $a_k^\dagger$ are called annihilation and creation operators, respectively.

Basis vectors $\langle n_1, n_2, \ldots, n_d|$ in dual space $\mathcal{F}^*$ are obtained from vectors (1.2.5) by Hermitian conjugation, and any vector $\langle\Phi| \in \mathcal{F}^*$ can be written as follows: $\langle\Phi| = \sum_{n_1,\ldots,n_d=0}^{\infty} \phi^*_{n_1,\ldots,n_d}\langle n_1,\ldots,n_d| = (|\Phi\rangle)^\dagger$, where $\phi_{n_1,\ldots,n_d} \in \mathbb{C}$.

Problem 1.2.1. Prove that vectors (1.2.5) are orthonormalized: $\langle k_1,\ldots,k_d| n_1,\ldots,n_d\rangle = \delta_{k_1 n_1}\cdots\delta_{k_d n_d}$. Prove that the Hamiltonian $\hat{\mathcal{H}}$ is Hermitian with respect to the scalar product $\langle\Phi|\Psi\rangle = \sum_{n_1,\ldots,n_d}(\phi^*_{n_1,\ldots,n_d}\psi_{n_1,\ldots,n_d})$. Check that the vectors (1.2.5) are eigenvectors of the Hamiltonian (1.2.3):

$$\hat{\mathcal{H}}|n_1,n_2,\ldots,n_d\rangle = \left(\sum_{k=1}^{d} n_k + \frac{1}{2}\right)|n_1,n_2,\ldots,n_d\rangle. \tag{1.2.7}$$

Problem 1.2.2. Find the wave function $\langle\vec{x}|0\rangle$ of the oscillator ground state (vacuum) in coordinate representation. Show that the eigenstates $|n_1,n_2,\ldots,n_d\rangle$ of the Hamiltonian have the following form in the coordinate representation: $\langle\vec{x}|n_1,\ldots,n_d\rangle = \left(\prod_{k=1}^{d} H_{n_k}(x_k)/(2^{n_k}n_k!)^{1/2}\right)\langle\vec{x}|0\rangle$, where $H_n(x) = (-1)^n e^{x^2}\frac{d^n}{dx^n}e^{-x^2}$ are Hermite polynomials.

The completeness relation (1.1.5) for vectors $|n_1,\ldots,n_d\rangle$ is written in the following way:

$$I = \sum_{n_1,\ldots,n_d=0}^{\infty} |n_1,\ldots,n_d\rangle\langle n_1,\ldots,n_d|, \tag{1.2.8}$$

where I is unit operator in the Fock space. Note that one can invert the relations (1.2.1) and obtain in this way the Fock representation of the Heisenberg algebra:

$$\hat{x}_k = \frac{1}{\sqrt{2}}(a_k + a_k^\dagger), \quad \hat{p}_k = \frac{1}{i\sqrt{2}}(a_k - a_k^\dagger).$$

Let us turn to holomorphic and antiholomorphic representations. In analogy to the eigenstates of coordinate and momentum (1.1.14), to obtain the antiholomorphic representation, we introduce the complete set of eigenvectors of the annihilation operators a_k:

$$a_k|z\rangle = z_k|z\rangle, \tag{1.2.9}$$

where $|z\rangle := |z_1, z_2, \ldots, z_d\rangle$, and z_k are complex eigenvalues. Vectors $|z\rangle$ are called coherent states, and they can be written as (hereafter summation over repeated indices k is assumed)

$$|z\rangle = \exp(z_k a_k^\dagger)|0\rangle, \tag{1.2.10}$$

where $|0\rangle$ is the vacuum Fock state. Equation (1.2.9) follows from (1.2.10) in view of (1.2.2). Making use of the representation (1.2.10), we obtain

$$a_k^\dagger |z\rangle = \frac{\partial}{\partial z_k}|z\rangle. \tag{1.2.11}$$

Problem 1.2.3. Find the wave functions $\langle \vec{x}|z\rangle$ of coherent states in coordinate representation.

We have similar formulas for conjugate vectors $\langle z^*| = (|z\rangle)^\dagger$

$$\langle z^*| = \langle 0|\exp(z_k^* a_k), \quad \langle z^*|a_k^\dagger = z_k^*\langle z^*|, \quad \langle z^*|a_k = \frac{\partial}{\partial z_k^*}\langle z^*|. \tag{1.2.12}$$

The convolution of two coherent states equals

$$\langle z^*|w\rangle = \langle 0|\exp(z_k^* a_k)|w\rangle = \langle 0|\exp(z_k^* w_k)|w\rangle = \exp(z_k^* w_k), \tag{1.2.13}$$

where $z, w \in \mathbb{C}$, and we employ Eq. (1.2.9) and the relation $\langle 0|w\rangle = 1$, $\forall w \in \mathbb{C}$. It is now clear that for $z \neq w$, coherent states $|w\rangle$ and $|z\rangle$ are not orthogonal. Nevertheless, they can be used for constructing new representations in quantum mechanics. The key point here is the following proposition.

Proposition 1.2.1. *The completeness condition* (1.2.8) *is written for coherent states as follows* (*cf.* (1.1.18))

$$\int d\mu(z, z^*)\exp(-z_k^* z_k)|z\rangle\langle z^*| = I, \tag{1.2.14}$$

where the measure $d\mu(z, z^*)$ on C^d is defined by[c]

$$d\mu(z, z^*) = \frac{1}{\pi^d}\prod_{k=1}^{d} d^2 z_k = \frac{1}{\pi^d}\prod_{k=1}^{d} dx_k dy_k, \quad z_k = x_k + iy_k. \tag{1.2.15}$$

[c]In mathematical literature (see, e.g., [17]), the often used notation (for measure on $\mathbb{C}$) is $d^2 z = \frac{i}{2} dz \wedge dz^*$, which is equivalent to our convention.

Proof. We prove (1.2.14) in one dimensional case $d = 1$; generalizing to arbitrary finite number of dimensions is trivial. We insert the definitions of coherent states (1.2.10), (1.2.12) into the left-hand side of (1.2.14) and write the integral in polar coordinates $z = \rho e^{i\phi}$:

$$\begin{aligned}
&\frac{1}{\pi}\int d^2z \exp(-|z|^2) \sum_{m,n=0}^{\infty} \frac{(za)^n}{n!}|0\rangle\langle 0|\frac{(z^*a^\dagger)^m}{m!} \\
&= \frac{1}{\pi}\int_0^\infty d\rho e^{-\rho^2} \sum_{m,n=0}^{\infty} \rho^{n+m+1} \int_0^{2\pi} d\phi e^{i(n-m)\phi} \frac{1}{\sqrt{n!m!}}|n\rangle\langle m| \\
&= \sum_{n=0}^{\infty} \frac{2}{n!}\left(\int_0^\infty d\rho \rho^{2n+1} e^{-\rho^2}\right)|n\rangle\langle n| = \sum_{n=0}^{\infty}|n\rangle\langle n| = I. \qquad (1.2.16)
\end{aligned}$$

This is the desired result. □

Now, like in the case of coordinate or momentum representation, we consider a function $\langle z^*|\Psi\rangle \equiv \Psi(z^*)$ that corresponds to a state vector $|\Psi\rangle$. This is the wave function in the *antiholomorphic representation.* In antiholomorphic representation, operators a_k and $a_k^\dagger$ act on a state $|\Psi\rangle$ as follows (see (1.2.12)):

$$\langle z^*|a_k|\Psi\rangle = \frac{\partial}{\partial z_k^*}\langle z^*|\Psi\rangle, \quad \langle z^*|a_k^\dagger|\Psi\rangle = z_k^*\langle z^*|\Psi\rangle, \qquad (1.2.17)$$

while the scalar product $\langle\Phi|\Psi\rangle$, in view of the completeness relation (1.2.14), reads

$$\langle\Phi|\Psi\rangle = \int d\mu(z,z^*)e^{-z_k^* z_k}\langle\Phi|z\rangle\langle z^*|\Psi\rangle = \int d\mu(z,z^*)e^{-z_k^* z_k}\left(\Phi(z^*)\right)^*\Psi(z^*). \qquad (1.2.18)$$

Thus, the formulas (1.2.17) define *antiholomorphic representation* $\bar{\rho}$ of the oscillator algebra a_k and $a_k^\dagger$ (cf. (1.1.19)):

$$\bar{\rho}(a_k) = \frac{\partial}{\partial z_k^*}, \quad \bar{\rho}(a_k^\dagger) = z_k^*, \qquad (1.2.19)$$

which acts in the Hilbert space of antiholomorphic functions $\Psi(z^*) = \langle z^*|\Psi\rangle$ with the scalar product (1.2.18). In literature, this representation is also known as Bargmann–Fock representation.

By performing the replacement $z_k^* \leftrightarrow z_k$, $(\forall k)$ in (1.2.12), (1.2.14), (1.2.17) and (1.2.18), we obtain yet another formulation of Bargmann–Fock representation denoted by ρ. It can be called holomorphic representation.

It acts in the Hilbert space of holomorphic functions $\langle z|\Psi\rangle = \Psi(z) \equiv \Psi(z_1, \ldots, z_d)$ with the scalar product

$$\langle\Phi|\Psi\rangle = \int d\mu(z, z^*) e^{-z_k^* z_k} \langle\Phi|z^*\rangle\langle z|\Psi\rangle = \int d\mu(z, z^*) e^{-z_k^* z_k} (\Phi(z))^* \Psi(z), \tag{1.2.20}$$

and the representation of the oscillator algebra (1.2.2) is given by[d]

$$\rho(a_k) = \frac{\partial}{\partial z_k}, \quad \rho(a_k^\dagger) = z_k. \tag{1.2.21}$$

The completeness condition (1.2.14) reads in the representation ρ as

$$\int d\mu(z, z^*) \exp(-z_k^* z_k) |z^*\rangle\langle z| = I, \tag{1.2.22}$$

while basis vectors (1.2.5) are

$$\langle z|n_1, \ldots, n_d\rangle = \frac{1}{\sqrt{n_1! \cdots n_d!}} z_1^{n_1} z_2^{n_2} \cdots z_d^{n_d}, \tag{1.2.23}$$

where we use $\langle z|a_k^\dagger = z_k\langle z|$; cf. (1.2.12).

Proposition 1.2.2. *The scalar product* (1.2.18), (1.2.20) *can be written as follows*:

$$\langle\Phi|\Psi\rangle = \Phi^*(\partial_1, \ldots, \partial_d) \cdot \Psi(z)|_{z_1 = \cdots = z_d = 0}, \tag{1.2.24}$$

where $\partial_k = \partial/\partial z_k$. *Here, if* $\Phi(z) = \sum C_{n_1,\ldots,d} z_1^{n_1} \ldots z_d^{n_d}$ *then* $\Phi^*(\partial_1, \ldots, \partial_d) = \sum C^*_{n_1,\ldots,n_d} \partial_1^{n_1} \ldots \partial_d^{n_d}$.

Proof. It suffices to prove (1.2.24) in the case when the vector $\Phi(z) = \langle z|\Phi\rangle$ is an arbitrary basis vector (1.2.23). In that case, we integrate by parts and obtain

$$\begin{aligned} \langle n_1, \ldots, n_d|\Psi\rangle &= \int d\mu(z, z^*) \langle n_1, \ldots, n_d|z^*\rangle\langle z|\Psi\rangle e^{-z_k^* z_k} \\ &= \int d\mu(z, z^*) \frac{z_1^{*n_1} \cdots z_d^{*n_d}}{\sqrt{n_1! \cdots n_d!}} \langle z|\Psi\rangle e^{-z_k^* z_k} \end{aligned}$$

[d]Sometimes in the literature one considers the representation $\rho'(a_k) = z_k$, $\rho'(a_k^\dagger) = -\frac{\partial}{\partial z_k}$ (cf. (1.2.21)) which acts in the space of holomorphic functions $\Psi(z)$ and also is called holomorphic representation. We will not discuss this possibility here.

$$= \int d\mu(z,z^*)(-1)^{\sum n_k}\langle z|\Psi\rangle \frac{\partial_1^{n_1}\cdots\partial_d^{n_d}}{\sqrt{n_1!\cdots n_d!}} e^{-z_k^* z_k}$$

$$= \int d\mu(z,z^*) e^{-z_k^* z_k} \frac{\partial_1^{n_1}\cdots\partial_d^{n_d}}{\sqrt{n_1!\cdots n_d!}} \langle z|\Psi\rangle$$

$$= \frac{\partial_1^{n_1}\cdots\partial_d^{n_d}}{\sqrt{n_1!\cdots n_d!}} \Psi(z)\Big|_{z_1=\cdots=z_d=0}. \tag{1.2.25}$$

Since an arbitrary holomorphic functions $\Phi(z)$ can be written as an expansion in basis monomials (1.2.23), then (1.2.25) gives (1.2.24). □

Remark 1. The representation (1.2.24) of the scalar product (1.2.20) is particularly convenient from technical viewpoint. We extensively use this representation below, in Sections 2.2.2, 2.4.4 and 2.5.4 (see, e.g., (2.2.28) and (2.4.86)).

Making use of (1.2.2), one can write any operator A in Fock space $\mathcal{F}$ in normal (Wick) form:

$$A = \sum_{\mu,\nu} A_{\mu,\nu} (a_1^\dagger)^{n_1}\cdot(a_2^\dagger)^{n_2}\cdots(a_d^\dagger)^{n_d}(a_1)^{m_1}\cdot(a_2)^{m_2}\cdots(a_d)^{m_d}, \tag{1.2.26}$$

where μ and ν are multi-indices, $\mu = (m_1,\dots,m_d)$ and $\nu = (n_1,\dots,n_d)$, and $A_{\mu,\nu} \in \mathbb{C}$. In this form, all annihilation operators in every term in (1.2.26) are on the right of the creation operators. Antiholomorphic representation is convenient for determining the *Wick symbols* of operators A in $\mathcal{F}$. To this end, we write the action of the operator (1.2.26) on a vector $|\Psi\rangle$ in antiholomorphic representation:

$$\langle w^*|A|\Psi\rangle = \int d\mu(z,z^*) e^{-z_k^* z_k}\langle w^*|A|z\rangle\langle z^*|\Psi\rangle, \tag{1.2.27}$$

where $\langle w^*|A|z\rangle$ is the kernel of the operator A. The latter is computed straightforwardly by making use of (1.2.9) and (1.2.12):

$$\langle w^*|A|z\rangle = A(w^*,z)e^{w_k^* z_k},$$

$$A(w^*,z) = \sum_{\mu\nu} A_{\mu,\nu}(w_1^*)^{n_1}\cdot(w_2^*)^{n_2}\cdots(w_d^*)^{n_d}(z_1)^{m_1}\cdot(z_2)^{m_2}\cdots(z_d)^{m_d}. \tag{1.2.28}$$

The function $A(w^*,z)$ is dubbed *Wick (normal) symbol* of the operator A. The only difference of the Wick symbol from the normal form is that the former involves complex numbers w_i^* and z_j instead of operators $a_i^\dagger$ and a_j.

Wick symbols are convenient for writing a product of two operators A_1 and A_2 in normal form (1.2.26).

Proposition 1.2.3. *Let $A_1(w^*, z)$ and $A_2(w^*, z)$ be the Wick symbols of operators A_1 and A_2, then the Wick symbol of the operator $A_1 \cdot A_2$ is given by*

$$(A_1{\cdot}A_2)(w^*, v) = \int d\mu(z, z^*) e^{-(z_k^* - w_k^*)(z_k - v_k)} A_1(w^*, z) A_2(z^*, v), \quad (1.2.29)$$

where the measure $d\mu(z, z^*)$ is defined in (1.2.15).

Proof. In accordance with (1.2.28), we have

$$\begin{aligned}(A_1 \cdot A_2)(w^*, v) &= \langle w^* | A_1 \cdot A_2 | v \rangle e^{-w_k^* v_k} \\ &= \int d\mu(z, z^*) \langle w^* | A_1 | z \rangle \langle z^* | A_2 | v \rangle e^{-z_k^* z_k - w_k^* v_k} \\ &= \int d\mu(z, z^*) A_1(w^*, z) A_2(z^*, v) e^{w_k^* z_k + z_k^* v_k} e^{-z_k^* z_k - w_k^* v_k},\end{aligned}$$

which coincides with (1.2.29). □

Thus, Eq. (1.2.29) gives an explicit formula for the Wick symbol of $A_1 \cdot A_2$, which straightforwardly translates to the normal form of $A_1 \cdot A_2$.

Remark 2. d-dimensional quantum oscillator algebra (1.2.2) is used for constructing the realizations of the generators of Lie algebras of classic series $s\ell(d, \mathbb{C})$, $so(d, \mathbb{C})$ and $sp(d, \mathbb{C})$. As an example, the generators $E_{ij} \in s\ell(d, \mathbb{C})$ can be represented as

$$E_{ij} = a_i^\dagger a_j - \frac{\delta_{ij}}{d} \sum_k a_k^\dagger a_k, \quad \sum_k E_{kk} = 0. \quad (1.2.30)$$

Indeed, it is straightforward to check that commutation relations of the operators (1.2.30) coincide with the defining relations for algebra $s\ell(d, \mathbb{C})$ (see Section **I-3.3.1** of the first book)

$$[E_{ij}, E_{km}] = \delta_{jk} E_{im} - \delta_{im} E_{kj}. \quad (1.2.31)$$

Likewise, the generators $L_{ij} \in so(d, \mathbb{C})$ and $M_{AB} \in sp(2d, \mathbb{C})$ have the following representations:

$$L_{ij} = a_i^\dagger a_j - a_j^\dagger a_i, \quad M_{AB} = \frac{1}{2}(\mathsf{z}_A \mathsf{z}_B + \mathsf{z}_B \mathsf{z}_A), \quad (1.2.32)$$

where $A, B, \ldots = 1, 2, \ldots, 2d$ and $(\mathsf{z}_1, \ldots, \mathsf{z}_{2d}) = (a_1^\dagger, \ldots, a_d^\dagger, a_1, \ldots, a_d)$. Commutation relations of the operators (1.2.32) coincide with the defining relations of algebras $so(d,\mathbb{C})$ and $sp(2d,\mathbb{C})$ (see Section **I-3.3.1** of the first book)

$$[L_{ij}, L_{k\ell}] = \delta_{jk}L_{i\ell} + \delta_{j\ell}L_{ki} + \delta_{ik}L_{\ell j} + \delta_{i\ell}L_{jk}, \tag{1.2.33}$$

$$[M_{AB}, M_{CD}] = J^{-1}_{BC}M_{AD} + J^{-1}_{AC}M_{BD} + J^{-1}_{AD}M_{CB} + J^{-1}_{BD}M_{CA}, \tag{1.2.34}$$

where $||J^{-1}_{AC}||$ is inverse of the matrix of symplectic metric

$$J = \begin{pmatrix} 0 & I_d \\ -I_d & 0 \end{pmatrix} = -J^{-1}, \quad [\mathrm{z}_A, \mathrm{z}_B] = J^{-1}_{AB}.$$

The representations (1.2.30) and (1.2.32) are called *Jordan–Schwinger representations* of Lie algebras $s\ell(d,\mathbb{C})$, $so(d,\mathbb{C})$ and $sp(d,\mathbb{C})$. The generators (1.2.30) and (1.2.32) can be written in holomorphic representation (1.2.21) and antiholomorphic representation (1.2.19). As an example, the holomorphic representation of the generators (1.2.30) of the Lie algebra $s\ell(d,\mathbb{C})$ is

$$E_{ij} = z_i\frac{\partial}{\partial z_j} - \frac{\delta_{ij}}{d}\sum_k z_k\frac{\partial}{\partial z_k}. \tag{1.2.35}$$

We use this representation below in Section 2.2.2, where we discuss the representations with highest weight for the differential realization of the algebra $s\ell(2,\mathbb{C})$.

1.3. Clifford Algebra and Free Fermion Algebra

In this section, we consider another important case where Dirac notations are particularly convenient. This example is used below in Sections 5.4 and 6.1 for constructing spinor representations of Lie algebra $so(N,\mathbb{C})$ and its real forms, as well as for describing spinor Lie groups.

Somewhat different presentation of the material of this section can be found in books [15–17].

1.3.1. Clifford algebra

Definition 1.3.1. Associative algebra $C\ell_N$ over the field of complex numbers $\mathbb{C}$ with unit element I and generators Γ_m $(m = 1, 2, \ldots, N)$ obeying the defining relations

$$\Gamma_m\Gamma_n + \Gamma_n\Gamma_m = 2\delta_{mn}I \tag{1.3.1}$$

is called complex *Clifford algebra*.[e]

[e]More general definition of Clifford algebras is given in Section 6.1.1.

The defining relations (1.3.1) stay intact under any permutation of generators $\Gamma_m \in \mathcal{C}\ell_N$, so the numbering of generators in algebra $\mathcal{C}\ell_N$ can be chosen arbitrarily.

Complex algebra $\mathcal{C}\ell_N$ has several inequivalent real forms $\mathcal{C}\ell_{(p,q)}$ (where $p, q \in \mathbb{Z}_{\geq 0}$ and $p + q = N$), i.e., real algebras with the set of generators Γ_m $(m = 1, \ldots, N)$ obeying defining relations

$$\Gamma_m \Gamma_n + \Gamma_n \Gamma_m = 2\eta_{mn} I, \quad \|\eta_{mn}\| = \mathrm{diag}(\underbrace{1, \ldots, 1}_{p}, \underbrace{-1, \ldots, -1}_{q}), \tag{1.3.2}$$

where η_{mn} is metric of (pseudo-)Euclidean space $\mathbb{R}^{p,q}$. Clearly, complexifications of real algebras $\mathcal{C}\ell_{(p,q)}$ are isomorphic to $\mathcal{C}\ell_N$. We discuss in detail and make use of real forms $\mathcal{C}\ell_{(p,q)}$ of Clifford algebra below in Chapter 6.

Proposition 1.3.1. *Clifford algebra $\mathcal{C}\ell_N$ with finite number N of its generators is finite dimensional, and its complex dimension is*

$$\dim(\mathcal{C}\ell_N) = 2^N. \tag{1.3.3}$$

Proof. Elements of algebra $\mathcal{C}\ell_N$ are linear combinations of monomials $\Gamma_{m_1} \cdot \Gamma_{m_2} \cdots \Gamma_{m_k}$. Each of these monomials can be decomposed into symmetric and antisymmetric parts in pairs of indices from the set $\{m_1, m_2, \ldots, m_k\}$. In accordance with (1.3.1), (partially) symmetrized combinations of products of generators Γ_m are reduced to polynomials of the order $(k - 2)$ and less. Therefore, the basis of algebra $\mathcal{C}\ell_N$ consists of unit element I and all *totally antisymmetric* products of generators Γ_m:

$$\Gamma_{[m_1 \ldots m_k]} = \frac{1}{k!} \sum_{\sigma \in S_k} (-1)^{\mathrm{p}(\sigma)} \Gamma_{\sigma(m_1)} \cdot \Gamma_{\sigma(m_2)} \cdots \Gamma_{\sigma(m_k)} \quad (k = 1, \ldots, N), \tag{1.3.4}$$

where summation runs over all permutations $\sigma \in S_k$ of indices $\{m_1, \ldots, m_k\}$, and $\mathrm{p}(\sigma)$ denotes the parity of permutation σ. Note that the set (1.3.4) with $k = 1$ contains all generators Γ_m, while due to antisymmetry of $\Gamma_{[m_1 \ldots m_k]}$ in indices $\{m_1, \ldots, m_k\}$ we have $\Gamma_{[m_1 \ldots m_k]} = 0$ for $k > N$. Thus, the number of basis elements $\Gamma_{[m_1 \ldots m_k]}$ is finite, so algebra $\mathcal{C}\ell_N$ is finite dimensional. Making use of defining relations (1.3.1) one reduces the set of basis elements to the elements of the form

$$\begin{aligned} \Gamma_A &= \{I, \Gamma_{m_1}, \Gamma_{m_1}\Gamma_{m_2}, \Gamma_{m_1}\Gamma_{m_2}\Gamma_{m_3}, \ldots, \Gamma_1\Gamma_2 \cdots \Gamma_N\} \\ &\equiv \{I, \Gamma_{m_1}, \Gamma_{m_1 m_2}, \Gamma_{m_1 m_2 m_3}, \ldots, \Gamma_{12\ldots N}\}, \end{aligned} \tag{1.3.5}$$

$$\Gamma_{m_1 m_2 \ldots m_k} \equiv \Gamma_{m_1}\Gamma_{m_2} \cdots \Gamma_{m_k},$$

where the product of generators Γ_{m_i} in each monomial $\Gamma_{m_1 m_2 \dots m_k}$ is ordered:

$$1 \leq m_1 < m_2 < m_3 < \cdots \leq N.$$

Let us find the number of independent elements (1.3.5) and in this way calculate dimension of algebra $\mathcal{C}\ell_N$. The number of kth-order monomials $\Gamma_{m_1 \dots m_k}$ in (1.3.5) is given by

$$C_N^k = \frac{N!}{k!(N-k)!},$$

so the dimension of Clifford algebra $\mathcal{C}\ell_N$ equals

$$\dim(\mathcal{C}\ell_N) = \sum_{k=0}^{N} \frac{N!}{k!(N-k)!} = (1+1)^N = 2^N, \tag{1.3.6}$$

which establishes (1.3.3). □

Let us consider the last, longest basis element in (1.3.5) which can be written either as an ordered product of all generators $\Gamma_m \in \mathcal{C}\ell_N$ or as their antisymmetric product:

$$\Gamma_{1 \dots N} = \Gamma_1 \cdot \Gamma_2 \cdots \Gamma_N = \frac{1}{N!} \varepsilon_{i_1 \dots i_N} \Gamma_{i_1} \cdots \Gamma_{i_N}, \tag{1.3.7}$$

where we use totally antisymmetric tensor (see Section **I-1.2.2** of the first book):

$$\varepsilon_{\dots i_k \dots i_\ell \dots} = -\varepsilon_{\dots i_\ell \dots i_k \dots}, \quad \varepsilon_{12 \dots N} = +1. \tag{1.3.8}$$

In accordance with (1.3.1), element (1.3.7) obeys the identity

$$(\Gamma_{1 \dots N})^2 = (-1)^{N(N-1)/2} (\Gamma_N \cdots \Gamma_2 \cdot \Gamma_1)(\Gamma_1 \cdot \Gamma_2 \cdots \Gamma_N) = (-1)^\nu I, \tag{1.3.9}$$

where $\nu = \left[\frac{N}{2}\right]$ is integer part of $\frac{N}{2}$, i.e., ν is nonnegative integer, and $N = 2\nu$ or $N = 2\nu + 1$. It is convenient to define normalized element

$$\Gamma_{N+1} = e^{-i\frac{\pi}{2}\nu} \Gamma_{1 \dots N} = e^{-i\frac{\pi}{2}\nu} \Gamma_1 \cdot \Gamma_2 \cdots \Gamma_N \in \mathcal{C}\ell_N, \tag{1.3.10}$$

such that

$$\Gamma_{N+1}^2 = I. \tag{1.3.11}$$

Let us now consider two cases.

1. The number of generators of $\mathcal{C}\ell_N$ is even, $N = 2\nu$. In this case we have, due to (1.3.1), the relation

$$\Gamma_{N+1} \cdot \Gamma_k = -\Gamma_k \cdot \Gamma_{N+1} \quad (\forall k = 1, \dots, N).$$

Therefore, using (1.3.11), we find that the elements $\Gamma_1, \ldots, \Gamma_N$ together with Γ_{N+1} obey relations (1.3.1) for odd number of generators $(N+1) = 2\nu + 1$ of algebra $\mathcal{C}\ell_{2\nu+1}$. We see that algebra $\mathcal{C}\ell_{2\nu}$ can be considered as quotient of algebra $\mathcal{C}\ell_{2\nu+1}$ with respect to the relation $\Gamma_{2\nu+1} = e^{-i\pi\nu/2}\Gamma_1 \cdots \Gamma_{2\nu}$ (or $\Gamma_{2\nu+1} = -e^{-i\pi\nu/2}\Gamma_1 \cdots \Gamma_{2\nu}$; recall that element $\Gamma_{N+1} \in \mathcal{C}\ell_{2\nu}$ is defined in (1.3.10) modulo sign). On the other hand, this means that there exist irreversible homomorphisms $\rho_\pm \colon \mathcal{C}\ell_{2\nu+1} \to \mathcal{C}\ell_{2\nu}$, defined by

$$\Gamma'_i \to \Gamma_i \ (i = 1, \ldots, 2\nu), \quad \pm\Gamma'_{2\nu+1} \to e^{-i\pi\nu/2}\Gamma_1 \cdots \Gamma_{2\nu},$$

where $\{\Gamma'_1, \ldots, \Gamma'_{2\nu}, \Gamma'_{2\nu+1}\}$ generate algebra $\mathcal{C}\ell_{2\nu+1}$. Under these homomorphisms, two elements $e^{-i\pi\nu/2}\Gamma'_1 \cdots \Gamma'_{2\nu} \in \mathcal{C}\ell_{2\nu+1}$ and $\Gamma'_{2\nu+1} \in \mathcal{C}\ell_{2\nu+1}$ (or $-\Gamma'_{2\nu+1} \in \mathcal{C}\ell_{2\nu+1}$) are mapped to one element $e^{-i\pi\nu/2}\Gamma_1 \cdots \Gamma_{2\nu} \in \mathcal{C}\ell_{2\nu}$.

2. The number of generators of algebra $\mathcal{C}\ell_N$ is odd, $N = 2\nu + 1$. Then we make use of (1.3.1) and find for the element (1.3.10) that

$$\Gamma_{N+1} \cdot \Gamma_k = \Gamma_k \cdot \Gamma_{N+1} \quad (\forall k = 1, \ldots, N = 2\nu + 1), \tag{1.3.12}$$

so Γ_{N+1} belongs to the center of algebra $\mathcal{C}\ell_N$. It follows from (1.3.11) that the central element Γ_{N+1} has only two eigenvalues ± 1. Consider now, instead of the standard set of generators $\Gamma_1, \ldots, \Gamma_{2\nu+1} \in \mathcal{C}\ell_{2\nu+1}$, a new set $\{\Gamma_1, \ldots, \Gamma_{2\nu}, \Gamma_{N+1}\}$, in which the element $\Gamma_{2\nu+1}$ is substituted by the central element (1.3.10). The new set is indeed a set of generators: any element of the complex algebra $\mathcal{C}\ell_{2\nu+1}$ can be written in terms of the generators from the new set, once one uses (1.3.10) and writes

$$\Gamma_{2\nu+1} \equiv \Gamma_N \to e^{-i\frac{\pi}{2}\nu}\Gamma_{N+1} \cdot \Gamma_1 \cdots \Gamma_{N-1}. \tag{1.3.13}$$

Note that the elements $\{\Gamma_1, \ldots, \Gamma_{N-1=2\nu}\}$ form subalgebra $\mathcal{C}\ell_{2\nu} \subset \mathcal{C}\ell_{2\nu+1}$, so algebra $\mathcal{C}\ell_{2\nu+1}$ with new generators $\{\Gamma_1, \ldots, \Gamma_{2\nu}, \Gamma_{N+1}\}$ can be viewed as central extension of $\mathcal{C}\ell_{2\nu}$. Then one recalls that the central element (1.3.10) obeys the identity $\Gamma_{N+1}^2 = I$ and writes any element $a \in \mathcal{C}\ell_{2\nu+1}$ as $a = a_1 + \Gamma_{N+1} \cdot a_2$, where $a_1, a_2 \in \mathcal{C}\ell_{2\nu}$. Therefore, the algebra $\mathcal{C}\ell_{2\nu+1}$ can be decomposed as follows:

$$\mathcal{C}\ell_{2\nu+1} = I \cdot \mathcal{C}\ell_{2\nu} + \Gamma_{N+1} \cdot \mathcal{C}\ell_{2\nu} = \frac{(I + \Gamma_{N+1})}{2} \cdot \mathcal{C}\ell_{2\nu} + \frac{(I - \Gamma_{N+1})}{2} \cdot \mathcal{C}\ell_{2\nu}, \tag{1.3.14}$$

where $\frac{1}{2}(I + \Gamma_{N+1})$ and $\frac{1}{2}(I - \Gamma_{N+1})$ are central projectors in algebra $\mathcal{C}\ell_{2\nu+1}$, which project the whole algebra $\mathcal{C}\ell_{2\nu+1}$ to eigenspaces of the element Γ_{N+1}

with $\Gamma_{N+1} = +I$ and $\Gamma_{N+1} = -I$, respectively. The formula (1.3.14) shows that there is isomorphism of complex Clifford algebras

$$\mathcal{C}\ell_{2\nu+1} = \mathcal{C}\ell_{2\nu} \oplus \mathcal{C}\ell_{2\nu}, \tag{1.3.15}$$

whose explicit form is

$$\begin{aligned} \Gamma_i &\to \begin{pmatrix} \Gamma'_i & 0 \\ 0 & \Gamma''_i \end{pmatrix}, \quad i = 1, \ldots, 2\nu, \\ \Gamma_{2\nu+1=N} &\to \begin{pmatrix} e^{-i\frac{\pi}{2}\nu}\Gamma'_1 \cdots \Gamma'_{2\nu} & 0 \\ 0 & -e^{-i\frac{\pi}{2}\nu}\Gamma''_1 \cdots \Gamma''_{2\nu} \end{pmatrix}, \\ \Gamma_{N+1} &\to \begin{pmatrix} I & 0 \\ 0 & -I \end{pmatrix}. \end{aligned} \tag{1.3.16}$$

Here $\Gamma_i, \Gamma_{2\nu+1}, \Gamma_{N+1} \in \mathcal{C}\ell_{2\nu+1}$, while Γ'_i and Γ''_i generate the first and second algebras $\mathcal{C}\ell_{2\nu}$ in (1.3.15); explicit form of the generator $\Gamma_{2\nu+1}$ follows from (1.3.13) (or from the definition (1.3.10)). Thus, we see that the study of the "odd" Clifford algebra $\mathcal{C}\ell_{2\nu+1}$ is reduced to the study of "even" algebra $\mathcal{C}\ell_{2\nu}$.

Let us consider Clifford algebra $\mathcal{C}\ell_N$ with even $N = 2\nu$ and divide the set of all generators $\Gamma_m \in \mathcal{C}\ell_N$ into two groups containing generators with even and odd m. Let us also introduce the set of operators $\{z_\alpha, \bar{z}_\alpha\}$ ($\alpha = 1, \ldots, \nu = N/2$):

$$z_\alpha = \frac{1}{2}(\Gamma_{2\alpha-1} + i\Gamma_{2\alpha}), \quad \bar{z}_\alpha = \frac{1}{2}(\Gamma_{2\alpha-1} - i\Gamma_{2\alpha}), \tag{1.3.17}$$

so that

$$\Gamma_{2\alpha-1} = z_\alpha + \bar{z}_\alpha, \quad \Gamma_{2\alpha} = i(\bar{z}_\alpha - z_\alpha). \tag{1.3.18}$$

Due to the property (1.3.1), these operators obey

$$[z_\alpha, z_\beta]_+ = 0, \quad [\bar{z}_\alpha, \bar{z}_\beta]_+ = 0, \quad [z_\alpha, \bar{z}_\beta]_+ = \delta_{\alpha\beta} I \quad (\forall \alpha, \beta), \tag{1.3.19}$$

where bracket $[\cdot, \cdot]_+$ denotes anticommutator $[a, b]_+ \equiv a \cdot b + b \cdot a$. Note that the operators z_α and $\bar{z}_\alpha$ are *nilpotent*, $z_\alpha^2 = 0$, $\bar{z}_\alpha^2 = 0$ (an operator is called nilpotent if it squares to zero).

Definition 1.3.2. Complex algebra $\mathcal{A}_\nu$ with unity I and generators $\{z_\alpha, \bar{z}_\alpha\}$ ($\alpha = 1, \ldots, \nu$) which obey the relations (1.3.19) is called *algebra of ν free fermions* or *fermion oscillator algebra.*

The change $\{\Gamma_m\} \to \{z_\alpha, \bar{z}_\beta\}$ is merely the change of the generator basis in complex algebra $\mathcal{C}\ell_{2\nu}$. In accordance with (1.3.18), this change is invertible, so it defines isomorphism $\mathcal{C}\ell_{2\nu} = \mathcal{A}_\nu$.

Let us set $(\Gamma_1, \ldots, \Gamma_{2\nu}) = \sqrt{2}(z_1, \ldots, z_\nu, \bar{z}_\nu, \ldots, \bar{z}_1)$. Then Eqs. (1.3.1) and (1.3.19) are written as follows (cf. (1.3.2)):

$$\Gamma_m \Gamma_n + \Gamma_n \Gamma_m = 2\eta_{mn} I, \quad \eta_{nm} = \delta_{n,N-m+1}, \tag{1.3.20}$$

where $||\eta_{mn}||$ is antidiagonal matrix, whose elements are equal to zero except for elements in the antidiagonal, which are equal to 1. This form of Clifford algebra is used below in Section 5.4.

Let ρ be an exact matrix representation of algebra $\mathcal{C}\ell_{2\nu}$, in which all basis elements (1.3.5) are linear independent. The minimal complex matrix space with $2^{2\nu}$ independent basis elements has dimension $2^{2\nu}$ — this space is $\mathrm{Mat}_{2^\nu}(\mathbb{C})$, whose complex dimension is $2^\nu \times 2^\nu = 2^{2\nu}$. Therefore, the matrices in the representation ρ must be at least $2^\nu \times 2^\nu$, i.e., the space of the representation ρ must have dimension not less than 2^ν. In the following section, we construct exact representation of algebra $\mathcal{C}\ell_{2\nu}$ whose space has dimension 2^ν. This points toward the isomorphism of complex Clifford algebra $\mathcal{C}\ell_{2\nu}$ and matrix algebra $\mathrm{Mat}_{2^\nu}(\mathbb{C})$. We establish this isomorphism in Chapter 6, Section 6.1.2.

1.3.2. Fock representation for free fermion algebra

Due to isomorphism $\mathcal{C}\ell_{2\nu} = \mathcal{A}_\nu$, the problem of constructing exact irreducible representation ρ of Clifford algebra $\mathcal{C}\ell_{2\nu}$ is the same as the problem of constructing exact irreducible representation of free fermion algebra (1.3.19). The latter problem is solved by introducing the *Fock space* $\mathcal{F}$ for free fermion algebra (1.3.19). Let us describe the Fock space $\mathcal{F}$ making use of the Dirac notations. We define the vacuum (lowest) vector $|0\rangle \in \mathcal{F}$ as a vector which is annihilated by all operators z_α:

$$z_\alpha|0\rangle = 0 \quad (\forall \alpha = 1, \ldots, \nu). \tag{1.3.21}$$

Operators $\bar{z}_\alpha$ act on $|0\rangle$ as creation operators, i.e., they generate, by acting on $|0\rangle$, all vectors of the Fock space $\mathcal{F}$. Since the operator $\bar{z}_\alpha$ is nilpotent, it gives zero when acting twice. Therefore, the entire Fock space $\mathcal{F}$, which, by construction, is the space of representation of algebra (1.3.19), consists of linear combinations of nonzero[f] basis vectors ($k = 0, 1, \ldots, \nu$)

$$|\alpha_1, \ldots, \alpha_k\rangle = \bar{z}_{\alpha_1} \cdots \bar{z}_{\alpha_k}|0\rangle \quad (1 \le \alpha_1 < \alpha_2 < \cdots < \alpha_k \le \nu). \tag{1.3.22}$$

[f]If any of the vectors (1.3.22) vanished, we would act on it by an appropriate combination of annihilation operators $z_{\alpha_1} \cdots z_{\alpha_k}$ and get $|0\rangle = 0$.

Thus, any vector $|\Psi\rangle \in \mathcal{F}$ can be written as

$$|\Psi\rangle = \sum_{k=0}^{\nu} \sum_{\alpha_1 < \cdots < \alpha_k} \psi_{\alpha_1,\ldots,\alpha_k} |\alpha_1, \ldots, \alpha_k\rangle, \quad \psi_{\alpha_1,\ldots,\alpha_k} \in \mathbb{C}. \tag{1.3.23}$$

Instead of ordering of indices $(\alpha_1, \ldots, \alpha_k)$ in (1.3.23), a useful convention is that components $\psi_{\alpha_1,\ldots,\alpha_k}$ are antisymmetric under all permutations of indices $(\alpha_1, \ldots, \alpha_k)$. The number of independent vectors (1.3.22) is calculated in the same way as the dimension of the Clifford algebra (see the chain of equalities (1.3.6)), and we obtain

$$\dim(\mathcal{F}) = \sum_{k=0}^{\nu} \frac{\nu!}{k!(\nu - k)!} = 2^{\nu} = 2^{N/2}. \tag{1.3.24}$$

This coincides with the minimal dimension of the space of an exact matrix representation of algebra $\mathcal{C}\ell_{2\nu}$. The generators of Clifford algebra $\Gamma_m \in \mathcal{C}\ell_{2\nu}$ are related to fermion operators z_{α_i} and $\bar{z}_{\beta_i}$ via (1.3.17), (1.3.18). Therefore, action of any element of Clifford algebra $\mathcal{C}\ell_{2\nu}$ in the Fock space $\mathcal{F}$ is determined by the action of fermion operators:

$$\begin{aligned}
z_{\beta}|\alpha_1, \ldots, \alpha_k\rangle &= \sum_{i=1}^{k} (-1)^{i-1} \delta_{\beta\alpha_i} |\alpha_1, \ldots, \alpha_{i-1}, \alpha_{i+1}, \ldots, \alpha_k\rangle, \\
\bar{z}_{\beta}|\alpha_1, \ldots, \alpha_k\rangle &= (-1)^i |\alpha_1, \ldots, \alpha_i, \beta, \alpha_{i+1}, \ldots, \alpha_k\rangle \quad (\alpha_i < \beta < \alpha_{i+1}), \\
\bar{z}_{\beta}|\alpha_1, \ldots, \alpha_k\rangle &= 0 \quad (\beta = \alpha_i).
\end{aligned} \tag{1.3.25}$$

These formulas, together with (1.3.18), define 2^{ν}-dimensional representation ρ of Clifford algebra $\mathcal{C}\ell_{2\nu}$. Explicit matrix form of generators $\rho(\Gamma_m)$ is discussed below in Section 6.1.2.

Proposition 1.3.2. *Complex 2^{ν}-dimensional representation ρ of Clifford algebra $\mathcal{C}\ell_{2\nu}$ defined by* (1.3.18) *and* (1.3.25), *is irreducible. All complex irreducible representations of Clifford algebra $\mathcal{C}\ell_{2\nu}$ are 2^{ν}-dimensional and equivalent to ρ.*

Proof. Any basis vector $|\alpha_1, \ldots, \alpha_k\rangle \in \mathcal{F}$ can be moved to any other basis vector $|\beta_1, \ldots, \beta_m\rangle \in \mathcal{F}$ by multiple action of operators $\{z_{\alpha_i}\}$ and $\{\bar{z}_{\beta_j}\}$. Therefore, there are no invariant subspaces in the space $\mathcal{F}$ of representation ρ, so the 2^{ν}-dimensional representation ρ is irreducible. Let $\mathcal{F}'$ be the space of another *irreducible* 2^{ν}-dimensional representation ρ' of algebra (1.3.19). Then there always exists a nonzero vector $|0\rangle'$ in space $\mathcal{F}'$ which obeys (1.3.21). Indeed, let $|\omega_0\rangle \in \mathcal{F}'$ be an arbitrarily chosen nonzero vector. Let

$z_1|\omega_0\rangle \neq 0$, then $z_1(z_1|\omega_0\rangle) = 0$, and we take vector $|\omega_1\rangle \equiv z_1|\omega_0\rangle$ instead of $|\omega_0\rangle$. Let $z_2|\omega_1\rangle \neq 0$, then $z_2(z_2|\omega_1\rangle) = 0$ and we switch from vector $|\omega_1\rangle$ to vector $|\omega_2\rangle \equiv z_2 z_1|\omega_0\rangle$. We continue this procedure until we get at step k the nonzero vector $|\omega_k\rangle \equiv |0\rangle' \in \mathcal{F}'$ which is annihilated by all z_i, i.e., which obeys (1.3.21). Now, by acting on the vector $|0\rangle' \in \mathcal{F}'$ by creation operators $\bar{z}_\beta$, we generate Fock space $\mathcal{F}''$ with basis vectors of the form (1.3.22):

$$|\alpha_1, \dots, \alpha_k\rangle' = \bar{z}_{\alpha_1} \cdots \bar{z}_{\alpha_k}|0\rangle' \quad (1 \leq \alpha_1 < \alpha_2 < \cdots < \alpha_k \leq \nu). \tag{1.3.26}$$

By construction, the space $\mathcal{F}''$ is a nontrivial invariant subspace in $\mathcal{F}'$. Since the representation ρ' is irreducible, the invariant subspace $\mathcal{F}''$ coincides with $\mathcal{F}'$. So, the space $\mathcal{F}'$ of irreducible representation ρ' contains all vectors (1.3.26); these vectors are nonzero and form basis in $\mathcal{F}'$. We conclude that mapping

$$|\alpha_1, \dots, \alpha_k\rangle' \;\rightarrow\; |\alpha_1, \dots, \alpha_k\rangle$$

is one-to-one, and this mapping establishes equivalence of representations ρ and ρ'. □

To define scalar product in the space $\mathcal{F}$, we introduce (see Section 1.1) dual Fock space $\mathcal{F}^*$ which consists of linear combinations of the following basis vectors:

$$\langle\beta_\ell, \dots, \beta_2, \beta_1| = \langle 0|z_{\beta_\ell} \cdots z_{\beta_1} \quad (\nu \geq \beta_\ell > \ldots > \beta_2 > \beta_1 \geq 1). \tag{1.3.27}$$

These are generated from dual vacuum vector $\langle 0| \in \mathcal{F}^*$ defined by the relations

$$\langle 0|\bar{z}_\beta = 0, \quad \forall \beta = 1, 2, \dots, \nu, \tag{1.3.28}$$

and normalized in such a way that $\langle 0|0\rangle = 1$. Using this normalization and Eqs. (1.3.25), one calculates the convolution of basis vectors in $\mathcal{F}$ and $\mathcal{F}^*$:

$$\langle\beta_\ell, \dots, \beta_2, \beta_1|\alpha_1, \alpha_2, \dots, \alpha_k\rangle = \delta_{k,\ell}(\delta_{\beta_1\alpha_1} \cdots \delta_{\beta_\ell\alpha_\ell}). \tag{1.3.29}$$

We can now define (see Section 1.1) Hermitian conjugation † of vectors in $\mathcal{F}$ and $\mathcal{F}^*$ as mappings $\mathcal{F} \rightarrow \mathcal{F}^*$ and $\mathcal{F}^* \rightarrow \mathcal{F}$:

$$\begin{gathered} |\alpha_1, \alpha_2, \dots, \alpha_k\rangle^\dagger = \langle\alpha_k, \dots, \alpha_2, \alpha_1|, \quad \langle\alpha_k, \dots, \alpha_2, \alpha_1|^\dagger = |\alpha_1, \alpha_2, \dots, \alpha_k\rangle, \\ \langle\Psi| = |\Psi\rangle^\dagger = \sum_{k=0}^{\nu} \sum_{\alpha_1 < \cdots < \alpha_k} \psi^*_{\alpha_1,\dots,\alpha_k}\langle\alpha_k, \dots, \alpha_2, \alpha_1|, \quad |\Psi\rangle = \langle\Psi|^\dagger. \end{gathered} \tag{1.3.30}$$

Then formula (1.3.29) defines Hermitian scalar product in $\mathcal{F}$:

$$\langle\Phi|\Psi\rangle = \sum_{k=0}^{\nu} \sum_{\alpha_1<\cdots<\alpha_k} \phi^*_{\alpha_1,\ldots,\alpha_k}\psi_{\alpha_1,\ldots,\alpha_k}. \tag{1.3.31}$$

Also, one identifies $\bar{z}_\alpha = z^\dagger_\alpha$, $\bar{z}^\dagger_\alpha = z_\alpha$.

Problem 1.3.1. Let vectors $|\alpha_1,\ldots,\alpha_k\rangle'$ and $\langle\beta_\ell,\ldots,\beta_1|'$ be defined by (1.3.22) and (1.3.27), but without ordering of parameters $(\alpha_1,\ldots,\alpha_k)$ and $(\beta_\ell,\ldots,\beta_1)$. Show that instead of (1.3.29) one has determinant formula

$$\begin{aligned}\langle\beta_\ell,\ldots,\beta_1|'|\alpha_1,\ldots,\alpha_k\rangle' &= \frac{\delta_{k,\ell}}{(\nu-k)!}\varepsilon_{\alpha_1\ldots\alpha_k\gamma_{k+1}\ldots\gamma_\nu}\cdot\varepsilon_{\beta_1\ldots\beta_k\gamma_{k+1}\ldots\gamma_\nu} \\ &= \delta_{k,\ell}\det(||\delta_{\alpha_i\beta_j}||_{i,j=1,\ldots,k}).\end{aligned} \tag{1.3.32}$$

Remark. In analogy to d-dimensional oscillator algebra (see Remark 2 in Section 1.2), Clifford algebra $\mathcal{C}\ell_N$ and free fermion algebra $\mathcal{A}_\nu$ (isomorphic to $\mathcal{C}\ell_N$ for even $N = 2\nu$) may be used for constructing realizations of Lie algebras of classical series $s\ell(\nu,\mathbb{C})$ and $so(N,\mathbb{C})$. As an example, the generators $E_{\alpha\beta} \in s\ell(\nu,\mathbb{C})$ can be represented as (cf. (1.2.30))

$$E_{\alpha\beta} = \bar{z}_\alpha z_\beta - \frac{\delta_{\alpha\beta}}{\nu}\sum_\gamma \bar{z}_\gamma z_\gamma, \quad \sum_\alpha E_{\alpha\alpha} = 0. \tag{1.3.33}$$

Likewise, we have for generators $L_{mn} \in so(N,\mathbb{C})$ the representation

$$L_{mn} \equiv \frac{1}{4}(\Gamma_m\Gamma_n - \Gamma_n\Gamma_m) = \frac{1}{2}\Gamma_{[mn]}, \tag{1.3.34}$$

where Γ_m are generators of Clifford algebra $\mathcal{C}\ell_N$.

Problem 1.3.2. Show that commutation relations for the operators (1.3.33) and (1.3.34) coincide with the defining relations (1.2.31) and (1.2.33) of Lie algebras $s\ell(\nu,\mathbb{C})$ and $so(N,\mathbb{C})$.

The realization (1.3.34) is used below in Section 5.4 for constructing spinor representations of algebras $so(N,\mathbb{C})$.

1.3.3. Grassmann algebra

Grassmann algebra is the fermionic analog of algebra of polynomials $\mathbb{C}[x_1,\ldots,x_d]$, where x_i are coordinates of vectors $\vec{x}$ in space $\mathbb{R}^d$, which was used for defining the coordinate representation (1.1.17) and, in particular, for constructing coordinate representation of bosonic oscillators (1.2.1).

Definition 1.3.3. Complex algebra Gr_d with unit element I and generators θ_α ($\alpha = 1, 2, \ldots, d$) obeying defining relations

$$\theta_\alpha \theta_\beta + \theta_\beta \theta_\alpha = 0, \quad \forall \alpha, \beta, \tag{1.3.35}$$

is called *Grassmann algebra* with d generators.

Example. Subalgebra of fermion oscillator algebra (1.3.19) whose generators are $\{z_\alpha\}$ where $\alpha = 1, \ldots, \nu$, is isomorphic to Grassmann algebra Gr_ν.

Basis in Grassmann algebra Gr_d is made of monomials (cf. (1.3.5))

$$\{I, \theta_{\alpha_1}, \theta_{\alpha_1}\theta_{\alpha_2}, \theta_{\alpha_1}\theta_{\alpha_2}\theta_{\alpha_3}, \ldots, \theta_{\alpha_1}\theta_{\alpha_2}\cdots\theta_{\alpha_d} \equiv \theta_1\theta_2\cdots\theta_d\},$$
$$1 \le \alpha_1 < \alpha_2 < \alpha_3 < \cdots \le d. \tag{1.3.36}$$

Thus, algebra Gr_d is finite dimensional, and its dimension is equal to 2^d (the calculation is similar to (1.3.24)). One introduces grading gr in algebra Gr_d: for basis monomials (1.3.36) one defines $\mathrm{gr}(\theta_{\beta_1}\cdots\theta_{\beta_k}) = k$. Then algebra Gr_d is expanded in the sum $\mathsf{Gr}_d = \oplus_{k=1}^d \mathsf{Gr}_d^{(k)}$, where $\mathsf{Gr}_d^{(k)}$ are vector spaces whose elements $a \in \mathsf{Gr}_d$ have $\mathrm{gr}(a) = k$, and the following relation holds: $\mathsf{Gr}_d^{(k)} \cdot \mathsf{Gr}_d^{(\ell)} = \mathsf{Gr}_d^{(k+\ell)}$.

Let us introduce partial derivative $\frac{\partial}{\partial\theta_\alpha}$ in Grassmann algebra Gr_d, which is a linear operator $\mathsf{Gr}_d \to \mathsf{Gr}_d$ defined on basis monomials (1.3.36) as follows:

$$\frac{\partial}{\partial\theta_\alpha}(\theta_{\beta_1}\cdots\theta_{\beta_k}) = \sum_{i=1}^k (-1)^{i+1}\delta^\alpha_{\beta_i}(\theta_{\beta_1}\cdots\theta_{\beta_{i-1}}\cdot\theta_{\beta_{i+1}}\cdots\theta_{\beta_k}), \quad \frac{\partial}{\partial\theta_\alpha}(I) = 0 \tag{1.3.37}$$

(the derivative $\partial/\partial\theta_\alpha$ anticommutes with θ_β for $\beta \neq \alpha$, i.e., $(\partial/\partial\theta_\alpha)\,\theta_\beta = -\theta_\beta(\partial/\partial\theta_\alpha)$, while acting on θ_α this derivative gives $(\partial/\partial\theta_\alpha)\theta_\alpha = 1$). This definition implies the graded Leibniz rule

$$\partial^\alpha_\theta(a\cdot b) = \partial^\alpha_\theta(a)\cdot b + (-1)^{\mathrm{gr}(a)}a\cdot\partial^\alpha_\theta(b), \tag{1.3.38}$$

so we assign grading to the derivative, $\mathrm{gr}(\partial^\alpha_\theta) = -1$. By comparing two consecutive actions (1.3.37) of derivatives ∂^α_θ and ∂^β_θ on basis monomials, we get

$$\partial^\alpha_\theta\partial^\beta_\theta(\theta_{\beta_1}\cdots\theta_{\beta_k}) = -\partial^\beta_\theta\partial^\alpha_\theta(\theta_{\beta_1}\cdots\theta_{\beta_k}) \;\Rightarrow\; \partial^\alpha_\theta\partial^\beta_\theta + \partial^\beta_\theta\partial^\alpha_\theta = 0. \tag{1.3.39}$$

Therefore, partial derivatives ∂_θ^α ($\alpha = 1, \ldots, d$) anticommute with each other and generate algebra isomorphic to Grassmann algebra Gr_d. Leibniz rule (1.3.38) can be also interpreted as the commutation rule of derivative ∂_θ^α with an element $a \in \mathsf{Gr}_d$:

$$\partial_\theta^\alpha \cdot a = \partial_\theta^\alpha(a) + (-1)^{\mathrm{gr}(a)} a \cdot \partial_\theta^\alpha. \tag{1.3.40}$$

In particular, one sets $a = \theta_\beta$ and recalls (1.3.37) to obtain

$$\partial_\theta^\alpha \cdot \theta_\beta = \delta_\beta^\alpha - \theta_\beta \cdot \partial_\theta^\alpha. \tag{1.3.41}$$

We compare the relations (1.3.35), (1.3.39) and (1.3.41) with defining relations (1.3.19) and arrive at the following.

Proposition 1.3.3. *Graded algebra $\mathcal{B}_d$ with generators $\{\theta_\alpha, \partial_\theta^\beta\}$, where $\alpha, \beta = 1, \ldots, d$, is isomorphic to free fermion algebra $\mathcal{A}_d$. The isomorphism is given by any of the two mappings $z_\alpha \leftrightarrow \theta_\alpha$, $\bar{z}_\beta \leftrightarrow \partial_\theta^\beta$, or $z_\alpha \leftrightarrow \partial_\theta^\alpha$, $\bar{z}_\beta \leftrightarrow \theta_\beta$.*

Remark. The operation $\frac{\partial}{\partial\theta_\alpha}$ defined by (1.3.37) is called left derivative. One can also define right partial derivative $\frac{\overleftarrow{\partial}}{\partial\theta_\alpha}$ by (cf. (1.3.37))

$$(\theta_{\beta_1} \cdots \theta_{\beta_k}) \frac{\overleftarrow{\partial}}{\partial\theta_\alpha} = \sum_{i=1}^{k} (-1)^{k-i} \delta_{\beta_i}^\alpha (\theta_{\beta_1} \cdots \theta_{\beta_{i-1}} \cdot \theta_{\beta_{i+1}} \cdots \theta_{\beta_k}).$$

The consideration of right derivative is completely analogous to that of left derivative. In particular, there is an analog of graded Leibniz rule for right derivative and analog of Proposition 1.3.3.

Proposition 1.3.3 implies that free fermion algebra $\mathcal{A}_d$ admits grading, just like algebra $\mathcal{B}_d$ with generators $\{\theta_\alpha, \partial_\theta^\beta\}$. To this end, we assign the following grading to generators of algebra $\mathcal{A}_d$:

$$\mathrm{gr}(z_\alpha) = +1, \quad \mathrm{gr}(\bar{z}_\alpha) = -1.$$

Considering algebras $\mathcal{A}_d$ and $\mathcal{B}_d$ together, we require that elements $a \in \mathcal{A}_d$ and $b \in \mathcal{B}_d$ commute in accordance to the following rule:

$$a \cdot b = (-1)^{\mathrm{gr}(a) \cdot \mathrm{gr}(b)} b \cdot a \tag{1.3.42}$$

(i.e., even elements commute with all elements, while odd elements a and b anticommute). We are now ready to construct analogs of (anti)holomorphic representations for fermion oscillator algebra. We introduce the set of eigenvectors for fermionic annihilation operators z_α (cf. (1.2.9)):

$$z_\alpha|\theta\rangle = \theta_\alpha|\theta\rangle, \tag{1.3.43}$$

where $|\theta\rangle := |\theta_1, \dots, \theta_d\rangle$, and θ_α are elements of Grassmann algebra $\mathsf{Gr}_d \subset \mathcal{B}_d$, which we call Grassmann variables. Vectors $|\theta\rangle$ are coherent states of fermion oscillators. They are constructed in the following way (cf. (1.2.10)):

$$|\theta\rangle = \exp(\bar{z}_\alpha\theta_\alpha)|0\rangle = \exp(-\theta_\alpha\bar{z}_\alpha)|0\rangle, \tag{1.3.44}$$

where $|0\rangle$ is the fermionic Fock space vacuum defined by (1.3.21). The fact that the vector (1.3.44) obeys the relation (1.3.43) follows from (1.3.19) and (1.3.42). Making use of the representation (1.3.44) and formulas (1.3.19), (1.3.42), we obtain

$$\bar{z}_\alpha|\theta\rangle = -\frac{\partial}{\partial\theta_\alpha}|\theta\rangle. \tag{1.3.45}$$

Dual coherent states

$$\langle\theta| = \langle 0|\exp(\theta_\alpha z_\alpha) = \langle 0|\exp(-z_\alpha\theta_\alpha), \tag{1.3.46}$$

with the dual vacuum $\langle 0|$ defined by (1.3.28), obey similar relations:

$$\langle\theta|\bar{z}_\alpha = \theta_\alpha\langle\theta|, \quad \langle\theta|z_\alpha = \frac{\partial}{\partial\theta_\alpha}\langle\theta|. \tag{1.3.47}$$

Problem 1.3.3. Check consistency of defining relations for free fermion algebra (1.3.19) and Eqs. (1.3.43), (1.3.45), (1.3.47).

Convolution of two coherent states (1.3.44) and (1.3.46) equals

$$\langle\theta|\bar{\theta}\rangle = \langle 0|\exp(\theta_\alpha z_\alpha)|\bar{\theta}\rangle = \langle 0|\exp(\theta_\alpha\bar{\theta}_\alpha)|\bar{\theta}\rangle = \exp(\theta_\alpha\bar{\theta}_\alpha), \tag{1.3.48}$$

where $\bar{\theta}_\alpha$ is another set of Grassmann variables, such that

$$\bar{\theta}_\alpha \cdot \theta_\beta = -\theta_\beta \cdot \bar{\theta}_\alpha, \quad \bar{\theta}_\alpha \cdot a = (-1)^{\mathrm{gr}(a)} a \cdot \bar{\theta}_\alpha, \ \forall a \in \mathcal{A}_d, \ \forall \alpha, \beta.$$

In other words, variables $\{\theta_1, \dots, \theta_d, \bar{\theta}_1, \dots, \bar{\theta}_d\}$ make Grassmann algebra Gr_{2d}. When deriving (1.3.48), we use Eq. (1.3.43) and the relation $\langle 0|\bar{\theta}\rangle = 1$.

We make use of fermionic coherent states $|\bar{\theta}\rangle$ and $\langle\theta|$ to construct representations of free fermion algebra analogous to (anti)holomorphic representation in bosonic case. To this end, we define wave functions (functions of Grassmann variables θ_α and $\bar{\theta}_\beta$) for a vector (1.3.23) from the Fock space $\mathcal{F}$ and dual vector $\langle\Psi| \in \mathcal{F}^*$:

$$\Psi(\theta) \equiv \langle\theta|\Psi\rangle = \langle\theta| \sum_{k=0}^{d} \sum_{\alpha_1<\cdots<\alpha_k} \psi_{\alpha_1,\ldots,\alpha_k}|\alpha_1,\ldots,\alpha_k\rangle$$

$$= \sum_{k=0}^{d} \sum_{\alpha_1<\cdots<\alpha_k} \psi_{\alpha_1,\ldots,\alpha_k}\theta_{\alpha_1}\cdots\theta_{\alpha_k}, \tag{1.3.49}$$

$$\bar{\Psi}(\bar{\theta}) \equiv \langle\Psi|\bar{\theta}\rangle = \sum_{k=0}^{d} \sum_{\alpha_1<\cdots<\alpha_k} \psi^*_{\alpha_1,\ldots,\alpha_k}\bar{\theta}_{\alpha_k}\cdots\bar{\theta}_{\alpha_1}, \tag{1.3.50}$$

where we make use of (1.3.22), (1.3.27), (1.3.30), with $\nu = d$, and also (1.3.44) and (1.3.47). Equation (1.3.47) shows that in terms of the wave functions (1.3.49), the operators z_α and $\bar{z}_\beta$ act on state $|\Psi\rangle$ as follows (cf. (1.2.17)):

$$\langle\theta|z_\alpha|\Psi\rangle = \frac{\partial}{\partial\theta_\alpha}\langle\theta|\Psi\rangle, \quad \langle\theta|\bar{z}_\alpha|\Psi\rangle = \theta_\alpha\langle\theta|\Psi\rangle. \tag{1.3.51}$$

These formulas define the representation $\bar{\rho}$ of fermion oscillators z_α and $\bar{z}_\alpha$ (cf. 1.2.19)):

$$\bar{\rho}(z_\alpha) = \frac{\partial}{\partial\theta_\alpha}, \quad \bar{\rho}(\bar{z}_\alpha) = \theta_\alpha, \tag{1.3.52}$$

which acts in the space of functions $\Psi(\theta) = \langle\theta|\Psi\rangle$ of Grassmann variables θ_α.

Problem 1.3.4. Show that the dimension of space of functions $\Psi(\theta)$ coincides with the dimension $\dim(\mathcal{F}) = 2^d$ of the Fock space of algebra $\mathcal{A}_d$ (see (1.3.24)).

We have shown (see Proposition 1.3.3) that mapping $\bar{\rho}$ is isomorphism of algebras $\mathcal{A}_d$ and $\mathcal{B}_d$; this means that the representation $\bar{\rho}$ is exact.

Problem 1.3.5. Making use of (1.3.18) and (1.3.52), construct representation $\bar{\rho}$ of the generators of Clifford algebra $\mathcal{C}\ell_N$.

Finally, the scalar product $\langle\Phi|\Psi\rangle$ is written in representation $\bar{\rho}$ as follows (cf. (1.2.18)):

$$\langle\Phi|\Psi\rangle = \int\left(\prod_\alpha d\theta_\alpha d\bar{\theta}_\alpha\right) e^{-\theta_\alpha\bar{\theta}_\alpha}\langle\Phi|\bar{\theta}\rangle\langle\theta|\Psi\rangle$$
$$= \int\left(\prod_\alpha d\theta_\alpha d\bar{\theta}_\alpha\right) e^{-\theta_\alpha\bar{\theta}_\alpha}\bar{\Phi}(\bar{\theta})\Psi(\theta). \tag{1.3.53}$$

Here $\int\left(\prod_\alpha d\theta_\alpha d\bar{\theta}_\alpha\right)$ is Berezin integral over Grassmann variables $\{\theta_1,\ldots,\theta_d,\bar{\theta}_1,\ldots,\bar{\theta}_d\}\in \mathsf{Gr}_{2d}$, whose definition is

$$\int\left(\prod_\alpha d\theta_\alpha d\bar{\theta}_\alpha\right)\Psi(\theta,\bar{\theta}) = \frac{\partial}{\partial\theta_d}\cdot\frac{\partial}{\partial\bar{\theta}_d}\cdots\frac{\partial}{\partial\theta_1}\cdot\frac{\partial}{\partial\bar{\theta}_1}\Psi(\theta,\bar{\theta}). \tag{1.3.54}$$

This expression can be viewed as multiple Berezin integral

$$\int\left(\prod_\alpha d\theta_\alpha d\bar{\theta}_\alpha\right)\Psi(\theta,\bar{\theta}) = \int d\theta_d\int d\bar{\theta}_d\cdots\int d\theta_1\int d\bar{\theta}_1\Psi(\theta,\bar{\theta}), \tag{1.3.55}$$

where each integration over θ_α or $\bar{\theta}_\beta$ is understood as partial derivative $\frac{\partial}{\partial\theta_\alpha}$ or $\frac{\partial}{\partial\bar{\theta}_\beta}$. In other words, Berezin integral over single variable is defined by

$$\int d\theta = 0, \quad \int d\theta\,\theta = 1 \tag{1.3.56}$$

(and, likewise, $\int d\bar{\theta} = 0$, $\int d\bar{\theta}\ \bar{\theta} = 1$), while the symbols $d\theta_\alpha$, $d\bar{\theta}_\beta$ anticommute with each other and with Grassmann variables θ_α, $\bar{\theta}_\beta$.

Problem 1.3.6. * Show that the integral over Grassmann variables on the right-hand side of (1.3.53) gives (1.3.31).

Berezin integral is a convenient tool which is used, e.g., for defining Feynman path integral over fermionic variables. The key point is that determinant of $d\times d$ matrix $||A_{\alpha\beta}||$ can be written in terms of the integral over Grassmann variables:

$$\det(A) = \int\left(\prod_\alpha d\theta_\alpha d\bar{\theta}_\alpha\right)\exp(\bar{\theta}_\alpha A_{\alpha\beta}\theta_\beta)$$
$$= \frac{1}{d!}\varepsilon_{\alpha_1\ldots\alpha_d}A_{\alpha_1\beta_1}\cdots A_{\alpha_d\beta_d}\varepsilon_{\beta_1\ldots\beta_d}. \tag{1.3.57}$$

The right-hand side of (1.3.57) is the formula for determinant that we discussed in the first book (see Section **I-1.2.2**). Let $d = 2n$ be even, and

$d \times d$ matrix A be antisymmetric, $A_{\alpha\beta} = -A_{\beta\alpha}$. Then

$$\int \left(\prod_{\alpha} d\theta_{\alpha}\right) \exp\left(\frac{1}{2}\theta_{\alpha}A_{\alpha\beta}\theta_{\beta}\right) = \frac{1}{2^n n!}\varepsilon_{\alpha_1\ldots\alpha_{2n}} A_{\alpha_1\alpha_2}\cdots A_{\alpha_{2n-1}\alpha_{2n}}$$
$$\equiv \mathrm{Pf}(A), \qquad (1.3.58)$$

where $\mathrm{Pf}(A)$ is Pfaffian of matrix A (see Section **I-1.2.2** of the first book). Here, we use the relation $\theta_{\alpha_1}\cdots\theta_{\alpha_d} = \varepsilon_{\alpha_1\ldots\alpha_d}\theta_1\cdots\theta_d$.

Problem 1.3.7. * Making use of integral representations (1.3.57) and (1.3.58), prove the following identity for antisymmetric $(2n \times 2n)$ matrix A: $\det(A) = (\mathrm{Pf}(A))^2$.

To end up this section we point out that Grassmann algebras are central in supersymmetry theory, and in particular in the theory of Lie supergroups and Lie superalgebras. This important topic is beyond the scope of this book.

Chapter 2

Finite-Dimensional Representations of Lie Algebras $su(2)$ and $s\ell(2,\mathbb{C})$ and Lie Groups $SU(2)$ and $SL(2,\mathbb{C})$

In this chapter, we study complex finite-dimensional representations of Lie algebras $s\ell(2,\mathbb{C})$, $su(2)$ and Lie group $SU(2)$, and also a restricted set of representations of Lie group $SL(2,\mathbb{C})$. Unless the opposite is stated, the representations that we analyze in this and following chapters are *finite dimensional.* The representations that we consider are in one-to-one correspondence with each other: any representation of algebra $s\ell(2,\mathbb{C})$ is at the same time a representation of algebra $su(2)$ and is lifted to representation of group $SU(2)$. These representations of $SU(2)$ are conveniently studied by relating them to certain representations of Lie group $SL(2,\mathbb{C})$, namely, to those representations which are contained in tensor products of its defining representation T. Clearly, this class of representations does not exhaust all finite-dimensional representations of $SL(2,\mathbb{C})$. As an example, it does not include sub-representations of the tensor products $T^{\otimes r_1} \otimes (T^*)^{\otimes r_2}$, where T^* is complex conjugate to defining representation.

2.1. Finite-Dimensional Representations of Lie Algebras $su(2)$ and $s\ell(2,\mathbb{C})$

Before reading this section, the reader is recommended to study Section I-4.7.3 of the first book. We have seen there that the complex Lie algebra $s\ell(2,\mathbb{C})$ (and its real form $su(2)$) has infinite number of inequivalent finite-dimensional irreducible representations $T^{(j)}$, which are labeled by nonnegative integer and half-integer numbers $j = 0, \frac{1}{2}, 1, \frac{3}{2}, 2, \dots$ (spins) and have dimensions $(2j+1)$. We have constructed these representations explicitly; these are representations with highest weights. We have also seen that there are no other finite-dimensional irreducible representations of Lie algebras $su(2)$ and $s\ell(2,\mathbb{C})$.

Recall the main steps in constructing the highest weight representations of the Lie algebra $s\ell(2,\mathbb{C})$. The algebra $s\ell(2,\mathbb{C})$ is a three-dimensional complex vector space whose basis elements can be chosen as vectors $\{e_+, e_-, \mathsf{h}\}$ with defining commutation relations

$$[e_+, e_-] = 2\mathsf{h}, \quad [\mathsf{h}, e_+] = e_+, \quad [\mathsf{h}, e_-] = -e_-. \tag{2.1.1}$$

These relations, in the enveloping algebra $\mathcal{U}(s\ell(2,\mathbb{C}))$, can be written as

$$e_+\, e_- = 2\,\mathsf{h} + e_-\, e_+, \quad \mathsf{h}\, e_\pm = e_\pm(\mathsf{h} \pm 1). \tag{2.1.2}$$

The irreducible representation $T^{(j)}$ of the algebra $s\ell(2,\mathbb{C})$ with highest weight j acts in the space $\mathcal{V}$ whose basis is generated starting from the highest weight vector $v_0 \in \mathcal{V}$:

$$\mathsf{h}\, v_0 = j\, v_0, \quad e_+\, v_0 = 0. \tag{2.1.3}$$

The representation $T^{(j)}$ is irreducible and finite dimensional if and only if the highest weight equals $j = n/2$ where $n \in \mathbb{Z}_{\geq 0}$ is integer and nonnegative. The highest weight j is called spin, and the space $\mathcal{V}$ of the representation $T^{(j)}$ has dimension $(2j+1)$. All basis vectors v_k in $\mathcal{V}$ are obtained by the multiple action of the lowering operator e_- on the highest weight vector $v_0 \in \mathcal{V}$:

$$v_k = \frac{1}{k!} e_-^k v_0, \quad k = 0, 1, \ldots, 2j. \tag{2.1.4}$$

The generators of $s\ell(2,\mathbb{C})$ in the representation $T^{(j)}$ act on the basis vectors $v_k \in \mathcal{V}$ as follows:

$$\mathsf{h}\, v_k = (j-k) v_k, \quad e_-\, v_k = (k+1) v_{k+1}, \quad e_+\, v_{k+1} = (2j-k) v_k \quad (k \geq 0). \tag{2.1.5}$$

The latter equality yields $e_+\, v_{2j+1} = 0$, which, for the irreducible representation, is equivalent to $v_{2j+1} = 0$. Formulas (2.1.5) follow from the relations (2.1.2) and definition (2.1.3) of the highest weight vector v_0. In particular, the last equality in (2.1.5) follows from the identity

$$e_+\, e_-^k = e_-^k\, e_+ + k\, e_-^{k-1}(2\,\mathsf{h} - k + 1), \quad e_\pm, \mathsf{h} \in \mathcal{U}(s\ell(2,\mathbb{C})), \tag{2.1.6}$$

which we use in what follows.

Let us recall the definition of the matrix $||T_{ij}||$ of a linear operator T acting in space $\mathcal{V}$. With the basis $\vec{e}_i \in \mathcal{V}$, this definition is

$$T \cdot \vec{e}_i = \vec{e}_k \, T_{ki}, \tag{2.1.7}$$

and the coordinates of a vector $\vec{x} = x_i \vec{e}_i$ transform under the linear transformation $\vec{x} \to \vec{x}\,' = T \cdot \vec{x}$ as

$$x_i' = (T \cdot \vec{x})_i = T_{ik} \, x_k. \tag{2.1.8}$$

In our case, we have

$$\mathsf{h} \cdot v_k = v_\ell \, \widetilde{T}_{\ell k}(\mathsf{h}), \quad e_\pm \cdot v_k = v_\ell \, \widetilde{T}_{\ell k}(e_\pm), \tag{2.1.9}$$

where $\ell, k = 0, 1, \ldots, 2j$ and, according to (2.1.5), the matrices of the generators $\{e_+, e_-, \mathsf{h}\}$ read (we mark them with tilde and reserve the notation without tilde for representations of $e_\pm, \mathsf{h}$ in different basis of $\mathcal{V}$, see below)

$$\begin{aligned} \widetilde{T}_{\ell k}(\mathsf{h}) &= (j-k)\delta_{\ell k}, \quad \widetilde{T}_{\ell k}(e_+) = (2j-k+1)\delta_{\ell, k-1}, \\ \widetilde{T}_{\ell k}(e_-) &= (k+1)\delta_{\ell, k+1}, \end{aligned} \tag{2.1.10}$$

where we make use of the convention $\delta_{\ell,-1} = 0 = \delta_{\ell,n+1}$. In the simplest nontrivial case $j = 1/2$ and two-dimensional space $\mathcal{V}$, we reproduce the defining matrix representation $T^{(1/2)}$:

$$e_+ = \begin{pmatrix} 0 & 1 \\ 0 & 0 \end{pmatrix}, \quad e_- = \begin{pmatrix} 0 & 0 \\ 1 & 0 \end{pmatrix}, \quad \mathsf{h} = \frac{1}{2}\begin{pmatrix} 1 & 0 \\ 0 & -1 \end{pmatrix}. \tag{2.1.11}$$

Problem 2.1.1. Show that any element $A \in SL(2, \mathbb{C})$ can be written as the product of exponentials

$$A = \exp(a_- e_-) \cdot \exp(a_0 \mathsf{h}) \cdot \exp(a_+ e_+), \tag{2.1.12}$$

where $a_\pm, a_0 \in \mathbb{C}$. Hint: write this element explicitly, using (2.1.11),

$$A = \begin{pmatrix} e^{\frac{a_0}{2}} & a_+ \, e^{\frac{a_0}{2}} \\ a_- \, e^{\frac{a_0}{2}} & e^{-\frac{a_0}{2}} + a_- \, a_+ \, e^{\frac{a_0}{2}} \end{pmatrix}. \tag{2.1.13}$$

Making use of the representation (2.1.13) show that the group $SL(2, \mathbb{C})$ is simply connected.

Problem 2.1.2. Construct finite-dimensional irreducible representation of the Lie algebra $s\ell(2,\mathbb{C})$ with the lowest weight μ, i.e., the representation generated from the lowest weight vector v:

$$\mathsf{h}\, v = \mu\, v, \quad e_-\, v = 0. \tag{2.1.14}$$

Show that the lowest weight μ must be nonpositive integer or half-integer, $\mu = -n/2$, where $n \in \mathbb{Z}_{\geq 0}$. Show that $(2j+1)$-dimensional representation of $s\ell(2,\mathbb{C})$ with the highest weight $j = n/2$ is equivalent to the representation with lowest weight $\mu = -n/2 = -j$.

Finite-dimensional irreducible representations $T^{(j)}$ of the Lie algebra $s\ell(2,\mathbb{C})$, which we have just discussed, at the same time determine finite-dimensional irreducible representations of the Lie algebra $su(2)$ (see Section I-4.7.3 in the first book). The latter representations of $su(2)$ are also labeled by the highest weight (spin) $j \in \mathbb{Z}_{\geq 0}/2$. They are obtained by making use of the relationship between the basis $\{e_\pm, \mathsf{h}\}$ in $s\ell(2,\mathbb{C})$ and the basis $\{\tau_\alpha\}$ of $su(2)$:

$$\begin{gathered} e_\pm = i\,\tau_1 \mp \tau_2, \quad \mathsf{h} = i\,\tau_3, \\ \tau_1 = -\frac{i}{2}(e_- + e_+), \quad \tau_2 = \frac{1}{2}(e_- - e_+), \quad \tau_3 = -i\,\mathsf{h}. \end{gathered} \tag{2.1.15}$$

As an example, one inserts the defining representation (2.1.11) of the generators of $s\ell(2,\mathbb{C})$ into (2.1.15) and obtains that the generators of $su(2)$ in the same representation $j = 1/2$ are given by $\tau_\alpha = -\frac{i}{2}\sigma_\alpha$, where σ_α are Pauli matrices

$$\sigma_1 = \begin{pmatrix} 0 & 1 \\ 1 & 0 \end{pmatrix}, \quad \sigma_2 = \begin{pmatrix} 0 & -i \\ i & 0 \end{pmatrix}, \quad \sigma_3 = \begin{pmatrix} 1 & 0 \\ 0 & -1 \end{pmatrix}. \tag{2.1.16}$$

Then the defining relations (2.1.1) of $s\ell(2,\mathbb{C})$ give the defining relations of $su(2)$:

$$[\tau_i,\, \tau_j] = \varepsilon_{ijk}\, \tau_k. \tag{2.1.17}$$

We know that every representation of a compact Lie algebra, such as the representation $\widetilde{T}$ of $su(2)$ corresponding to (2.1.10), is equivalent to anti-Hermitian representation T, for which $T_{\ell k}(\tau_\alpha) = -T^*_{k\ell}(\tau_\alpha)$, for all α. Then Eq. (2.1.15) implies that the matrices $T(e_\pm)$ of the generators $e_\pm$ of

$s\ell(2,\mathbb{C})$ obey

$$T_{\ell k}(e_-) = (T_{k\ell}(e_+))^*. \tag{2.1.18}$$

However, the matrices $\widetilde{T}(e_\pm)$ given by (2.1.10) do not have this property for $j > 1/2$; the problem has to do with the normalization of the basis vectors. Let us construct the analogs of matrices (2.1.10) which satisfy (2.1.18). To this end, we change the basis in the space $\mathcal{V}$ of the representation $T^{(j)}$, namely, instead of v_k $(k = 0,\ldots,2j)$, we define new basis vectors

$$|j,m\rangle := \sqrt{(j+m)!(j-m)!}\, v_{j-m} = \sqrt{\frac{(j+m)!}{(j-m)!}}\; e_-^{j-m}\cdot v_0, \quad m = -j,-j+1,\ldots,j-1,j. \tag{2.1.19}$$

Hereafter we use Dirac notations (see Chapter 1). In particular, the highest weight vector is now denoted by $|j,j\rangle := \sqrt{(2j)!}\cdot v_0$.

In accordance with (2.1.5), we have for the new basis vectors

$$\begin{gathered} \mathsf{h}\cdot|j,m\rangle = m\,|j,m\rangle, \quad e_\pm\cdot|j,m\rangle = \sqrt{(j\pm m+1)(j\mp m)}|j,m\pm 1\rangle, \\ J^2\cdot|j,m\rangle = j(j+1)\,|j,m\rangle, \\ m = -j,-j+1,\ldots,j-1,j, \end{gathered} \tag{2.1.20}$$

where we introduce the quadratic Casimir operator J^2:

$$J^2 = \frac{1}{2}(e_-e_+ + e_+e_-) + \mathsf{h}^2 = e_-e_+ + \mathsf{h}(\mathsf{h}-1) \in \mathcal{U}(s\ell(2,\mathbb{C})). \tag{2.1.21}$$

Making use of the relations (2.1.20) and the rule (2.1.7) we obtain the matrix form for the generators $e_\pm$ and h in the representation $T^{(j)}$ (cf. (2.1.9) and (2.1.10)):

$$\mathsf{h}\cdot|j,m\rangle = |j,m'\rangle\, T_{m'm}(\mathsf{h}), \quad e_\pm\cdot|j,m\rangle = |j,m'\rangle\, T_{m'm}(e_\pm), \tag{2.1.22}$$

where

$$\begin{aligned} T_{m'm}(\mathsf{h}) &= m\,\delta_{m'm}, \\ T_{m'm}(e_+) &= \sqrt{(j+m+1)(j-m)}\,\delta_{m',m+1}, \\ T_{m'm}(e_-) &= \sqrt{(j+m)(j-m+1)}\,\delta_{m',m-1}. \end{aligned} \tag{2.1.23}$$

We now have $T_{m'm}(e_+) = T_{mm'}(e_-)$, i.e., these matrices, being real, obey (2.1.18), so the representation (2.1.23) of $s\ell(2,\mathbb{C})$ yields an anti-Hermitian representation of $su(2)$ defined by (2.1.15). The representation (2.1.22),

(2.1.23) is equivalent to the representation defined by (2.1.9) and (2.1.10), since it is obtained by changing the basis in $\mathcal{V}$.

We now define scalar product of two vectors $|\alpha\rangle = \sum_m \alpha_m|j,m\rangle$ and $|\beta\rangle = \sum_m \beta_m|j,m\rangle$ $(\alpha_m, \beta_m \in \mathbb{C})$ in the complex space $\mathcal{V}$ of representation $T^{(j)}$:

$$\langle \beta \,|\, \alpha \rangle := \sum_m \beta_m^* \alpha_m, \quad \langle \alpha \,|\, \beta \rangle = \langle \beta \,|\, \alpha \rangle^*. \tag{2.1.24}$$

Basis vectors $|j,m\rangle$ are orthonormalized in this scalar product,

$$\langle j,m' \,|j,m\rangle = \delta_{mm'}, \tag{2.1.25}$$

while the matrix elements (2.1.23) are given by (see (2.1.22))

$$\langle j,m' \,|\, \mathsf{h}|j,m\rangle = T_{m'm}(\mathsf{h}), \quad \langle j,m' \,|\, e_\pm|j,m\rangle = T_{m'm}(e_\pm). \tag{2.1.26}$$

Let us now make use of the general construction given in Chapter 1, and introduce the space $\mathcal{V}^*$ dual to $\mathcal{V}$. We denote the basis vectors in $\mathcal{V}^*$ by $\langle j,m\,|$ (here $m = -j,\ldots,j-1,j$) and choose them in such a way that Eq. (2.1.25) is satisfied. Thus, $\mathcal{V}^*$ is a complex $(2j+1)$-dimensional vector space with vectors $\langle\alpha| = \sum_m \langle j,m\,|\,\alpha_m^*$. We define the Hermitian conjugation operation $\dagger$:

$$|j,m\rangle^\dagger = \langle j,m\,|, \quad |\alpha\rangle^\dagger = \langle\alpha| = \sum_m \langle j,m\,|\,\alpha_m^*, \quad (|\alpha\rangle^\dagger)^\dagger = |\alpha\rangle. \tag{2.1.27}$$

According to (2.1.24), (2.1.25) and (2.1.27), the scalar product $\langle\beta\,|\,\alpha\rangle$ can be considered as convolution of vector $|\,\alpha\rangle \in \mathcal{V}$ and vector $\langle\beta\,| \in \mathcal{V}^*$. The action of an element $a \in s\ell(2,\mathbb{C})$ in the dual space $\mathcal{V}^*$ is defined as $(\langle\beta\,|\cdot a)\,|\,\alpha\rangle := \langle\beta\,|\,(a\cdot|\,\alpha\rangle)$. Then, making use of (2.1.22), (2.1.25) and (2.1.26) we have for the basis vectors $\langle j,m| \in \mathcal{V}^*$

$$\langle j,m|\cdot\mathsf{h} = \langle j,m|\,m, \quad \langle j,m|\cdot e_\pm = \langle j,m\mp 1|\,\sqrt{(j\pm m)(j\mp m+1)}. \tag{2.1.28}$$

Hermitian conjugate element $a^\dagger$ of an element $a \in s\ell(2,\mathbb{C})$ is defined according to the rule $(a\cdot|\,\alpha\rangle)^\dagger = \langle\alpha\,|\cdot a^\dagger$, so, by comparing (2.1.20) and (2.1.28), we have $e_\pm^\dagger = e_\mp$ and $\mathsf{h}^\dagger = \mathsf{h}$. Now, the last relation in (2.1.28) gives

$$\langle j,m| = \sqrt{\frac{(j+m)!}{(j-m)!(2j)!}}\;\; \langle j,j|\cdot e_+^{j-m}. \tag{2.1.29}$$

Here $\langle j,j| = |j,j\rangle^\dagger$ is the dual highest weight vector, for which we have from (2.1.28) (cf. (2.1.3))

$$\langle j,j|\cdot e_- = 0, \quad \langle j,j|\cdot \mathsf{h} = j\,\langle j,j|. \tag{2.1.30}$$

Since the lowest weight vector in $\mathcal{V}$ is defined by the relations (see (2.1.14))

$$e_-\cdot|j,-j\rangle = 0, \quad \mathsf{h}\cdot|j,-j\rangle = -j\,|j,-j\rangle, \tag{2.1.31}$$

the dual lowest weight vector obeys

$$\langle j,-j|\cdot e_+ = 0, \quad \langle j,-j|\cdot \mathsf{h} = -j\langle j,-j|. \tag{2.1.32}$$

Note that orthonormality of vectors (2.1.25) is consistent with (2.1.22), (2.1.28) and (2.1.29). The latter property can be seen directly:

$$\begin{aligned}\langle j,m'|j,m\rangle &= \sqrt{\frac{(j+m')!}{(j-m')!(2j)!}}\;\langle j,j|\cdot e_+^{j-m'}\cdot|j,m\rangle \\ &= \delta_{m',m}\langle j,j|j,j\rangle = \delta_{m',m}.\end{aligned} \tag{2.1.33}$$

Basis vectors (2.1.19) in the space of representation $T^{(j)}$ are labeled by two numbers: spin j (which is a characteristic of the representation) and weight m, the eigenvalue of the operator $\mathsf{h} = i\,\tau_3$ ("spin projection on the third axis"), where, according to (2.1.20), the weight can take values $m = j, j-1, \ldots, 1-j, -j$. The operator of spin vector squared equals in this representation

$$T^{(j)}(J^2) = -T^{(j)}(\tau_\alpha)T^{(j)}(\tau_\alpha) = j(j+1)I_{2j+1}, \tag{2.1.34}$$

which follows from (2.1.15), (2.1.20) and (2.1.21).

Remark 1. Representations of $s\ell(2,\mathbb{C})$ with highest weight j (or lowest weight $\mu = -j$), where $j \notin \mathbb{Z}_{\geq 0}/2$, are infinite dimensional. Lie algebra $s\ell(2,\mathbb{C})$ also has infinite-dimensional representations which are neither highest weight nor lowest weight, i.e., no vector in those representations obeys (2.1.3) or (2.1.31).

Remark 2. At first sight, it appears puzzling that the Lie algebra $s\ell(2,\mathbb{C})$ has only one two-dimensional irreducible representation, namely, that of the highest weight $j = 1/2$, while the Lie *group* $SL(2,\mathbb{C})$ has the defining representation $\tilde{T}$ which is not equivalent to its complex conjugate[a] $\tilde{T}^*$ (see Section I-4.1.3 in the first book). In fact, there is no discrepancy here. In the first place, complex conjugate representation can be defined for *real* Lie algebra only, see Section I-4.1.2 of the first book, while algebra $s\ell(2,\mathbb{C})$ is complex. Therefore, the notion of complex conjugate representations has nothing to do with algebra $s\ell(2,\mathbb{C})$. On the other hand, Lie algebra $\mathcal{A}(G)$ of Lie group G (tangent space to real manifold G) is always real algebra; in particular, Lie algebra of Lie group $SL(2,\mathbb{C})$ is a six-dimensional real Lie algebra $\mathcal{A}(SL(2,\mathbb{C})) = \mathcal{A}^6_{\mathbb{R}} = so(1,3)$. The latter is the realification of the complex Lie algebra $s\ell(2,\mathbb{C})$. It is this real algebra that has inequivalent representations, $\tilde{T}$ and $\tilde{T}^*$. Complex representations of Lie algebra $\mathcal{A}^6_{\mathbb{R}} = so(1,3)$ are at the same time representations of its complexification $\mathcal{A}^6_{\mathbb{C}} = so(4,\mathbb{C})$. The latter is isomorphic to direct sum $s\ell(2,\mathbb{C}) \oplus s\ell(2,\mathbb{C})$ (see Proposition I-3.3.5 in Section I-3.3.3 of the first book). So, it has two two-dimensional inequivalent irreducible representations denoted by $(1/2,0)$ and $(0,1/2)$, where the first and second number in each pair refers to spin of representation of the first and second algebra in direct sum $s\ell(2,\mathbb{C}) \oplus s\ell(2,\mathbb{C})$. These two representations correspond precisely to the defining representation of group $SL(2,\mathbb{C})$ and complex conjugate to this representation.

Remark 3. In Chapters 2 and 4, we are primarily interested in complex finite-dimensional representations of Lie algebras $s\ell(2,\mathbb{C})$, $su(2)$ and Lie group $SU(2)$, as well as Lie algebras $s\ell(N,\mathbb{C})$, $su(N)$ and Lie group $SU(N)$. As we know, these representations are in one-to-one correspondence with each other: any representation of $s\ell(2,\mathbb{C})$ is at the same time representation of $su(2)$ and is lifted to representation of group $SU(2)$, and vice versa. One can associate with all representations of $s\ell(2,\mathbb{C})$ the limited class of representations of Lie group $SL(2,\mathbb{C})$, namely, if element $g \in SL(2,\mathbb{C})$ is a

[a]It is worth emphasizing that we treat all Lie groups G including $SL(2,\mathbb{C})$ as *real* manifolds. Alternatively, group $SL(2,\mathbb{C})$ may be treated as complex manifold. In the latter approach, $s\ell(2,\mathbb{C})$ is its complex Lie algebra. The real analog of the group $SL(2,\mathbb{C})$ is then the group $\mathrm{Spin}(1,3)$, whose Lie algebra is $\mathcal{A}(\mathrm{Spin}(1,3)) = so(1,3)$. We do not make use of the approach in which group $SL(2,\mathbb{C})$ is interpreted as complex manifold, and identify groups $\mathrm{Spin}(1,3)$ and $SL(2,\mathbb{C})$. Groups $\mathrm{Spin}(p,q)$ are considered in Section 6.2.

product (see (2.1.12))

$$g = \exp(a_- e_-) \cdot \exp(a_0 \mathsf{h}) \cdot \exp(a_+ e_+),$$

where $e_\pm$, h are generators of $s\ell(2,\mathbb{C})$ and $a_\pm$, a_0 are complex parameters, then representations $T(g)$ of group $SL(2,\mathbb{C})$ we are talking about are related to representations $T(A)$ of algebra $s\ell(2,\mathbb{C})$ via

$$T(g) = \exp(a_- T(e_-)) \cdot \exp(a_0 T(\mathsf{h})) \cdot \exp(a_+ T(e_+)) \tag{2.1.35}$$

(here we deal with exponentials of complex matrices). Obviously, these representations of $SL(2,\mathbb{C})$ are in one-to-one correspondence with representations of Lie algebras $s\ell(2,\mathbb{C})$, $su(2)$ and Lie group $SU(2)$. We will see below that these (and only these) representations of group $SL(2,\mathbb{C})$ are subrepresentations in the tensor product of certain number of its defining representations.[b] Similar observation applies in the case of $SL(N,\mathbb{C})$ considered in Chapter 4. Our remark shows that the class of representations of groups $SL(2,\mathbb{C})$ and $SL(N,\mathbb{C})$ under consideration does not exhaust all finite-dimensional irreducible representations of these groups, so in Chapters 2 and 4 we do not pretend to give complete theory of representations of $SL(2,\mathbb{C})$ and $SL(N,\mathbb{C})$.

2.2. Differential Realization of Lie Algebra $s\ell(2,\mathbb{C})$ and Highest Weight Representations

2.2.1. Realization of Lie algebra $s\ell(2,\mathbb{C})$ in terms of differential operators

1. Group $SL(2,\mathbb{C})$ acts in two-dimensional complex vector space $\mathcal{V}_2$. Let us define this action in slightly unconventional way, namely, let us define $\mathcal{V}_2$ as two-dimensional space of all linear combinations of two complex variables $s = u^1$ and $t = u^2$,

$$\alpha_1 \, s + \alpha_2 \, t = \alpha_\beta u^\beta, \quad \alpha_\beta \in \mathbb{C}. \tag{2.2.1}$$

The monomials $s = u^1$ and $t = u^2$ serve as basis in this space, and the coefficients α_β serve as coordinates of a vector in $\mathcal{V}_2$. Now we define the

---

[b] In particular, we do not consider complex conjugated representations of $SL(2,\mathbb{C})$.

action of group $SL(2,\mathbb{C})$ in the space of homogeneous polynomials (2.2.1) by postulating its action on basis monomials $s = u^1$ and $t = u^2$:

$$T(g)\cdot(s,t) = (s,t)g = (sA_1^{\ 1} + tA_2^{\ 1}, sA_1^{\ 2} + tA_2^{\ 2}) = (s',t'), \qquad T(g)\cdot u^\alpha = u^\beta A_\beta{}^\alpha, \tag{2.2.2}$$

where $g = \begin{pmatrix} A_1^{\ 1} & A_1^{\ 2} \\ A_2^{\ 1} & A_2^{\ 2} \end{pmatrix}$ is an element of group $SL(2,\mathbb{C})$ ($A_\beta{}^\alpha \in \mathbb{C}$ and $\det(g) = 1$). Representation T in (2.2.2) is nothing but defining representation. Let the element $g \in SL(2,\mathbb{C})$ be cast in the exponential form (2.1.12), (2.1.13). Then, taking derivatives of the two sides of equality (2.2.2) over each of the parameters $a_0, a_\pm$ and setting then $a_0 = a_+ = a_- = 0$, we obtain three relations

$$T(e_+)\cdot u = u\cdot e_+ = (0,s), \quad T(e_-)\cdot u = u\cdot e_- = (t,0), \qquad T(\mathsf{h})\cdot u = u\cdot \mathsf{h} = \frac{1}{2}(s,-t), \tag{2.2.3}$$

where 2×2 matrices $\mathsf{h}, e_\pm \in s\ell(2,\mathbb{C})$ are given in (2.1.11) and we use shorthand notation u for vector $(u^1,u^2) = (s,t)$. It follows from (2.2.3) that $T(e_+)$, $T(e_-)$ and $T(\mathsf{h})$ can be realized as differential operators

$$\hat{e}_+ := T(e_+) = s\,\partial_t, \quad \hat{e}_- := T(e_-) = t\,\partial_s, \quad \hat{\mathsf{h}} := T(\mathsf{h}) = \frac{1}{2}(s\partial_s - t\partial_t) \tag{2.2.4}$$

(cf. the differential realization (1.2.35) of algebra $s\ell(d,\mathbb{C})$). By construction, these operators obey the commutation relations (2.1.1) for generators $\mathsf{h}, e_\pm$ of the Lie algebra $s\ell(2,\mathbb{C})$.

Problem 2.2.1. Check that the differential operators (2.2.4) obey (2.1.1).

The defining representation (2.2.2) of group $SL(2,\mathbb{C})$ admits natural generalization. To this end, consider space $\mathcal{F}$ of smooth functions $f(s,t) \equiv f(u)$ and define the action of group $SL(2,\mathbb{C})$ in this space:

$$[T(g)\cdot f](u) = f(u\cdot g), \quad \forall g \in SL(2,\mathbb{C}), \tag{2.2.5}$$

(cf. the right regular representation in Section I-4.5 of the first book, and Eq. (2.2.21) below). Since the space $\mathcal{F}$ is infinite dimensional, the representation T defined in (2.2.5) is infinite dimensional. We show below that this representation is reducible.

Problem 2.2.2. Check that mapping T from group $SL(2,\mathbb{C})$ to the space of linear operators in $\mathcal{F}$, defined by (2.2.5), is homomorphism.

We take g in the form (2.1.12) and evaluate the derivatives of (2.2.5) with respect to a_i $(i = 0, \pm)$ at $a_i = 0$. In this way, we again obtain the differential representation (2.2.4) of the generators of $s\ell(2,\mathbb{C})$:

$$\begin{aligned}
[T(\mathsf{h}) \cdot f](u) &= \frac{\partial}{\partial a_0} f(u \cdot g)\Big|_{a_i=0} = \frac{1}{2}(s\partial_s - t\partial_t) f(u), \\
[T(e_+) \cdot f](u) &= \frac{\partial}{\partial a_+} f(u \cdot g)\Big|_{a_i=0} = s\,\partial_t f(u), \\
[T(e_-) \cdot f](u) &= \frac{\partial}{\partial a_-} f(u \cdot g)\Big|_{a_i=0} = t\,\partial_s f(u)
\end{aligned} \tag{2.2.6}$$

(cf. the differential realizations of Lie algebras in regular representation, see Section I-7.5.3 in the first book). Thus, the differential realization (2.2.4) of the defining representation of the Lie algebra $s\ell(2,\mathbb{C})$ can be extended to the representation in infinite-dimensional functional space $\mathcal{F}$, whose elements are functions of two complex variables.

2. Consider a vector-column $\bar{u} = \begin{pmatrix} \bar{s} \\ \bar{t} \end{pmatrix}$ whose components are $\bar{u}_1 = \bar{s}$, $\bar{u}_2 = \bar{t}$ and define complex vector space $\mathcal{V}_2$ of first-order polynomials in variables $\bar{s}$ and $\bar{t}$:

$$\bar{\alpha}^1\,\bar{s} + \bar{\alpha}^2\,\bar{t} = \bar{\alpha}^\beta\,\bar{u}_\beta, \quad \bar{\alpha}^\beta \in \mathbb{C}. \tag{2.2.7}$$

The basis vectors in this space are monomials $\bar{s}$ and $\bar{t}$. We define the action of an element g of group $SL(2,\mathbb{C})$ on basis vectors as follows (cf. (2.2.2)):

$$\begin{aligned}
\overline{T}(g) \cdot \bar{u}_\alpha = (g^{-1} \cdot \bar{u})_\alpha = (A^{-1})_\alpha^{\ \beta} \bar{u}_\beta \Leftrightarrow \overline{T}(g) \cdot \begin{pmatrix} \bar{s} \\ \bar{t} \end{pmatrix} \\
= \begin{pmatrix} A_2^{\ 2}\bar{s} - A_1^{\ 2}\bar{t} \\ A_1^{\ 1}\bar{t} - A_2^{\ 1}\bar{s} \end{pmatrix} = \begin{pmatrix} \bar{s}' \\ \bar{t}' \end{pmatrix},
\end{aligned} \tag{2.2.8}$$

where $g^{-1} = ||(A^{-1})_\alpha^{\ \beta}|| = ||\varepsilon^{\beta\beta'} A_{\beta'}^{\ \alpha'} \varepsilon_{\alpha'\alpha}||$ is the matrix inverse to element $g = ||A_\alpha^{\ \beta}|| \in SL(2,\mathbb{C})$ and

$$||\varepsilon_{\alpha\beta}|| = \begin{pmatrix} 0 & 1 \\ -1 & 0 \end{pmatrix}, \quad ||\varepsilon^{\alpha\beta}|| = \begin{pmatrix} 0 & -1 \\ 1 & 0 \end{pmatrix}, \quad \varepsilon_{\alpha\beta}\varepsilon^{\beta\gamma} = \delta_\alpha^\gamma. \tag{2.2.9}$$

Formula (2.2.8) defines the representation $\overline{T}$ of group $SL(2,\mathbb{C})$, which is contragredient to the defining representation (2.2.2) (contragredient representation is introduced in item 3 of Section I-4.1.1 of the first book).

We choose the element g in the form (2.1.12) in (2.2.8), differentiate the latter equation over $a_0, a_\pm$, and set $a_0 = a_\pm = 0$. We obtain

$$\overline{T}(e_+) \cdot \bar{u} = -e_+ \cdot \bar{u} = -\begin{pmatrix} \bar{t} \\ 0 \end{pmatrix}, \quad \overline{T}(e_-) \cdot \bar{u} = -e_- \cdot \bar{u} = -\begin{pmatrix} 0 \\ \bar{s} \end{pmatrix},$$
$$\overline{T}(\mathsf{h}) \cdot \bar{u} = -\mathsf{h} \cdot \bar{u} = -\frac{1}{2} \cdot \begin{pmatrix} \bar{s} \\ \bar{t} \end{pmatrix}. \tag{2.2.10}$$

These define the generators $\overline{T}(e_\pm)$ and $\overline{T}(\mathsf{h})$ in the form of differential operators,

$$\bar{e}_+ := \overline{T}(e_+) = -\bar{t}\,\partial_{\bar{s}}, \quad \bar{e}_- := \overline{T}(e_-) = -\bar{s}\,\partial_{\bar{t}},$$
$$\bar{\mathsf{h}} := \overline{T}(\mathsf{h}) = \frac{1}{2}\,(\bar{t}\,\partial_{\bar{t}} - \bar{s}\,\partial_{\bar{s}}). \tag{2.2.11}$$

These operators, in analogy to (2.2.4), make basis in the Lie algebra $s\ell(2,\mathbb{C})$ in the representation $\overline{T}$, i.e., they obey defining relations (2.1.1). Like in the previous case, this representation is generalized to infinite-dimensional representation acting in the space of functions $f(\bar{s},\bar{t})$ of two complex variables, and the generators still have the form (2.2.11).

Problem 2.2.3. Check that the differential representation (2.2.11) of the Lie algebra $s\ell(2,\mathbb{C})$ corresponds to the representation of the Lie group $SL(2,\mathbb{C})$:

$$[\overline{T}(g) \cdot f](\bar{u}) = f(g^{-1} \cdot \bar{u}). \tag{2.2.12}$$

Compare (2.2.12) with the formula for left regular representation from Section I-4.5 of the first book (see also (2.2.22) below).

Note that the transformations (2.2.8) and (2.2.2), performed together, leave invariant the convolution of basis vectors

$$(s,t)\begin{pmatrix} \bar{s} \\ \bar{t} \end{pmatrix} = (s\bar{s} + t\bar{t}) = u^\beta\,\bar{u}_\beta \quad (\beta = 1,2). \tag{2.2.13}$$

The function (2.2.13) is a homogeneous first-order polynomial in s and t, and at the same time a homogeneous first-order polynomial in $\bar{s}$ and $\bar{t}$, so it can be considered as an element of a direct product of vector spaces (2.2.1) and (2.2.7).

3. We now construct generalizations of the above representations. To this end, we consider space $\mathcal{F}_{(1,\bar{1})}$ of smooth functions $f(u,\bar{u})$ (we consider variables $u = (s,t)$ and $\bar{u} = (\bar{s},\bar{t})$ as independent) and define, in this space, representation $T_{(1,\bar{1})}$ of group $SL(2,\mathbb{C})$:

$$[T_{(1,\bar{1})}(g) \cdot f](u,\bar{u}) = f(u \cdot g, g^{-1} \cdot \bar{u}), \quad \forall g \in SL(2,\mathbb{C}). \tag{2.2.14}$$

It is straightforward to see that this is homomorphism. In the same manner as above, we obtain the following expressions for the generators of algebra $s\ell(2,\mathbb{C})$ in representation $T_{(1,\bar{1})}$:

$$T_{(1,\bar{1})}(e_\pm) = (\hat{e}_\pm + \bar{e}_\pm), \quad T_{(1,\bar{1})}(\mathsf{h}) = (\hat{\mathsf{h}} + \bar{\mathsf{h}}), \tag{2.2.15}$$

where operators $\{\hat{e}_\pm, \hat{\mathsf{h}}\}$ and $\{\bar{e}_\pm, \bar{\mathsf{h}}\}$ are given by (2.2.4) and (2.2.11). Thus, for every element $a \in s\ell(2,\mathbb{C})$, we have $T_{(1,\bar{1})}(a) = T(a) + \bar{T}(a)$, where representations T and $\bar{T}$ are defined in (2.2.5) and (2.2.12). This reflects the fact that the representation $T_{(1,\bar{1})}$ is a direct product of representations T and $\bar{T}$ (direct product of representations of Lie groups and Lie algebras is defined in Section I-4.3 of the first book).

Problem 2.2.4. Check that the differential operators (2.2.11), and also operators (2.2.15), acting in space of functions $f(u,\bar{u})$, obey the commutation relations (2.1.1). Show that any function $f(u^\beta \bar{u}_\beta) \in \mathcal{F}_{(1,\bar{1})}$ that depends only on the combination $u^\beta \bar{u}_\beta = s\bar{s} + t\bar{t}$, obey differential equations

$$T_{(1,\bar{1})}(e_\pm) \cdot f(u^\beta \bar{u}_\beta) = 0, \quad T_{(1,\bar{1})}(\mathsf{h}) \cdot f(u^\beta \bar{u}_\beta) = 0. \tag{2.2.16}$$

Hint: derive Eq. (2.2.16) for the simplest function (2.2.13).

Remark 1. Transformations (2.2.8), (2.2.10) can be obtained from (2.2.2), (2.2.3) by identifying

$$\bar{s} = t, \quad \bar{t} = -s \Leftrightarrow \bar{u}_\alpha = \varepsilon_{\alpha\beta}\, u^\beta, \tag{2.2.17}$$

where $\varepsilon_{\alpha\beta}$ is antisymmetric second rank tensor (2.2.9). Accordingly, the generators (2.2.4) and (2.2.11) transform into each other under the transformation (2.2.17). This fact means that *defining representation of the Lie algebra $s\ell(2,\mathbb{C})$ (of the group $SL(2,\mathbb{C})$) and its contragredient representation are equivalent*, so for classification purposes it is sufficient to consider one of them. Nevertheless, we are going to use both representations.

Problem 2.2.5. Construct differential representations of the generators τ_α of the Lie algebra $su(2)$ in representation $T_{(1,\bar{1})}$. Hint: make use of the relations (2.1.15).

Remark 2. Let us make a change of variables (cf. (2.2.17)) in Eq. (2.2.14), which defines the representation $T_{(1,\bar{1})} = T \otimes \bar{T}$ of group $SL(2,\mathbb{C})$:

$$s = s_1, \quad t = t_1, \quad \bar{s} = t_2, \quad \bar{t} = -s_2, \tag{2.2.18}$$

and denote $u^\alpha = (s_1, t_1)$, $v^\alpha = (s_2, t_2)$. Equation (2.2.14) takes the form

$$[T_{(2,\bar{0})}(g) \cdot f](u,v) = f(u \cdot g, v \cdot g), \tag{2.2.19}$$

and defines representations $T_{(2,\bar{0})}$ of group $SL(2,\mathbb{C})$, which is nothing but the direct product of representations (2.2.5), i.e., $T_{(2,\bar{0})} = T \otimes T$. It acts in the space $\mathcal{F}_{(2,\bar{0})}$ of functions $f(u^\alpha, v^\beta)$. The constructions (2.2.14) and (2.2.19) are straightforwardly generalized to the direct product $T_{(n,\bar{m})}$ of any number of representations T and $\bar{T}$, namely, let vectors $u_{(k)} = (s_k, t_k)$, where $k = 1, \ldots, n$, transform as in (2.2.2), whereas vectors $\bar{u}_{(i)} = (\bar{s}_i, \bar{t}_i)$, where $i = 1, \ldots, m$, transform as written in (2.2.8). Then representations $T_{(n,\bar{m})} = T^{\otimes n} \otimes \bar{T}^{\otimes m}$ act in spaces $\mathcal{F}_{(n,\bar{m})}$ of functions $f(u_{(k)}; \bar{u}_{(i)}) := f(u_{(1)}, \ldots, u_{(n)}; \bar{u}_{(1)}, \ldots, \bar{u}_{(m)})$ and these representations are given by

$$[T_{(n,\bar{m})}(g) \cdot f](u_{(k)}; \bar{u}_{(i)}) = f(u_{(k)} \cdot g;\, g^{-1} \cdot \bar{u}_{(i)}). \tag{2.2.20}$$

We make use of representations $T_{(n,\bar{m})}$ below in Sections 2.4.4 and 2.5.4 where we derive explicit expressions for Clebsch–Gordan coefficients and for 6-j symbols.

Problem 2.2.6. Recall (see Section I-4.5 in the first book) that right and left regular representations of group G, $\tilde{T}_R$ and T_R, are defined by

$$[\tilde{T}_R(g) \cdot f](h) = f(h \cdot g), \tag{2.2.21}$$

$$[T_R(g) \cdot f](h) = f(g^{-1} \cdot h), \tag{2.2.22}$$

where $g, h \in G$ and f are functions on group G. Find additional conditions on coordinates of vectors $u_{(i)}$ and $\bar{u}_{(i)}$ in (2.2.20), such that the representations $T_{(2,\bar{0})}$ and $T_{(0,\bar{2})}$ of group $SL(2,\mathbb{C})$ reduce to the regular representations (2.2.21), (2.2.22).

2.2.2. Highest weight representations of differential realizations of $s\ell(2,\mathbb{C})$

The generators of algebra $s\ell(2,\mathbb{C})$ realized as operators (2.2.4) act in infinite-dimensional space $\mathcal{F} = \mathcal{F}_{(1,0)}$ of functions $f(s,t)$ of two variables

s and t, as given by (2.2.6). This representation of algebra $s\ell(2,\mathbb{C})$ is reducible. Let us extract, in space $\mathcal{F}$, finite-dimensional invariant subspaces of irreducible representations of $s\ell(2,\mathbb{C})$. To this end, we make use of the highest weight representation technique.

Recall that the highest weight vector $v_0(s,t) \in \mathcal{F}$ obeys (2.1.3), so we have

$$\hat{e}_+ \cdot v_0 = 0, \quad \hat{\mathsf{h}} \cdot v_0 = jv \Rightarrow s\,\partial_t\, v_0 = 0, \quad (s\partial_s - t\partial_t)v_0 = 2\,j\,v_0, \quad (2.2.23)$$

where j is the highest weight. The solution to these differential equations, modulo a normalization constant, is

$$v_0 = s^{2j} \in \mathcal{F}. \quad (2.2.24)$$

This is precisely the highest weight vector in $\mathcal{F}$. We now follow the standard procedure, act on this function by lowering operators $\hat{e}_- = t\partial_s$ and obtain the sequence of vectors, analogous to (2.1.4):

$$\begin{aligned} \hat{e}_- \cdot v_0 = 2j\, s^{2j-1}\, t \Rightarrow \hat{e}_-^2 \cdot v_0 = 2j\,(2j-1)\, s^{2j-2}\, t^2 \\ \Rightarrow \cdots \Rightarrow \hat{e}_-^k \cdot v_0 = 2j(2j-1)\cdots(2j-k+1)s^{2j-k}\, t^k, \ldots. \end{aligned} \quad (2.2.25)$$

At each step, the power of s decreases by 1, and the power of t increases by 1, so all vectors $\hat{e}_-^k \cdot v_0$ are monomials of one and the same degree $2j$ in (s,t). If $j = \frac{n}{2}$, where $n \in \mathbb{Z}_{\geq 0}$ is nonnegative integer, then the sequence of vectors $\hat{e}_-^k \cdot v_0$ terminates at $k = 2j+1$, since in view of (2.2.25) we have $\hat{e}_-^{2j+1}\, v_0 = 0$. In this case we generate, out of the highest weight vector v_0, finite number $(2j+1)$ of vectors (cf. (1.2.23))

$$\begin{aligned} T^j_m(s,t) = \frac{1}{\sqrt{(j+m)!(j-m)!}}\, s^{j+m}\, t^{j-m} =: \langle s,t\,|j,m\rangle, \\ m = -j, -j+1, \ldots, j. \end{aligned} \quad (2.2.26)$$

They form basis in $(2j+1)$-dimensional space $\mathcal{V}_{2j+1}$ of irreducible representation $T^{(j)}$ of algebras $s\ell(2,\mathbb{C})$ and $su(2)$. Note that index m in (2.2.26) takes integer values for integer j and half-integer values for half-integer j. We introduce in (2.2.26) notation $\langle s,t\,|j,m\rangle$, instead of $|j,m\rangle$, for monomials $T^j_m(s,t)$ to distinguish this particular form from the vectors (2.1.19) used in abstract construction of highest weight representations. These are standard Dirac notations introduced in Chapter 1.

The space $\mathcal{V}_{2j+1}$ with basis vectors (2.2.26) consists of all homogeneous polynomials in two variables (s, t) of common degree $2j$ (cf. (2.2.1)):

$$P_{2j}(s,t) = \sum_{m=-j}^{j} \alpha_m T^j_m(s,t), \quad \alpha_m \in \mathbb{C}. \tag{2.2.27}$$

At the same time, it is the space of irreducible representation $T^{(j)}$ of the Lie group $SL(2,\mathbb{C})$ (and $SU(2)$). This follows from the action (2.2.5) of this group in the representation T: this action preserves the degree of a homogeneous polynomial.

The normalization in (2.2.26) is chosen in such a way that vectors (2.2.26) form orthonormal basis with respect to the scalar product (cf. (1.2.24))

$$(P(s,t), P'(s,t)) = P^*(\partial_s, \partial_t) \cdot P'(s,t)|_{s=t=0}, \tag{2.2.28}$$

where $P(s,t)$ and $P'(s,t)$ are arbitrary polynomials in $\mathcal{F}$, and for $P(s,t) = \sum_{n_s,n_t} C_{n_s,n_t} s^{n_s} t^{n_t}$ we have, by definition, $P^*(\partial_s, \partial_t) = \sum_{n_s,n_t} C^*_{n_s,n_t} \partial_s^{n_s} \partial_t^{n_t}$. The scalar product (2.2.28) is equivalent to the scalar product defined in (2.1.33), since these scalar products coincide for basis vectors in the pertinent subspaces of irreducible representations.

Remark 1. *Let us clarify the relationship between the scalar product (2.2.28) emerging in $s\ell(2,\mathbb{C})$ context, and scalar product (1.2.24) introduced (see Proposition 1.2.2)) in the context of the (anti)holomorphic representation in quantum mechanics. This relationship emerges when the generators $E_{ij} \in s\ell(d,\mathbb{C})$ are written in (1.2.30) in terms of quadratic combinations of oscillators (Jordan–Schwinger representation). In the (anti)holomorphic representation of oscillators, the basis elements E_{ij} of algebra $s\ell(d,\mathbb{C})$ is given by differential operators (1.2.35), which, for $d = 2$ and upon the change of notation $(z_1, z_2) = (s,t)$, coincide with operators (2.2.4). Recall that the holomorphic representation is defined in the space of functions with basis (1.2.23) and scalar product (1.2.24). This agrees with the construction in this section for $d = 2$, i.e., with our basis (2.2.26) and scalar product (2.2.28).*

As an example, we give the functions $T^j_m(s,t)$ for $j = \frac{1}{2}, 1$:

(1) $j = 1/2 (m = \pm 1/2) \Rightarrow T^{1/2}_{-1/2} = t$, $T^{1/2}_{1/2} = s$,
(2) $j = 1 (m = -1, 0, 1) \Rightarrow T^1_{-1} = t^2/\sqrt{2}$, $T^1_0 = st$, $T^1_1 = s^2/\sqrt{2}$.

In the case $j = 1/2$, functions $T^{1/2}_m$ coincide with the basis vectors $u = (s,t)$ of defining representation. For $j = 1$, the monomials $T^j_m(s,t)$ form basis in three-dimensional space of quadratic polynomials, which is the space of adjoint representation of $s\ell(2,\mathbb{C})$.

Problem 2.2.7. Construct finite-dimensional irreducible highest weight contragredient representations $\overline{T}^{(j)}$ of the Lie algebra $s\ell(2,\mathbb{C})$ and Lie group $SL(2,\mathbb{C})$, making use of (2.2.11) and (2.2.12). Show that basis vectors in the spaces of these representations are homogeneous monomials (cf. (2.2.26)):

$$\overline{T}^j_m = \frac{(-1)^{j+m}}{\sqrt{(j+m)!(j-m)!}}\, \bar{s}^{j-m}\, \bar{t}^{j+m}. \tag{2.2.29}$$

Hint: choose the highest weight vector as follows: $(-\bar{t})^{2j} \in \mathcal{F}_{(0,\bar{1})}$.

Note that under the identification (2.2.17) we have $\overline{T}^j_m = T^j_m$, as expected.

By construction, all vectors (2.2.26) are eigenvectors of the differential operator $\hat{\mathsf{h}}$ given by (2.2.4)

$$\hat{\mathsf{h}}\, T^j_m = m\, T^j_m \Leftrightarrow \hat{\mathsf{h}}\, |j,m\rangle = m\, |j,m\rangle \tag{2.2.30}$$

(hereafter in this section we omit part $\langle s,t|$ in the notation $\langle s,t|j,m\rangle$ for vectors (2.2.26)). Also, using explicit formulas (2.2.4) for generators, we get the relations

$$\begin{aligned} \hat{e}_+\, |j,m\rangle &= \sqrt{(j+m+1)(j-m)}\, |j,m+1\rangle, \\ \hat{e}_-\, |j,m\rangle &= \sqrt{(j-m+1)(j+m)}\, |j,m-1\rangle. \end{aligned} \tag{2.2.31}$$

Formulas (2.2.30) and (2.2.31) coincide with (2.1.22) and (2.1.23), while the basis vectors (2.2.26), as we have seen, are orthonormalized with respect to the scalar product (2.2.28). Thus, we have reproduced the results of Section 2.1 in the formalism of differential realizations.

The same formulas (2.2.30) and (2.2.31) are valid also for contragredient representations (2.2.11) and (2.2.29). This follows from the equivalence (2.2.17).

Problem 2.2.8. Derive the generalization of (2.2.31):

$$\hat{e}^k_\pm\, |j,m\rangle = \sqrt{\frac{(j\pm m+k)!(j\mp m)!}{(j\pm m)!(j\mp m-k)!}}\, |j,m\pm k\rangle. \tag{2.2.32}$$

Consider direct product $T^{(j)} \otimes \overline{T}^{(j)}$ of two irreducible representations $T^{(j)}$ and $\overline{T}^{(j)}$ of the same highest weight $j = n/2$, $n \in \mathbb{Z}_{\geq 0}$. The two representations act in functional spaces with bases (2.2.26) and (2.2.29), respectively. The basis in the space of representation $T^{(j)} \otimes \overline{T}^{(j)}$ is made of monomials $T^j_m(u) \cdot \overline{T}^j_{m'}(\bar{u})$, where indices m, m' take values $-j, \dots, j-1, j$. Representation $T^{(j)} \otimes \overline{T}^{(j)}$ is reducible and (in its decomposition in irreducible representations) contains one-dimensional trivial representation $T^{(0)}$. Since we use this trivial representation for calculating Clebsch–Gordan coefficients (see Section 2.4.4), we extract representation $T^{(0)}$ from $T^{(j)} \otimes \overline{T}^{(j)}$ explicitly. To this end, we have to find a vector $f(u, \bar{u})$ in the space of representation $T^{(j)} \otimes \overline{T}^{(j)}$, which is invariant under the transformations (2.2.14). Consider function $(u^\beta \bar{u}_\beta)^{2j} = (s\bar{s} + t\bar{t})^{2j}$, which is a homogeneous polynomial of degree $2j$ in variables (s, t), as well as in variables $(\bar{s}, \bar{t})$. This function, therefore, belongs to the space of representation $T^{(j)} \otimes \overline{T}^{(j)}$. It is invariant under transformations (2.2.14), which is clear from (2.2.16). So, the function $(s\bar{s} + t\bar{t})^{2j}$ is the desired vector which determines the one-dimensional subspace of trivial representation in the space of representation $T^{(j)} \otimes \overline{T}^{(j)}$. Finally, we expand

$$(s\bar{s} + t\bar{t})^{2j} = (2j)! \sum_{m=-j}^{j} \frac{(s\bar{s})^{j+m}(t\bar{t})^{j-m}}{(j+m)!(j-m)!}$$

$$= (2j)! \sum_{m=-j}^{j} (-1)^{j-m} \overline{T}^j_{-m}(\bar{u}) T^j_m(u). \qquad (2.2.33)$$

Due to invariance of $(s\bar{s} + t\bar{t})^{2j}$, the right-hand side here is interpreted as invariant convolution of basis vectors T^j_m and $\overline{T}^j_m$, given by (2.2.26) and (2.2.29).

Remark 2. *Any homogeneous polynomial $P_{2j}(s,t)$ of degree $2j$ can be written as*

$$P_{2j}(s,t) = t^{2j} \phi_{2j}(z), \qquad (2.2.34)$$

where we change variables $s, t \to z = s/t, t$ and define degree-$2j$ polynomial $\phi_{2j}(z)$ in one variable $z = s/t$. In terms of new variables (z, t), operators $\hat{\mathsf{e}}_\pm, \hat{\mathsf{h}}$ (2.2.4) have the following form:

$$\hat{\mathsf{e}}_+ = zt\partial_t - z^2\partial_z, \quad \hat{\mathsf{e}}_- = \partial_z, \quad \hat{\mathsf{h}} = z\partial_z - \frac{1}{2}t\partial_t. \qquad (2.2.35)$$

Since the action of operators $\hat{\mathsf{e}}_\pm, \hat{\mathsf{h}}$ does not modify the degree of a polynomial $P_{2j}(s,t)$, we make use of (2.2.34) and (2.2.35) to define these operators in the space of polynomials $\phi_{2j}(z)$:

$$\hat{\mathsf{e}}_\pm \cdot (t^{2j}\, \phi_{2j}(z)) = t^{2j}\, T_z(\mathsf{e}_\pm) \cdot \phi_{2j}(z), \;\; \hat{\mathsf{h}} \cdot (t^{2j}\, \phi_{2j}(z)) = t^{2j}\, T_z(\mathsf{h}) \cdot \phi_{2j}(z), \qquad (2.2.36)$$

which give

$$T_z(e_+) = 2\,j\,z - z^2\partial_z, \quad T_z(e_-) = \partial_z, \quad T_z(\mathsf{h}) = z\partial_z - j. \tag{2.2.37}$$

Formulas (2.2.37) define representations of algebra $s\ell(2,\mathbb{C})$ in the space of polynomials in one variable z of degree $2j$. The corresponding representation of group $SL(2,\mathbb{C})$ follows from (2.2.2), (2.2.5), namely, the action of an element $g \in SL(2,\mathbb{C})$ on a homogeneous polynomial $P_{2j}(s,t)$ reads

$$[T(g)\cdot P_{2j}](s,t) = P_{2j}(s',t') = (t')^{2j}\;\phi_{2j}(s'/t') = (t)^{2j}\;(t'/t)^{2j}\;\phi_{2j}(s'/t'), \tag{2.2.38}$$

where the transformed variables s', t' and function $\phi_{2j}(s/t)$ are defined in (2.2.5) and (2.2.34). By making the change of variables (2.2.34) in (2.2.38), we obtain the representation of group $SL(2,\mathbb{C})$ in the space of polynomials $\phi_{2j}(z)$:

$$[T_z(g)\cdot\phi_{2j}](z) = ({A_1}^2\,z + {A_2}^2)^{2j}\;\phi_{2j}\left(\frac{{A_1}^1\,z + {A_2}^1}{{A_1}^2\,z + {A_2}^2}\right). \tag{2.2.39}$$

Representations (2.2.37) and (2.2.39) are considered in Section 2.2.3 from somewhat different viewpoint.

Problem 2.2.9. *Show that operators (2.2.37) make basis in the Lie algebra $s\ell(2,\mathbb{C})$. Obtain the realization (2.2.37) of $s\ell(2,\mathbb{C})$ starting from the realization (2.2.39) of group $SL(2,\mathbb{C})$. Calculate Casimir operator J^2 for realization (2.2.37).*

Problem 2.2.10. *Find the highest weight vector v_0 for realization (2.2.37) of the Lie algebra $s\ell(2,\mathbb{C})$:*

$$e_+\,v_0 = 0, \quad \mathsf{h}\,v_0 = \lambda\,v_0,$$

show that $\lambda = j$. Construct representation T of $s\ell(2,\mathbb{C})$ with the highest weight $\lambda = j$. Show that T is finite dimensional, if and only if $2j$ is nonnegative integer. Describe in the latter case the space of representation T and determine its dimension. Describe infinite-dimensional representations of $s\ell(2,\mathbb{C})$ with the highest weight in the case $2j \neq \mathbb{Z}_{\geq 0}$.

2.2.3. Coherent states for Lie algebra $s\ell(2,\mathbb{C})$

In this Section, we construct coherent states for Lie algebras $s\ell(2,\mathbb{C})$ and $su(2)$, which are somewhat similar to coherent states of the oscillator algebra (see Section 1.2). We show that the differential realization (2.2.37) of algebra $s\ell(2,\mathbb{C})$ and representation (2.2.39) of group $SL(2,\mathbb{C})$ arise naturally as representations in the space of coherent states.

Let us choose the standard basis $\{e_\pm, \mathsf{h}\}$ in algebra $s\ell(2,\mathbb{C})$. As we pointed out in Section 2.1, anti-Hermiticity $\tau_\alpha^\dagger = -\tau_\alpha$ of generators of $su(2)$ (the real form of $s\ell(2,\mathbb{C})$) in any of its representations imposes the following relations for the generators of $s\ell(2,\mathbb{C})$ (see (2.1.18)):

$$\mathsf{h}^\dagger = \mathsf{h}, \quad e_+^\dagger = e_-, \quad e_-^\dagger = e_+. \tag{2.2.40}$$

Let $T_{(-j)}$ be representation of $s\ell(2,\mathbb{C})$ with the lowest weight $(-j)$. The lowest weight vector $|-j\rangle$ in the space $\mathcal{V}$ of representation $T_{(-j)}$ obeys

$$e_-\,|-j\rangle = 0, \quad \mathsf{h}|-j\rangle = -j\,|-j\rangle, \tag{2.2.41}$$

where $j \in \mathbb{R}$. In analogy to Fock representation (see Section 1.2), we refer to the lowest weight vector $|-j\rangle$ as vacuum. All basis vectors $|m\rangle \in \mathcal{V}$ are obtained by multiple action of raising operators e_+ on vacuum $|-j\rangle$:

$$|m\rangle := c_m \, e_+^{j+m} \, |-j\rangle \Rightarrow \mathsf{h} \, |m\rangle = m \, |m\rangle, \tag{2.2.42}$$

where $m = -j, -j+1, \ldots$, i.e., $(j+m)$ is nonnegative integer. Recall that the representation $T_{(-j)}$ is finite dimensional and irreducible if and only if $j \in \mathbb{Z}_{\geq 0}/2$ is nonnegative integer or half-integer. In that case, the representation $T_{(-j)}$ is equivalent to the representation with the highest weight j. However, we start with the generic case of $j \in \mathbb{R}$.

Let the space $\mathcal{V}$ of representation $T_{(-j)}$ be equipped with Hermitian scalar product $\langle \Psi \mid \Phi \rangle \in \mathbb{C}$ (see Sections 1.1 and 2.1):

$$\langle \Phi \mid \Psi \rangle = (\langle \Psi \mid \Phi \rangle)^*. \tag{2.2.43}$$

Hermitian conjugate vectors $\langle \Psi \mid = \mid \Psi \rangle^\dagger$ form the space $\mathcal{V}^*$ which is dual to $\mathcal{V}$. We choose the basis in $\mathcal{V}^*$ that consists of vectors $\langle m| = |m\rangle^\dagger$ (cf. (2.1.27)), where $|m\rangle$ are basis vectors (2.2.42) in space $\mathcal{V}$. In view of properties (2.2.40), (2.2.41), dual lowest weight vector $\langle -j|$ is defined by

$$\langle -j| e_+ = 0, \quad \langle -j| \mathsf{h} = (-j) \, \langle -j|, \tag{2.2.44}$$

and all basis vectors $\langle m| \in \mathcal{V}^*$ are obtained by the right action of operators e_- on the dual lowest vector:

$$\langle m| = |m\rangle^\dagger = (c_m \, e_+^{j+m} \, |-j\rangle)^\dagger = c_m^* \, \langle -j| e_-^{j+m}, \quad m = -j, -j+1, \ldots.$$

The calculation of convolution of two basis vectors gives

$$\langle m'|m\rangle = c_{m'}^* \, c_m \, \langle -j| \, e_-^{j+m'} \, e_+^{j+m} |-j\rangle = \delta_{j+m',j+m} \, |c_m|^2 (j+m)! \, (2j)_m,$$

where we many times use the formula (2.1.6) rewritten as

$$e_-^k \, e_+ = e_+ \, e_-^k - k(2\,\mathsf{h} + k - 1) e_-^{k-1},$$

set $\langle -j| - j\rangle = 1$ and introduce the notation

$$(2j)_m = (2j) \cdot (2j-1) \cdots (j-m+1) = \frac{\Gamma(2j+1)}{\Gamma(j-m+1)}, \tag{2.2.45}$$

where $\Gamma(x)$ is Euler Gamma-function. To ensure orthonormality of the system of vectors (2.2.42)

$$\langle m'|m\rangle = \delta_{j+m',j+m}, \tag{2.2.46}$$

we set, modulo phase factor (cf. normalization coefficient in (2.1.29)),

$$c_m = ((j+m)! \, (2j)_m)^{-1/2} = \left(\frac{\Gamma(2j+1)\Gamma(j+m+1)}{\Gamma(j-m+1)} \right)^{-1/2}. \tag{2.2.47}$$

Unit operator I in space $\mathcal{V}$ is expanded in terms of the complete set of basis vectors (2.2.42) in a standard way (see (1.1.5)):

$$I = \sum_{m=-j}^{\infty} |m\rangle\langle m| = |-j\rangle\langle -j| + |1-j\rangle\langle 1-j| + |2-j\rangle\langle 2-j| + \cdots. \tag{2.2.48}$$

We recall that here j is an arbitrary real number. Accordingly, any vector $|\Psi\rangle \in \mathcal{V}$ is represented as linear combination of vectors $|m\rangle$:

$$|\Psi\rangle = \sum_{k=0}^{\infty} |k-j\rangle\, \langle k-j|\Psi\rangle, \tag{2.2.49}$$

where $\langle m|\Psi\rangle \in \mathbb{C}$ are coordinates of vector $|\Psi\rangle \in \mathcal{V}$ in the chosen basis.

Consider now generating function for basis vectors (2.2.42):

$$|z\rangle := \exp(ze_+)\,|-j\rangle = \sum_{m=-j}^{\infty} z^{j+m} \left(\frac{\Gamma(2j+1)}{\Gamma(m+j+1)\Gamma(j-m+1)}\right)^{1/2} |m\rangle, \tag{2.2.50}$$

which is called coherent state of the Lie algebra $s\ell(2,\mathbb{C})$ in representation $T_{(-j)}$. We stress that the sum in (2.2.50) runs over $m = -j + k$ $(k = 0, 1, 2, \dots)$, where j does not have to be integer.

Proposition 2.2.1. *Coherent state* (2.2.50) *obeys the following equations:*

$$e_+\,|z\rangle = \partial_z\,|z\rangle, \quad \mathsf{h}\,|z\rangle = (z\,\partial_z - j)\,|z\rangle, \quad e_-\,|z\rangle = (2\,j\,z - z^2\,\partial_z)\,|z\rangle. \tag{2.2.51}$$

Proof. *The first relation in (2.2.51) is obvious. The other two are obtained by making use of the relations*

$$e^{-ze_+}\,\mathsf{h}\,e^{ze_+} = \mathsf{h} + z\,e_+, \quad e^{-ze_+}\,e_-\,e^{ze_+} = e_- - 2\,z\,\mathsf{h} - z^2\,e_+,$$

which follow from the identity

$$e^A \cdot B \cdot e^{-A} = \sum_{k=0}^{\infty} \frac{1}{k!} \underbrace{[A, [A, \dots\, [A}_{k}, B] \dots]],$$

where A and B are arbitrary operators. □

Upon Hermitian conjugation of Eq. (2.2.50), we obtain dual coherent state in space $\mathcal{V}^*$:

$$\langle z| := \langle -j|\exp(ze_-). \tag{2.2.52}$$

Hermitian conjugation of (2.2.51) gives

$$\langle z|e_- = \partial_z \langle z|, \quad \langle z|\,\mathsf{h} = (z\,\partial_z - j)\,\langle z|, \quad \langle z|\,e_+ = (2\,j\,z - z^2\,\partial_z)\,\langle z|. \tag{2.2.53}$$

We now specify to finite-dimensional irreducible representations $T_{(-j)}$ with $j \in \mathbb{Z}_{\geq 0}/2$ and $m = -j, -j+1, \dots, j$. In that case, the space $\mathcal{V}$ of the representation has dimension $(2j+1)$, and unit operator (2.2.48) reads

$$I = \sum_{m=-j}^{j} |m\rangle\langle m|. \tag{2.2.54}$$

For finite-dimensional representations $T_{(-j)}$, coherent states form complete set of vectors in $\mathcal{V}$. This follows from the fact that the unit operator (2.2.48) can be decomposed in coherent state basis.

Proposition 2.2.2. *The unit operator in space $\mathcal{V}$ is written in terms of the coherent states* (2.2.50) *as follows:*

$$I = \int d^2z\, \mu(z,\bar{z})\, |z\rangle\langle\bar{z}|, \tag{2.2.55}$$

where $z = x + iy \in \mathbb{C}$, $\bar{z} := z^* = x - iy$ *and* $d^2z := dx\,dy$, *while the measure* $\mu(z,\bar{z})$ *is given by*

$$\mu(z,\bar{z}) = \frac{(2j+1)}{\pi}\,\frac{1}{(1+z\,\bar{z})^{2j+2}}. \tag{2.2.56}$$

Proof. *Let us show that the operator (2.2.55) commutes with all generators* $\mathsf{e}_\pm, \mathsf{h} \in s\ell(2,\mathbb{C})$ *in representation* $T_{(-j)}$:

$$\mathsf{e}_\pm \cdot I = I \cdot \mathsf{e}_\pm, \quad \mathsf{h} \cdot I = I \cdot \mathsf{h}. \tag{2.2.57}$$

Indeed, we insert in (2.2.57) the expression (2.2.55) for operator I *and make use of Eqs. (2.2.51) and (2.2.53) to obtain a system of equations for* $\mu(z,\bar{z})$:

$$(\partial_{\bar{z}}\bar{z}^2 + 2\bar{z}j)\mu = -\partial_z\,\mu, \quad (\partial_z z^2 + 2zj)\mu = -\partial_{\bar{z}}\,\mu, \quad \bar{z}\,\partial_{\bar{z}}\,\mu = z\,\partial_z\,\mu.$$

The last equation implies that $\mu(z,\bar{z})$ *depends only on product* $x = z\bar{z}$. *Then the first and second equations are reduced to one and the same equation*

$$(2(1+j) + (1+x)\partial_x)\mu = 0, \tag{2.2.58}$$

whose general solution is $\mu(x) = c(1+x)^{-2j-2}$ *where* c *is a constant. Inserting this solution into (2.2.55), we obtain an operator*

$$I = c\int \frac{d^2z}{(1+z\bar{z})^{2+2j}}|z\rangle\langle\bar{z}|, \tag{2.2.59}$$

which commutes with all generators of $s\ell(2,\mathbb{C})$ *in irreducible representation* $T_{(-j)}$. *Due to Schur's lemma,* I *is proportional to unit operator. The constant* c *is obtained from the requirement that* $|w\rangle = I \cdot |w\rangle$, *where* $|w\rangle$ *is an arbitrarily chosen coherent state (2.2.50):*

$$\begin{aligned}|w\rangle = I \cdot |w\rangle &= c\int \frac{d^2z}{(1+z\,\bar{z})^{2+2j}}\,|z\rangle\langle\bar{z}|w\rangle = c\int \frac{d^2z}{(1+z\,\bar{z})^{2+2j}}\,|z\rangle\langle\bar{z}|e^{w\,\mathsf{e}_+}|-j\rangle \\ &= c\int \frac{d^2z}{(1+z\,\bar{z})^{2+2j}}\,e^{z\,\mathsf{e}_+}|w\rangle\,\langle\bar{z}|-j\rangle = \left(c\int \frac{d^2z}{(1+z\,\bar{z})^{2+2j}}\right)|w\rangle,\end{aligned} \tag{2.2.60}$$

where we use (2.2.57), identity $e^{w\,\mathsf{e}_+}|z\rangle = e^{z\,\mathsf{e}_+}|w\rangle$, *condition* $\langle\bar{z}|-j\rangle = 1$, *and also relation* $\int d^2z\mu(z\bar{z})z^k \sim \delta_{k,0}$. *As a result, we obtain*

$$c^{-1} = \int \frac{d^2z}{(1+z\,\bar{z})^{2+2j}} = \frac{\pi}{2j+1}, \tag{2.2.61}$$

which finally gives the form (2.2.55), (2.2.56) of unit operator in $\mathcal{V}$. □

Problem 2.2.11. *Show by direct calculation that the right-hand sides of Eqs. (2.2.54) and (2.2.55) coincide identically. Hint: calculate the integral*[c]

$$\int d^2z\,\frac{z^{m+j}\,\bar{z}^{k+j}}{(1+z\,\bar{z})^{2+2j}} = \pi\,\delta_{k,m}\,\frac{(j+m)!(j-m)!}{(2j+1)!}. \tag{2.2.62}$$

[c]This integral generalizes (2.2.61), and also reinstates the relation $\int d^2z\mu(z\bar{z})z^k \sim \delta_{k,0}$ which we used in deriving (2.2.60).

Making use of the completeness relation (2.2.55), we can expand any vector $|\Psi\rangle \in \mathcal{V}$ in coherent state basis:

$$|\Psi\rangle = \int d^2 z\, \mu(z,\bar{z})\, |z\rangle\langle\bar{z}|\Psi\rangle = \int d^2 z\, \mu(z,\bar{z})\, |\bar{z}\rangle\langle z|\Psi\rangle \tag{2.2.63}$$

(we changed integration variables $z \leftrightarrow \bar{z}$ to obtain the last equality). Thus, the role of the coordinates of vector $|\Psi\rangle \in \mathcal{V}$ in the coherent state basis is played by wave functions $\Psi(z) := \langle z|\Psi\rangle$, and, according to (2.2.53), the generators $\{\mathsf{e}_\pm, \mathsf{h}\}$ act on these wave functions as follows:

$$\langle z|\, \mathsf{e}_-|\Psi\rangle = \partial_z\, \langle z|\Psi\rangle = \partial_z\, \Psi(z), \quad \langle z|\, \mathsf{h}|\Psi\rangle = (z\,\partial_z - j)\,\langle z|\Psi\rangle = (z\,\partial_z - j)\Psi(z),$$
$$\langle z|\, \mathsf{e}_+|\Psi\rangle = (2\,j\,z - z^2\,\partial_z)\,\langle z|\Psi\rangle = (2\,j\,z - z^2\,\partial_z)\Psi(z). \tag{2.2.64}$$

Note that the right-hand sides here involve differential realization of the generators of $s\ell(2,\mathbb{C})$, which we have already obtained (see (2.2.37)) from another viewpoint. Finally, we use the expression (2.2.59) for unit operator and write the Hermitian scalar product (2.2.43) in terms of the wave functions $\Psi(z) = \langle z|\Psi\rangle$:

$$\langle\Psi|\Phi\rangle = c\int \frac{d^2 z}{(1+z\,\bar{z})^{2+2j}}\,\langle\Psi|\bar{z}\rangle\langle z|\Phi\rangle = c\int \frac{\overline{\Psi(z)}\Phi(z)}{(1+|z|^2)^{2+2j}} dz d\bar{z}, \tag{2.2.65}$$

where the constant c is given by (2.2.61).

In the coherent state basis, vector $|m\rangle$ defined in (2.2.42) has the wave function

$$\langle z|m\rangle = a_m\, z^{j+m}, \qquad a_m := \left(\frac{(2j)!}{(j+m)!(j-m)!}\right)^{1/2}. \tag{2.2.66}$$

Thus, the basis vector $|m\rangle$ has the wave function which is a monomial in z of degree $(j+m)$. Any vector $|\Psi\rangle \in \mathcal{V}$ is expanded in the basis vectors $|m\rangle$, so its wave function is a polynomial in z of degree $2j$,

$$\Psi(z) = \langle z|\Psi\rangle = \sum_{m=-j}^{j} \langle z|m\rangle\,\langle m|\Psi\rangle = \sum_{m=-j}^{j} \Psi_m\, z^{j+m},$$

where $\Psi_m = a_m\,\langle m|\Psi\rangle \in \mathbb{C}$. Thus, irreducible finite-dimensional representation $T_{(-j)}$ of algebra $s\ell(2,\mathbb{C})$ is acting in the space $\mathcal{P}_{2j}$ of polynomials in variable z of degree $2j$. The action of the $s\ell(2,\mathbb{C})$ generators in the space $\mathcal{P}_{2j}$ is given by (2.2.64) and it transforms polynomials belonging to $\mathcal{P}_{2j}$ to polynomials belonging to $\mathcal{P}_{2j}$. We have already considered this representation of algebra $s\ell(2,\mathbb{C})$ in the end of Section 2.2.2.

Problem 2.2.12. *Check the orthonormality (2.2.46) of basis wave functions (2.2.66). Hint: make use of the representation (2.2.65) and recall (2.2.62).*

The action of the generators of the Lie algebra $s\ell(2,\mathbb{C})$, realized as differential operators (2.2.64), is directly extended to infinite-dimensional space $\mathcal{F}$ of smooth functions of z. The latter representation $T_z^{\{j\}}$ (here the notation reflects the fact that the differential operators involve j as a parameter) is reducible, as $\mathcal{F}$ contains invariant subspace of polynomials of degree $2j$, which is the subspace of finite-dimensional representation $T_{(-j)}$. The representation $T_z^{\{j\}}$ of the Lie algebra $s\ell(2,\mathbb{C})$ corresponds to some infinite-dimensional representation $T_z^{\{j\}}$ of Lie group $SL(2,\mathbb{C})$ in space $\mathcal{F}$. Let us find explicit form of the action of group $SL(2,\mathbb{C})$ in representation $T_z^{\{j\}}$.

Proposition 2.2.3. *Let representation $T_z^{\{j\}}$ of Lie group $SL(2,\mathbb{C})$ in space $\mathcal{F}$ correspond to differential representation (2.2.64) of Lie algebra $s\ell(2,\mathbb{C})$. The element $g = \left(\begin{smallmatrix} a & b \\ c & d \end{smallmatrix}\right) \in SL(2,\mathbb{C})$, in representation $T_z^{\{j\}}$, acts on functions $\Psi(z) \in \mathcal{F}$ in the following way (cf. (2.2.39)):*

$$T_z(g) \cdot \Psi(z) = (b\,z + d)^{2j}\,\Psi\left(\frac{a\,z + c}{b\,z + d}\right). \tag{2.2.67}$$

Proof. *An arbitrary element $g \in SL(2,\mathbb{C})$ can be written as (see Problem 2.1.1)*

$$g = \exp(a_- e_-) \cdot \exp(a_0 \mathsf{h}) \cdot \exp(a_+ e_+), \tag{2.2.68}$$

where $a_\pm, a_0$ are complex parameters. In defining representation, this is, explicitly,

$$g = \begin{pmatrix} e^{\frac{a_0}{2}} & a_+\, e^{\frac{a_0}{2}} \\ a_-\, e^{\frac{a_0}{2}} & e^{-\frac{a_0}{2}} + a_-\, a_+\, e^{\frac{a_0}{2}} \end{pmatrix} \equiv \begin{pmatrix} a & b \\ c & d \end{pmatrix}. \tag{2.2.69}$$

In representation $T_z^{\{j\}}$, the generators $e_\pm, \mathsf{h}$ act as given by (2.2.64), and the following relations hold:

$$\begin{aligned} T_z(\exp(a_- e_-)) \cdot \Psi(z) &= e^{a_- \partial_z} \cdot \Psi(z) = \Psi(z + a_-), \\ T_z(\exp(a_0 \mathsf{h})) \cdot \Psi(z) &= e^{a_0 (z\partial_z - j)} \cdot \Psi(z) = e^{-j\,a_0}\, \Psi(e^{a_0}\, z), \end{aligned} \tag{2.2.70}$$

$$T_z(\exp(a_+ e_+)) \cdot \Psi(z) = e^{a_+ (-z^2\,\partial_z + 2\,j\,z)} \cdot \Psi(z) = (1 + a_+ z)^{2j}\, \Psi\left(\frac{z}{1 + a_+ z}\right). \tag{2.2.71}$$

Relations (2.2.70) are fairly straightforward to check; we leave this check for the reader. To check (2.2.71), we write its left-hand side in the form

$$e^{\lambda e_+} \cdot \Psi(z) = \Psi(e^{\lambda e_+}\, z\, e^{-\lambda e_+}) e^{\lambda e_+} \cdot 1, \tag{2.2.72}$$

where we write $e^{\lambda e_+}$ instead of $T_z(e^{\lambda e_+})$ for brevity. The formula (2.2.71) then follows from equalities

$$e^{\lambda e_+}\, z\, e^{-\lambda e_+} = \frac{z}{(1 + \lambda z)}, \qquad e^{\lambda e_+} \cdot 1 = (1 + \lambda z)^{2j}. \tag{2.2.73}$$

The first formula in (2.2.73) is understood as operator equality, while dot in the second formula denotes the action of $e^{\lambda e_+}$ on function (say, $\partial_z \cdot 1 = 0$, etc.).

Problem 2.2.13. * *Derive the identities (2.2.72) and (2.2.73). Making use of these identities, obtain Eq. (2.2.71).*

Finally, we make use of (2.2.70), (2.2.71) and find the action of an element (2.2.68) on function $\Psi(z)$:

$$\begin{aligned} T_z(g) \cdot \Psi(z) &= T_z(e^{a_- e_-}) \cdot T_z(e^{a_0 \mathsf{h}}) \cdot T_z(e^{a_+ e_+}) \cdot \Psi(z) \\ &= T_z(e^{a_- e_-}) \cdot T_z(e^{a_0 \mathsf{h}}) \cdot (1 + a_+ z)^{2j}\, \Psi\left(\frac{z}{1 + a_+ z}\right) \\ &= T_z(e^{a_- e_-}) \cdot e^{-j\,a_0}\, (1 + a_+ e^{a_0} z)^{2j}\, \Psi\left(\frac{e^{a_0}\, z}{1 + a_+ e^{a_0}\, z}\right) \end{aligned}$$

$$= (e^{-\frac{a_0}{2}} + a_+ e^{\frac{a_0}{2}} (z + a_-))^{2j}\ \Psi\left(\frac{e^{a_0}(z + a_-)}{1 + a_+ e^{a_0}(z + a_-)}\right)$$

$$= (b\,z + d)^{2j}\ \Psi\left(\frac{a\,z + c}{b\,z + d}\right),$$

where coefficients a, b, c, d are defined in (2.2.69). □

Problem 2.2.14. *Show that for $j \in \mathbb{Z}_{\geq 0}/2$, the action (2.2.67) of group $SL(2,\mathbb{C})$ moves polynomials $\Psi(z) \in \mathcal{P}_{2j}$ to polynomials belonging to $\mathcal{P}_{2j}$.*

2.3. Finite-Dimensional Representations of Groups $SU(2)$ and $SL(2,\mathbb{C})$

2.3.1. Parameterizations of Group $SU(2)$

Parameterizations of groups $U(n)$ in terms of $\frac{n(n-1)}{2}$ hyperspherical angles and $\frac{n(n+1)}{2}$ phases were constructed in Section I-7.2 of the first book (see formula (I-7.2.38)). Using these parameterizations, one can construct parameterizations of groups $SU(n)$. In particular, we introduced in Section I-7.2 of the first book the parameterization of group $SU(2)$ in terms of Euler angles (see formula (I-7.2.37)). In this section, we discuss in detail several parameterizations of group $SU(2)$, some of which we use in what follows. Recall that the group manifold of $SU(2)$ is sphere S^3, so the parameterizations we discuss are valid for S^3 as well; they are widely used in physics applications.

Let us recall (see Section I-2.1.2 of the first book) that any element U of Lie group $SU(2)$ can be written as follows:

$$U = \begin{pmatrix} \alpha & \beta \\ -\beta^* & \alpha^* \end{pmatrix}, \quad |\alpha|^2 + |\beta|^2 = 1, \tag{2.3.1}$$

where $\alpha,\ \beta \in \mathbb{C}$. We set $\alpha = x_0 + ix_3$, $\beta = x_2 + ix_1$ ($x_m \in \mathbb{R}$), then the formula (2.3.1) becomes

$$U = I_2\, x_0 + \sum_{k=1}^{3} i\,\sigma_k\, x_k, \quad \sum_{m=0}^{3} x_m x_m = 1, \tag{2.3.2}$$

where σ_k are Pauli matrices (2.1.16), and I_2 is unit 2×2 matrix. Formula (2.3.2) reiterates that the group manifold of $SU(2)$ is 3-sphere S^3, and also shows that U is quaternion of unit length (the definition of quaternions is given in Section I-3.3.3 of the first book).

Any element A of Lie algebra $su(2)$ is a linear combination of three generators $\tau_k = -i\sigma_k/2$:

$$A = \Psi \cdot (n_k \tau_k) = -i\frac{\Psi}{2}(n_1\sigma_1 + n_2\sigma_2 + n_3\sigma_3), \tag{2.3.3}$$

where $\Psi \in \mathbb{R}$ and $\vec{n} = (n_1, n_2, n_3)$ is unit vector in $\mathbb{R}^3$. Components of vector $\vec{n}$ can be defined in terms of spherical angles $\varphi \in [0, 2\pi)$ and $\vartheta \in [0, \pi]$:

$$n_1 = \sin\vartheta\cos\varphi, \quad n_2 = \sin\vartheta\sin\varphi, \quad n_3 = \cos\vartheta. \tag{2.3.4}$$

We make use of the exponential mapping and construct an element of group $SU(2)$ in terms of element $A \in su(2)$:

$$\exp(A) = \exp(\Psi \cdot (n_k \tau_k)) = \exp\left(-i\frac{\Psi}{2}(n_k\sigma_k)\right). \tag{2.3.5}$$

As a side remark, we recall (see Section I-3.5 of the first book) that group $SU(2)$ covers twice group $SO(3)$: two elements $\pm\exp(\Psi(n_k \tau_k))$ of $SU(2)$ correspond to one element of $SO(3)$:

$$O(\Psi, \vec{n}) = \exp(\Psi \cdot (n_k S_k)), \tag{2.3.6}$$

where matrices

$$S_1 = \begin{pmatrix} 0 & 0 & 0 \\ 0 & 0 & -1 \\ 0 & 1 & 0 \end{pmatrix}, \quad S_2 = \begin{pmatrix} 0 & 0 & 1 \\ 0 & 0 & 0 \\ -1 & 0 & 0 \end{pmatrix}, \quad S_3 = \begin{pmatrix} 0 & -1 & 0 \\ 1 & 0 & 0 \\ 0 & 0 & 0 \end{pmatrix}, \tag{2.3.7}$$

realize adjoint (spin $j = 1$) representation of generators τ_1, τ_2, τ_3 of $SU(2)$ and generate rotations in $\mathbb{R}^3$ around axes $\vec{e}_1 = (1, 0, 0)$, $\vec{e}_2 = (0, 1, 0)$ and $\vec{e}_3 = (0, 0, 1)$, respectively. We now expand the exponential in (2.3.6) and, making use of the identity $(n_k S_k)^3 = -(n_k S_k)$, obtain

$$\exp(\Psi \cdot (n_k S_k)) = I_3 + (n_k S_k)\sin\Psi + (n_k S_k)^2(1 - \cos\Psi). \tag{2.3.8}$$

This element corresponds to the rotation in $\mathbb{R}^3$ by angle Ψ around the axis directed along vector $\vec{n}$, so for group $SO(3)$ one chooses the range of parameter Ψ as $\Psi \in [-\pi, \pi)$, or one can shift this interval and set $\Psi \in [0, 2\pi)$.

Turning back to $SU(2)$, we parameterize its elements as

$$U_{\vec{n}}(\Psi) = \exp\left(i\frac{\Psi}{2}(n_i\sigma_i)\right). \tag{2.3.9}$$

Pauli matrices obey $\sigma_k\sigma_j = \delta_{kj}I_2 + i\varepsilon_{kji}\sigma_i$, which gives

$$(n_k\,\sigma_k)^2 = \vec{n}^{\,2}I_2 = I_2 \Rightarrow (i\,n_k\,\sigma_k)^2 = -I_2. \tag{2.3.10}$$

Therefore, we make use of the operator version of Euler formula,

$$\exp(\phi\, J) = (I\,\cos\phi + J\,\sin\phi),$$

where ϕ is a parameter, I is unit operator and J is any operator such that $J^2 = -I$. As a result, we obtain the matrix expression

$$\begin{aligned} U_{\vec{n}}(\Psi) &= \exp\left(i(n_i\sigma_i)\frac{\Psi}{2}\right) = I_2\cos\left(\frac{\Psi}{2}\right) + i(n_i\sigma_i)\sin\left(\frac{\Psi}{2}\right) \\ &= \begin{pmatrix} \cos\frac{\Psi}{2} + in_3\sin\frac{\Psi}{2}, & (n_2+in_1)\sin\frac{\Psi}{2} \\ -(n_2-in_1)\sin\frac{\Psi}{2}, & \cos\frac{\Psi}{2} - in_3\sin\frac{\Psi}{2} \end{pmatrix}. \end{aligned} \tag{2.3.11}$$

It agrees with (2.3.1) provided the range of Ψ is $\Psi \in [0, 2\pi)$. Indeed, one compares (2.3.2) and (2.3.11) and finds

$$x_0 = \cos\frac{\Psi}{2}, \qquad x_i = n_i\sin\frac{\Psi}{2}, \tag{2.3.12}$$

which, together with (2.3.4), give the standard parameterization of 3-sphere

$$\begin{aligned} x_0 &= \cos\chi, \quad x_1 = \sin\chi\sin\vartheta\cos\varphi, \\ x_2 &= \sin\chi\sin\vartheta\sin\varphi, \quad x_3 = \sin\chi\cos\vartheta, \end{aligned} \tag{2.3.13}$$

where

$$\chi = \frac{\Psi}{2}. \tag{2.3.14}$$

The range of spherical angles is

$$\chi \in [0,\pi), \qquad \vartheta \in [0,\pi], \qquad \varphi \in [0, 2\pi),$$

which agrees with the parameterization of sphere S^n studied in Section I-7.2 (item 3) of the first book. We have explicitly

$$U = I_2\,x_0 + i\sigma_k\,x_k = \begin{pmatrix} \cos\chi + i\sin\chi\cos\vartheta, & i\sin\chi\sin\vartheta\,\mathrm{e}^{-i\varphi} \\ i\sin\chi\sin\vartheta\,\mathrm{e}^{i\varphi}, & \cos\chi - i\sin\chi\cos\vartheta \end{pmatrix}. \tag{2.3.15}$$

Note that there is a similarity transformation that takes any element $U_{\vec{n}}(\Psi) \in SU(2)$, parameterized as in (2.3.9), to diagonal form:

$$U_{\vec{n}}(\Psi) = U_{\vartheta,\varphi}^{-1} \cdot \exp\left(i\frac{\Psi}{2}\sigma_3\right) \cdot U_{\vartheta,\varphi}, \tag{2.3.16}$$

where $U_{\vartheta,\varphi} = U_{\vec{e}_2}(\vartheta) \cdot U_{\vec{e}_3}(\varphi)$ and (see (2.3.11))

$$U_{\vec{e}_2}(\vartheta) = \exp\left(i\frac{\vartheta}{2}\sigma_2\right) = \begin{pmatrix} \cos\frac{\vartheta}{2} & \sin\frac{\vartheta}{2} \\ -\sin\frac{\vartheta}{2} & \cos\frac{\vartheta}{2} \end{pmatrix},$$
$$U_{\vec{e}_3}(\varphi) = \exp\left(i\frac{\varphi}{2}\sigma_3\right) = \begin{pmatrix} e^{i\frac{\varphi}{2}} & 0 \\ 0 & e^{-i\frac{\varphi}{2}} \end{pmatrix}. \tag{2.3.17}$$

Formula (2.3.16) follows from the relation $(n_i\sigma_i) = U_{\vartheta,\varphi}^{-1} \cdot \sigma_3 \cdot U_{\vartheta,\varphi}$, which is straightforward to check.

Problem 2.3.1. Show that with parameterization (2.3.11), (2.3.4), the invariant measure $d\mu(U_{\vec{n}}(\Psi))$ on group $SU(2)$ reads

$$d\mu(U_{\vec{n}}(\Psi)) = \frac{1}{2}\sin(\vartheta)\sin^2(\Psi/2)d\Psi\, d\vartheta\, d\varphi. \tag{2.3.18}$$

Hint: make use of the expression for the Haar measure on group $SU(2)$ (see Section I-2.2.2 of the first book)

$$d\mu(g) = -\mathrm{Tr}\left(\frac{\partial g}{\partial y_{i_1}}g^{-1} \cdot \frac{\partial g}{\partial y_{i_2}}g^{-1} \cdot \frac{\partial g}{\partial y_{i_3}}g^{-1}\right) \epsilon_{i_1 i_2 i_3} dy_1 dy_2 dy_3,$$

where y_i are coordinates on group manifold.

In what follows, we use another parameterization of group $SU(2)$, which is obtained as follows. Complex parameters α and β in (2.3.1) can be written as $\alpha = \rho_1 e^{i\xi_1}$ and $\beta = \rho_2 e^{i\xi_2}$, where $\rho_1^2 + \rho_2^2 = 1$. Then we make a change of variables

$$\rho_1 = \cos\frac{\theta}{2}, \quad \rho_2 = \sin\frac{\theta}{2}, \quad \xi_1 = \frac{1}{2}(\phi + \psi), \quad \xi_2 = \frac{1}{2}(\phi - \psi),$$

and obtain the following parameterization of group $SU(2)$:

$$U = \begin{pmatrix} \cos\left(\frac{\theta}{2}\right) e^{i\frac{1}{2}(\psi+\phi)}, & \sin\left(\frac{\theta}{2}\right) e^{i\frac{1}{2}(\phi-\psi)} \\ -\sin\left(\frac{\theta}{2}\right) e^{i\frac{1}{2}(\psi-\phi)}, & \cos\left(\frac{\theta}{2}\right) e^{-i\frac{1}{2}(\psi+\phi)} \end{pmatrix} \tag{2.3.19}$$

$$= U_{\vec{e}_3}(\phi) \cdot U_{\vec{e}_2}(\theta) \cdot U_{\vec{e}_3}(\psi)$$

$$= \begin{pmatrix} e^{i\frac{\phi}{2}} & 0 \\ 0 & e^{-i\frac{\phi}{2}} \end{pmatrix} \begin{pmatrix} \cos\frac{\theta}{2} & \sin\frac{\theta}{2} \\ -\sin\frac{\theta}{2} & \cos\frac{\theta}{2} \end{pmatrix} \begin{pmatrix} e^{i\frac{\psi}{2}} & 0 \\ 0 & e^{-i\frac{\psi}{2}} \end{pmatrix}. \tag{2.3.20}$$

Formula (2.3.20) is an analog of the parameterization of group $SO(3)$ in terms of Euler angles (Item 2 in Section I-7.2 of the first book).

Matrix (2.3.19), (2.3.20) coincides with (2.3.1), once the ranges of angles ϕ, θ and ψ are chosen appropriately. This point is relevant, in particular, when one integrates over the whole group $SU(2)$ in coordinates ϕ, θ and ψ. Recall that in the case of group $SO(3)$ the range of Euler angles is

$$SO(3): \quad \psi \in [-\pi, \pi), \quad \theta \in [0, \pi], \quad \phi \in [0, 2\pi). \tag{2.3.21}$$

To cover the whole group $SU(2)$ (sphere S^3) with the standard ranges of angles ϕ, θ given by (2.3.21) we have to set

$$SU(2): \quad \psi \in [-2\pi, 2\pi), \quad \theta \in [0, \pi], \quad \phi \in [0, 2\pi), \tag{2.3.22}$$

(compare the range of ψ in (2.3.21) and (2.3.22)). With this range of angles, group $SU(2)$ indeed covers twice group $SO(3)$ and elements $U_{\vec{e}_3}(\psi)$, right factors in the product (2.3.20), make the whole subgroup $U(1)$ in $SU(2)$. We use the latter fact below, when discussing spherical functions on homogeneous space $SU(2)/U(1)$.

We emphasize that the parameterization (2.3.11) of $SU(2)$ (say, in terms of spherical angles (2.3.15)) should not be confused with the parameterization of $SU(2)$ in terms of Euler angles, Eq. (2.3.19). In the Euler parameterization, the relation to parameters x_m in (2.3.2) reads (cf. (2.3.12) and (2.3.13))

$$\begin{aligned} x_0 &= \cos\frac{\theta}{2}\cos\frac{\phi+\psi}{2}, \quad x_1 = \sin\frac{\theta}{2}\sin\frac{\phi-\psi}{2}, \\ x_2 &= \sin\frac{\theta}{2}\cos\frac{\phi-\psi}{2}, \quad x_3 = \cos\frac{\theta}{2}\sin\frac{\phi+\psi}{2}. \end{aligned} \tag{2.3.23}$$

2.3.2. Finite-dimensional representations of groups $SU(2)$, $SL(2,\mathbb{C})$ and $SO(3)$. Tensor representations — Wigner functions

Let us discuss in more detail representations of group $SU(2)$, and also $SO(3)$. As before, it is convenient to consider also special representations of group $SL(2,\mathbb{C})$, specified in Remark 3 in the end of Section 2.1, which are in one-to-one correspondence with representations of $SU(2)$.

Making use of the notations $s = u^1, t = u^2$ introduced in (2.2.1) and the expression (2.2.26) for basis vectors $T^j_m(s,t)$:

$$T^j_m(s,t) = ((j+m)!(j-m)!)^{-1/2}(u^1)^{j+m}(u^2)^{j-m}, \tag{2.3.24}$$

we write an arbitrary homogeneous polynomial (2.2.27) as

$$P_{2j}(s,t) = \psi_{\alpha_1,\alpha_2,\ldots,\alpha_{2j}}\, u^{\alpha_1}\, u^{\alpha_2}\cdots u^{\alpha_{2j}}, \tag{2.3.25}$$

where components $\psi_{\alpha_1,\ldots,\alpha_{2j}} \in \mathbb{C}$ make symmetric (in view of symmetry of the product $u^{\alpha_1}\cdots u^{\alpha_{2j}}$) rank-$2j$ tensor. Therefore, the space $\mathcal{V}_{2j+1}$ of homogeneous polynomials (2.2.27) (which is the space of irreducible representation $T^{(j)}$) is equivalent to space of rank-$2j$ symmetric tensors. In this regard, we recall that the general definition of tensor is given in Section I-4.3.1 of the first book.

Problem 2.3.2. Show that components $\psi_{\alpha_1,\ldots,\alpha_{2j}}$ in (2.3.25) are related to parameters α_m in (2.2.27) as follows: $\psi_{\underbrace{1,\ldots,1}_{j+m},\underbrace{2,\ldots,2}_{j-m}} = \sqrt{(j+m)!(j-m)!}\,\frac{\alpha_m}{(2j)!}$.

In accordance with (2.2.2) and (2.2.5), action of element $g = ||A_\beta{}^\alpha||$ of group $SL(2,\mathbb{C})$ (and group $SU(2)$) on basis monomials $u^{\alpha_1}\cdots u^{\alpha_{2j}} \in \mathcal{V}_{2j+1}$ reads

$$u^{\alpha_1}\cdots u^{\alpha_{2j}} \to T(g)\cdot[u^{\alpha_1}\cdots u^{\alpha_{2j}}] = u^{\beta_1}\cdot u^{\beta_2}\cdots u^{\beta_{2j}}\, A_{\beta_1}{}^{\alpha_1}\, A_{\beta_2}{}^{\alpha_2}\cdots A_{\beta_{2j}}{}^{\alpha_{2j}}. \tag{2.3.26}$$

This action is extended to linear combinations of these monomials, i.e., to arbitrary homogeneous polynomials (2.3.25), vectors in space $\mathcal{V}_{2j+1}$ of irreducible representation $T^{(j)}$. Following the standard procedure (2.1.7) and (2.1.8), action (2.3.26) is written as transformation of coordinates of vectors (2.3.25), i.e., components $\psi_{\alpha_1,\dots,\alpha_{2j}}$ of rank-$2j$ symmetric tensor:

$$\psi_{\alpha_1,\dots,\alpha_{2j}} \to [T^{(j)}(g)\cdot\psi]_{\alpha_1,\dots,\alpha_{2j}} = A_{\alpha_1}^{\ \beta_1}\cdots A_{\alpha_{2j}}^{\ \beta_{2j}}\,\psi_{\beta_1,\dots,\beta_{2j}}. \tag{2.3.27}$$

We now recall the definition of direct product of representations (see Section I-4.3.1 of the first book) and find that the space $\mathcal{V}_{2j+1}$ of rank-$2j$ symmetric tensors is a subspace of space $\underbrace{\mathcal{V}_2\otimes\mathcal{V}_2\otimes\cdots\otimes\mathcal{V}_2}_{2j}$, where $\mathcal{V}_2$ is space of two-dimensional defining representation $T^{(1/2)}$. The representation $T^{(j)}(g)$ of an element $g\in SL(2,\mathbb{C})$ acts in $\mathcal{V}_{2j+1}$ as operator written as the tensor product of two-dimensional operators (see (2.3.26) and (2.3.27)):

$$\underbrace{A\otimes\cdots\otimes A}_{2j} = \underbrace{T^{(1/2)}(g)\otimes\cdots\otimes T^{(1/2)}(g)}_{2j} \equiv (T^{(1/2)}(g))^{\otimes 2j}. \tag{2.3.28}$$

Thus, the irreducible representation $T^{(j)}$ is embedded in representation $(T^{(1/2)})^{\otimes 2j}$. In other words, representation $(T^{(1/2)})^{\otimes 2j}$ is reducible, and its expansion in irreducible representations contains representation $T^{(j)}$ which acts in space $\mathcal{V}_{2j+1}\subset\mathcal{V}_2^{\otimes 2j}$ of symmetric rank-$2j$ tensors (we discuss this fact in more detail in Sections 2.4.2 and 2.4.3).

Likewise, any homogeneous polynomial of degree $2j$ in variables $(\bar{u}_1,\bar{u}_2)=(\bar{s},\bar{t})$ is a linear combination of dual basis monomials $\tilde{T}^j_m(\bar{s},\bar{t})$, which, modulo sign factors, coincide with (2.2.29) (cf. (2.3.24)):

$$\tilde{T}^j_m(\bar{s},\bar{t}) \equiv (-1)^{j-m}\,\overline{T}^j_{-m}(\bar{s},\bar{t}) = ((j+m)!(j-m)!)^{-1/2}(\bar{u}_1)^{j+m}(\bar{u}_2)^{j-m}. \tag{2.3.29}$$

Any such polynomial is written as

$$\overline{P}_{2j}(\bar{s},\bar{t}) = \bar{u}_{\alpha_1}\bar{u}_{\alpha_2}\cdots\bar{u}_{\alpha_{2j}}\,\overline{\psi}^{\,\alpha_1,\dots,\alpha_{2j}}.$$

The components $\overline{\psi}^{\,\alpha_1,\dots,\alpha_{2j}}\in\mathbb{C}$ again make symmetric rank-$2j$ tensor. In accordance with (2.2.8), analogs of transformations (2.3.27) are

$$\overline{\psi}^{\,\alpha_1,\dots,\alpha_{2j}} \to [T^{(j)}(g)\cdot\overline{\psi}]^{\alpha_1,\dots,\alpha_{2j}} = \overline{\psi}^{\beta_1,\dots,\beta_{2j}}\,(A^{-1})_{\beta_1}^{\ \alpha_1}\cdots(A^{-1})_{\beta_{2j}}^{\ \alpha_{2j}}. \tag{2.3.30}$$

Thus, the symmetric tensor $\overline{\psi}$ transforms in contragredient way with respect to transformations (2.3.27) of tensor ψ, i.e., the space of symmetric tensors $\overline{\psi}$ is the space of representation $\overline{T}^{(j)}$ which is contragredient to representation $T^{(j)}$. Representations of the types (2.3.27) and (2.3.30) are sometimes called tensor, or spin-tensor. These representations are further discussed in Section 4.4.

Remark. Irreducible representations $T^{(j)}$ and their co-representations $\overline{T}^{(j)}$ are equivalent for both $SL(2,\mathbb{C})$ and $SU(2)$. The equivalence relation is

$$\varepsilon_{\alpha_1\beta_1}\cdots\varepsilon_{\alpha_{2j}\beta_{2j}}\overline{\psi}^{\beta_1,\dots,\beta_{2j}} = \psi_{\alpha_1,\dots,\alpha_{2j}}, \tag{2.3.31}$$

which follows from Eq. (2.2.17). Therefore, we concentrate in this section on representations $T^{(j)}$ only. We emphasize that the situation with complex conjugate representations $T^{(j)*}$ is different. For unitary matrices $A \in SU(2)$ we have $A^* = (A^{\mathsf{T}})^{-1}$. Therefore, in the case of $SU(2)$, complex conjugate representation $T^{(j)*}$ (whose matrices are complex conjugate to (2.3.28)) coincides with contragredient representation $\overline{T}^{(j)}$, which is equivalent to $T^{(j)}$. This means that for group $SU(2)$, representation $T^{(j)}$ is equivalent to its complex conjugate representation $T^{(j)*}$. However, this is not the case for group $SL(2,\mathbb{C})$ which we treat as real six-dimensional group: representations $T^{(j)}$ and $T^{(j)*}$, like representations $T^{(\frac{1}{2})}$ and $T^{(\frac{1}{2})*}$, are inequivalent (see Remark 2 in the end of Section 2.1). Inequivalent representations $T^{(j)}$ and $T^{(j)*}$ of group $SL(2,\mathbb{C})$ and their tensor products are important for physics applications.

We now find explicit form of representation $T^{(j)}(g)$ of element g of group $SL(2,\mathbb{C})$ (or group $SU(2)$). Operators $T^{(j)}(g)$ act in space $\mathcal{V}_{2j+1}$ whose basis elements are monomials (2.3.24). According to (2.2.2), (2.2.5) and (2.3.26), the element $g = ||A^{\alpha}_{\beta}|| \in SL(2,\mathbb{C})$ transforms the monomial $T^j_m(s,t)$ into homogeneous polynomial of degree $2j$, namely

$$[T^{(j)}(g)\cdot T^j_m](s,t) = T^j_m(s',t') = \frac{(sA_1^1 + tA_2^1)^{j+m}(sA_1^2 + tA_2^2)^{j-m}}{\sqrt{(j+m)!(j-m)!)}} \tag{2.3.32}$$

$$= \sum_{k=0}^{j+m}\sum_{k'=0}^{j-m} \frac{\sqrt{(j+m)!(j-m)!}}{k!k'!(j+m-k)!(j-m-k')!}(t^{k+k'}s^{2j-k-k'})$$
$$\times (A_1^1)^{j+m-k}(A_1^2)^{j-m-k'}(A_2^1)^k(A_2^2)^{k'}, \tag{2.3.33}$$

where in the last equality we use binomial expansion. The next step is to write the polynomial (2.3.33) as a linear combination of basis monomials

$T^j_{m'}(s,t)$. It is convenient to define

$$n! = \infty \quad \text{for} \quad n < 0, \quad 0! = 1, \tag{2.3.34}$$

which agree with analytic continuation of factorial as Gamma-function: $n! = \Gamma(n+1)$. With this definition we can extend the limits of summation in (2.3.33) to $\pm\infty$, since the factorials in the denominators are equal to infinity beyond summation limits. One sets $m' = j - k - k'$, then m' takes all integer values for integer j and all half-integer values for half-integer j. Now we extract basis monomials $T^j_{m'}(s,t)$ on the right-hand side of (2.3.33) and obtain

$$[T^{(j)}(g) \cdot T^j_m](s,t) = \sum_{m'=-j}^{j} T^j_{m'}(s,t)\mathcal{D}^{(j)}_{m'm}(||A^\alpha_\beta||), \tag{2.3.35}$$

where $g = ||A^\alpha_\beta||$ and $(2j+1) \times (2j+1)$ matrix $\mathcal{D}^{(j)}$ has elements

$$\mathcal{D}^{(j)}_{m'm}(g) = \sum_k \frac{\sqrt{(j+m)!(j-m)!(j+m')!(j-m')!}}{k!(j-k-m')!(j+m-k)!(k-m+m')!} \times (A_1^{\ 1})^{j+m-k}(A_1^{\ 2})^{k-m+m'}(A_2^{\ 1})^k(A_2^{\ 2})^{j-k-m'}. \tag{2.3.36}$$

This matrix is precisely the matrix of element g in irreducible representation $T^{(j)}$. These matrices obey

$$\mathcal{D}^{(j)}_{m_1m_2}(||A_\beta^{\ \alpha}B_\alpha^{\ \gamma}||) = \sum_{m_3} \mathcal{D}^{(j)}_{m_1m_3}(||A_\beta^{\ \alpha}||)\mathcal{D}^{(j)}_{m_3m_2}(||B_\beta^{\ \alpha}||), \tag{2.3.37}$$

which reflects the fact that the mapping (representation) $T^{(j)}$ is homomorphism. Note that given the form (2.3.36) of matrices $\mathcal{D}^{(j)}_{m'm}$, the property (2.3.37) is actually a set of nontrivial identities.

Finite-dimensional representation $T^{(j)}$ of group $SL(2,\mathbb{C})$, constructed in (2.3.35), is nonunitary. This can be seen explicitly, by noticing that matrix $||\mathcal{D}^{(j)}_{m'm}(g)||$ is not unitary for some $g = ||A^\alpha_\beta|| \in SL(2,\mathbb{C})$. Indeed, Hermitian convolution of vectors $T^j_m(s,t)$ (cf. the first equality in (2.2.33))

$$\sum_{m=-j}^{j} T^j_m(s,t)(T^j_m(s,t))^* = \frac{(|s|^2+|t|^2)^{2j}}{(2j)!} \tag{2.3.38}$$

must be invariant under the transformations (2.3.35) with unitary matrix $||\mathcal{D}^{(j)}_{m'm}(g)||$. However, it is clear that the right-hand side of (2.3.38) changes under transformations corresponding to nonunitary matrices $g = ||A^\alpha_\beta|| \in SL(2,\mathbb{C})$ (nonunitary two-dimensional transformations (2.2.2) do not conserve the quadratic form, $|s|^2 + |t|^2 \neq |s'|^2 + |t'|^2$). The fact that matrices $||\mathcal{D}^{(j)}_{m'm}||$ are nonunitary, is no surprise: a finite-dimensional representation of noncompact $SL(2,\mathbb{C})$ is always nonunitary. However, matrices $||\mathcal{D}^{(j)}_{m'm}||$ with elements (2.3.36) still have a number of remarkable properties.

Proposition 2.3.1. *Matrices $||\mathcal{D}^{(j)}_{m'm}(g)||$ of finite-dimensional representation of group $SL(2,\mathbb{C})$, which are defined in (2.3.36), are equivalent to pseudo-orthogonal matrices for integer j and symplectic matrices*[d] *for half-integer j.*

Proof. Using the identification (2.2.17), we write invariant (2.2.33) as follows:

$$(s_1 t_2 - s_2 t_1)^{2j} = (2j)! \sum_{m=-j}^{j} (-1)^{j-m}\, T^j_m(s_1,t_1)\, T^j_{-m}(s_2,t_2)$$
$$= (2j)! \sum_{m,m'} T^j_m(s_1,t_1)\eta^{(j)}_{m,m'}\, T^j_{m'}(s_2,t_2), \qquad (2.3.39)$$

where we define metric[e]

$$\eta^{(j)}_{m,m'} = (-1)^{j-m}\delta_{m,-m'}, \qquad (2.3.40)$$

which obeys

$$\eta^{(j)}_{m',m} = (-1)^{2m}\eta^{(j)}_{m,m'} = (-1)^{2j}\eta^{(j)}_{m,m'}.$$

So, this metric is symmetric $\eta^{(j)}_{m,m'} = \eta^{(j)}_{m',m}$ for integer j and antisymmetric $\eta^{(j)}_{m,m'} = -\eta^{(j)}_{m',m}$ for half-integer j. Since the convolution (2.3.39) is invariant under group action, we find that metric (2.3.40) has the following invariance property:

$$\mathcal{D}^{(j)}_{km}(g)\mathcal{D}^{(j)}_{k'm'}(g)\,\eta^{(j)}_{m,m'} = \eta^{(j)}_{k,k'}$$
$$\Rightarrow \mathcal{D}^{(j)}(g)\cdot\eta^{(j)}\cdot(\mathcal{D}^{(j)}(g))^{\mathsf{T}} = \eta^{(j)}, \quad \forall g \in SL(2,\mathbb{C}), \qquad (2.3.41)$$

[d]Definitions of pseudo-orthogonal and symplectic matrices are given in Section I-1.2.5 in the first book.

[e]Recall that $(j-m)$ is always integer, so nonzero elements of metric $\eta^{(j)}_{m,m'}$ are equal to ± 1.

where matrices $||\mathcal{D}^{(j)}_{m'm}(g)||$, given by (2.3.36), define representation $T^{(j)}$ of $SL(2,\mathbb{C})$. Square of metric matrix reads

$$\eta^{(j)} \cdot \eta^{(j)} = (-1)^{2j} I_{2j+1}. \tag{2.3.42}$$

For integer j we have $(\eta^{(j)})^2 = I_{2j+1}$, so eigenvalues of symmetric matrix $||\eta^{(j)}_{m,m'}||$ are equal to (± 1), and this matrix can be diagonalized, and takes the standard form of pseudo-Euclidean metric (see Section I-1.2.5 in the first book):

$$I_{p,q} = \begin{pmatrix} I_p & 0 \\ 0 & -I_q \end{pmatrix}, \quad p + q = 2j + 1. \tag{2.3.43}$$

Here zeroes denote zero blocks $(p \times q)$ and $(q \times p)$. Then Eq. (2.3.41) determines odd-dimensional matrices $||\mathcal{D}^{(j)}_{m'm}(g)||$, which are equivalent to pseudo-orthogonal matrices, and for even j we have $p = j + 1$, $q = j$, while for odd j we have $p = j$, $q = j + 1$.

Problem 2.3.3. Derive these values of parameters p and q. Hint: find all eigenvectors and eigenvalues of matrix (2.3.40) for integer j.

For half-integer j, Eq. (2.3.42) shows that antisymmetric and even-dimensional matrix $||\eta^{(j)}_{m,m'}||$ obeys $(\eta^{(j)})^2 = -I_{2j+1}$, so by equivalence transformation it reduces to the standard form of symplectic matrix

$$J = \begin{pmatrix} 0 & I_r \\ -I_r & 0 \end{pmatrix}, \quad 2r = 2j + 1, \tag{2.3.44}$$

where 0 denotes zero $(r \times r)$ matrix. In this case, Eq. (2.3.41) is equivalent to defining relation $\mathcal{D} \cdot J \cdot \mathcal{D}^{\mathsf{T}} = J$ for symplectic matrices $\mathcal{D}$. □

Corollary 2.3.2. *Given the metric* (2.3.40), *invariant* (2.2.33) *takes the form similar to* (2.3.39):

$$\begin{aligned}(s\bar{s} + t\bar{t})^{2j} &= (2j)! \sum_{m=-j}^{j} T^j_m(s,t)\eta^{(j)}_{m,m'}\overline{T}^j_{m'}(\bar{s},\bar{t}) \\ &= (2j)! \sum_{m=-j}^{j} T^j_m(s,t)\tilde{T}^j_m(\bar{s},\bar{t}), \end{aligned} \tag{2.3.45}$$

where $\tilde{T}^j_m(\bar{s},\bar{t}) = \eta^{(j)}_{m,m'}\overline{T}^j_{m'}(\bar{s},\bar{t})$. *Hence, the action of group* $SL(2,\mathbb{C})$ *on vector* $\overline{T}^j_{m'}(\bar{s},\bar{t})$ *must agree*[f] *with the action* (2.3.35) *of this group on vector* $T^j_m(s,t)$, *so in view of* (2.3.41) *we have*

$$[T^{(j)}(g)\cdot\overline{T}^j_m](\bar{s},\bar{t}) = \overline{T}^j_m(\bar{s}',\bar{t}') = \sum_{m'=-j}^{j}\overline{T}^j_{m'}(\bar{s},\bar{t})\mathcal{D}^{(j)}_{m'm}(g). \qquad (2.3.46)$$

Here the transformed variables $(\bar{s}',\bar{t}')$ *are defined in* (2.2.8). *The invariance of the function* (2.3.45) *under transformation Eq.* (2.3.35) *leads to the transformation law for monomials* $\tilde{T}^j_m(\bar{s},\bar{t})$:

$$[T^{(j)}(g)\cdot\tilde{T}^j_m](\bar{s},\bar{t}) = \tilde{T}^j_m(\bar{s}',\bar{t}') = \sum_{m'=-j}^{j}\mathcal{D}^{(j)}_{mm'}(g^{-1})\,\tilde{T}^j_{m'}(\bar{s},\bar{t}). \qquad (2.3.47)$$

This means that $\tilde{T}^j_m(\bar{s},\bar{t})$ *are dual basis vectors with respect to the basis vectors* $T^j_m(s,t)$. *We use formula* (2.3.47) *in what follows.*

We now point out that the above construction of finite-dimensional irreducible representations of group $SL(2,\mathbb{C})$ can be directly translated to its compact subgroup $SU(2)$. To this end, matrices $g = ||A^\alpha_\beta||$ in (2.2.2), (2.3.32)–(2.3.36)) must be considered as unitary matrices U belonging to the subgroup $SU(2)\subset SL(2,\mathbb{C})$. These matrices have the form (2.3.1). The transformations (2.2.2) read in this case

$$(s',t') = (s,t)U = (s\,\alpha - t\,\beta^*, s\,\beta + t\,\alpha^*), \qquad (2.3.48)$$

while Eqs. (2.3.35) and (2.3.36) are written as follows:

$$T^j_m(s',t') = [T^{(j)}(U)\cdot T^j_m](s,t) = \sum_{m'} T^j_{m'}(s,t)\mathcal{D}^{(j)}_{m'm}(\alpha,\beta), \qquad (2.3.49)$$

where

$$\begin{aligned}\mathcal{D}^{(j)}_{m'm}(\alpha,\beta) &= \mathcal{D}^{(j)}_{m'm}(U) = \langle j,m'|\,U\,|j,m\rangle \\ &= \sum_k (-1)^k\frac{\sqrt{(j+m)!(j-m)!(j+m')!(j-m')!}}{k!(j-k-m')!(j+m-k)!(k-m+m')!} \\ &\quad\times \alpha^{j+m-k}\beta^{k-m+m'}(\beta^*)^k(\alpha^*)^{j-k-m'}, \qquad (2.3.50)\end{aligned}$$

$|\alpha|^2 + |\beta|^2 = 1$ and summation runs over integer k.

---

[f] This property reflects the fact that representations $T^{(j)}$ are equivalent to their contragredient representations.

Problem 2.3.4. * Show that elements $\mathcal{D}^{(j)}_{m'm}(\alpha,\beta)$ in (2.3.50) can be written in compact form

$$\mathcal{D}^{(j)}_{m'm}(\alpha,\beta) = A^{(j)}_{m'm}(\alpha,\beta)\, (\partial_y)^{j+m'} (y-1)^{j+m}\, (y+1)^{j-m}\Big|_{y=2\alpha\alpha^*-1},$$
$$A^{(j)}_{m'm}(\alpha,\beta) \equiv \frac{2^{m'-j}\sqrt{(j-m')!}}{\sqrt{(j+m')!(j-m)!(j+m)!}} \cdot \alpha^{m'+m}\beta^{m'-m}, \tag{2.3.51}$$

where $\partial_y = \frac{\partial}{\partial y}$ and $|\alpha|^2 + |\beta|^2 = 1$.

When writing (2.3.50), we use Dirac notations $\langle j,m'|U|j,m\rangle$ for elements of matrix $||\mathcal{D}^{(j)}_{m'm}(U)||$, which is an element $U \in SU(2)$ in representation $T^{(j)}$. In these notations, the relations (2.3.49) and (2.3.47) for $g = U \in SU(2)$ read

$$\langle s',t'|j,m\rangle = \langle s,t|U\,|j,m\rangle = \sum_{m'} \langle s,t|j,m'\rangle\, \langle j,m'|\,U\,|j,m\rangle, \tag{2.3.52}$$

$$\langle j,m|\bar{s}',\bar{t}'\rangle = \langle j,m|U^{-1}\,|\bar{s},\bar{t}\,\rangle = \sum_{m'} \langle j,m|\,U^{-1}\,|j,m'\rangle\langle j,m'|\bar{s},\bar{t}\rangle, \tag{2.3.53}$$

where, in accordance with (2.2.26), (2.3.48) and (2.2.8), we set[g]

$$\langle s',t'| = \langle s,t|U, \quad \langle s,t|j,m\rangle := T^{(j)}_m(s,t), \tag{2.3.54}$$
$$|\bar{s}',\bar{t}'\rangle = U^{-1}|\bar{s},\bar{t}\,\rangle, \quad \langle j,m|\bar{s},\bar{t}\rangle := \tilde{T}^{(j)}_m(\bar{s},\bar{t}). \tag{2.3.55}$$

Note that the second equalities in (2.3.52) and (2.3.53) can be obtained from universal and compact formulas,

$$U\,|j,m\rangle = \sum_{m'} |j,m'\rangle\, \langle j,m'|\,U\,|j,m\rangle, \tag{2.3.56}$$

$$\langle j,m|\,U^{-1} = \sum_{m'} \langle j,m|\,U^{-1}\,|j,m'\rangle\langle j,m'|, \tag{2.3.57}$$

where the operator U is understood as an abstract element of $SU(2)$ rather than (2×2) matrix, while vectors $|j,m\rangle$ and $\langle j,m|$ in (2.3.52), (2.3.53), (2.3.56) and (2.3.57) should be also treated as abstract vectors[h] introduced in (2.1.19) and (2.1.29). Relations (2.3.49), (2.3.52) are obtained from (2.3.56) by the convolution of the two sides of the latter formula with the

[g] Upon the change of variables $s/t = z$, polynomials $t^{-2j}\langle s,t|j,m\rangle$ are straightforwardly related to coherent states (2.2.66).

[h] Space of these vectors is called module in mathematical literature.

dual vector $\langle s,t|$ that obeys (2.3.54). Likewise, Eq. (2.3.57) yields (2.3.47), (2.3.53). Formulas (2.3.52) and (2.3.56) are analogs of Eq. (1.1.7), which we obtained in Section 1.1. Formula (2.3.37) can also be written in Dirac notations:

$$\langle j,m_1|U\cdot U'|j,m_2\rangle = \sum_{m_3}\langle j,m_1|U|j,m_3\rangle\ \langle j,m_3|U'|j,m_2\rangle,$$

$$\forall U,U'\in SL(2,\mathbb{C}). \tag{2.3.58}$$

Finally, it follows from (2.3.56), (2.3.57) and (2.3.58) that the states $|j,m\rangle$ must form a complete set of basis vectors in the space of irreducible representation $T^{(j)}$ (cf. (1.1.5), (2.2.54)):

$$\sum_{m=-j}^{j} |j,m\rangle\ \langle j,m| = I.$$

Clearly, Proposition 2.3.1 is valid not only for matrices of representation of group $SL(2,\mathbb{C})$ but also for matrices $||\mathcal{D}^{(j)}_{m'm}(\alpha,\beta)||$ of representation of unitary group $SU(2)$. Furthermore, unlike in the case of $SL(2,\mathbb{C})$, finite-dimensional representations (2.3.49), (2.3.50) of group $SU(2)$ are unitary,

$$\sum_m \mathcal{D}^{(j)}_{mm'}(\alpha,\beta)\mathcal{D}^{(j)\,*}_{mm''}(\alpha,\beta) = \delta_{m'm''},$$

since the Hermitian convolution (2.3.38) is, obviously, invariant under transformations (2.3.49) (because of the invariance of the bilinear form $|s|^2+|t|^2$, entering (2.3.38), under all two-dimensional unitary transformations (2.3.48)).

The explicit formula (2.3.50) shows that

$$\mathcal{D}^{(j)}_{m'm}(-\alpha,-\beta) = (-1)^{2j}\ \mathcal{D}^{(j)}_{m'm}(\alpha,\beta). \tag{2.3.59}$$

For integer $j=\ell$, elements (2.3.50) do not change under the transformation $U\to -U$, so they are single-valued functions on group $SO(3)$, whose elements are in one-to-one correspondence with pairs $\pm U\in SU(2)$. This means that for integer $j=\ell$, irreducible representations (2.3.49), (2.3.50) of $SU(2)$ are at the same time irreducible representations of $SO(3)$, and that these representations exhaust all finite-dimensional representations of $SO(3)$. The latter statement can be understood as follows. Group $SU(2)$ is universal covering of group $SO(3)$, so all finite-dimensional irreducible representations of $SO(3)$ belong to the set of finite-dimensional irreducible representations of $SU(2)$. It remains to be seen that representations (2.3.49), (2.3.50) of $SU(2)$ with half-integer j are not representations of $SO(3)$. The latter fact is obvious from the property of functions $\mathcal{D}^{(j)}_{m'm}(\alpha,\beta)$ with half-integer j, which, according to (2.3.59), change sign under the transformation $U \to -U$ and hence cannot be representations of $SO(3)$ (in conventional sense[i]).

Let us parameterize the matrix elements in (2.3.1) as given by (2.3.11), (2.3.4) (see also (2.3.15)):

$$\alpha = \cos\chi + i\sin\chi\cdot\cos\vartheta, \quad \beta = i\sin\chi\cdot\sin\vartheta\cdot e^{-i\varphi}.$$

Then coefficients (2.3.50) take the form

$$\begin{aligned}\mathcal{D}^{(j)}_{m'm}(\chi,\vec{n}) &= e^{i\varphi(m-m')}\frac{(1-\sin^2\vartheta\sin^2\chi)^j}{(-1)^{m'}(\sin\vartheta)^{m-m'}}\frac{(\cos\vartheta - i\cot\chi)^m}{(\cos\vartheta + i\cot\chi)^{m'}} \\ &\quad\times\sum_k(-1)^k\frac{\sqrt{(j+m)!(j-m)!(j+m')!(j-m')!}}{k!(j-k-m')!(j-k+m)!(k-m+m')!} \\ &\quad\times\left(\frac{\sin^2\vartheta\sin^2\chi}{1-\sin^2\vartheta\sin^2\chi}\right)^k. \end{aligned} \tag{2.3.60}$$

If the matrix elements $U \in SU(2)$ are parameterized in terms of Euler angles (2.3.19),

$$\alpha = \cos(\theta/2)e^{i\frac{1}{2}(\psi+\phi)}, \quad \beta = \sin(\theta/2)e^{i\frac{1}{2}(\phi-\psi)}, \tag{2.3.61}$$

we have

$$\begin{aligned}\mathcal{D}^{(j)}_{m'm}(\phi,\theta,\psi) &= \langle j,m'\,|\,U(\phi,\theta,\psi)\,|\,j,m\rangle = e^{im'\phi+im\psi}d^{(j)}_{m'm}(\theta) \\ &= e^{im'\phi+im\psi}\sum_k\frac{(-1)^k\sqrt{(j+m)!(j-m)!(j+m')!(j-m')!}}{k!(j-k-m')!(j+m-k)!(k-m+m')!} \\ &\quad\times\left(\tan\left(\frac{\theta}{2}\right)\right)^{2k-m+m'}\cos^{2j}\left(\frac{\theta}{2}\right). \end{aligned} \tag{2.3.62}$$

[i]Representations $T^{(j)}$ with half-integer spins j are sometimes called in physics literature double-valued representations of $SO(3)$.

Here elements $U(\phi,\theta,\psi)$ are given by (2.3.20), and we introduce functions

$$d^{(j)}_{m'm}(\theta) = \langle j, m' \,|U_{\vec{e}_2}(\theta)|\, j, m\rangle$$
$$= \cos^{2j}\left(\frac{\theta}{2}\right) \frac{\sqrt{(j+m)!(j-m)!(j+m')!(j-m')!}}{\left(\tan\left(\frac{\theta}{2}\right)\right)^{m-m'}}$$
$$\times \sum_k \frac{(-1)^k \tan^{2k}\left(\frac{\theta}{2}\right)}{k!(j-k-m')!(j+m-k)!(k-m+m')!}, \qquad (2.3.63)$$

which are dubbed Wigner D-functions. Functions (2.3.62) simplify in some cases. As an example, for $\theta=\phi=0$ in (2.3.62), we have $\tan(\frac{\theta}{2})=0$ and, in view of factorials in the denominator, the entire sum over k has the only nonvanishing term with $2k=m-m'=0$. As a result, we get

$$\langle j, m' |\, U_{\vec{e}_3}(\psi)\, | j, m\rangle = \mathcal{D}^{(j)}_{m'm}(0,0,\psi) = e^{im\psi}\delta_{m,m'},$$
$$m', m = -j, -j+1, \ldots, j. \qquad (2.3.64)$$

We show in what follows that spherical harmonics, associated Legendre functions, Chebyshev and Legendre polynomials are all expressed in terms of Wigner D-functions.

Problem 2.3.5. Making use of (2.3.62), obtain the following equalities:

$$\mathcal{D}^{(j)}_{jm} = \left(\frac{(2j)!}{(j+m)!(j-m)!}\right)^{\frac{1}{2}} \tan^{j-m}\left(\frac{\theta}{2}\right)\cos^{2j}\left(\frac{\theta}{2}\right) e^{i(j\phi+m\psi)},$$

$$\mathcal{D}^{(j)}_{mj} = (-1)^{j-m}\left(\frac{(2j)!}{(j+m)!(j-m)!}\right)^{\frac{1}{2}}$$
$$\times \tan^{j-m}\left(\frac{\theta}{2}\right)\cos^{2j}\left(\frac{\theta}{2}\right) e^{i(m\phi+j\psi)}, \qquad (2.3.65)$$

$$\mathcal{D}^{(j)}_{m'm}(\phi,\theta,\psi) = \mathcal{D}^{(j)}_{-m\,-m'}(-\psi,\theta,-\phi). \qquad (2.3.66)$$

2.3.3. Spherical functions on $S^2 = SU(2)/U(1)$. Laplace operators on $SU(2)$ and $SU(2)/U(1)$

Material of this subsection can be thought of as an illustration of general constructions introduced in Sections I-2.2.2, I-4.6.2 *and* I-7.5 *of the first book.*

In Section 2.3.2 *we have found the explicit form* (2.3.50) *of elements* $\mathcal{D}^{(j)}_{m'm}(U) = \langle j, m' \,|\, U \,|\, j, m\rangle$ *of matrices of irreducible representations* $T^{(j)}$ *of group* $SU(2)$ *for all* j. *In accordance with Peter–Weyl theorem (Theorem* I-4.6.7, *Section* I-4.6.2 *of the first book), elements* (2.3.50) *form complete set of functions on group* $SU(2)$, *i.e., on sphere* S^3.

If we restrict ourselves with elements (2.3.50) with integer nonnegative $j = \ell$*, then we get complete set of functions on group manifold* $SO(3)$*, i.e., on* $\mathbb{RP}^3$ *(recall, see Section 2.3.2, that representations* $T^{(\ell)}$ *with integer* $\ell \geq 0$ *exhaust all finite-dimensional irreducible representations of group* $SO(3)$*).*

In this subsection, we use parameterization (2.3.19), (2.3.20) of elements $U \in SU(2)$ *in terms of Euler angles* ϕ, θ, ψ*. Elements* $U(\phi, \theta, \psi) \in SU(2)$ *act on basis vectors* $|j, m\rangle$ *of the space* $\mathcal{V}_{2j+1}$ *of representation* $T^{(j)}$ *as written in (2.3.56):*

$$|j, m\rangle \to U(\phi, \theta, \psi)\, |j, m\rangle = \sum_{m'=-j}^{j} |j, m'\rangle\, \mathcal{D}^{(j)}_{m'm}(\phi, \theta, \psi), \tag{2.3.67}$$

where the functions $\mathcal{D}^{(j)}_{m'm}(\phi, \theta, \psi) = \langle j, m'|U(\phi, \theta, \psi)|j, m\rangle$ *are defined in (2.3.62). We reiterate that these functions form complete set on the manifold* $SU(2)$ *(sphere* S^3*).*

Let us now construct complete set of functions on coset space $S^2 = SU(2)/U(1)$*. We follow the general approach outlined in Section I-7.5.5 of the first book. Group* $SU(2)$ *contains subgroup* $H = U(1)$ *of diagonal matrices*

$$U_{\vec{e}_3}(\psi) = \begin{pmatrix} e^{i\frac{\psi}{2}} & 0 \\ 0 & e^{-i\frac{\psi}{2}} \end{pmatrix}, \quad \psi \in [-2\pi, 2\pi). \tag{2.3.68}$$

Elements $U_{\vec{e}_3}(\psi) \in H \subset SU(2)$ *in representation* $T^{(j)}$ *are* $(2j+1)$*-dimensional diagonal matrices* $||\mathcal{D}^{(j)}_{m'm}(0, 0, \psi)||$ *given by (2.3.64), and for* $j = 1/2$ *matrix (2.3.64) coincides with (2.3.68). For integer* $j = \ell$*, representations (2.3.64) of subgroup* $U(1)$ *are at the same time complex representations of subgroup* $SO(2) \subset SO(3)$*, which involves rotations (2.3.6) about the axis* $\vec{n} = \vec{e}_3 = (0, 0, 1)$*:*

$$T_3(-\psi) = O(-\psi, \vec{e}_3) = \exp(-\psi\, S_3), \quad \psi \in [-\pi, \pi).$$

We insert (2.3.64) into (2.3.67) and obtain the action of element $U = U_{\vec{e}_3}(\psi)$ *on basis vectors:*

$$U_{\vec{e}_3}(\psi)\, |j, m\rangle = \sum_{m'=-j}^{j} |j, m'\rangle\, \langle j, m'|U_{\vec{e}_3}(\psi)|j, m\rangle = e^{im\psi}\, |j, m\rangle, \qquad m \in [-j, -j+1, \ldots, j-1, j] \tag{2.3.69}$$

(hereafter we write for brevity $U_{\vec{e}_k}(\psi)|j, m\rangle$ *instead of* $T^{(j)}(U_{\vec{e}_k}(\psi))|j, m\rangle$*). Thus, the representation* $T^{(j)}$ *of the Abelian subgroup* $H = U(1) \in SU(2)$ *is reducible and is expanded in direct sum of one-dimensional representations, whose number is* $(2j+1)$ *(in complete agreement with Corollary I-4.4.3 to Schur's Lemma). For integer spin* $j = \ell$*, index* m *takes integer values* $m = -\ell, -\ell+1, \ldots, \ell$*, and, in particular, takes zero value. Therefore, the set of basis vectors* $|\ell, m\rangle$ *contains a special vector* $|\ell, 0\rangle$*, which, in view of (2.3.69) is stationary (invariant) under subgroup* $H = U(1)$*:*

$$U_{\vec{e}_3}(\psi)|\ell, 0\rangle = |\ell, 0\rangle, \quad \forall \psi. \tag{2.3.70}$$

Making use of this fact, we obtain for any $U_{\vec{e}_3}(\psi) \in H$ *and* $U \in SU(2)$

$$\mathcal{D}^{(\ell)}_{m0}(U \cdot U_{\vec{e}_3}(\psi)) = \langle \ell, m|U \cdot U_{\vec{e}_3}(\psi)|\ell, 0\rangle = \langle \ell, m|U|\ell, 0\rangle = \mathcal{D}^{(\ell)}_{m0}(U). \tag{2.3.71}$$

So, components $\mathcal{D}^{(\ell)}_{m0}(U)$ *are the same for all elements in a given coset of subgroup* $U(1)$ *in* $SU(2)$*, and these components are actually functions on* $SU(2)/U(1) = SO(3)/SO(2)$*.*

In Euler angle parameterization, components $\mathcal{D}^{(\ell)}_{m0}(U)$ are

$$\mathcal{D}^{(\ell)}_{m0}(\phi,\theta,\psi) = \langle \ell, m|U_{\vec{e}_3}(\phi)\cdot U_{\vec{e}_2}(\theta)\cdot U_{\vec{e}_3}(\psi)|\ell,\, 0\rangle$$

$$= \langle \ell, m|U_{\vec{e}_3}(\phi)\cdot U_{\vec{e}_2}(\theta)|\ell,\, 0\rangle = \mathcal{D}^{(\ell)}_{m0}(\phi,\theta,0). \tag{2.3.72}$$

They depend only on angles ϕ and θ, which parameterize the coset space $SU(2)/U(1) = SO(3)/SO(2)$, i.e., sphere S^2. Functions (2.3.72) deserve special notation:

$$Y^{(\ell)}_m(\phi,\theta) := \mathcal{D}^{(\ell)}_{m0}(\phi,\theta,0) = e^{im\phi}d^{(\ell)}_{m0}(\theta), \tag{2.3.73}$$

and these are called spherical harmonics on S^2. From the viewpoint of general nomenclature introduced in Section I-7.5.5 of the first book, functions (2.3.71), (2.3.72) are adjoint spherical functions of the representation $T^{(\ell)}$. In what follows, we use both terms "spherical harmonics" and "adjoint spherical functions of the representation $T^{(\ell)}$", when it comes to group $SU(2)$ and sphere $S^2 = SU(2)/U(1)$.

Making use of Eqs. (2.3.51) and (2.3.61), Wigner function $d^{(\ell)}_{m0}(\theta)$ given by (2.3.63) and entering $Y^{(\ell)}_m(\phi,\theta)$, can be written as

$$d^{(\ell)}_{m0}(\theta) = \frac{\sqrt{(\ell-m)!}}{\sqrt{(\ell+m)!}}\cdot P^{(\ell)}_m(\cos\theta), \tag{2.3.74}$$

where

$$P^{(\ell)}_m(y) = \frac{1}{2^\ell \ell!}(1-y^2)^{m/2}\left(\frac{\partial}{\partial y}\right)^{\ell+m}(y^2-1)^\ell \tag{2.3.75}$$

is the associated Legendre function. Among the set of functions $Y^{(\ell)}_m(\phi,\theta)$, there is one, $Y^{(\ell)}_0(\theta) = d^{(\ell)}_{00}(\theta)$, which is independent of ϕ and called zonal spherical harmonic. From (2.3.74) and (2.3.75) we find $Y^{(\ell)}_0(\theta) = P^{(\ell)}(\cos\theta)$, where

$$P^{(\ell)}(y) = P^{(\ell)}_0(y) = \frac{1}{2^\ell \ell!}\left(\frac{\partial}{\partial y}\right)^{\ell}(y^2-1)^\ell \tag{2.3.76}$$

are Legendre polynomials.

Problem 2.3.6. * *Making use of formulas (2.3.58), (2.3.66), (2.3.72), (2.3.73) and (2.3.74), prove the addition theorem for spherical harmonics*

$$\sum_m Y^{(\ell)}_{-m}(-\phi_1,\theta_1)Y^{(\ell)}_m(\phi_2,\theta_2) = \sum_m e^{im(\phi_1+\phi_2)}P^{(\ell)}_{-m}(\cos\theta_1)P^{(\ell)}_m(\cos\theta_2)$$

$$= P^{(\ell)}(\cos\omega_{12}), \tag{2.3.77}$$

where $\cos\omega_{12} = \cos\theta_1\cos\theta_2 - \sin\theta_1\sin\theta_2\cos(\phi_1+\phi_2)$.

It is pointed out in Section I-7.5.5 of the first book that if group G is compact and its subgroup $H\subset G$ is such that all irreducible representations $T^{(\lambda)}$ of G are class 1 representations with respect to H, then all adjoint spherical functions on coset space G/H form complete system on G/H. Recall that the representation $T^{(\lambda)}$ of group G is class 1 with respect to Abelian subgroup H if there is a vector $|0\rangle$ in the space of this representation which is invariant under H: $H|0\rangle = |0\rangle$. Then the adjoint spherical functions that form complete set on G/H are written as matrix elements $\langle m|T^{(\lambda)}(g)|0\rangle$, where $g\in G$, while $\langle m|$ are basis vectors in dual space $\mathcal{V}^*_\lambda$. It follows from (2.3.70) that in our case the vectors $|0\rangle$ are $|\ell,\,0\rangle$, so all representations $T^{(\ell)}$ of group $SO(3)$ are class 1 *with respect to subgroup $SO(2)$. Thus, in our case the matrix elements $\langle m|T^{(\lambda)}(g)|0\rangle$ are spherical harmonics $Y_m^{(\ell)}(\phi,\theta) = \langle \ell, m|U(\phi,\theta,0)|\ell, 0\rangle$, and they form complete system of functions on sphere $S^2 = SO(3)/SO(2)$.*

Let us now construct two infinite-dimensional representations of group $SU(2)$, one of which acts on the space of functions on manifold $S^3 = SU(2)$, and another acts in the space of functions on coset space $S^2 = SU(2)/U(1)$. To this end, consider the left action of an element $U \in SU(2)$ on element $U(\phi,\theta,\psi) \in SU(2)$, parameterized by Euler angles:

$$U \cdot U(\phi,\theta,\psi) = U \cdot U_{\vec{e}_3}(\phi)U_{\vec{e}_2}(\theta)U_{\vec{e}_3}(\psi) = U_{\vec{e}_3}(\phi')U_{\vec{e}_2}(\theta')U_{\vec{e}_3}(\psi'), \tag{2.3.78}$$

where the result is again written in Euler angle parameterization. We multiply the two sides of (2.3.78) on the right by $U_{\vec{e}_3}(-\psi)$ and obtain

$$U \cdot U_{\vec{e}_3}(\phi)U_{\vec{e}_2}(\theta) = U_{\vec{e}_3}(\phi')U_{\vec{e}_2}(\theta')U_{\vec{e}_3}(\psi''), \tag{2.3.79}$$

where $\psi'' = \psi' - \psi$. The relation (2.3.79) can be thought of as the left action of an element $U \in SU(2)$ on coset space $SU(2)/U(1)$, whose points are labeled (see (2.3.72)) by representatives $U_{\vec{e}_3}(\phi)U_{\vec{e}_2}(\theta)$, i.e., a point on the sphere $S^2 = SU(2)/U(1)$ with coordinates (θ,ϕ) is moved by the element $U \in SU(2)$ to the point with coordinates (θ',ϕ').

Let an element $U = \exp(\frac{i}{2}t^k\,\sigma_k) = \exp(-t^k\,\tau_k)$ of group $SU(2)$ be close to unity, i.e., parameters t^i be small. Then functions ϕ',θ',ψ' in (2.3.78), (2.3.79), to the first order in t^i, are

$$\begin{aligned} \phi' &= \phi + (t^1\cos(\phi) - t^2\sin(\phi))\cot(\theta) + t^3 + \cdots, \\ \theta' &= \theta + t^1\sin(\phi) + t^2\cos(\phi) + \cdots, \\ \psi' &= \psi + (-t^1\cos(\phi) + t^2\sin(\phi))\sin^{-1}(\theta) + \cdots, \end{aligned} \tag{2.3.80}$$

where dots denote terms of second and higher orders in variables t^1, t^2, t^3.

‖ **Problem 2.3.7.** * *Derive the expressions (2.3.80).*

We now introduce the notations $\vec{x} = (\phi,\theta,\psi)$ and $\vec{F}(\vec{t},\vec{x}) = (\phi',\theta',\psi')$, similar to ones used in Section I-2.2.2 in the first book. Then formulas (2.3.80) can be written as $F^i(\vec{t},\vec{x}) = x^i + t^k\,R_k^i(\vec{x}) + \cdots$, where we introduce matrix

$$R_i^j(\vec{x}) = \partial_{t^i}F^j(\vec{t},\vec{x})\big|_{\vec{t}=0} = \begin{pmatrix} \partial_1\phi', & \partial_1\theta', & \partial_1\psi' \\ \partial_2\phi', & \partial_2\theta', & \partial_2\psi' \\ \partial_3\phi', & \partial_3\theta', & \partial_3\psi' \end{pmatrix}$$

$$= \begin{pmatrix} \cos\phi\cot\theta, & \sin\phi, & -\cos\phi/\sin\theta \\ -\sin\phi\cot\theta, & \cos\phi, & \sin\phi/\sin\theta \\ 1, & 0, & 0 \end{pmatrix}. \tag{2.3.81}$$

By definition, this matrix is equal to matrix of right-invariant vector field $R_i^{\,j}\partial_j$ on S^3.

It follows from general theory (see Section I-7.3 of the first book) that starting from the left action (2.3.78) of group $SU(2)$ on itself, and left action (2.3.79) of $SU(2)$ on coset space $SU(2)/U(1)$, one can construct two infinite-dimensional representations of group $SU(2)$, namely, using Eq. (2.3.78) (which determines the transformation of parameters

$(\phi,\theta,\psi) \to (\phi',\theta',\psi')$ on $SU(2) = S^3$), one constructs left regular representation ρ_R of $SU(2)$ (cf. representation T_R in (2.2.22)):

$$[\rho_R(U^{-1}) \cdot f](\phi,\theta,\psi) = f'(\phi,\theta,\psi) \equiv f(\phi',\theta',\psi'), \tag{2.3.82}$$

which acts in space of functions $f(\phi,\theta,\psi)$ on $SU(2) = S^3$. The second infinite-dimensional representation ρ of group $SU(2)$ is constructed by making use of (2.3.79):

$$[\rho(U^{-1}) \cdot f](\phi,\theta) = f'(\phi,\theta) \equiv f(\phi',\theta'), \quad \forall U \in SU(2). \tag{2.3.83}$$

This representation acts in space of functions $f(\phi,\theta)$ on coset space $S^2 = SU(2)/U(1)$.

Problem 2.3.8. *Show that mappings ρ_R and ρ, defined by (2.3.82) and (2.3.83), are homomorphisms.*

We again insert $U = \exp(-t^i \tau_i)$ into (2.3.82), (2.3.83), take derivative with respect to t^i, use (2.3.78), (2.3.79) and (2.3.80), and then set $t^i = 0$. As a result, we obtain representations ρ_R and ρ for generators τ_i of algebra $su(2)$ in terms of differential operators. We have for regular representation ρ_R, corresponding to (2.3.82):

$$\begin{aligned} \rho_R(\tau_1) &= \cos(\phi)\cot(\theta)\partial_\phi + \sin(\phi)\partial_\theta - \cos(\phi)\sin^{-1}(\theta)\partial_\psi, \\ \rho_R(\tau_2) &= -\sin(\phi)\cot(\theta)\partial_\phi + \cos(\phi)\partial_\theta + \sin(\phi)\sin^{-1}(\theta)\partial_\psi, \quad \rho_R(\tau_3) = \partial_\phi. \end{aligned} \tag{2.3.84}$$

These operators act on functions on $SU(2) = S^3$ and, by construction (see Sections I-2.2.2 and I-7.5.3) are[j] *right-invariant vector fields $\rho_R(\tau_k) = R^j_k \partial_j$ on $SU(2)$. Likewise, by differentiating (2.3.83) over parameters t^i, and then setting $t^i = 0$ we obtain representation ρ of the $SU(2)$ generators:*

$$\begin{aligned} \hat{\tau}_1 &= \rho(\tau_1) = \cos(\phi)\cot(\theta)\partial_\phi + \sin(\phi)\partial_\theta, \\ \hat{\tau}_2 &= \rho(\tau_2) = -\sin(\phi)\cot(\theta)\partial_\phi + \cos(\phi)\partial_\theta, \quad \hat{\tau}_3 = \rho(\tau_3) = \partial_\phi. \end{aligned} \tag{2.3.85}$$

These act in the space of functions on coset space $SU(2)/U(1) = S^2$. Note that representation (2.3.85) can be obtained from (2.3.84) by setting in the latter $\partial_\psi = 0$ and thus considering functions on manifold $SU(2)$, which are constant along the orbits of the subgroup $U(1)$.

Problem 2.3.9. *Check that operators (2.3.84) and (2.3.85) make basis in algebra $su(2)$, i.e., that they obey commutation relations (2.1.17).*

Problem 2.3.10. *Show that quadratic Casimir operator $\hat{\tau}_k \hat{\tau}_k$ (Laplace operator on sphere $S^3 = SU(2)$), with generators (2.3.84) is given by*

$$\rho_R(\tau_k\,\tau_k) = \left(\frac{1}{\sin^2\theta}\partial^2_\phi + \cot\theta\partial_\theta + \partial^2_\theta\right) + \frac{1}{\sin^2\theta}(\partial_\psi - 2\cos\theta\,\partial_\phi)\partial_\psi \equiv \Delta_{SU(2)}. \tag{2.3.86}$$

[j]If we make a change of parameters $t^k \to -t^k$, then, according to (2.3.80), (2.3.81) all elements R^j_k flip sign, $R^j_k \to -R^j_k$. Then right-invariant vector fields are written as $(-R^j_k \partial_j)$. It is the latter convention that is used in Sections I-2.2.2 and I-7.5 of the first book.

Problem 2.3.11. * *Laplacian (2.3.86) on sphere S^3 (group $SU(2)$) is written in Euler angle parameterization (2.3.23). Show that Laplacian (2.3.86) modulo factor 4, is equal to the standard Laplacian on sphere S^3, written in spherical angle parameterization (2.3.13):*

$$\Delta'_{SU(2)} = \frac{1}{\sin^2\chi}\partial_\chi(\sin^2\chi\cdot\partial_\chi) + \frac{1}{\sin^2\chi}\frac{1}{\sin\vartheta}\partial_\vartheta(\sin\vartheta\cdot\partial_\vartheta) + \frac{1}{\sin^2\chi}\frac{1}{\sin^2\vartheta}\partial^2_\varphi. \tag{2.3.87}$$

The precise relation is $4\cdot\Delta_{SU(2)} = \Delta'_{SU(2)}$.

Quadratic Casimir operator $\hat{\tau}_k\hat{\tau}_k$ with generators (2.3.85) (Laplace operator on 2-sphere S^2) is obtained from (2.3.86) by setting $\partial_\psi = 0$. In this way, the Casimir operator of $su(2)$ in representation (2.3.85) takes the standard form of Laplace operator on sphere $S^2 = SU(2)/U(1)$:

$$\hat{\tau}_k\hat{\tau}_k = \left(\frac{1}{\sin^2\theta}\partial^2_\phi + \cot\theta\,\partial_\theta + \partial^2_\theta\right) \equiv \Delta_{SU(2)/U(1)}. \tag{2.3.88}$$

The infinite-dimensional representations ρ_R and ρ of group $SU(2)$, introduced above, are reducible. As an example, we extract finite-dimensional irreducible representations from the representation ρ given by (2.3.83). To this end, we sandwich the operator equality (2.3.79) between vector $\langle\ell,m|$ on the left, and vector $|\ell,0\rangle$ on the right, and make use of (2.3.58), where we set $m_3 = m'$, $j = \ell$ and $U' = U(\phi,\theta,\psi)$. We also use (2.3.70), (2.3.72) and (2.3.73). As a result, we obtain

$$\langle\ell,m|\,U\cdot U_{\vec{e}_3}(\phi)\cdot U_{\vec{e}_2}(\theta)|\ell,0\rangle = \mathcal{D}^{(\ell)}_{mm'}(U)Y^{(\ell)}_{m'}(\phi,\theta) = Y^{(\ell)}_m(\phi',\theta'), \tag{2.3.89}$$

where U is an arbitrary element of $SU(2)$, and angles ϕ', θ' depend on U, ϕ and θ and are uniquely determined by Eq. (2.3.79). In accordance with general theory given in Section I-7.5.5 of the first book, relations (2.3.89) define a representation $\rho^{(\ell)}$ of group $SU(2)$:

$$[\rho^{(\ell)}(U^{-1})\cdot Y^{(\ell)}_m](\phi,\theta) = \mathcal{D}^{(\ell)}_{mm'}(U)Y^{(\ell)}_{m'}(\phi,\theta) = Y^{(\ell)}_m(\phi',\theta'). \tag{2.3.90}$$

It follows from the first equality in (2.3.90) that spherical harmonics $Y^{(\ell)}_m(\phi,\theta)$ form basis in the space of finite-dimensional irreducible representation $\rho^{(\ell)}$ of group $SU(2)$. These representations are labeled by integers $\ell = 0,1,2,\ldots$. They are equivalent to representations $T^{(\ell)}$ and are embedded as subrepresentations, into infinite-dimensional representation ρ, defined in (2.3.83) (this is clear from the second equality in (2.3.90)). Recall that representations (2.3.90) of $SU(2)$ are at the same time irreducible representations of $SO(3)$ (since ℓ is integer).

Let us find representation of algebra $su(2)$ that corresponds to representation $\rho^{(\ell)}$ of group $SU(2)$. We begin with inserting the exponential form $U = \exp(-t^i\tau_i)$ in (2.3.90) and, with standard manipulation, obtain the following formula for action of generators τ_k in differential representation (2.3.85) on spherical harmonics:

$$\hat{\tau}_k\cdot Y^{(\ell)}_m(\phi,\theta) = -\sum_{m'}\mathcal{D}^{(\ell)}_{mm'}(\tau_k)Y^{(\ell)}_{m'}(\phi,\theta).$$

Here $||\mathcal{D}^{(\ell)}_{mm'}(\tau_k)||$ are matrices of generators τ_k of Lie algebra $su(2)$ in representation $T^{(\ell)}$. The same formula in Dirac notations reads

$$\hat{\tau}_k\langle\ell,m\,|\,\phi,\theta\rangle = -\langle\ell,m\,|\,\tau_k\,|\,\phi,\theta\rangle = -\sum_{m'}\langle\ell,m\,|\,\tau_k|\ell,\,m'\rangle\langle\ell,m'\,|\,\phi,\theta\rangle, \tag{2.3.91}$$

where we set

$$\begin{gathered}|\phi,\theta\rangle := U(\phi,\theta,0)\,|\ell,\,0\rangle, \quad \mathcal{D}^{(\ell)}_{mm'}(\tau_k) = \langle \ell, m\,|\,\tau_k|\ell,m'\rangle, \\ Y^{(\ell)}_m(\phi,\theta) = \langle \ell, m\,|\,\phi,\theta\rangle = \langle \ell, m\,|\,U(\phi,\theta,0)\,|\ell,\,0\rangle. \end{gathered} \tag{2.3.92}$$

Note that the relation (2.3.91) between matrix representations $||\mathcal{D}^{(\ell)}_{mm'}(\tau_k)||$ and their differential realizations (2.3.85) shows that the action of differential operator $\hat{\tau}_k$ on spherical harmonics $Y^{(\ell)}_m(\phi,\theta)$ is equivalent to action of the generator $(-\tau_k)$ on dual vectors $\langle \ell, m|$ on the right. Let us use the generators $\mathsf{e}_\pm, \mathsf{h}$ of algebra $s\ell(2,\mathbb{C})$, see (2.1.15), in differential realization (2.3.85):

$$\hat{\mathsf{e}}_\pm = i\,\hat{\tau}_1 \mp \hat{\tau}_2 = e^{\mp i\phi}\left(\mp\partial_\theta + i\cot\theta\partial_\phi\right), \quad \hat{\mathsf{h}} = i\,\hat{\tau}_3 = i\,\partial_\phi. \tag{2.3.93}$$

In accordance with (2.3.91), action of differential generators (2.3.93) on functions $Y^{(\ell)}_m(\phi,\theta) = \langle \ell, m\,|\,\phi,\theta\rangle$ is determined by the action of generators $(-\mathsf{e}_\pm)$, $(-\mathsf{h})$ in representation $T^{(\ell)}$ on dual vectors $\langle \ell, m|$ on the right. Making use of (2.1.28), (2.3.91), we obtain[k]

$$\begin{aligned} (-\hat{\mathsf{h}})Y^{(\ell)}_m(\phi,\theta) &= \langle \ell, m\,|\,\mathsf{h}\,|\,\phi,\theta\rangle = m\,Y^{(\ell)}_m(\phi,\theta), \\ (-\hat{\mathsf{e}}_-)Y^{(\ell)}_m(\phi,\theta) &= \langle \ell, m\,|\,\mathsf{e}_-\,|\,\phi,\theta\rangle = \sqrt{(\ell+m+1)(\ell-m)}\,Y^{(\ell)}_{m+1}(\phi,\theta), \\ (-\hat{\mathsf{e}}_+)\,Y^{(\ell)}_m(\phi,\theta) &= \langle \ell, m\,|\,\mathsf{e}_+\,|\,\phi,\theta\rangle = \sqrt{(\ell-m+1)(\ell+m)}\,Y^{(\ell)}_{m-1}(\phi,\theta). \end{aligned} \tag{2.3.94}$$

Since multiple action of generators e_+ (and e_-) generates (via (2.1.28), (2.1.29)) all basis vectors $\langle m, \ell|$ of dual space $\mathcal{V}^*_{2\ell+1}$, starting from dual highest weight (lowest weight) vectors $\langle \ell, \ell|$ ($\langle \ell, -\ell|$), the differential operators $\hat{\mathsf{e}}_\pm$, given by (2.3.93), generate all spherical harmonics $Y^{(\ell)}_m(\phi,\theta)$ starting from $Y^{(\ell)}_\ell(\phi,\theta)$ and $Y^{(\ell)}_{-\ell}(\phi,\theta)$, which correspond to vectors $\langle \ell, \ell|$ and $\langle \ell, -\ell|$. We set $j = \ell$ in (2.3.62) and then $m' = \ell$, $m = 0$, or $m' = -\ell$, $m = 0$ and obtain

$$\begin{aligned} Y^{(\ell)}_\ell(\phi,\theta) &= \frac{\sqrt{(2\ell)!}}{2^\ell(\ell!)}e^{i\,\ell\phi}\sin^\ell(\theta) = \langle \ell, \ell\,|\,\phi,\theta\rangle, \\ Y^{(\ell)}_{-\ell}(\phi,\theta) &= (-1)^\ell\frac{\sqrt{(2\ell)!}}{2^\ell(\ell!)}e^{-i\,\ell\phi}\sin^\ell(\theta) = \langle \ell, -\ell\,|\,\phi,\theta\rangle. \end{aligned} \tag{2.3.95}$$

Problem 2.3.12. *Show that functions $Y^{(\ell)}_m(\phi,\theta)$ given by (2.3.73), (2.3.74) and (2.3.75), obey (2.3.94). Check, in particular, that functions $Y^{(\ell)}_\ell(\phi,\theta)$ and $Y^{(\ell)}_{-\ell}(\phi,\theta)$ given by (2.3.95) obey the equations for highest weight and lowest weight dual vectors (see (2.1.32), (2.1.30)): $\hat{\mathsf{e}}_-\,Y^{(\ell)}_\ell(\phi,\theta) = 0$, $\hat{\mathsf{e}}_+\,Y^{(\ell)}_{-\ell}(\phi,\theta) = 0$. Derive the representation (2.3.73), (2.3.74) and (2.3.75) for $Y^{(\ell)}_\ell(\phi,\theta)$, by acting by the operators $\hat{\mathsf{e}}^m_+$ and $\hat{\mathsf{e}}^m_-$ on $Y^{(\ell)}_\ell(\phi,\theta)$ and $Y^{(\ell)}_{-\ell}(\phi,\theta)$, respectively.*

[k]We recall that there is automorphism of algebra $s\ell(2,\mathbb{C})$: $\hat{\mathsf{e}}_\pm \to -\hat{\mathsf{e}}_\mp$, $\hat{\mathsf{h}} \to -\hat{\mathsf{h}}$. We can change variables in (2.3.93) and (2.3.94) accordingly, and write these equalities in more conventional form familiar from quantum mechanics.

Problem 2.3.13. *Making use of (2.3.94), show that spherical harmonics $Y_m^{(\ell)}$ are eigenfunctions of Laplace operator (2.3.88) on 2-sphere:*

$$\Delta_{SU(2)/U(1)} Y_m^{(\ell)}(\phi,\theta) = \ell(\ell+1) Y_m^{(\ell)}(\phi,\theta), \quad \forall m,$$

i.e., check the validity of (2.1.34) for this representation (cf. the general formula for the action of Laplace operator $\Delta_{G/H}$ on adjoint spherical functions, Section I-7.5.5 *of the first book).*

Problem 2.3.14. *Using the representation (2.3.84) and formula (2.3.62), show that functions $\mathcal{D}^{(j)}_{mm'}(\phi,\theta,\psi)$ are eigenfunctions of the Laplace operator (2.3.86) on $SU(2) = S^3$,*

$$\Delta_{SU(2)} \mathcal{D}^{(j)}_{mm'}(\phi,\theta,\psi) = j(j+1)\mathcal{D}^{(j)}_{mm'}(\phi,\theta,\psi), \quad \forall m, m',$$

where j are integer and half-integer nonnegative numbers. Show that the spectral problem for Laplace operator (2.3.87) on S^3 is written as

$$\Delta'_{SU(2)} \Psi^{(j)}(\varphi,\vartheta,\chi) = n(n+2)\Psi^{(j)}(\varphi,\vartheta,\chi),$$

where n are nonnegative integers.

Let us find Haar measure for group $SU(2)$. To this end, we calculate the determinant of matrix (2.3.81):

$$\det(||R^j_k||) = \sin^{-1}(\theta). \tag{2.3.96}$$

Recall (see Section I-2.2.2 of the first book) that Haar measure on n-dimensional compact group G in a local patch near the point $g(\vec{x}) \in G$ (where $\vec{x} = (x_1, \dots, x_n)$ are local coordinates) is determined through matrix $||R^j_k||$ of right vector fields:

$$d\mu(g) = c \cdot \det{}^{-1}(||R^j_k||) dx_1 \cdots dx_n.$$

We insert here the result (2.3.96) and obtain Haar measure on group $SU(2)$ in Euler angle parameterization:

$$d\mu(U) = \frac{1}{16\pi^2} \sin(\theta) d\phi\, d\theta\, d\psi, \tag{2.3.97}$$

where constant $c = \frac{1}{16\pi^2}$ is chosen in such a way that the total volume of $SU(2)$ is equal to 1 (the ranges of Euler angles are given in (2.3.22)). The orthogonality condition (see Section I-4.6.2 of the first book) for matrix elements $\mathcal{D}^{(j)}_{m'm}(U)$ of irreducible representation $T^{(j)}$, given by (2.3.62), reads

$$\int d\mu(U) \mathcal{D}^{(j)}_{m'm}(U) (\mathcal{D}^{(j')}_{k'k}(U))^* = c(j)\delta^{jj'}\delta_{m'k'}\delta_{mk}, \tag{2.3.98}$$

where the constant $c(j)$ can be calculated by choosing $j = j'$, $k = m = j$, $k' = m' = j$ and using (2.3.62) and (2.3.65). In this case, the integral (2.3.98) reduces to

$$\frac{1}{2}\int_0^\pi d\theta \sin(\theta) d^{(j)}_{jj}(\theta) d^{(j)}_{jj}(\theta) = \frac{1}{2}\int_0^\pi d\theta \sin(\theta)\cos^{4j}(\theta/2) = c(j), \tag{2.3.99}$$

and we obtain

$$c(j) = \frac{1}{2j+1}. \tag{2.3.100}$$

Formula (2.3.98) also yields the orthogonality property (see Section I-7.5.5 of the first book) of spherical harmonics:

$$\frac{1}{4\pi}\int_0^{2\pi} d\phi \int_0^{\pi} d\theta \sin(\theta) Y_m^{(\ell)}(\phi,\theta)\big(Y_{m'}^{(\ell')}(\phi,\theta)\big)^* = c(\ell)\delta^{\ell\ell'}\delta_{mm'}, \tag{2.3.101}$$

which in turn implies orthogonality properties of associated Legendre functions (2.3.75) and Legendre polynomials (2.3.76).

Problem 2.3.15. *Using (2.3.100), (2.3.73) and (2.3.74), show that the formula (2.3.101) yields the following orthogonality property of associated Legendre functions:*

$$\int_{-1}^{1} dx\, P_m^{(\ell)}(x) P_m^{(\ell')}(x) = \frac{2}{(2\ell+1)}\frac{(\ell+m)!}{(\ell-m)!}\,\delta^{\ell\ell'}.$$

2.4. Tensor Product of Representations of $SU(2)$ and Clebsch–Gordan Series

2.4.1. Clebsch–Gordan expansion

Let $T^{(1)}$ and $T^{(2)}$ be irreducible representations of group G, and $\mathcal{V}_1$ and $\mathcal{V}_2$ be spaces of these representations. Tensor product of these representations, $T^{(1)} \otimes T^{(2)}$ is, generally speaking, reducible. Let group G be compact. Then Proposition I-4.5.1 (Section I-4.5 of the first book) guarantees that $T^{(1)} \otimes T^{(2)}$ can always be decomposed into *direct sum* of irreducible representations $T^{(\alpha)}$ of group G, where index α labels these representations. Representations $T^{(\alpha)}$ act in spaces $\mathcal{V}_\alpha$. Thus, we have the decomposition

$$\mathcal{V}_1 \otimes \mathcal{V}_2 = \bigoplus_\alpha m_\alpha \mathcal{V}_\alpha, \tag{2.4.1}$$

where m_α is the multiplicity of the pertinent representation (number of times the representation equivalent to $T^{(\alpha)}$ enters $T^{(1)} \otimes T^{(2)}$). More precisely, the right-hand side of (2.4.1) is the sum $\oplus_\alpha \oplus_{\kappa=1}^{m_\alpha} \mathcal{V}_{\kappa;\alpha}$, where index κ labels representations equivalent to $T^{(\alpha)}$. In other words, there exists an invertible matrix C which transforms all matrices of the representation $T^{(1)} \otimes T^{(2)}$ into block-diagonal form

$$C(T^{(1)}(g) \otimes T^{(2)}(g))C^{-1} = \bigoplus_\alpha \bigoplus_{\kappa=1}^{m_\alpha} T_\kappa^{(\alpha)}(g), \tag{2.4.2}$$

for all $g \in G$. Matrix C transforms the basis of space $\mathcal{V}_1 \otimes \mathcal{V}_2$, which consists of tensor products of basis vectors in $\mathcal{V}_1$ and $\mathcal{V}_2$, into the natural basis in space $\oplus_\alpha m_\alpha \mathcal{V}_\alpha$, which consists of all basis vectors of all of the subspaces $\mathcal{V}_\alpha$. Thus, we have

$$\sum_{d,b} C^{(1,d),(2,b)}_{(\alpha,c)} |d\rangle_1 \otimes |b\rangle_2 = |c\rangle_\alpha, \tag{2.4.3}$$

where $|d\rangle_1$, $|b\rangle_2$ and $|c\rangle_\alpha$ are normalized basis vectors of spaces $\mathcal{V}_1$, $\mathcal{V}_2$ and $\mathcal{V}_\alpha \subset \mathcal{V}_1 \otimes \mathcal{V}_2$, respectively (we omit index κ). Here $C^{(1,d),(2,b)}_{(\alpha,c)}$ are *Clebsch–Gordan coefficients*, and (2.4.3) is Clebsch–Gordan expansion (series).

The problem of explicit evaluation of the Clebsch–Gordan series is one of the most difficult problems in group representation theory, whose solution is known only for limited classes of groups and certain types of representations.

In this section, we study Clebsch–Gordan series for finite-dimensional representations $T^{(j)}$ of group $SU(2)$ with spin j. These representations are at the same time representations of Lie algebras $su(2)$ and $s\ell(2,\mathbb{C})$.

In the first place, we study the multiplicities m_α in expansions (2.4.1) and (2.4.2). Let $T^{(j_1)}$ and $T^{(j_2)}$ be two irreducible representations of group $SU(2)$ with spins j_1 and j_2. Recall that representation $T^{(j)}$ acts in space $\mathcal{V}_{2j+1}$ with basis vectors $|j,m\rangle$, see (2.1.19), and elements of $SU(2)$ are represented by $(2j+1)$-dimensional matrices (2.3.50). In our case, equality (2.4.2) reads

$$C(T^{(j_1)}(U) \otimes T^{(j_2)}(U))C^{-1} = \bigoplus_j \bigoplus_{\kappa=1}^{m_j} T^{(j)}_\kappa(U), \quad \forall U \in SU(2). \tag{2.4.4}$$

Let us choose parameterization (2.3.9) for an element $U \in SU(2)$, $U = U_{\vec{n}}(\psi)$ (we use the notation ψ instead of Ψ). Equation (2.4.4) yields the relation for characters:

$$\chi^{(j_1)}(\psi) \cdot \chi^{(j_2)}(\psi) = \sum_j m_j \chi^{(j)}(\psi), \tag{2.4.5}$$

where

$$\chi^{(j)}(\psi) = \mathrm{Tr}_{2j+1}(T^{(j)}(U_{\vec{n}}(\psi))) = \sum_{m=-j}^{j} \langle j,m|\, U_{\vec{n}}(\psi)\, |j,m\rangle, \tag{2.4.6}$$

and Tr_{2j+1} denotes trace in space $\mathcal{V}_{2j+1}$ of the representation $T^{(j)}$. On the right-hand side of (2.4.6), we omit symbol $T^{(j)}$ for brevity.

Proposition 2.4.1. *Character* (2.4.6) *is independent of the direction* $\vec{n}$, *and is given by*

$$\chi^{(j)}(\psi) = \frac{\sin(\psi(j+1/2))}{\sin(\psi/2)}. \tag{2.4.7}$$

Here, the angle $\psi \in [-2\pi, 2\pi)$ *labels conjugacy classes in* $SU(2)$ *which consist of all elements* $U_{\vec{n}}(\psi)$, *given by* (2.3.9), *with a fixed* ψ *and arbitrary* $\vec{n}$.

Proof. By similarity transformation (2.3.16), any element $U_{\vec{n}}(\psi)$ given by (2.3.9) with arbitrary $\vec{n}$ reduces to element $U_{\vec{e}_3}(\psi)$, which can be considered as a representative of the pertinent conjugacy class. Using this fact, character of an element $U_{\vec{n}}(\psi) \in SU(2)$ in representation $T^{(j)}$ is

$$\begin{aligned}\chi^{(j)}(\psi) &= \mathrm{Tr}_{2j+1}(T^{(j)}(U_{\vec{n}}(\psi))) = \mathrm{Tr}_{2j+1}(T^{(j)}(U^{-1}_{\vartheta,\varphi} \cdot U_{\vec{e}_3}(\psi) \cdot U_{\vartheta,\varphi}))\\ &= \mathrm{Tr}_{2j+1}(T^{(j)}(U_{\vec{e}_3}(\psi))) = \sum_m \langle j,m|\, U_{\vec{e}_3}(\psi)\,|j,m\rangle.\end{aligned} \tag{2.4.8}$$

It is indeed independent of $\vec{n}$ and depends on ψ only. Making use of (2.4.8) and explicit form (2.3.64) of matrix $U_{\vec{e}_3}(\psi)$ in representation $T^{(j)}$, we calculate

$$\begin{aligned}\chi^{(j)}(\psi) &= \sum_{m=-j}^{j} \langle j,m|\, U_{\vec{e}_3}(\psi)\,|j,m\rangle = \sum_{m=-j}^{j} e^{im\psi}\\ &= \frac{e^{i\psi(j+1/2)} - e^{-i\psi(j+1/2)}}{e^{i\psi/2} - e^{-i\psi/2}},\end{aligned} \tag{2.4.9}$$

which gives (2.4.7). □

Explicit expressions (2.4.7), (2.4.9) for characters $\chi^{(j)}$, together with completeness and orthogonality of functions $\chi^{(j)}(\psi)$, which hold for characters of all compact groups (see Section I-4.6.2 of the first book) are sufficient to obtain the multiplicities m_j. To this end, we insert (2.4.7) into the left-hand side of (2.4.5) and make use of the identity

$$\sin(a)\sin(b) = \sin(a+b-c)\sin(c) + \sin(a-c)\sin(b-c),$$

valid for all a, b, c. We set $a = \psi(j_1 + \frac{1}{2})$, $b = \psi(j_2 + \frac{1}{2})$ and $c = \psi/2$. Then we write the left-hand side of (2.4.5) as follows:

$$\begin{aligned}\chi^{(j_1)}(\psi)\cdot\chi^{(j_2)}(\psi) &= \frac{\sin\left(\psi\left(j_1+\frac{1}{2}\right)\right)\sin\left(\psi\left(j_2+\frac{1}{2}\right)\right)}{\sin^2(\psi/2)} \\ &= \frac{\sin\left(\psi\left(j_1+j_2+\frac{1}{2}\right)\right)}{\sin(\psi/2)} + \frac{\sin(\psi\, j_1)\sin(\psi\, j_2)}{\sin^2(\psi/2)} \\ &= \chi^{(j_1+j_2)}(\psi) + \chi^{(j_1-\frac{1}{2})}(\psi)\cdot\chi^{(j_2-\frac{1}{2})}(\psi). \qquad (2.4.10)\end{aligned}$$

The second term on the right-hand side of (2.4.10) is again rewritten by using (2.4.10), and we have

$$\chi^{(j_1)}(\psi)\cdot\chi^{(j_2)}(\psi) = \chi^{(j_1+j_2)}(\psi) + \chi^{(j_1+j_2-1)}(\psi) + \chi^{(j_1-1)}(\psi)\cdot\chi^{(j_2-1)}(\psi).$$

We again apply (2.4.10) to the last term here, and continue the procedure. Without loss of generality, we consider the case $j_1 \geq j_2$, then after $(2\,j_2)$ steps we find that one of the factors in the last, quadratic in $\chi^{(j)}$ term equals 1 (since $\chi^{(0)}(\psi) = 1$), the procedure terminates, and we arrive at the desired expansion

$$\chi^{(j_1)}(\psi)\cdot\chi^{(j_2)}(\psi) = \chi^{(j_1+j_2)}(\psi) + \chi^{(j_1+j_2-1)}(\psi) + \cdots + \chi^{(j_1-j_2)}(\psi). \qquad (2.4.11)$$

By comparing this with expansion (2.4.5), we see that $m_j = 1$ for $|j_1 - j_2| \leq j \leq j_1 + j_2$ and $m_j = 0$ for other j. Thus, we have proven the following proposition.

Proposition 2.4.2. *Expansion* (2.4.4) *of tensor product of two irreducible finite-dimensional representations* $T^{(j_1)}$ *and* $T^{(j_2)}$ *of group* $SU(2)$ *in irreducible representations has the following form:*

$$(T^{(j_1)} \otimes T^{(j_2)}) = \bigoplus_{j=|j_1-j_2|}^{j_1+j_2} T^{(j)}. \qquad (2.4.12)$$

This formula means that tensor product of representations with spins j_1 and j_2 is decomposed into direct sum of representations with spins j in the range $|j_1 - j_2| \leq j \leq j_1 + j_2$ with step 1, and each of the latter spins appears precisely once. This result is well known in quantum mechanics and is dubbed spin addition rule there. Note that the formula (2.4.12) agrees with counting of dimensions of the representations, since

$$(2j_1+1)(2j_2+1) = \sum_{j=|j_1-j_2|}^{j_1+j_2} (2j+1).$$

Remark. Formula (2.4.11) for $j_1 = j$ and $j_2 = 1/2$ gives

$$\chi^{(j)} \cdot \chi^{(\frac{1}{2})} = \chi^{(j+\frac{1}{2})} + \chi^{(j-\frac{1}{2})}, \tag{2.4.13}$$

where

$$\chi^{(0)} = 1, \quad \chi^{(\frac{1}{2})} = 2\cos(\psi/2). \tag{2.4.14}$$

In accordance with the identity (2.4.13) considered as recurrence relation with the initial data (2.4.14), all functions $\chi^{(j)}(\psi)$, given by (2.4.7), are degree $2j$ polynomials $P_{2j}(x)$ of variable $x = \cos(\psi/2)$. Polynomials $P_{2j}(x)$ are called Chebyshev polynomials of the second kind. Their orthogonality,

$$\int_{-1}^{1} P_n(x) P_m(x)(1-x^2)^{1/2} dx = \frac{\pi}{2}\delta_{nm}, \tag{2.4.15}$$

can be understood in terms of the general orthogonality property of characters of irreducible representations of any compact group, see Section I-4.6.2 of the first book.

Problem 2.4.1. Derive the formula (2.4.15) by considering the integral of the product of characters $\chi^{(\frac{n}{2})}(\psi) \cdot \chi^{(\frac{m}{2})}(\psi)$ over the group $SU(2)$ with measure (2.3.18).

Problem 2.4.2. Show that the generating function for polynomials $P_{2j}(x)$ is

$$\sum_{n=0}^{\infty} P_n(x) t^n = \frac{1}{1 - 2tx + t^2}. \tag{2.4.16}$$

Hint: multiply both parts of (2.4.16) by $(1 - 2tx + t^2)$, use (2.4.13), written in the form $2xP_n(x) = P_{n+1}(x) + P_{n-1}(x)$, and take into account initial data (2.4.14), namely, $P_0 = 1$, $P_1 = 2x$.

Thus, we have found the multiplicities in (2.4.2). Our next problem is to extract explicitly irreducible representations from tensor product $T^{(j_1)} \otimes T^{(j_2)}$ and calculate the Clebsch–Gordan coefficients in (2.4.3). We consider various methods of solving the latter problems in the next subsections.

2.4.2. Highest weight representations in $T^{(j_1)} \otimes T^{(j_2)}$

Here, we describe the method of extracting irreducible components from tensor product of two representations, which is based on the construction of highest weight representations (see Section 2.1 and Section I-4.7.3 of the first book). This method is convenient for not very high spins.

We choose, as it is common in quantum mechanics, the generators of algebra $su(2)$ in the form of Hermitian operators S_α, which, in the defining representation, are given by

$$S_\alpha = i\tau_\alpha = \frac{1}{2}\sigma_\alpha.$$

They obey commutation relations that follow from (2.1.17):

$$[S_\alpha, S_\beta] = \mathrm{i}\,\varepsilon_{\alpha\beta\gamma}S_\gamma \quad (\alpha, \beta, \gamma = 1, 2, 3). \tag{2.4.17}$$

We write in what follows $S_x = S_1$, $S_y = S_2$, $S_z = S_3$ and refer to these operators as spin components.

Let $T^{(\frac{1}{2})}$ be two-dimensional defining representation of algebra $su(2)$ and group $SU(2)$, and $\vec{e}^A = (\vec{e}^1, \vec{e}^2)$ be basis in two-dimensional space $\mathcal{V}_2 \equiv \mathcal{V}$ of this representation. Any vector $v \in \mathcal{V}$ is expanded in this basis, $v = v_A\vec{e}^A$, and can be written in coordinate form as a column $v_A = \binom{v_1}{v_2}$. For basis vectors, the coordinate representation reads

$$\vec{e}^A = (\vec{e}^A)_B\vec{e}^B, \quad (\vec{e}^A)_B = \delta^A_B \tag{2.4.18}$$

$$\Rightarrow \ |\uparrow\rangle := \begin{pmatrix}1\\0\end{pmatrix} = (\vec{e}^1)_A, \quad |\downarrow\rangle := \begin{pmatrix}0\\1\end{pmatrix} = (\vec{e}^2)_A. \tag{2.4.19}$$

In this subsection, we identify, wherever possible, basis vectors $\vec{e}^A$ and their coordinates $(\vec{e}^A)_B$. The action of operator S_α on vector $v \in \mathcal{V}$ is defined in the standard way

$$v \to v' = S_\alpha \cdot v = v_A(S_\alpha \cdot \vec{e}^A) = v_B\vec{e}^A(S_\alpha)_A{}^B \Rightarrow v_A \to v'_A = (S_\alpha)_A{}^B v_B. \tag{2.4.20}$$

This yields the rule for the action of S_α on basis vectors in coordinate form:

$$(\vec{e}^A)_B \to (S_\alpha \cdot \vec{e}^A)_B = (S_\alpha)_B{}^C(\vec{e}^A)_C = (\vec{e}^C)_B(S_\alpha)_C{}^A, \tag{2.4.21}$$

where we use (2.4.18) in the last inequality. According to this rule, arrows in notations $|\uparrow\rangle, |\downarrow\rangle$ show the direction of z-component of spin (up or down), or, in other words, the sign of eigenvalue $m = +1/2$ or $m = -1/2$

of the operator $S_z = 1/2\sigma_3$:

$$S_z \,|\uparrow\rangle = \frac{1}{2}\,|\uparrow\rangle, \quad S_z \,|\downarrow\rangle = -\frac{1}{2}\,|\downarrow\rangle.$$

Now, for basis vectors

$$\vec{e}^{A_1} \otimes \vec{e}^{A_2} \otimes \cdots \otimes \vec{e}^{A_n} \in \mathcal{V}^{\otimes n} \tag{2.4.22}$$

(here $\vec{e}^A$ are basis vectors in $\mathcal{V}_2$), we use the notations

$$\begin{gathered} |\uparrow\uparrow\rangle = \begin{pmatrix}1\\0\end{pmatrix} \otimes \begin{pmatrix}1\\0\end{pmatrix} = \vec{e}^1 \otimes \vec{e}^1, \quad |\downarrow\uparrow\rangle = \begin{pmatrix}0\\1\end{pmatrix} \otimes \begin{pmatrix}1\\0\end{pmatrix} = \vec{e}^2 \otimes \vec{e}^1, \\ |\uparrow\downarrow\rangle = \begin{pmatrix}1\\0\end{pmatrix} \otimes \begin{pmatrix}0\\1\end{pmatrix} = \vec{e}^1 \otimes \vec{e}^2, \quad |\downarrow\downarrow\rangle = \begin{pmatrix}0\\1\end{pmatrix} \otimes \begin{pmatrix}0\\1\end{pmatrix} = \vec{e}^2 \otimes \vec{e}^2, \end{gathered} \tag{2.4.23}$$

$$\begin{gathered} |\uparrow\uparrow\uparrow\rangle = \vec{e}^1 \otimes \vec{e}^1 \otimes \vec{e}^1, \quad |\downarrow\uparrow\uparrow\rangle = \vec{e}^2 \otimes \vec{e}^1 \otimes \vec{e}^1, \\ |\downarrow\uparrow\downarrow\rangle = \vec{e}^2 \otimes \vec{e}^1 \otimes \vec{e}^2, \ldots . \end{gathered} \tag{2.4.24}$$

In quantum mechanics, vectors (2.4.22) describe states of n particles of spin 1/2.

As usual, the operators of spin components S_x, S_y, S_z act on states from space of representation $T^{(j_1)} \otimes T^{(j_2)}$ as follows:

$$T(S_\alpha)(v \otimes u) = (T^{(j_1)}(S_\alpha)v) \otimes u + v \otimes (T^{(j_2)}(S_\alpha)u), \tag{2.4.25}$$

where v and u are vectors from spaces $\mathcal{V}_{2j_1+1}$ and $\mathcal{V}_{2j_2+1}$ of representations $T^{(j_1)}$ and $T^{(j_2)}$, respectively. So, we have

$$T(S_\alpha) = T^{(j_1)}(S_\alpha) \otimes I_{2j_2+1} + I_{2j_1+1} \otimes T^{(j_2)}(S_\alpha), \tag{2.4.26}$$

where I_n is unit operator in $\mathcal{V}_n$.

In particular, for tensor product of two defining representations $j_1 = j_2 = 1/2$ we have the representation $T \equiv \Delta$ which acts in $\mathcal{V} \otimes \mathcal{V}$ (one can consider Δ as a representation of enveloping algebras $\mathcal{U}(su(2))$)

$$\Delta(S_\alpha) = S_\alpha \otimes I + I \otimes S_\alpha, \tag{2.4.27}$$

where $I = I_2$ is unit operator in $\mathcal{V}$ and operators S_α act on vectors from $\mathcal{V}$ in defining representation $T^{(\frac{1}{2})}$. The action of $\Delta(S_\alpha)$ on states $(v \otimes u)$

where $v, u \in \mathcal{V}$ is

$$\Delta(S_\alpha)(v \otimes u) = (S_\alpha \otimes I + I \otimes S_\alpha)(v \otimes u) = (S_\alpha v) \otimes u + v \otimes (S_\alpha u). \tag{2.4.28}$$

The action of the generators S_α on states (2.4.24) in $\mathcal{V} \otimes \mathcal{V} \otimes \mathcal{V}$ is given by

$$\Delta^2(S_\alpha) = S_\alpha \otimes I \otimes I + I \otimes S_\alpha \otimes I + I \otimes I \otimes S_\alpha, \tag{2.4.29}$$

where Δ^2 denotes representation of $\mathcal{U}(su(2))$ in the space $\mathcal{V} \otimes \mathcal{V} \otimes \mathcal{V}$, etc. In general, action of spin operators S_α on states in the tensor product of n defining representations reads (cf. (2.4.27), (2.4.29))

$$\Delta^{n-1}(S_\alpha) = \sum_{k=1}^{n} S_\alpha^{(k)}, \tag{2.4.30}$$

where

$$S_\alpha^{(k)} = I^{\otimes(k-1)} \otimes S_\alpha \otimes I^{\otimes(n-k)}. \tag{2.4.31}$$

In other words, let

$$\psi = \psi_{A_1 A_2 \cdots A_n}\, \vec{e}^{A_1} \otimes \vec{e}^{A_2} \otimes \cdots \otimes \vec{e}^{A_n} \in \mathcal{V}^{\otimes n}, \quad (\psi_{A_1 A_2 \cdots A_n} \in \mathbb{C}), \tag{2.4.32}$$

be a state in the tensor product of n defining representations, which describes n particles of spin 1/2. Then the action of spin operator S_α on this state is

$$\begin{aligned} \Delta^{n-1}(S_\alpha)\psi = \sum_{k=1}^{n} &\psi_{A_1 A_2 \cdots A_n} \vec{e}^{A_1} \otimes \cdots \otimes \vec{e}^{A_{k-1}} \otimes \vec{e}^{B_k} (S_\alpha)_{B_k}^{\ A_k} \\ &\otimes \vec{e}^{A_{k+1}} \otimes \cdots \otimes \vec{e}^{A_n}. \end{aligned} \tag{2.4.33}$$

Vectors (2.4.22) form basis in 2^n-dimensional space $\mathcal{V}^{\otimes n}$. To extract irreducible components from $\mathcal{V}^{\otimes n}$ we employ the construction of the highest weight representation. We introduce raising and lowering operators $S_\pm$ (analogs of generators $e_\pm \in s\ell(2, \mathbb{C})$ given by (2.1.2)):

$$S_+ = S_x + iS_y, \quad S_- = S_x - iS_y,$$
$$[S_+, S_-] = 2S_z, \quad [S_z, S_\pm] = \pm S_\pm.$$

Here S_z is an analog of generator h in (2.1.2). We have in defining representation

$$S_+ = \begin{pmatrix} 0 & 1 \\ 0 & 0 \end{pmatrix}, \quad S_- = \begin{pmatrix} 0 & 0 \\ 1 & 0 \end{pmatrix},$$

and according to (2.4.21) we find

$$\begin{aligned} &S_+|\uparrow\rangle = 0, \quad S_+|\downarrow\rangle = |\uparrow\rangle, \\ &S_-|\uparrow\rangle = |\downarrow\rangle, \quad S_-|\downarrow\rangle = 0. \end{aligned} \tag{2.4.34}$$

We see that in this representation, operators S_+ and S_- either flip spin, or annihilate basis vectors.

We first consider the simplest case of four-dimensional space $\mathcal{V}^{\otimes 2}$ with basis vectors (2.4.23). It is convenient to arrange these vectors in a tower, in accordance with their eigenvalues m of the operator S_z in representation Δ (spin projection on z-axis):

$$\begin{aligned} &|\uparrow\uparrow\rangle && m = 1, \\ &|\uparrow\downarrow\rangle, |\downarrow\uparrow\rangle && m = 0, \\ &|\downarrow\downarrow\rangle && m = -1. \end{aligned} \tag{2.4.35}$$

According to definition (2.1.3) one of the highest weight vectors in space $\mathcal{V}^{\otimes 2}$ is the state $|\uparrow\uparrow\rangle$ whose spin projection $m = 1$ is maximal. Indeed, this eigenstate of S_z is annihilated by raising operator $\Delta(S_+) = S_+ \otimes I + I \otimes S_+$. To construct the whole space of irreducible representation out of the highest weight vector $|\uparrow\uparrow\rangle$, we use the standard procedure and act (several times) by lowering operator $\Delta(S_-) = S_- \otimes I + I \otimes S_- \equiv \Delta S_-$ on the vector $|\uparrow\uparrow\rangle$. As a result, we get

$$\Delta(S_-) \cdot |\uparrow\uparrow\rangle = |\downarrow\uparrow\rangle + |\uparrow\downarrow\rangle, \quad [\Delta(S_-)]^2 \cdot |\uparrow\uparrow\rangle = 2|\downarrow\downarrow\rangle,$$
$$[\Delta(S_-)]^3 \cdot |\uparrow\uparrow\rangle = 0.$$

Thus, three symmetric (under spin transpositions) vectors

$$\begin{aligned} |\uparrow\uparrow\rangle \quad & m = 1, \\ (\Delta S_-) \cdot |\uparrow\uparrow\rangle = |\downarrow\uparrow\rangle + |\uparrow\downarrow\rangle \quad & m = 0, \\ (\Delta S_-)^2 \cdot |\uparrow\uparrow\rangle = 2|\downarrow\downarrow\rangle \quad & m = -1 \end{aligned} \tag{2.4.36}$$

form basis in invariant three-dimensional space embedded in four-dimensional space $\mathcal{V}^{\otimes 2}$. This three-dimensional space is the space of irreducible representation of $su(2)$ with spin $j = 1$. The latter fact follows, e.g., from the calculation of Casimir operator (2.1.21) equal to $j(j+1)$ in irreducible representation with spin j. Indeed, the Casimir operator reads

$$J^2 = \frac{1}{2}(S_+S_- + S_-S_+) + S_z^2 = S_-S_+ + (1 + S_z)S_z, \tag{2.4.37}$$

and its value on the highest weight vector is

$$\Delta(J^2)|\uparrow\uparrow\rangle = (1 + \Delta(S_z))\Delta(S_z)|\uparrow\uparrow\rangle = 2|\uparrow\uparrow\rangle.$$

Recall that the eigenvalue of the Casimir operator is the same on all vectors of a given irreducible representation. So, we have $j(j+1) = 2$, i.e., $j = 1$. Using Dirac notations $|j,m\rangle$, we write normalized basis vectors (2.4.36) as follows:

$$\begin{aligned} &|1,1\rangle = |\uparrow\uparrow\rangle = \vec{e}^{\,1} \otimes \vec{e}^{\,1}, \quad |1,-1\rangle = |\downarrow\downarrow\rangle = \vec{e}^{\,2} \otimes \vec{e}^{\,2}, \\ &|1,0\rangle = \frac{1}{\sqrt{2}}(|\downarrow\uparrow\rangle + |\uparrow\downarrow\rangle) = \frac{1}{\sqrt{2}}(\vec{e}^{\,2} \otimes \vec{e}^{\,1} + \vec{e}^{\,1} \otimes \vec{e}^{\,2}). \end{aligned} \tag{2.4.38}$$

Thus far, we have extracted, from the space $\mathcal{V}^{\otimes 2}$ with basis (2.4.35), three-dimensional invariant subspace $\mathcal{V}_3 \subset \mathcal{V}^{\otimes 2}$ with basis (2.4.38). Since $\mathcal{V}^{\otimes 2}$ is four-dimensional, there exists one-dimensional invariant subspace $\mathcal{V}_1 \subset \mathcal{V}^{\otimes 2}$ which does not belong to $\mathcal{V}_3$. By comparing (2.4.35), with (2.4.38), we see that the vector generating the subspace $\mathcal{V}_1$ has spin projection $m = 0$. It can be extracted from $\mathcal{V}^{\otimes 2}$ by imposing again the highest weight requirement, but now on a state with $m = 0$:

$$\Delta(S_+) \cdot (a_1|\downarrow\uparrow\rangle + a_2|\uparrow\downarrow\rangle) = 0.$$

This gives $a_1 = -a_2$. Thus, the last normalized basis vector in $\mathcal{V}^{\otimes 2}$, in addition to the three vectors (2.4.38), is

$$|0,0\rangle = \frac{1}{\sqrt{2}}(|\downarrow\uparrow\rangle - |\uparrow\downarrow\rangle) = \frac{1}{\sqrt{2}}(\vec{e}^{\,2} \otimes \vec{e}^{\,1} - \vec{e}^{\,1} \otimes \vec{e}^{\,2}). \tag{2.4.39}$$

It is antisymmetric under spin permutations and indeed generates (trivial) one-dimensional irreducible representation of spin $j = 0$. The latter fact is seen from the property that all operators $\Delta(S_+)$, $\Delta(S_-)$ and $\Delta(S_z)$ vanish on this vector. Thus, we have obtained explicit decomposition $\mathcal{V}^{\otimes 2} = \mathcal{V}_3 + \mathcal{V}_1$, where $\mathcal{V}_3$ and $\mathcal{V}_1$ are spaces of irreducible representations $T^{(1)}$ and $T^{(0)}$ of algebra $su(2)$ and group $SU(2)$. Explicitly,

$$\psi = \psi_{A_1A_2}\vec{e}^{\,A_1} \otimes \vec{e}^{\,A_2} = \psi_1 + \psi_0,$$

where

$$\psi_1 = \psi_{(A_1A_2)}\vec{e}^{A_1} \otimes \vec{e}^{A_2} = \psi_{11}\,|1,1\rangle + \sqrt{2}\,\psi_{(12)}\,|1,0\rangle + \psi_{22}\,|1,-1\rangle \in \mathcal{V}_3,$$
$$\psi_0 = \psi_{[A_1A_2]}\vec{e}^{A_1} \otimes \vec{e}^{A_2} = \tfrac{1}{\sqrt{2}}\left(\varepsilon^{A_1A_2}\,\psi_{[A_1A_2]}\right)|0,0\rangle \in \mathcal{V}_1,$$
$$(2.4.40)$$

where $\varepsilon^{A_1A_2}$ are components of antisymmetric tensor ($\varepsilon^{A_1A_2} = -\varepsilon^{A_2A_1}$; $\varepsilon^{12} = -1$), brackets $(\ldots)$ and $[\ldots]$ denote symmetrization and antisymmetrization of indices, respectively, and we use (2.4.38), (2.4.39) and identities

$$\psi_{[A_1A_2]} = \tfrac{1}{2}\varepsilon_{A_1A_2}(\varepsilon^{B_1B_2}\psi_{[B_2B_1]}),$$
$$\vec{e}^{A_1} \otimes \vec{e}^{A_2} - \vec{e}^{A_2} \otimes \vec{e}^{A_1} = \varepsilon^{A_1A_2}\left(\varepsilon_{B_1B_2}\,\vec{e}^{B_2} \otimes \vec{e}^{B_1}\right) = \sqrt{2}\,\varepsilon^{A_1A_2}\,|0,0\rangle.$$
$$(2.4.41)$$

Here $\varepsilon_{AB}\varepsilon^{BC} = \delta_A^C$, i.e., $\varepsilon^{AB} = -\varepsilon_{AB}$.

Problem 2.4.3. Making use of (2.4.33), check that vector ψ_0 given by (2.4.40) is invariant under $su(2)$:

$$\Delta(S_\alpha)\psi_{[A_1A_2]}\vec{e}^{A_1} \otimes \vec{e}^{A_2} = 0 \quad (\alpha = 1,2,3).$$

Problem 2.4.4. Show that vectors $|\Phi\rangle = \phi_{B[AC]}\vec{e}^{A} \otimes \vec{e}^{B} \otimes \vec{e}^{C}$ and $|\Psi\rangle = \psi_{A[BC]}\vec{e}^{A} \otimes \vec{e}^{B} \otimes \vec{e}^{C}$ generate two spaces of spin-1/2 representation of $su(2)$. Hint: make use of (2.4.33) and the result of the previous problem, and calculate the Casimir operator on these vectors.

Now we apply similar procedure for tensor product of two irreducible representations $T^{(j_1)}$ and $T^{(j_2)}$ with arbitrary j_1 and j_2. Basis in space $\mathcal{V}_{2j_1+1} \otimes \mathcal{V}_{2j_2+1}$ consists of vectors $|j_1, m_1\rangle \otimes |j_2, m_2\rangle$, where $m_i = -j_i, -j_i + 1, \ldots, j_i - 1, j_i$. The action of operators S_α on these vectors is given by (2.4.25). Vector $|v_0\rangle = |j_1, j_1\rangle \otimes |j_2, j_2\rangle$ obeys the equation for the highest weight vector of representation $T^{(j_1+j_2)}$:

$$T(S_z)|v_0\rangle = (j_1 + j_2)|v_0\rangle, \quad T(S_+)|v_0\rangle = 0,$$

thus

$$|j_1 + j_2, j_1 + j_2\rangle = |j_1, j_1\rangle \otimes |j_2, j_2\rangle. \qquad (2.4.42)$$

We act on both sides of this relation by lowering operator $T(S_-)$ and make use of (2.4.25), (2.1.22), (2.2.31). For left- and right-hand sides, we have

$$
\begin{aligned}
T(S_-)|j_1+j_2, j_1+j_2\rangle &= \sqrt{2(j_1+j_2)}|j_1+j_2, j_1+j_2-1\rangle, \\
T(S_-)(|j_1,j_1\rangle \otimes |j_2,j_2\rangle) &= (T^{(j_1)}(S_-)|j_1,j_1\rangle) \otimes |j_2,j_2\rangle + |j_1,j_1\rangle \\
&\quad \otimes (T^{(j_2)}(S_-)|j_2,j_2\rangle) \\
&= \sqrt{2j_1}\,|j_1,j_1-1\rangle \otimes |j_2,j_2\rangle + \sqrt{2j_2}\,|j_1,j_1\rangle \\
&\quad \otimes |j_2,j_2-1\rangle.
\end{aligned}
$$

Therefore, we get

$$
\begin{aligned}
|j_1+j_2, j_1+j_2-1\rangle = \frac{\sqrt{j_1}}{\sqrt{j_1+j_2}}|j_1,j_1-1\rangle \otimes |j_2,j_2\rangle \\
+ \frac{\sqrt{j_2}}{\sqrt{j_1+j_2}}|j_1,j_1\rangle \otimes |j_2,j_2-1\rangle. \quad (2.4.43)
\end{aligned}
$$

This vector belongs to the space of irreducible representation $T^{(j_1+j_2)}$ and is a linear combination of two eigenvectors of the operator $T(S_z)$ with eigenvalue $m = (j_1+j_2-1)$. We again act by lowering operator $T(S_-)$, now on state (2.4.43), and obtain vector with eigenvalue $m = (j_1+j_2-2)$, etc. As a result, we obtain all basis vectors of space $\mathcal{V}_{2(j_1+j_2)+1}$ of irreducible representation $T^{(j_1+j_2)}$.

For $j_1 \geq 1/2$, $j_2 \geq 1/2$, the space $\mathcal{V}_{2j_1+1} \otimes \mathcal{V}_{2j_2+1}$ contains two linearly independent eigenvectors of operator $T(S_z)$ with eigenvalue $m = (j_1+j_2-1)$. One of them is given by (2.4.43). We search the second vector $|v_1\rangle$ in the following form:

$$
|v_1\rangle = \alpha_1\,|j_1,j_1\rangle \otimes |j_2,j_2-1\rangle + \alpha_2\,|j_1,j_1-1\rangle \otimes |j_2,j_2\rangle. \quad (2.4.44)
$$

This vector must be orthogonal to (2.4.43) and hence it is annihilated by raising operator $T(S_+)$.

Problem 2.4.5. Show that orthogonality property $\langle j_1+j_2, j_1+j_2-1|v_1\rangle = 0$ yields $T(S_+)|v_1\rangle = 0$.

Thus, vector $|v_1\rangle$ obeys

$$
T(S_z)|v_1\rangle = (j_1+j_2-1)|v_1\rangle, \quad T(S_+)|v_1\rangle = 0,
$$

so it is the highest weight vector $|v_1\rangle = |j_1+j_2-1, j_1+j_2-1\rangle$ of representation $T^{(j_1+j_2-1)}$. Equation $T(S_+)|v_1\rangle = 0$, normalization condition $\langle v_1|v_1\rangle = 1$ and reality of coefficients α_1 and α_2 defines vector (2.4.44) in unique way (modulo sign, which is a matter of convention):

$$|v_1\rangle = |j_1+j_2-1, j_1+j_2-1\rangle = \frac{\sqrt{j_1}}{\sqrt{j_1+j_2}}|j_1,j_1\rangle \otimes |j_2,j_2-1\rangle - \frac{\sqrt{j_2}}{\sqrt{j_1+j_2}}|j_1,j_1-1\rangle \otimes |j_2,j_2\rangle. \tag{2.4.45}$$

Acting on vector (2.4.45) many times by lowering operator $T(S_-)$, we construct all basis vectors in space $\mathcal{V}_{2(j_1+j_2)-1}$ of representation $T^{(j_1+j_2-1)}$.

Consider now subspace $\mathcal{V}'$ in $\mathcal{V}_{2j_1+1} \otimes \mathcal{V}_{2j_2+1}$ of all vectors $|j_1,m_1\rangle \otimes |j_2,m_2\rangle$ with eigenvalue $m = m_1+m_2 = j_1+j_2-2$ of the operator $T(S_z)$. If $j_1 \geq 1$, $j_2 \geq 1$, then the subspace $\mathcal{V}'$ is three dimensional, and by solving equation $T(S_+)|v_2\rangle = 0$ one finds unique highest weight vector $|v_2\rangle \in \mathcal{V}'$ with weight (j_1+j_2-2). It is orthogonal to two other vectors $|j_1+j_2, j_1+j_2-2\rangle$ and $|j_1+j_2-1, j_1+j_2-2\rangle$ from $\mathcal{V}'$. Then one acts on highest weight vector $|v_2\rangle$ by lowering operator $T(S_-)$ and obtains invariant subspace of irreducible representation $T^{(j_1+j_2-2)}$. Continuing with this procedure, one obtains expansion (2.4.12) of tensor product $T^{(j_1)} \otimes T^{(j_2)}$ into direct sum of representations $T^{(j)}$. The pertinent formulas, analogous to (2.4.43) and (2.4.45), have the following general form (cf. Clebsch–Gordan series (2.4.3)):

$$|j_3, m\rangle = \sum_{\substack{m_1,m_2 \\ m_1+m_2=m}} |j_1\, m_1\rangle \otimes |j_2\, m_2\rangle \, \langle j_1\, m_1, j_2\, m_2|j_3\, m\rangle, \tag{2.4.46}$$

where spin j_3 belongs to the range $|j_1-j_2| \leq j_3 \leq j_1+j_2$, while $|j_3,m\rangle$, $|j_1\, m_1\rangle$ and $|j_2\, m_2\rangle$ are *orthonormalized* basis vectors in the spaces of representations $T^{(j_3)}$, $T^{(j_1)}$ and $T^{(j_2)}$, respectively. Here, we introduce the notation

$$\langle j_1\, m_1, j_2\, m_2|j_3\, m\rangle := (\langle j_1\, m_1| \otimes \langle j_2, m_2|)\; |j_3\, m\rangle. \tag{2.4.47}$$

for Clebsch–Gordan coefficients.

Example. *Let us calculate Clebsch–Gordan coefficients in the expansion of tensor product* $T^{(j_1)} \otimes T^{(\frac{1}{2})}$ *of irreducible representations, one of which is defining. In accordance with Proposition 2.4.2, we have* $T^{(j_1)} \otimes T^{(\frac{1}{2})} = T^{(j_1+\frac{1}{2})} \oplus T^{(j_1-\frac{1}{2})}$. *The highest weight vector of representation* $T^{(j_1+\frac{1}{2})}$ *is*

$$|j_1+1/2, j_1+1/2\rangle = |j_1, j_1\rangle \otimes |1/2, 1/2\rangle. \tag{2.4.48}$$

We act on both sides of (2.4.48) by operator

$$T^{(j_1+\frac{1}{2})}(S_-^k) = \left(T^{(j_1)}(S_-)\otimes I + I\otimes T^{(\frac{1}{2})}(S_-)\right)^k, \tag{2.4.49}$$

and obtain ($k \le j_1$)

$$S_-^k\left|j_1+\frac{1}{2}, j_1+\frac{1}{2}\right\rangle = S_-^k|j_1,j_1\rangle\otimes\left|\frac{1}{2},\frac{1}{2}\right\rangle + kS_-^{k-1}|j_1,j_1\rangle\otimes S_-\left|\frac{1}{2},\frac{1}{2}\right\rangle, \tag{2.4.50}$$

where we use shorthand notation $S_-^k|j,m\rangle \equiv T^{(j)}(S_-^k)|j,m\rangle$. *Now, making use of (2.2.32), we rewrite Eq. (2.4.50) as follows:*

$$\left|j_1+\frac{1}{2}, j_1+\frac{1}{2}-k\right\rangle = \sqrt{\frac{(2j_1+1-k)}{(2j_1+1)}}|j_1,j_1-k\rangle\otimes\left|\frac{1}{2},\frac{1}{2}\right\rangle$$
$$+\sqrt{\frac{k}{(2j_1+1)}}|j_1,j_1-k+1\rangle\otimes\left|\frac{1}{2},-\frac{1}{2}\right\rangle. \tag{2.4.51}$$

We search for the highest weight vector $|v_1\rangle = |j_1-\frac{1}{2}, j_1-\frac{1}{2}\rangle$ *of representation* $T^{(j_1-\frac{1}{2})}$ *in the form (cf. (2.4.44)):*

$$\left|j_1-\frac{1}{2}, j_1-\frac{1}{2}\right\rangle = \alpha_1|j_1,j_1\rangle\otimes\left|\frac{1}{2},-\frac{1}{2}\right\rangle + \alpha_2|j_1,j_1-1\rangle\otimes\left|\frac{1}{2},\frac{1}{2}\right\rangle, \tag{2.4.52}$$

where coefficients α_1, α_2 *are obtained in the same way as in the case of vector (2.4.44) and are given by (2.4.45) with* $j_2 = 1/2$:

$$\alpha_1 = \sqrt{\frac{2j_1}{(2j_1+1)}}, \quad \alpha_2 = -\sqrt{\frac{1}{(2j_1+1)}}.$$

We again act on the two sides of (2.4.52) by lowering operator (2.4.49). Using (2.2.32), we obtain for the left- and right-hand sides

$$S_-^k\left|j_1-\frac{1}{2}, j_1-\frac{1}{2}\right\rangle = \sqrt{\frac{k!(2j_1-1)!}{(2j_1-1-k)!}}\left|j_1-\frac{1}{2}, j_1-\frac{1}{2}-k\right\rangle,$$

$$\alpha_2 S_-^k|j_1,j_1-1\rangle\otimes\left|\frac{1}{2},\frac{1}{2}\right\rangle + \alpha_2 kS_-^{k-1}|j_1,j_1-1\rangle\otimes S_-\left|\frac{1}{2},\frac{1}{2}\right\rangle$$
$$+\alpha_1 S_-^k|j_1,j_1\rangle\otimes\left|\frac{1}{2},-\frac{1}{2}\right\rangle$$
$$=\alpha_2\sqrt{\frac{(k+1)!(2j_1-1)!}{(2j_1-1-k)!}}|j_1,j_1-1-k\rangle\otimes\left|\frac{1}{2},\frac{1}{2}\right\rangle$$
$$+\alpha_2 k\sqrt{\frac{k!(2j_1-1)!}{(2j_1-k)!}}|j_1,j_1-k\rangle\otimes\left|\frac{1}{2},-\frac{1}{2}\right\rangle$$
$$+\alpha_1\sqrt{\frac{k!(2j_1)!}{(2j_1-k)!}}|j_1,j_1-k\rangle\otimes\left|\frac{1}{2},-\frac{1}{2}\right\rangle$$
$$=-\sqrt{\frac{(k+1)!(2j_1-1)!}{(2j_1+1)(2j_1-1-k)!}}|j_1,j_1-1-k\rangle\otimes\left|\frac{1}{2},\frac{1}{2}\right\rangle$$
$$+\sqrt{\frac{(2j_1-k)k!(2j_1-1)!}{(2j_1+1)(2j_1-k-1)!}}|j_1,j_1-k\rangle\otimes\left|\frac{1}{2},-\frac{1}{2}\right\rangle.$$

Finally, we have

$$\left| j_1 - \frac{1}{2}, j_1 - \frac{1}{2} - k \right\rangle$$
$$= -\sqrt{\frac{(k+1)}{(2j_1+1)}} |j_1, j_1 - 1 - k\rangle \otimes \left| \frac{1}{2}, \frac{1}{2} \right\rangle + \sqrt{\frac{(2j_1 - k)}{(2j_1+1)}} |j_1, j_1 - k\rangle \otimes \left| \frac{1}{2}, -\frac{1}{2} \right\rangle. \tag{2.4.53}$$

By convolution of relations (2.4.51) and (2.4.53) with vectors $\langle j_1, m_1| \otimes \langle \frac{1}{2}, m_2|$ *we find, making use of (2.4.47), the following table for Clebsch–Gordan coefficients* $\langle j_1, m_1; \frac{1}{2}, m_2 | j, m\rangle$*:*

	$m_2 = 1/2$	$m_2 = -1/2$
$j = j_1 + \frac{1}{2}$	$\left(\frac{j_1+m+1/2}{2j_1+1}\right)^{1/2}$	$\left(\frac{j_1-m+1/2}{2j_1+1}\right)^{1/2}$
$j = j_1 - \frac{1}{2}$	$-\left(\frac{j_1-m+1/2}{2j_1+1}\right)^{1/2}$	$\left(\frac{j_1+m+1/2}{2j_1+1}\right)^{1/2}$

This result is actually well known and quoted in textbooks on quantum mechanics.

2.4.3. Heisenberg spin chain

Let $\mathcal{V}$ *again denote the space* $\mathcal{V}_2$ *of defining representation* $T^{(\frac{1}{2})}$ *of algebra* $su(2)$*. The problem of extracting irreducible representations of* $su(2)$ *from the tensor product* $(T^{(\frac{1}{2})})^{\otimes n}$ *has applications in theories of various quantum systems. In particular, the space* $\mathcal{V}^{\otimes n}$ *of representation* $(T^{(\frac{1}{2})})^{\otimes n}$ *can be viewed as the space of quantum states of the Heisenberg spin chain model. This model has* n *nodes (each factor* $\mathcal{V}$ *in* $\mathcal{V}^{\otimes n}$ *corresponds to chain node) and is described by the Hamiltonian*

$$\mathcal{H}_n = \sum_{k=1}^{n} \sum_{\alpha} S_\alpha^{(k)} \cdot S_\alpha^{(k+1)} = \sum_{k=1}^{n} \left(\frac{1}{2} S_+^{(k)} \cdot S_-^{(k+1)} + \frac{1}{2} S_-^{(k)} \cdot S_+^{(k+1)} + S_z^{(k)} \cdot S_z^{(k+1)} \right), \tag{2.4.54}$$

which acts in the space $\mathcal{V}^{\otimes n}$*. Summation here runs over all nodes of the chain, and nodes with* $k = 1$ *and* $k = n + 1$ *are identified (closed chain). Each node* k *contains an atom, whose degrees of freedom are described by spin operator* $S_\alpha^{(k)}$ *defined in (2.4.31). The problem is to find all eigenvalues and all eigenstates of the Hamiltonian* $\mathcal{H}_n$*, i.e., to find all energy levels and corresponding quantum states of quantum Heisenberg spin chain.*

Hamiltonian $\mathcal{H}_n$ *commutes with components of total spin vector, i.e., it commutes with the generators* S_α *in representation (2.4.30):*

$$\left[\mathcal{H}_n, \sum_k S_\alpha^{(k)} \right] = 0, \quad \alpha = x, y, z. \tag{2.4.55}$$

> **Problem 2.4.6.** *Prove the relation (2.4.55). Hint: show that the term* $\sum_\alpha S_\alpha^{(k)} \cdot S_\alpha^{(k+1)}$ *in (2.4.54) commutes with* $(S_\beta^{(k)} + S_\beta^{(k+1)})$ *(the term* $\sum_\alpha S_\alpha^{(k)} \cdot S_\alpha^{(k+1)}$ *is polarized Casimir operator of algebra* $su(2)$*, see Section* I-4.7.4 *of the first book).*

The existence of the operators $\sum_k S_\alpha^{(k)}$ *commuting with the Hamiltonian (2.4.54) yields the degeneracy of the energy spectrum. Subspaces in* $\mathcal{V}^{\otimes n}$ *that correspond to irreducible* *representations of algebra $su(2)$ (i.e., subspaces invariant under the action of operators $\sum_k S_\alpha^{(k)}$) can be chosen in such a way that their vectors $|\Psi\rangle \in \mathcal{V}^{\otimes n}$ have one and the same eigenvalue E of the Hamiltonian (2.4.54),*

$$\mathcal{H}_n\,|\Psi\rangle = E\,|\Psi\rangle. \tag{2.4.56}$$

It is clear that the Hamiltonian, the total spin squared (the Casimir operator $\sum_\alpha(\sum_k S_\alpha^{(k)})^2$) and total spin projection $\sum_k S_z^{(k)}$ can be diagonalized simultaneously.

Let us sketch the procedure of extracting the invariant subspaces from $\mathcal{V}^{\otimes n}$, which correspond to irreducible representations of $su(2)$ and at the same time have definite energy E (eigenvalue of the Hamiltonian). This procedure is a slight modification of the method given in Section 2.4.2. Let us consider a tower of basis vectors in $\mathcal{V}^{\otimes n}$:

$$\begin{array}{lll}
(1) & |\underbrace{\uparrow\uparrow\cdots\uparrow}_{n}\rangle & m = n/2 \\
(n) & |\underbrace{\downarrow\uparrow\cdots\uparrow}_{n}\rangle,\ |\underbrace{\uparrow\downarrow\uparrow\cdots\uparrow}_{n}\rangle\ldots,|\underbrace{\uparrow\uparrow\ldots\uparrow\downarrow}_{n}\rangle & m = n/2-1 \\
(\frac{n(n-1)}{2}) & |\underbrace{\downarrow\downarrow\uparrow\cdots\uparrow}_{n}\rangle,\ |\underbrace{\downarrow\uparrow\downarrow\uparrow\cdots\uparrow}_{n}\rangle\ldots,|\underbrace{\uparrow\uparrow\cdots\uparrow\downarrow\downarrow}_{n}\rangle & m = n/2-2 \\
\vdots & \vdots & \vdots \\
C_n^k & |\underbrace{\downarrow\cdots\downarrow}_{k}\underbrace{\uparrow\cdots\uparrow}_{n-k}\rangle,\ldots\ldots,|\underbrace{\uparrow\cdots\uparrow}_{n-k}\underbrace{\downarrow\cdots\downarrow}_{k}\rangle & m = n/2-k \\
\vdots & \vdots & \vdots \\
(n) & |\underbrace{\uparrow\downarrow\cdots\downarrow}_{n}\rangle,\ |\underbrace{\downarrow\uparrow\downarrow\cdots\downarrow}_{n}\rangle\ldots,|\underbrace{\downarrow\downarrow\cdots\downarrow\uparrow}_{n}\rangle & m = -n/2+1 \\
(1) & |\underbrace{\downarrow\downarrow\cdots\downarrow}_{n}\rangle & m = -n/2
\end{array} \tag{2.4.57}$$

We show in the right column the eigenvalues m of operator S_z in representation (2.4.30), and in the left column the dimensions of the pertinent subspaces: subspace with $m = n/2 - k$ has dimension C_n^k.

The state $|\uparrow\uparrow\cdots\uparrow\rangle$, with maximal projection $m = n/2$ (all spins are directed upwards) is the highest weight vector. Indeed, we have

$$\Delta^{n-1}(S_z)\,|\uparrow\cdots\uparrow\rangle = \frac{n}{2}|\uparrow\cdots\uparrow\rangle, \quad \Delta^{n-1}(S_+)\,|\uparrow\cdots\uparrow\rangle = 0,$$

where, in accordance with (2.4.30), raising operator is

$$\Delta^{n-1}(S_+) = S_+^{(1)} + S_+^{(2)} + \cdots + S_+^{(n)}. \tag{2.4.58}$$

Furthermore, it is straightforward to see that this vector is an eigenvector of the Hamiltonian (2.4.54) with eigenvalue $E = n/4$. We act on the highest weight vector $|\uparrow\cdots\uparrow\rangle$ by lowering operator

$$\Delta^{n-1}(S_-) = S_-^{(1)} + S_-^{(2)} + \cdots + S_-^{(n)}, \tag{2.4.59}$$

and, in view of (2.4.34), obtain a vector with $m = n/2 - 1$, which is symmetric linear combination of all vectors with one spin down,

$$|n/2, n/2-1\rangle = \frac{1}{\sqrt{n}}(|\downarrow\uparrow\cdots\uparrow\rangle + |\uparrow\downarrow\uparrow\cdots\uparrow\rangle + \cdots + |\uparrow\cdots\uparrow\downarrow\rangle). \tag{2.4.60}$$

Here we use standard notation $|j,m\rangle$ *for basis vectors in the space of representation* $T^{(j)}$ *and choose the normalization factor in such a way that the vector* $|n/2, n/2-1\rangle$ *has unit length. At the next step, we apply to this vector the operator* $\Delta^{n-1}(S_-)$ *and continue the procedure. Upon action by* $(\Delta^{n-1}(S_-))^k$*, we obtain symmetric vector with projection* $m = n/2 - k$*, i.e., with* k *spins down. The procedure terminates at* n*th step, when the action of* $\Delta^{n-1}(S_-)$ *on the last state* $|\downarrow \ldots \downarrow\rangle$ *gives zero. The space* $\mathcal{V}_{n+1}$ *constructed in this way is the space of irreducible representation* $T^{(n/2)}$*; its dimension is* $(n+1)$*, and all its vectors are eigenvectors of the Hamiltonian (2.4.54) with one and the same eigenvalue* $E = n/4$*. The latter fact follows from (2.4.55). Note that the space* $\mathcal{V}_{n+1}$ *contains one and only one vector from subspace of a given* m *(one and only one linear combination of basis vectors from every line in (2.4.57)).*

Now, we consider a linear combination of vectors in $\mathcal{V}^{\otimes n}$ *with spin projection* $m = n/2 - 1$ *(second line in (2.4.57)):*

$$|n/2-1, n/2-1\rangle = a_1|\downarrow\uparrow\cdots\uparrow\rangle + a_2|\uparrow\downarrow\uparrow\cdots\uparrow\rangle + \cdots + a_n|\uparrow\cdots\uparrow\downarrow\rangle, \tag{2.4.61}$$

and impose the highest weight condition

$$\Delta^{n-1}(S_+)|n/2-1, n/2-1\rangle = 0. \tag{2.4.62}$$

This gives

$$a_1 + a_2 + \cdots + a_n = 0, \tag{2.4.63}$$

which is equivalent to orthogonality of the vectors (2.4.61) and (2.4.60). A new point is that one requires that the states (2.4.61) be eigenstates of the Hamiltonian (2.4.54). This requirement gives extra equations for parameters a_i*. Their solutions determine* $(n-1)$ *eigenstates of the Hamiltonian with* $j = m = n/2 - 1$*. Each of these states is called single-magnon state and is used as the highest weight vector for standard construction of invariant subspaces of representations* $T^{(n/2-1)}$*. We do not give explicit forms of the pertinent equations and single-magnon states, as this material goes beyond the scope of this book, see, e.g., monograph [33].*

To sketch the continuation of the procedure, we do not require that the states considered are eigenstates of the Hamiltonian. So, we can choose any $(n-1)$ *states (2.4.61), (2.4.63). In particular, we can take* $a_k = -a_{k+1}$ *and* $a_i = 0$ $(\forall i \neq k, k+1)$*, where* $k = 1, 2, \ldots, n-1$*. In this way, we obtain* $(n-1)$ *independent states with the highest weight* $m = n/2 - 1$ *and spin* $j = n/2 - 1$*. Starting with these highest weight vectors, we generate, by applying the lowering operator (2.4.59), equivalent* $(n-1)$*-dimensional representations* $T^{(n/2-1)}$*. Indeed, as we pointed out in Section 2.4.2, antisymmetric combination of two spins* $(\downarrow\uparrow - \uparrow\downarrow)$ *is a scalar (singlet). As an example, vector (2.4.61) with* $a_k = -a_{k+1}$ *and* $a_i = 0$ $(\forall i \neq k, k+1)$*, which can be written as*

$$\frac{1}{\sqrt{2}}(S_-^k - S_-^{k+1})|\underbrace{\uparrow\ldots\uparrow}_{n}\rangle = \frac{1}{\sqrt{2}}(|\underbrace{\uparrow\cdots\uparrow}_{k-1}\downarrow\uparrow\uparrow\ldots\rangle - |\underbrace{\uparrow\ldots\uparrow}_{k-1}\uparrow\downarrow\uparrow\ldots\rangle), \tag{2.4.64}$$

is naturally the highest weight vector $|\underbrace{\uparrow\ldots\uparrow}_{n-2}\rangle$*. As described above, this vector generates* $(n-1)$*-dimensional subspace with spin* $j = n/2-1$*. After that we again find in the space* $\mathcal{V}^{\otimes n}$ *the highest weight vectors, but now with* $m = n/2-2$ *(third line in (2.4.57)). These are orthogonal to vectors from subspaces of irreducible representations* $T^{(\frac{n}{2})}$*,* $T^{(\frac{n}{2}-1)}$ *constructed starting with the states (2.4.60), (2.4.61), (2.4.63). The number of the new highest weight vectors is* $(n-1)n/2 - n = (n-3)n/2$*. Eigenstates of the Hamiltonian in this sector are double-magnon states, since they are linear combinations of the states with* *two arrows down. By acting on these states by lowering operators, we generate $(n-3)$-dimensional invariant subspaces of representations $T^{(\frac{n}{2}-2)}$ with spins $j = (n/2-2)$. The complete set of relevant states is obtained by continuing this procedure. At the level $m = n/2-k+1$, there are C_n^{k-1} states, which either belong to representations $T^{(j)}$ with $j > \frac{n}{2}-k+1$ or are the highest weight vectors of representation $T^{(\frac{n}{2}-k+1)}$. Therefore, at kth step $(k < n/2)$ at the level $m = n/2-k$ we have C_n^{k-1} vectors that belong to representations $T^{(j)}$ with $j > \frac{n}{2}-k$, and the remaining states are the highest weight vectors of representation $T^{(\frac{n}{2}-k)}$. The number of the latter vectors is*

$$m(n,k) = C_n^k - C_n^{k-1} = C_n^k \left(\frac{n-2k+1}{n-k+1}\right), \quad k < n/2. \tag{2.4.65}$$

Thus, by making use of the above procedure, one finds invariant $(2j+1)$-dimensional subspaces $\mathcal{V}_{2j+1} \in \mathcal{V}^{\otimes n}$ of representations $T^{(j)}$ with spins $j = n/2, n/2-1, \ldots, 0$ or $j = n/2, n/2-1, \ldots, 1/2$, for even or odd n. The general formula for expansion of $\mathcal{V}^{\otimes n}$ in invariant subspaces $\mathcal{V}_{2j+1}$ reads

$$\mathcal{V}^{\otimes n} = \mathcal{V}_{n+1} + (n-1)\mathcal{V}_{n-1} + \frac{n(n-3)}{2}\mathcal{V}_{n-3} + \cdots = \mathcal{V}_{n+1} + \sum_{k=1}^{[n/2]} m(n,k)\mathcal{V}_{n-2k+1}, \tag{2.4.66}$$

where $[n/2]$ is integer part of $n/2$. In other words, the expansion (analog of (2.4.1)) is

$$(T^{(1/2)})^{\otimes n} = T^{(n/2)} + \sum_{k=1}^{[n/2]} m(n,k) T^{(n/2-k)}. \tag{2.4.67}$$

Formula (2.4.65) for multiplicities $m(n,k)$ of irreducible representations $T^{(n/2-k)}$ in the expansion of representation $(T^{(1/2)})^{\otimes n}$ gives the number of independent k-magnon states of the Heisenberg spin chain. It gives the number of eigenstates of the Hamiltonian (2.4.54) at the level $m = n/2-k$, whose energies may differ from the energies of magnon states constructed starting from the highest weight states at previous levels $m > n/2-k$. The formula (2.4.65) has been obtained by H. Bethe in 1931 in his famous paper [34] and sometimes it is called Bethe formula.

Problem 2.4.7. *Check the consistency of formula (2.4.66) by inserting the dimensions of the pertinent spaces* $\dim(\mathcal{V}^{\otimes n}) = 2^n$ *and* $\dim(\mathcal{V}_m) = m$.

Problem 2.4.8. *Derive the formula (2.4.67) starting from (2.4.12). Hint: make use of induction in n.*

2.4.4. Calculating Clebsch–Gordan coefficients

In this subsection, we give a method of calculating explicitly the elements of matrix C in (2.4.2) and (2.4.3) for finite-dimensional representations of group $SU(2)$. In other words, we are going to calculate Clebsch–Gordan coefficients $\langle j_1 m_1, j_2 m_2 | j_3 m\rangle$ in (2.4.46). This method was suggested by van der Waerden in [20], see also [10]. As before, it is convenient to think of representations of $SU(2)$ as embedded in representations of $SL(2,\mathbb{C})$ from the special class discussed in Remark 3 in the end of Section 2.1. We stress that we consider here representations of $SL(2,\mathbb{C})$ from this class only.

Let $T^{(j_1)}$ and $T^{(j_2)}$ be two irreducible representations of group $SL(2,\mathbb{C})$ (or $SU(2)$). We consider them in the differential formalism, so the basis vectors in spaces of these representations are $T^{j_1}_m$ and $T^{j_2}_m$ as given by (2.2.26). These are monomials of degrees $(2j_1)$ and $(2j_2)$ of variables $u^\alpha = (s_1,t_1)$ and $v^\alpha = (s_2,t_2)$, respectively. Tensor product $T^{(j_1)} \otimes T^{(j_2)}$ acts in space $\mathcal{P}_{(j_1,j_2)} \subset \mathcal{F}_{(2,\bar{0})}$, whose basis vectors are monomials $T^{j_1}_m(s_1,t_1)\cdot T^{j_2}_{m'}(s_2,t_2)$. We begin by noticing that for $j_1 = j_2 = j$, the representation $T^{(j)} \otimes T^{(j)}$ contains trivial one-dimensional representation whose basis vector is an invariant $(s_1t_2 - s_2t_1)^{2j} \in \mathcal{P}_{(j,j)}$. Indeed, function $\varepsilon_{\alpha\beta}u^\alpha v^\beta = (s_1t_2 - s_2t_1)$ is invariant under transformations (2.2.19), so any power of this function is an invariant. In accordance with (2.3.39), expansion of the function $(s_1t_2-s_2t_1)^{2j}$ in series in variables s_i and t_i gives an invariant (by construction) convolution of basis vectors $T^j_m(s_1,t_1)$ and $T^j_m(s_2,t_2)$. The convolution (2.3.39) has been written in terms of invariant metric (2.3.40), which is convenient to write as

$$\eta^{(j)}_{m,-m'} = (-1)^{j-m}\delta_{m,m'}. \tag{2.4.68}$$

We make use of this metric in Section 2.5.

Let us come back to the general case of two representations $T^{(j_1)}$ and $T^{(j_2)}$. We consider yet another co-representation $\overline{T}^{(j_3)}$ of $SL(2,\mathbb{C})$ which acts in the space of homogeneous polynomials of degree $2j_3$ of dual variables $(\bar{s}_3,\bar{t}_3)$, with basis (2.2.29). Representation $T^{(j_1)} \otimes T^{(j_2)} \otimes \overline{T}^{(j_3)}$ in space $\mathcal{P}_{(j_1,j_2,j_3)} \subset \mathcal{F}_{(2,\bar{1})}$ with basis monomials $T^{j_1}_{m_1}(s_1,t_1)\cdot T^{j_2}_{m_2}(s_2,t_2)\cdot \bar{T}^{j_3}_m(\bar{s}_3,\bar{t}_3)$ is nothing but representation $T_{(2,\bar{1})}$, in accordance with the general formula (2.2.20). In representation $T^{(j_1)} \otimes T^{(j_2)} \otimes \overline{T}^{(j_3)}$, there is a trivial one-dimensional sub-representation whose basis vector is a function $\mathcal{I}$ of six variables $\{s_1,t_1,s_2,t_2,\bar{s}_3,\bar{t}_3\}$:

$$\mathcal{I} = (s_1t_2 - s_2t_1)^{n_3}(s_1\bar{s}_3 + t_1\bar{t}_3)^{n_2}(s_2\bar{s}_3 + t_2\bar{t}_3)^{n_1}, \tag{2.4.69}$$

where n_1, n_2, n_3 are nonnegative integers,

$$n_3 = j_1 + j_2 - j_3, \quad n_2 = j_3 + j_1 - j_2, \quad n_1 = j_3 + j_2 - j_1. \tag{2.4.70}$$

Indeed, the degree of this function in every pair of variables (s_1,t_1), (s_2,t_2) and $(\bar{s}_3,\bar{t}_3)$ equals $n_3 + n_2 = 2j_1$, $n_3 + n_1 = 2j_2$ and $n_1 + n_2 = 2j_3$, respectively, hence the vector (2.4.69) belongs to the space

$\mathcal{P}_{(j_1,j_2,j_3)}$ of representation $T^{(j_1)} \otimes T^{(j_2)} \otimes \overline{T}^{(j_3)}$. Furthermore, this vector is invariant under the transformations (2.2.20), which follows from invariance of functions (2.2.33) and (2.3.39), so it is indeed a basis vector of one-dimensional representation of $SL(2,\mathbb{C})$.

Problem 2.4.9. Making use of the differential realizations (2.2.4) and (2.2.11), check the invariance property of $\mathcal{I}$, namely,

$$(\hat{e}^{(1)}_{\pm} + \hat{e}^{(2)}_{\pm} + \bar{e}^{(3)}_{\pm})\mathcal{I} = 0, \quad (\hat{\mathsf{h}}^{(1)} + \hat{\mathsf{h}}^{(2)} + \bar{\mathsf{h}}^{(3)})\mathcal{I} = 0. \tag{2.4.71}$$

Since the invariant $\mathcal{I}$ given by (2.4.69) is a homogeneous polynomial in $\bar{s}_3$, $\bar{t}_3$ of degree $2j_3$, it can be expanded in basis vectors $\overline{T}^{j_3}_m(\bar{s}_3,\bar{t}_3)$, and written as a convolution

$$\mathcal{I} = \sum_{m=-j_3}^{j_3} (-1)^{j_3-m} \quad W^{j_3}_m(s_1,t_1;s_2,t_2)\cdot \overline{T}^{j_3}_{-m}(\bar{s}_3,\bar{t}_3). \tag{2.4.72}$$

Since this convolution is invariant, the polynomials $W^{j_3}_m$ transform under $SL(2,\mathbb{C})$ in the same way as $T^{j_3}_m$. At the same time, these polynomials belong to the space $\mathcal{P}_{(j_1,j_2)}$ of representation $T^{(j_1)} \otimes T^{(j_2)}$. Thus, the functions $W^{j_3}_m$ are basis vectors of the space of irreducible representation, equivalent to $T^{(j_3)}$ and embedded in $T^{(j_1)} \otimes T^{(j_2)}$.

Let us figure out the expansion of vector $W^{j_3}_m(s_1,t_1;s_2,t_2)$ in basis monomials $T^{(j_1)}_{m_1}(s_1,t_1) \cdot T^{(j_2)}_{m_2}(s_2,t_2)$ in space $\mathcal{P}_{(j_1,j_2)}$ of representation $T^{(j_1)} \otimes T^{(j_2)}$. To this end, we expand the invariant (2.4.69) in all variables $(s_1,t_1,s_2,t_2,\bar{s}_3,\bar{t}_3)$:

$$\begin{aligned}\mathcal{I} = &\sum_{k,k_1,k_2} (-1)^k C^k_{n_3} C^{k_1}_{n_1} C^{k_2}_{n_2} (s_1 t_2)^{n_3-k} (s_2 t_1)^k (s_1 \bar{s}_3)^{n_2-k_2} \\ &\times (t_1 \bar{t}_3)^{k_2} (s_2 \bar{s}_3)^{n_1-k_1} (t_2 \bar{t}_3)^{k_1} \\ = &\sum_{k,k_1,k_2} (-1)^k C^k_{n_3} C^{k_1}_{n_1} C^{k_2}_{n_2} (s_1^{n_3+n_2-k-k_2}\, t_1^{k+k_2})(s_2^{n_1-k_1+k}\, t_2^{n_3-k+k_1}) \\ &\times (\bar{s}_3^{n_1+n_2-k_1-k_2}\, \bar{t}_3^{k_1+k_2}),\end{aligned} \tag{2.4.73}$$

where summation runs over $0 \le k \le n_3$, $0 \le k_1 \le n_1$, $0 \le k_2 \le n_2$. In fact, we can extend the summation over k, k_1, k_2 in (2.4.73) to infinite range, since, in accordance with (2.3.34), the binomial coefficients vanish outside the original range of summation. To write (2.4.73) in terms of the basis vectors of spaces of representations $T^{(j_1)}$, $T^{(j_2)}$ and $\overline{T}^{(j_3)}$ (see (2.2.26) and

(2.2.29)), we write n_1, n_2 and n_3 in terms of j_1, j_2 and j_3 via (2.4.70), and, instead of indices $k_{1,2}$ introduce indices $m_{1,2}$:

$$\begin{aligned} k + k_2 &= j_1 - m_1 &&\Rightarrow 0 \le j_1 - m_1 \le n_3 + n_2 \Rightarrow -j_1 \le m_1 \le j_1, \\ n_3 - k + k_1 &= j_2 - m_2 &&\Rightarrow 0 \le j_2 - m_2 \le n_3 + n_1 \Rightarrow -j_2 \le m_2 \le j_2. \end{aligned}$$

We have, in particular,

$$k_1 + k_2 = j_3 - (m_1 + m_2).$$

After these changes of variables, summation over k in (2.4.73) involves the binomial coefficients only. Therefore, the expansion (2.4.73) reads

$$\begin{aligned} \mathcal{I} &= \sum_{m_1,m_2} \left(\sum_k (-1)^k C^k_{n_3} C^{j_2-n_3-m_2+k}_{2j_2-n_3} C^{j_1-m_1-k}_{2j_1-n_3} \right) \\ &\quad \times (s_1^{j_1+m_1} t_1^{j_1-m_1})(s_2^{j_2+m_2} t_2^{j_2-m_2})(\bar{s}_3^{j_3+m} \bar{t}_3^{j_3-m}) \\ &= \mathcal{D}(j_1,j_2,j_3) \sum_{m_1,m_2} \langle\langle j_1, m_1; j_2, m_2 | j_3, m \rangle\rangle (-1)^{j_3-m} T^{j_1}_{m_1} T^{j_2}_{m_2} \overline{T}^{j_3}_{-m}, \end{aligned} \tag{2.4.74}$$

where $m = m_1 + m_2$, $\sum_{m_1,m_2} = \sum_{m_1=-j_1}^{j_1} \sum_{m_2=-j_2}^{j_2}$ and we extract the factor

$$\mathcal{D}(j_1,j_2,j_3) = (j_2 + j_3 - j_1)!(j_1 + j_3 - j_2)!(j_2 + j_1 - j_3)!. \tag{2.4.75}$$

Another object in (2.4.74), denoted by $\langle\langle j_1\, m_1, j_2\, m_2 | j\, m\rangle\rangle$, is

$$\begin{aligned} &\langle\langle j_1\, m_1, j_2\, m_2 | j_3\, m\rangle\rangle \\ &\quad = \sum_k \left[(-1)^k \frac{\sqrt{(j_1+m_1)!(j_1-m_1)!(j_2+m_2)!}}{(j_1-m_1-k)!(j_1-n_3+m_1+k)!(j_2+m_2-k)!} \right. \\ &\qquad \left. \times \frac{\sqrt{(j_2-m_2)!(j_3+m)!(j_3-m)!}}{(j_2-n_3-m_2+k)!k!(n_3-k)!} \right] \\ &\quad = \sum_k \left[(-1)^k \frac{\sqrt{(j_1+m_1)!(j_1-m_1)!(j_2+m_2)!}}{(j_1-m_1-k)!(j_3-j_2+m_1+k)!(j_2+m_2-k)!} \right. \\ &\qquad \left. \times \frac{\sqrt{(j_2-m_2)!(j_3+m)!(j_3-m)!}}{(j_3-j_1-m_2+k)!k!(j_1+j_2-j_3-k)!} \right]. \end{aligned} \tag{2.4.76}$$

We find below that this object, modulo overall normalization, equals the Clebsch–Gordan coefficient.

We can again extend summation over k in (2.4.76) to $\pm\infty$, since in view of convention (2.3.34), the ratio on the right-hand side of (2.4.76) vanishes, if any of the factorials in the denominator, say, $k!$ or $(j_1 - m_1 - k)!$ diverges. By comparing the invariant (2.4.74) with invariant convolution (2.4.72), we obtain

$$W^{j_3}_m(s_1,t_1,s_2,t_2) = \mathcal{D}(j_1,j_2,j_3) \sum_{\substack{m_1,m_2\\ m_1+m_2=m}} T^{j_1}_{m_1} T^{j_2}_{m_2} \langle\langle j_1,m_1;j_2,m_2|j_3,m\rangle\rangle. \tag{2.4.77}$$

Now, vector $W^{j_3}_m$ is not normalized with respect to scalar product (2.2.28). So, we introduce the normalized vector

$$T^{j_3}_m(s_1,t_1,s_2,t_2) = \frac{\mu}{\mathcal{D}} W^{j_3}_m = \sum_{m_1,m_2} T^{j_1}_{m_1} T^{j_2}_{m_2} \langle j_1,m_1;j_2,m_2|j_3,m\rangle, \tag{2.4.78}$$

where $\mu(j_1,j_2,j_3)$ is normalization factor that we determine below (see Proposition 2.4.3), and Clebsch–Gordan coefficients are

$$\langle j_1\, m_1, j_2\, m_2|j_3\, m\rangle = \mu(j_1,j_2,j_3)\,\langle\langle j_1\, m_1, j_2\, m_2|j_3\, m\rangle\rangle \delta_{m,m_1+m_2}. \tag{2.4.79}$$

The fact that these are indeed Clebsch–Gordan coefficients follows from comparison of (2.4.78) and (2.4.46).

In view of (2.4.78), invariant (2.4.69), (2.4.72) takes the form

$$\mathcal{I} = \frac{\mathcal{D}}{\mu} \sum_{m=-j_3}^{j_3} (-1)^{j_3-m}\, T^{j_3}_m(s_1,t_1;s_2,t_2)\cdot \overline{T}^{j_3}_{-m}(\bar{s}_3,\bar{t}_3) \tag{2.4.80}$$

$$\begin{aligned} &= \frac{\mathcal{D}}{\mu} \sum_{m_1,m_2,m_3} (-1)^{j_3-m_3} T^{j_1}_{m_1}(s_1,t_1) T^{j_2}_{m_2}(s_2,t_2) \\ &\quad \times \langle j_1,m_1;j_2,m_2|j_3,m_3\rangle \overline{T}^{j_3}_{-m_3}(\bar{s}_3,\bar{t}_3). \end{aligned} \tag{2.4.81}$$

Because of the invariance of function $\mathcal{I}$ under $SL(2,\mathbb{C})$ transformations (2.2.20), and the transformation rule (2.3.35) for $T^j_m(s,t)$, we find that the coefficients $\langle j_1,m_1;j_2,m_2|j_3,m_3\rangle$ obey

$$\begin{aligned} &\mathcal{D}^{(j_1)}_{m_1'm_1}(g)\mathcal{D}^{(j_2)}_{m_2'm_2}(g)\,\langle j_1,m_1;j_2,m_2|j_3,m_3\rangle\,\mathcal{D}^{(j_3)}_{m_3m_3'}(g^{-1}) \\ &\quad = \langle j_1,m_1';j_2,m_2'|j_3,m_3'\rangle, \quad \forall g\in SL(2,\mathbb{C}). \end{aligned} \tag{2.4.82}$$

Thus, Clebsch–Gordan coefficients can be treated as coefficients of invariant third rank tensor (see Definition I-4.3.2 in the first book).

Proposition 2.4.3. *Explicit expression for Clebsch–Gordan coefficient* (2.4.79) *is*

$$\begin{aligned}\langle j_1\, m_1, j_2\, m_2 | j_3\, m\rangle \\ &= \delta_{m,m_1+m_2}\, \mu(j_1, j_2, j_3) \\ &\times \sum_k \left[(-1)^k \frac{\sqrt{(j_1+m_1)!(j_1-m_1)!(j_2+m_2)!}}{(j_1-m_1-k)!(j_3-j_2+m_1+k)!(j_2+m_2-k)!}\right. \\ &\left.\times \frac{\sqrt{(j_2-m_2)!(j_3+m)!(j_3-m)!}}{(j_3-j_1-m_2+k)!k!(j_1+j_2-j_3-k)!}\right], \end{aligned} \tag{2.4.83}$$

where the normalization factor $\mu(j_1, j_2, j_3)$ *equals*

$$\mu(j_1, j_2, j_3) = \sqrt{2j_3+1}\Delta(j_1, j_2, j_3), \tag{2.4.84}$$

and the combination

$$\Delta(j_1, j_2, j_3) = \sqrt{\frac{(j_1+j_2-j_3)!(j_1+j_3-j_2)!(j_2+j_3-j_1)!}{(j_1+j_2+j_3+1)!}} \tag{2.4.85}$$

is called triangle coefficient.

Proof. In view of (2.4.76) and (2.4.79), it remains to calculate the normalization coefficient $\mu(j_1, j_2, j_3)$. Consider invariant vector (2.4.69). In accordance with (2.2.28), its norm is given by

$$(\partial_{s_1}\partial_{t_2} - \partial_{s_2}\partial_{t_1})^{n_3} \cdot (\partial_{s_1}\partial_{\bar{s}_3} + \partial_{t_1}\partial_{\bar{t}_3})^{n_2} \cdot (\partial_{s_2}\partial_{\bar{s}_3} + \partial_{t_2}\partial_{\bar{t}_3})^{n_1} \cdot \mathcal{I}. \tag{2.4.86}$$

To calculate this norm, we make use of remarkable identities

$$\begin{aligned}&(\partial_{s_2}\partial_{\bar{s}_3} + \partial_{t_2}\partial_{\bar{t}_3}) \cdot (s_1t_2 - s_2t_1)^p(s_1\bar{s}_3 + t_1\bar{t}_3)^q(s_2\bar{s}_3 + t_2\bar{t}_3)^r \\ &\quad = r(r+p+q+1)(s_1t_2 - s_2t_1)^p(s_1\bar{s}_3 + t_1\bar{t}_3)^q(s_2\bar{s}_3 + t_2\bar{t}_3)^{r-1}, \end{aligned} \tag{2.4.87}$$

$$(\partial_{s_1}\partial_{t_2} - \partial_{t_1}\partial_{s_2}) \cdot (s_1t_2 - s_2t_1)^p = p(p+1)(s_1t_2 - s_2t_1)^{p-1}, \tag{2.4.88}$$

which are derived by straightforward calculation. Similar identities hold for operators $(\partial_{s_1}\partial_{t_2} - \partial_{s_2}\partial_{t_1})$ and $(\partial_{s_1}\partial_{\bar{s}_3} + \partial_{t_1}\partial_{\bar{t}_3})$ acting on invariant $\mathcal{I}$.

We make use of (2.4.87) and (2.4.88) to calculate the norm (2.4.86):

$$
\begin{aligned}
&(\partial_{s_1}\partial_{t_2}-\partial_{s_2}\partial_{t_1})^{n_3}(\partial_{s_1}\partial_{\bar{s}_3}+\partial_{t_1}\partial_{\bar{t}_3})^{n_2}(\partial_{s_2}\partial_{\bar{s}_3}+\partial_{t_2}\partial_{\bar{t}_3})^{n_1}\mathcal{I} \\
&=\frac{n_1!(n_1+n_2+n_3+1)!}{(n_2+n_3+1)!}(\partial_{s_1}\partial_{t_2}-\partial_{s_2}\partial_{t_1})^{n_3}(\partial_{s_1}\partial_{\bar{s}_3}+\partial_{t_1}\partial_{\bar{t}_3})^{n_2} \\
&\times(s_1t_2-s_2t_1)^{n_3}(s_1\bar{s}_3+t_1\bar{t}_3)^{n_2} \\
&=\frac{n_1!(n_1+n_2+n_3+1)!}{(n_2+n_3+1)!}\cdot\frac{n_2!(n_2+n_3+1)!}{(n_3+1)!}(\partial_{s_1}\partial_{t_2}-\partial_{s_2}\partial_{t_1})^{n_3} \\
&\times(s_1t_2-s_2t_1)^{n_3} \\
&=\frac{n_1!(n_1+n_2+n_3+1)!}{(n_2+n_3+1)!}\cdot\frac{n_2!(n_2+n_3+1)!}{(n_3+1)!}\cdot n_3!(n_3+1)! \\
&=n_1!n_2!n_3!(n_1+n_2+n_3+1)!.
\end{aligned}
$$

So, the normalized invariant vector is

$$
\widetilde{\mathcal{I}}=\frac{\mathcal{I}}{\sqrt{n_1!n_2!n_3!(n_1+n_2+n_3+1)!}}, \tag{2.4.89}
$$

and its expansion in basis polynomials $\bar{T}^{j_3}_{m_3}(\bar{s}_3,\bar{t}_3)$ reads

$$
\widetilde{\mathcal{I}}=\frac{1}{\sqrt{2j_3+1}}\sum_{m_3=-j_3}^{j_3}(-1)^{j_3-m_3}\bar{T}^{j_3}_{-m_3}(\bar{s}_3,\bar{t}_3)T^{j_3}_{m_3}(s_1,t_1,s_2,t_2). \tag{2.4.90}
$$

We remind that $T^{j_3}_{m_3}(s_1,t_1,s_2,t_2)$ are normalized basis vectors in the subspace of representation $T^{(j_3)}$ embedded into the space of representation $T^{j_1}\otimes T^{j_2}$. The overall coefficient $(2j_3+1)^{-1/2}$ on the right-hand side of (2.4.90) is determined by the requirement that this convolution must be normalized vector in the space of representation $T^{(j_1)}\otimes T^{(j_2)}\otimes \overline{T}^{(j_3)}$.

Problem 2.4.10. Making use of (2.4.88), show that the norm of vector $\frac{1}{(2j)!}(s\bar{s}+t\bar{t})^{2j}=\sum_{m=-j}^{j}(-1)^{j-m}\bar{T}^j_{-m}(\bar{s},\bar{t})T^j_m(s,t)$ is equal to $(2j+1)$.

By comparing (2.4.90) and (2.4.80), we obtain the coefficient $\mu(j_1,j_2,j_3)$ given by (2.4.84).

□

Together with Clebsch–Gordan coefficients, one often uses coefficients $\begin{pmatrix} j_1 & j_2 & j_3 \\ m_1 & m_2 & m_3 \end{pmatrix}$, known as Wigner 3-j symbols. They are related to Clebsch–Gordan coefficients as follows:

$$\begin{pmatrix} j_1 & j_2 & j_3 \\ m_1 & m_2 & m_3 \end{pmatrix} = \frac{(-1)^{j_2-j_1+m_3}}{\sqrt{2j_3+1}} \langle j_1\, m_1,\, j_2\, m_2 | j_3, -m_3 \rangle. \tag{2.4.91}$$

Like Clebsch–Gordan coefficients, they do not vanish only if $m_1 + m_2 + m_3 = 0$. Wigner 3-j symbols obey simple symmetry properties, see Sec. 2.4.5.

2.4.5. Properties of Clebsch–Gordan coefficients and 3-j symbols

Making use of Dirac notations, we rewrite (2.4.78) as follows (see (2.4.46)):

$$|j_3, m\rangle = \sum_{m_1,m_2} |j_1\, m_1\rangle \otimes |j_2\, m_2\rangle \langle j_1\, m_1, j_2\, m_2 | j_3\, m\rangle, \tag{2.4.92}$$

where we remind that j_3 is any spin from the range $|j_1 - j_2| \le j_3 \le j_1 + j_2$, and $|j_3, m\rangle$, $|j_1\, m_1\rangle$ and $|j_2\, m_2\rangle$ are orthonormalized basis vectors in the spaces of representations $T^{(j_3)}$, $T^{(j_1)}$ and $T^{(j_2)}$, respectively. Due to the orthonormality of basis vectors, this gives the formal expression for Clebsch–Gordan coefficients:

$$\langle j_1\, m_1, j_2\, m_2 | j_3\, m\rangle := \left(\langle j_1\, m_1| \otimes \langle j_2, m_2|\right) |j_3\, m\rangle. \tag{2.4.93}$$

Note that the invariants (2.4.69) have the property

$$\mathcal{I}_{j_1,j_2,j_3}(s_1, t_1; s_2, t_2; \bar{s}_3, \bar{t}_3) = (-1)^{j_1+j_2-j_3} \mathcal{I}_{j_2,j_1,j_3}(s_2, t_2; s_1, t_1; \bar{s}_3, \bar{t}_3),$$

which yields the following transformation law of Clebsch–Gordan coefficients (2.4.93) under the permutation of vectors $\langle j_1\, m_1|$ and $\langle j_2, m_2|$ on the right-hand side of (2.4.93):

$$\langle j_1\, m_1, j_2\, m_2 | j_3\, m\rangle = (-1)^{j_1+j_2-j_3} \langle j_2\, m_2, j_1\, m_1 | j_3\, m\rangle. \tag{2.4.94}$$

We also note that it is sometimes convenient to denote the basis vector $|j_3\ m\rangle$ on the left-hand side of (2.4.92) and right-hand side of (2.4.93) in more informative way, $|j_1\, j_2; j_3\ m\rangle$, where spins of the original states j_1 and j_2 are explicitly indicated. Basis in the space of representation $T^{(j_1)} \otimes T^{(j_2)}$ composed of vectors $|j_1\, j_2; j_3\ m\rangle$, where $|j_1 - j_2| \le j_3 \le j_1 + j_2$, is called basis of coupled states, or simply coupled basis.

We take Hermitian conjugation of both sides of (2.4.92) and make use of (2.1.27) to obtain the equality

$$\langle j_1, j_2\, ; j_3, m| = \sum_{m_1,m_2} \langle j_3\, m | j_1\, m_1, j_2\, m_2\rangle \langle j_1\, m_1| \otimes \langle j_2\, m_2|, \tag{2.4.95}$$

where, due to real-valuedness of Clebsch–Gordan coefficients (2.4.79), we have

$$\langle j_3\, m | j_1\, m_1, j_2\, m_2\rangle := \langle j_1\, m_1, j_2\, m_2 | j_3\, m\rangle^* = \langle j_1\, m_1, j_2\, m_2 | j_3\, m\rangle. \tag{2.4.96}$$

By convoluting the two sides of (2.4.95) with vector $|j_1, j_2\, ; j_4, m_4\rangle$ and using the representation (2.4.93), one finds (with substitution $m \to m_3$) the orthogonality property of Clebsch–Gordan coefficients

$$\delta_{j_3,j_4} \delta_{m_3,m_4} = \sum_{m_1,m_2} \langle j_3\, m_3 | j_1\, m_1, j_2\, m_2\rangle \langle j_1\, m_1; j_2\, m_2 | \, j_4, m_4\rangle. \tag{2.4.97}$$

Now, consider operator $\sum_{j_3=|j_1-j_2|}^{j_1+j_2} |j_1, j_2\,;\, j_3, m\rangle\langle j_1, j_2\,;\, j_3, m|$, *which acts in the space* $\mathcal{V} = \mathcal{V}_1 \otimes \mathcal{V}_2$ *of representation* $T^{(j_1)} \otimes T^{(j_2)}$ *and, in accordance with Proposition 2.4.2, serves as a projector in* $\mathcal{V}$ *on subspace of all eigenvectors of the operator* h *with eigenvalue* m. *One can define another operator acting in* $\mathcal{V}$:

$$\sum_{\substack{m_1,m_2\\ m_1+m_2=m}} (|j_1\, m_1\rangle \otimes |j_2\, m_2\rangle)(\langle j_1\, m_1| \otimes \langle j_2\, m_2|)$$

$$= \sum_{\substack{m_1,m_2\\ m_1+m_2=m}} |j_1\, m_1\rangle\langle j_1\, m_1| \otimes |j_2\, m_2\rangle\langle j_2\, m_2|.$$

It is also projector on subspace of all eigenvectors of the operator h *with eigenvalue* m, *so the two operators are, in fact, equal to each other,*

$$\sum_{j_3=|j_1-j_2|}^{j_1+j_2} |j_1, j_2\,;\, j_3, m\rangle\langle j_1, j_2\,;\, j_3, m| = \sum_{\substack{m_1',m_2'\\ m_1'+m_2'=m}} |j_1\, m_1'\rangle\langle j_1\, m_1'| \otimes |j_2\, m_2'\rangle\langle j_2\, m_2'|. \tag{2.4.98}$$

This relation enables one to invert Clebsch–Gordan expansion (2.4.92). Indeed, one multiplies (2.4.98) on the right by vector $|j_1\, m_1\rangle \otimes |j_2\, m_2\rangle$. *Then, using representation conjugate to (2.4.93),*

$$\langle j_3\; m|j_1\, m_1, j_2\, m_2\rangle := \langle j_3\; m|(|j_1\, m_1\rangle \otimes |j_2\, m_2\rangle), \tag{2.4.99}$$

one finds inverted Clebsch–Gordan expansion

$$\sum_{j_3=|j_1-j_2|}^{j_1+j_2} |j_1, j_2; j_3\, m\rangle\; \langle j_3\, m|j_1\, m_1, j_2\, m_2\rangle = |j_1\, m_1\rangle \otimes |j_2\, m_2\rangle, \tag{2.4.100}$$

where we still have $m = m_1 + m_2$. *Then, by taking convolution of the two sides of (2.4.100) with vector* $\langle j_1\, m_1'| \otimes \langle j_2\, m_2'|$ *one obtains another orthonormalization property of Clebsch–Gordan coefficients*

$$\sum_{j_3=|j_1-j_2|}^{j_1+j_2} \langle j_1\, m_1', j_2\, m_2'|j_3\, m\rangle\; \langle j_3\, m|j_1\, m_1, j_2\, m_2\rangle = \delta_{m_1 m_1'}\delta_{m_2 m_2'},$$

where $m = m_1 + m_2 = m_1' + m_2'$.

Remark 1. *In view of definition (2.4.70) and identification* $\bar{t}_3 = -s_3$, $\bar{s}_3 = t_3$ *considered in (2.2.17), invariant (2.4.69) can be written in symmetric form*

$$(-1)^{j_2-j_1-j_3}\mathcal{I} = (s_1t_2 - s_2t_1)^{j_1+j_2-j_3}(s_3t_1 - s_1t_3)^{j_3+j_1-j_2}(s_2t_3 - s_3t_2)^{j_3+j_2-j_1}. \tag{2.4.101}$$

Then, upon normalizing (2.4.89), we have

$$\frac{(-1)^{j_2-j_1-j_3}\mathcal{I}}{\sqrt{n_1!n_2!n_3!(n_1+n_2+n_3+1)!}}$$

$$= \sum_{m_1,m_2,m_3} \begin{pmatrix} j_1 & j_2 & j_3 \\ m_1 & m_2 & m_3 \end{pmatrix} T^{j_1}_{m_1}(s_1t_1)\, T^{j_2}_{m_2}(s_2t_2)\, T^{j_3}_{m_3}(s_3t_3), \tag{2.4.102}$$

where $\begin{pmatrix} j_1 & j_2 & j_3 \\ m_1 & m_2 & m_3 \end{pmatrix}$ *are actually Wigner 3-j symbols (2.4.91). The latter fact follows from comparison of (2.4.102) and (2.4.81), where one substitutes* $\bar{t}_3 \to -s_3$, $\bar{s}_3 \to t_3$ *and uses (2.4.70), (2.4.75) and (2.4.84).*

Since the expression (2.4.101) does not change (modulo sign) under any permutation of the three indices $\{1,2,3\}$, *and also under the replacement* $s_i \leftrightarrow t_i$, *the 3-j symbol (2.4.91) has the same symmetry property. In particular, we have*

$$(-1)^{j_1+j_2+j_3}\begin{pmatrix} j_1 & j_2 & j_3 \\ m_1 & m_2 & m_3 \end{pmatrix} = \begin{pmatrix} j_2 & j_1 & j_3 \\ m_2 & m_1 & m_3 \end{pmatrix} = \begin{pmatrix} j_1 & j_3 & j_2 \\ m_1 & m_3 & m_2 \end{pmatrix}, \tag{2.4.103}$$

where, in view of (2.4.70), the sum $(j_1 + j_2 + j_3)$ *is always integer. Thus, 3-j symbols (2.4.91) have the following properties.*

1. *They do not change under even permutations of columns:*

$$\begin{pmatrix} j_1 & j_2 & j_3 \\ m_1 & m_2 & m_3 \end{pmatrix} = \begin{pmatrix} j_3 & j_1 & j_2 \\ m_3 & m_1 & m_2 \end{pmatrix} = \begin{pmatrix} j_2 & j_3 & j_1 \\ m_2 & m_3 & m_1 \end{pmatrix},$$

and acquire sign factor $(-1)^{j_1+j_2+j_3}$ *under odd permutations of columns.*

2. *Under replacement* $s_i \leftrightarrow t_i$ $(i = 1,2,3)$, *we have* $T^{j_i}_{m_i}(s_i,t_i) \to T^{j_i}_{-m_i}(s_i,t_i)$, *and formula (2.4.101) tells that under this replacement, the invariant transforms as* $\mathcal{I} \to (-1)^{j_1+j_2+j_3}\mathcal{I}$. *Therefore, in view of (2.4.102), 3-j symbols acquire sign factor* $(-1)^{j_1+j_2+j_3}$ *when signs of all projections* m_i *are flipped:*

$$\begin{pmatrix} j_1 & j_2 & j_3 \\ m_1 & m_2 & m_3 \end{pmatrix} = (-1)^{j_1+j_2+j_3}\begin{pmatrix} j_1 & j_2 & j_3 \\ -m_1 & -m_2 & -m_3 \end{pmatrix}. \tag{2.4.104}$$

3. *In accordance with (2.4.79) we have*

$$m_1 + m_2 + m_3 \neq 0 \Longrightarrow \begin{pmatrix} j_1 & j_2 & j_3 \\ m_1 & m_2 & m_3 \end{pmatrix} = 0. \tag{2.4.105}$$

Problem 2.4.11. *By making use of (2.4.76), (2.4.79), (2.4.84) and (2.4.91), derive the formula*

$$\begin{pmatrix} j_1 & j_2 & j_3 \\ j_1 & m_2 & m_3 \end{pmatrix} = \frac{(-1)^{j_2-j_1+m_3}\Delta(j_1,j_2,j_3)}{(j_1+j_2-j_3)!(j_1+j_3-j_2)!} \times\sqrt{\frac{(2j_1)!(j_2-m_2)!(j_3-m_3)!}{(j_2+m_2)!(j_3+m_3)!}}\,\delta_{j_1,-m_2-m_3}. \tag{2.4.106}$$

Using this result, show that

$$\begin{pmatrix} 0 & j_2 & j_3 \\ 0 & m_2 & m_3 \end{pmatrix} = \frac{(-1)^{j_2-m_2}}{\sqrt{(j_2+j_3+1)}}\,\delta_{m_2,-m_3}\,\delta^{j_2,j_3}. \tag{2.4.107}$$

Remark 2. *In terms of homogeneous coordinates* $z_i = s_i/t_i$, *the symmetric invariant (2.4.101) is written as*

$$\mathcal{I} = t_1^{2j_1}t_2^{2j_2}t_3^{2j_3}(z_1-z_2)^{j_1+j_2-j_3}(z_1-z_3)^{j_3+j_1-j_2}(z_2-z_3)^{j_3+j_2-j_1},$$

while invariance properties (2.4.71) read

$$\sum_{i=1}^{3}\partial_{z_i}\mathcal{I}=0,\quad \sum_{i=1}^{3}(z_it_i\partial_{t_i}-z_i^2\partial_{z_i})\mathcal{I}=0,\quad \sum_{i=1}^{3}\left(z_i\partial_{z_i}-\frac{1}{2}t_i\partial_{t_i}\right)\mathcal{I}=0. \tag{2.4.108}$$

Problem 2.4.12. *Derive Eq. (2.4.108) from Eq. (2.4.71), by making change of variables* $\bar{t}_3=-s_3$, $\bar{s}_3=t_3$ *and inserting* $\hat{e}_+^{(i)}=s_i\,\partial_{t_i}$, $\hat{e}_-^{(i)}=t_i\,\partial_{s_i}$ *and* $\hat{h}^{(i)}=\frac{1}{2}(s_i\,\partial_{s_i}-t_i\,\partial_{t_i})$.

Upon appropriate renormalization, the invariant function $\mathcal{I}^{-1}$ *is proportional to holomorphic part*

$$G_{j_1,j_2,j_3}(z_1,z_2,z_3)=\frac{1}{(z_1-z_2)^{j_1+j_2-j_3}(z_1-z_3)^{j_3+j_1-j_2}(z_2-z_3)^{j_3+j_2-j_1}} \tag{2.4.109}$$

of 3-point correlation function in two-dimensional conformal field theory

$$\langle\,\phi_1(z_1,\bar{z}_1)\phi_2(z_2,\bar{z}_2)\phi_3(z_3,\bar{z}_3)\,\rangle=G_{j_1,j_2,j_3}(z_1,z_2,z_3)\cdot G_{\bar{j}_1,\bar{j}_2,\bar{j}_3}(\bar{z}_1,\bar{z}_2,\bar{z}_3). \tag{2.4.110}$$

The parameters $\Delta_a=j_a+\bar{j}_a$ *and* $s_a=j_a-\bar{j}_a$ *determine conformal dimensions and spins of 2-dimensional fields* $\phi_a(z_a,\bar{z}_a)$*, respectively. This follows from the form of conformal transformations* $\phi_a(z,\bar{z})\to\lambda^{j_a}\bar{\lambda}^{\bar{j}_a}\,\phi_a(\lambda z,\bar{\lambda}\bar{z})$*, which leave the correlation function (2.4.110) invariant. The relationship to the correlation function is, in fact, not accidental: it has to do with the fact that in two-dimensional conformal field theories, holomorphic 3-point functions* $G_{j_1,j_2,j_3}(z_1,z_2,z_3)\equiv G(z_1,z_2,z_3)$ *obey Ward identities*

$$\sum_{i=1}^{3}\partial_{z_i}G=0,\quad \sum_{i=1}^{3}(z_i^2\partial_{z_i}+z_i\Delta_i)G=0,\quad \sum_{i=1}^{3}\left(z_i\partial_{z_i}+\frac{\Delta_i}{2}\right)G=0. \tag{2.4.111}$$

These are equivalent to invariance conditions (2.4.108) with $\Delta_i=2j_i$ *(compare differential operators in (2.4.111) and operators in (2.2.37)).*

Remark 3. *In practice, it is often sufficient to calculate Clebsch–Gordan coefficients for relatively small values of at least one of the spins. In this situation, one uses the following approach, based on the formulas of tensor product of representations, Eqs. (I-4.3.12), (I-4.7.73), (2.4.25). One writes*

$$T(e_\pm)=T^{(j_1)}(e_\pm)\otimes 1+1\otimes T^{(j_2)}(e_\pm),\quad T(\mathsf{h})=T^{(j_1)}(\mathsf{h})\otimes 1+1\otimes T^{(j_2)}(\mathsf{h}). \tag{2.4.112}$$

Consider now the following chain of equalities

$$\begin{aligned}\langle j_1m_1|\otimes\langle j_2m_2|(T^{(j)}(e_\pm)\,|jm\rangle)&=(\langle j_1m_1|\otimes\langle j_2m_2|\,T(e_\pm))|jm\rangle\\&=(\langle j_1m_1|\otimes\langle j_2m_2|\,[T^{(j_1)}(e_\pm)\otimes 1+1\otimes T^{(j_2)}(e_\pm)])|jm\rangle\\&=(\langle j_1m_1|\,T^{j_1}(e_\pm)\otimes\langle j_2m_2\,|)|\,jm\rangle+(\langle j_1m_1|\otimes\langle j_2m_2\,|T^{j_2}(e_\pm)|)|jm\rangle,\end{aligned} \tag{2.4.113}$$

where we use (2.4.112). By comparing left- and right-hand side of (2.4.113), and using the formula for action of generators $e_\pm$ *on dual vectors (2.1.28), we obtain the recurrence relation*

$$\sqrt{(j \pm m+1)(j \mp m)}\,\langle j_1,m_1;j_2,m_2\,|j,m\pm 1\rangle$$
$$= \sqrt{(j_1 \mp m_1+1)(j_1 \pm m_1)}\,\langle j_1,m_1 \mp 1;j_2,m_2|\,j,m\rangle$$
$$+ \sqrt{(j_2 \mp m_2+1)(j_2 \pm m_2)}\,\langle j_1,m_1;j_2,m_2 \mp 1|\,j,m\rangle. \tag{2.4.114}$$

Note that if we consider operator h *instead of* $e_\pm$ *in (2.4.113) we get*

$$(m_1+m_2-m)\langle j_1,m_1;j_2,m_2\,|j,m\rangle = 0, \tag{2.4.115}$$

which reiterates that nonvanishing Clebsch–Gordan coefficients have $m = m_1+m_2$.

Problem 2.4.13. *Consider matrix element* $\langle jm|\,T^{(j)}(e_\mp)(|j_1m_1\rangle \otimes |j_2m_2\rangle)$ *and derive equations for inverted Clebsch–Gordan coefficients (2.4.99):*

$$\sqrt{(j \pm m+1)(j \mp m)}\,\langle j,m\pm 1\,|j_1,m_1;j_2,m_2\rangle$$
$$= \sqrt{(j_1 \mp m_1+1)(j_1 \pm m_1)}\,\langle j,m|\,j_1,m_1 \mp 1;j_2,m_2\rangle \tag{2.4.116}$$
$$+ \sqrt{(j_2 \mp m_2+1)(j_2 \pm m_2)}\,\langle j,m|\,j_1,m_1;j_2,m_2 \mp 1\rangle,$$

$$(m_1+m_2-m)\langle j,m\,|j_1,m_1;j_2,m_2\rangle = 0. \tag{2.4.117}$$

Compare these equations with Eqs. (2.4.114), (2.4.115); make use of (2.4.96) for that matter.

Now, take Eq. (2.4.114) with upper sign, and set $m = j$ *there. Then the left-hand side of (2.4.114) vanishes, and we obtain a set of relations which enable us to find Clebsch–Gordan coefficients* $\langle j_1,m_1;j_2,m_2\,|j,j\rangle$ *(modulo common factor determined by normalization and the convention that* $\langle j_1,j_1;j_2,j-j_1\,|j,j\rangle$ *is real and positive). By choosing then the lower sign in (2.4.114), we express* $\langle j_1,m_1;j_2,m_2\,|j,m-1\rangle$ *through* $\langle j_1,m_1;j_2,m_2\,|j,m\rangle$. *Going down in* m, *we find all Clebsch–Gordan coefficients in this way.*

2.5. Tensor Operators and 3n-j Symbols

2.5.1. Tensor operators and Wigner–Eckart theorem

Let T be an n-dimensional representation of group G in space $\mathcal{V}_n(\mathbb{K})$. Let $\vec{e}_k$ $(k = 1,\dots,n)$ be basis in $\mathcal{V}_n(\mathbb{K})$. Every element $g \in G$ is mapped to $n \times n$ matrix with elements $T_{kl}(g) \in \mathbb{K}$ (see (2.1.7)):

$$T(g)\cdot\vec{e}_k = \vec{e}_\ell\, T_{\ell k}(g).$$

Let ρ be unitary representation of group G in complex vector space $\mathcal{V}$ with Hermitian scalar product.

Definition 2.5.1. A set of linear operators $\{\hat{\mathsf{T}}_k^{(T)}\}$ $(k = 1,\dots,n)$, acting in space $\mathcal{V}$ of representation ρ of group G and obeying

$$\rho(g)\cdot\hat{\mathsf{T}}_k^{(T)}\cdot\rho(g^{-1}) = \hat{\mathsf{T}}_\ell^{(T)}T_{\ell k}(g), \quad \forall g \in G. \tag{2.5.1}$$

is called *tensor operator* for group G in representation T. If the representation T is reducible (irreducible), then tensor operator $\{\hat{\mathsf{T}}_k^{(T)}\}$ is reducible (irreducible).

Problem 2.5.1. Let $\{\hat{\mathsf{T}}_k^{(T)}\}$ $(k = 1, \dots, n)$ be a set of linear independent operators in $\mathcal{V}$, and ρ be a representation of G in $\mathcal{V}$. Let the relation (2.5.1) hold. Show that matrices $||T_{\ell k}(g)||$ define mapping $T : G \to GL(n, \mathbb{K})$ which is a representation of G.

We point out that in certain cases it is possible to relax the conditions that the representation T is finite-dimensional and the representation ρ is unitary. It is sometimes sufficient to require that the pertinent space $\mathcal{V}$ is Hilbert space (see Definition 1.1.1 in Section 1.1). This comment is unimportant for our purposes.

Let G be Lie group and $\mathcal{A}(G)$ its Lie algebra. Let (2.5.1) hold. By considering a curve $g(t) = \exp(tA)$, where $A \in \mathcal{A}(G)$, near $t = 0$ we obtain in the standard manner

$$[\rho(A),\ \hat{\mathsf{T}}_k^{(T)}] = \hat{\mathsf{T}}_\ell^{(T)}\, T_{\ell k}(A), \quad \forall A \in \mathcal{A}(G). \tag{2.5.2}$$

This relation can be viewed as Lie algebraic definition of tensor operator.

Tensor operators are of importance in quantum mechanics, and especially in quantum field theory. As an example, momentum and spin operators in nonrelativistic quantum mechanics are tensor (vector) operators with respect to group of rotations of three-dimensional space. Quantum fields in relativistic quantum field theory are tensor operators with respect to Poincaré group.

Consider irreducible tensor operator $\{\hat{\mathsf{T}}_m^j\}$ $(m = -j, \dots, j-1, j)$ for group $SU(2)$ in representation $T^{(j)}$, i.e., a set of operators that transform under adjoint action (2.5.2) of Lie algebra $su(2)$, or its complexification $s\ell(2,\mathbb{C})$, in the same way as basis vectors (2.2.26) in the space of irreducible representation $T^{(j)}$. In accordance with (2.1.22), (2.2.30), (2.2.31) and (2.5.2), we have

$$[\rho(\mathsf{h}),\ \hat{\mathsf{T}}_m^j] = m\hat{\mathsf{T}}_m^j, \tag{2.5.3}$$

$$[\rho(\hat{e}_\pm),\ \hat{\mathsf{T}}_m^j] = \sqrt{(j \pm m + 1)(j \mp m)}\,\hat{\mathsf{T}}_{m\pm1}^j. \tag{2.5.4}$$

A concrete example of tensor operator with respect to group $SU(2)$ are spherical harmonics $\{Y_m^{(\ell)}(\theta,\phi)\}$ $(m = -\ell, \dots, \ell)$ treated as operators that act in space $\mathcal{V}$ of functions $f(\theta,\phi)$ on a sphere S^2 as follows (see

Section 2.3.3):

$$f(\theta,\phi) \to f'(\theta,\phi) = Y_m^{(\ell)}(\theta,\phi)\cdot f(\theta,\phi).$$

With this understanding, the left-hand sides of (2.3.94) can be written as commutators of operators $Y_m^{(\ell)}$ and generators $\hat{\mathsf{h}} = -\rho(\mathsf{h})$ and $\hat{e}_\pm = -\rho(e_\mp)$, defined in terms of differential operators (2.3.93). Then the relations (2.3.94) become identical to (2.5.3) and (2.5.4).

Problem 2.5.2. Show that monomials (2.2.26) can be viewed as tensor operator in the space of functions $f(s,t)$. Write Eq. (2.2.31) in terms of commutators (2.5.3) and (2.5.4).

In the example with spherical harmonics, ρ is unitary infinite-dimensional representation of $SU(2)$, which acts in the space $\mathcal{V}$ of all (square-integrable) functions $f(\theta,\phi)$ on a sphere S^2. An important point here is that tensor operators $\hat{\mathsf{T}}_m^{(\ell)} = Y_m^{(\ell)}(\theta,\phi)$ for all $\ell \in \mathbb{Z}_{\geq 0}$ act in one and the same space $\mathcal{V}$, and they can be multiplied by each other. In other words, $\hat{\mathsf{T}}_m^{(j)}$ make algebra of tensor operators. Let us consider this point in more detail.

By definition (see (2.5.3) and (2.5.4)), tensor operators $\hat{\mathsf{T}}_m^j$, like vectors $|j,m\rangle$, define finite-dimensional representation $T^{(j)}$ of algebra $su(2)$. Product of irreducible tensor operators $\hat{\mathsf{T}}_{m_1}^{j_1}\cdot\hat{\mathsf{T}}_{m_2}^{j_2}$ is a new tensor operator (reducible, generally speaking) which transforms under $su(2)$ in the same way as tensor product $|j_1,m_1\rangle\otimes|j_2,m_2\rangle$. Thus, products of irreducible tensor operators obey analogs of Clebsch–Gordan expansions (2.4.78) and (2.4.100), namely

$$\hat{\mathsf{X}}_m^{j_3} = \sum_{\substack{m_1,m_2\\ m_1+m_2=m}} \hat{\mathsf{T}}_{m_1}^{j_1}\cdot\hat{\mathsf{U}}_{m_2}^{j_2}\cdot\langle j_1\,m_1, j_2\,m_2|j_1,j_2;j_3\,m\rangle, \tag{2.5.5}$$

$$\hat{\mathsf{T}}_{m_1}^{j_1}\cdot\hat{\mathsf{U}}_{m_2}^{j_2} = \sum_{j_3=|j_1-j_2|}^{j_1+j_2} \hat{\mathsf{X}}_m^{j_3}\cdot\langle j_1,j_2;j_3\,m|j_1\,m_1,j_2\,m_2\rangle, \tag{2.5.6}$$

where $\hat{\mathsf{X}}_m^{j_3}$, $\hat{\mathsf{T}}_{m_1}^{j_1}$ and $\hat{\mathsf{U}}_{m_2}^{j_2}$ are tensor operators obeying (2.5.3), (2.5.4). As an example, let us set $j_1 = j_2 = j$ and $m=0, j_3=0$ in (2.5.5) and make use of the following relation that follows from (2.4.83) (cf. (2.4.107) and expression for metric (2.4.68))

$$\langle j\,m_1, j\,m_2|0,0\rangle = \frac{(-1)^{j-m_1}}{\sqrt{2j+1}}\delta_{m_1,-m_2}.$$

As a result, Eq. (2.5.5) takes the following form:

$$\hat{\mathsf{X}}^0_0 = \frac{1}{\sqrt{2j+1}} \sum_{m=-j}^{j} (-1)^{j-m} \hat{\mathsf{T}}^j_m \, \hat{\mathsf{U}}^j_{-m}, \tag{2.5.7}$$

which is invariant convolution, with metric (2.4.68), of two tensor operators $\hat{\mathsf{T}}^j_m$ and $\hat{\mathsf{U}}^j_m$ (compare this convolution with (2.3.39)). For spherical harmonics with integer $j = \ell$, acting in *different* factors $\mathcal{V}_1 = \{f(\theta_1, \phi_1)\}$ and $\mathcal{V}_2 = \{f(\theta_2, \phi_2)\}$ of the space $\mathcal{V}_1 \otimes \mathcal{V}_2$, Eq. (2.5.7) corresponds to well-known addition theorem (2.3.77), which, upon the change of coordinate $\phi_1 \to \pi - \phi_1$ is written as follows:

$$(-1)^\ell P^{(\ell)}(\cos w_{12}) = \sum_{m=-\ell}^{\ell} (-1)^{\ell-m} Y^{(\ell)}_m(\theta_1, \phi_1) Y^{(\ell)}_{-m}(\theta_2, \phi_2),$$

where $\cos w_{12} = \cos\theta_1 \cos\theta_2 + \sin\theta_1 \sin\theta_2 \cos(\phi_1 - \phi_2)$ is invariant under adjoint action of generators (2.3.85):

$$[\tau_\alpha(\theta_1, \phi_1) + \tau_\alpha(\theta_2, \phi_2), \cos w_{12}] = 0. \tag{2.5.8}$$

In quantum mechanics, one often calculates matrix elements $\langle j_3\, m_3 | \hat{\mathsf{T}}^{j_2}_{m_2} | j_1\, m_1 \rangle$ of tensor operators $\hat{\mathsf{T}}^{j_2}_{m_2}$ which act in space $\mathcal{V}$ containing all spaces $\mathcal{V}_{2j+1}$ of irreducible representations $T^{(j)}$. Let us show that this problem has a solution in terms of Clebsch–Gordan coefficients. Let us sandwich (2.5.3) and (2.5.4) between the states $\langle j_3\, m_3|$ and $|j_1\, m_1\rangle$ to obtain

$$(m_3 - m_1)\langle j_3\, m_3 | \hat{\mathsf{T}}^j_m | j_1\, m_1 \rangle = m\, \langle j_3\, m_3 | \hat{\mathsf{T}}^j_m | j_1\, m_1 \rangle, \tag{2.5.9}$$

$$\langle j_3\, m_3 | [\rho(\hat{e}_\pm), \hat{\mathsf{T}}^j_m] | j_1\, m_1 \rangle = \sqrt{(j \pm m + 1)(j \mp m)}\, \langle j_3\, m_3 | \hat{\mathsf{T}}^j_{m\pm1} | j_1\, m_1 \rangle. \tag{2.5.10}$$

The former equation gives the selection rule: matrix element $\langle j_3\, m_3 | \hat{\mathsf{T}}^j_m | j_1\, m_1 \rangle$ vanishes for all $m_3 \neq m_1 + m$. In view of (2.2.31), Eq. (2.5.10) gives

$$\begin{aligned} &\sqrt{(j_3 \mp m_3 + 1)(j_3 \pm m_3)}\langle j_3\, m_3 \mp 1 | \hat{\mathsf{T}}^j_m | j_1\, m_1 \rangle \\ &\quad = \sqrt{(j \pm m + 1)(j \mp m)}\langle j_3\, m_3 | \hat{\mathsf{T}}^j_{m\pm1} | j_1\, m_1 \rangle \\ &\qquad + \sqrt{(j_1 \pm m_1 + 1)(j_1 \mp m_1)}\langle j_3\, m_3 | \hat{\mathsf{T}}^j_m | j_1\, m_1 \pm 1 \rangle. \end{aligned} \tag{2.5.11}$$

Equations (2.5.9) and (2.5.11) coincide with Eqs. (2.4.116) and (2.4.117) which define Clebsch–Gordan coefficients modulo overall normalization.

Therefore, matrix elements of tensor operators are proportional to Clebsch–Gordan coefficients:

$$\langle j_3\, m_3|\hat{\mathsf{T}}^{j_2}_{m_2}|j_1\, m_1\rangle = \frac{(j_3||\hat{\mathsf{T}}^{j_2}||j_1)}{\sqrt{2j_3+1}} \cdot \langle j_3\, m_3\,|\, j_1\, m_1, j_2\, m_2\rangle, \tag{2.5.12}$$

where the normalization coefficient $(j_3||\hat{\mathsf{T}}^{j_2}||j_1)$ is independent of m_1, m_2 and m_3 and is called in physics literature *reduced matrix element* (factor $\frac{1}{\sqrt{2j_3+1}}$ is introduced for convenience). Formula (2.5.12) which relates matrix elements of tensor operator $\hat{\mathsf{T}}^{j_2}_{m_2}$ with Clebsch–Gordan coefficients is known as *Wigner–Eckart theorem.*

2.5.2. Racah coefficients and 3n-j symbols

Clebsch–Gordan coefficients (2.4.79), matrix elements of tensor operators (2.5.12) and also 3-j symbols (2.4.91) can be treated as components of third rank tensors, e.g.,

$$\begin{aligned} &t^{(j_1,j_2,\overline{j_3})} \\ &\quad = \langle\, j_1\, m_1, j_2\, m_2|j_3\, m_3\,\rangle|m_1\rangle_{j_1} \otimes |m_2\rangle_{j_2} \otimes \langle m_3|_{j_3} \in \mathcal{V}_{(1)} \otimes \mathcal{V}_{(2)} \otimes \mathcal{V}^*_{(3)}, \\ &t^{(j_1,j_2,j_3)} = \begin{pmatrix} j_1 & j_2 & j_3 \\ m_1 & m_2 & m_3 \end{pmatrix} |m_1\rangle_{j_1} \otimes |m_2\rangle_{j_2} \otimes |m_3\rangle_{j_3} \in \mathcal{V}_{(1)} \otimes \mathcal{V}_{(2)} \otimes \mathcal{V}_{(3)}, \end{aligned} \tag{2.5.13}$$

where summation over indices m_1, m_2, m_3 is assumed in the range $m_k = (-j_k, -j_k+1, \ldots, j_k-1, j_k)$, and we use shorthand notations $|m\rangle_j = |j\, m\rangle$, $\langle m|_j = \langle j\, m|$, $\mathcal{V}_{(k)} = \mathcal{V}_{2j_k+1}$. Recall that rth rank tensor is an object that is transformed as an element of tensor product of r vector spaces, see Definition I-4.3.1 in Section I-4.3.1 of the first book.

Action of an element $g \in SU(2)$ on basis vectors $|j\, m\rangle \in \mathcal{V}_{2j+1}$ and $\langle j\, m| \in \mathcal{V}^*_{2j+1}$ in spaces of representations $T^{(j)}$ and $\overline{T}^{(j)}$ is given by (2.3.56) and (2.3.57). This action is written as the transformation of tensor components (2.5.13):

$$\begin{aligned} &\langle j_1 m_1, j_2 m_2|j_3 m_3\rangle \to \mathcal{D}^{(j_1)}_{m_1 m_1'}(g)\mathcal{D}^{(j_2)}_{m_2 m_2'}(g)\langle j_1 m_1', j_2 m_2'|j_3 m_3'\rangle \mathcal{D}^{(j_3)}_{m_3' m_3}(g^{-1}), \\ &\begin{pmatrix} j_1 & j_2 & j_3 \\ m_1 & m_2 & m_3 \end{pmatrix} \to \mathcal{D}^{(j_1)}_{m_1 m_1'}(g)\mathcal{D}^{(j_2)}_{m_2 m_2'}(g)\; \mathcal{D}^{(j_3)}_{m_3 m_3'}(g) \begin{pmatrix} j_1 & j_2 & j_3 \\ m_1' & m_2' & m_3' \end{pmatrix}, \end{aligned} \tag{2.5.14}$$

(summation over repeated indices m_k' is assumed). We have already pointed out (see (2.4.82)) that Clebsch–Gordan coefficients are invariant under

$SU(2)$-transformations (2.5.14), i.e., the right-hand side of the first line in Eq. (2.5.14) coincides with its left-hand side. Likewise, the invariance of the convolution in (2.4.102) under transformations (2.3.35) with $g \in SU(2)$ yields invariance of 3-j symbols:

$$\mathcal{D}^{(j_1)}_{m_1 m'_1}(g)\, \mathcal{D}^{(j_2)}_{m_2 m'_2}(g)\, \mathcal{D}^{(j_3)}_{m_3 m'_3}(g) \begin{pmatrix} j_1 & j_2 & j_3 \\ m'_1 & m'_2 & m'_3 \end{pmatrix} = \begin{pmatrix} j_1 & j_2 & j_3 \\ m_1 & m_2 & m_3 \end{pmatrix}. \tag{2.5.15}$$

Thus, in accordance with Definition I-4.3.2 from Section I-4.3.1 of the first book, both Clebsch–Gordan coefficients and 3-j symbols are components of invariant third rank tensors. Clearly, one can also define convolution of these tensors in indices m_k that correspond to one and the same representation (with one and the same spin j_k), with invariant metric $\eta^{(j_k)}$ defined in (2.3.40), (2.4.68). These convolutions again give invariant tensors (generally, of another rank), including invariant scalars.

Following this rule, we consider, as an example, the right-hand side of orthogonality condition (2.4.97) as invariant convolution of two Clebsch–Gordan coefficients, giving second rank invariant tensor. Making use of (2.4.91) and symmetry properties (2.4.103), (2.4.104) we write the orthogonality relation (2.4.97) in terms of 3-j symbols as

$$\begin{aligned} \sum_{m_1,m_2} (-1)^{j_1+j_2-m_1-m_2} &\begin{pmatrix} j_1 & j_2 & j_3 \\ m_1 & m_2 & m_3 \end{pmatrix} \begin{pmatrix} j_1 & j_2 & j_4 \\ -m_1 & -m_2 & m_4 \end{pmatrix} \\ &= \frac{(-1)^{j_3-m_3}}{2j_3+1} \delta_{m_3,-m_4} \delta^{j_3,j_4}, \end{aligned} \tag{2.5.16}$$

where the right-hand side is second rank tensor proportional to invariant metric $||\eta^{(j_3)}_{m_3,m_4}||$. If we set $j_4 = j_3$ and $m_4 = -m_3$ in (2.5.16), multiply by $(-1)^{j_3-m_3}$ and sum over m_3, then we obtain, on the left-hand side, the complete invariant convolution of two 3-j symbols, which, in fact, is equal to 1:

$$\begin{aligned} &\sum_{m_1,m_2,m_3} \sum_{m'_1,m'_2,m'_3} \eta^{(j_1)}_{m_1,m'_1} \eta^{(j_2)}_{m_2,m'_2} \eta^{(j_3)}_{m_3,m'_3} \begin{pmatrix} j_1 & j_2 & j_3 \\ m_1 & m_2 & m_3 \end{pmatrix} \begin{pmatrix} j_1 & j_2 & j_3 \\ m'_1 & m'_2 & m'_3 \end{pmatrix} \\ &= \sum_{m_1,m_2,m_3} (-1)^{\sum_{k=1}^{3}(j_k-m_k)} \begin{pmatrix} j_1 & j_2 & j_3 \\ m_1 & m_2 & m_3 \end{pmatrix} \begin{pmatrix} j_1 & j_2 & j_3 \\ -m_1 & -m_2 & -m_3 \end{pmatrix} \\ &= \sum_{m_1,m_2,m_3} \begin{pmatrix} j_1 & j_2 & j_3 \\ m_1 & m_2 & m_3 \end{pmatrix}^2 = 1. \end{aligned} \tag{2.5.17}$$

Here, we make use of (2.4.104) and the property $\sum_{k=1}^{3} m_k = 0$. Thus, there is an invariant convolution of two 3-j symbols, but it is trivial.

Clearly, there is no scalar invariant convolution of three 3-j symbols, since there is no way to contract pairwise their nine indices m_k. Therefore, we consider the next possible case of scalar invariant convolution of four 3-j symbols. We show below that there is only one such contraction which cannot be made trivial by using identities (2.5.16). In terms of Clebsch–Gordan coefficients, this convolution is

$$\begin{aligned} \mathbf{I} \equiv \sum_{m_1,\dots,m_6} & \langle j_1\, m_1, j_2\, m_2 | j_3\, m_3 \rangle \langle j_5\, m_5 | j_1\, m_1, j_6\, m_6 \rangle \\ & \times \langle j_6\, m_6 | j_2\, m_2, j_4\, m_4 \rangle \langle j_4\, m_4, j_3\, m_3 | j_5\, m_5 \rangle. \end{aligned} \tag{2.5.18}$$

It is invariant under $SU(2)$ transformations and gives a scalar. Its invariance follows from relation (2.4.82) and its dual

$$\begin{aligned} & \mathcal{D}^{(j_3)}_{m_3' m_3}(g)\, \langle j_3, m_3 | j_1, m_1; j_2, m_2 \rangle\, \mathcal{D}^{(j_1)}_{m_1 m_1'}(g^{-1})\, \mathcal{D}^{(j_2)}_{m_2 m_2'}(g^{-1}) \\ & = \langle j_3, m_3' | j_1, m_1'; j_2, m_2' \rangle. \end{aligned} \tag{2.5.19}$$

The latter is obtained by complex conjugation of (2.4.82) in view of real-valuedness of Clebsch–Gordan coefficients and unitarity of matrices $||\mathcal{D}^{(j)}_{m'm}(g)||$ for $g \in SU(2)$). Due to (2.4.91), (2.4.103) and (2.4.104), invariant (2.5.18) can be written, modulo an overall factor depending on spins j_k only, as an invariant convolution, over all spin projections m_i, of four 3-j symbols

$$\begin{aligned} & \frac{(-1)^{2j_4 - j_1 - j_2 - j_3}}{(2j_5+1)\sqrt{(2j_3+1)(2j_6+1)}} \mathbf{I} \\ & = \sum_{m_1,\dots,m_6} (-1)^{j_6 - j_1 - j_2 + 2m_5 + m_3 + m_6} \begin{pmatrix} j_1 & j_2 & j_3 \\ m_1 & m_2 & -m_3 \end{pmatrix} \begin{pmatrix} j_1 & j_6 & j_5 \\ m_1 & m_6 & -m_5 \end{pmatrix} \\ & \quad \times \begin{pmatrix} j_2 & j_4 & j_6 \\ m_2 & m_4 & -m_6 \end{pmatrix} \begin{pmatrix} j_4 & j_3 & j_5 \\ m_4 & m_3 & -m_5 \end{pmatrix} \\ & = \sum_{m_1,\dots,m_6} (-1)^{\sum_{i=1}^{6}(j_i - m_i)} \begin{pmatrix} j_1 & j_2 & j_3 \\ -m_1 & -m_2 & -m_3 \end{pmatrix} \begin{pmatrix} j_1 & j_5 & j_6 \\ m_1 & -m_5 & m_6 \end{pmatrix} \end{aligned}$$

$$\times \begin{pmatrix} j_4 & j_2 & j_6 \\ m_4 & m_2 & -m_6 \end{pmatrix} \begin{pmatrix} j_4 & j_5 & j_3 \\ -m_4 & m_5 & m_3 \end{pmatrix}$$

$$\equiv \begin{Bmatrix} j_1 & j_2 & j_3 \\ j_4 & j_5 & j_6 \end{Bmatrix}, \tag{2.5.20}$$

where we make use of (2.4.96) to obtain the first equality, rename summation indices and use (2.4.103), (2.4.104), (2.4.105) to obtain the third line. Also, we employ the conditions

$$2(j_1 \pm j_2 \pm j_3) = 0 \bmod(2), \quad 2(j_k \pm m_k) = 0 \bmod(2) \tag{2.5.21}$$

(upper and lower signs are uncorrelated), which are valid for any 3-j symbol $\begin{pmatrix} j_1 & j_2 & j_3 \\ m_1 & m_2 & m_3 \end{pmatrix}$. All contractions over m_k in the second sum in (2.5.20) are taken with invariant metric (2.4.68). Scalar invariants introduced in (2.5.20) are denoted in a standard way, $\begin{Bmatrix} j_1 & j_2 & j_3 \\ j_4 & j_5 & j_6 \end{Bmatrix}$ and are dubbed as *Wigner 6-j symbols*. The nomenclature "6-j" has to do with the fact that invariants (2.5.20) are functions of six spins $j_1, \ldots, j_6$. Explicit expression for them is (see, e.g., [21, 24]):

$$\begin{Bmatrix} j_1 \, j_2 \, j_3 \\ j_4 \, j_5 \, j_6 \end{Bmatrix} = \Delta(j_1, j_2, j_3)\Delta(j_1, j_5, j_6)\Delta(j_4, j_2, j_6)\Delta(j_4, j_5, j_3)$$

$$\times \sum_k \left[(-1)^k \, (k+1)! \, \frac{[(k - j_1 - j_2 - j_3)!(k - j_1 - j_5 - j_6)!]^{-1}}{(j_1 + j_2 + j_4 + j_5 - k)!(j_1 + j_3 + j_4 + j_6 - k)!} \right.$$

$$\left. \times \frac{[(k - j_4 - j_2 - j_6)!(k - j_4 - j_5 - j_3)!]^{-1}}{(j_2 + j_3 + j_5 + j_6 - k)!} \right]. \tag{2.5.22}$$

Here, we make use of triangle coefficients $\Delta(j_1, j_2, j_3)$ defined in (2.4.85). We obtain the formula (2.5.22) below in Section 2.5.4.

In spectroscopy, instead of Wigner 6-j symbols one often uses *Racah coefficients* $W(j_1 j_2 j_5 j_4; j_3 j_6)$ which were introduced independently by G. Racah,

$$W(j_1 j_2 j_5 j_4; j_3 j_6) = (-1)^{j_1 + j_2 + j_4 + j_5} \begin{Bmatrix} j_1 & j_2 & j_3 \\ j_4 & j_5 & j_6 \end{Bmatrix}.$$

An important aspect of representation theory of group $SU(2)$ is construction of all invariant contractions of any number of 3-j symbols (analogous to contractions (2.5.20)). The study of these invariant contractions conveniently involves diagrammatic techniques, in which 3-j symbols are represented by vertices with three outgoing lines (three-valence vertices), each line labeled by a pair of indices (j_k, m_k) where $k = 1, 2, 3$. Invariant contraction (with metric (2.4.68)) of two 3-j symbols over index m_k is represented by an oriented line connecting vertices representing these 3-j symbols. As an example, convolution of two 3-j symbols (cf. (2.5.17)) over a pair of indices m_3, m_3',

$$\sum_{m_3,m_3'} \eta^{(j_3)}_{m_3,m_3'} \begin{pmatrix} j_1 & j_2 & j_3 \\ m_1 & m_2 & m_3 \end{pmatrix} \begin{pmatrix} j_1' & j_2' & j_3 \\ m_1' & m_2' & m_3' \end{pmatrix}$$

$$= \sum_{m_3} (-1)^{j_3 - m_3} \begin{pmatrix} j_1 & j_2 & j_3 \\ m_1 & m_2 & m_3 \end{pmatrix} \begin{pmatrix} j_1' & j_2' & j_3 \\ m_1' & m_2' & -m_3 \end{pmatrix} \tag{2.5.23}$$

is represented by two-vertex graph with one internal oriented line and four external lines

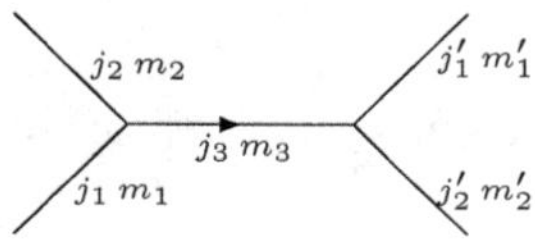

In what follows, we often do not write index of summation m_3 at internal oriented line. This diagrammatic technique[1] is given in detail in Figs. 2.5.1 and 2.5.2, which show the graph for 3-j symbol (2.4.91) (Fig. 2.5.1) and graph for invariant metric (2.3.40), (2.4.68) (Fig. 2.5.2). To perform the diagrammatic calculations, one also needs graphical representation (Fig. 2.5.3) of convolution of two external legs of a graph with invariant metric. We note that similar diagrammatic technique has been used in Ref. [24].

When working with this diagrammatic technique one should keep track of the orientations of diagrams: reflection of a diagram, which is inequivalent to rotation, may lead to flip of sign (see the second line in Fig. 2.5.1). An example of diagrammatic technique is given by graphic representation of identities (2.5.16), (2.5.17) shown in Fig. 2.5.4.

Clearly, there is no complete invariant convolution of odd number of 3-j symbols, since pairwise contraction of odd number of indices m_k is

[1]In literature, a dual diagrammatic technique is sometimes used, with 3-j symbol represented by a triangle (rather than three-valence vertex), and pairs (j_k, m_k) labeling 3-j symbol attached to the sides of the triangle (rather than outgoing lines). The two techniques are equivalent. We make use of the dual technique in Section 2.5.3.

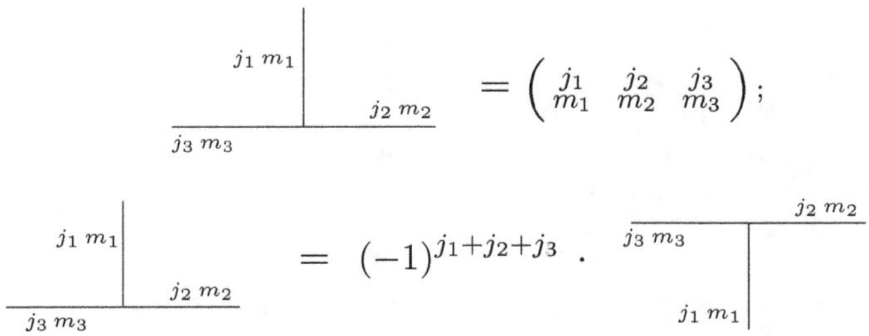

Fig. 2.5.1. Graphic representation of 3-j symbol.

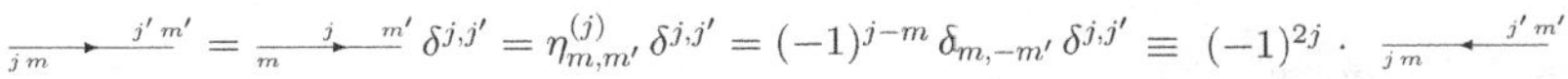

Fig. 2.5.2. Graphic representation of invariant metric.

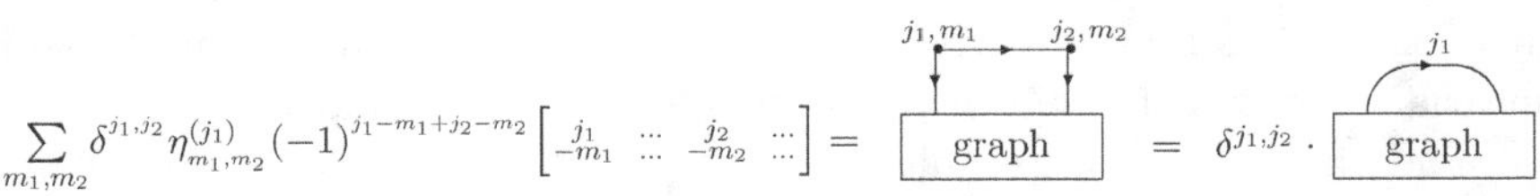

Fig. 2.5.3. Convolution of two incoming lines of a graph with invariant metric.

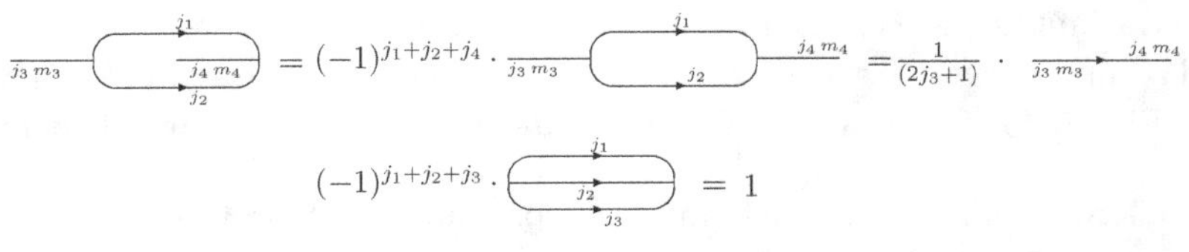

Fig. 2.5.4. Diagrammatic representation of identities (2.5.16) and (2.5.17).

impossible. Consider, therefore, a product of even number of 3-j symbols, with all their projections m_k pairwise summed over with invariant metric, in accordance to the above rules (see, e.g., (2.5.17), (2.5.20) and (2.5.23)). The pertinent convolution of 3-j symbols is a scalar invariant. It is shown graphically as a closed graph with three-valence vertices corresponding to 3-j symbols and internal lines which are oriented, labeled by their spins j_k and correspond to contractions in m_k with invariant metric. There is a relation $3V = 2E$ for such a closed graph, where V is the number of vertices (number of 3-j symbols), and E is the number of internal lines (number of pairings, equal to the number of spins $j_1, j_2, \ldots$). This relation between

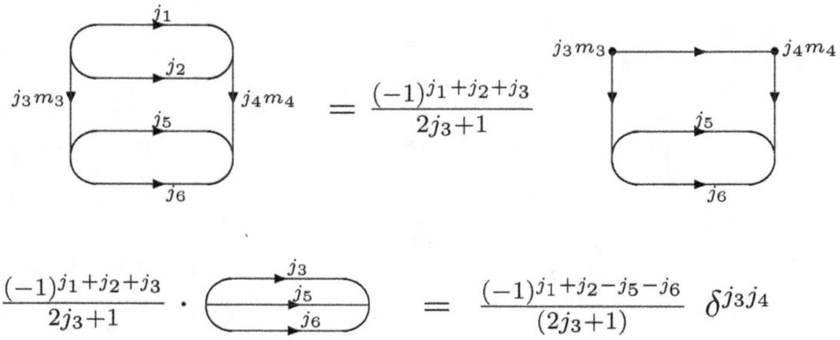

Fig. 2.5.5. Trivial convolution of four 3-j symbols.

V and E shows that V is *even*, say, $V = 2n$, while $E = 3n$ is divisible by 3. Therefore, as a result of complete contraction of $2n$ 3-j symbols, we obtain the scalar invariants depending on $3n$ spins $j_1, j_2, \ldots, j_{3n}$. In accordance with this observation, these scalar invariants are dubbed 3n-j symbols (unless they are trivialized, i.e., are factorized into a product of invariants constructed out of smaller numbers of 3-j symbols).

The diagrammatic technique enables one to represent graphically convolutions of 3-j symbols and make various manipulations with these convolutions, by making use of the identities shown in Figs. 2.5.3 and 2.5.4. Let us demonstrate how this technique works by calculating (see Fig. 2.5.5) an invariant scalar convolution of four 3-j symbols. Here, we use the first and second identities shown in Fig. 2.5.4, and also make use of the formula (see Fig. 2.5.3) for convolution of two external lines with invariant metric.

Problem 2.5.3. Check the formula for convolution of four 3-j symbols given in Fig. 2.5.5, without employing the diagrammatic technique. Hint: make use of the identities (2.5.16) and (2.5.17).

The result shown in Fig. 2.5.5 is that the particular scalar convolution of four 3-j symbols is trivial. By the analysis of all connected graphs with four three-valence vertices one finds that the only nontrivial convolution of four 3-j symbols is given by Eq. (2.5.20) which defines 6-j symbol. In accordance with the diagrammatic technique, the convolution (2.5.20) is represented by a tetrahedron, see Fig. 2.5.6. This representation immediately shows that 6-j symbol is symmetric under specific permutations of spins j_k which correspond to rotations and reflections of the tetrahedron. This again demonstrates the advantage of the diagrammatic technique. We consider these symmetries below in Problem 2.5.4.

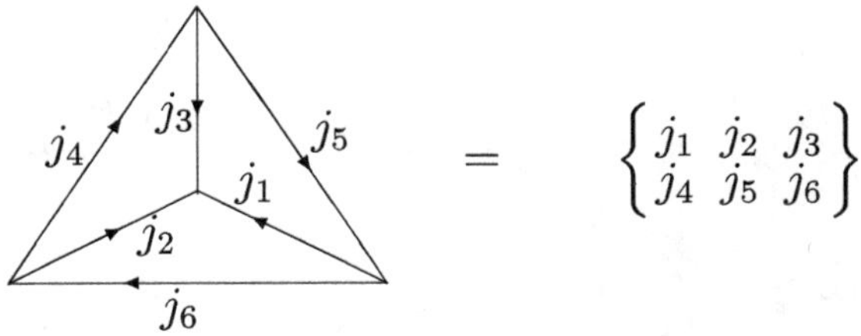

Fig. 2.5.6. Diagrammatic representation of 6-j symbol.

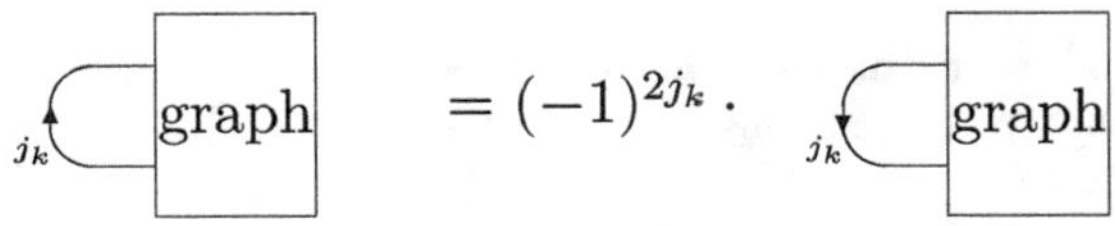

Fig. 2.5.7. Flip of the orientation of a line.

Note that orientation of lines in graphs determining 3n-j symbols is important. As an example, the closed graph with reversed orientation of the line labeled by spin j_k is related to the original graph by a sign factor, see Fig. 2.5.7.

Problem 2.5.4. * Show that 6-j symbol (2.5.20) represented by tetrahedron in Fig. 2.5.6 is symmetric under the permutations of spins j_k which correspond to rotations and reflections of this tetrahedron:

(1) 6-j symbol is invariant under permutations of columns:

$$\begin{Bmatrix} j_1 & j_2 & j_3 \\ j_4 & j_5 & j_6 \end{Bmatrix} = \begin{Bmatrix} j_2 & j_1 & j_3 \\ j_5 & j_4 & j_6 \end{Bmatrix} = \begin{Bmatrix} j_1 & j_3 & j_2 \\ j_4 & j_6 & j_5 \end{Bmatrix}. \tag{2.5.24}$$

(2) 6-j symbol is invariant under permutation of the first and second rows in the first two columns:

$$\begin{Bmatrix} j_1 & j_2 & j_3 \\ j_4 & j_5 & j_6 \end{Bmatrix} = \begin{Bmatrix} j_4 & j_5 & j_3 \\ j_1 & j_2 & j_6 \end{Bmatrix}. \tag{2.5.25}$$

Show that all compositions of transformations (2.5.24) and (2.5.25) make a group of symmetries of tetrahedron which is isomorphic to symmetric group (group of permutations) S_4.

The only nontrivial invariant made of six 3-j symbols and called 9-j symbol is given by a nonplanar graph shown in Fig. 2.5.8. The fact that we do not indicate orientation of lines in this graph is reflected by an arbitrary

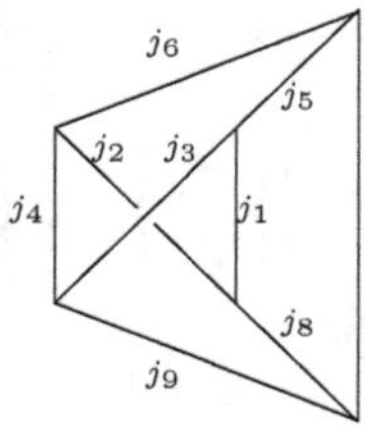

$$= \quad (-1)^{B''} \cdot \begin{Bmatrix} j_3 & j_1 & j_5 \\ j_6 & j_8 & j_7 \\ j_4 & j_2 & j_9 \end{Bmatrix}$$

Fig. 2.5.8. Graphic representation of 9-j symbol: tetrahedron with a pair of opposite edges connected by another edge j_1.

sign factor $(-1)^{B''}$ in the right-hand and side. An example of trivial scalar invariant made of six 3-j symbols is given below in Problem 2.5.9 (see Fig. 2.5.15).

Problem 2.5.5. Describe symmetries of 9-j symbol that follow from its diagrammatic representation shown in Fig. 2.5.8.

2.5.3. 6-j symbols and associativity of product of representations

Remarkably, 6-j symbols are deeply connected with associativity in tensor algebra of representations of $SU(2)$:

$$(T^{(j_1)} \otimes T^{(j_2)}) \otimes T^{(j_3)} = T^{(j_1)} \otimes (T^{(j_2)} \otimes T^{(j_3)}). \qquad (2.5.26)$$

To see this, we consider representation $T = T^{(j_1)} \otimes T^{(j_2)} \otimes T^{(j_3)}$ in space $\mathcal{V} = \mathcal{V}_{2j_1+1} \otimes \mathcal{V}_{2j_2+1} \otimes \mathcal{V}_{2j_3+1}$, and construct coupled basis in the vector space $\mathcal{V}$ via Clebsch–Gordan expansion. There are two ways to construct such a basis in terms of basis vectors $|j_1, m_1\rangle \otimes |j_2, m_2\rangle \otimes |j_3, m_3\rangle \in \mathcal{V}$. The first way is to apply Clebsch–Gordan formula (2.4.100) first to the product $|j_1, m_1\rangle \otimes |j_2, m_2\rangle$ and thus expand this product in basis vectors $|j_1, j_2; j_4, m_4\rangle$, where $m_4 = m_1 + m_2$ and $|j_1 - j_2| \le j_4 \le j_1 + j_2$, and then apply Clebsch–Gordan formula to the product $|j_1, j_2; j_4, m_4\rangle \otimes |j_3, m_3\rangle$. In this way, we obtain

$$\begin{aligned} &|j_4, j_3;\, j_5, m_5\rangle \\ &\quad = \sum_{m_1, m_2, m_3} (|j_1, m_1\rangle \otimes |j_2, m_2\rangle) \otimes |j_3, m_3\rangle \langle j_1, m_1; j_2, m_2 | j_4, m_4\rangle \\ &\quad\quad \times \langle j_4, m_4; j_3, m_3 | j_5, m_5\rangle, \end{aligned} \qquad (2.5.27)$$

where $m_4 = m_1 + m_2$ and $m_5 = m_1 + m_2 + m_3$. Importantly, there may exist several basis vectors with quantum numbers j_5, m_5, these vectors differ by the value of intermediate spin j_4. Likewise, the second way is to apply Clebsch–Gordan formula (2.4.100) first to the product $|j_2,m_2\rangle \otimes |j_3,m_3\rangle$, and then to the product of vector $|j_1,m_1\rangle$ and intermediate "coupled state" $|j_2,j_3;j_6,m_6\rangle$:

$$\begin{aligned} &|j_1,j_6;\, j_5,m_5\rangle \\ &\quad = \sum_{m_1,m_2,m_3} |j_1,m_1\rangle \otimes (|j_2,m_2\rangle \otimes |j_3,m_3\rangle)\,\langle j_2,m_2;j_3,m_3|j_6,m_6\rangle \\ &\qquad \times \langle j_1,m_1;j_6,m_6|j_5,m_5\rangle, \end{aligned} \tag{2.5.28}$$

where $m_6 = m_2+m_3$ and $m_5 = m_1+m_2+m_3$. It is sometimes convenient to denote intermediate spins by $j_4 = j_{12}$ and $j_6 = j_{23}$, respectively. On the left-hand sides of (2.5.27) and (2.5.28) and in what follows we use more informative notations $|j_4,j_3;\, j_5,m_5\rangle$ and $|j_1,j_6;\, j_5,m_5\rangle$ for vectors of coupled basis. This is useful for distinguishing (generally speaking, different) vectors $|j_4,j_3;\, j_5,m_5\rangle$ and $|j_1,j_6;\, j_5,m_5\rangle$ which correspond to different intermediate states.

Let us illustrate the fact that the states $|j_4,j_3;\, j_5,m_5\rangle$ and $|j_1,j_6;\, j_5,m_5\rangle$ indeed can be different. To this end, consider the product of three representations with spins $j_1=1/2$, $j_2=3$ and $j_3=1$:

$$T^{(1/2)}\otimes T^{(3)}\otimes T^{(1)} \begin{array}{c} \nearrow\ (T^{(5/2)}\oplus T^{(7/2)})\otimes T^{(1)}\ \searrow \\ \searrow\ T^{(1/2)}\otimes(T^{(2)}\oplus T^{(3)}\oplus T^{(4)})\ \nearrow \end{array} T^{(3/2)}\oplus 2T^{(5/2)}\oplus 2T^{(7/2)}\oplus T^{(9/2)}. \tag{2.5.29}$$

In accordance with the right-hand side of this diagram, there are two vectors in the coupled basis with quantum numbers $j=5/2$, $m=m_5$, and two vectors with quantum numbers $j=7/2$, $m=m_5$. Therefore, the vectors of the types $|5/2,m_5\rangle$ and $|7/2,m_5\rangle$ that are constructed in the first approach (2.5.27) (upper path in (2.5.29)) do not have to coincide with vectors of the same types obtained in the second approach (2.5.28) (lower path in (2.5.29)); rather, these are linearly related. As an example, two vectors of the type $|5/2,m_5\rangle$ obtained in the first approach are linear combinations of two similar vectors found in the second approach:

$$\begin{aligned} |5/2,1;5/2,m_5\rangle &= w_1\,|1/2,2;5/2,m_5\rangle + w_2\,|1/2,3;5/2,m_5\rangle, \\ |7/2,1;5/2,m_5\rangle &= w_3\,|1/2,2;5/2,m_5\rangle + w_4\,|1/2,3;5/2,m_5\rangle. \end{aligned} \tag{2.5.30}$$

Analogous formulas are valid for vectors of the type $|7/2,m_5\rangle$. We show below that coefficients w_α in (2.5.30) depend only on initial spins j_1, j_2,

j_3 and spins of coupled states j_4, j_5 and j_6, while they are independent of the projection m_5. Moreover, we show that coefficients w_α, modulo simple functions of spins j_k, are given by 6-j symbols (2.5.22).

Let us come back to the general case. Like in our example (2.5.30), basis vectors (2.5.27) and (2.5.28) with equal quantum numbers j_5 and m_5 are linearly related to each other:

$$|j_4, j_3;\, j_5, m_5\rangle = \sum_{j_6} \begin{bmatrix} j_1 & j_2 & j_4 \\ j_3 & j_5 & j_6 \end{bmatrix} |j_1, j_6; j_5, m_5\rangle, \tag{2.5.31}$$

where the coefficients (analogs of w_α) can be written as matrix elements (see (2.5.31))

$$\langle j_1, j_{23}; j_5, m_5 | j_{12}, j_3; j_5, m_5'\rangle = \begin{bmatrix} j_1 & j_2 & j_4 \\ j_3 & j_5 & j_6 \end{bmatrix} \delta_{m_5, m_5'}, \tag{2.5.32}$$

where $j_{12} = j_4$ and $j_{23} = j_6$.

Proposition 2.5.1. *Coefficients* $\left[\begin{smallmatrix} j_1 & j_2 & j_4 \\ j_3 & j_5 & j_6 \end{smallmatrix}\right]$ *in* (2.5.31) *and* (2.5.32) *are independent of projection* m_5.

Proof. We make use of the fact that

$$|j_{12}, j_3;\, j_5, j_5 - 1\rangle = (2j_5)^{-1/2} e_- |j_{12}, j_3;\, j_5, j_5\rangle,$$

and write

$$\begin{aligned} &\langle j_1, j_{23};\, j_5, j_5 - 1 | j_{12}, j_3;\, j_5, j_5 - 1\rangle \\ &\quad = (2j_5)^{-1} \langle j_1, j_{23};\, j_5, j_5 | e_+ \cdot e_- | j_{12}, j_3;\, j_5, j_5\rangle \\ &\quad = (2j_5)^{-1} \langle j_1, j_{23};\, j_5, j_5 |\, 2\,\mathsf{h}\, | j_{12}, j_3;\, j_5, j_5\rangle \\ &\quad = \langle j_1, j_{23};\, j_5, j_5 |\, j_{12}, j_3;\, j_5, j_5\rangle. \end{aligned}$$

So, the left-hand sides of (2.5.32) are the same for $m_5 = j_5$ and $m_5 = j_5 - 1$. Proof for lower m_5 is obtained by induction. □

In view of the relations (2.5.27) and (2.5.28), and making use of the fact that vectors $|j_1, m_1\rangle \otimes |j_2, m_2\rangle \otimes |j_3, m_3\rangle$ form basis in $\mathcal{V}$, we write (2.5.31) as follows:

$$\sum_{m_4} \langle j_1, m_1; j_2, m_2 | j_4, m_4 \rangle \, \langle j_4, m_4; j_3, m_3 | j_5, m_5 \rangle$$
$$= \sum_{j_6} \begin{bmatrix} j_1 & j_2 & j_4 \\ j_3 & j_5 & j_6 \end{bmatrix} \sum_{m_6} \langle j_2, m_2; j_3, m_3 | j_6, m_6 \rangle \langle j_1, m_1; j_6, m_6 | j_5, m_5 \rangle. \tag{2.5.33}$$

Problem 2.5.6. Making use of (2.4.94), derive the following graphic representation of the relation (2.5.33) (this is a special case of Yang–Baxter equation [25])

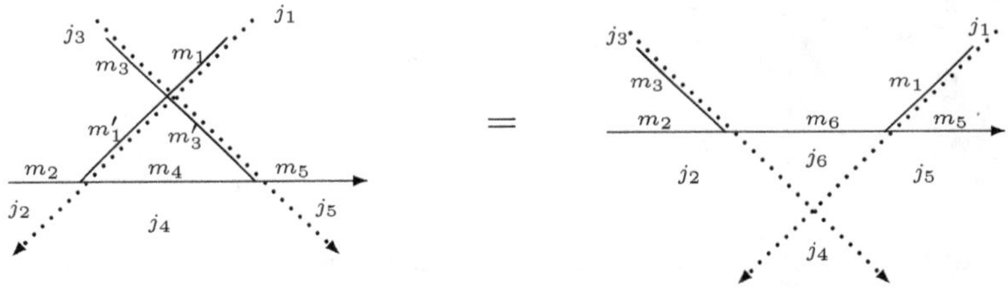

where indices m_i,m_i' are assigned to solid lines (there is a summation over internal lines with indices m_1',m_3',m_4,m_6), indices j_1,j_3 are assigned to dashed lines, indices j_2, j_4, j_5, j_6 are assigned to two-dimensional facets (there is a summation over internal facet with index j_6), and

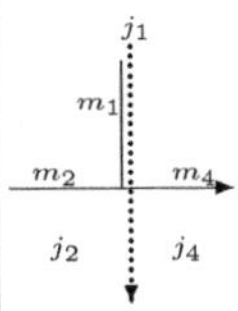

$= \langle j_1, m_1; j_2, m_2 | j_4, m_4 \rangle$,

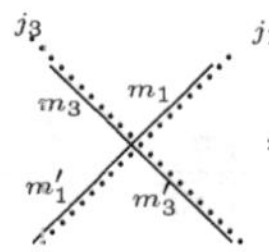

$= R(j_1, j_3)^{m_1 m_3}_{m_1' m_3'} = \delta^{m_1}_{m_1'} \, \delta^{m_3}_{m_3'}$,

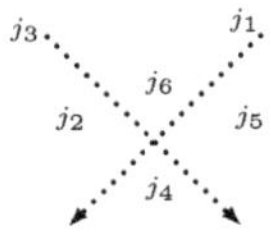

$= (-1)^{j_2+j_5-j_4-j_6} \begin{bmatrix} j_1 & j_2 & j_4 \\ j_3 & j_5 & j_6 \end{bmatrix} =$

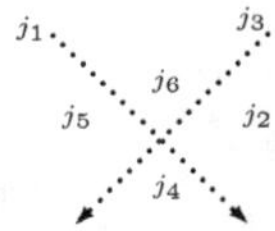

.

Under $SU(2)$ transformations, the left-hand side and all terms in the sum over j_6 in the right-hand side of the identity (2.5.33) behave as components of invariant fourth rank tensors. So, their convolution with basis vectors $(-1)^{j_5-m_5} T^{j_5}_{-m_5} \cdot T^{j_1}_{m_1} \cdot T^{j_2}_{m_2} \cdot T^{j_3}_{m_3}$ is invariant under $SU(2)$. The relation

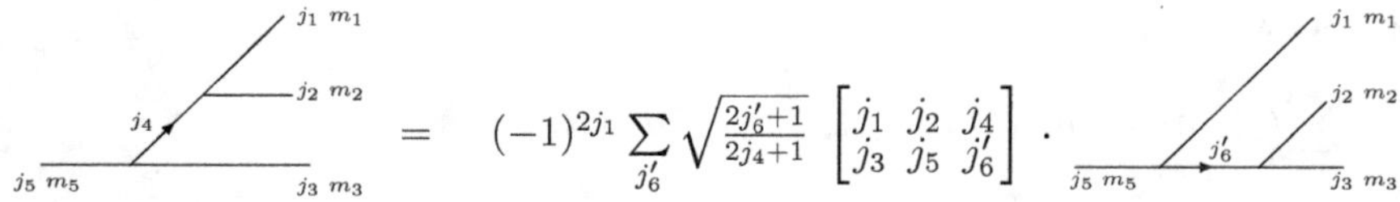

Fig. 2.5.9. Diagrammatic representation of associativity relation (2.5.34).

(2.5.33) between fourth rank tensors is conveniently written in terms of 3-j symbols (2.4.91):

$$\sum_{m_4}(-1)^{j_4-m_4}\begin{pmatrix} j_1 & j_2 & j_4 \\ m_1 & m_2 & -m_4 \end{pmatrix}\begin{pmatrix} j_4 & j_3 & j_5 \\ m_4 & m_3 & m_5 \end{pmatrix}$$

$$= (-1)^{2j_1}\sum_{j_6}\sqrt{\frac{2j_6+1}{2j_4+1}}\begin{bmatrix} j_1 & j_2 & j_4 \\ j_3 & j_5 & j_6 \end{bmatrix}$$

$$\times\sum_{m_6}(-1)^{j_6-m_6}\begin{pmatrix} j_2 & j_3 & j_6 \\ m_2 & m_3 & -m_6 \end{pmatrix}\begin{pmatrix} j_1 & j_6 & j_5 \\ m_1 & m_6 & m_5 \end{pmatrix}. \quad (2.5.34)$$

To evaluate the sign factor here, we make use of (2.5.21). With the diagrammatic technique of Figs. 2.5.1 and 2.5.2, this relation is represented as shown in Fig. 2.5.9 (where we rename the index of summation $j_6 \to j_6'$ for future convenience).

Proposition 2.5.2. *Coefficients* $\begin{bmatrix} j_1 & j_2 & j_3 \\ j_4 & j_5 & j_6 \end{bmatrix}$ *in associativity relations (2.5.33) and (2.5.34) are equal to 6-j symbols, modulo a simple factor that depends on spins* j_k*:*

$$\begin{Bmatrix} j_1 & j_2 & j_4 \\ j_3 & j_5 & j_6 \end{Bmatrix} = \frac{(-1)^{j_1+j_2+j_3+j_5}}{\sqrt{(2j_4+1)(2j_6+1)}}\cdot\begin{bmatrix} j_1 & j_2 & j_4 \\ j_3 & j_5 & j_6 \end{bmatrix}$$

$$= \frac{(-1)^{j_1+j_2+j_3+j_5}}{\sqrt{(2j_4+1)(2j_6+1)}}\cdot\langle j_1, j_6;\, j_5, m_5 | j_4, j_3;\, j_5, m_5\rangle. \quad (2.5.35)$$

Proof. We start with the diagram for fourth rank tensor on the right-hand side of Fig. 2.5.9, return back to notation $j_6' \to j_6$ and reflect this diagram with respect to the vertical axis. As a result, we obtain diagrammatic representation for dual tensor, see Fig. 2.5.10. We then contract (by means of invariant metric) the tensors in both sides of Fig. 2.5.9 with the dual

$$\sum_{m_6}(-1)^{j_6-m_6}\begin{pmatrix} j_3 & j_2 & j_6 \\ m_3 & m_2 & -m_6 \end{pmatrix}\begin{pmatrix} j_1 & j_5 & j_6 \\ m_1 & m_5 & m_6 \end{pmatrix} =$$

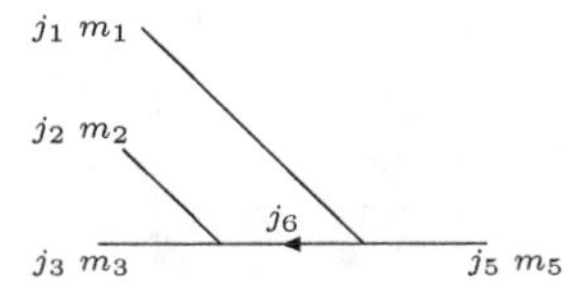

Fig. 2.5.10. Diagrammatic representation of dual tensor.

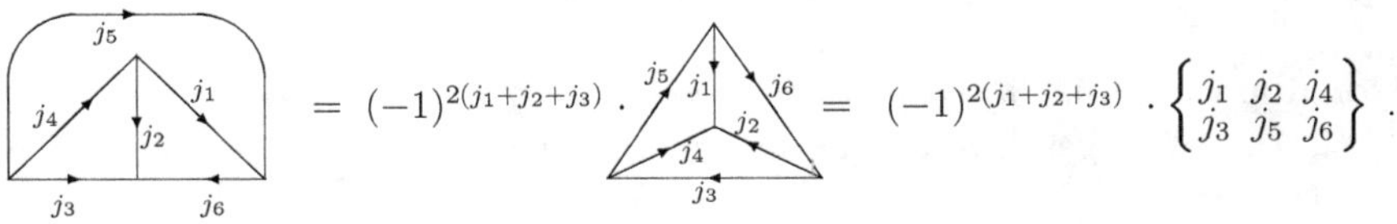

Fig. 2.5.11. Invariant convolution of the left-hand side of Fig. 2.5.9.

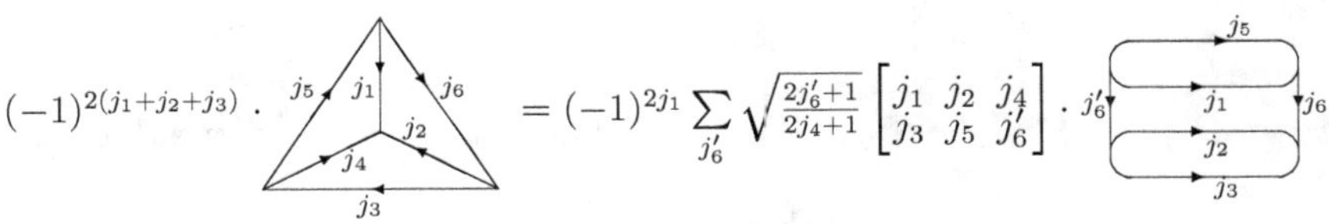

Fig. 2.5.12. Invariant scalar form of the relation shown in Fig. 2.5.9.

tensor shown in Fig. 2.5.10, in such a way that all external lines in the dual tensor are incoming.

Then on the left-hand side, as shown in Figs. 2.5.11 and 2.5.12, one obtains convolution of four 3-j symbols whose diagrammatic representation is tetrahedron. In accordance to Fig. 2.5.6, this convolution is just 6-j symbol (take note of arrangement of spins j_k and orientations of edges). The right-hand side (see Fig. 2.5.12) involves the sum of coefficients $\begin{bmatrix} j_1 & j_2 & j_4 \\ j_3 & j_5 & j'_6 \end{bmatrix}$ multiplied by scalar invariants shown on the left-hand side of Fig. 2.5.5.

Thus, upon convolution with dual tensor, given in Fig. 2.5.10 the relation shown in Fig. 2.5.9 is written as an equality of scalar invariants, as shown in Fig. 2.5.12.

The invariant convolution of four 3-j symbols entering the right-hand side of Fig. 2.5.12, has been evaluated in Fig. 2.5.5. We insert this result into the right-hand side of Fig. 2.5.12 and replace the tetrahedron on the left-hand side by 6-j symbol (see Fig. 2.5.11). In this way, we finally write

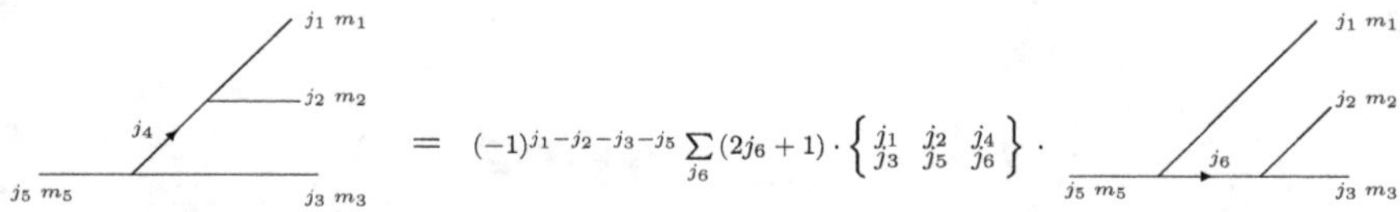

Fig. 2.5.13. Associativity condition (2.5.34) (see also Fig. 2.5.9), in terms of 6-j symbols.

the identity shown in Fig. 2.5.12 as the first equality in (2.5.35). The second equality in (2.5.35) is a consequence of (2.5.32). □

Corollary 2.5.3. *In view of Proposition 2.5.1, the right-hand side of (2.5.35) is independent of* m_5. *Therefore, 6-j symbol can be defined as follows:*

$$\begin{Bmatrix} j_1 & j_2 & j_4 \\ j_3 & j_5 & j_6 \end{Bmatrix} = \frac{(-1)^{B'}}{\sqrt{(2j_4+1)(2j_6+1)(2j_5+1)}} \times \sum_{m_5} \langle j_1, j_6;\ j_5, m_5 | j_4, j_3;\ j_5, m_5 \rangle, \qquad (2.5.36)$$

where $B' = j_1 + j_2 + j_3 + j_5$. *We use this definition in the next subsection.*

Problem 2.5.7. Check the formula (2.5.35) by direct calculation of scalar product $\langle j_1, j_6;\ j_5, m_5 | j_4, j_3;\ j_5, m_5 \rangle$ of vectors (2.5.27) and (2.5.28).

Due to relation (2.5.35), associativity property (2.5.34) (see also Fig. 2.5.9) can be written in terms of 6-j symbols. In other words, 6-j symbols completely determine coefficients in the associativity condition (2.5.34) for tensor product (2.5.26) of three representations. The associativity condition given by Fig. 2.5.9 is presented in terms of 6-j symbols as shown in Fig. 2.5.13.

Problem 2.5.8. Making use of the procedure employed for passing from Fig. 2.5.9 to Fig. 2.5.12, prove the identity shown in Fig. 2.5.14, where $B'' = (j_1 + j_2 + j_3) + 2(j_4 + j_5 + j_6)\ \mathrm{mod}(2)$. Hint: direct all external lines on the left-hand side toward the triangle, which gives a factor $(-1)^{B''}$, and then take convolution of both sides with 3-j symbol.

Problem 2.5.9. Show that invariant constructed out of six 3-j symbols and given in Fig. 2.5.15 is not a nontrivial 9-j invariant, as it is equal to product of two 6-j symbols. Hint: make use of the identity shown in Fig. 2.5.14.

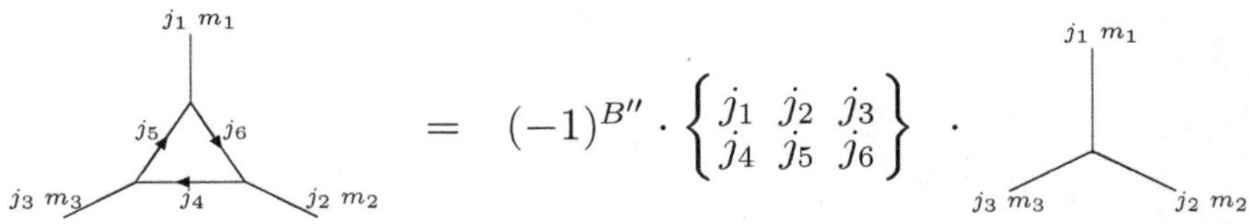

Fig. 2.5.14. Identity for convolution of three 3-j symbols.

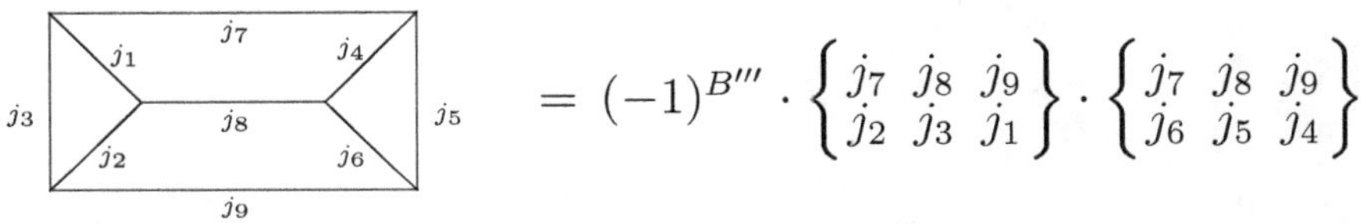

Fig. 2.5.15. Convolution of six 3-j symbols giving "trivial" 9-j invariant (here the sign factor $(-1)^{B'''}$ depends on the orientation of lines on the left-hand side).

Remark 1. Identity shown in Fig. 2.5.14 can be used to express 6-j symbol in a compact form, in terms of a sum over one variable. To this end, one sets $m_1 = j_1$ in this identity and makes use of explicit form of 3-j symbol as given by (2.4.106).

Remark 2. Nontrivial 9-j symbol shown in Fig. 2.5.8 is related to the associativity condition for tensor product of four representations (cf. (2.5.26))

$$[T^{(j_1)} \otimes (T^{(j_2)} \otimes T^{(j_3)}]_{13} \otimes T^{(j_4)})_{24} = (T^{(j_1)} \otimes T^{(j_2)}) \otimes (T^{(j_3)} \otimes T^{(j_4)}). \tag{2.5.37}$$

Here the brackets $[\ldots]_{13}$ and $(\ldots)_{24}$ denote the application of Clebsch–Gordan formula (2.4.100) to tensor products of representations $T^{(j_1)} \otimes T^{(j_3)}$ and $T^{(j_2)} \otimes T^{(j_4)}$, respectively. This associativity condition gives an identity for 3-j symbols whose diagrammatic counterpart is shown in Fig. 2.5.16 (this is an analog of Fig. 2.5.9). The right-hand side of Fig. 2.5.16 involves a linear combination of 9-j symbols, and $f(j_1, \ldots, j_9)$ is a function of spins j_k, which we do not write here.

Let us see why the coefficients on the right-hand side of Fig. 2.5.16 can be expressed through the 9-j symbols shown in Fig. 2.5.8. We make use of the same procedure that is used to obtain the relation in Fig. 2.5.12 from that shown in Fig. 2.5.9, i.e., we define dual tensor (see Fig. 2.5.17) to tensor on the right-hand side of Fig. 2.5.16. We then make a convolution

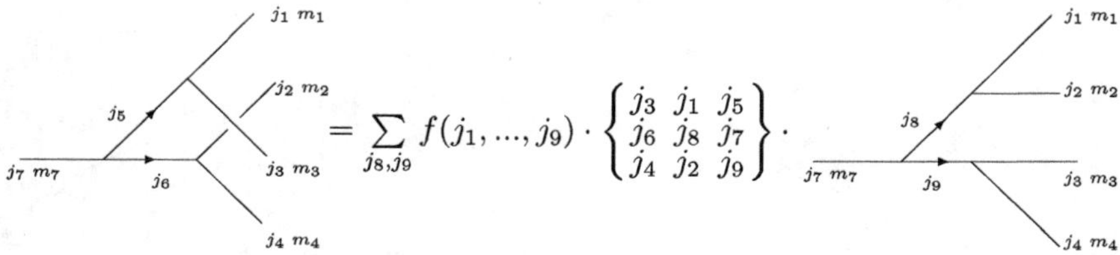

Fig. 2.5.16. Diagrammatic representation of associativity condition that involves 9-j symbols.

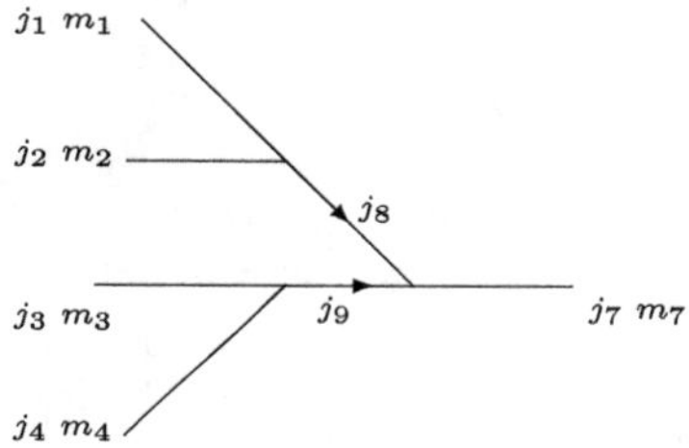

Fig. 2.5.17. Dual tensor to the tensor on the right-hand side of Fig. 2.5.16.

of this dual tensor with the two sides of Fig. 2.5.16 and find that the left-hand side is the graph for 9-j symbol (see Fig. 2.5.8), while the right-hand side is a linear combination of 9-j symbols whose coefficients are invariant convolutions of six 3-j symbols which are calculated by employing formulas shown in Figs. 2.5.4 and 2.5.5:

$$\sum_{j_8',j_9'} f(j_1,\ldots,j_7,j_8',j_9')\cdot\begin{Bmatrix} j_3 & j_1 & j_5 \\ j_6 & j_8' & j_7 \\ j_4 & j_2 & j_9' \end{Bmatrix}\cdot$$

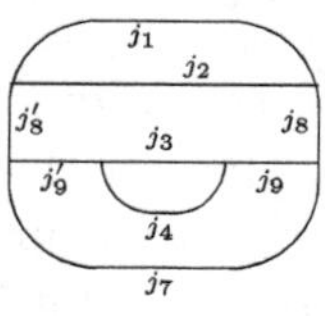

$$= \begin{Bmatrix} j_3 & j_1 & j_5 \\ j_6 & j_8 & j_7 \\ j_4 & j_2 & j_9 \end{Bmatrix}\cdot\frac{f(j_1,\ldots,j_9)\cdot(-1)^B}{(2j_8+1)(2j_9+1)}\,. \tag{2.5.38}$$

Here $(-1)^B$ is a sign factor that depends on orientation of lines in the graphs in the left- and right-hand sides of Fig. 2.5.16. We then compare (2.5.38) with 9-j symbol (see Fig. 2.5.8) that results from the convolution of the left-hand side of Fig. 2.5.16 and dual tensor in Fig. 2.5.17, and in this way obtain equation that defines the function $f(j_1,\ldots,j_9) = (-1)^B(2j_8+1)(2j_9+1)$.

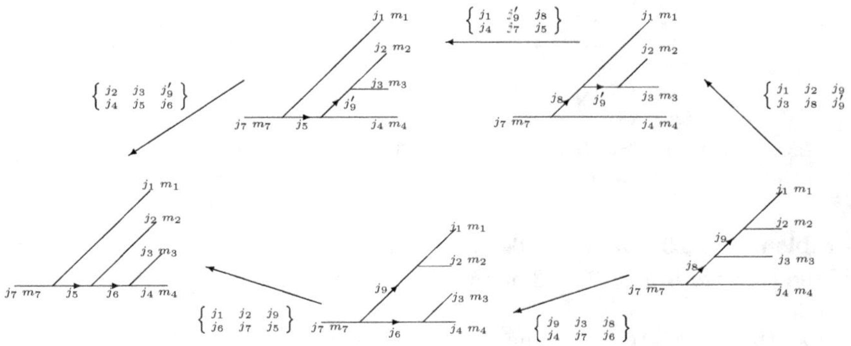

Fig. 2.5.18. Pentagonal relation following from associativity of product of four representations.

Remark 3. Since 6-j symbol is involved in the associativity condition for three representations, as shown in Fig. 2.5.13, the 6-j symbols obey pentagonal relation which follows from the equivalence of two paths going from tensor product of four representations $((T^{(j_1)}\otimes T^{(j_2)})\otimes T^{(j_3)})\otimes T^{(j_4)}$ to tensor product of the same four representations, but with the opposite order of fusion: $T^{(j_1)} \otimes (T^{(j_2)} \otimes (T^{(j_3)} \otimes T^{(j_4)}))$. Here parentheses show the order in which the fusion of representations is performed, or the order in which the Clebsch–Gordan formula is applied. This equivalence is graphically represented in Fig. 2.5.18.

Vertices of this diagram are fifth rank invariant tensors $t^{(j_1,j_2,j_3,j_4,j_7)}_{m_1,m_2,m_3,m_4,m_7}$ constructed as contractions of three 3-j symbols. These tensors are visualized graphically as "trees" which show the order how the representations are fused, or the order of application of Clebsch–Gordan formula. Each step shown in Fig. 2.5.18 by arrow corresponds to applying associativity condition (which is given in Fig. 2.5.13) to tensor product of three representations. At each step the "tree" is rearranged. As a result of two different paths from the utmost right tensor in Fig. 2.5.18 to utmost left tensor, there emerge two different linear combinations with coefficients made of 6-j symbols. By equating these two linear combinations, we obtain pentagonal identity for five 6-j symbols, called *Biedenharn–Elliott identity*:

$$\begin{Bmatrix} j_1 & j_2 & j_9 \\ j_6 & j_7 & j_5 \end{Bmatrix} \begin{Bmatrix} j_9 & j_3 & j_8 \\ j_4 & j_7 & j_6 \end{Bmatrix} = \sum_{j_9'} (-1)^R \, (2j_9' + 1) \begin{Bmatrix} j_2 & j_3 & j_9' \\ j_4 & j_5 & j_6 \end{Bmatrix} \begin{Bmatrix} j_1 & j_2 & j_9 \\ j_3 & j_8 & j_9' \end{Bmatrix} \begin{Bmatrix} j_1 & j_9' & j_8 \\ j_4 & j_7 & j_5 \end{Bmatrix}, \tag{2.5.39}$$

where $(-1)^R$ is a certain sign factor.

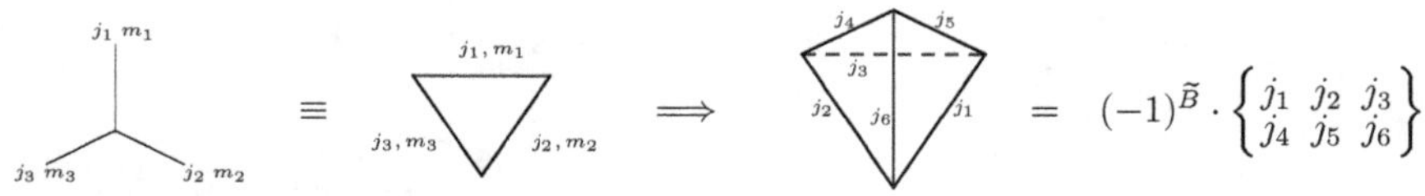

Fig. 2.5.19. Dual representation of 3-j and 6-j symbols. In the second equality the tetrahedron is obtained by gluing four 3-j symbols (four triangle faces).

Problem 2.5.10. Derive Biedenharn–Elliott identity (2.5.39) by making use of the diagram of Fig. 2.5.18 and calculate the sign factor $(-1)^R$.

Identity (2.5.39) has remarkable geometric interpretation, once one recalls that 6-j symbols are represented in Fig. 2.5.6 by tetrahedrons. It is convenient, however, to represent 6-j symbol as tetrahedron dual to the original one (shown in Fig. 2.5.6). The dual representation is given in Fig. 2.5.19. In the dual diagrammatic technique,[m] 3-j symbol (which is three-valence vertex in the original diagrammatic technique) is represented by a triangle, as shown in Fig. 2.5.19, and these triangles make dual tetrahedron representing 6-j symbol. We emphasize that the convolution of 3-j symbols over projection m_k corresponds, in dual technique, to gluing together the relevant edges of pertinent triangles.

Like in the original diagrammatic technique, one has to adopt a convention concerning orientation of edges. This convention determines the sign factor $(-1)^{\widetilde{B}}$ in Fig. 2.5.19.

Note that the representation of 3-j symbol, in dual diagram technique, by a triangle (rather than vertex as in the original diagram technique) is more natural. Indeed, 3-j symbol does not vanish only if its spins j_1, j_2 and j_3 obey triangle inequalities

$$j_1 + j_2 \geq j_3, \quad j_2 + j_3 \geq j_1, \quad j_3 + j_1 \geq j_2.$$

Thus, in dual technique, the spins j_1, j_2, j_3 are naturally associated with lengths of the edges of the triangle.

Remark 4. Making use of dual technique, we represent the associativity condition given in Fig. 2.5.13 in the form shown in Fig. 2.5.20. Then, the dual analog of the diagram in Fig. 2.5.18 takes the form of the diagram in Fig. 2.5.21, and we graphically represent Biedenharn–Elliott identity (2.5.39) as shown in Fig. 2.5.22. This graphical relation may

[m]Dual diagrammatic technique for contraction of 3-j symbols is used in the context of 6-j symbols, e.g., in book [21].

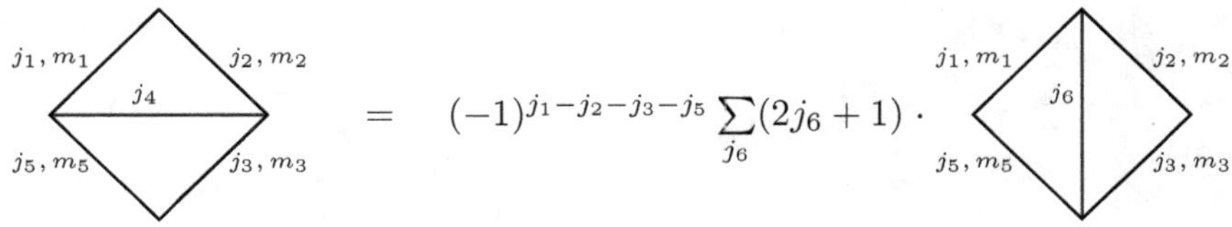

Fig. 2.5.20. Associativity condition (2.5.34), shown in Fig. 2.5.13, is represented with the help of dual technique in the form of rearrangements of planar triangulation.

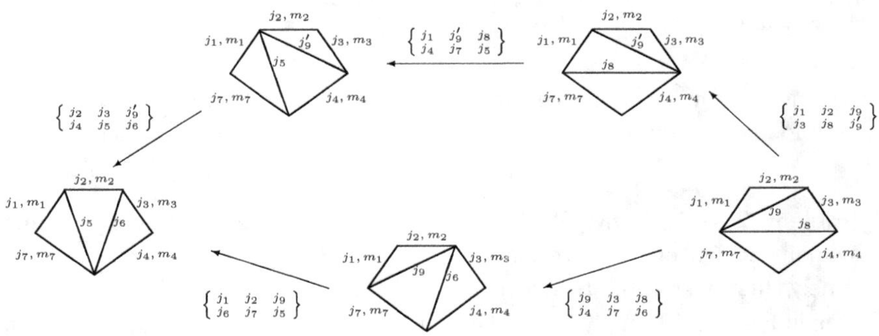

Fig. 2.5.21. Pentagon relation for associators (see Fig. 2.5.18), given in the dual representation.

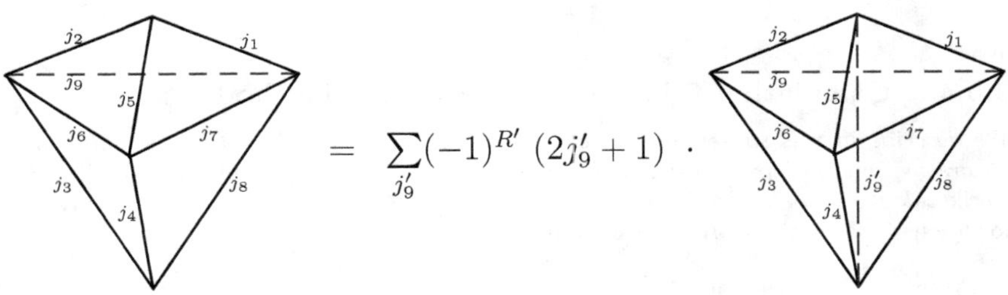

Fig. 2.5.22. Diagrammatic representation of Biedenharn–Elliott identity (2.5.39).

be interpreted as one of the rearrangements of triangulations of three-dimensional space.[n] This geometric property of 6-j symbols makes them

[n]Gluing together two tetrahedrons which have a common face in left part of Fig. 2.5.22 (product of two 6-j symbols on the left-hand side of (2.5.39)) is represented as gluing together three tetrahedrons which have a common edge j_9', with summation over j_9', on the right-hand side of Fig. 2.5.22 (convolution of three 6-j symbols on the right-hand side of (2.5.39)).

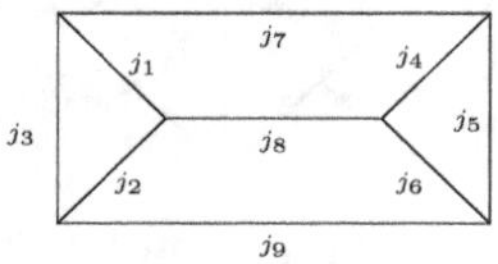

$$= \sum_{j_9'} (-1)^{B^{(5)}} (2j_9' + 1) \begin{Bmatrix} j_1 & j_3 & j_7 \\ j_5 & j_4 & j_9' \end{Bmatrix} \cdot$$

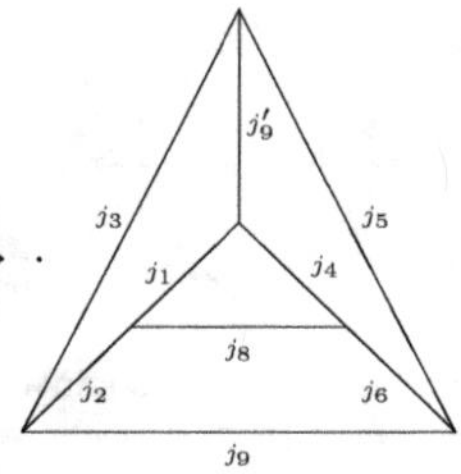

Fig. 2.5.23. Transformation of diagram given in Fig. 2.5.15 by making use of associativity condition.

instrumental for constructing models of gravity based on triangulations of manifolds (Ponzano–Regge model [35], Turaev–Viro model [36], etc.).

Problem 2.5.11. Derive Biedenharn–Elliott identity (2.5.39) by calculating invariant shown on the left-hand side of Fig. 2.5.15 in an alternative way, namely, first transform this diagram making use of associativity condition given in Fig. 2.5.13 (apply it to subgraph involving spins j_1, j_3, j_7, j_5 and j_4 on the left-hand side of Fig. 2.5.23) and obtain the relation shown in Fig. 2.5.23. Then calculate the right-hand side of the latter by making use of the identity shown in Fig. 2.5.14. Finally, employ the symmetries of 6-j symbols.

2.5.4. Calculating 6-j symbols. Schwinger method

We describe in this subsection the calculation of 6-j symbol based on the method of generating functions proposed by J. Schwinger [37] (see also [38]). Similar method applies to the calculation of 9-j symbol [37]. Our purpose is to obtain compact explicit expression for 6-j symbol which involves the smallest possible number of summations.

1. Functional representation

Let us write tensor $\sum_{m_4} \langle j_1, m_1; j_2, m_2 | j_4, m_4\rangle \langle j_4, m_4; j_3, m_3 | j_5, m_5\rangle$ on the left-hand side of (2.5.33) as follows:

$$\sum_{m_4, m_4'} \langle j_1, m_1; j_2, m_2 | j_4, m_4\rangle \langle j_4, m_4 | j_4, m_4'\rangle \langle j_4, m_4'; j_3, m_3 | j_5, m_5\rangle,$$

multiply it by vectors

$$T^{j_1}_{m_1}(s_1, t_1),\ T^{j_2}_{m_2}(s_2, t_2),\ T^{j_3}_{m_3}(s_3, t_3),\ (-1)^{j_5 - m_5} \overline{T}^{j_5}_{-m_5}(\bar{S}, \bar{T}), \tag{2.5.40}$$

make use of the formula (2.2.28) to write the scalar product of basis vectors

$$\langle j_4, m_4 | j_4, m_4'\rangle = (-1)^{j_4 - m_4} \overline{T}^{j_4}_{-m_4}(\partial_\sigma, \partial_\tau) \cdot T^{j_4}_{m_4'}(\sigma, \tau)\Big|_{\sigma=\tau=0} = \delta_{m_4 m_4'}, \tag{2.5.41}$$

and sum over projections $m_1, m_2, m_3, m_4, m_4', m_5$. We also make use of the relation (2.4.81):

$$(s_1 t_2 - s_2 t_1)^{j_1+j_2-j_3} (s_1 \bar{s}_3 + t_1 \bar{t}_3)^{j_1+j_3-j_2} (s_2 \bar{s}_3 + t_2 \bar{t}_3)^{j_2+j_3-j_1}$$
$$= \frac{\mathcal{D}}{\mu} \sum_{m_1, m_2, m_3} (-1)^{j_3-m_3} T^{j_1}_{m_1}(s_1, t_1) T^{j_2}_{m_2}(s_2, t_2)$$
$$\times \langle j_1, m_1; j_2, m_2 | j_3, m_3 \rangle \overline{T}^{j_3}_{-m_3}(\bar{s}_3, \bar{t}_3). \tag{2.5.42}$$

As a result, we obtain, in view of (2.4.81), the invariant polynomial in variables $s_1, t_1, s_2, t_2, s_3, t_3, \bar{S}, \bar{T}$ (modulo overall normalization):

$$\mathcal{I}_{12} = (s_1 t_2 - t_1 s_2)^{n_{12}} (s_1 \partial_\sigma + t_1 \partial_\tau)^{n_{1\sigma}} (s_2 \partial_\sigma + t_2 \partial_\tau)^{n_{2\sigma}}$$
$$\times (\sigma t_3 - \tau s_3)^{n_{3\sigma}} (\sigma \bar{S} + \tau \bar{T})^{n_{4\sigma}} (s_3 \bar{S} + t_3 \bar{T})^{n_{34}} \big|_{\sigma=\tau=0}, \tag{2.5.43}$$

where ($j_{12} := j_4$)

$$\begin{aligned} &n_{12} = j_1 + j_2 - j_{12}, \quad n_{1\sigma} = j_{12} + j_1 - j_2, \quad n_{2\sigma} = j_{12} + j_2 - j_1, \\ &n_{3\sigma} = j_{12} + j_3 - j_5, \quad n_{4\sigma} = j_5 + j_{12} - j_3, \quad n_{34} = j_5 + j_3 - j_{12}. \end{aligned} \tag{2.5.44}$$

By construction, invariant (2.5.43) is homogeneous polynomial in pairs of variables (s_1, t_1), (s_2, t_2), (s_3, t_3) of degree $2j_1$, $2j_2$, $2j_3$, respectively, so for given $\bar{S}$, $\bar{T}$ it belongs to $\mathcal{V}_{(1)} \otimes \mathcal{V}_{(2)} \otimes \mathcal{V}_{(3)}$, where $\mathcal{V}_{(k)} \equiv \mathcal{V}_{2j_k+1}$ is the space of representation $T^{(j_k)}$. This invariant has the same transformation properties with respect to the variables (s_1, t_1), (s_2, t_2) as the invariant polynomial

$$\mathcal{I} = (s_1 t_2 - t_1 s_2)^{n_{12}} (s_1 \bar{s} + t_1 \bar{t})^{n_{1\sigma}} (s_2 \bar{s} + t_2 \bar{t})^{n_{2\sigma}}.$$

The latter belongs to space $\mathcal{V}_{(12)} \subset \mathcal{V}_{(1)} \otimes \mathcal{V}_{(2)}$ of representation with spin j_{12}, since the degree of $\mathcal{I}$ in dual variables $\bar{s}, \bar{t}$ equals $2j_{12}$; see Section 2.4.4. Finally, the invariant polynomial (2.5.43) has degree $2j_5$ in variables $(\bar{S}, \bar{T})$, so it is expanded as follows:

$$\mathcal{I}_{12} = N_{12} \sum_{m_5} (-1)^{j_5-m_5} \bar{T}^{j_5}_{-m_5}(\bar{S}, \bar{T}) T^{j_5, j_{12}}_{m_5}(s_1, \ldots, t_3), \tag{2.5.45}$$

where $\bar{T}^{j_5}_{-m_5}(\bar{S}, \bar{T})$ and $T^{j_5, j_{12}}_{m_5}(s_1, \ldots, t_3)$ are normalized basis vectors in spaces $\bar{\mathcal{V}}_{(5)}$ and $\mathcal{V}_{(5)}$ embedded in $\mathcal{V}_{(12)} \otimes \mathcal{V}_{(3)}$, and N_{12} is a normalization coefficient. Modulo sign, the latter is determined by the normalization condition

$$(2j_5 + 1) N_{12}^2 = (\mathcal{I}_{12}, \mathcal{I}_{12}) = \mathcal{I}_{12}(\hat{s}_1, \ldots \hat{\bar{T}}) \cdot \mathcal{I}_{12}(s_1, \ldots \bar{T}) \big|_{s_i = t_i = \bar{S} = \bar{T} = 0}, \tag{2.5.46}$$

where $\hat{s} = \partial_s$, etc. The factor $(2j_5 + 1)$ on the left-hand side of (2.5.46) has to do with the fact that the vector (2.5.45) modulo factor N_{12} is equal to the sum of $(2j_5 + 1)$ vectors $(-1)^{j_5-m_5} \bar{T}^{j_5}_{-m_5} T^{j_5, j_{12}}_{m_5} \in \bar{\mathcal{V}}_{(5)} \otimes \mathcal{V}_{(5)}$, each of which is normalized to unity.

Similarly, we take a convolution of tensor

$$\sum_{m_6, m_6'} \langle j_2, m_2; j_3, m_3 | j_6, m_6 \rangle \langle j_6, m_6 | j_6, m_6' \rangle \langle j_1, m_1; j_6, m_6' | j_5, m_5 \rangle,$$

entering the right-hand side of (2.5.33), with vectors (2.5.40), make use of (2.5.41) and construct invariant polynomial

$$\mathcal{I}_{23} = (s_2 t_3 - t_2 s_3)^{r_{23}} (s_2 \partial_\sigma + t_2 \partial_\tau)^{r_{2\sigma}} (s_3 \partial_\sigma + t_3 \partial_\tau)^{r_{3\sigma}}$$
$$\times (s_1 \tau - t_1 \sigma)^{r_{1\sigma}} (\sigma \bar{S} + \tau \bar{T})^{r_{4\sigma}} (s_1 \bar{S} + t_1 \bar{T})^{r_{14}} \big|_{\sigma=\tau=0}, \tag{2.5.47}$$

where $(j_{23} := j_6)$

$$\begin{aligned} &r_{23} = j_2 + j_3 - j_{23}, \quad r_{2\sigma} = j_{23} + j_2 - j_3, \quad r_{1\sigma} = j_{23} + j_1 - j_5, \\ &r_{3\sigma} = j_{23} + j_3 - j_2, \quad r_{4\sigma} = j_5 + j_{23} - j_1, \quad r_{14} = j_5 + j_1 - j_{23}. \end{aligned} \tag{2.5.48}$$

In the same way as in (2.5.45), polynomial (2.5.47) is expanded as follows:

$$\mathcal{I}_{23} = N_{23} \sum_{m_5} (-1)^{j_5 - m_5}\, \bar{T}^{j_5}_{-m_5}(\bar{S}, \bar{T})\, T^{j_5, j_{23}}_{m_5}(s_1, \dots, t_3), \tag{2.5.49}$$

where $\bar{T}^{j_5}_{-m_5}(\bar{S}, \bar{T})$ and $T^{j_5,j_{23}}_{m_5}(s_1, \dots, t_3)$ are normalized basis vectors in spaces $\bar{\mathcal{V}}_{(5)}$ and $\mathcal{V}_{(5)} \subset \mathcal{V}_{(1)} \otimes \mathcal{V}_{(23)}$, while normalization coefficient N_{23} is found from the relation analogous to (2.5.46):

$$(2j_5 + 1) N_{23}^2 = (\mathcal{I}_{23}, \mathcal{I}_{23}) = \mathcal{I}_{23}(\hat{s}_1, \dots \hat{\bar{T}}) \cdot \mathcal{I}_{23}(s_1, \dots \bar{T})|_{s_i = t_i = \bar{S} = \bar{T} = 0}. \tag{2.5.50}$$

Expansions (2.5.45) and (2.5.49) imply that

$$\begin{aligned} (\mathcal{I}_{23}, \mathcal{I}_{12}) &= N_{23}\, N_{12} \sum_{m_5, m_5'} (-1)^{2j_5 - m_5 - m_5'} (\bar{T}^{j_5}_{-m_5'}, \bar{T}^{j_5}_{-m_5})(T^{j_5,j_{23}}_{m_5'}, T^{j_5,j_{12}}_{m_5}) \\ &= N_{23}\, N_{12} \sum_{m_5} (T^{j_5,j_{23}}_{m_5}, T^{j_5,j_{12}}_{m_5}). \end{aligned} \tag{2.5.51}$$

This is used, with account of (2.5.46) and (2.5.50), to write the formula (2.5.36) in the following form:

$$\begin{aligned} \begin{Bmatrix} j_1 & j_2 & j_{12} \\ j_3 & j_5 & j_{23} \end{Bmatrix} &= \frac{(-1)^{B'}}{\sqrt{(2j_{12}+1)(2j_{23}+1)(2j_5+1)}} \sum_{m_5} (T^{j_5,j_{23}}_{m_5}, T^{j_5,j_{12}}_{m_5}) \\ &= \frac{(-1)^{B'}}{\sqrt{(2j_{12}+1)(2j_{23}+1)}} \frac{(\mathcal{I}_{23}, \mathcal{I}_{12})}{\sqrt{(\mathcal{I}_{12}, \mathcal{I}_{12})(\mathcal{I}_{23}, \mathcal{I}_{23})}}, \end{aligned} \tag{2.5.52}$$

where the scalar products $(,)$ are understood as in (2.5.46), (2.5.50). It is worth noting that the following relations hold:

$$\begin{aligned} 2j_{12} &= n_{1\sigma} + n_{2\sigma} = n_{3\sigma} + n_{4\sigma}, \quad 2j_{23} = r_{1\sigma} + r_{4\sigma} = r_{2\sigma} + r_{3\sigma}, \\ B' &= j_1 + j_2 - j_3 - j_5 = n_{12} - n_{34} = r_{1\sigma} - r_{3\sigma} = r_{2\sigma} - r_{4\sigma}. \end{aligned} \tag{2.5.53}$$

Therefore, in accordance with (2.5.52), and (2.5.43), (2.5.47), 6-j symbol is a function of 12 parameters (2.5.44) and (2.5.48), which are not, however, independent. It is convenient for what follows to make change of variables $\bar{S} \to t_4$, $\bar{T} \to -s_4$ and consider polynomials

$$\begin{aligned} \mathcal{I}_{12} = {} & (s_1 t_2 - t_1 s_2)^{n_{12}} (s_1 \partial_\sigma + t_1 \partial_\tau)^{n_{1\sigma}} (s_2 \partial_\sigma + t_2 \partial_\tau)^{n_{2\sigma}} \\ & \times (\sigma t_3 - \tau s_3)^{n_{3\sigma}} (\sigma t_4 - \tau s_4)^{n_{4\sigma}} (s_3 t_4 - t_3 s_4)^{n_{34}}|_{\sigma=\tau=0}, \end{aligned} \tag{2.5.54}$$

$$\begin{aligned} \mathcal{I}_{23} = {} & (s_2 t_3 - t_2 s_3)^{r_{23}} (s_2 \partial_\sigma + t_2 \partial_\tau)^{r_{2\sigma}} (s_3 \partial_\sigma + t_3 \partial_\tau)^{r_{3\sigma}} \\ & \times (\tau s_1 - \sigma t_1)^{r_{1\sigma}} (\sigma t_4 - \tau s_4)^{r_{4\sigma}} (s_1 t_4 - t_1 s_4)^{r_{14}}|_{\sigma=\tau=0}. \end{aligned} \tag{2.5.55}$$

Proposition 2.5.3. *Norms of invariant vectors (2.5.54) and (2.5.55) are*

$$(\mathcal{I}_{12}, \mathcal{I}_{12}) = \mu_{12} \frac{(1 + n_{12} + n_{1\sigma} + n_{2\sigma})!(1 + n_{34} + n_{3\sigma} + n_{4\sigma})!}{(2j_{12} + 1)}, \tag{2.5.56}$$

$$(\mathcal{I}_{23}, \mathcal{I}_{23}) = \mu_{23} \frac{(1 + r_{23} + r_{2\sigma} + r_{3\sigma})!(1 + r_{14} + r_{1\sigma} + r_{4\sigma})!}{(2j_{23} + 1)}, \tag{2.5.57}$$

where

$$\mu_{12} = n_{12}!\,n_{1\sigma}!\,n_{2\sigma}!\,n_{3\sigma}!\,n_{4\sigma}!\,n_{34}!, \quad \mu_{23} = r_{23}!\,r_{1\sigma}!\,r_{2\sigma}!\,r_{3\sigma}!\,r_{4\sigma}!\,r_{14}!. \tag{2.5.58}$$

Proof. *Formula (2.5.57) is obtained from (2.5.56) by appropriate replacements, so we concentrate on (2.5.56). We write by definition*

$$\begin{aligned}(\mathcal{I}_{12},\mathcal{I}_{12}) &= \mathcal{I}_{12}(\hat s_1,\ldots,\hat s_4,\hat t_1,\ldots,\hat t_4)\cdot\mathcal{I}_{12}(s_1,\ldots,s_4,t_1,\ldots,t_4)\\ &= (\hat s_1\hat t_2-\hat t_1\hat s_2)^{n_{12}}(\hat s_1\partial_\zeta+\hat t_1\partial_\theta)^{n_{1\sigma}}(\hat s_2\partial_\zeta+\hat t_2\partial_\theta)^{n_{2\sigma}}\\ &\quad\times(\zeta\hat t_3-\theta\hat s_3)^{n_{3\sigma}}(\zeta\hat t_4-\theta\hat s_4)^{n_{4\sigma}}(\hat s_3\hat t_4-\hat t_3\hat s_4)^{n_{34}}\\ &\quad\times(s_1t_2-t_1s_2)^{n_{12}}(s_1\partial_\sigma+t_1\partial_\tau)^{n_{1\sigma}}(s_2\partial_\sigma+t_2\partial_\tau)^{n_{2\sigma}}\\ &\quad\times(\sigma t_3-\tau s_3)^{n_{3\sigma}}(\sigma t_4-\tau s_4)^{n_{4\sigma}}(s_3t_4-t_3s_4)^{n_{34}},\end{aligned}$$

where it is implicit that one sets all variables $\theta,\sigma,s_1,t_1,\ldots$ equal to zero after evaluating all derivatives $\partial_\theta,\partial_\sigma,\hat s_1,\hat t_1,\ldots$. Let us apply operators $(\hat s_1\hat t_2-\hat t_1\hat s_2)^{n_{12}}$ and $(\hat s_3\hat t_4-\hat t_3\hat s_4)^{n_{34}}$ to $\mathcal{I}_{12}(s_1,\ldots,s_4,t_1,\ldots,t_4)$. In accordance with (2.4.87), we get

$$\begin{aligned}&(\hat s_3\hat t_4-\hat t_3\hat s_4)^{n_{34}}(\hat s_1\hat t_2-\hat t_1\hat s_2)^{n_{12}}\cdot\mathcal{I}_{12}\\ &=\frac{n_{12}!\,(1+n_{12}+n_{1\sigma}+n_{2\sigma})!}{(1+n_{1\sigma}+n_{2\sigma})!}\cdot\frac{n_{34}!\,(1+n_{34}+n_{3\sigma}+n_{4\sigma})!}{(1+n_{3\sigma}+n_{4\sigma})!}\\ &\quad\times(s_1\partial_\sigma+t_1\partial_\tau)^{n_{1\sigma}}(s_2\partial_\sigma+t_2\partial_\tau)^{n_{2\sigma}}(\sigma t_3-\tau s_3)^{n_{3\sigma}}(\sigma t_4-\tau s_4)^{n_{4\sigma}}.\end{aligned}\tag{2.5.59}$$

It remains to calculate

$$\begin{aligned}&(\hat s_1\partial_\zeta+\hat t_1\partial_\theta)^{n_{1\sigma}}(\hat s_2\partial_\zeta+\hat t_2\partial_\theta)^{n_{2\sigma}}(\zeta\hat t_3-\theta\hat s_3)^{n_{3\sigma}}(\zeta\hat t_4-\theta\hat s_4)^{n_{4\sigma}}\\ &\quad\times(s_1\partial_\sigma+t_1\partial_\tau)^{n_{1\sigma}}(s_2\partial_\sigma+t_2\partial_\tau)^{n_{2\sigma}}(\sigma t_3-\tau s_3)^{n_{3\sigma}}(\sigma t_4-\tau s_4)^{n_{4\sigma}}\\ &=(\hat s_1\partial_\zeta+\hat t_1\partial_\theta)^{n_{1\sigma}}(s_1\partial_\sigma+t_1\partial_\tau)^{n_{1\sigma}}(\hat s_2\partial_\zeta+\hat t_2\partial_\theta)^{n_{2\sigma}}(s_2\partial_\sigma+t_2\partial_\tau)^{n_{2\sigma}}\\ &\quad\times(\zeta\hat t_3-\theta\hat s_3)^{n_{3\sigma}}(\sigma t_3-\tau s_3)^{n_{3\sigma}}(\zeta\hat t_4-\theta\hat s_4)^{n_{4\sigma}}(\sigma t_4-\tau s_4)^{n_{4\sigma}}.\end{aligned}\tag{2.5.60}$$

To evaluate derivatives in (2.5.60) over $s_1,t_1,\ldots,s_4,t_4$, we write (2.5.60) as follows:

$$\begin{aligned}&[(\hat s_1\partial_\zeta+\hat t_1\partial_\theta)^{n_{1\sigma}},(s_1\partial_\sigma+t_1\partial_\tau)^{n_{1\sigma}}]\cdot[(\hat s_2\partial_\zeta+\hat t_2\partial_\theta)^{n_{2\sigma}},(s_2\partial_\sigma+t_2\partial_\tau)^{n_{2\sigma}}]\\ &\times[(\zeta\hat t_3-\theta\hat s_3)^{n_{3\sigma}},(\sigma t_3-\tau s_3)^{n_{3\sigma}}]\cdot[(\zeta\hat t_4-\theta\hat s_4)^{n_{4\sigma}},(\sigma t_4-\tau s_4)^{n_{4\sigma}}],\end{aligned}$$

and make use of the identities

$$[\hat s_k\partial_\zeta+\hat t_k\partial_\theta,s_k\partial_\sigma+t_k\partial_\tau]=(\partial_\zeta\partial_\sigma+\partial_\theta\partial_\tau),\quad[\zeta\hat t_k-\theta\hat s_k,\sigma t_k-\tau s_k]=(\zeta\sigma+\theta\tau).$$

Then (2.5.60) becomes

$$n_{1\sigma}!\,n_{2\sigma}!\,n_{3\sigma}!\,n_{4\sigma}!\times(\partial_\zeta\partial_\sigma+\partial_\theta\partial_\tau)^{n_{1\sigma}+n_{2\sigma}}(\zeta\sigma+\theta\tau)^{n_{3\sigma}+n_{4\sigma}},\tag{2.5.61}$$

where, according to (2.5.44), we have $n_{1\sigma}+n_{2\sigma}=n_{3\sigma}+n_{4\sigma}=2j_{12}$. Finally, we make use of (2.4.88) and obtain

$$\begin{aligned}&(\partial_\zeta\partial_\sigma+\partial_\theta\partial_\tau)^{n_{1\sigma}+n_{2\sigma}}(\zeta\sigma+\theta\tau)^{n_{3\sigma}+n_{4\sigma}}=(n_{3\sigma}+n_{4\sigma})!\,(1+n_{3\sigma}+n_{4\sigma})!\\ &=\frac{1}{2j_{12}+1}(1+n_{1\sigma}+n_{2\sigma})!\,(1+n_{3\sigma}+n_{4\sigma})!.\end{aligned}\tag{2.5.62}$$

Collecting all factors in (2.5.59), (2.5.61), (2.5.62), we arrive at (2.5.56). □

We insert (2.5.56), (2.5.57) in (2.5.52) and get

$$\begin{Bmatrix} j_1 & j_2 & j_{12} \\ j_3 & j_5 & j_{23} \end{Bmatrix} = (-1)^{B'} \Delta(j_1, j_2, j_{12})\Delta(j_2, j_3, j_{23})\Delta(j_1, j_{23}, j_5)\Delta(j_{12}, j_3, j_5)$$

$$\times \frac{1}{\mu_{12}\,\mu_{23}} (\mathcal{I}_{23}, \mathcal{I}_{12}), \tag{2.5.63}$$

where $\Delta(p, q, r)$ are triangle coefficients (2.4.85), while factors μ_{12} and μ_{23} in denominator are explicitly defined in (2.5.58). Note that upon recalling the identification $j_4 \equiv j_{12}$, $j_6 \equiv j_{23}$ and replacing $j_3 \leftrightarrow j_4$, the product of Δ-factors in (2.5.63) coincides with the product entering (2.5.22).

Remark. *From symmetry viewpoint, it is convenient to understand 6-j symbol as function of 12 nonnegative integer parameters defined in (2.5.44), (2.5.48), rather than as function of six spins j_k. The former parameters are arranged into a table 3×4 introduced by Shelepin [39] and called R-symbol:*

$$\begin{Vmatrix} n_{34} & r_{23} & r_{14} & n_{12} \\ r_{3\sigma} & n_{3\sigma} & n_{1\sigma} & r_{1\sigma} \\ r_{4\sigma} & n_{2\sigma} & n_{4\sigma} & r_{2\sigma} \end{Vmatrix} = \begin{Vmatrix} R_{11} & R_{12} & R_{13} & R_{14} \\ R_{21} & R_{22} & R_{23} & R_{24} \\ R_{31} & R_{32} & R_{33} & R_{34} \end{Vmatrix} = \begin{Bmatrix} j_1 & j_2 & j_{12} \\ j_3 & j_5 & j_{23} \end{Bmatrix}.$$

As an example, it has been shown that 6-j symbol is invariant under arbitrary permutation of columns or rows in R-symbol.

Problem 2.5.12. *Write all compositions of permutations of rows and columns in R-symbol that correspond to symmetries of tetrahedron (2.5.24) and (2.5.25).*

Problem 2.5.13. *Show that the invariance of R-symbol under transposition of the last columns is equivalent to symmetry of 6-j symbol [40]*

$$\begin{Bmatrix} j_1 & j_2 & j_4 \\ j_3 & j_5 & j_6 \end{Bmatrix} = \begin{Bmatrix} j_1 & \frac{1}{2}(j_2 + j_5 + j_4 - j_6) & \frac{1}{2}(j_2 + j_6 + j_4 - j_5) \\ j_3 & \frac{1}{2}(j_2 + j_5 + j_6 - j_4) & \frac{1}{2}(j_6 + j_5 + j_4 - j_2) \end{Bmatrix}.$$

Problem 2.5.14. *Check the identities obeyed by the parameters $R_{\alpha p}$:*

$$R_{\alpha p} + R_{\beta q} = R_{\alpha q} + R_{\beta p}, \quad (\alpha, \beta = 1, 2, 3;\ \ p, q = 1, 2, 3, 4).$$

Find all relations analogous to $2j_1 = R_{13} + R_{24} = R_{14} + R_{23}$ (see (2.5.53)), which express spins j_k through coefficients $R_{\alpha p}$.

2. Method of generating functions

To derive explicit formula for 6-j symbol, we have to calculate the scalar product $(\mathcal{I}_{23}, \mathcal{I}_{12})$ entering (2.5.52). We perform this calculation using the method of generating functions proposed by J. Schwinger [37].

Consider functions

$$\mathcal{K}_{12}(z_{12}, z_{i\sigma}, z_{34}) = \exp\left[z_{12}(s_1 t_2 - t_1 s_2) + z_{1\sigma}(s_1\partial_\sigma + t_1\partial_\tau) + z_{2\sigma}(s_2\partial_\sigma + t_2\partial_\tau)\right]$$
$$\times \exp\left[z_{3\sigma}(\sigma t_3 - \tau s_3) + z_{4\sigma}(\sigma t_4 - \tau s_4) + z_{34}(s_3 t_4 - t_3 s_4)\right]\big|_{\sigma,\tau=0},$$

$$\mathcal{K}_{23}(w_{23}, w_{i\sigma}, w_{14}) = \exp\left[w_{23}(s_2t_3 - t_2s_3) + w_{2\sigma}(s_2\partial_\zeta + t_2\partial_\theta) + w_{3\sigma}(s_3\partial_\zeta + t_3\partial_\theta)\right] \times \exp\left[w_{1\sigma}(\theta s_1 - \zeta t_1) + w_{4\sigma}(\zeta t_4 - \theta s_4) + w_{14}(s_1t_4 - t_1s_4)\right]|_{\zeta,\theta=0}. \quad (2.5.64)$$

Their expansion coefficients in monomials

$$\frac{(z_{12}^{n_{12}} \cdot z_{1\sigma}^{n_{1\sigma}} \cdot z_{2\sigma}^{n_{2\sigma}} \cdot z_{3\sigma}^{n_{3\sigma}} \cdot z_{4\sigma}^{n_{4\sigma}} \cdot z_{34}^{n_{34}})}{n_{12}!\,n_{1\sigma}!\,n_{2\sigma}!\,n_{3\sigma}!\,n_{4\sigma}!\,n_{34}!}, \qquad \frac{(w_{23}^{r_{23}} \cdot w_{1\sigma}^{r_{1\sigma}} \cdot w_{2\sigma}^{r_{2\sigma}} \cdot w_{3\sigma}^{r_{3\sigma}} \cdot w_{4\sigma}^{r_{4\sigma}} \cdot w_{14}^{r_{14}})}{r_{23}!\,r_{1\sigma}!\,r_{2\sigma}!\,r_{3\sigma}!\,r_{4\sigma}!\,r_{14}!} \quad (2.5.65)$$

coincide, respectively, with vectors $\mathcal{I}_{12}$ and $\mathcal{I}_{23}$ given by (2.5.54) and (2.5.55). So, the functions (2.5.65) are generating functions for vectors $\mathcal{I}_{12}$ and $\mathcal{I}_{23}$. Note that denominators in (2.5.65) are equal to coefficients μ_{12} and μ_{23} introduced in (2.5.58). We evaluate the derivatives with respect to ζ, θ, σ, τ in (2.5.64) by making use of the standard formulas like $e^{a\partial_\theta}F(\theta) = F(\theta + a)e^{a\partial_\theta}$. As a result, we obtain

$$\mathcal{K}_{12} = \exp\left[\sum_{p<q} z_{pq}(s_pt_q - t_ps_q)\right], \quad \mathcal{K}_{23} = \exp\left[\sum_{i<j} w_{ij}(s_it_j - t_is_j)\right], \quad (2.5.66)$$

where $i, j, p, q = 1, 2, 3, 4$ and the sets of variables are

$$\begin{aligned} &z\colon\ (z_{12};\ z_{13} = z_{1\sigma}z_{3\sigma};\ z_{14} = z_{1\sigma}z_{4\sigma};\ z_{23} = z_{2\sigma}z_{3\sigma};\ z_{24} = z_{2\sigma}z_{4\sigma};\ z_{34}),\\ &w\colon\ (w_{12} = w_{1\sigma}w_{2\sigma};\ w_{23};\ w_{24} = w_{2\sigma}w_{4\sigma};\ w_{13} = w_{1\sigma}w_{3\sigma};\ w_{14};\ w_{34} = w_{3\sigma}w_{4\sigma}). \end{aligned} \quad (2.5.67)$$

Now, the expression

$$\frac{1}{\mu_{12}\mu_{23}}(\mathcal{I}_{23}, \mathcal{I}_{12}) \quad (2.5.68)$$

(where μ_{12}, μ_{23} are defined in (2.5.58)) is a coefficient of monomial

$$(w_{23}^{r_{23}} \cdot w_{1\sigma}^{r_{1\sigma}} \cdot w_{2\sigma}^{r_{2\sigma}} \cdot w_{3\sigma}^{r_{3\sigma}} \cdot w_{4\sigma}^{r_{4\sigma}} \cdot w_{14}^{r_{14}}) \cdot (z_{12}^{n_{12}} \cdot z_{1\sigma}^{n_{1\sigma}} \cdot z_{2\sigma}^{n_{2\sigma}} \cdot z_{3\sigma}^{n_{3\sigma}} \cdot z_{4\sigma}^{n_{4\sigma}} \cdot z_{34}^{n_{34}}), \quad (2.5.69)$$

in the expansion of the function

$$\mathcal{K}(w_{ij}, z_{pq}) = \left\{\exp\left[\sum_{i<j} w_{ij}(\hat{s}_i\hat{t}_j - \hat{t}_i\hat{s}_j)\right] \times \exp\left[\sum_{p<q} z_{pq}(s_pt_q - t_ps_q)\right]\right\}_{s_i=t_i=0}. \quad (2.5.70)$$

Thus, the function $\mathcal{K}(w_{ij}, z_{pq})$ is the generating function for 6-j symbols. Here we keep using the notation $\hat{s}_i = \partial_{s_i}$, $\hat{t}_i = \partial_{t_i}$. It is convenient for what follows to define z_{ij} and w_{ij} for $i > j$ in such a way that matrices $||w_{ij}||$ and $||z_{ij}||$ are antisymmetric, $z_{ij} = -z_{ji}$, $w_{ij} = -w_{ji}$.

Proposition 2.5.4. *Consider d-dimensional analog of function (2.5.70), in which indices i, j, p, q take values $\{1, 2, \ldots, d\}$, while variables z_{ij} and w_{ij} define antisymmetric d-dimensional matrices $w = ||w_{ij}||$ and $z = ||z_{ij}||$. Then the d-dimensional function (2.5.70) has determinant representation*

$$\mathcal{K}(w_{ij}, z_{pq}) = \frac{1}{\det(I_d + w \cdot z)}, \quad (2.5.71)$$

where I_d is unit d-dimensional matrix.

Proof. *To begin with, we write d-dimensional function $\mathcal{K}(w_{ij}, z_{pq})$ given by (2.5.70), as matrix element*

$$\mathcal{K}(w_{ij}, z_{pq}) = \langle K_w|K_z\rangle, \tag{2.5.72}$$

where

$$|K_z\rangle = \exp\left(\sum_{i,j=1}^{d} z_{ij}\, a_i^\dagger\, b_j^\dagger\right)|0\rangle, \quad \langle K_w| = \langle 0|\exp\left(\sum_{i,j=1}^{d} w_{ij}\, a_i\, b_j\right), \tag{2.5.73}$$

and $\{a_i, a_i^\dagger\}$ and $\{b_j, b_j^\dagger\}$ are two independent sets of annihilation and creation operators of d-dimensional oscillators (see Section 1.2). We use completeness relation (1.2.22) for coherent states $|\alpha^*, \beta^*\rangle$ in holomorphic representation

$$a_i\,|\alpha^*, \beta^*\rangle = \alpha_i^*|\alpha^*, \beta^*\rangle, \quad b_j\,|\alpha^*, \beta^*\rangle = \beta_j^*|\alpha^*, \beta^*\rangle,$$

and write the matrix element (2.5.72) as

$$\begin{aligned}\langle K_w|K_z\rangle &= \int d\mu(\alpha, \alpha^*)\, d\mu(\beta, \beta^*)\, e^{-\alpha_i\alpha_i^* - \beta_i\beta_i^*}\langle K_w|\alpha^*, \beta^*\rangle\,\langle\alpha, \beta|K_z\rangle \\ &= \int d\mu(\alpha, \alpha^*)\, d\mu(\beta, \beta^*)\exp(w_{ij}\alpha_i^*\,\beta_j^* + z_{ij}\alpha_i\,\beta_j)\,\exp\left[-\alpha_i\alpha_i^* - \beta_i\beta_i^*\right],\end{aligned} \tag{2.5.74}$$

where summation over repeated indices is implicit, and measure $d\mu(\alpha, \alpha^*)$ is defined in (1.2.15). Gaussian integral (2.5.74) is calculated by the change of integration variables $\beta_i \to \gamma_i + \alpha_k^* w_{ki}$, $\beta_i^* \to \bar{\gamma}_i + \alpha_k z_{ki}$. As a result, we get

$$\langle K_w|K_z\rangle = \left(\int d\mu(\alpha, \alpha^*)\exp(-\alpha_i^*(\delta_{ij} + w_{ik}z_{kj})\alpha_j)\right)\left(\int d\mu(\gamma, \bar{\gamma})e^{-\gamma_i\bar{\gamma}_i}\right), \tag{2.5.75}$$

where the first integral over α_i and α_i^* is equal to $\det^{-1}(I + w\,z)$, while the integral over γ_i and $\bar{\gamma}_i$ is equal to 1. □

Determinant on the right-hand side of (2.5.71) can be calculated explicitly. We note, to this end, that there is an identity

$$\det(I_d + w \cdot z) = \left[\mathrm{Pf}\begin{pmatrix} z & -I_d \\ I_d & w \end{pmatrix}\right]^2, \tag{2.5.76}$$

where Pf *denotes Pfaffian of antisymmetric, even-dimensional matrix. Recall (see (1.3.58)) that Pfaffian of an antisymmetric $2n \times 2n$ matrix $Z = ||Z_{ij}||$ is defined by*

$$\mathrm{Pf}(Z) = \frac{1}{2^n n!}\sum_{i_1,\ldots,i_{2n}} \varepsilon_{i_1\,i_2\,i_3\,i_4\cdots i_{2n-1}\,i_{2n}} Z_{i_1 i_2} Z_{i_3 i_4}\ldots Z_{i_{2n-1}i_{2n}}. \tag{2.5.77}$$

Making use of this definition, one writes Pfaffian on the right hand side of (2.5.76) as

$$\mathrm{Pf}\begin{pmatrix} z & -I_d \\ I_d & w \end{pmatrix} = (-1)^{d(d+1)/2}\left(1 + \sum_{k=1}^{[\frac{d}{2}]}\frac{(-1)^k}{(d-2k)!}\widetilde{w}_{i_1\ldots i_{d-2k}} \cdot \widetilde{z}_{i_1\ldots i_{d-2k}}\right), \tag{2.5.78}$$

where $[\frac{d}{2}]$ is an integer part of $\frac{d}{2}$, and the expression

$$\widetilde{z}_{i_1\ldots i_{d-2k}} = \frac{1}{2^k\,k!}\sum_{j_1,\ldots,j_{2k}} \varepsilon_{i_1\ldots i_{d-2k}\,j_1 j_2 j_3 j_4\cdots j_{2k}}(z_{j_1 j_2} z_{j_3 j_4}\cdots z_{j_{2k-1}j_{2k}}) \tag{2.5.79}$$

is naturally called incomplete Pfaffian.

Problem 2.5.15. * *Derive the expression (2.5.76) for determinant* $\det(I_d + w\cdot z)$. *Obtain the equality (2.5.78).*

Let us give explicit expressions for $\det(I_d + w\cdot z)$ *for small* d. *We substitute (2.5.78) into (2.5.76) and for* $d=2$ *and* $d=3$ *(i.e., for* 2×2 *or* 3×3 *matrices* w *and* z*) we find, respectively,*

$$\det(I_2 + wz) = [1 - z_{12}w_{12}]^2, \quad \det(I_3 + wz) = \Big[1 - \sum_{i<j} z_{ij}w_{ij}\Big]^2 .$$

These formulas are straightforward to check by explicit calculations. For 4×4 *matrices by using (2.5.76) and (2.5.78) we find*

$$\det(I_4 + wz) = \Big[1 - \sum_{i<j} z_{ij}w_{ij} + \mathrm{Pf}(z)\cdot\mathrm{Pf}(w)\Big]^2 , \tag{2.5.80}$$

where, in accordance with its definition (2.5.77), Pfaffian $\mathrm{Pf}(z)$ *equals*

$$\mathrm{Pf}(z) = \frac{1}{8}\varepsilon_{ijkm}z_{ij}z_{km} = (z_{12}z_{34} - z_{13}z_{24} + z_{23}z_{14}). \tag{2.5.81}$$

Finally, for 5×5 *matrices* w *and* z *we have* $\det(I_5 + wz) = [1 - \sum_{i<j} z_{ij}w_{ij} + \widetilde{z}_\ell\,\widetilde{w}_\ell]^2$, *where* $\widetilde{z}_\ell = \frac{1}{8}\varepsilon_{ijkm\ell}\, z_{ij}\, z_{km}$. *As an example,* $\widetilde{z}_1 = (z_{23}z_{45} - z_{24}z_{35} + z_{34}z_{25})$, *cf. (2.5.81).*

3. Explicit formula for 6-j-symbol

To finish the calculation of 6-j symbol, we use the expression for determinant in (2.5.71) in the case $d=4$. *In that case, by making use of (2.5.80) and (2.5.81), we obtain for generating function (2.5.70) the following formula*

$$\mathcal{K} = \Big[1 - \frac{1}{2}z_{ij}w_{ij} + (z_{12}z_{34} - z_{13}z_{24} + z_{23}z_{14})(w_{12}w_{34} - w_{13}w_{24} + w_{23}w_{14})\Big]^{-2} . \tag{2.5.82}$$

The sets of variables z_{ij} *and* w_{ij} *are special, as they come from (2.5.67). The formula (2.5.82) is therefore simplified, since* $z_{13}z_{24} = z_{14}z_{23}$, $w_{12}w_{34} = w_{13}w_{24}$ *and we have*

$$\mathcal{K}(z_{ij}, w_{ij}) = \frac{1}{\Big[1 - \Big(\sum_{i<j} z_{ij}\, w_{ij}\Big) + z_{12}\, z_{34}\, w_{23}\, w_{14}\Big]^2}. \tag{2.5.83}$$

This is the final formula for the generating function of 6-j symbols. We are interested in the coefficient of monomial (2.5.69) in the expansion of (2.5.83). We write

$$\begin{aligned}\mathcal{K}(z_{ij}, w_{ij}) &= \sum_{Q=0}^{\infty}(Q+1)\left[\Big(\sum_{i<j} z_{ij}\, w_{ij}\Big) - z_{12}\, z_{34}\, w_{23}\, w_{14}\right]^Q \\ &= \sum_{Q=0}^{\infty}(Q+1)!\left\{\frac{1}{Q!}\left[\Big(\sum_{i<j} z_{ij}\, w_{ij}\Big) - z_{12}\, z_{34}\, w_{23}\, w_{14}\right]^Q\right\},\end{aligned} \tag{2.5.84}$$

and expand the function in curly braces by making use of generalized binomial formula

$$[x_1 + \cdots + x_k]^Q = \sum_{\substack{a_1,\ldots,a_k \\ a_1+\ldots+a_k=Q}} \frac{Q!}{a_1!\cdots a_k!}\, x_1^{a_1}\cdots x_k^{a_k}.$$

Then the general term in the expansion of the curly braces in (2.5.84) reads

$$(-1)^b \cdot \left(\prod_{i<j} \frac{(w_{ij} z_{ij})^{a_{ij}}}{a_{ij}!} \right) \cdot \frac{(w_{23}\, w_{14}\, z_{12}\, z_{34})^b}{b!},$$

where a_{ij} $(i < j)$ and b are integers obeying

$$\sum_{i<j} a_{ij} + b = Q. \tag{2.5.85}$$

We now recall the relationships (2.5.67) and write this general term explicitly

$$(-1)^b \cdot \frac{(w_{1\sigma} w_{2\sigma} z_{12})^{a_{12}}}{a_{12}!} \cdot \frac{(w_{1\sigma} w_{3\sigma} z_{1\sigma} z_{3\sigma})^{a_{13}}}{a_{13}!} \cdot \frac{(w_{14} z_{1\sigma} z_{4\sigma})^{a_{14}}}{a_{14}!}$$

$$\times \frac{(w_{23} z_{2\sigma} z_{3\sigma})^{a_{23}}}{a_{23}!} \cdot \frac{(w_{2\sigma} w_{4\sigma} z_{2\sigma} z_{4\sigma})^{a_{24}}}{a_{24}!} \cdot \frac{(w_{3\sigma} w_{4\sigma} z_{34})^{a_{34}}}{a_{34}!} \cdot \frac{(w_{23} w_{14} z_{12} z_{34})^b}{b!}.$$

We compare this general term with monomial (2.5.69) and make use of (2.5.85). In this way, we obtain a set of equations for parameters a_{ij}, b whose solution, in view of (2.5.44) and (2.5.48), reads

$$\begin{aligned} b &= j_1 + j_2 + j_3 + j_5 - Q, \\ a_{12} &= Q - j_3 - j_{12} - j_5, \\ a_{13} &= j_1 + j_3 + j_{12} + j_{23} - Q, \\ a_{14} &= Q - j_2 - j_3 - j_{23}, \\ a_{23} &= Q - j_1 - j_{23} - j_5, \\ a_{24} &= j_2 + j_{12} + j_{23} + j_5 - Q, \\ a_{34} &= Q - j_1 - j_2 - j_{12}. \end{aligned} \tag{2.5.86}$$

This gives for the scalar product (2.5.68) the final result

$$\frac{1}{\mu_{12}\, \mu_{23}} (\mathcal{I}_{23}, \mathcal{I}_{12}) = (-1)^{j_1+j_2+j_3+j_5} \sum_Q (-1)^Q \frac{(Q+1)!}{\left(\prod_{i<j} a_{ij}!\right) b!}.$$

Accordingly, we obtain the expression for 6-j symbol (2.5.63)

$$\begin{aligned} \begin{Bmatrix} j_1 & j_2 & j_{12} \\ j_3 & j_5 & j_{23} \end{Bmatrix} &= (-1)^{B'+j_1+j_2+j_3+j_5} \Delta(j_1, j_2, j_{12}) \Delta(j_2, j_3, j_{23}) \Delta(j_1, j_{23}, j_5) \\ &\times \Delta(j_3, j_{12}, j_5) \sum_Q (-1)^Q \frac{(Q+1)!}{(\prod_{i<j} a_{ij}!) b!}. \end{aligned} \tag{2.5.87}$$

In view of the convention (2.5.36) for sign factor, $(-1)^{B'+j_1+j_2+j_3+j_5} = 1$, this result coincides with (2.5.22) (see also Ref. [21]), modulo replacement $j_3 \leftrightarrow j_{12}$ (recall that $j_{12} = j_4$, $j_{23} = j_6$) and $Q \to k$.

Chapter 3

Representations of Simple Lie Algebras. Weight Theory

This chapter builds on material presented in Refs. [5]–[8]. Also, we heavily use the results given in the first book in Chapter I-6 "Root systems and classification of simple Lie algebras" (Sections I-6.1 and I-6.2). To avoid confusion, we emphasize again that we study finite-dimensional representations, unless we state the opposite.

3.1. Root Systems of Simple Lie Algebras

3.1.1. Root systems of Lie algebra $s\ell(n, \mathbb{C})$, $so(n, \mathbb{C})$ and $sp(2r, \mathbb{C})$

Here, we summarize the properties of root systems of Lie algebras of classical series $s\ell(r+1, \mathbb{C})$ (A_r-series), $so(2r+1, \mathbb{C})$ (B_r-series), $sp(2r, \mathbb{C})$ (C_r-series) and $so(2r, \mathbb{C})$ (D_r-series). These root systems are of great interest by themselves; also, they are used in what follows for general constructions of weight theory and representation theory of simple Lie algebras. Detailed account of root systems of simple Lie algebras is given in Section I-6.2 of the first book.

Before turning to concrete algebras, let us summarize some facts concerning root systems.

Consider a simple complex Lie algebra $\mathcal{A}$ of rank r and its Cartan–Weyl basis $\{H_i, E_\alpha\}$. Here $i = 1, \ldots, r$, and roots $\alpha = (\alpha_1, \ldots, \alpha_r)$ are vectors in real r-dimensional space $\mathfrak{V}_r$, dubbed root space. The set of all roots (root system) of algebra $\mathcal{A}$ is denoted by $\Phi(\mathcal{A})$. The defining commutation relations are

$$[H_i, H_k] = 0, \quad [H_k, E_\alpha] = \alpha_k \, E_\alpha, \tag{3.1.1}$$

$$[E_\alpha,\, E_\beta] = N_{\alpha,\beta}\, E_{\alpha+\beta},\;\; N_{\alpha,\beta} \neq 0,\;\; \text{if } (\alpha+\beta) \text{ is a root},$$
$$[E_\alpha,\, E_\beta] = 0,\;\; \text{if } (\alpha+\beta) \text{ is not a root and } \alpha+\beta \neq 0\,, \tag{3.1.2}$$

$$[E_\alpha,\, E_{-\alpha}] = \alpha^k\, H_k\,, \quad \alpha^k = g^{kj}\, \alpha_j\,. \tag{3.1.3}$$

We choose the normalization of Cartan generators H_i and root generators E_α in such a way that Killing metric is

$$g_{kj} = \mathrm{Tr}(\mathrm{ad}(H_k)\cdot \mathrm{ad}(H_j)) = \sum_\alpha \alpha_k\, \alpha_j, \tag{3.1.4}$$

$$\mathrm{Tr}(\mathrm{ad}(H_k)\cdot \mathrm{ad}(E_\alpha)) = 0, \quad \mathrm{Tr}(\mathrm{ad}(E_\alpha)\cdot \mathrm{ad}(E_\beta)) = \delta_{\alpha,-\beta}. \tag{3.1.5}$$

Elements g^{jk} in (3.1.3) make inverse matrix to $||g_{jk}||$. The summation in (3.1.4) runs over all roots α of Lie algebra $\mathcal{A}$. Making use of matrices $||g^{jk}||$ and $||g_{jk}||$ we can raise and lower indices of components of vectors belonging to root space $\mathfrak{V}_r$.

Basis $\{H_i\}$ in Cartan subalgebra of any simple complex algebra $\mathcal{A}$ can be chosen in such a way that it coincides with Cartan basis in compact real form of algebra $\mathcal{A}$. Then, upon the replacement $H_k \to i\, H_k$ and equivalence transformation, basis elements of Cartan subalgebra in any finite-dimensional representation are Hermitian matrices (see Remark 2 in Section I-6.1.2 of the first book). With this convention, the roots α are real vectors and metric g_{ij}, in accordance with (3.1.4), is a Euclidean metric in real root space $\mathfrak{V}_r$. Indeed, the scalar product

$$(x, y) = x^i\, g_{ij}\, y^j = x^i\, y_i \quad (\forall\, x, y \in \mathfrak{V}_r), \tag{3.1.6}$$

is positive definite, since $(x,x) = (x^i g_{ij} x^j) = \sum_\alpha (x,\alpha)^2 > 0$ for all non vanishing real vectors x. We use this convention for the basis of Cartan subalgebra in the rest of the book.

If α is a root of Lie algebra $\mathcal{A}$, then $(-\alpha)$ is also a root of this algebra. Therefore (see Section I-6.2.2 of the first book), upon choosing some vector $x \in \mathfrak{V}_r$ such that $(x,\alpha) \neq 0$ $(\forall \alpha)$, one divides the set of roots into two subsets of positive roots with $(x,\alpha) > 0$ and negative roots with $(x,\alpha) < 0$. Positive roots that cannot be represented as a sum of two or more positive roots are dubbed *simple*. The number of simple roots is equal to the rank r of algebra $\mathcal{A}$. Simple roots $\alpha^{(i)}$ $(i = 1, \ldots, r)$ make basis in the root space $\mathfrak{V}_r$. All other roots α are linear combinations of simple roots, $\alpha = \sum_i \alpha^{(i)} n_i$, with integer coefficients n_i; for positive (negative) roots one has $n_i \geq 0$ $(n_i \leq 0)$ for all i.

In what follows, we use the standard notation $\Phi(\mathcal{A})$ for root system of algebra $\mathcal{A}$. For positive and negative roots α, we write $\alpha > 0$ and $\alpha < 0$, respectively.

1. Root system of algebra $s\ell(r+1,\mathbb{C})$ (A_r-series)
In Lie algebra $s\ell(r+1,\mathbb{C})$ (A_r-series), Cartan–Weyl basis can be chosen to consist of the following elements (see Section I-6.2.3 of the first book)

$$E_{\alpha^{(ij)}} = e_{ij}, \quad E_{-\alpha^{(ij)}} = e_{ji} \quad (i<j), \tag{3.1.7}$$

$$H_i \equiv e_{ii} - 1/(r+1)\cdot I_{r+1} \quad (i=1,\ldots,r+1), \tag{3.1.8}$$

where e_{ij} are $(r+1)\times(r+1)$ matrix units (matrices with elements $(e_{ij})_{kl} = \delta_{ik}\delta_{jl}$), $I_{r+1} = \sum_i e_{ii}$ is unit $(r+1)$-dimensional matrix and $\mathrm{Tr}(H_i) = 0$. Elements H_i obey

$$\sum_{i=1}^{r+1} H_i = 0, \tag{3.1.9}$$

and form basis in Cartan subalgebra of $s\ell(r+1,\mathbb{C})$, while $E_{\alpha^{(ij)}}$ and $E_{-\alpha^{(ij)}}$ are root generators. Elements (3.1.7) and (3.1.8) obey commutation relations $(k<l)$:

$$[H_i, E_{\alpha^{(kl)}}] = \alpha_i^{(kl)}\, E_{\alpha^{(kl)}}, \quad [H_i, E_{-\alpha^{(kl)}}] = -\alpha_i^{(kl)}\, E_{-\alpha^{(kl)}}, \tag{3.1.10}$$

where $\alpha_i^{(kl)}$ are components of $(r+1)$-dimensional root vectors

$$\alpha^{(kl)} = e^{(k)} - e^{(l)} \quad (k,l = 1,\ldots,r+1;\ k\neq l). \tag{3.1.11}$$

Hereafter $e^{(k)}$ denote the Cartesian basis vectors in vector space, whose coordinates are

$$e_i^{(k)} = \delta_i^k.$$

Note that all roots (3.1.11) are orthogonal to vector

$$u = \sum_k e^{(k)} = (1,1,\ldots,1),$$

so they belong to r-dimensional hyperplane P_u orthogonal to vector u (the latter fact is no accident, as it follows from (3.1.9)). Nondegenerate Euclidean metric in hyperplane P_u is defined as follows:

$$g_{ij} = \frac{1}{\kappa}\mathrm{Tr}(H_i H_j) = \frac{1}{\kappa}\Big(\delta_{ij} - \frac{u_i u_j}{(r+1)}\Big), \tag{3.1.12}$$

so we have $(\alpha^{(ik)}, \alpha^{(ik)}) = \alpha_\ell^{(ik)} \alpha_s^{(ik)} g^{\ell s} = \kappa \cdot |e^{(i)} - e^{(k)}|^2 = 2\kappa$. The overall scale κ is determined by comparing (3.1.4) and (3.1.12); in fact, it cancels out in all relations below, so we can set it equal to 1 for brevity. We choose the vector x defining positive ($(\alpha, x) > 0$) and negative ($(\alpha, x) < 0$) roots in the following way: $x = (x^1, x^2, \dots, x^{r+1}) \in P_u$, where $\sum_i x^i = 0$, and $x^i > x^j$ for $i < j$. Then roots $(e^{(k)} - e^{(l)})$ of the root system $\Phi(s\ell(r+1, \mathbb{C}))$ with $k < l$ form the system of positive roots $\{\alpha_+^{(kl)}\}$, while for $k > l$ we have the system of negative roots $\{\alpha_-^{(kl)}\}$:

$$\alpha_+^{(kl)} = (e^{(k)} - e^{(l)}), \quad \alpha_-^{(kl)} = -(e^{(k)} - e^{(l)}), \quad k < l. \tag{3.1.13}$$

The r simple roots, that form basis in root space $\mathfrak{V}_r = P_u$, are

$$\alpha^{(i)} = \alpha^{(i\,i+1)} = (e^{(i)} - e^{(i+1)}), \quad i = 1, \dots, r. \tag{3.1.14}$$

Indeed, all roots (3.1.13) are linear combinations of the simple roots (3.1.14) with integer coefficients n_i, where $n_i \geq 0$ for positive roots $\alpha_+^{(kl)}$ and $n_i \leq 0$ for negative roots $\alpha_-^{(kl)}$.

2. Root systems of algebras $so(2r+1, \mathbb{C})$ (B_r-series) and $so(2r, \mathbb{C})$ (D_r-series)

Algebra $so(n, \mathbb{C})$ is defined as a set of complex $n \times n$ matrices A obeying

$$A^{\mathsf{T}} \eta + \eta A = 0, \tag{3.1.15}$$

where η is symmetric nondegenerate real metric whose explicit form is a matter of convenience (algebras with different η's are isomorphic). In the first book (item 6 of Section I-3.3.1), we have chosen $\eta = I_n$. Here it is convenient to choose metric η as antidiagonal matrix whose nonzero matrix elements are equal to 1 and are placed at antidiagonal:

$$||\eta_{ij}|| = \begin{pmatrix} 0\,0 \dots 0\,1 \\ 0\,0 \dots 1\,0 \\ \vdots \; \vdots \;\; \vdots \;\; \vdots \; \vdots \\ 0\,1 \dots 0\,0 \\ 1\,0 \dots 0\,0 \end{pmatrix}, \quad \eta_{ij} = \delta_{i, n-j+1}. \tag{3.1.16}$$

With this metric, the condition (3.1.15) reads

$$A_{ij} = -A_{j'i'}, \tag{3.1.17}$$

where

$$i' = n - i + 1.$$

This condition says that matrix A is antisymmetric under reflection with respect to antidiagonal. The convenience of the definition of algebra $so(n,\mathbb{C})$ with metric (3.1.16) stems from the fact that its Cartan subalgebra is generated by diagonal matrices

$$H_k = e_{kk} - e_{k'k'}, \quad k = 1, \ldots, \left[\tfrac{n}{2}\right], \tag{3.1.18}$$

while the root generators of Cartan–Weyl basis in $so(n,\mathbb{C})$ read

$$E_{ij} = e_{ij} - e_{j'i'}, \quad E_{ij'} = e_{ij'} - e_{ji'}, \quad 1 \le i < j \le \left[\tfrac{n}{2}\right], \tag{3.1.19}$$

$$E_{ji} = e_{ji} - e_{i'j'}, \quad E_{j'i} = e_{j'i} - e_{i'j}, \quad 1 \le i < j \le \left[\tfrac{n}{2}\right], \tag{3.1.20}$$

where $[y]$ is integer part of y. For $n = 2r+1$ one adds extra generators to the set (3.1.19), (3.1.20):

$$E_{i,r+1} = e_{i,r+1} - e_{r+1,i'}, \quad E_{r+1,i} = e_{r+1,i} - e_{i',r+1} \quad (i = 1, \ldots, r). \tag{3.1.21}$$

Generators (3.1.18), (3.1.19), (3.1.20) and (3.1.21) fill in (in view of antisymmetry with respect to antidiagonal) the matrix of generators in a way schematically shown below:

$A \;=$ 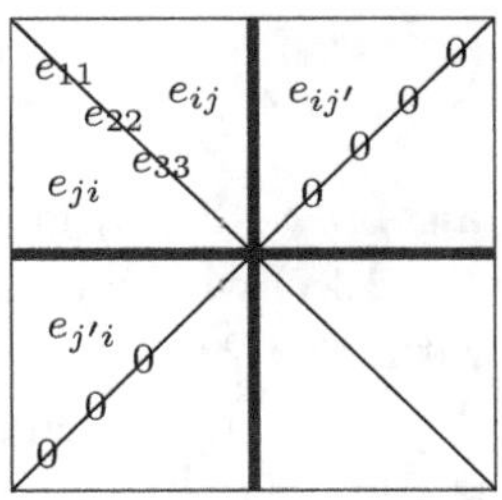

We emphasize that the places in matrix A shown here by two thick lines are absent for $n = 2r$, while in the case $n = 2r+1$ these places are filled with extra generators (3.1.21) (with exception of central element $A_{r+1,r+1} = 0$ on intersection of main diagonal and antidiagonal).

Commutation relations for generators (3.1.18), (3.1.19) and (3.1.20) read

$$[H_k, E_{ij}] = (e_k^{(i)} - e_k^{(j)})\, E_{ij}, \quad [H_k, E_{ij'}] = (e_k^{(i)} + e_k^{(j)})\, E_{ij'}, \quad i < j, \tag{3.1.22}$$

$$[H_k, E_{ji}] = -(e_k^{(i)} - e_k^{(j)})\, E_{ji}, \quad [H_k, E_{j'i}] = -(e_k^{(i)} + e_k^{(j)})\, E_{j'i}, \quad i < j. \tag{3.1.23}$$

They define the root vectors. For $n = 2r+1$ we have additionally

$$[H_k, E_{i,r+1}] = e_k^{(i)} E_{i,r+1}, \quad [H_k, E_{r+1,i}] = -e_k^{(i)} E_{r+1,i} \quad (i = 1, \dots, r). \tag{3.1.24}$$

Thus, in accordance with (3.1.22) and (3.1.23), roots α of algebra $so(2r, \mathbb{C})$ are as follows (see also Section I-6.2.3 of the first book):

$$\alpha = \{\alpha_{\pm}^{(i,j)} = (e^{(i)} \pm e^{(j)}), \quad -\alpha_{\pm}^{(i,j)} = -(e^{(i)} \pm e^{(j)})\}, \quad 1 \le i < j \le r. \tag{3.1.25}$$

Invariant metric in root space $\mathfrak{V}_r$ is defined in a standard way,

$$g_{ij} = \frac{1}{2\kappa} \mathrm{Tr}(H_i H_j) = \frac{1}{\kappa} \delta_{ij}, \tag{3.1.26}$$

so the root vectors (3.1.25) have one and the same length equal to $|\alpha_{\pm}^{(i,j)}| = \sqrt{2\kappa}$ (the scale κ is irrelevant for what follows). Roots $e^{(i)} \pm e^{(j)}$ $(i < j)$ are positive, while roots $-(e^{(i)} \pm e^{(j)})$ $(i < j)$ are negative. Accordingly, elements (3.1.19) are positive, and elements (3.1.20) are negative root generators. Simple roots that form basis in the root space of algebra $so(2r, \mathbb{C})$ are chosen as follows:

$$\alpha^{(i)} = e^{(i)} - e^{(i+1)} \quad (i = 1, \dots, r-1), \quad \alpha^{(r)} = e^{(r-1)} + e^{(r)}, \tag{3.1.27}$$

so that all roots of algebra $so(2r, \mathbb{C})$ given by (3.1.25) are linear combinations of simple roots (3.1.27) with integer coefficients.

In accordance with (3.1.24), in the case of algebra $so(2r+1, \mathbb{C})$ (B_r-series), the roots (3.1.25) are supplemented by root vectors $\pm e^{(k)}$ $(k = 1, \dots, r)$. Thus, the complete root system of $so(2r+1, \mathbb{C})$ consists of vectors (see also Section I-6.2.3 of the first book)

$$\alpha = \{\pm e^{(k)}, \alpha_{\pm}^{(i,j)} = (e^{(i)} \pm e^{(j)}), -\alpha_{\pm}^{(i,j)} = -(e^{(i)} \pm e^{(j)})\}, \; 1 \le i < j \le r, \tag{3.1.28}$$

where the lengths of roots are equal to $|\alpha_{\pm}^{(i,j)}| = \sqrt{2\kappa}$ and $|e^{(i)}| = \sqrt{\kappa}$ (we again use metric (3.1.26); the scale κ is again irrelevant). Simple roots of system (3.1.28) are

$$\alpha^{(i)} = e^{(i)} - e^{(i+1)} \quad (i = 1, \dots, r-1), \quad \alpha^{(r)} = e^{(r)}, \tag{3.1.29}$$

so all root vectors (3.1.28) of algebra $so(2r+1, \mathbb{C})$ are linear combinations of simple roots (3.1.29) with integer coefficients.

3. Root system of algebra $sp(2r, \mathbb{C})$ ($\mathbf{C}_r$-series)

Lie algebra $sp(2r, \mathbb{C})$ is defined (see Section I-6.2.3 of the first book) as a set of complex matrices A, obeying the condition (3.1.15), but now with *antisymmetric* (symplectic) nondegenerate real metric

$$\eta = \begin{pmatrix} 0 & I_r \\ -I_r & 0 \end{pmatrix}, \tag{3.1.30}$$

where 0 denotes zero $(r \times r)$ block, and I_r is unit $(r \times r)$ matrix. We write A in block form

$$A = \begin{pmatrix} X & Y \\ Z & W \end{pmatrix}, \tag{3.1.31}$$

where X, Y, Z, W are complex $r \times r$ matrices. Then the condition (3.1.15) with metric (3.1.30) gives

$$X = -W^{\mathsf{T}}, \quad Y^{\mathsf{T}} = Y, \quad Z^{\mathsf{T}} = Z, \tag{3.1.32}$$

so the Cartan–Weyl basis in $sp(2r, \mathbb{C})$ can be chosen as follows:

$$\begin{gathered} e_{i,i} - e_{r+i,r+i} \equiv H_i \ (i = 1, \ldots, r), \ e_{i,j} - e_{r+j,r+i} \equiv X_{ij} \ (i \neq j), \\ e_{i,r+j} + e_{j,r+i} \equiv Y_{ij}, \ e_{r+i,j} + e_{r+j,i} \equiv Z_{ij}, \quad (i, j = 1, \ldots, r). \end{gathered} \tag{3.1.33}$$

Diagonal elements $H_i (i = 1, \ldots, r)$ generate Cartan subalgebra in $sp(2r, \mathbb{C})$.

Making use of (3.1.33), we obtain commutation relations that determine root vectors:

$$\begin{gathered} [H_i,\, X_{kl}] = (e^{(k)} - e^{(l)})_i\, X_{kl}, \quad [H_i,\, Y_{kl}] = (e^{(k)} + e^{(l)})_i\, Y_{kl}, \\ [H_i,\, Z_{kl}] = -(e^{(k)} + e^{(l)})_i\, Z_{kl}. \end{gathered} \tag{3.1.34}$$

Thus, the root system of $sp(2r, \mathbb{C})$ is (see also section I-6.2.3 of the first book)

$$\pm(e^{(k)} - e^{(l)}) \ (k < l), \quad \pm(e^{(k)} + e^{(l)}) \ (k < l), \quad \pm 2\, e^{(k)}, \quad k, l = 1, \ldots, r. \tag{3.1.35}$$

Metric in root space $\mathfrak{V}_r$ reads

$$g_{ij} = \frac{1}{2\kappa}\, \mathrm{Tr}(H_i\, H_j) = \frac{1}{\kappa}\, \delta_{ij}. \tag{3.1.36}$$

Then the lengths of root vectors (3.1.35) are $|(e^{(k)} \pm e^{(l)})| = \sqrt{2\kappa}$ and $|2e^{(k)}| = 2\sqrt{\kappa}$ (the scale κ is again irrelevant for what follows). Roots

$$(e^{(k)} - e^{(l)}) \quad k < l, \quad (e^{(k)} + e^{(l)}) \quad k \leq l \tag{3.1.37}$$

are positive. The basis of simple roots is

$$\alpha^{(i)} = e^{(i)} - e^{(i+1)} \, (i = 1, \ldots, r-1), \, \alpha^{(r)} = 2\, e^{(r)}. \tag{3.1.38}$$

Problem 3.1.1. Check that all roots (3.1.35) are linear combinations of simple roots (3.1.38) with integer coefficients, which are nonnegative for positive roots (3.1.37).

Remark. As we discuss in Section I-6.2.4 of the first book, system of simple roots $\alpha^{(i)}$ $(i = 1, \ldots, r)$ of simple rank-r Lie algebra as well as its Cartan matrix

$$K_{ij} = 2\frac{(\alpha^{(i)}, \alpha^{(j)})}{(\alpha^{(j)})^2} \tag{3.1.39}$$

are graphically represented by Dynkin diagram (see Fig. 3.1.1). The ith vertex of diagram encodes simple root $\alpha^{(i)}$; arrow $>$ (or $<$) points to root of smaller length (if the vertices are connected by one line, their roots have equal lengths, and there is no arrow between these vertices); the number of lines between vertices i and j equals $(K_{ij} \cdot K_{ji})$. Explicit forms of Cartan matrices of Lie algebras $s\ell(r+1, \mathbb{C})$, $so(n, \mathbb{C})$ and $sp(2r, \mathbb{C})$ are given in Section I-6.2.3 of the first book.

3.1.2. Root systems of exceptional Lie algebras

In this chapter, we mostly concentrate on Lie algebras of classical series, leaving it to the reader to consider exceptional Lie algebras. Nevertheless, we give here a description of root systems of exceptional Lie algebras G_2, F_4, E_6, E_7 and E_8. Detailed analysis of the latter root systems is given in Section I-6.2.5 of the first book, and also in Refs. [2–4, 9]).

1. Root system of Lie algebra G_2

Rank of lie algebra G_2 equals 2, and its Dynkin diagram is shown in Fig. 3.1.1. This diagram encodes the Cartan matrix (3.1.39), which reads

$$||K_{ij}|| = \begin{pmatrix} 2 & -1 \\ -3 & 2 \end{pmatrix}. \tag{3.1.40}$$

Two-dimensional root space of algebra G_2 is conveniently chosen in the same way as in the case of algebra $s\ell(3, \mathbb{C})$, namely, it is a plane P_u in

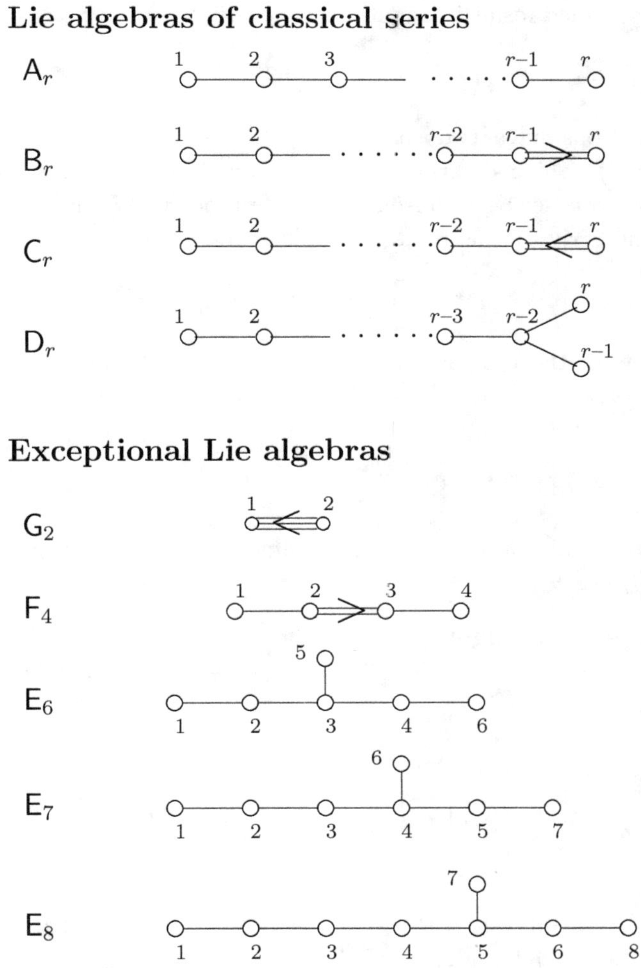

Fig. 3.1.1. Dynkin diagrams for simple, finite-dimensional Lie algebras.

three-dimensional Euclidean space, normal to vector $u = e^{(1)} + e^{(2)} + e^{(3)} = (1,1,1)$. Let us choose the short simple root of G_2 as follows: $\alpha^{(1)} = (e^{(1)} - e^{(2)})$. We search for the second simple root in the form

$$\alpha^{(2)} = \beta_1 e^{(1)} + \beta_2 e^{(2)} - (\beta_1 + \beta_2) e^{(3)} \in P_u.$$

Making use of the Cartan matrix, (3.1.40) and the property $(\alpha^{(1)})^2 = 2$, we find that vector $\alpha^{(2)}$ must obey two equations: $(\alpha^{(1)}, \alpha^{(2)}) = -3$ and

$(\alpha^{(2)})^2 = 6$, which result in equations for coefficients β_1, β_2:

$$\beta_2 = \beta_1 + 3, \quad \beta_1^2 + 3\beta_1 + 2 = 0.$$

These equations have two solutions ($\beta_1 = -2$, $\beta_2 = 1$) and ($\beta_1 = -1$, $\beta_2 = 2$), which determine two equivalent root systems (related by mirror reflection, see Section I-6.2.5 of the first book). We choose the former solution and arrive at the basis of simple roots of G_2:

$$\alpha^{(1)} = e^{(1)} - e^{(2)}, \quad \alpha^{(2)} = -2e^{(1)} + e^{(2)} + e^{(3)}. \tag{3.1.41}$$

All 12 roots of the root system $\Phi(\mathsf{G}_2)$ are obtained from (3.1.41) in the standard way, as admissible sums of simple roots (3.1.41) with integer coefficients,

$$\Phi(\mathsf{G}_2) = \{\pm(e^{(j)} - e^{(k)}), \ \pm(3e^{(i)} - e^{(1)} - e^{(2)} - e^{(3)})\}, \tag{3.1.42}$$

where $1 \leq j < k \leq 3$ and $i = 1, 2, 3$. Dimension of Lie algebra G_2 equals $\dim(\Phi(\mathsf{G}_2)) + 2 = 14$.

2. Root system of algebra F_4

Dynkin diagram shown in Fig. 3.1.1 determines Cartan matrix (3.1.39) of algebra F_4:

$$||K_{ij}|| = \begin{pmatrix} 2 & -1 & 0 & 0 \\ -1 & 2 & -2 & 0 \\ 0 & -1 & 2 & -1 \\ 0 & 0 & -1 & 2 \end{pmatrix}. \tag{3.1.43}$$

Note that by erasing the vertex 4 in diagram F_4, one obtains the diagram B_3 (see Fig. 3.1.1) for algebra $so(7,\mathbb{C})$. Therefore, algebra $so(7,\mathbb{C})$ is a subalgebra of algebra F_4, and root system $\Phi(\mathsf{F}_4)$ contains, as a subsystem, the root system $\Phi(so(7,\mathbb{C}))$ given by (3.1.28) and (3.1.29) with $r = 3$. As the first three simple roots of algebra F_4 we take the simple roots of algebra $so(7,\mathbb{C})$,

$$\alpha^{(1)} = e^{(1)} - e^{(2)}, \quad \alpha^{(2)} = e^{(2)} - e^{(3)}, \quad \alpha^{(3)} = e^{(3)},$$

and search for the fourth simple root of F_4 as vector $\alpha^{(4)} = \sum_{i=1}^{4} a_i e^{(i)} \in \mathbb{R}^4$, which must obey the following relations dictated by the Cartan matrix (3.1.43):

$$(\alpha^{(i)}, \alpha^{(4)}) = 0 \ (i = 1, 2), \quad (\alpha^{(4)})^2 = 1,$$
$$K_{34} = -1 \ \Rightarrow \ (\alpha^{(3)}, \alpha^{(4)}) = -1/2.$$

We obtain $a_1 = a_2 = a_3 = -1/2$, $a_4 = \pm 1/2$ and postulate the positive sign of a_4 (negative sign yields the root system which is mirror image of positive sign system with respect to $P_{e^{(4)}}$). In this way, we arrive at the following simple roots of algebra F_4:

$$\begin{gathered} \alpha^{(1)} = e^{(1)} - e^{(2)}, \quad \alpha^{(2)} = e^{(2)} - e^{(3)}, \quad \alpha^{(3)} = e^{(3)}, \\ \alpha^{(4)} = \tfrac{1}{2}(-e^{(1)} - e^{(2)} - e^{(3)} + e^{(4)}). \end{gathered} \tag{3.1.44}$$

The complete root system $\Phi(\mathsf{F}_4)$ consists of roots (3.1.28) (with $r = 3$) of algebra $so(7, \mathbb{C})$ and root vectors involving the simple root $\alpha^{(4)}$. As a result, complete root system of algebra F_4 is (see Section I-6.2.5 of the first book)

$$\Phi(\mathsf{F}_4) = \{\pm e^{(i)}; \ \pm e^{(i)} \pm e^{(j)} \ (i \neq j); \ \frac{1}{2}(\pm e^{(1)} \pm e^{(2)} \pm e^{(3)} \pm e^{(4)})\}, \tag{3.1.45}$$

where $i, j = 1, 2, 3, 4$ and signs $\pm$ are uncorrelated. In view of dimension of Cartan subalgebra, equal to 4, dimension of algebra F_4 is $\dim(\Phi(\mathsf{F}_4)) + 4 = (8 + 24 + 2^4) + 4 = 52$.

3. Root system of Lie algebra E_6

Dynkin diagram of algebra E_6 determines its Cartan matrix:

$$||K_{ij}|| = \begin{pmatrix} 2 & -1 & 0 & 0 & 0 & 0 \\ -1 & 2 & -1 & 0 & 0 & 0 \\ 0 & -1 & 2 & -1 & -1 & 0 \\ 0 & 0 & -1 & 2 & 0 & -1 \\ 0 & 0 & -1 & 0 & 2 & 0 \\ 0 & 0 & 0 & -1 & 0 & 2 \end{pmatrix}. \tag{3.1.46}$$

In analogy to the previous case, by erasing vertex 6 in the diagram E_6 shown in Fig. 3.1.1 one obtains diagram D_5 of algebra $so(10, \mathbb{C})$. So, algebra $so(10, \mathbb{C})$ is a subalgebra in E_6, and the root system $\Phi(\mathsf{E}_6)$ contains the subsystem $\Phi(so(10, \mathbb{C}))$ given by (3.1.25) and (3.1.27) with $r = 5$. We take simple roots (3.1.27) of $so(10, \mathbb{C})$ as the first five roots of E_6, and add the simple sixth root $\alpha^{(6)} = \sum_{i=1}^{6} a_i e^{(i)}$. In accordance with Cartan matrix (3.1.46), this root obeys $(\alpha^{(6)}, \alpha^{(i)}) = 0$ $(i = 1, 2, 3, 5)$, $(\alpha^{(6)}, \alpha^{(4)}) = -1$

and $(\alpha^{(6)})^2 = 2$. By choosing one of the two solutions for $\alpha^{(6)}$ (the other solution is again equivalent via reflection to the chosen one), we obtain the following simple roots of algebra E_6:

$$\alpha_i = e^{(i)} - e^{(i+1)} \ (i = 1, \ldots, 4), \ \alpha_5 = e^{(4)} + e^{(5)},$$

$$\alpha_6 = -\frac{e'}{2} + e^{(5)} + \beta e^{(6)}, \tag{3.1.47}$$

where $e' = \sum_{i=1}^{6} e^{(i)}$ and $\beta = (1+\sqrt{3})/2$. The complete root system $\Phi(\mathsf{E}_6)$ is (see Section I-6.2.5 of the first book):

$$\Big\{ \pm(e^{(i)} - e^{(j)}); \ \pm(e^{(i)} + e^{(j)}); \ \pm\left(\frac{e'}{2} + (\beta - 1)e^{(6)} - e^{(i)} - e^{(j)}\right);$$

$$\pm\left(\frac{e'}{2} + (\beta - 1)e^{(6)}\right); \ \pm\left(-\frac{e'}{2} + e^{(i)} + \beta e^{(6)}\right)\Big\}, \quad 1 \le i < j \le 5. \tag{3.1.48}$$

The number of all roots (3.1.48) equals to $\dim(\Phi(\mathsf{E}_6)) = 2(3\,C_5^2 + 1 + 5) = 72$. Therefore, $\dim(\mathsf{E}_6) = \dim(\Phi(\mathsf{E}_6)) + \mathrm{rank}(\mathsf{E}_6) = 78$.

4. Root systems of algebras E_7 and E_8

We construct first the root system $\Phi(\mathsf{E}_8)$ and then obtain from it the root system $\Phi(\mathsf{E}_7)$.

Let us find simple roots of algebra E_8. By erasing vertex 8 in diagram E_8 shown in Fig. 3.1.1 one obtains diagram D_7 for algebra $so(14, \mathbb{C})$. So, it is natural to choose the first seven simple roots $\alpha^{(1)}, \ldots, \alpha^{(7)}$ of algebra E_8 as simple roots (3.1.27) of $so(14, \mathbb{C})$: $\alpha^{(i)} = e^{(i)} - e^{(i+1)}$, $(i = 1, \ldots, 6)$ and $\alpha^{(7)} = e^{(6)} + e^{(7)}$. The remaining simple root is written as $\alpha^{(8)} = \sum_{i=1}^{8} a_i\, e^{(i)}$. Cartan matrix of E_8, as determined by its Dynkin diagram, is

$$||K_{ij}|| = \begin{pmatrix} 2 & -1 & 0 & 0 & 0 & 0 & 0 & 0 \\ -1 & 2 & -1 & 0 & 0 & 0 & 0 & 0 \\ 0 & -1 & 2 & -1 & 0 & 0 & 0 & 0 \\ 0 & 0 & -1 & 2 & -1 & 0 & 0 & 0 \\ 0 & 0 & 0 & -1 & 2 & -1 & -1 & 0 \\ 0 & 0 & 0 & 0 & -1 & 2 & 0 & -1 \\ 0 & 0 & 0 & 0 & -1 & 0 & 2 & 0 \\ 0 & 0 & 0 & 0 & 0 & -1 & 0 & 2 \end{pmatrix}. \tag{3.1.49}$$

Coefficients K_{i8} of matrix (3.1.49) dictate the following equations for the root $\alpha^{(8)}$:

$$(\alpha^{(i)}, \alpha^{(8)}) = 0 \quad (i = 1, \dots, 5, 7), \quad (\alpha^{(6)}, \alpha^{(8)}) = -1, \quad (\alpha^{(8)})^2 = 2. \tag{3.1.50}$$

These lead to equations for coefficients a_i:

$$a_1 = \dots = a_6 \equiv a,\ a_6 = -a_7,\ a_6 - a_7 = -1,\ 7\,a^2 + a_8^2 = 2.$$

They have two solutions: $a_i = -1/2$ for $i = 1, \dots, 6$, $a_7 = 1/2$ and $a_8 = \pm 1/2$. The conventional choice is $a_8 = +1/2$, see, e.g., Refs. [3, 4]. The choice $a_8 = -1/2$ is equivalent to $a_8 = +1/2$, since it corresponds to mirror reflection with respect to hyperplane normal to vector $e^{(8)}$. Thus, simple roots of algebra E_8 are

$$\begin{aligned} &\alpha^{(i)} = e^{(i)} - e^{(i+1)}\ (i = 1, \dots, 6),\ \alpha^{(7)} = e^{(6)} + e^{(7)}, \\ &\alpha^{(8)} = \frac{1}{2}\left[-(e^{(1)} + \dots + e^{(6)}) + (e^{(7)} + e^{(8)})\right] = -\frac{e}{2} + e^{(7)} + e^{(8)}, \end{aligned} \tag{3.1.51}$$

where $e = \sum_{i=1}^{8} e^{(i)}$. Now, by forming admissible linear combinations of simple root $\alpha^{(8)}$ given by (3.1.51) and all roots (3.1.25) of $so(14, \mathbb{C})$ one constructs the complete root system $\Phi(\mathsf{E}_8)$ (see Section I-6.2.5 of the first book):

$$\begin{aligned} \Phi(\mathsf{E}_8) = \Big\{ &\pm (e^{(i)} - e^{(j)});\ \pm(e^{(i)} + e^{(j)}); \pm\frac{e}{2};\ \pm\Big(\frac{e}{2} - (e^{(i)} + e^{(j)})\Big); \\ &\Big(\frac{e}{2} - (e^{(n)} + e^{(s)} + e^{(k)} + e^{(l)})\Big)\Big\},\ 1 \le i < j \le 8, \\ &1 \le n < s < k < l \le 8. \end{aligned} \tag{3.1.52}$$

The number of all roots (3.1.52) of algebra E_8 is

$$\dim(\Phi(\mathsf{E}_8)) = 2(C_8^2 + C_8^2 + 1 + C_8^2) + C_8^4 = 2(3 \cdot 28 + 1) + 70 = 240,$$

therefore $\dim(\mathsf{E}_8) = 240 + \mathrm{rank}(\mathsf{E}_8) = 248$.

Now, let us turn to the root system $\Phi(\mathsf{E}_7)$. We note that Dynkin diagram E_7 is obtained by removing vertex 1 from Dynkin diagram E_8 (see Fig. 3.1.1). We thus remove simple root $\alpha^{(1)}$ that corresponds to vertex 1 from the set (3.1.51) and obtain the set of simple roots of algebra E_7:

$$\alpha^{(i)} = e^{(i+1)} - e^{(i+2)},\ \alpha^{(6)} = e^{(6)} + e^{(7)},\ \alpha^{(7)} = -\frac{e}{2} + e^{(7)} + e^{(8)}, \tag{3.1.53}$$

where $i = 1, \dots, 5$ and $e = \sum_{k=1}^{8} e^{(k)}$. Here, simple roots of algebra E_7 are embedded in eight-dimensional space $\mathbb{R}^8$ and belong to seven-dimensional hyperplane V normal to vector $v = e^{(1)} + e^{(8)}$. Then we can either make use of a remarkable property $\Phi(\mathsf{E}_7) = \Phi(\mathsf{E}_8) \cap V$ (see Ref. [4]) or proceed in the same way as in the case of algebra E_8. In either way, we arrive at the following complete root system $\Phi(\mathsf{E}_7) \subset \mathbb{R}^8$ (see Section I-6.2.5 of the first book):

$$\begin{aligned}\Phi(\mathsf{E}_7) = \Big\{ & \pm(e^{(8)} - e^{(1)});\ \pm(e^{(i)} - e^{(j)}),\ \pm(e^{(i)} + e^{(j)}),\ 2 \le i < j \le 7; \\ & \pm\left(\frac{e}{2} - e^{(k)} - e^{(8)}\right),\ \pm\left(\frac{e}{2} - e^{(1)} - e^{(k)}\right),\ k = 2, \ldots, 7; \\ & \pm\left(\frac{e}{2} - (e^{(i)} + e^{(j)} + e^{(k)} + e^{(8)})\right),\quad 2 \le i < j < k \le 7 \Big\}. \end{aligned} \tag{3.1.54}$$

In accordance with (3.1.54), the number of roots of algebra E_7 equals

$$\dim(\Phi(\mathsf{E}_7)) = 2(C_6^2 + C_6^2 + 1 + 6 + 6 + C_6^3) = 2(2 \cdot 15 + 1 + 12 + 20) = 126,$$

and the dimension of this algebra is $\dim(\mathsf{E}_7) = 126 + 7 = 133$.

Problem 3.1.2. Show that by removing the root $\alpha^{(1)}$ from the set (3.1.53) of simple roots of algebra E_7, one obtains the set of simple roots of algebra E_6:

$$\alpha^{(i)} = e^{(i+2)} - e^{(i+3)},\ \alpha^{(5)} = e^{(6)} + e^{(7)},\ \alpha^{(6)} = -\frac{e}{2} + e^{(7)} + e^{(8)}, \tag{3.1.55}$$

where $i = 1, \ldots, 4$. Extract the roots system $\Phi(\mathsf{E}_6)$ from the root system $\Phi(\mathsf{E}_7)$, by making use of the property $\Phi(\mathsf{E}_6) = \Phi(\mathsf{E}_7) \cap V'$ (see Ref: [4]), where $V' \subset \mathbb{R}^8$ is the vector space normal to vectors $v = e^{(1)} + e^{(8)}$ and $v' = e^{(1)} + e^{(2)} + 2e^{(8)}$.

3.1.3. Weyl group. Dual root systems

1. Weyl group

Consider the system $\Phi(\mathcal{A})$ of all roots of a simple Lie algebra $\mathcal{A}$. For two vectors α, β belonging to root space $\mathfrak{V}_r$ we define reflection of vector β with respect to plane P_α normal to vector α:

$$\sigma_\alpha(\beta) = \beta - \frac{2\,(\beta, \alpha)}{(\alpha, \alpha)}\,\alpha. \tag{3.1.56}$$

In accordance with the general theory (see Section I-6.2.1 of the first book), for two roots $\alpha, \beta \in \Phi(\mathcal{A})$ the combination $\frac{2\,(\beta,\alpha)}{(\alpha,\alpha)}$ is an integer, and the reflected vector $\sigma_\alpha(\beta)$ is again a root. Thus, the root system $\Phi(\mathcal{A})$ is invariant under all reflections (3.1.56), and operators σ_α, with $\alpha \in \Phi(\mathcal{A})$, are called Weyl reflections and generate a group W. Since the root system is finite, group W is finite.

Definition 3.1.1. Group W generated by all reflections (3.1.56) with $\alpha \in \Phi(\mathcal{A})$ is called *Weyl group.*

Hyperplanes P_α normal to roots α, divide r-dimensional root space $\mathfrak{V}_r$ into nonintersecting cones, each of which is called *Weyl chamber.* Weyl group W takes one Weyl chamber to another, so W acts in the set of Weyl chambers. This action is transitive; the latter fact follows from Corollary 3.4.4 (see below).

Problem 3.1.3. Show that the reflections (3.1.56) obey an identity $\sigma_\alpha \sigma_\beta \sigma_\alpha = \sigma_{\sigma_\alpha(\beta)}$.

We show below that the whole Weyl group W is generated, in fact, by reflections $\sigma_i := \sigma_{\alpha^{(i)}}$, where $\alpha^{(i)}$ are simple roots. This can be understood as follows. From the relation $\sigma_\alpha \sigma_\beta \sigma_\alpha = \sigma_{\sigma_\alpha(\beta)}$ and Definition 3.1.1 it results that Weyl group contains all reflections σ_α, where vectors α are obtained by reflections σ_i of simple roots $\alpha^{(j)}$:

$$\alpha = \sigma_{i_1} \cdot \sigma_{i_2} \cdots \sigma_{i_k}(\alpha^{(i)}). \tag{3.1.57}$$

These vectors form the whole root system $\Phi(\mathcal{A})$, and not its part, otherwise algebra $\mathcal{A}$ would not be simple. The latter fact is not obvious, so we give another proof of completeness of the set of generators σ_i in group W, which is based on explicit representation of W as group of operators in root space $\mathfrak{V}_r$.

Proposition 3.1.1. *Weyl group W of root system $\Phi(\mathcal{A})$ is generated by reflections $\sigma_i := \sigma_{\alpha^{(i)}}$ associated with simple roots $\alpha^{(i)}$ of algebra $\mathcal{A}$. Generators σ_i of group W obey the following relations:*

$$\sigma_i^2 = e, \quad (\sigma_i \cdot \sigma_j)^{m_{ij}} = e, \tag{3.1.58}$$

where e is unit element in W and exponents m_{ij} are determined by Dynkin diagram $\mathcal{D}$ of algebra $\mathcal{A}$ (see Fig. 3.1.1) : if vertices $\alpha^{(i)}$ and $\alpha^{(j)}$ of diagram $\mathcal{D}$ are connected by $0, 1, 2, 3$ lines, then $m_{ij} = 2, 3, 4, 6$, respectively.

Proof. We give the proof for Lie algebras of classical series, and leave to the reader to consider the case of exceptional algebras.

(1) We begin with Weyl group of root system $\Phi(s\ell(r+1,\mathbb{C}))$. Any vector μ belonging to the root space $P_u \subset \mathbb{R}^{r+1}$ of algebra $s\ell(r+1,\mathbb{C})$ (A_r-series, see item 1 of Section 3.1.1) can be expanded in basis vectors $e^{(s)} \in \mathbb{R}^{r+1}$:

$$\mu = \sum_{s=1}^{r+1} \mathsf{m}_s \, e^{(s)}, \;\; (\mu, u) = 0 \;\Rightarrow\; \sum_{s=1}^{r+1} \mathsf{m}_s = 0. \tag{3.1.59}$$

Making use of the formula (3.1.56) and expression (3.1.11) for roots $\alpha^{(ik)} \in \Phi(s\ell(r+1,\mathbb{C}))$, we find reflected vector

$$\sigma_{\alpha^{(ik)}}(\mu) = \mu - (\mathsf{m}_i - \mathsf{m}_k)(e^{(i)} - e^{(k)}) = \sum_{\substack{s=1 \\ s \neq k,i}}^{r+1} \mathsf{m}_s \, e^{(s)} + \mathsf{m}_k e^{(i)} + \mathsf{m}_i e^{(k)}. \tag{3.1.60}$$

By comparing (3.1.59) and (3.1.60), one finds that vectors μ and $\sigma_{\alpha^{(ik)}}(\mu)$ differ by transposition of coordinates m_i and m_k, so group W generated by reflections $\sigma_{\alpha^{(ik)}}$ is isomorphic to symmetric group S_{r+1} of permutations of components m_i. In view of (3.1.60), reflections $\sigma_i = \sigma_{\alpha^{(i)}}$ ($i = 1,\dots,r$) associated with simple roots $\alpha^{(i)} = e^{(i)} - e^{(i+1)}$ are neighboring (adjacent) transpositions

$$\sigma_i \cdot (\mathsf{m}_1, \dots, \mathsf{m}_i, \mathsf{m}_{i+1}, \dots, \mathsf{m}_{r+1}) = (\mathsf{m}_1, \dots, \mathsf{m}_{i+1}, \mathsf{m}_i, \dots, \mathsf{m}_{r+1}). \tag{3.1.61}$$

Neighboring transpositions σ_i generate the whole group $W = S_{r+1}$ (see Section I-1.1.4 of the first book); these are *Coxeter generators*. In terms of generators σ_i, defining relations for group S_{r+1} follow from (3.1.61):

$$\begin{gathered} \sigma_i \cdot \sigma_{i+1} \cdot \sigma_i = \sigma_{i+1} \cdot \sigma_i \cdot \sigma_{i+1} \quad (i = 1,\dots,r-1), \\ \sigma_i \cdot \sigma_j = \sigma_j \cdot \sigma_i, \quad (|i-j| > 1), \quad \sigma_i^2 = e, \quad (i = 1,\dots,r), \end{gathered} \tag{3.1.62}$$

where e is identical permutation. The relations (3.1.62) coincide with relations (3.1.58), written for Lie algebras of A_r-series.

(2) Root system of algebra $so(2r+1,\mathbb{C})$ (B_r-series) consists of vectors (3.1.28). Any vector μ in r-dimensional root space $\mathfrak{V}_r = \mathbb{R}^r$ is expanded in

basis vectors $e^{(s)}$:

$$\mu = \sum_{s=1}^{r} \mathsf{m}_s\, e^{(s)} \in \mathfrak{V}_r. \tag{3.1.63}$$

Action of elements $\sigma_\alpha \in W$ associated with roots $\alpha = (e^{(i)} - e^{(k)}) \in \Phi(so(2r+1,\mathbb{C}))$ is given by the formula analogous to (3.1.60), while for other roots $\{\pm(e^{(i)} + e^{(k)}), \pm e^{(i)}\} \in \Phi(so(2r+1,\mathbb{C}))$ we find that

$$\begin{aligned} \sigma_{\pm(e^{(i)}+e^{(k)})}(\mu) &= \mu - (\mathsf{m}_i + \mathsf{m}_k)(e^{(i)} + e^{(k)}) \\ &= \sum_{\substack{s=1 \\ s\neq k,i}}^{r} \mathsf{m}_s\, e^{(s)} - \mathsf{m}_k e^{(i)} - \mathsf{m}_i e^{(k)}, \end{aligned} \tag{3.1.64}$$

$$\sigma_{\pm e^{(i)}}(\mu) = \mu - 2\,\mathsf{m}_i\, e^{(i)} = \sum_{\substack{s=1 \\ s\neq i}}^{r} \mathsf{m}_s\, e^{(s)} - \mathsf{m}_i e^{(i)}. \tag{3.1.65}$$

Thus, Weyl group W of algebra $so(2r+1,\mathbb{C})$ is isomorphic to extension of group S_r of permutations of components m_i by sign flips of any of the coordinates m_i. Reflections $\sigma_i = \sigma_{\alpha^{(i)}}$ associated with first $(r-1)$ simple roots $\alpha^{(i)} = e^{(i)} - e^{(i+1)}$ transpose neighboring coordinates m_i and m_{i+1},

$$\begin{gathered} \sigma_i \cdot (\mathsf{m}_1, \ldots, \mathsf{m}_i, \mathsf{m}_{i+1}, \ldots, \mathsf{m}_r) = (\mathsf{m}_1, \ldots, \mathsf{m}_{i+1}, \mathsf{m}_i, \ldots, \mathsf{m}_r), \\ i = 1, \ldots, r-1, \end{gathered} \tag{3.1.66}$$

and hence generate subgroup $S_r \subset W$, while reflection $\sigma_r := \sigma_{\alpha^{(r)}}$ associated with the last simple root $\alpha^{(r)} = e^{(r)}$, in view of (3.1.65), flips sign of the last component m_r:

$$\sigma_r \cdot (\mathsf{m}_1, \ldots, \mathsf{m}_{r-1}, \mathsf{m}_r) \; = \; (\mathsf{m}_1, \ldots, \mathsf{m}_{r-1}, -\mathsf{m}_r). \tag{3.1.67}$$

Combined with transformations (3.1.66), the latter operation flips signs of any number of components μ_i. Thus, generators σ_i associated with all simple roots of algebra $so(2r+1,\mathbb{C})$ indeed generate the whole Weyl group W. Structure relations in terms of generators σ_i follow from (3.1.66) and (3.1.67); they are given by (3.1.62) for the first $(r-1)$ generators σ_i $(i = 1, \ldots, r-1)$, while the last generator $\sigma_r \in W$ obeys

$$\sigma_r \cdot \sigma_{r-1} \cdot \sigma_r \cdot \sigma_{r-1} \; = \; \sigma_{r-1} \cdot \sigma_r \cdot \sigma_{r-1} \cdot \sigma_r, \quad \sigma_r^2 = e. \tag{3.1.68}$$

Relations (3.1.68) and (3.1.62) coincide with relations (3.1.58) written for Lie algebras of B_r-series.

(3) Root system of algebra $sp(2r,\mathbb{C})$ (C_r-series) consists of vectors (3.1.35). We compare root systems $\Phi(so(2r+1,\mathbb{C})$ and $\Phi(sp(2r,\mathbb{C}))$ and find that the action, on vector (3.1.63), of reflections σ_α associated with roots $\alpha = \pm(e^{(i)} \pm e^{(k)}) \in \Phi(sp(2r,\mathbb{C}))$ transforms components m_i in the same way as in (3.1.60) and (3.1.64), while reflections σ_α associated with roots $\alpha = \pm 2e^{(i)}$ coincide with operations (3.1.65). Thus, Weyl group of algebra $sp(2r,\mathbb{C})$ (C_r-series) is isomorphic to Weyl group of algebra $so(2r+1,\mathbb{C})$ (B_r-series). Generators $\sigma_i = \sigma_{\alpha^{(i)}}$ associated with simple roots (3.1.38) act on μ in accordance with (3.1.66), (3.1.67), and again generate the whole Weyl group of algebra $sp(2r,\mathbb{C})$. Structure relations for generators σ_i are the same as in the case of $so(2r+1,\mathbb{C})$; they have the form (3.1.58) specified to Lie algebras of C_r-series.

(4) Root systems of algebras $so(2r,\mathbb{C})$ (D_r-series) is given in (3.1.25). For roots $\alpha = (e^{(i)} - e^{(k)}) \in \Phi(so(2r,\mathbb{C}))$, reflection of vector μ given by (3.1.63) is analogous to (3.1.60), while for roots $\alpha = \pm(e^{(i)} + e^{(k)}) \in \Phi(so(2r,\mathbb{C}))$ the reflections are given by (3.1.64). Therefore, Weyl group W consists of all permutations of components m_i plus sign flips of even number of components m_i. For $i = 1,\dots,(r-1)$, reflections $\sigma_i = \sigma_{\alpha^{(i)}}$, where $\alpha^{(i)}$ are simple roots (3.1.27), give neighboring transpositions (3.1.66) and generate subgroup $S_r \subset W$, while formula (3.1.64) shows that the reflection $\sigma_r := \sigma_{\alpha^{(r)}}$, associated with the last simple root $\alpha^{(r)} = e^{(r-1)} + e^{(r)}$, gives

$$\sigma_r \cdot (\mathsf{m}_1,\dots,\mathsf{m}_{r-2},\mathsf{m}_{r-1},\mathsf{m}_r) = (\mathsf{m}_1,\dots,\mathsf{m}_{r-2},-\mathsf{m}_{r-1},-\mathsf{m}_r). \quad (3.1.69)$$

Combined with transpositions (3.1.66), the latter flips signs of any even number of components m_i. So, reflections σ_i associated with simple roots (3.1.27) generate the whole Weyl group W of algebra $so(2r,\mathbb{C})$.

Structure relations for generators σ_i follow from (3.1.66) and (3.1.69), they are given by (3.1.62) for the first $(r-1)$ generators σ_i ($i = 1,\dots,r-1$), while for the last generator $\sigma_r \in W$ we have

$$\sigma_r \cdot \sigma_{r-1} = \sigma_{r-1} \cdot \sigma_r, \quad \sigma_r \cdot \sigma_{r-2} \cdot \sigma_r = \sigma_{r-2} \cdot \sigma_r \cdot \sigma_{r-2}, \quad \sigma_r^2 = e. \quad (3.1.70)$$

Relations (3.1.70) and (3.1.62) coincide with (3.1.58) written for algebras of D_r-series.

□

Proposition 3.1.2. *Reflection $\sigma_{\alpha^{(i)}}$ which acts on roots α as given by (3.1.56), namely,*

$$\alpha \to \sigma_{\alpha^{(i)}}(\alpha) = \alpha - \frac{2(\alpha, \alpha^{(i)})}{(\alpha^{(i)}, \alpha^{(i)})}\alpha^{(i)}, \tag{3.1.71}$$

permutes positive roots α different from simple root $\alpha^{(i)}$ whereas for $\alpha = \alpha^{(i)}$ one has $\sigma_{\alpha^{(i)}}(\alpha^{(i)}) = -\alpha^{(i)}$.

Proof. Let root $\alpha = \sum_{j=1}^{r} m_j \alpha^{(j)}$ be positive, so that $m_j \geq 0$ for all j. Then

$$\sigma_{\alpha^{(i)}}(\alpha) = \left(m_i - \frac{2(\alpha, \alpha^{(i)})}{(\alpha^{(i)}, \alpha^{(i)})}\right)\alpha^{(i)} + \sum_{j \neq i} m_j \alpha^{(j)}. \tag{3.1.72}$$

Since $\alpha^{(j)}$ are simple roots, all coefficients in the expansion of root $\sigma_{\alpha^{(i)}}(\alpha)$ in $\alpha^{(j)}$ must have one and the same sign (see discussion after (3.1.6)). If $m_i \geq \frac{2(\alpha,\alpha^{(i)})}{(\alpha^{(i)},\alpha^{(i)})}$, then all coefficients in the expansion of root $\sigma_{\alpha^{(i)}}(\alpha)$ are nonnegative, and hence the root $\sigma_{\alpha^{(i)}}(\alpha)$ is positive. In this case, the operation $\sigma_{\alpha^{(i)}}$ transforms positive root to positive root, i.e., it is a permutation of positive roots. If, on the other hand, $m_i < \frac{2(\alpha,\alpha^{(i)})}{(\alpha^{(i)},\alpha^{(i)})}$, then all other coefficients must obey $m_k \leq 0$, and because the root α is positive, we have $m_k = 0$ ($\forall k \neq i$). Then $\alpha = \alpha^{(i)}$ ($m_i = 1$ since there are no multiple roots) and we obtain from (3.1.71) that $\sigma_{\alpha^{(i)}}(\alpha^{(i)}) = -\alpha^{(i)}$. □

2. Dual root systems. Co-roots

Given a root system $\Phi(\mathcal{A})$ of a simple Lie algebra $\mathcal{A}$, one defines another system $\Phi(\mathcal{A}^\vee)$, whose vectors are

$$\alpha^\vee = \frac{\alpha}{\alpha^2} \quad \Leftrightarrow \quad \alpha = \frac{\alpha^\vee}{(\alpha^\vee)^2}, \tag{3.1.73}$$

where $\alpha \in \Phi(\mathcal{A})$, $\alpha^2 = (\alpha, \alpha)$. Vectors $\alpha^\vee$ are called co-roots of Lie algebra $\mathcal{A}$. The set of vectors $\alpha^\vee$ satisfies all requirements for root system. Indeed, for any $\alpha^\vee, \beta^\vee \in \Phi(\mathcal{A}^\vee)$, we have

$$2\frac{(\alpha^\vee, \beta^\vee)}{(\beta^\vee)^2} = 2\frac{(\alpha, \beta)}{(\alpha)^2} \in \mathbb{Z}. \tag{3.1.74}$$

Also, since directions of vectors α and $\alpha^\vee$ coincide, the Weyl chambers of systems $\Phi(\mathcal{A})$ and $\Phi(\mathcal{A}^\vee)$ also coincide. In fact, Weyl reflections of the mutually dual systems agree:

$$\alpha \to \alpha - 2\frac{(\alpha,\beta)}{(\beta^2)}\beta \quad \Leftrightarrow \quad \alpha^\vee \to \alpha^\vee - 2\frac{(\alpha^\vee,\beta^\vee)}{(\beta^{\vee 2})}\beta^\vee,$$

so Weyl groups of systems $\Phi(\mathcal{A})$ and $\Phi(\mathcal{A}^\vee)$ are isomorphic. We now show that the dual system $\Phi(\mathcal{A}^\vee)$ is actually a root system of some Lie algebra $\mathcal{A}^\vee$, which is called *dual Lie algebra*. In other words, co-roots of Lie algebra $\mathcal{A}$ are roots of dual Lie algebra $\mathcal{A}^\vee$, and vice versa. Since $(\alpha^\vee)^\vee = \alpha$, we have $(\mathcal{A}^\vee)^\vee = \mathcal{A}$. Concretely, the following relationships hold.

Proposition 3.1.3. *The duality relations between simple Lie algebras are*

$$s\ell(r+1,\mathbb{C})^\vee = s\ell(r+1,\mathbb{C}),$$

$$so(2r,\mathbb{C})^\vee = so(2r,\mathbb{C}), \tag{3.1.75}$$

$$so(2r+1,\mathbb{C})^\vee = sp(2r,\mathbb{C}), \tag{3.1.76}$$

$$\mathsf{G}_2^\vee = \mathsf{G}_2, \quad \mathsf{F}_4^\vee = \mathsf{F}_4, \quad \mathsf{E}_r^\vee = \mathsf{E}_r \quad (r = 6,7,8), \tag{3.1.77}$$

i.e., most of simple Lie algebras are selfdual, with the exception of algebras $sp(2r,\mathbb{C})$ and $so(2r+1,\mathbb{C})$, which are dual to each other. Simple roots $\alpha^{(i)}$ $(i = 1,\ldots,r)$ of Lie algebra $\mathcal{A}$ are related to simple roots $\alpha^{(i)\vee}$ of dual algebra $\mathcal{A}^\vee$ by duality transformation

$$(\alpha^{(i)})^\vee = \frac{\alpha^{(i)}}{(\alpha^{(i)})^2}. \tag{3.1.78}$$

Proof. If all roots α of algebra $\mathcal{A}$ have one and the same length, the co-roots (3.1.73) also have the same length, and the root system of algebra $\mathcal{A}^\vee$ is obtained from root system of algebra $\mathcal{A}$ by scaling (multiplication of all roots by one and the same factor $1/(\alpha)^2$). Hence, in this case, Lie algebra $\mathcal{A}^\vee$ is identical to algebra $\mathcal{A}$. One can show that Lie algebras $s\ell(r+1,\mathbb{C})$, $so(2r,\mathbb{C})$, E_6, E_7 and E_8, which have simply laced Dynkin diagrams, have one and the same length of all roots and therefore these algebras are self-dual.

It remains to consider algebras whose roots have different lengths. Let us check duality relation (3.1.76). All roots of $so(2r,\mathbb{C})$ are given in (3.1.28), so the co-roots (3.1.73) are

$$\pm\frac{1}{\kappa}e^{(i)}, \quad \alpha_\pm^{(i,j)} = \frac{1}{2\kappa}(e^{(i)} \pm e^{(j)}), \quad -\alpha_\pm^{(i,j)} = -\frac{1}{2\kappa}(e^{(i)} \pm e^{(j)}), \quad i<j. \tag{3.1.79}$$

Modulo scale factor $\frac{1}{2\kappa}$ (which can always be redefined by renormalization of generators), this root system coincides with root system (3.1.35) of algebra

$sp(2r, \mathbb{C})$, therefore $so(2r+1, \mathbb{C})^\vee = sp(2r, \mathbb{C})$. We leave it to the reader to check selfduality (3.1.77) of exceptional Lie algebras G_2 and F_4, which have roots of different lengths (root systems of exceptional Lie algebras are given in Section 3.1.2 and Section I-6.2.5 of the first book).

Let us come to relation (3.1.78) for simple roots of systems $\Phi(\mathcal{A})$ and $\Phi(\mathcal{A}^\vee)$. We do not give the general proof of this relation (see Ref. [9]). It suffices to check (3.1.78) explicitly, by making use of concrete expressions (3.1.14), (3.1.27), (3.1.29) and (3.1.38) for simple roots of algebras $s\ell(r+1, \mathbb{C})$, $so(2r, \mathbb{C})$, $so(2r+1, \mathbb{C})$ and $sp(2r, \mathbb{C})$. This check, as well as the corresponding check for exceptional Lie algebras, is left to the reader. □

Corollary 3.1.4. *Since all simple roots* $(\alpha^{(i)})^\vee$ *of dual algebra* $\mathcal{A}^\vee$ *are given by* (3.1.78), *all roots* $\alpha^\vee$ *of algebra* $\mathcal{A}^\vee$ *are linear combinations*

$$\alpha^\vee = \frac{\alpha}{(\alpha)^2} = \sum_{i=1}^{r} \mathsf{n}_i \frac{\alpha^{(i)}}{(\alpha^{(i)})^2} \tag{3.1.80}$$

with integer coefficients $\mathsf{n}_i \in \mathbb{Z}$.

Corollary 3.1.5. *In accordance with* (3.1.78), *we find for Cartan matrix* (3.1.39) *of the dual algebra* $\mathcal{A}^\vee$

$$K^\vee_{ij} = 2\frac{(\alpha^{(i)\vee}, \alpha^{(j)\vee})}{(\alpha^{(j)\vee})^2} = 2\frac{(\alpha^{(i)}, \alpha^{(j)})}{(\alpha^{(i)})^2} = K_{ji}. \tag{3.1.81}$$

Thus, Cartan matrix of dual algebra $\mathcal{A}^\vee$ is transposed Cartan matrix of original algebra $\mathcal{A}$. This enables one to identify dual Lie algebras by making use of Dynkin diagrams of Fig. 3.1.1, Namely, we have

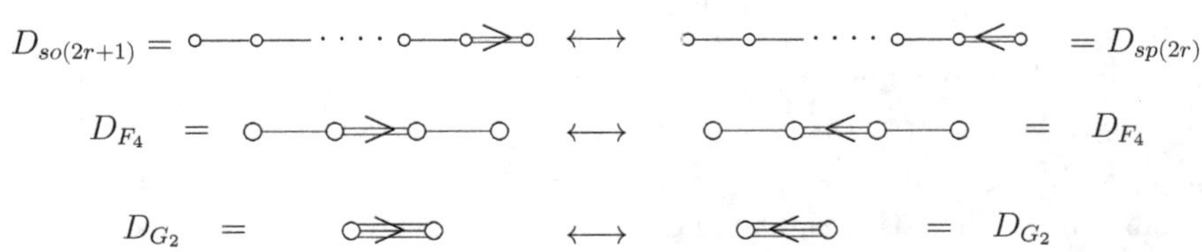

This reiterates the result summarized in Proposition 3.1.3.

3.2. Representations and Weights

Let H_i $(i = 1, \ldots, r)$ be Cartan generators of simple complex Lie algebra $\mathcal{A}$ of rank r, and T be a complex representation of algebra $\mathcal{A}$ in space $\mathcal{V}$.

Consider an eigenvector $|\mu\rangle \in \mathcal{V}$ of the commuting set of operators $T(H_i)$:

$$T(H_i)\,|\mu\rangle = \mu_i|\mu\rangle \quad (i = 1, \ldots, r). \tag{3.2.1}$$

r-component (generally, complex) vector $\mu = (\mu_1, \mu_2, \ldots, \mu_r)$ is called *weight* of representation T, and also weight of vector $|\mu\rangle$. In what follows, we omit symbol T and understand Cartan–Weyl generators $\{H_i, E_\alpha\}$ as operators acting in vector space $\mathcal{V}$ of representation T.

To develop the theory of irreducible representations of simple Lie algebras, we have to understand the properties of weights. Let us study these properties.

A. Every representation has at least one weight. Indeed, operator H_1 in space $\mathcal{V}$ has at least one eigenvector whose eigenvalue we denote by μ_1. We denote the subspace of eigenvectors $|\mu_1\rangle$: $H_1|\mu_1\rangle = \mu_1|\mu_1\rangle$ with eigenvalue μ_1 by $\mathcal{V}_1$. Since

$$H_1 H_2|\mu_1\rangle = H_2 H_1|\mu_1\rangle = \mu_1 H_2|\mu_1\rangle, \tag{3.2.2}$$

we have $H_2|\mu_1\rangle \in \mathcal{V}_1$. Operator H_2 has at least one eigenvector $|\mu_1, \mu_2\rangle \in \mathcal{V}_1$ such that $H_2|\mu_1, \mu_2\rangle = \mu_2|\mu_1, \mu_2\rangle$. We now have a subspace in $\mathcal{V}_1$ of eigenvectors of H_2 with eigenvalue μ_2, which we denote by $\mathcal{V}_2$. By continuing this process, we arrive at subspace $\mathcal{V}_r$ of all eigenvectors $|\mu\rangle$ of the set of operators $H_1, H_2, \ldots, H_r$, which have a set of eigenvalues $\mu = (\mu_1, \mu_2, \ldots, \mu_r)$. This is a subspace associated with weight μ.

B. Eigenvectors $|\mu\rangle$ of Cartan operators $H_1, H_2, \ldots, H_r$, belonging to the space $\mathcal{V}$ of representation T, are linear independent for different weights μ. To see this, we consider eigenvectors $|u_1\rangle, |u_2\rangle, \ldots, |u_\ell\rangle$:

$$H_i|u_k\rangle = \mu_i^{(k)}|u_k\rangle.$$

Let $|u\rangle$ be yet another eigenvector of weight $\mu \neq \mu^{(k)}$ $(\forall k)$ such that $|u\rangle = \sum_k |u_k\rangle$. We act on the latter equality by the operator $\prod_k \sum_i \xi^i(H_i - \mu_i^{(k)}) = (\sum_i \xi^i(H_i - \mu_i^{(1)})) \cdot (\sum_i \xi^i(H_i - \mu_i^{(2)})) \ldots$, where ξ^i are arbitrary coefficients. Then the right-hand side vanishes, while the left-hand side is zero for all ξ^i if and only if $|u\rangle = 0$. This proves linear independence of eigenvectors of different weights. Note that the latter property implies that the maximum number of different weights in a given representation is $N = \dim \mathcal{V}$.

It is worth emphasizing that weights may be degenerate, i.e., there may exist several linear independent eigenvectors with one and the same weight. In that case, it is useful to keep in mind the following property.

Let $|u_0\rangle, |u_1\rangle, \ldots, |u_\ell\rangle \in \mathcal{V}$ be eigenvectors of operators $\{H_i\}$ with weights $\mu^{(0)}, \mu^{(1)}, \ldots, \mu^{(\ell)}$, and let them be linear dependent:

$$|u_0\rangle = |u_1\rangle + |u_2\rangle + \cdots + |u_\ell\rangle. \tag{3.2.3}$$

Then, without loss of generality, every vector on the right-hand side has the same weight $\mu^{(0)}$ as vector $|u_0\rangle$, $\mu^{(1)} = \mu^{(2)} = \cdots = \mu^{(0)}$. Indeed, since vectors with different weights are linear independent, there are vectors on the right-hand side of (3.2.3) whose weight is $\mu^{(0)}$. Let all these vectors be $|u_1\rangle, |u_2\rangle, \ldots, |u_k\rangle$, where $k \leq \ell$. We move them to the left-hand side of (3.2.3):

$$|u_0\rangle - |u_1\rangle - \cdots - |u_k\rangle = |u_{k+1}\rangle + |u_{k+2}\rangle + \cdots + |u_\ell\rangle. \tag{3.2.4}$$

The left-hand side is now a new vector $|\mu^{(0)}\rangle \equiv |u_0\rangle - \sum_{i=1}^{k} |u_i\rangle$ of weight $\mu^{(0)}$. Since there are no longer vectors with weight $\mu^{(0)}$ on the right-hand side of (3.2.4), linear independence of vectors with different weights guarantees that the both sides in (3.2.4) vanish, so we have

$$|\mu^{(0)}\rangle = 0 \Rightarrow |u_0\rangle = |u_1\rangle + |u_2\rangle + \cdots + |u_k\rangle, \tag{3.2.5}$$

where $|u_1\rangle, |u_2\rangle, \ldots, |u_k\rangle$ have the same weight $\mu^{(0)}$ as vector $|u_0\rangle$, as desired.

In what follows, we limit ourselves to *finite-dimensional irreducible* representations of *simple complex Lie algebras* $\mathcal{A}$. These representations are at the same time representations of compact real forms of Lie algebras $\mathcal{A}$. Making use of this fact, we have shown in Section I-6.1.2 of the first book (see Remark 2 there) that any finite-dimensional representation of Lie algebra $\mathcal{A}$ is equivalent to representation T with Cartan–Weyl basis elements (normalized in accordance with structure relations (3.1.1)–(3.1.3)) represented by operators $T(H_i)$, $T(E_\alpha)$ (we temporarily restore the notation T) such that

$$[T(E_\alpha)]^\dagger = T(E_{-\alpha}), \quad [T(H_j)]^\dagger = T(H_j). \tag{3.2.6}$$

This means that the space $\mathcal{V}$ of representation T is equipped with Hermitian scalar product:

$$\langle\psi|\phi\rangle \in \mathbb{C}, \quad \langle\psi| = (|\psi\rangle)^\dagger, \quad |\psi\rangle, |\phi\rangle \in \mathcal{V} \tag{3.2.7}$$

and that the following relations hold:

$$\big(\langle\psi|T(E_\alpha)|\phi\rangle\big)^* = \langle\phi|T(E_{-\alpha})|\psi\rangle, \quad \big(\langle\psi|T(H_j)|\phi\rangle\big)^* = \langle\phi|T(H_j)|\psi\rangle. \tag{3.2.8}$$

We emphasize that Hermitian conjugation formulas in (3.2.6) and (3.2.8) are consistent with structure relations in (3.1.1)–(3.1.3) in the case of real root vectors $\alpha = (\alpha_1, \dots, \alpha_r)$.

Since in the case we consider, Cartan generators in representation T are Hermitian, the properties **A** and **B** are simple consequences of general properties of systems of commuting Hermitian operators. Moreover, common eigenvectors of operators H_i (we again write H_i instead of $T(H_i)$) form basis in space $\mathcal{V}$, and weights $\mu = (\mu_1, \dots, \mu_r)$ $(k = 1, \dots, \dim\,\mathcal{V})$ are real.

Note that root vectors $\alpha = (\alpha_1, \dots, \alpha_r)$ of simple Lie algebra $\mathcal{A}$ are weights of adjoint representation of this algebra. In this regard, it is convenient to consider all weights of algebra $\mathcal{A}$ (i.e., weights of all representations of this algebra) as vectors belonging to root space $\mathfrak{V}_r$ of algebra $\mathcal{A}$.

C. If $|\mu\rangle \in \mathcal{V}$ is vector of weight μ, then vector $E_\alpha|\mu\rangle$, if nonzero, has definite weight $\mu + \alpha$. Indeed, we have

$$H_i\, E_\alpha|\mu\rangle = [H_i,\, E_\alpha]\,|\mu\rangle + E_\alpha H_i|\mu\rangle = (\alpha_i + \mu_i)\, E_\alpha|\mu\rangle. \tag{3.2.9}$$

Since representation T is irreducible, by acting many times on vector $|\mu\rangle$ by root operators E_α:

$$\dots\, E_\delta E_\gamma E_\beta E_\alpha|\mu\rangle,$$

and making linear combinations of these vectors, we obtain *all* vectors of space $\mathcal{V}$ (in this way, we construct an invariant subspace which must coincide with $\mathcal{V}$ for irreducible representation T). In general, a vector $|\mu\rangle \in \mathcal{V}$ which is used for generating the whole representation space $\mathcal{V}$ in the way described just now is called *cyclic vector*, and the representation itself is called cyclic. So, for irreducible representation, any vector from the space of representation is cyclic. This reasoning tells also that *all weights of irreducible representation T are obtained from each other by adding roots*, i.e., for any two weights μ and λ of T we have

$$\mu = \lambda + \sum_\alpha m_\alpha\, \alpha,$$

where summation runs over all roots α, and the coefficients are integer, $m_\alpha \in \mathbb{Z}_{\geq 0}$.

We emphasize again that weights of representation T may be degenerate. As an example, let $(\alpha + \beta + \gamma) = 0$, then vectors $|\mu\rangle$ and $E_\alpha E_\beta E_\gamma|\mu\rangle$ (if the latter does not vanish) are, generally speaking, different, but both have weight μ.

Definition 3.2.1. The number of linear independent vectors in $\mathcal{V}$ with one and the same weight is called *multiplicity* of this weight. If the multiplicity is equal to 1, the weight is called *simple.*

Example. For Lie algebra $s\ell(2,\mathbb{C})$, weights $m = -j, -j+1, \ldots, j$ of finite-dimensional irreducible representation with spin j are simple (see Section 2.1).

> **Problem 3.2.1.** Show that for adjoint representation of simple Lie algebra of rank $r > 1$, zero weight $\mu = 0$ is always degenerate, and its multiplicity equals r.

D. The key property of weights is as follows.

Theorem 3.2.1. *For any weight μ of finite-dimensional irreducible representation T of Lie algebra $\mathcal{A}$ and for any root α of $\mathcal{A}$, the value of $2(\mu,\alpha)/(\alpha,\alpha)$ is integer:*

$$\frac{2(\mu,\alpha)}{(\alpha,\alpha)} \in \mathbb{Z}, \tag{3.2.10}$$

and vector

$$\sigma_\alpha(\mu) = \mu - \frac{2\,(\mu,\alpha)}{(\alpha,\alpha)}\alpha, \tag{3.2.11}$$

obtained by Weyl reflection of weight vector μ with respect to plane P_α normal to root α, is again a weight of representation T.

Recall that the scalar product in space $\mathfrak{V}_r$ is defined by (3.1.6) with metric (3.1.4), and that this scalar product is positive-definite.

Proof of Theorem 3.2.1. The proof almost completely repeats the proof of Theorem I-6.2.1 of the first book (see Section I-6.2.1), which applies to the case when μ are roots of simple Lie algebras. The new aspect in Theorem I-6.2.1 is that weights, unlike roots, can have multiplicities greater than one.

For any root α of Lie algebra $\mathcal{A}$, three elements

$$H_\alpha = \frac{1}{\alpha^2}\,(\alpha^i\,H_i), \quad \mathcal{E}_{+\alpha} = \sqrt{\frac{2}{\alpha^2}}\,E_\alpha, \quad \mathcal{E}_{-\alpha} = \sqrt{\frac{2}{\alpha^2}}\,E_{-\alpha} \in \mathcal{A}, \tag{3.2.12}$$

where $\alpha^2 \equiv (\alpha, \alpha)$, generate Lie subalgebra $s\ell(2, \mathbb{C}) \subset \mathcal{A}$ with defining relations (2.1.1):

$$[H_\alpha,\, \mathcal{E}_{\pm\alpha}] = \pm\,\mathcal{E}_{\pm\alpha}, \quad [\mathcal{E}_{+\alpha},\, \mathcal{E}_{-\alpha}] = 2\,H_\alpha. \tag{3.2.13}$$

The latter equations follow from (3.1.1), (3.1.2), (3.1.3). For weight vector $|\mu\rangle$ belonging to space $\mathcal{V}$ of representation T, one has

$$H_\alpha\,|\mu\rangle = \frac{1}{\alpha^2}\,(\alpha^i\,\mu_i)\,|\mu\rangle. \tag{3.2.14}$$

As shown in Section 2.1, for finite-dimensional representations, eigenvalues of element $H_\alpha \in s\ell(2, \mathbb{C})$ are integer or half-integer. Therefore, $(\alpha,\, \mu)/\alpha^2 \in \mathbb{Z}/2$, which gives (3.2.10).

To prove that (3.2.11) is a weight, we consider weight vector $|u_0\rangle \in \mathcal{V}$ with weight μ', such that $E_\alpha|u_0\rangle = 0$. Such a vector exists, otherwise there would be an infinite number of weight (in view of (3.2.9)) vectors $(E_\alpha)^k|u_0\rangle$, which is impossible for finite-dimensional representation. We construct a chain of vectors

$$|u_1\rangle = E_{-\alpha}|u_0\rangle, \quad |u_2\rangle = E_{-\alpha}|u_1\rangle, \ldots, \quad |u_{k+1}\rangle = E_{-\alpha}|u_k\rangle, \ldots, \tag{3.2.15}$$

and assume that $|u_1\rangle = E_{-\alpha}|u_0\rangle \neq 0$ (otherwise, we deal with the trivial case: $H_\alpha|u_0\rangle = [E_\alpha,\, E_{-\alpha}]\,|u_0\rangle = 0$, so that according to (3.2.14) we have $(\alpha,\, \mu') = 0$, vector α is normal to μ' and $\sigma_\alpha(\mu') = \mu'$ is a weight). Clearly, at kth step in (3.2.15) we have

$$H_i|u_k\rangle = (\mu'_i - k\alpha_i)|u_k\rangle, \tag{3.2.16}$$

i.e., weight of vector $|u_k\rangle$ is $\mu^{(k)} = \mu' - k\alpha$. Now, let us prove, by induction, that

$$E_\alpha|u_{k+1}\rangle = m_{k+1}|u_k\rangle \tag{3.2.17}$$

(where the coefficient m_{k+1} is found below). The latter relation is not completely obvious because weights can have multiplicities greater than one.

Let us assume that (3.2.17) is valid for some $(k-1)$, then we make use of (3.2.16) to obtain

$$E_\alpha|u_{k+1}\rangle = E_\alpha E_{-\alpha}|u_k\rangle = [E_\alpha, E_{-\alpha}]|u_k\rangle + E_{-\alpha}E_\alpha|u_k\rangle$$
$$= (\alpha^i H_i)|u_k\rangle + m_k E_{-\alpha}|u_{k-1}\rangle = ((\mu', \alpha) - k(\alpha, \alpha) + m_k)\,|u_k\rangle.$$

Thus, relation (3.2.17) is valid for k, and we have

$$m_{k+1} = (\mu', \alpha) - k(\alpha, \alpha) + m_k. \tag{3.2.18}$$

Since $E_\alpha|u_0\rangle = 0$, relation (3.2.17) is valid for initial value (base of induction) $k=-1$ for which $m_0 = 0$, so it is valid for all $k > -1$.

Since the representation is finite dimensional, the procedure (3.2.15) terminates at some step j, i.e., $|u_{j+1}\rangle = 0$. Then it follows from (3.2.17) that $m_{j+1} = 0$. So, recursion relation (3.2.18) with boundary conditions $m_0 = 0 = m_{j+1}$ gives

$$m_k = k(\mu',\, \alpha) - \frac{k(k-1)}{2}(\alpha,\, \alpha), \quad \frac{2(\mu',\, \alpha)}{(\alpha,\, \alpha)} = j, \tag{3.2.19}$$

where the second equality is obtained from the first one for $k = j+1$. Thus, we have shown that if α is a root, and weight vector $|u_0\rangle$ has weight μ' and obeys $E_\alpha|u_0\rangle = 0$, then for $(\alpha,\, \mu') \neq 0$ the chain of vectors (3.2.15) defines sequence (string) of weights

$$\mu', \quad \mu' - \alpha, \ldots, \; \mu' - j\,\alpha = \mu' - 2\frac{(\alpha,\, \mu')}{(\alpha,\, \alpha)}\,\alpha, \tag{3.2.20}$$

where the last weight is determined by the second relation in (3.2.19). For $(\alpha,\, \mu') = 0$ the weight string consists of a single weight μ'. The last weight $(\mu' - j\alpha)$ in (3.2.20) coincides with $\sigma_\alpha(\mu')$, where the reflection σ_α is defined in (3.2.11). Since weight $(\mu' - j\alpha)$ is obtained from the first weight μ' by reflection with respect to plane P_α normal to root α, the entire string (3.2.20) is symmetric under this reflection, and if weight μ belongs to string (3.2.20), then weight $\sigma_\alpha(\mu)$ also belongs to this string.

It remains to be shown that every weight μ of representation T belongs to one of strings (3.2.20). Let μ be an arbitrary weight, and $|\mu\rangle$ be the pertinent weight vector. Multiple action by operator E_α on $|\mu\rangle$ gives new weight vectors. Since the representation is finite dimensional, there exists q such that $(E_\alpha)^q|\mu\rangle = |u_0\rangle \neq 0$ and $(E_\alpha)^{q+1}|\mu\rangle = E_\alpha|u_0\rangle = 0$. Weight string constructed starting from vector $|u_0\rangle$ with weight $\mu' = \mu + q\alpha$ contains weight μ; the pertinent weight vector is[a] $(E_{-\alpha})^q|u_0\rangle$ (this vector with weight μ, generally speaking, may not coincide with $|\mu\rangle$, see the following item E). This string of weights is symmetric under reflections σ_α. Thus, vector $\sigma_\alpha(\mu)$ in root space obtained by reflection (3.2.11) from vector μ belongs to this string and hence is a weight. □

Corollary 3.2.2. *In view of* (3.2.19), *formula* (3.2.17) *is written as follows*:

$$E_\alpha E_{-\alpha}|u_k\rangle = \frac{(\alpha,\,\alpha)}{2}(k+1)(j-k)\,|u_k\rangle \quad (k=0,\ldots,j), \tag{3.2.21}$$

where $|u_k\rangle = (E_{-\alpha})^k|u_0\rangle$, $E_\alpha|u_0\rangle = 0 = E_{-\alpha}|u_j\rangle$, *and weights of vectors* $|u_k\rangle$ *are given by* (3.2.16). *We emphasize that formula* (3.2.21) *is valid only for weight vectors* $|u_k\rangle$ *that are obtained by multiple action of operator* $E_{-\alpha}$ *on vector* $|u_0\rangle$ *which obeys* $E_\alpha|u_0\rangle = 0$.

Remark. The property (3.2.10) is a necessary condition for a vector μ of the root space to be a weight of finite-dimensional irreducible representation of Lie algebra $\mathcal{A}$. We call vectors μ obeying (3.2.10) *admissible weights.* In fact, this condition is sufficient for μ to be a weight of some representation T of algebra $\mathcal{A}$. Below, in Sections 3.4.1 and 3.4.2, we prove the latter fact for Lie algebras of classical series $s\ell(n,\mathbb{C})$, $so(n,\mathbb{C})$ and $sp(2r,\mathbb{C})$.

E. Consider the weight string of representation T along a root α, which corresponds to the set of eigenvectors (3.2.15). Let the initial weight $\mu' = \mu^{(0)}$ of weight vector $|u_0\rangle$ be such that $(\mu'+\alpha)$ is not a weight. Let some weights $\mu^{(k)}$ of vectors $|u_k\rangle$ given by (3.2.15) with $k>0$ have nontrivial multiplicities (i.e., these weights are not simple). Let us take the smallest k for which weight $\mu^{(k)}$ has multiplicity greater than 1. Then, besides vector $|u_k\rangle$, there exists another vector $|v_k\rangle$ with the same weight $\mu^{(k)}$. The latter vector obeys $E_\alpha|v_k\rangle = 0$, otherwise $E_\alpha|v_k\rangle$, which is linear independent of $|u_{k-1}\rangle$, would have weight $\mu^{(k-1)}$, and weight $\mu^{(k-1)}$ would not be simple, which would contradict our choice of k. For vector $|v_k\rangle$, we can repeat the procedure developed when proving Theorem 3.2.1 and construct additional string of weight vectors along the root α; this string begins with weight $\mu^{(k)}$, it is symmetric under reflection with respect to plane P_α and the last weight in the string is $\sigma_\alpha(\mu^{(k)})$. This procedure can be repeated for all other vectors with weights $\mu^{(k)}$, and in this way it demonstrates complete

[a]Vector $(E_{-\alpha})^q|u_0\rangle$ is nonzero, since $\langle\mu|(E_{-\alpha})^q|u_0\rangle = \langle u_0|u_0\rangle \neq 0$.

symmetry of system of weights (with their multiplicities included) under reflection with respect to plane P_α. This means, in particular, that weights $\mu^{(k)}$ and $\sigma_\alpha(\mu^{(k)}) = \mu^{(k)} - \frac{2(\mu^{(k)},\alpha)}{(\alpha,\alpha)}$ have equal multiplicities.

Problem 3.2.2. Show that any weight vector $|\mu\rangle$ of representation T is a linear combination of the form

$$|\mu\rangle = \sum_b (E_{-\alpha})^{k_b} |u_0^{(b)}\rangle, \tag{3.2.22}$$

where vectors $|u_0^{(b)}\rangle$ (analogs of weight vectors $|v_k\rangle$, see item E above) obey $E_\alpha |u_0^{(b)}\rangle = 0$. Hint: to begin with, show that all root vectors in strings, constructed above, can be chosen orthogonal to each other.

Definition 3.2.2. Weights of representation T related by reflections (3.2.11) from group W are *equivalent*.

An important notion in weight theory is *a set of all weights of all irreducible finite-dimensional representations* of a given complex simple Lie algebra $\mathcal{A}$. We illustrate below in examples, and further discuss in Section 3.3, that this set has a remarkable property: all weights of all finite-dimensional representations of simple Lie algebra $\mathcal{A}$ belong to a lattice Λ in root space $\mathfrak{V}_r$. This lattice is invariant under transformations (3.2.11) from Weyl group of reflections.

In the following examples, we consider all weights of all irreducible finite-dimensional representations of concrete Lie algebras.

Example 1. Let us consider weight system of Lie algebra $s\ell(r+1,\mathbb{C})$ (A_r-series). Since Cartan elements (3.1.8) in any representation obey $\sum_{i=1}^{r+1} H_i = 0$, the weights μ of *all* finite-dimensional representations belong to root space P_u which is r-dimensional hyperplane in $\mathbb{R}^{r+1}$ normal to vector $u = (1,\ldots,1) \in \mathbb{R}^{r+1}$ (see item 1 in Section 3.1.1). Thus, weights can be expanded in basis vectors $e^{(s)} \in \mathbb{R}^{r+1}$:

$$\mu = \sum_{s=1}^{r+1} \mathsf{m}_s \, e^{(s)}, \tag{3.2.23}$$

where $\sum_{s=1}^{r+1} \mathsf{m}_s = 0$. Furthermore, weights (3.2.23) must obey (3.2.10), which, with explicit expression (3.1.11) for roots, gives extra condition

$$\frac{2(\mu, e^{(i)} - e^{(k)})}{|e^{(i)} - e^{(k)}|^2} = \mathsf{m}_i - \mathsf{m}_k \in \mathbb{Z}, \quad \forall i \neq k. \tag{3.2.24}$$

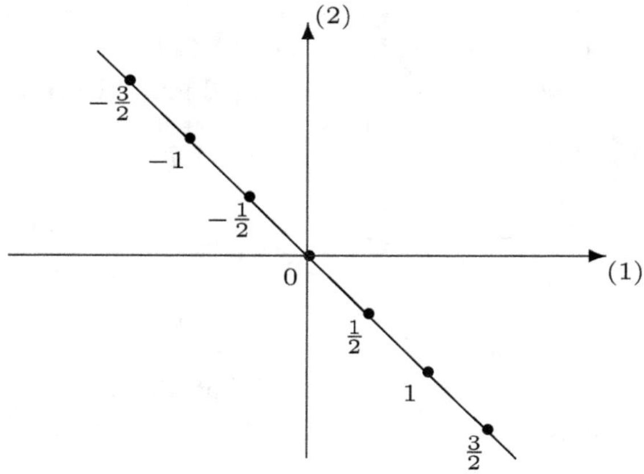

Fig. 3.2.1. Weight of Lie algebra $s\ell(2,\mathbb{C})$. Numbers $\ldots,-\frac{1}{2},0,\frac{1}{2},1,\frac{3}{2},\ldots$ are values of m.

We have shown in item 1 of the proof of Proposition 3.1.1 (see Section 3.1.3) that Weyl group $W = S_{r+1}$ of algebra $s\ell(r+1,\mathbb{C})$ permutes components m_i of weight vector (3.2.23), which is consistent with (3.2.24). Thus, Weyl group W is a symmetry of the lattice formed by weights (3.2.23), (3.2.24) of algebra $s\ell(r+1,\mathbb{C})$.

In the simplest case $r = 1$, Lie algebra $s\ell(2,\mathbb{C})$ has one simple root $\alpha = (e^{(1)} - e^{(2)})$, all weights (3.2.23) obey $\mathsf{m}_1 + \mathsf{m}_2 = 0$, and hence they are written as $\mu = \mathsf{m}(e^{(1)} - e^{(2)})$, where according to (3.2.24) we have $\mathsf{m} \in \mathbb{Z}/2$. So, all weights of all finite-dimensional representation of algebra $s\ell(2,\mathbb{C})$ form one-dimensional lattice and belong to a straight line directed along the root α, as shown in Fig. 3.2.1. The transformation (3.1.61) from Weyl group W of algebra $s\ell(2,\mathbb{C})$ is simply reflection $\mu \to -\mu$, so the Weyl group is isomorphic to $S_2 = \mathbb{Z}_2$, as anticipated.

The case $r = 2$ corresponds to Lie algebra $s\ell(3,\mathbb{C})$. Weight systems of some representations of algebra $s\ell(3,\mathbb{C})$ and weight lattice of all representations are studied in Section 3.3.

Example 2. Root system of algebra $so(2r+1,\mathbb{C})$ (B_r-series) consists of vectors (3.1.28) (see item 2 in Section 3.1.1). We expand weights μ in basis vectors $e^{(s)}$ of r-dimensional root space

$$\mu = \sum_{s=1}^{r} \mathsf{m}_s \, e^{(s)} \in \mathfrak{V}_r \tag{3.2.25}$$

and insert this into the condition (3.2.10) where root vectors α are given in (3.1.28). In this way, we find that $(\mathsf{m}_i \pm \mathsf{m}_k) \in \mathbb{Z}$ and $2(\mu, e^{(i)}) = 2\mathsf{m}_i \in \mathbb{Z}$. So, components of any weight (3.2.25) must be integer or half-integer,

$$\mathsf{m}_i \in \mathbb{Z}/2. \tag{3.2.26}$$

We have shown in item 2 of the proof of Proposition 3.1.1 that the Weyl group of $\Phi(so(2r+1,\mathbb{C}))$ is isomorphic to extension of the group S_r of permutations of m_i by any number of sign flips $\mathsf{m}_i \to -\mathsf{m}_i$. All these transformations are consistent with (3.2.26), so Weyl group W is a symmetry group of the lattice formed by weights (3.2.25), (3.2.26) of algebra $so(2r+1,\mathbb{C})$.

Example 3. Root system for algebra $sp(2r,\mathbb{C})$ (C_r-series) is given by (3.1.35) (see Section 3.1.1, item 3), so weights (3.2.25) obey $(\mathsf{m}_i \pm \mathsf{m}_k) \in \mathbb{Z}$ and also

$$2(\mu, 2e^{(i)})/(2e^{(i)},\, 2e^{(i)}) = \mathsf{m}_i \in \mathbb{Z}. \tag{3.2.27}$$

Making use of Eq. (3.2.11), it is straightforward to show that Weyl group for algebras of C_r-series is the same as Weyl group for algebras of B_r-series (see item 3 in the proof of Proposition 3.1.1, Section 3.1.3). Transformations from this group preserve the relation (3.2.27), so it is a symmetry group of weight lattice of algebra $sp(2r,\mathbb{C})$.

Example 4. Root system of algebra $so(2r,\mathbb{C})$ (D_r-series) is given by (3.1.25) (see Section 3.1.1, item 2). The condition (3.2.10) for weights (3.2.25) gives $\mathsf{m}_i \pm \mathsf{m}_k \in \mathbb{Z}$, where $i \neq k$, which is equivalent to (3.2.26). So, the weights are the same as for algebra $so(2r+1,\mathbb{C})$ (B_r-series). However, Weyl group of $so(2r,\mathbb{C})$ consists of permutations of components m_i plus even number of sign flips of these components (see item 4 in the proof of Proposition 3.1.1, Section 3.1.3). These transformations preserve (3.2.26), so the Weyl group is a symmetry group of weight lattice of algebra $so(2r,\mathbb{C})$.

Remark. Weyl reflections from group W can be realized on elements of Lie algebra $\mathcal{A}$. We define operators

$$T_3(\alpha) = H_\alpha = \frac{\alpha^i H_i}{\alpha^2}, \quad T_2(\alpha) = \frac{1}{2i}(\mathcal{E}_\alpha - \mathcal{E}_{-\alpha}), \quad T_1(\alpha) = \frac{1}{2}(\mathcal{E}_\alpha + \mathcal{E}_{-\alpha}), \tag{3.2.28}$$

which, in accordance with (3.2.13), form angular momentum algebra,

$$[T_k(\alpha),\, T_j(\alpha)] = i\varepsilon_{kjn} T_n(\alpha).$$

Consider rotation by angle π about the second axis, which is associated with operator $e^{i\pi T_2(\alpha)}$ and leads to reflection of the third axis $T_3(\alpha) \to -T_3(\alpha)$:

$$e^{i\pi T_2(\alpha)}\, \alpha^i H_i \, e^{-i\pi T_2(\alpha)} = -\alpha^i H_i. \tag{3.2.29}$$

The operator $e^{i\pi T_2(\alpha)}$ realizes the Weyl reflection: we show momentarily that

$$e^{i\pi T_2(\alpha)}\, H_i \, e^{-i\pi T_2(\alpha)} = [\sigma_\alpha(H)]_i, \tag{3.2.30}$$

where reflection σ_α is defined in (3.2.11). Due to the latter property, the transformed algebra elements $e^{i\pi T_2(\alpha)}\, H_i \, e^{-i\pi T_2(\alpha)}$ have eigenvalues $[\sigma_\alpha(\mu)]_i$, which precisely means that automorphism $\mathcal{A} \to e^{i\pi T_2(\alpha)}\, \mathcal{A}\, e^{-i\pi T_2(\alpha)}$ realizes Weyl reflection.

To prove (3.2.30), we note that if vector ξ is normal to α, i.e., $(\xi, \alpha) = 0$, then $[\mathcal{E}_{\pm\alpha},\, \xi^i H_i] = (\xi, \alpha)\mathcal{E}_{\pm\alpha} = 0$, and we have

$$e^{i\pi T_2(\alpha)}\, (\xi, H) \, e^{-i\pi T_2(\alpha)} = (\xi, H) \quad (\forall \xi \perp \alpha). \tag{3.2.31}$$

An arbitrary vector x in Euclidean space $\mathfrak{V}_r$ is a linear superposition of vector α and vector ξ normal to α. Then (3.2.29) and (3.2.31) give

$$e^{i\pi T_2(\alpha)}\, (x, H) \, e^{-i\pi T_2(\alpha)} = (\sigma_\alpha(x), H) = (x, \sigma_\alpha(H)), \tag{3.2.32}$$

which gives (3.2.30), since x is arbitrary. The second equality in (3.2.32) makes use of the fact that

$$(\sigma_\alpha(\beta),\, \gamma) = (\beta,\, \sigma_\alpha(\gamma)) \tag{3.2.33}$$

for any reflection (3.2.11) from the Weyl group W and any two vectors $\beta, \gamma \in \mathfrak{V}_r$. The latter equality means also that Weyl group generators σ_α are orthogonal operators with respect to scalar product $(\beta,\, \gamma) = \beta_i\, g^{ij}\, \gamma_j$. Therefore, for any element $w \in W$, which is, by definition, a product of generators σ_α, we have

$$(w \cdot \beta,\, \gamma) = \left(\beta,\, w^{-1} \cdot \gamma\right). \tag{3.2.34}$$

So, for metric g we have $w^{\mathsf{T}} \cdot g \cdot w = g$ and there are two options

$$\det(w) = +1, \quad \det(w) = -1, \tag{3.2.35}$$

depending on whether $w \in W$ is a product of even or odd reflections σ_α.

Problem 3.2.3. Show that any root generator $E_\beta \in \mathcal{A}$ obeys

$$e^{i\pi T_2(\alpha)}\, E_\beta \, e^{-i\pi T_2(\alpha)} \propto E_{\sigma_\alpha(\beta)}. \tag{3.2.36}$$

3.3. Weight Lattice

In Section 3.2, we have seen that the quantity $2(\mu,\alpha)/\alpha^2$ is integer for all roots α of simple Lie algebra $\mathcal{A}$ and weights μ of its representations, see (3.2.10). This property is not only necessary but also sufficient for μ to be a weight of some finite-dimensional representation of $\mathcal{A}$. We do not give here the general proof of the latter statement for *all* simple Lie algebras, but rather show its validity for algebras of classical series $s\ell(n,\mathbb{C})$, $so(n,\mathbb{C})$ and $sp(2r,\mathbb{C})$, namely, for any vector μ obeying (3.2.10) we construct explicitly (see Sections 3.4.1 and 3.4.2) a representation which has μ as a weight. Here we take for granted that all admissible weights, i.e., vectors μ obeying (3.2.10), are actually weights of some representations.

Let $\{\alpha^{(i)}\}$ be simple roots of algebra $\mathcal{A}$. In view of (3.2.10), for any admissible weight μ we have

$$\frac{2(\mu,\alpha^{(i)})}{(\alpha^{(i)})^2} = \mathsf{m}_i \in \mathbb{Z}. \tag{3.3.1}$$

Among vectors obeying (3.3.1), there are vectors $\lambda_{(i)}$ $(i=1,\ldots,r)$ such that

$$\frac{2(\lambda_{(i)},\alpha^{(j)})}{(\alpha^{(j)})^2} = \delta_i^j. \tag{3.3.2}$$

Vectors $\lambda_{(i)}$ are admissible weights, since they obey (3.2.10) for any root α. Indeed, in accordance with (3.1.80) and (3.3.2), we obtain for any root α

$$\frac{2(\lambda_{(i)},\alpha)}{(\alpha)^2} = \sum_{j=1}^{r} \mathsf{n}_j \frac{2(\lambda_{(i)},\alpha^{(j)})}{(\alpha^{(j)})^2} = \mathsf{n}_i \in \mathbb{Z}.$$

Thus, every simple root $\alpha^{(i)}$ has its weight counterpart $\lambda_{(i)}$, which is defined by (3.3.2) and is called *fundamental weight.* Making use of (3.3.1) and (3.3.2), we find for any weight μ

$$\left(\mu - \sum_{i=1}^{r} \mathsf{m}_i\lambda_{(i)},\ \alpha^{(j)}\right) = 0 \tag{3.3.3}$$

Since simple roots $\{\alpha^{(i)}\}$ form complete system of vectors in the root space $\mathfrak{V}_r$, and metric g_{ij} in $\mathfrak{V}_r$ is Euclidean, one finds from Eq. (3.3.3) that

$$\mu = \sum_{i=1}^{r} \mathsf{m}_i\lambda_{(i)}, \quad \mathsf{m}_i \in \mathbb{Z}. \tag{3.3.4}$$

So, any weight μ is expanded in fundamental weights with integer coefficients. This "quantized" spectrum of weights is a remarkable property of finite-dimensional representations of simple complex Lie algebras and their compact real forms.

The property (3.3.4) means that all admissible weights μ are some sites of a lattice Λ determined by fundamental weights $\lambda_{(i)}$. This lattice is called *weight lattice.* The following inverse statement is also true.

Proposition 3.3.1. *Any site of weight lattice Λ is an admissible weight. In other words, for any set of integers ℓ_i, $i = 1, \ldots, r$, vector*

$$\mu = \sum_{i=1}^{r} \ell_i \, \lambda_{(i)}, \tag{3.3.5}$$

is admissible weight.

Proof. We have to show that vector (3.3.5) obeys (3.2.10). Let us consider co-roots $\alpha^\vee = \frac{\alpha}{(\alpha)^2}$. Recall that co-roots are linear combinations (3.1.80) of simple co-roots $\frac{\alpha^{(i)}}{(\alpha^{(i)})^2}$ with integer coefficients. Then we have for any root α

$$\frac{2(\mu, \alpha)}{(\alpha)^2} = \sum_{i,j=1}^{r} \mathsf{n}_i \ell_j \, \frac{(2\lambda_{(j)}, \alpha^{(i)})}{(\alpha^{(i)})^2} = \sum_{i=1}^{r} \mathsf{n}_i \ell_i \, . \tag{3.3.6}$$

The right-hand side of (3.3.6) is integer, since $\mathsf{n}_i, \ell_j \in \mathbb{Z}$, so vector μ is admissible weight. □

Since roots of Lie algebra $\mathcal{A}$ are weights of adjoint representation of $\mathcal{A}$, they belong to weight lattice Λ. Lattice Λ_r generated by simple roots $\alpha^{(i)}$, whose sites are

$$\alpha = \sum_{i} \alpha^{(i)} \, \mathsf{n}_i, \tag{3.3.7}$$

where $\mathsf{n}_i \in \mathbb{Z}$, is *root lattice.* Lattice Λ_r is sublattice in Λ, $\Lambda_r \subset \Lambda$. Lattice Λ can be viewed as Abelian group of translations in r-dimensional space, and its sublattice Λ_r is its normal subgroup. In fact, the quotient group Λ/Λ_r is isomorphic to center $Z(\widetilde{G})$ of simply connected (universal covering) group $\widetilde{G}$, whose Lie algebra is $\mathcal{A}$. We do not prove the latter fact, and merely illustrate it by examples below. Since root lattice Λ_r is sublattice in Λ, all roots, including simple roots, are expanded in fundamental weights with integer coefficients. Also, since simple roots $\alpha^{(i)}$ make basis in root space $\mathfrak{V}_r$, fundamental weights $\lambda_{(k)}$ also make basis in $\mathfrak{V}_r$.

Problem 3.3.1. ★ Prove that expansion of simple roots $\alpha^{(i)}$ in fundamental weights $\lambda_{(k)}$ reads $\alpha^{(i)} = \sum_{k=1}^{r} K_{ik}\,\lambda_{(k)}$, where $K_{ik} = 2(\alpha^{(i)}, \alpha^{(k)})/(\alpha^{(k)})^2$ are elements of Cartan matrix (see Section I-6.2.2 of the first book). Show that order of group Λ/Λ_r is equal to $\det ||K_{ik}||$.

For irreducible, and hence cyclic, representation T in space $\mathcal{V}$, one can start with any basis vector $|\mu\rangle \in \mathcal{V}$ and construct any other vector in $\mathcal{V}$ by acting on $|\mu\rangle$ by root generators E_α and making linear combinations of these vectors. In accordance with property C from Section 3.2, vector $E_\gamma \dots E_\beta E_\alpha|\mu\rangle \in \mathcal{V}$ has weight $(\mu + \alpha + \beta + \cdots + \gamma)$. So, all weights of irreducible representation T differ from the chosen weight μ of this representation by a linear combination of roots with integer coefficients; in other words, all weights *of a given representation* T belong to one and the same element in Λ/Λ_r.

Example 1. Root space of algebra $s\ell(2, \mathbb{C})$ is one dimensional (see Example 1 of Section 3.2), and we can identify it with real axis $\mathbb{R}$. One makes the following choice of the only simple root: $\alpha = 1 \in \mathbb{R}$. Root lattice Λ_r is the set of integers, $\Lambda_r = \mathbb{Z}$:

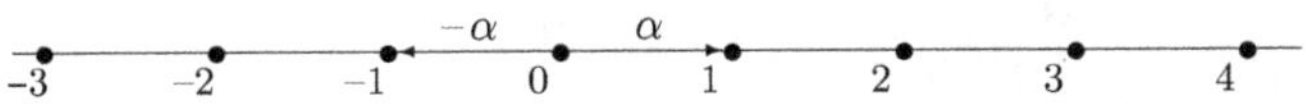

Fundamental weight λ obeys (3.3.2):

$$\frac{2(\lambda, \alpha)}{(\alpha)^2} = 1 \Rightarrow \lambda = \frac{1}{2}. \tag{3.3.8}$$

So, weight lattice has half-integer spacing, $\Lambda = \mathbb{Z}/2$, see Fig. 3.3.1. This weight lattice has been already constructed in Example 1 in Section 3.2 (see Fig. 3.2.1). Clearly, $\Lambda_r \subset \Lambda$ and $\Lambda/\Lambda_r = \mathbb{Z}_2$. The latter property is clear from the fact that there are only two cosets $\bar{0} = \{0 + \Lambda_r\} = \{\Lambda_r\}$ and $\bar{\frac{1}{2}} = \{\frac{1}{2} + \Lambda_r\}$, with multiplication table identical to that of group $\mathbb{Z}_2$. On the other hand, group $\mathbb{Z}_2 = \{I_2, -I_2\}$ is center in $SU(2)$, therefore, center of $SU(2)$ is indeed isomorphic to quotient group Λ/Λ_r.

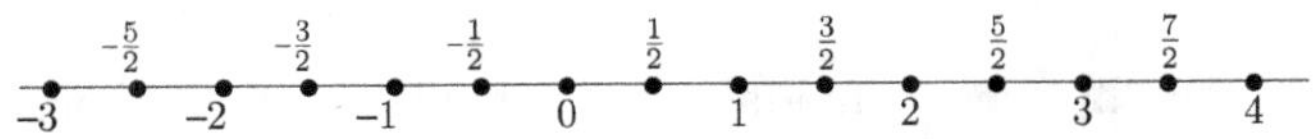

Fig. 3.3.1. Weight lattice Λ of Lie algebra $s\ell(2, \mathbb{C})$.

Example 2. Let us turn to Lie algebra $s\ell(3,\mathbb{C})$. We choose Cartan–Weyl basis in defining representation as follows (see Section I-6.2.1 of the first book):

$$H_1 = e_{11} - e_{22}, \quad H_2 = \frac{1}{\sqrt{3}}(e_{11} + e_{22} - 2e_{33}),$$

$$E_{\alpha^{(1)}} = \frac{1}{\sqrt{6}}e_{12}, \quad E_{\alpha^{(2)}} = \frac{1}{\sqrt{6}}e_{23}, \quad E_{\alpha^{(1)}+\alpha^{(2)}} = \frac{1}{\sqrt{6}}e_{13}, \tag{3.3.9}$$

$$E_{-\alpha^{(1)}} = \frac{1}{\sqrt{6}}e_{21}, \quad E_{-\alpha^{(2)}} = \frac{1}{\sqrt{6}}e_{32}, \quad E_{-(\alpha^{(1)}+\alpha^{(2)})} = \frac{1}{\sqrt{6}}e_{31},$$

where e_{ij} are (3×3) matrix units, H_1 and H_2 are Cartan generators and $E_{\alpha^{(1)}}$, $E_{\alpha^{(2)}}$ are generators that correspond to simple roots

$$\alpha^{(1)} = (2, 0), \quad \alpha^{(2)} = (-1, \sqrt{3}), \tag{3.3.10}$$

so that $[H_i,\, E_{\alpha^{(k)}}] = \alpha_i^{(k)}\, E_{\alpha^{(k)}}$. Normalization of root generators (3.3.9) is chosen in such a way that the second relation in (3.1.5) is satisfied. Root space $\mathfrak{V}_2$ of Lie algebra $s\ell(3,\mathbb{C})$ is two dimensional, and root lattice is constructed out of the two simple roots in accordance with (3.3.7). The fundamental weights are obtained by solving Eq. (3.3.2):

$$2\frac{(\alpha^{(i)},\, \lambda_{(j)})}{(\alpha^{(i)})^2} = \delta^i_j \Rightarrow (\alpha^{(i)},\, \lambda_{(j)}) = 2\,\delta^i_j. \tag{3.3.11}$$

Here to define scalar product we use metric $g^{ij} = \kappa\delta^{ij}$ in $\mathfrak{V}_2$, which is proportional to $\mathrm{Tr}(H_i H_j) = 2\delta_{ij}$, and since the left-hand sides of (3.2.10) and (3.3.2) are independent of the parameter κ, we choose $\kappa = 1$ for definiteness. We find the fundamental weights by solving (3.3.11) with simple roots (3.3.10):

$$\lambda_{(1)} = \left(1, \frac{1}{\sqrt{3}}\right), \quad \lambda_{(2)} = \left(0, \frac{2}{\sqrt{3}}\right). \tag{3.3.12}$$

The part of weight system is shown in Fig. 3.3.2, where $\alpha^{(1)}$ and $\alpha^{(2)}$ are simple roots, and $\lambda_{(1)}$, $\lambda_{(2)}$ are fundamental weights.

Note that weights

$$\lambda_{(1)},\ \mu_{(3)} = \lambda_{(1)} - \alpha^{(1)} = (-1, 1/\sqrt{3}),$$

$$\mu_{(5)} = \lambda_{(1)} - \alpha^{(1)} - \alpha^{(2)} = (0, -2/\sqrt{3})$$

shown in Fig. 3.3.2 refer to fundamental (defining) triplet representation, while weights

$$\lambda_{(2)},\ \mu_{(6)} = \lambda_{(2)} - \alpha^{(2)},\ \mu_{(4)} = \lambda_{(2)} - \alpha^{(1)} - \alpha^{(2)}$$

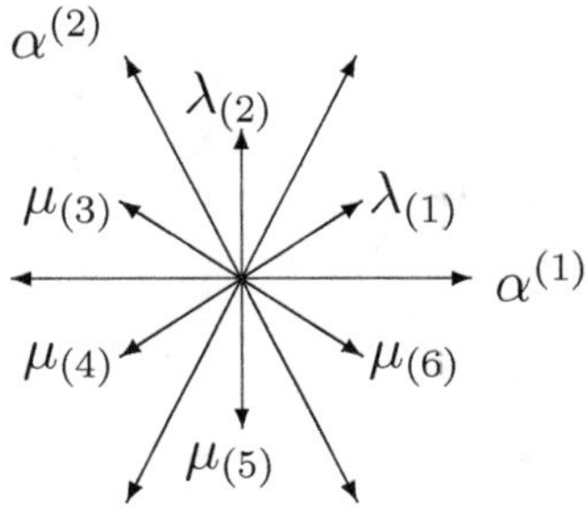

Fig. 3.3.2. Simple roots and weight systems for defining and co-defining representations of Lie algebra $s\ell(3, \mathbb{C})$.

refer to antitriplet, i.e., to the representation, contragredient to defining one (co-defining). Indeed, by making use of explicit form (3.3.9) of root generators E_α, all basis vectors in space $\mathcal{V}_3$ of defining representation can be written as

$$|\mu\rangle = \begin{pmatrix} 1 \\ 0 \\ 0 \end{pmatrix}, \quad |\mu'\rangle = E_{-\alpha^{(1)}}\,|\mu\rangle \sim \begin{pmatrix} 0 \\ 1 \\ 0 \end{pmatrix}, \quad |\mu''\rangle = E_{-\alpha^{(2)}}\,E_{-\alpha^{(1)}}\,|\mu\rangle \sim \begin{pmatrix} 0 \\ 0 \\ 1 \end{pmatrix}. \tag{3.3.13}$$

We emphasize that the first vector $|\mu\rangle$ has an important property

$$E_{\alpha^{(i)}}|\mu\rangle = 0, \tag{3.3.14}$$

i.e., it is annihilated by simple root generators. Vectors (3.3.13) indeed have weights $\mu = \lambda_{(1)}$, $\mu' = \mu_{(3)}$ and $\mu'' = \mu_{(5)}$. This can be checked directly by acting on them Cartan elements H_i in defining representation (3.3.9).

Problem 3.3.2. Show that weights $\lambda_{(2)}$, $\mu_{(6)}$ and $\mu_{(4)}$ correspond to antitriplet, i.e., to three basis vectors in the space of co-defining representation.

Fundamental weights $\lambda_{(1)}$ and $\lambda_{(2)}$ generate the whole weight lattice, which is shown in Fig. 3.3.3. This figure shows, in particular, that root lattice Λ_r is sublattice of weight lattice Λ, and that quotient group is $\Lambda/\Lambda_r = \mathbb{Z}_3$. Indeed, there are three cosets $0 + \Lambda_r$, $\overline{\lambda}_{(1)} = \lambda_{(1)} + \Lambda_r$, $\overline{\lambda}_{(2)} = \lambda_{(2)} + \Lambda_r$, and their multiplication table coincides with that of $\mathbb{Z}_3$. The latter fact follows from (3.3.10) and (3.3.12), which give $2\lambda_{(1)} = \lambda_{(2)} + \alpha^{(1)} \in \overline{\lambda}_{(2)}$,

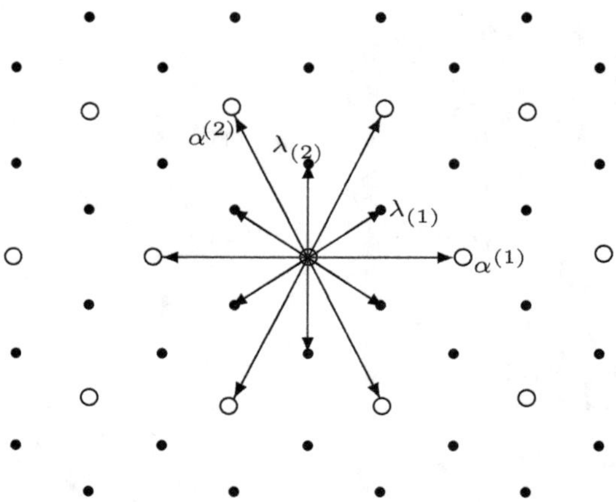

Fig. 3.3.3. Weight lattice Λ of Lie algebra $s\ell(3,\mathbb{C})$. Open circles are sites of root sublattice Λ_r.

$3\lambda_{(1)} = 2\alpha^{(1)}+\alpha^{(2)} \in \Lambda_r$, $\lambda_{(1)}+\lambda_{(2)} = \alpha^{(1)}+\alpha^{(2)} \in \Lambda_r$, $2\lambda_{(2)} = \lambda_{(1)}+\alpha^{(2)} \in \overline{\lambda}_{(1)}$, $3\lambda_{(2)} = \alpha^{(1)} + 2\alpha^{(2)} \in \Lambda_r$, i.e.,

$$\overline{\lambda}_{(1)} + \overline{\lambda}_{(1)} = \overline{\lambda}_{(2)}, \quad \overline{\lambda}_{(2)} + \overline{\lambda}_{(2)} = \overline{\lambda}_{(1)},$$
$$\overline{\lambda}_{(1)} + \overline{\lambda}_{(1)} + \overline{\lambda}_{(1)} = \overline{\lambda}_{(2)} + \overline{\lambda}_{(2)} + \overline{\lambda}_{(2)} = \overline{\lambda}_{(1)} + \overline{\lambda}_{(2)} = \Lambda_r.$$

Recall that center of group $SU(3)$ has three elements $\{I_3,\, e^{i\frac{2\pi}{3}}I_3,\, e^{i\frac{4\pi}{3}}I_3\}$, and it is isomorphic to $\mathbb{Z}_3$, i.e., to Λ/Λ_r, as anticipated.

Remark. It is sometimes convenient to consider weight lattices for dual Lie algebras $\mathcal{A}^\vee$ defined toward the end of Section 3.1. Fundamental weights $\lambda^\vee_{(i)}$ of dual algebra $\mathcal{A}^\vee$ are determined by

$$2\,\frac{(\lambda^\vee_{(i)}, \alpha^{(j)\vee})}{(\alpha^{(j)\vee})^2} = \delta^i_j \Rightarrow 2\,(\lambda^\vee_{(i)}, \alpha^{(j)}) = \delta^i_j. \tag{3.3.15}$$

By comparing (3.3.2) with the second relation in (3.3.15), we find

$$\lambda^\vee_{(i)} = \frac{\lambda_{(i)}}{(\alpha^{(i)})^2}. \tag{3.3.16}$$

Weights (3.3.16) are called fundamental co-weights of Lie algebra $\mathcal{A}$.

3.4. Classification of Finite-Dimensional Irreducible Representations

3.4.1. Highest weight representations

Definition 3.4.1. Weight λ of representation T of simple Lie algebra $\mathcal{A}$ is called *highest weight* if $\lambda + \alpha$ is not a weight of representation T for any positive root α of algebra $\mathcal{A}$. Representation that has highest weight is called *highest weight representation.*

Vector $|\lambda\rangle$ in the space of representation T with highest weight λ is called highest weight vector. A necessary condition for vector $|\lambda\rangle$ to be highest weight vector is $E_\alpha|\lambda\rangle = 0$ for all positive roots α. If highest weight λ is simple (multiplicity 1) — and we will see shortly that this is indeed the case for irreducible representations — this condition is also sufficient. So, Definition 3.4.1 is equivalent to the statement that $|\lambda\rangle$ is highest weight vector if and only if $E_\alpha|\lambda\rangle = 0$ for all positive roots α. Importantly, *any finite-dimensional representation T of Lie algebra $\mathcal{A}$ is highest weight representation.* Indeed, since representation T is finite dimensional, there must exist weight λ such that $(\lambda + \alpha)$ is not a weight for all positive roots α.

Problem 3.4.1. Prove that weight λ of irreducible representation T is highest weight (i.e., $E_\alpha|\lambda\rangle = 0 \ \forall\alpha > 0$) if $\lambda + \alpha^{(i)}$ is not a weight of this representation for any simple root $\alpha^{(i)}$ of algebra $\mathcal{A}$. Hint: make use of the fact that generator $E_\alpha \in \mathcal{A}$, where α is an arbitrary positive root, is a multiple commutator of generators $E_{\alpha^{(i)}}$.

Example. It follows from (3.3.14) that vector $|\mu\rangle$, given by (3.3.13), is highest weight vector of defining representation of algebra $s\ell(3,\mathbb{C})$. Its highest weight is the fundamental weight $\lambda_{(1)}$.

Theorem 3.4.1. (1) *If finite-dimensional representation T is irreducible, then its highest weight is simple.* (2) *For a given irreducible representation T, the highest weight is unique.*

Proof. (1) Let $|\mu^{(0)}\rangle$ be highest weight vector with highest weight $\mu^{(0)}$ in space $\mathcal{V}$ of irreducible representation T. Since any vector in $\mathcal{V}$, including $|\mu^{(0)}\rangle$, is cyclic, then, in accordance with items B and C of Section 3.2, we have to show that any vector of $\mathcal{V}$ which has the form

$$\dots E_\delta E_\gamma E_\beta E_\alpha |\mu^{(0)}\rangle \tag{3.4.1}$$

and the same weight $\mu^{(0)}$ is equal to $k\,|\mu^{(0)}\rangle$, where k is a constant.[b] It follows from item C of Section 3.2 that $\dots + \delta + \gamma + \beta + \alpha = 0$, so at least one of the roots must be positive. Let the utmost right positive root be, say, γ. We move E_γ to the right by using $E_\gamma E_\beta = [E_\gamma,\, E_\beta] + E_\beta E_\gamma$ until E_γ hits $|\mu^{(0)}\rangle$ and annihilates it (recall that $E_\gamma|\mu^{(0)}\rangle = 0$ for $\gamma > 0$). In this way, we obtain a sum of vectors with the same weight $\mu^{(0)}$ but smaller number of generators E as compared with (3.4.1) (since $[E_\gamma,\, E_\beta]$ is a linear combination of E_α and H_i). We continue this process until all operators E with positive roots disappear. At this point, we are left with a sum of products H_i acting on $|\mu^{(0)}\rangle$, which is equal to $|\mu^{(0)}\rangle$ times a polynomial of components of vector $\mu^{(0)}$.

(2) Let the highest weight of an irreducible representation T be not unique, i.e., let there exist two highest weight vectors $|\mu^{(0)}\rangle$ and $|\lambda^{(0)}\rangle$, whose highest weights $\mu^{(0)}$ and $\lambda^{(0)}$ are not equal to each other. Any vector of irreducible representation is cyclic, so, in accordance with items B and C of Section 3.2, there exists a sequence of roots $\delta, \gamma, \dots, \alpha$ such that[c]

$$|\lambda^{(0)}\rangle = E_\delta E_\gamma \cdots E_\alpha |\mu^{(0)}\rangle, \tag{3.4.2}$$

where $\delta + \cdots + \alpha \neq 0$. In accordance with the discussion in part (1) of this proof, we can, without loss of generality, assume that all roots $\delta, \gamma, \dots, \alpha$ are negative. Let us apply positive root generator $E_{-\delta}$ to both sides of (3.4.2). As a result, the left-hand side vanishes, while the right-hand side does not. The latter property follows from the fact that if $E_\delta|\psi\rangle \neq 0$, then $E_{-\delta}E_\delta|\psi\rangle \neq 0$, since, in view of (3.2.6), $\langle\psi|E_{-\delta}E_\delta|\psi\rangle = \langle\psi|E_\delta^\dagger E_\delta|\psi\rangle > 0$. □

Remark. When proving the first part of proposition, we have seen that if $|\mu^{(0)}\rangle \in \mathcal{V}$ is the highest weight vector with highest weight $\mu^{(0)}$, then any vector in space $\mathcal{V}$ can be written as linear combination of vectors (3.4.1) *with negative roots* $\dots, \delta, \gamma, \beta, \alpha$. In other words, any weight $\mu \neq \mu^{(0)}$ of

[b]This would ensure that any linear combination of vectors (3.4.1) with weight $\mu^{(0)}$ equals $k'\,|\mu^{(0)}\rangle$ with constant k'.

[c]In principle, vector $|\lambda^{(0)}\rangle$ in (3.4.2) may be equal to a linear combination of vectors $E_\delta \cdots E_\alpha|\mu^{(0)}\rangle$, each of which (due to item B) has weight $\lambda^{(0)}$. However, all these vectors, because of part (1) of this proposition, are proportional to each other, and we are back to (3.4.2).

representation T is obtained from the highest weight by subtracting positive roots from the highest weight $\mu^{(0)}$, so we have

$$\mu = \mu^{(0)} - \sum_i \mathsf{n}_i \alpha^{(i)}, \qquad \mathsf{n}_i \in \mathbb{Z}_{\geq 0}, \tag{3.4.3}$$

where $\alpha^{(i)}$ are simple roots of algebra $\mathcal{A}$.

Theorem 3.4.2. *Two finite-dimensional irreducible representations T and T' of Lie algebra $\mathcal{A}$ are equivalent if and only if their highest weights are equal.*

Proof. To distinguish between the objects related to representations T and T', we use prime in the latter case, e.g., $\mathcal{V}$ and $\mathcal{V}'$ are spaces of representations T and T'. The fact that equality of highest weights is necessary for equivalence of representations is shown in the following way. Equivalence means that there exists invertible operator $S : \mathcal{V} \to \mathcal{V}'$ such that $T'(X) = S \cdot T(X) \cdot S^{-1}$ ($\forall X \in \mathcal{A}$). So, highest weight vector $|u_0\rangle'$ is related to $|u_0\rangle$ by $|u_0\rangle' = S|u_0\rangle$ and has the same weight.

Let us show that equality of highest weights is sufficient for equivalence of representations. Let $|u_0\rangle$ and $|u_0\rangle'$ be highest weight vectors from $\mathcal{V}$ and $\mathcal{V}'$ with the same highest weight $\mu^{(0)}$. All vectors in $\mathcal{V}$ and $\mathcal{V}'$ are linear combinations of vectors

$$E_{-\alpha_k} E_{-\alpha_{k-1}} \cdots E_{-\alpha_1} |u_0\rangle, \quad E'_{-\alpha_k} E'_{-\alpha_{k-1}} \cdots E'_{-\alpha_1} |u_0\rangle', \tag{3.4.4}$$

respectively. Here we use shorthand notations E_α and E'_α for $T(E_\alpha)$ and $T'(E_\alpha)$. In accordance with Remark to Theorem 3.4.1, it is sufficient to consider only *positive* roots α_i in (3.4.4). To prove equivalence between T and T', we have to show that any linear relation for nonprimed vectors corresponds to the same linear relation (with the same coefficients) for primed vectors, and vice versa: this would precisely mean that there is one-to-one correspondence between spaces $\mathcal{V}$ and $\mathcal{V}'$. In other words, let vectors in space $\mathcal{V}$ obey

$$\sum_k \sum_\alpha \gamma_{\alpha_1,\dots\,\alpha_k} E_{-\alpha_k} E_{-\alpha_{k-1}} \cdots E_{-\alpha_1} |u_0\rangle = 0. \tag{3.4.5}$$

We have to show that the same relation, but with primes, is also valid,

$$\sum_k \sum_\alpha \gamma_{\alpha_1,\dots,\alpha_k} E'_{-\alpha_k} E'_{-\alpha_{k-1}} \cdots E'_{-\alpha_1} |u_0\rangle' = 0,$$

and vice versa. Without loss of generality, all vectors in the sum (3.4.5) have one and the same weight, otherwise the relation (3.4.5) would be a set of independent relations. It is worth noting that the sum (3.4.5) is finite, since we consider finite-dimensional representations.

Before giving the general proof, it is instructive to consider an example. Let $|u_0\rangle \in \mathcal{V}$ and $|u_0\rangle' \in \mathcal{V}'$ be highest weight vectors. Consider vectors $E_{-\alpha}|u_0\rangle \in \mathcal{V}$ and $E'_{-\alpha}|u_0\rangle' \in \mathcal{V}'$ where α is a positive root. These vectors, if nonzero, have the same weight $(\mu^{(0)} - \alpha)$. Let us show that if $E'_{-\alpha}|u_0\rangle' \neq 0$ then $E_{-\alpha}|u_0\rangle \neq 0$, and vice versa. For nonzero vector $E'_{-\alpha}|u_0\rangle'$ we have $\langle u_0|' E'_{\alpha} E'_{-\alpha}|u_0\rangle' \neq 0$ so that

$$E'_{\alpha}\, E'_{-\alpha}\, |u_0\rangle' = [E'_{\alpha},\, E'_{-\alpha}]\, |u_0\rangle' = \alpha^i\, H'_i\, |u_0\rangle' = (\alpha,\, \mu^{(0)})\, |u_0\rangle' \neq 0,$$

i.e., $(\alpha,\, \mu^{(0)}) \neq 0$. The same calculation for nonprimed objects gives the relation $E_{\alpha} E_{-\alpha}|u_0\rangle = (\alpha,\, \mu^{(0)})|u_0\rangle$, and since $(\alpha,\, \mu^{(0)}) \neq 0$ we have $E_{-\alpha}|u_0\rangle \neq 0$, as desired. This establishes correspondence between (nonzero) vectors $E_{-\alpha}|u_0\rangle$ and $E'_{-\alpha}|u_0\rangle'$, where α is any positive root. In principle, this procedure can be continued and used to establish one-to-one correspondence between spaces $\mathcal{V}$ and $\mathcal{V}'$.

We do not follow this route and give general proof *ad absurdum*. Let (3.4.5) be satisfied, but

$$\sum_k \sum_\alpha \gamma_{\alpha_1,\ldots,\alpha_k} E'_{-\alpha_k} E'_{-\alpha_{k-1}} \cdots E'_{-\alpha_1} |u_0\rangle' \equiv |\Psi\rangle' \neq 0. \tag{3.4.6}$$

Since representation T' is irreducible, vector $|\Psi\rangle'$, like any other vector, is cyclic in $\mathcal{V}'$. In particular, there exists element $\mathcal{O}'$ of enveloping algebra $\mathcal{U}(\mathcal{A})$ in representation T', which moves $|\Psi\rangle'$ to original highest weight vector $|u_0\rangle'$. In other words, we have

$$|u_0\rangle' = \mathcal{O}'|\Psi\rangle',$$

where $\mathcal{O}'$ is a linear combination of products of generators of algebra $\mathcal{A}$ in representation T'. In analogy to discussion in the proof of the first part of Theorem 3.4.1, the action of operator $\mathcal{O}'$ on the left-hand side of (3.4.6) is reduced to operations of commutation, which enable one to move Cartan generators and positive root generators to the right. Cartan operators and positive root generators acting on $|u_0\rangle'$ give $\mu_i^{(0)}|u_0\rangle'$ and zero, respectively. It is important here that we only use commutation relations of algebra $\mathcal{A}$ and the fact that $|u_0\rangle'$ is the highest weight vector with highest weight $\mu^{(0)}$. A finite number of these operations give vector $|u_0\rangle'$.

We take now operator $\mathcal{O}$ which is different from $\mathcal{O}'$ by omission of primes only. We apply this operator to the left-hand side of (3.4.5) and do the same manipulations as in the primed case. This gives vector $|u_0\rangle$, which contradicts the fact that (3.4.5) vanishes. We see that (3.4.5) implies that the left-hand side of (3.4.6) is zero, so there is one-to-one correspondence between spaces $\mathcal{V}$ and $\mathcal{V}'$. The action of generators on vectors (3.4.4), as well as their linear combinations, is completely determined by commutation relations in algebra $\mathcal{A}$ and the highest weight $\mu^{(0)}$. So, this correspondence is isomorphism of representations. □

Recall that vector x belongs to positive Weyl chamber (including its boundary hyperplanes) if $(x, \alpha) \geq 0$ for all positive roots α. Let us introduce partial ordering of weights. We choose, once and for all, some vector x in positive Weyl chamber. Let μ and λ be two weights of representation T. We say that $\lambda \geq \mu$ if $(\lambda, x) \geq (\mu, x)$. If the difference of two weights $\lambda - \mu = \sum_i \alpha_i$ is a sum of positive roots $\alpha_i > 0$, then $\lambda \geq \mu$ irrespectively of the choice of vector x. If λ is highest weight of representation T, and μ is any other weight of this representation, then, in view of (3.4.3), we have $\lambda \geq \mu$. We use this ordering of weights, in particular, in Section 3.5.

For given representation T, weight μ' is *dominant* if it belongs to positive Weyl chamber. Recall that Weyl group W acts on the set of Weyl chambers.

Proposition 3.4.3. *For any weight μ, there exists element $w \in W$ such that the equivalent weight $\mu' = w(\mu)$ is dominant.*

Proof. We choose vector x inside the positive Weyl chamber (x does not belong to any of its boundary hyperplanes), i.e., $(x, \alpha) > 0$ for all positive root α. Let us order weights with respect to this vector x. Let weight μ be nondominant (does not belong to positive Weyl chamber), i.e., there exists positive root α such that $(\mu, \alpha) < 0$. Then equivalent weight $\sigma_\alpha(\mu) = (\mu - 2\frac{(\mu,\alpha)}{\alpha^2}\alpha)$ obeys $(\sigma_\alpha(\mu), x) > (\mu, x)$, i.e., $\sigma_\alpha(\mu) > \mu$. Let now the weight $\sigma_\alpha(\mu)$ be nondominant too, then there exists positive root β such that $(\sigma_\alpha(\mu), \beta) < 0$, and $\sigma_\beta(\sigma_\alpha(\mu)) > \sigma_\alpha(\mu) > \mu$. Since Weyl group is finite, this procedure must terminate, and in its end we get that for some $w \in W$, the equivalent weight $w(\mu) = \mu'$ obeys $(\mu', \alpha) \geq 0$ for all $\alpha > 0$, meaning that the weight μ' is dominant. □

Corollary 3.4.4. *Since group W permutes Weyl chambers, the element $w \in W$, constructed in the above proof, moves Weyl chamber containing* *weight μ to positive Weyl chamber. In fact, the arguments used in the proof of Proposition* 3.4.3 *are valid for arbitrary vector μ (not necessarily weight vector). This shows that for any Weyl chamber, there exists an element $w \in W$ which moves it to positive Weyl chamber. An inverse element $w^{-1} \in W$ moves positive Weyl chamber into a given Weyl chamber. Hence, Weyl group W acts transitively in the set of Weyl chambers.*

Problem 3.4.2. Show that if μ' is dominant weight, then $\mu' \geq \sigma(\mu')$ for any element $\sigma \in W$ and any choice of the reference vector x in positive Weyl chamber.

Proposition 3.4.5. *Highest weight λ of finite-dimensional irreducible representation T is always dominant.*

Proof. Let λ be highest weight of representation T, then, in accordance with Theorem 3.2.1, its Weyl reflection

$$\sigma_\alpha(\lambda) = \lambda - \frac{2(\lambda, \alpha)}{(\alpha, \alpha)}\alpha$$

is also weight of representation T. For positive root α, we have

$$\frac{2(\lambda, \alpha)}{(\alpha, \alpha)} \geq 0, \tag{3.4.7}$$

otherwise weight $\sigma_\alpha(\lambda)$ would be a sum of weight λ and positive roots α, which would contradict the definition of highest weight. Therefore, we have $(\lambda, \alpha) \geq 0$ for all positive roots α, so λ belongs to positive Weyl chamber. □

We consider simple root $\alpha^{(i)}$ in (3.4.7) and write for highest weight λ

$$\frac{2(\lambda, \alpha^{(i)})}{(\alpha^{(i)}, \alpha^{(i)})} = n_i \geq 0, \tag{3.4.8}$$

where n_i are nonnegative integers. Making use of (3.3.1) and (3.3.4), we obtain the expansion of highest weight of any finite-dimensional irreducible representation in fundamental weights:

$$\lambda = \sum_{i=1}^{r} n_i \lambda_{(i)} \quad (n_i \in \mathbb{Z}_{\geq 0}). \tag{3.4.9}$$

Subset in Λ with weights (3.4.9) is denoted by Λ^+. In what follows, we call the subset Λ^+ a *sublattice*, even though this is not completely adequate from mathematical viewpoint.

All of the above leads us to the following theorem.

Theorem 3.4.6. *Every weight λ of simple Lie algebra $\mathcal{A}$, which belongs to positive Weyl chamber, is the highest weight of one and only one (modulo equivalence) finite-dimensional irreducible representation. In other words, finite-dimensional irreducible representations of $\mathcal{A}$ are labeled by sites of sublattice $\Lambda^+ = \{\lambda \in \Lambda | (\lambda, \alpha) \geq 0$ for all positive roots $\alpha\}$.*

Proof. It is sufficient to show that any fundamental weight $\lambda_{(f)}$ ($f = 1, \ldots, r$) is highest weight of some finite-dimensional irreducible representation $T_{(f)}$ in space $\mathcal{V}_{(f)}$. We prove this below. Once this fact is established, then, as we see immediately, irreducible representation $T^{(n_1,\ldots,n_r)}$ with any given highest weight (3.4.9) is constructed as subrepresentation in tensor product of representations $T_{(f)}$,

$$\tilde{T} := T_{(1)}^{\otimes n_1} \otimes T_{(2)}^{\otimes n_2} \otimes \cdots \otimes T_{(r)}^{\otimes n_r}, \quad n_i \in \mathbb{Z}_{\geq 0}. \tag{3.4.10}$$

Tensor product (3.4.10) acts in space

$$\tilde{\mathcal{V}} := \mathcal{V}_{(1)}^{\otimes n_1} \otimes \mathcal{V}_{(2)}^{\otimes n_2} \otimes \cdots \otimes \mathcal{V}_{(r)}^{\otimes n_r}. \tag{3.4.11}$$

Generally speaking, representation (3.4.10) is reducible. Elements $X \in \mathcal{A}$ in representation (3.4.10) are operators

$$\tilde{T}(X) = \sum_{f=1}^{r} \Big(I_{(1)}^{\otimes n_1} \otimes \cdots \otimes I_{(f-1)}^{\otimes n_{f-1}} \otimes T_{(f)}^{\otimes n_f}(X) \otimes I_{(f+1)}^{\otimes n_{f+1}} \otimes \cdots \otimes I_{(r)}^{\otimes n_r} \Big), \tag{3.4.12}$$

which act in space (3.4.11), where operators

$$T_{(f)}^{\otimes n_f}(X) = \sum_{k=1}^{n_f} \Big(I_{(f)}^{\otimes k-1} \otimes T_{(f)}(X) \otimes I_{(f)}^{\otimes (n_f - k)} \Big)$$

act in $\mathcal{V}_{(f)}^{\otimes n_f}$ and $I_{(f)}$ are unit operators in $\mathcal{V}_{(f)}$. Let $|\lambda_{(f)}\rangle$ be highest weight vectors of representations $T_{(f)}$ in spaces $\mathcal{V}_{(f)}$ whose weights are fundamental,

$$T_{(f)}(H_i)|\lambda_{(f)}\rangle = (\lambda_{(f)})_i \, |\lambda_{(f)}\rangle, \quad T_{(f)}(E_\alpha)|\lambda_{(f)}\rangle = 0 \quad (\forall \alpha > 0). \tag{3.4.13}$$

We construct a special vector in space (3.4.11),

$$|\vec{n}\,\rangle = |\lambda_{(1)}\rangle^{\otimes n_1} \otimes |\lambda_{(2)}\rangle^{\otimes n_2} \otimes \cdots \otimes |\lambda_{(r)}\rangle^{\otimes n_r}. \tag{3.4.14}$$

Here, we use shorthand notation $\vec{n} = (n_1, \ldots, n_r)$. Making use of (3.4.12) and the second relation in (3.4.13) we find for all positive roots

α that

$$\tilde{T}(E_\alpha)\,|\vec{n}\,\rangle = 0. \tag{3.4.15}$$

Also, again using (3.4.12) and now the first relation in (3.4.13), we find that vector (3.4.14) is eigenvector of Cartan generators $\tilde{T}(H_i)$:

$$\tilde{T}(H_i)\,|\vec{n}\,\rangle = \lambda_i^{(\vec{n})}\,|\vec{n}\,\rangle, \tag{3.4.16}$$

and its weight is $\lambda^{(\vec{n})} = \sum_{i=1}^{r} n_i \lambda_{(i)}$, where by construction we have $n_i \in \mathbb{Z}_{\geq 0}$. Thus, vector $|\vec{n}\,\rangle$ is highest weight vector of some irreducible subrepresentation $T^{(\vec{n})}$ in representation (3.4.10), and its (highest) weight is given by (3.4.9). Invariant subspace $\mathcal{V}^{(\vec{n})} \subset \overline{\mathcal{V}}$ of irreducible representation $T^{(\vec{n})}$ in $\tilde{T}$ is, in accordance with item C of Section 3.2, a space of linear combinations of vectors

$$\cdots\, \tilde{T}(E_\gamma)\cdot\tilde{T}(E_\beta)\cdot\tilde{T}(E_\alpha)\,|\vec{n}\,\rangle, \tag{3.4.17}$$

where $\alpha, \beta, \gamma, \ldots$ are negative roots of $\mathcal{A}$ (see discussion in the proof of the first part of Theorem 3.4.1).

To finish the proof of Theorem 3.4.6, we have to show that every fundamental weight $\lambda_{(f)}$ is indeed the highest weight of some finite-dimensional representation $T_{(f)}$. We do not give a general proof of this fact, and rather construct explicitly in Section 3.4.2 these representations $T_{(f)}$ for Lie algebras of classical series $s\ell(n,\mathbb{C})$, $so(n,\mathbb{C})$ and $sp(2r,\mathbb{C})$ and in this way establish the validity of Theorem 3.4.6 for these algebras. □

Corollary 3.4.7. *Let w be an element of Weyl group W. In accordance with Theorem* 3.2.1, *if μ is weight, then vector $w(\mu)$ is also weight. It follows from Theorem* 3.4.6 *that all vectors* (3.4.9) *in positive Weyl chamber are highest weight vectors. Since the Weyl group acts in the set of Weyl chambers transitively (i.e., every Weyl chamber V can be obtained by acting on positive Weyl chamber by some element $w \in W$), every vector* (3.3.5) *in the entire weight lattice Λ (that belongs to some V) can be obtained by the action of element $w \in W$ on some highest weight. Thus, all admissible weights μ (which obey* (3.2.10)) *are indeed weights, so the property* (3.2.10) *is indeed a sufficient condition for μ to be a weight.*

Corollary 3.4.8. *It follows from Theorem* 3.4.6 *and formula* (3.4.9) *that highest weights of finite-dimensional irreducible representations $T^{(\vec{n})}$ of simple Lie algebra $\mathcal{A}$ of rank r are labeled by a set of r nonnegative integers n_i $(i = 1, \ldots, r)$ which are multiplicities of weights $\lambda_{(i)}$ in the sum* (3.4.9). *Since fundamental weights $\lambda_{(i)}$ are in one-to-one correspondence with* *simple roots $\alpha_{(i)}$, the numbers n_i can be assigned to vertices of Dynkin diagram of algebra $\mathcal{A}$. Such a weighted Dynkin diagram encodes an irreducible representation of algebra $\mathcal{A}$ with highest weight* (3.4.9).

As an example, any irreducible representation of algebra $s\ell(3,\mathbb{C})$ is encoded by diagram $\overset{n_1}{\circ}\!\!-\!\!-\!\!\overset{n_2}{\circ}$. In particular, defining and co-defining representations of $s\ell(3,\mathbb{C})$, discussed in Example 2 of Section 3.3 (see Fig. 3.3.2) correspond to diagrams

$$\underset{\alpha^{(1)}}{\overset{1}{\circ}}\!\!-\!\!-\!\!-\!\!-\!\!\underset{\alpha^{(2)}}{\overset{0}{\circ}}, \qquad \underset{\alpha^{(1)}}{\overset{0}{\circ}}\!\!-\!\!-\!\!-\!\!-\!\!\underset{\alpha^{(2)}}{\overset{1}{\circ}}. \tag{3.4.18}$$

Indeed, the highest weight for defining representation is fundamental weight $\lambda_{(1)}$, while for co-defining representation, the highest weight is fundamental weight $\lambda_{(2)}$. We note that nonnegative integers n_i determining the highest weight (3.4.9) and assigned to vertices of Dynkin diagram are often called *Dynkin indices.*

It is sometimes convenient to use the following notation for matrices of elements $X \in \mathcal{A}$ in irreducible representation $T^{(\vec{n})}$:

$$D^{(\lambda)}_{\mu,\mu'}(X) = D^{(n_1,\dots,n_r)}_{\mu,\mu'}(X) = \langle\mu|\, T^{(\vec{n})}(X)\,|\mu'\rangle, \tag{3.4.19}$$

where the highest weight λ is determined, in accordance with (3.4.9), by the set of Dynkin indices $(n_1,\dots,n_r)$, while μ and μ' label all basis vectors in the space of representation, i.e., they run through all possible weights (with account for their multiplicities) of the representation.

3.4.2. Fundamental weights and representations of Lie algebras $s\ell(n,\mathbb{C})$, $so(n,\mathbb{C})$ and $sp(2r,\mathbb{C})$

Here, we figure out the fundamental weights $\lambda_{(f)}$ for Lie algebras $s\ell(n,\mathbb{C})$, $so(n,\mathbb{C})$ and $sp(2r,\mathbb{C})$, and construct representations $T_{(f)}$ with the highest weights $\lambda_{(f)}$. In a certain sense (see proof of Theorem 3.4.5), representations $T_{(f)}$ are building blocks for other finite-dimensional irreducible representations, which can be used, in accordance with (3.4.15), (3.4.16) and (3.4.17), to construct any finite-dimensional irreducible representation with highest weight (3.4.9) as subrepresentation in a tensor product of representations $T_{(f)}$. So, knowledge of fundamental weights of simple complex Lie algebra $\mathcal{A}$ enables one to classify and construct all finite-dimensional irreducible representations, each of which has a highest weight (3.4.9) and corresponds to a site of sublattice Λ^+.

We emphasize again that we consider in this chapter *complex* finite-dimensional irreducible representations of simple Lie algebras.

1. Algebra $s\ell(2,\mathbb{C})$.
Weight lattice of algebra $s\ell(2,\mathbb{C})$ is a one-dimensional lattice $\Lambda = \mathbb{Z}/2$ shown in Fig. 3.3.1 (see Example 1 in Section 3.3). It follows from (3.3.8) that there is one fundamental weight $\lambda_{(1)} = \frac{1}{2}$, and, in accordance with (3.4.9), all highest weights are $\lambda = n\,\lambda_{(1)} = \frac{n}{2}$, where $n \in \mathbb{Z}_{\geq 0}$. They belong to sublattice $\Lambda^+ = \mathbb{Z}_{\geq 0}/2$. We choose a site $j \in \Lambda^+ = \mathbb{Z}_{\geq 0}/2$ and denote representation with this highest weight by $T^{(j)}$. Other weights $\mu_{(k)}$ of representation $T^{(j)}$ are obtained from the highest weight j by subtracting simple root $\alpha = 1$: $\mu_{(k)} = j - k$, where k are positive integers. Weight string must be symmetric under Weyl group of reflections $\mu \to -\mu$ (see Example 1 in Section 3.2). Thus, this string is $\{j, j-1, \dots, -j+1, -j\}$, and all weights of this string are simple (this follows from Corollary 3.2.2). So, we have $\dim(T^{(\lambda)}) = 2j+1$. In this way, we reproduce, in the framework of weight theory, known results concerning finite-dimensional irreducible representations of $s\ell(2,\mathbb{C})$ which were obtained by explicit construction in Section I-4.7.3 of the first book (see also Sections 2.1 and 2.2.1).

2. Algebras $s\ell(r+1,\mathbb{C})$, $r > 1$
Root system of Lie algebra $s\ell(r+1,\mathbb{C})$ (A_r-series) consists of real vectors (3.1.11) which belong to r-dimensional hyperplane $P_u \subset \mathbb{R}^{r+1}$ normal to vector $u = (1,1,\dots,1) \in \mathbb{R}^{r+1}$. The weight and root lattices of Lie algebra $s\ell(r+1,\mathbb{C})$ are studied in Example 1 in Section 3.2 (see formulas (3.2.23) and (3.2.24)). Dominant weights λ obey $(\alpha,\lambda) \geq 0$ for all positive roots α, and with the choice (3.1.13) of positive roots $\alpha_+^{(kl)}$ this means that components λ_i of dominant weights satisfy

$$\lambda_1 \geq \lambda_2 \geq \cdots \geq \lambda_r \geq \lambda_{r+1}. \tag{3.4.20}$$

Fundamental weights $\lambda_{(i)}$ are obtained by solving Eq. (3.3.2), where simple roots $\alpha^{(i)}$ are given in (3.1.14). Also, weights $\lambda_{(i)}$ belong to root space P_u, so they obey $(\lambda_{(i)}, u) = 0$. These requirements enable one to find fundamental weights $\lambda_{(i)} \in \mathbb{R}^{r+1}$ in a unique way:

$$\begin{aligned}
\lambda_{(1)} &= \frac{1}{r+1}(r,-1,\dots,-1),\\
\lambda_{(2)} &= \frac{1}{r+1}(r-1,r-1,-2,\dots,-2)\\
&\;\vdots\\
\lambda_{(r-1)} &= \frac{1}{r+1}(2,\dots,2,1-r,\,1-r),\\
\lambda_{(r)} &= \frac{1}{r+1}(1,1,\dots,1,-r).
\end{aligned} \tag{3.4.21}$$

Note that these weights obey dominance condition (3.4.20) automatically. It is convenient to write them in a unified way:

$$\lambda_{(f)} = (\underbrace{1,\dots,1}_{f},0,\dots,0) - \frac{f}{r+1}(1,1,\dots,1), \quad f = 1,\dots,r. \tag{3.4.22}$$

Problem 3.4.3. In the case of algebra $s\ell(3,\mathbb{C})$, find the basis in two-dimensional plane P_u in such a way that components of simple roots and fundamental weights, found in (3.1.14) and (3.4.21), take the form of two-dimensional vectors (3.3.10) and (3.3.12), respectively.

Now we prove that fundamental weights (3.4.21), (3.4.22) are indeed highest weights of certain finite-dimensional irreducible representations of algebra $s\ell(r+1,\mathbb{C})$. This is done by explicitly constructing these representations.

Let $T_{(1)}$ be defining representation of algebra $s\ell(r+1,\mathbb{C})$, and $\epsilon^{(i)}$ $(i = 1,\dots r+1)$ be basis vectors in $(r+1)$-dimensional complex space[d] $\mathcal{V}_{r+1}$ of representation $T_{(1)}$. Then tensor products $\epsilon^{(i_1)} \otimes \cdots \otimes \epsilon^{(i_f)}$ make basis in space $\mathcal{V}_{r+1}^{\otimes f}$ of representation $T_{(1)}^{\otimes f}$; elements of this space are rank-f tensors

$$\Psi = \Psi^{i_1\dots i_f} \cdot (\epsilon^{(i_1)} \otimes \cdots \otimes \epsilon^{(i_f)}) \in \mathcal{V}_{r+1}^{\otimes f}, \tag{3.4.23}$$

where $\Psi^{i_1\dots i_f} \in \mathbb{C}$ are components of tensor Ψ. Element X of algebra $s\ell(r+1,\mathbb{C})$ acts in $\mathcal{V}_{r+1}^{\otimes f}$ in the standard way

$$T_{(1)}^{\otimes f}(X) = \sum_{k=1}^{f} (I_{r+1})^{\otimes(k-1)} \otimes T_{(1)}(X) \otimes (I_{r+1})^{\otimes(f-k)}. \tag{3.4.24}$$

Note that representation $T_{(1)}^{\otimes f}$ is, generally speaking, reducible.

Consider now a subspace of $\mathcal{V}_{r+1}^{\otimes f}$ which consists of totally antisymmetric tensors $\Psi_{(a)}$, whose components $\Psi_{(a)}^{i_1\dots i_f}$ in expansion (3.4.23) are antisymmetric in all their indices. This space is denoted by

$$\mathcal{V}_{r+1}^{\wedge f} \equiv \underbrace{\mathcal{V}_{r+1} \wedge \mathcal{V}_{r+1} \wedge \cdots \wedge \mathcal{V}_{r+1}}_{f} \subset \mathcal{V}_{r+1}^{\otimes f}, \tag{3.4.25}$$

and its elements are written as

$$\Psi_{(a)} = \Psi_{(a)}^{i_1\dots i_f} \cdot \epsilon^{(i_1)} \wedge \epsilon^{(i_2)} \wedge \cdots \wedge \epsilon^{(i_f)}, \tag{3.4.26}$$

[d]To avoid confusion, we use different notations for basis vectors $\epsilon^{(i)}$ in the space of defining representation $\mathcal{V}_{r+1}$ and for basis vectors $e^{(i)}$ of $(r+1)$-dimensional real space where the root space $\mathfrak{V}_r = P_u$ is embedded.

where

$$\begin{aligned}\epsilon^{(i_1)} \wedge \epsilon^{(i_2)} \wedge \cdots \wedge \epsilon^{(i_f)} \; &:= \frac{1}{f!} \sum_{\sigma \in S_f} (-1)^{p(\sigma)} \epsilon^{(i_{\sigma_1})} \otimes \epsilon^{(i_{\sigma_2})} \otimes \cdots \otimes \epsilon^{(i_{\sigma_f})} \\ &= \epsilon^{(j_1)} \otimes \epsilon^{(j_2)} \otimes \cdots \otimes \epsilon^{(j_f)} \, A^{j_1 \ldots j_f}_{i_1 \ldots i_f} \\ &= A \cdot (\epsilon^{(i_1)} \otimes \cdots \otimes \epsilon^{(i_f)}). \end{aligned} \tag{3.4.27}$$

Summation here runs over all permutations $\sigma : (1, 2, \ldots, f) \to (\sigma_1, \sigma_2, \ldots, \sigma_f)$, $p(\sigma)$ is parity of permutation σ, and operator (projector) A is antisymmetrizer of rank f and its matrix has the form

$$A^{j_1 \ldots j_f}_{i_1 \ldots i_f} = \frac{1}{f!} \sum_{\sigma \in S_f} (-1)^{p(\sigma)} \delta^{j_1}_{i_{\sigma_1}} \ldots \delta^{j_f}_{i_{\sigma_f}} \;\Rightarrow\; A^2 = A. \tag{3.4.28}$$

Elements $\epsilon^{(i_1)} \wedge \ldots \wedge \epsilon^{(i_f)}$, given by (3.4.27), make basis in $\mathcal{V}^{\wedge f}_{r+1}$. It follows from (3.4.24) that any element X of algebra $s\ell(r+1, \mathbb{C})$ in representation $T^{\otimes f}_{(1)}$ commutes with antisymmetrizer A:

$$T^{\otimes f}_{(1)}(X) \cdot A = A \cdot T^{\otimes f}_{(1)}(X). \tag{3.4.29}$$

In other words, operator $T^{\otimes f}_{(1)}(X)$ takes totally antisymmetric tensor (3.4.26) to another totally antisymmetric tensor. Therefore, $\mathcal{V}^{\wedge f}_{r+1}$ is invariant subspace in $\mathcal{V}^{\otimes f}_{r+1}$. Moreover representation $T^{\otimes f}_{(1)}$, restricted to subspace (3.4.25) of totally antisymmetric tensors, is irreducible and is denoted by $T^{\wedge f}_{(1)}$ in what follows. Representations $T^{\wedge f}_{(1)}$ are discussed in detail in Chapter 4. We show there that representation $T^{\wedge f}_{(1)}$ of Lie algebra $s\ell(r+1, \mathbb{C})$ is irreducible, because the corresponding representation $T^{\wedge f}_{(1)}$ of Lie group $SL(r+1, \mathbb{C})$ is irreducible. In its turn, the fact that the representation $T^{\wedge f}_{(1)}$ of $SL(r+1, \mathbb{C})$ is irreducible, is a consequence of Proposition 4.2.5 (see Section 4.2), and also of the results in Section 4.4.1.

Problem 3.4.4. Show that $\dim(\mathcal{V}^{\wedge f}_{r+1}) = \frac{(r+1)!}{(r+1-f)!f!}$.

We are now ready to specify finite-dimensional irreducible representations $T_{(f)}$ of algebra $s\ell(r+1, \mathbb{C})$, whose highest weights are fundamental weights $\lambda_{(f)}$ given in (3.4.21).

Proposition 3.4.9. *Fundamental weights $\lambda_{(f)}$ in (3.4.22) are highest weights of finite-dimensional irreducible representations $T_{(f)} = T^{\wedge f}_{(1)}$ of algebra $s\ell(r+1, \mathbb{C})$. In particular, the first fundamental weight $\lambda_{(1)}$ is highest weight of defining representation $T_{(1)}$.*

Proof. To begin with, we consider the case $f = 1$ and show that fundamental weight $\lambda_{(1)}$ is indeed highest weight of defining representation. We choose orthonormalized basis in the space of defining representation $\mathcal{V}_{r+1}$ which consists of unit vectors $|\mu^{(k)}\rangle \equiv \epsilon^{(k)}$ with components $(\epsilon^{(k)})_j = \delta_{kj}$ $(k,j = 1,\ldots,r+1)$. Cartan generators H_i in defining representation are diagonal matrices (3.1.8), and we immediately find all weights of this representation, which we use later on:

$$H_i|\mu^{(k)}\rangle = \mu_i^{(k)}\,|\mu^{(k)}\rangle \Rightarrow \mu^{(k)} = e^{(k)} - \frac{1}{r+1}u, \quad k = 1,\ldots,r+1, \tag{3.4.30}$$

where $u = (1,\ldots,1) \in \mathbb{R}^{r+1}$ and we write H_i instead of $T_{(1)}(H_i)$. Weights $\mu^{(k)}$ in (3.4.30) and roots (3.1.11) are related by $\mu^{(k)} + \alpha^{(ik)} = \mu^{(i)}$. So, for all $k > i$, when root $\alpha^{(ik)}$ is positive, weight $\mu^{(k)}$ is not highest weight. Therefore, highest weight of defining representation $T_{(1)}$ is $\mu^{(i)}$ for $i = 1$:

$$\mu^{(1)} = e^{(1)} - \frac{1}{r+1}\,u = (1,0,\ldots,0) - \frac{1}{r+1}(1,\ldots,1) \equiv \lambda_{(1)}, \tag{3.4.31}$$

and it is indeed the first fundamental weight in (3.4.21); see also formula (3.4.22) for $f = 1$.

Highest weight vector with highest weight $\mu^{(1)} = \lambda_{(1)}$ is $\epsilon^{(1)} = |\mu^{(1)}\rangle \in \mathcal{V}_{r+1}$. One can see explicitly that this is indeed the case by checking the relations

$$E_{\alpha^{(i)}}|\mu^{(1)}\rangle = e_{i\,i+1}\cdot\epsilon^{(1)} = 0. \tag{3.4.32}$$

Here, we use defining representation (3.1.7) of root generators $E_{\alpha^{(i)}}$. Equation (3.4.32) is valid for all i, i.e., for all simple roots $\alpha^{(i)}$ given by (3.1.14).

Consider now representation $T_{(1)}^{\wedge f}$ of algebra $s\ell(r+1,\mathbb{C})$. We choose in space $\mathcal{V}_{r+1}^{\wedge f}$ a particular vector (cf. (3.4.27))

$$|\lambda_{(f)}\rangle = \frac{1}{f!}\sum_{\sigma\in S_f}(-1)^{p(\sigma)}\epsilon^{(\sigma_1)}\otimes\epsilon^{(\sigma_2)}\otimes\cdots\otimes\epsilon^{(\sigma_f)} = \epsilon^{(1)}\wedge\epsilon^{(2)}\wedge\cdots\wedge\epsilon^{(f)}, \tag{3.4.33}$$

where, like in (3.4.27), summation runs over all permutations σ and $p(\sigma)$ is parity of permutation σ. Vector (3.4.33) is antisymmetrized product of first f basis vectors $\{\epsilon^{(1)},\ldots,\epsilon^{(f)}\}$ from the set of all basis vectors

$\epsilon^{(1)}, \ldots, \epsilon^{(r+1)} \in \mathcal{V}_{r+1}$. This vector obeys

$$T_{(1)}^{\otimes f}\left(E_{\alpha^{(i)}}\right)|\lambda_{(f)}\rangle = 0 \tag{3.4.34}$$

for all simple roots $\alpha^{(i)}$ and, as we see below, it is an eigenvector of Cartan generators H_i in representation (3.4.24), i.e., operators $T_{(1)}^{\otimes f}(H_i)$.

‖ **Problem 3.4.5.**★ Prove that vector (3.4.33) obeys (3.4.34).

Thus, (3.4.33) is highest weight vector in invariant subspace $\mathcal{V}_{r+1}^{\wedge f} \subset \mathcal{V}_{r+1}^{\otimes f}$ of irreducible representation $T_{(1)}^{\wedge f}$. To obtain this highest weight, we act on vector (3.4.33) by operators $T_{(1)}^{\otimes f}(H_i)$ and make use of formula (3.4.30) for all weights $\mu^{(k)}$ of defining representation $T_{(1)}$. Then for all $f \leq r$ we obtain

$$\begin{aligned} T_{(1)}^{\otimes f}(H_i)\,|\lambda_{(f)}\rangle &= \left(\sum_{k=1}^{f} I_{r+1}^{\otimes(k-1)} \otimes T_{(1)}(H_i) \otimes I_{r+1}^{\otimes(f-k)}\right)|\lambda_{(f)}\rangle \\ &= \left(\sum_{k=1}^{f} \mu_i^{(k)}\right)|\lambda_{(f)}\rangle, \end{aligned} \tag{3.4.35}$$

where $i = 1, \ldots, (r+1)$. So, weight of the highest weight vector $|\lambda_{(f)}\rangle \in \mathcal{V}_{r+1}^{\wedge f}$ (i.e., highest weight of representation $T_{(f)} = T_{(1)}^{\wedge f}$) is

$$\sum_{k=1}^{f} \mu^{(k)} = \sum_{k=1}^{f} e^{(k)} - \frac{f}{r+1}\,u,$$

which is precisely the fundamental weight $\lambda_{(f)}$ in (3.4.22). □

The result of the Proposition 3.4.9 completes the proof of Theorem 3.4.6 for algebras $s\ell(r+1, \mathbb{C})$.

In accordance with Theorem 3.4.6 and analysis made in its proof, highest weights of all finite-dimensional irreducible representations of algebra $s\ell(r+1, \mathbb{C})$ are determined by (3.4.9) and belong to sublattice Λ^+ generated by fundamental weights (3.4.21) and (3.4.22). Every point in this sublattice is characterized by a vector $\vec{n} = (n_1, \dots, n_r)$ with $n_i \in \mathbb{Z}_{\geq 0}$, i.e.,

$$\lambda^{(\vec{n})} = \sum_{f=1}^{r} n_f \cdot \lambda_{(f)} = (\lambda_1, \dots, \lambda_r, \lambda_{r+1}) - \frac{n}{r+1}(1, 1, \dots, 1), \qquad (3.4.36)$$

where $n_f \in \mathbb{Z}_{\geq 0}$, $\lambda_{r+1} = 0$ and

$$\lambda_k - \lambda_{k+1} = n_k \;\Rightarrow\; \lambda_k = \sum_{f=k}^{r} n_f, \quad n = \sum_{k=1}^{r} \lambda_k = \sum_{f=1}^{r} f \cdot n_f \equiv |\lambda| \quad (3.4.37)$$

(see also [5]).

The pertinent irreducible representation $T^{(\vec{n})}$ is a subrepresentation in tensor product of representations $T_{(f)} = T_{(1)}^{\wedge f}$,

$$T_{(1)}^{\otimes n_1} \otimes T_{(2)}^{\otimes n_2} \otimes \cdots \otimes T_{(r)}^{\otimes n_r}, \quad n_i \in \mathbb{Z}_{\geq 0}. \qquad (3.4.38)$$

This tensor product acts in space

$$\tilde{\mathcal{V}} = \mathcal{V}_{r+1}^{\otimes n_1} \otimes (\mathcal{V}_{r+1}^{\wedge 2})^{\otimes n_2} \otimes \cdots \otimes (\mathcal{V}_{r+1}^{\wedge r})^{\otimes n_r}, \qquad (3.4.39)$$

and is reducible. The highest weight vector of $T^{(\vec{n})}$ in the space (3.4.39) is the tensor product of highest weight vectors of representations $T_{(f)}$,

$$|\vec{n}\,\rangle = |\lambda_{(1)}\rangle^{\otimes n_1} \otimes |\lambda_{(2)}\rangle^{\otimes n_2} \otimes \cdots \otimes |\lambda_{(r)}\rangle^{\otimes n_r}. \qquad (3.4.40)$$

The invariant subspace $\mathcal{V}^{(\vec{n})} \subset \tilde{\mathcal{V}}$ of irreducible representation $T^{(\vec{n})}$ is a set of linear combinations of vectors (see (3.4.17))

$$\cdots T^{(\vec{n})}(E_\gamma)\, T^{(\vec{n})}(E_\beta)\, T^{(\vec{n})}(E_\alpha)\, |\vec{n}\,\rangle, \qquad (3.4.41)$$

where $\alpha, \beta, \gamma, \dots$ are negative roots and operators $T^{(\vec{n})}(E_\alpha)$ are defined in accordance to the rule (3.4.12). All this is summarized as follows.

Proposition 3.4.10. *Any finite-dimensional irreducible representation* $T^{(\vec{n})}$ *of algebra* $s\ell(r+1, \mathbb{C})$ *is uniquely characterized by its highest weight* (3.4.36), *i.e., by the set of* r *nonnegative integers* n_f, *and is constructed as subrepresentation in tensor product* (3.4.38) *of representations* $T_{(f)}$ *whose highest weights are fundamental weights* (3.4.21) *and* (3.4.22).

As we have already pointed out, integers n_i are associated with fundamental weights $\lambda_{(i)}$ and hence simple roots $\alpha^{(i)}$. This is visualized by weighted Dynkin diagrams, as shown in Fig. 3.4.1 (cf. (3.4.18)).

Fig. 3.4.1. Weighted Dynkin diagram for representation $T^{(\vec{n})}$ of algebra $s\ell(r+1,\mathbb{C})$ with highest weight (3.4.36).

Since any fundamental representation[e] $T_{(f)} = T_{(1)}^{\wedge f}$ of $s\ell(r+1,\mathbb{C})$, in accordance with Proposition 3.4.9, can be viewed as subrepresentation in tensor product of f defining representations $T_{(1)}$, we have yet another important property which is used in Chapter 4.

Proposition 3.4.11. *Any finite-dimensional irreducible representation $T^{(\vec{n})}$ of algebra $s\ell(r+1,\mathbb{C})$ is a subrepresentation in tensor product $T_{(1)}^{\otimes n}$ of defining representations, where exponent n is $n = \sum_{f=1}^{r} f\cdot n_f$; see* (3.4.37).

Remark 1. Highest weight (3.4.36) of irreducible representation $T^{(\vec{n})}$ of algebra $s\ell(r+1,\mathbb{C})$ is determined by the set of integers λ_i $(i=1,\dots,r)$, which, in accordance with (3.4.37), are ordered, $\lambda_1 \geq \lambda_2 \geq \cdots \geq \lambda_r \geq 0$, and their sum equals n. A set $\Lambda = [\lambda_1,\dots,\lambda_r]$ of nonnegative integers with these properties is called *partition* of number n and is denoted by $\Lambda \vdash n$. So, highest weights $\lambda^{(\vec{n})} \in \Lambda^+$ of algebra $s\ell(r+1,\mathbb{C})$ (finite-dimensional irreducible representations $T^{(\vec{n})}$) are in one-to-one correspondence with partitions $\Lambda \vdash n$. A partition can be represented by Young diagram $\Lambda = [\lambda_1,\dots,\lambda_r]$ with n boxes, whose first row has λ_1 boxes, second row has λ_2 boxes, etc., and the last, rth row has λ_r boxes. We have met partitions and Young diagrams when discussing symmetric group in Section I-1.1.4 of the first book; they are key combinatorial objects in what follows. Integers $\lambda_k \geq 0$, defining Young diagrams, and integers $n_f \geq 0$, determining highest weight (3.4.9) (and assigned to vertices of Dynkin diagram in Fig. 3.4.1) are related by (3.4.37). In particular, for fundamental highest weights $\lambda_{(f)}$ we have, by comparing (3.4.22) and (3.4.36),

$$n_i = \delta_{if} \;\Leftrightarrow\; \Lambda = [\lambda_1,\dots,\lambda_r] = [\underbrace{1,\dots,1}_{f},0,\dots,0] \equiv [1^f].$$

[e]Here, we follow mathematical nomenclature, and call fundamental any representation whose highest weight is fundamental. In physics literature, fundamental representation is a synonym of defining representation, the latter being just one of "mathematical" fundamental representations.

Therefore, irreducible representation $T_{(f)} = T_{(1)}^{\wedge f}$ with highest weight equal to fundamental weight $\lambda_{(f)}$, which acts in space of rank-f totally antisymmetric tensors (3.4.26) is associated with Young diagram that has just one column of height f:

$$T_{(f)} = T_{(1)}^{\wedge f} \longleftrightarrow [1^f] = \left.\begin{array}{|c|}\hline \\ \hline \\ \hline \vdots \\ \hline \\ \hline \end{array}\right\} f. \tag{3.4.42}$$

Starting from the upper box, one assigns to the boxes indices $(i_1, \ldots, i_f)$ of antisymmetric components $\Psi_{(a)}^{i_1 \ldots i_k}$. The resulting Young tableaux is associated with totally antisymmetric tensors (3.4.26), whose linear space is the space of irreducible representation $T_{(1)}^{\wedge f}$.

Remark 2. Space $\mathcal{V}^{(\vec{n})}$ of irreducible representation $T^{(\vec{n})}$ of algebra $s\ell(r+1, \mathbb{C})$ with highest weight (3.4.36) is a subspace of tensors of rank $n = \sum_{f=1}^{r} f \cdot n_f$, see (3.4.37) and (3.4.38). Symmetry of these tensors is determined by Young diagram $\Lambda = [\lambda_1, \ldots, \lambda_r]$ with $n = \sum_{i=1}^{r+1} \lambda_i$ boxes. We study the latter property in detail in Section 4.4.1, and here we give preliminary discussion.

Let us write tensor product (3.4.38) in inverse order, and make use of graphic representation (3.4.42). As a result we have the representation shown in Fig. 3.4.2. Reducible representation $\tilde{T}$ shown in Fig. 3.4.2 acts in space (3.4.39) of tensors Ψ whose rank is $n = \sum_f f \cdot n_f$. We now put indices $(i_1, \ldots, i_n)$ of components $\Psi^{i_1 \ldots i_n}$ of tensor Ψ into the boxes of the diagrams in Fig. 3.4.2. Then we see that antisymmetry in indices in each of factors $T_{(f)}$ translates into antisymmetry in indices belonging to one and the same column. In the process of extraction of invariant subspace of irreducible representation $T^{(\vec{n})}$ from the space $\tilde{\mathcal{V}}$ (see (3.4.41)), this antisymmetry in indices belonging to one and the same column will not, in general, be preserved, and there will emerge additional *symmetrization* in indices belonging to different columns. As a result, Young diagram associated with

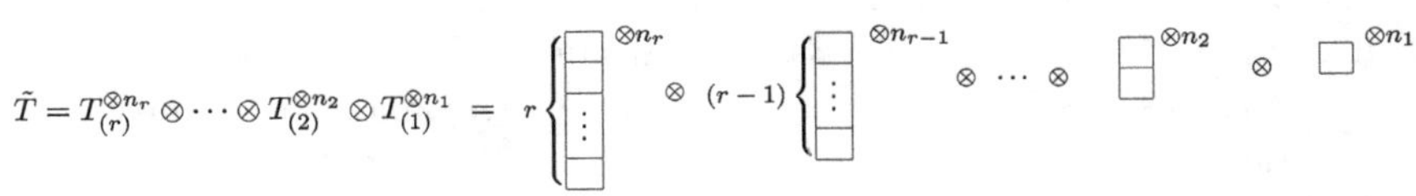

Fig. 3.4.2. Visualization of representation (3.4.38) of Lie algebra $s\ell(r+1, \mathbb{C})$.

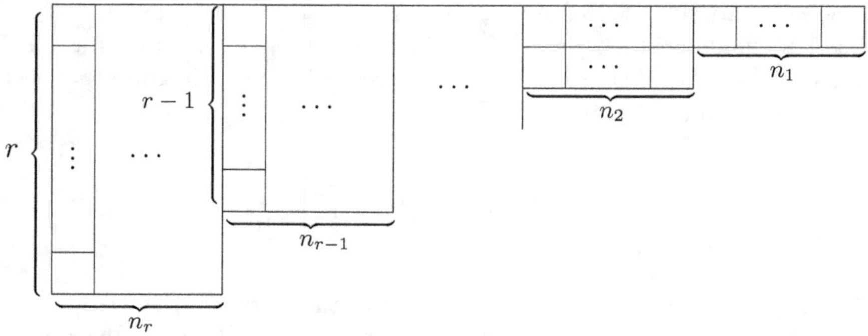

Fig. 3.4.3. Young diagram Λ made of n_r columns of height r, n_{r-1} columns of height $(r-1)$, etc. The first row has length $\lambda_1 = n_r + \cdots + n_2 + n_1$, the length of the second row is $\lambda_2 = n_r + \cdots + n_2$, etc.; the last row has length $\lambda_r = n_r$. This precisely corresponds to $\Lambda = [\lambda_1, \dots, \lambda_r]$.

the irreducible representation $T^{(\vec{n})}$ is obtained from diagram in Fig. 3.4.2 by gluing columns together, as shown in Fig. 3.4.3.

Young diagram obtained in this way, $\Lambda = [\lambda_1, \dots, \lambda_r]$, coincides with the Young diagram discussed in Remark 1 (see the first formulas in (3.4.37)). In fact, this diagram determines completely the symmetry properties of rank-n tensors Ψ which form the space of irreducible representation $T^{(\vec{n})}$. The precise relationship between Young diagrams and symmetry properties of tensors in spaces of irreducible representations of Lie algebra $s\ell(r+1,\mathbb{C})$ is studied in Section 4.4.1.

2.1. Algebras $g\ell(r+1,\mathbb{C})$

Below, in Section 3.5.2 where we derive explicit formulas for characters of linear groups, and in Section 4.6.5 where we discuss Gelfand–Tsetlin basis in the spaces of representations of Lie algebras, we make use of properties of highest weight representations of nonsimple algebras $g\ell(r+1,\mathbb{C})$ and $u(r+1)$.

The above discussion of irreducible complex representations of algebra $s\ell(r+1,\mathbb{C})$ is straightforwardly applied to algebra $g\ell(r+1,\mathbb{C})$ (and its real compact form $u(r+1)$). Basis in algebra $g\ell(r+1,\mathbb{C})$ is formed, in defining representation, by root generators (3.1.7) and Cartan elements (cf. (3.1.8))

$$H'_i \equiv e_{ii} \quad (i = 1, \dots, r+1), \tag{3.4.43}$$

which, however, no longer obey $s\ell$-condition (3.1.9). Therefore, Cartan subalgebra in $g\ell(r+1,\mathbb{C})$ is an extension (by addition of unit matrix) of Cartan subalgebra in $s\ell(r+1,\mathbb{C})$, so the *weight* space $\mathfrak{V}_{r+1}$ of algebra $g\ell(r+1,\mathbb{C})$ has dimension $(r+1)$, since this space is defined as the space of vectors $\mu = (\mu_1,\ldots,\mu_{r+1})$ whose components μ_i are determined by

$$H'_i\,|\mu\rangle = \mu_i\,|\mu\rangle \quad (i = 1,\ldots,r+1),$$

where $|\mu\rangle$ are vectors from some representation of algebra $g\ell(r+1,\mathbb{C})$.

Since root generators (3.1.7) of algebra $s\ell(r+1,\mathbb{C})$ coincide with root generators of algebra $g\ell(r+1,\mathbb{C})$, while Cartan subalgebra of $s\ell(r+1,\mathbb{C})$ is embedded in Cartan subalgebra $g\ell(r+1,\mathbb{C})$ (cf. (3.1.8) and (3.4.43)), root vectors $\alpha^{(kl)}$ of algebra $g\ell(r+1,\mathbb{C})$, which are again defined in terms of the commutation relations (3.1.10) (difference between elements H_i and H'_i is an operator proportional to unit operator), coincide with root vectors (3.1.11) of algebra $s\ell(r+1,\mathbb{C})$. Therefore, the *root* spaces of the two algebras are the same. Recall that the root space of algebra $s\ell(r+1,\mathbb{C})$ is r-dimensional hyperplane P_u in $\mathfrak{V}_{r+1}$, which is orthogonal to vector $u = (1,\ldots,1)$. So, unlike in the case of simple Lie algebras, root space P_u of Lie algebras $g\ell(r+1,\mathbb{C})$ and $u(r+1)$ does not coincide with their weight space $\mathfrak{V}_{r+1}$ (root space is a subspace in $\mathfrak{V}_{r+1}$).

The above discussion implies that (i) Weyl groups of algebras $s\ell(r+1,\mathbb{C})$ and $g\ell(r+1,\mathbb{C})$ coincide, (ii) if we leave aside one-dimensional representations, then highest weight vectors of irreducible representations of $s\ell(r+1,\mathbb{C})$ are at the same time highest weight vectors of irreducible representations of $g\ell(r+1,\mathbb{C})$, so for these algebras there is one-to-one correspondence (modulo multiplication by one-dimensional representations) between irreducible representations with highest weights. In particular, highest weight vectors $|\lambda_{(f)}\rangle$ of fundamental representations of algebra $s\ell(r+1,\mathbb{C})$, given by (3.4.33), are also highest weight vectors of fundamental representations $T_{(f)}$ of algebra $g\ell(r+1,\mathbb{C})$. Representations $T_{(f)}$ ($f = 1,\ldots,r,r+1$) of algebra $g\ell(r+1,\mathbb{C})$ act in the same spaces $\mathcal{V}_{r+1}^{\wedge f}$ of totally antisymmetric rank-f tensors as the fundamental representations of algebra $s\ell(r+1,\mathbb{C})$. Here, the space $\mathcal{V}_{r+1} = \mathbb{C}^{r+1}$ of defining representation of algebra $g\ell(r+1,\mathbb{C})$ is identified with the space of defining representation of algebra $s\ell(r+1,\mathbb{C})$. Note, however, that in $g\ell$-case, there exists an additional nontrivial one-dimensional representation $T_{(r+1)}$ (with highest weight $\lambda_{(r+1)} \equiv u = (1,\ldots,1)$) which is trivial in $s\ell$-case (its highest weight is equal to zero). Also, since the generators of Cartan subalgebras (3.1.8) and (3.4.43) are different, the highest weights $\lambda_{(f)}$ of representations $T_{(f)}$ of algebra $g\ell(r+1,\mathbb{C})$ associated with vectors $|\lambda_{(f)}\rangle$ are different from highest weights (3.4.22) of representations $T_{(f)}$ of algebra $s\ell(r+1,\mathbb{C})$, namely, the analog of Proposition 3.4.9 reads as follows.

Proposition 3.4.12. *Weights $\lambda_{(f)} \in \mathfrak{V}_{r+1}$ of finite-dimensional irreducible representations $T_{(f)} = T_{(1)}^{\wedge f}$ of algebra $g\ell(r+1,\mathbb{C})$ (algebra $u(r+1)$), which act in spaces $\mathcal{V}_{r+1}^{\wedge f}$ of totally antisymmetric rank-f tensors, are given by* (*cf.* (3.4.22))

$$\lambda_{(f)} = (\underbrace{1,\ldots,1}_{f},0,\ldots,0), \quad f = 1,\ldots,r+1. \tag{3.4.44}$$

Proof. We have to prove the relations $H'_i\,|\lambda_{(f)}\rangle = (\lambda_{(f)})_i\,|\lambda_{(f)}\rangle$, where $(\lambda_{(f)})_i$ are components of weights (3.4.44). These relations follow from (3.4.35) and the relation $H_i = H'_i - \frac{1}{r+1}I_{r+1}$ between the generators (3.1.8) and (3.4.43). □

In accordance with general theory, all finite-dimensional irreducible representations $T^{(\vec{n})}$ of algebra $g\ell(r+1,\mathbb{C})$ have highest weights (3.4.9):

$$\lambda^{(\vec{n})} = \sum_{f=1}^{r+1} n_f \lambda_{(f)} = (\lambda_1,\lambda_2,\ldots,\lambda_{r+1}) \in \mathfrak{V}_{r+1} \quad (n_f \in \mathbb{Z}_{\geq 0}), \tag{3.4.45}$$

where $\lambda_k = \sum_{f=k}^{r+1} n_f$ and so we have

$$\lambda_1 \geq \lambda_2 \geq \cdots \geq \lambda_{r+1} \geq 0. \tag{3.4.46}$$

This weight can be represented by Young diagram $\Lambda = [\lambda_1,\lambda_2,\ldots,\lambda_{r+1}]$, shown in Fig. 3.4.3, with n_{r+1} columns of height $(r+1)$ added on the left (this corresponds to multiplication of representation with diagram $[\lambda_1,\ldots,\lambda_r]$ by n_{r+1} one-dimensional representations $T_{(r+1)}$). Thus, finite-dimensional irreducible representations of algebra $g\ell(r+1,\mathbb{C})$ and its compact real form $u(r+1)$ are associated with Young diagrams with not more than $(r+1)$ rows.

3. Algebras $so(2r,\mathbb{C})$ (series $\mathbf{D}_r$)

Roots of Lie algebra $so(2r,\mathbb{C})$ are given in (3.1.25), and the system of simple roots is written in (3.1.27) (see Section 3.1). With simple roots (3.1.27),

Eq. (3.3.2) gives a system of linear equations on coordinates $(\lambda_{(i)})_j$ $(i,j = 1,\dots,r)$ of fundamental weights $\lambda_{(i)} \in \mathfrak{V}_r$. By solving this system, we find

$$\lambda_{(1)} = (1,0,0,\dots,0) = e^{(1)},\ \lambda_{(2)} = (1,1,0,\dots,0),\dots,$$
$$\lambda_{(r-2)} = (1,1,\dots,1,0,0),\ \lambda_{(r-1)} = \left(\frac{1}{2},\frac{1}{2},\dots,\frac{1}{2},-\frac{1}{2}\right),$$
$$\lambda_{(r)} = \left(\frac{1}{2},\frac{1}{2},\dots,\frac{1}{2}\right). \tag{3.4.47}$$

We consider the case $f = 1,\dots,r-2$ and construct finite-dimensional irreducible representations $T_{(f)}$ with highest weights $\lambda_{(f)}$ given by (3.4.47). We comment on representations with highest weights $\lambda_{(r-1)}$ and $\lambda_{(r)}$ toward the end of our discussion.

Let us show first that the highest weight of defining representation $T_{(1)}$ of algebra $so(2r,\mathbb{C})$ equals the fundamental weight $\lambda_{(1)} = (1,0,0,\dots,0)$. To this end, we choose, in the space $\mathcal{V}_{2r}$ of defining representation, orthonormalized basis $\epsilon^{(k)} = |\mu^{(k)}\rangle$ $(k = 1,\dots,2r)$, where coordinates of $\epsilon^{(k)}$ are again $\epsilon^{(k)}_\ell = \delta^k_\ell$. Given the explicit form (3.1.18) of Cartan generators H_i $(i = 1,\dots,r)$ in defining representation, the basis vectors $\epsilon^{(k)}$ have weights

$$\mu^{(k)} = e^{(k)} - e^{(k')} \in \mathfrak{V}_r, \quad k = 1,\dots,2r, \quad k' = 2r-k+1, \tag{3.4.48}$$

which follows from the relations $(m = 1,\dots,2r)$

$$(H_i\epsilon^{(k)})_m = (e_{ii} - e_{i'i'})_{m\ell}\epsilon^{(k)}_\ell = (\delta_{ik} - \delta_{i'k})\epsilon^{(k)}_m \ \Rightarrow\ \mu^{(k)}_i = e^{(k)}_i - e^{(k')}_i,$$

where $i = 1,\dots,r$, and we use $\epsilon^{(k)}_\ell = \delta^k_\ell$. Weights (3.4.48) and roots (3.1.25) of $so(2r,\mathbb{C})$ are related by

$$\mu^{(k)} - \mu^{(\ell)} = \alpha^{(k,\ell)}_- \quad (k,\ell \le r), \quad \mu^{(k)} - \mu^{(\ell)} = \alpha^{(k,\ell')}_+ \quad (k \le r,\ \ell > r).$$

For $k < \ell \le r$ and $k \le r < \ell$, roots $\alpha^{(k,\ell)}_-$ and $\alpha^{(k,\ell')}_+$ are positive (in both cases we have $k < \ell$). This shows that the highest weight of defining representation is

$$\mu^{(1)}_i = e^{(1)}_i - e^{(2r)}_i = \delta_{1i} \ \Rightarrow\ \mu^{(1)} = (1,0,\dots,0) \equiv \lambda_{(1)},$$

as promised.

Now, in complete analogy to algebra $s\ell(r+1,\mathbb{C})$, representation $T_{(f)}$ of algebra $so(2r,\mathbb{C})$ with highest weight equal to fundamental weight $\lambda_{(f)}$

$(f = 2, \ldots, r-2)$ given by (3.4.47) is constructed as the tensor product of representations $T_{(1)}$ in space

$$\underbrace{\mathcal{V}_{2r} \wedge \mathcal{V}_{2r} \wedge \cdots \wedge \mathcal{V}_{2r}}_{f} \equiv (\mathcal{V}_{2r})^{\wedge f}, \tag{3.4.49}$$

with highest weight vectors given by (3.4.33). The action of Cartan generators in representation (3.4.24) on vectors (3.4.33) gives weights equal to fundamental weights $\lambda_{(f)}$ $(f = 2, \ldots, r-2)$.

Problem 3.4.6. Prove the last statement.

Representations $T_{(r-1)}$ and $T_{(r)}$ with highest weights $\lambda_{(r-1)}$ and $\lambda_{(r)}$ from the list (3.4.47) cannot be subrepresentations in a tensor product of defining representations $T_{(1)}$ since the eigenvalues of operators H_i in defining representation are either $+1$ or -1, so one cannot obtain half-integer coordinates of weights. However, representations $T'_{(r-1)}$ and $T'_{(r)}$, $T''_{(r)}$ with highest weights

$$\begin{gathered} \lambda'_{(r-1)} = \lambda_{(r-1)} + \lambda_{(r)} = (1, 1, \ldots, 1, 0), \\ \lambda'_{(r)} = 2\,\lambda_{(r)} = (1, 1, \ldots, 1), \quad \lambda''_{(r)} = 2\,\lambda_{(r-1)} = (1, 1, \ldots, -1), \end{gathered} \tag{3.4.50}$$

can be constructed as antisymmetrized tensor products of defining representations $T_{(1)}$ acting in spaces (3.4.49), where $f = (r-1)$ for the case of representation $T'_{(r-1)}$ and $f = r$ for the cases of representations $T'_{(r)}$, $T''_{(r)}$. The latter fact indicates that the representation of algebra $so(2r, \mathbb{C})$ acting in the space $\mathcal{V}_{2r}^{\wedge r}$ of antisymmetric rank r tensors is reducible and decomposes into two irreducible representations with highest weights $\lambda'_{(r)}$ and $\lambda''_{(r)}$, which are called selfdual and anti-selfdual representations (we discuss these representations in Section 5.1.3).

Problem 3.4.7. Prove that the highest weight vectors of representations $T'_{(r-1)}$, $T'_{(r)}$ and $T''_{(r)}$ are $|\lambda'_{(r-1)}\rangle = \epsilon^{(1)} \wedge \cdots \wedge \epsilon^{(r-1)} \in \mathcal{V}_{2r}^{\wedge(r-1)}$, $|\lambda'_{(r)}\rangle = \epsilon^{(1)} \wedge \cdots \wedge \epsilon^{(r)} \in \mathcal{V}_{2r}^{\wedge r}$ and $|\lambda''_{(r)}\rangle = \epsilon^{(1)} \wedge \cdots \wedge \epsilon^{(r-1)} \wedge \epsilon^{(r+1)} \in \mathcal{V}_{2r}^{\wedge r}$. Hint: make use of the explicit form of Cartan elements (3.1.18) and root generators (3.1.19) that correspond to simple roots of algebra $so(2r, \mathbb{C})$

$$\begin{gathered} E_{12} = e_{12} - e_{2r-1,2r}, \quad E_{23} = e_{23} - e_{2r-2,2r-1}, \ldots, \\ E_{r-1,r} = e_{r-1,r} - e_{r+1,r+2}, \quad E_{r-1,r'} = e_{r-1,r+1} - e_{r,r+2}\,. \end{gathered}$$

Use also the result of Problem 3.4.5.

Representations $T_{(r-1)}$ and $T_{(r)}$ with highest weights $\lambda_{(r-1)}$ and $\lambda_{(r)}$ are called *spinor* representations of algebra $so(2r, \mathbb{C})$. These two representations are inequivalent; they are studied below in Section 5.4.1, and also in Sections 6.2.2 and 6.4.1. In particular, we show in Section 6.4.1 (see Problem 207) that defining representation $T_{(1)}$ of $so(2r, \mathbb{C})$ can be constructed as subrepresentation in tensor product $T_{(r-1)} \otimes T_{(r)}$ for even r and $T_{(r-1)} \otimes T_{(r-1)}$ (and also $T_{(r)} \otimes T_{(r)}$) for odd r. Recall that all finite-dimensional irreducible representations of simple algebra $\mathcal{A}$ can be constructed as subrepresentations in tensor product (3.4.10) of representations $T_{(f)}$ (see the proof of Theorem 3.4.6). The last two statements taken together ensure that all finite-dimensional irreducible representations of algebra $so(2r, \mathbb{C})$ can be constructed as subrepresentations in direct product of spinor representations $T_{(r-1)}^{\otimes N_1} \otimes T_{(r)}^{\otimes N_2}$ where $N_1, N_2 \in \mathbb{Z}_{\geq 0}$.

To end this item, we note that the root system and system of fundamental weights of algebra $so(2r, \mathbb{C})$ is invariant under transformation of simple roots $\alpha^{(i)} \to \alpha^{(i)}$ $(i = 1, \ldots, r-2)$, $\alpha^{(r-1)} \leftrightarrow \alpha^{(r)}$ and corresponding transformation of fundamental weights $\lambda_{(i)} \to \lambda_{(i)}$ $(i = 1, \ldots, r-2)$, $\lambda_{(r-1)} \leftrightarrow \lambda_{(r)}$. This transformation is obtained by the action of operator $A = \text{diag}(\underbrace{+1, \ldots +1, -1}_{r})$ in root space. Under this transformation, the weight (and root) lattice of algebra $so(2r, \mathbb{C})$ is transformed into itself. This invariance is due to the symmetry of Dynkin diagram of series D_r shown in Fig. 3.1.1. It implies that there is an automorphism of algebra $so(2r, \mathbb{C})$ which performs linear transformation of its generators (including Cartan elements) and, as we see below (see Remark 3 in Section 6.2.2), leads to interchange of spinor representations $T_{(r-1)}$ and $T_{(r)}$.

Problem 3.4.8. Prove that the automorphism of algebra $so(2r, \mathbb{C})$, that corresponds to the action of operator $A = I_{r-1,1}$ in root space, takes $L_{i,2r} \to -L_{i,2r}$ while leaving other L_{ij} $(i, j = 1, \ldots, 2r-1)$ intact, where L_{ij} are generators obeying defining relations (1.2.33). Hint: choose Cartan–Weyl basis in terms of L_{ij} in the same way as in the first book [1] (see Section I-6.2.3); take note that under this automorphism we have $H_r \to -H_r$, $E_{\alpha_\pm^{(i,r)}} \leftrightarrow E_{\alpha_\mp^{(i,r)}}$, $E_{-\alpha_\pm^{(i,r)}} \leftrightarrow E_{-\alpha_\mp^{(i,r)}}$, where $\alpha_\pm^{(i,j)}$ are positive roots of algebra $so(2r, \mathbb{C})$, while other generators stay intact.

4. Lie algebras $so(2r+1, \mathbb{C})$ (series B_r)

Simple roots of Lie algebra $so(2r+1, \mathbb{C})$ are given in (3.1.29) (see Section 3.1). Weight lattice Λ for all finite-dimensional representations and symmetry of this lattice under Weyl reflections are studied in Example 2 of Section 3.2; see Eqs. (3.2.25) and (3.2.26). Fundamental weights $\lambda_{(f)}$ $(f = 1, \ldots, r)$ are determined by Eq. (3.3.2), whose solution is

$$\lambda_{(1)} = (1, 0, 0, \ldots, 0),\ \lambda_{(2)} = (1, 1, 0, \ldots, 0),\ \ldots,\ \lambda_{(r-1)} = (1, 1, \ldots, 1, 0),$$
$$\lambda_{(r)} = (\tfrac{1}{2}, \tfrac{1}{2}, \ldots, \tfrac{1}{2}). \tag{3.4.51}$$

Leaving aside for a while the fundamental weight $\lambda_{(r)}$, one constructs, in the same way as in the case of Lie algebra $s\ell(r+1, \mathbb{C})$, representations $T_{(f)}$ $(f = 1, \ldots, r-1)$ with highest weights $\lambda_{(f)}$, as subrepresentations in tensor products of defining representations $T_{(1)}$, which act in spaces $\mathcal{V}_{2r+1}^{\wedge f} \subset \mathcal{V}_{2r+1}^{\otimes f}$ of antisymmetric rank-f tensors. Here $\mathcal{V}_{2r+1}$ is $(2r+1)$-dimensional space of defining representation $T_{(1)}$ of algebra $so(2r+1, \mathbb{C})$. In particular, representation with highest weight $\lambda_{(1)}$ is equivalent to defining representation $T_{(1)}$.

Problem 3.4.9. Let us choose the standard orthonormal basis in space $\mathcal{V}_{2r+1}$ with basis vectors $\epsilon^{(a)}$ $(a = 1, \ldots, 2r+1)$ having coordinates $(\epsilon^{(a)})_b = \delta_{ab}$. Making use of explicit formulas (3.1.18), (3.1.19), (3.1.20), (3.1.21), for generators H_i and E_α of algebra $so(2r+1, \mathbb{C})$ in defining representation, show that $\epsilon^{(1)}$ is the highest weight vector in space $\mathcal{V}_{2r+1}$ of defining representation $T_{(1)}$ whose highest weight is $\lambda_{(1)}$. Show that tensors (cf. (3.4.33))

$$|\lambda_{(f)}\rangle = \epsilon^{(1)} \wedge \epsilon^{(2)} \wedge \ldots \wedge \epsilon^{(f)}, \quad f = 1, \ldots, r-1, \tag{3.4.52}$$

are highest weight vectors in spaces $\mathcal{V}_{2r+1}^{\wedge f}$ of irreducible representations $T_{(f)}$ with highest weights equal to fundamental weights $\lambda_{(f)}$ $(f = 1, \ldots, r-1)$ from the list (3.4.51).

The last fundamental weight $\lambda_{(r)}$ in (3.4.51) is the highest weight of irreducible *spinor* representation $T_{(r)}$, which cannot be a subrepresentation in tensor product of defining representations. The reason is the same as in the case of algebra $so(2r, \mathbb{C})$. However, like in the case of algebra $so(2r, \mathbb{C})$, representation $T'_{(r)}$ of $so(2r+1, \mathbb{C})$ with highest weight

$$\lambda'_{(r)} = 2\,\lambda_{(r)} = (1, 1, \ldots, 1), \tag{3.4.53}$$

can be constructed as antisymmetrized subrepresentation in $T_{(1)}^{\otimes r}$, acting in space $\mathcal{V}_{2r+1}^{\wedge r}$. The highest weight vector of representation $T'_{(r)}$ is given by (3.4.52) with $f = r$.

Note that algebra $so(2r+1,\mathbb{C})$ has *one* spinor representation, unlike algebra $so(2r,\mathbb{C})$ with *two* inequivalent spinor representations with highest weights $\lambda_{(r-1)}$ and $\lambda_{(r)}$ (see (3.4.47)). We study spinor representation $T_{(r)}$ of algebra $so(2r+1,\mathbb{C})$ in Sections 5.4.2, 6.2.2 and 6.4.1. In particular, we show there that defining representation $T_{(1)}$ of algebra $so(2r+1,\mathbb{C})$ can be viewed as subrepresentation in tensor product $T_{(r)} \otimes T_{(r)}$ of two spinor representations. So, in accordance with general theory, we find that any finite-dimensional representation of algebra $so(2r+1,\mathbb{C})$ can be constructed as subrepresentation in tensor product $T_{(r)}^{\otimes M}$ of spinor representations, where $M = 2\sum_{f=1}^{r-1} f \cdot n_f + r \cdot n_r$.

Let us make here a general comment concerning representations of Lie algebras $so(2r,\mathbb{C})$ and $so(2r+1,\mathbb{C})$.

Remark 3. The same arguments as in the case of algebra $s\ell(r+1,\mathbb{C})$ (see the end of item 2 above) show that inequivalent representations of algebras $so(2r,\mathbb{C})$ and $so(2r+1,\mathbb{C})$, *which are subrepresentations in tensor products of defining representations*, are embedded in representations

$$\bar{T} = (T'_{(r)} \oplus T''_{(r)})^{\otimes n_r} \otimes T'^{\otimes n_{r-1}}_{(r-1)} \otimes T^{\otimes n_{r-2}}_{(r-2)} \otimes \cdots \otimes T^{\otimes n_2}_{(2)} \otimes T^{\otimes n_1}_{(1)}$$
$$(\text{for } so(2r,\mathbb{C})),$$
$$\bar{T} = T'^{\otimes n_r}_{(r)} \otimes T^{\otimes n_{r-1}}_{(r-1)} \otimes \cdots \otimes T^{\otimes n_2}_{(2)} \otimes T^{\otimes n_1}_{(1)}$$
$$(\text{for } so(2r+1,\mathbb{C}))$$

(note that representations $T'_{(r-1)}, T'_{(r)}, T''_{(r)}$ with highest weights (3.4.50) and (3.4.53) are not fundamental). Like in the case of algebra $s\ell(r+1,\mathbb{C})$, the highest weights of irreducible representations are

$$\lambda = \sum_{i=1}^{r-2} n_i \lambda_{(i)} + n_{r-1}\lambda'_{(r-1)} + n_r \mu_{(r)} \equiv (\lambda_1, \ldots, \lambda_r) \in \mathfrak{V}_r \quad (n_k, \lambda_k \in \mathbb{Z}_{\geq 0}), \tag{3.4.54}$$

where $\mu_{(r)} = \lambda'_{(r)}$ or $\mu_{(r)} = \lambda''_{(r)}$, and

$$\lambda = \sum_{i=1}^{r-1} n_i \lambda_{(i)} + n_r \lambda'_{(r)} \equiv (\lambda_1, \ldots, \lambda_r) \in \mathfrak{V}_r \quad (n_k, \lambda_k \in \mathbb{Z}_{\geq 0}), \tag{3.4.55}$$

for representations of $so(2r,\mathbb{C})$ and $so(2r+1,\mathbb{C})$, respectively. Highest weights (3.4.54) and (3.4.55) can be associated with Young diagrams (see Fig. 3.4.3) whose number of rows does not exceed r.

5. Algebras $sp(2r, \mathbb{C})$ (series $\mathbf{C}_r$)

Roots of Lie algebra $sp(2r, \mathbb{C})$ and its system of simple roots are given in (3.1.35) and (3.1.38), respectively (see Section 3.1). By solving Eq. (3.3.2), we obtain fundamental weights

$$\begin{aligned} &\lambda_{(1)} = (1, 0, 0, \dots, 0) = e^{(1)}, \quad \lambda_{(2)} = (1, 1, 0, \dots, 0), \dots, \\ &\lambda_{(r-1)} = (1, 1, \dots, 1, 0), \quad \lambda_{(r)} = (1, 1, \dots, 1, 1). \end{aligned} \tag{3.4.56}$$

All representations $T_{(f)}$ with highest weights equal to fundamental weights $\lambda_{(f)}$, listed in (3.4.56), are *antisymmetrized* tensor products of defining representation $T_{(1)}$ with weight $\lambda_{(1)}$; they act in spaces (3.4.49). Thus, all finite-dimensional irreducible representations of algebra $sp(2r, \mathbb{C})$ can be constructed as subrepresentations in tensor products of defining representations $T_{(1)}$. All these representations with highest weights given by (3.4.9),

$$\lambda = \sum_{i=1}^{r} n_i \lambda_{(i)} \equiv (\lambda_1, \dots, \lambda_r) \in \mathfrak{V}_r \quad (n_k, \lambda_k \in \mathbb{Z}_{\geq 0}), \tag{3.4.57}$$

can be constructed as irreducible subrepresentations in

$$\bar{T} = T_{(r)}^{\otimes n_r} \otimes T_{(r-1)}^{\otimes n_{r-1}} \otimes \cdots \otimes T_{(2)}^{\otimes n_r} \otimes T_{(1)}^{\otimes n_r},$$

and can be associated with Young diagram of Fig. 3.4.3, whose number of rows does not exceed rank r of algebra $sp(2r, \mathbb{C})$.

To summarize, we have shown that fundamental weights of algebras $s\ell(r+1, \mathbb{C})$, $so(n, \mathbb{C})$ and $sp(2r, \mathbb{C})$, are indeed highest weights of certain representations $T_{(f)}$ of these algebras. We have explicitly constructed these representations $T_{(f)}$, with the exception of spinor representations of algebras $so(n, \mathbb{C})$. We fill the latter gap in Section 5.4. Thus, we have completed the proof of Theorem 3.4.6 in the case of Lie algebras of classical series.

6. Examples

Here, we give three examples related to "accidental" isomorphisms (see Section I-3.3.3 of the first book): $so(4, \mathbb{C}) = s\ell(2, \mathbb{C}) + s\ell(2, \mathbb{C})$, $so(5, \mathbb{C}) = sp(4, \mathbb{C})$ and $so(6, \mathbb{C}) = s\ell(4, \mathbb{C})$, and discuss the relationships between fundamental representations of these algebras.

Example 1: $\mathbf{so(4, \mathbb{C}) = s\ell(2, \mathbb{C}) + s\ell(2, \mathbb{C})}$

In accordance with (3.4.47), we have the following fundamental weights of algebra $so(4, \mathbb{C})$: $\lambda_{(1)} = (1/2, -1/2)$ and $\lambda_{(2)} = (1/2, 1/2)$. As we pointed out in item 3, these are associated with two inequivalent spinor representations of $so(4,\mathbb{C})$. On the other hand, semisimple algebra $s\ell(2,\mathbb{C})+s\ell(2,\mathbb{C})$ has two-dimensional root space $\mathfrak{V}_2=\mathbb{R}^2$ (direct sum of one-dimensional root spaces of the two algebras $s\ell(2,\mathbb{C})$) and, as we see from Example 1 in Section 3.3 and item 1 in this section, there are two simple roots $(1,0)$ and $(0,1)$ and hence two fundamental weights $(\frac{1}{2},0)$ and $(0,\frac{1}{2})$. Since algebras $so(4,\mathbb{C})$ and $s\ell(2,\mathbb{C})+s\ell(2,\mathbb{C})$ are isomorphic, two irreducible representations of $so(4,\mathbb{C})$ with highest weights $\lambda_{(1)}=(1/2,-1/2)$ and $\lambda_{(2)}=(1/2,1/2)$ must be equivalent to two fundamental representations of $s\ell(2,\mathbb{C})+s\ell(2,\mathbb{C})$ with weights $\lambda'_{(1)}=(\frac{1}{2},0)$ and $\lambda'_{(2)}=(0,\frac{1}{2})$. The latter representations are thus spinor representations[f] of $so(4,\mathbb{C})$. Note that it is natural to denote irreducible representations of $s\ell(2,\mathbb{C})+s\ell(2,\mathbb{C})=so(4,\mathbb{C})$ as follows (cf. (3.4.9)):

$$(j_1,j_2)\equiv n_1\lambda'_{(1)}+n_2\lambda'_{(2)}=\left(\frac{n_1}{2},\frac{n_2}{2}\right),\quad n_1,n_2\in\mathbb{Z}_{\geq 0},$$

where $j_1,j_2\in\mathbb{Z}_{\geq 0}/2$ are highest weights (spins) of the representations of the first and second subalgebra $s\ell(2,\mathbb{C})$, respectively.

Example 2: $\mathbf{so(5,\mathbb{C})=sp(4,\mathbb{C})}$

In accordance with (3.4.51), algebra $so(5,\mathbb{C})$ has fundamental weights $\lambda_{(1)}=(1,0)$ and $\lambda_{(2)}=(1/2,1/2)$, which are highest weights of defining and (a single) spinor representations. Fundamental weights of algebra $sp(4,\mathbb{C})$ are $\lambda'_{(1)}=(1,0)$ and $\lambda'_{(2)}=(1,1)$, which are the highest weight of defining four-dimensional representation in space $\mathcal{V}_4$ and highest weight of antisymmetric *irreducible* subrepresentation in space $\mathcal{V}_4^{\wedge 2}$. The antisymmetric representation of algebra $sp(4,\mathbb{C})$ in space $\mathcal{V}_4^{\wedge 2}$ has one-dimensional subrepresentation, so the antisymmetric *irreducible* subrepresentation with highest weight $\lambda'_{(2)}=(1,1)$ has dimension $4{\cdot}3/2-1=5$. This coincides with the dimension of defining representation of algebra $so(5,\mathbb{C})$ with highest weight $\lambda_{(1)}=(1,0)$. For this reason, the representation with $\lambda'_{(2)}=(1,1)$ of algebra $sp(4,\mathbb{C})$ is to be identified with defining representation with $\lambda_{(1)}=(1,0)$ of algebra $so(5,\mathbb{C})$, while the remaining four-dimensional defining representation of $sp(4,\mathbb{C})$ (with highest weight $\lambda'_{(1)}=(1,0)$) is identified with spinor representation of $so(5,\mathbb{C})$ (with highest weight $\lambda_{(2)}=(1/2,1/2)$).

[f] In physics and mathematics literature, these spinors are called Weyl spinors and semi-spinors (half-spinors), respectively.

Example 3: $so(6, \mathbb{C}) = s\ell(4, \mathbb{C})$
Fundamental weights of algebras $so(6, \mathbb{C})$ and $s\ell(4, \mathbb{C})$ are

$$\begin{aligned} so(6,\mathbb{C})\colon \lambda_{(1)} &= (1,0,0), \lambda_{(2)} = (1/2,1/2,-1/2), \\ \lambda_{(3)} &= (1/2,1/2,1/2), \\ s\ell(4,\mathbb{C})\colon \lambda'_{(1)} &= (1,0,0,0) - \frac{1}{4}u, \ \lambda'_{(2)} = (1,1,0,0) - \frac{2}{4}u, \\ \lambda'_{(3)} &= (1,1,1,0) - \frac{3}{4}u. \end{aligned}$$

Antisymmetric representation of $s\ell(4, \mathbb{C})$ with highest weight $\lambda'_{(2)}$ (acting in space $\mathcal{V}_4^{\wedge 2}$) has dimension $4 \cdot 3/2 = 6$, which is equal to dimension of defining representation $T_{(1)}$ of algebra $so(6, \mathbb{C})$ with highest weight $\lambda_{(1)}$. So, antisymmetric representation $\lambda'_{(2)}$ of $s\ell(4, \mathbb{C})$ is identified with defining representation $\lambda_{(1)}$ of $so(6, \mathbb{C})$. The remaining two four-dimensional representations of $s\ell(4, \mathbb{C})$, defining one with highest weight $\lambda'_{(1)}$ and contragredient to defining representation with highest weight $\lambda'_{(3)}$ (which acts in space $\mathcal{V}_4^{\wedge 3}$), are identified with two inequivalent spinor representations of $so(6, \mathbb{C})$ with highest weights $\lambda_{(2)} = (1/2, 1/2, -1/2)$ and $\lambda_{(3)} = (1/2, 1/2, 1/2)$.

Problem 3.4.10. Making use of Eq. (3.3.2), find fundamental weights associated with systems of simple roots (3.1.41), (3.1.44), (3.1.51), (3.1.53), (3.1.55) of exceptional Lie algebras (and compare with weights in Ref. [4]):

$$\begin{aligned}
&G_2\colon \lambda_{(1)} = (0,-1,1), \ \lambda_{(2)} = (-1,-1,2); \\
&F_4\colon \lambda_{(1)} = (1,0,0,1), \ \lambda_{(2)} = (1,1,0,2), \ \lambda_{(3)} = \tfrac{1}{2}(1,1,1,3), \\
&\quad \lambda_{(4)} = (0,0,0,1); \\
&E_6\colon \lambda_{(1)} = (-\tfrac{1}{3},-\tfrac{1}{3},1,0,0,0,0,\tfrac{1}{3}), \ \lambda_{(2)} = (-\tfrac{2}{3},-\tfrac{2}{3},1,1,0,0,0,\tfrac{2}{3}), \\
&\quad \lambda_{(3)} = (-1,-1,1,1,1,0,0,1), \ \lambda_{(4)} = (-\tfrac{5}{6},-\tfrac{5}{6},\tfrac{1}{2},\tfrac{1}{2},\tfrac{1}{2},\tfrac{1}{2},-\tfrac{1}{2},\tfrac{5}{6}), \\
&\quad \lambda_{(5)} = \tfrac{1}{2}(-1,-1,1,1,1,1,1,1), \ \lambda_{(6)} = \tfrac{2}{3}(-1,-1,0,0,0,0,0,1); \\
&E_7\colon \lambda_{(1)} = (-\tfrac{1}{2},1,0,0,0,0,0,\tfrac{1}{2}), \ \lambda_{(2)} = (-1,1,1,0,0,0,0,1), \\
&\quad \lambda_{(3)} = (-\tfrac{3}{2},1,1,1,0,0,0,\tfrac{3}{2}), \ \lambda_{(4)} = (-2,1,1,1,1,0,0,2), \\
&\quad \lambda_{(5)} = \tfrac{1}{2}(-3,1,1,1,1,1,-1,3), \ \lambda_{(6)} = \tfrac{1}{2}(-2,1,1,1,1,1,1,2); \\
&\quad \lambda_{(7)} = (-1,0,0,0,0,0,0,1); \\
&E_8\colon \lambda_{(1)} = (1,0,0,0,0,0,0,1), \ \lambda_{(2)} = (1,1,0,0,0,0,0,2), \\
&\quad \lambda_{(3)} = (1,1,1,0,0,0,0,3), \ \lambda_{(4)} = (1,1,1,1,0,0,0,4), \\
&\quad \lambda_{(5)} = (1,1,1,1,1,0,0,5), \ \lambda_{(6)} = \tfrac{1}{2}(1,1,1,1,1,1,-1,7); \\
&\quad \lambda_{(7)} = \tfrac{1}{2}(1,1,1,1,1,1,1,5), \ \lambda_{(8)} = (0,0,0,0,0,0,0,2).
\end{aligned}$$

3.4.3. Quadratic Casimir operator

Let X_a $(a = 1, \ldots, \dim\mathcal{A})$ be basis elements of simple Lie algebra $\mathcal{A}$ with defining relations $[X_a, X_b] = C^d_{ab} X_d$, and g^{ab} be inverse of Killing metric in this basis. As we show in the first book (see Section I-4.7.2), operator $\hat{C}_2 = X_a \mathsf{g}^{ab} X_b$ commutes with all generators $X_a \in \mathcal{A}$ and is called quadratic Casimir operator. We choose normalization of elements $\{H_i, E_\alpha\}$ of the Cartan–Weyl basis in such a way that Killing metric has the form (3.1.4), (3.1.5), then quadratic Casimir operator reads

$$\hat{C}_2 = X_a \mathsf{g}^{ab} X_b = \sum_{i,j=1}^{r} H_i g^{ij} H_j + \sum_{\alpha} E_\alpha E_{-\alpha}, \tag{3.4.58}$$

where summation in $\sum_\alpha$ runs over all roots α.

Problem 3.4.11. Show that Casimir operator, written in the form (3.4.58), commutes with all generators $\{H_i, E_\alpha\}$. Hint: make use of identity $C^f_{ab} \mathsf{g}_{fd} = -C^f_{ad} \mathsf{g}_{fb}$, that leads, in Cartan–Weyl basis, to $N_{\alpha-\beta,\beta} = -N_{-\alpha,\beta}$ (see (3.1.2) for definition of $N_{\alpha,\beta}$).

Due to Schur's lemma, Casimir operator $\hat{C}_2$ is proportional to unit operator in any irreducible complex representation T. Proportionality coefficient, denoted by $c_2^{(T)}$, depends on the representation. So, for any vector $|u\rangle$ in the space of irreducible representation $T^{(\lambda)}$ with highest weight λ, we have $\hat{C}_2|u\rangle = c_2^{(\lambda)}|u\rangle$. In particular, for highest weight vector $|\lambda\rangle$, this relation gives

$$\begin{aligned} \hat{C}_2|\lambda\rangle &= \left(\sum_{i,j=1}^{r} H_i g^{ij} H_j + \sum_{\alpha>0} (E_\alpha E_{-\alpha} + E_{-\alpha} E_\alpha) \right) |\lambda\rangle \\ &= \left(\sum_{i,j=1}^{r} \lambda_i g^{ij} \lambda_j + \sum_{\alpha>0} [E_\alpha, E_{-\alpha}] \right) |\lambda\rangle \\ &= \left((\lambda, \lambda) + \left(\lambda, \sum_{\alpha>0} \alpha \right) \right) |\lambda\rangle, \end{aligned} \tag{3.4.59}$$

where we make use of the fact that $E_\alpha|\lambda\rangle = 0$ for all positive roots α (generators H_i, E_α in (3.4.59) are understood as operators $T^{(\lambda)}(H_i), T^{(\lambda)}(E_\alpha)$ acting in space of representation $T^{(\lambda)}$). Thus, the element $\hat{C}_2$ in representation $T^{(\lambda)}$ is proportional to unit operator, and the proportionality coefficient is

$$c_2^{(\lambda)} = (\lambda,\, \lambda + 2\,\delta) = (\lambda + \delta,\, \lambda + \delta) - (\delta,\, \delta), \tag{3.4.60}$$

where we use convenient notation

$$\delta = \frac{1}{2} \sum_{\alpha > 0} \alpha. \tag{3.4.61}$$

Example. Lie algebra $s\ell(2,\mathbb{C})$ (and its real form $su(2)$) has one positive root $\alpha = 1$ (in normalization where weights are integers or half-integers), so we have $\delta = 1/2$ and eigenvalue $c_2^{(\lambda)}$ of quadratic Casimir operator equals $\lambda(\lambda+1)$, where $\lambda \equiv j \in \mathbb{Z}_+/2$ is highest weight (spin) of representation (see item 1 of Section 3.4.2).

We see that highest weight λ determines the value of quadratic Casimir operator in representation $T^{(\lambda)}$. In general, the opposite is not true[g]: highest weight $\lambda = (\lambda_1, \ldots, \lambda_r)$ cannot be found by knowing the value of $c_2^{(\lambda)}$ only. In other words, knowledge of the value of quadratic Casimir operator is insufficient to identity irreducible representation of algebra $\mathcal{A}$. As an example, the value (3.4.60) of quadratic Casimir operator is the same for representation $T^{(\lambda)}$ and its contragredient representation $\bar{T}^{(\lambda)}$ for which $\bar{T}^{(\lambda)}(X_a) = -T^{(\lambda)}(X_a)^{\mathsf{T}}$. To find the highest weight $\lambda = (\lambda_1, \ldots, \lambda_r)$ and thus identify irreducible representation, one defines a set of Casimir operators (central elements commuting with all generators, see Section I-4.7.2 of the first book)

$$\hat{C}_k = d^{a_1 \ldots a_k}\, X_{a_1} \cdots X_{a_k}, \quad d^{a_1 \ldots a_k} = \mathsf{g}^{a_1 b_1} \cdots \mathsf{g}^{a_k b_k}\, \mathrm{Tr}(T_{b_1} \cdots T_{b_k}), \tag{3.4.62}$$

where $k = 2, 3, \ldots$ and in the second formula we use matrices $(T_a)^b_d = C^b_{ad} = \mathrm{ad}(X_a)^b_d$ of generators $X_a \in \mathcal{A}$ in adjoint representation. The number of *independent* Casimir operators in this set is equal to rank r of simple Lie algebra $\mathcal{A}$. So, once one knows the values of r independent Casimir operators, one has r equations for r unknown components of highest weight λ. Note though (see [5]) that eigenvalues $c_k^{(\lambda)}$ of operators (3.4.62) are functions of vectors $\lambda + \delta$, so they determine highest weights λ

[g]One of exceptions is algebra $s\ell(2,\mathbb{C})$.

modulo Weyl reflections $(\lambda+\delta)\to\sigma_\alpha(\lambda+\delta)$. Indeed, Weyl reflections are realized in Lie algebra as equivalence transformations (3.2.32), (3.2.36), which do not change the eigenvalues of operators $\hat{C}_k$ in (3.4.62). To obtain unique solution, one imposes additionally the dominance condition on highest weights: Weyl reflections move dominant weights into nondominant.

Remark 1. Let us choose the normalization of generators X_a as determined by the second equality in (3.4.58). Let Tr_λ denote trace in space of representation $T^{(\lambda)}$. The following equality holds for simple Lie algebras (see Section I-4.7.2 of the first book):

$$\mathrm{Tr}_\lambda(T^{(\lambda)}(X_a)T^{(\lambda)}(X_b)) = \mathsf{c}_\lambda\, \mathsf{g}_{ab}, \tag{3.4.63}$$

where g_{ab} is Killing metric and constant c_λ is called *index* of representation $T^{(\lambda)}$. Let us calculate the index. To this end, we note that

$$\mathrm{Tr}_\lambda(T^{(\lambda)}(X_a)T^{(\lambda)}(X_b))\, \mathsf{g}^{ab} = \mathrm{Tr}_\lambda(T^{(\lambda)}(C_2)) = (\lambda,\, \lambda+2\,\delta)\cdot \dim(T^{(\lambda)}), \tag{3.4.64}$$

where we use (3.4.60). On the other hand, Eq. (3.4.63) gives

$$\mathrm{Tr}_\lambda(T^{(\lambda)}(X_a)T^{(\lambda)}(X_b))\, \mathsf{g}^{ab} = \mathsf{c}_\lambda\cdot \dim(\mathcal{A}).$$

By comparing these two relations, we find that index of representation $T^{(\lambda)}$ is given by

$$\mathsf{c}_\lambda = \frac{\dim(T^{(\lambda)})}{\dim(\mathcal{A})}\,(\lambda,\, \lambda+2\,\delta).$$

Remark 2. Vector δ defined in (3.4.61) obeys remarkable identity, which we employ later on:

$$\frac{1}{2}\sum_{\alpha>0}\alpha = \sum_{f=1}^{r}\lambda_{(f)} \;\Rightarrow\; \delta = \sum_{f=1}^{r}\lambda_{(f)}, \tag{3.4.65}$$

where $\lambda_{(f)}$ are fundamental weights of algebra $\mathcal{A}$. To prove (3.4.65) we consider Weyl reflection $\sigma_{\alpha^{(i)}}$ of vector $\delta=\frac{1}{2}\sum_{\alpha>0}\alpha$ determined by any given simple root $\alpha^{(i)}$. In accordance with Proposition 3.1.2, reflection $\sigma_{\alpha^{(i)}}$ permutes all positive roots with the exception of root $\alpha^{(i)}$, whose reflection is $\sigma_{\alpha^{(i)}}(\alpha^{(i)}) = -\alpha^{(i)}$. We therefore obtain

$$\sigma_{\alpha^{(i)}}(\delta) = \frac{1}{2}\sigma_{\alpha^{(i)}}\left(\sum_{\alpha>0}\alpha - \alpha^{(i)}\right) + \frac{1}{2}\sigma_{\alpha^{(i)}}(\alpha^{(i)}) = \frac{1}{2}\left(\sum_{\alpha>0}\alpha - \alpha^{(i)}\right) - \frac{1}{2}\alpha^{(i)} = \delta - \alpha^{(i)}. \tag{3.4.66}$$

Since fundamental weights $\lambda_{(f)}$ form basis in root space, we can write $\delta = \sum_{f=1}^{r} \mathsf{n}_f \lambda_{(f)}$ and then calculate its reflection:

$$\sigma_{\alpha^{(i)}}(\delta) = \sum_f \mathsf{n}_f \sigma_{\alpha^{(i)}}(\lambda_{(f)}) = \sum_f \mathsf{n}_f(\lambda_{(f)} - \delta_{fi}\alpha^{(i)}) = \delta - \mathsf{n}_i\,\alpha^{(i)}, \tag{3.4.67}$$

where we make use of (3.2.11) and (3.3.2). By comparing right-hand sides of (3.4.66) and (3.4.67), we find $\mathsf{n}_i = 1$ for all $i = 1, \ldots, r$, which gives (3.4.65), as promised.

Finally, we stress that according to (3.4.9) and (3.4.65) vector $\delta \in \Lambda^+$ is a highest (and dominant) weight of certain representation T of the Lie algebra $\mathcal{A}$.

Problem 3.4.12. Check explicitly the identity (3.4.65) for simple algebras $s\ell(r+1, \mathbb{C})$, $so(n, \mathbb{C})$, $sp(2r, \mathbb{C})$ by making use of expressions for roots (3.1.13), (3.1.25), (3.1.28), (3.1.35) and fundamental weights (3.4.22), (3.4.47), (3.4.51), (3.4.56).

3.5. Weyl Character Formula for Representations of Compact Simple Lie Groups

3.5.1. Weyl denominator and Weyl character formula

Irreducible representations of compact Lie groups G are related to irreducible representations of their Lie algebras $\mathcal{A}(G)$. This correspondence is one-to-one for universal covering groups, see Sections I-3.5 and I-5.3 of the first book. In the theory of compact Lie groups, a lot of information on their irreducible representations $T^{(\lambda)}$ (with highest weight λ) is encoded in characters, which are functions on group G:

$$\hat{\chi}_\lambda(g) = \mathrm{Tr}_\lambda\left(T^{(\lambda)}(g)\right) = \sum_\mu D^{(\lambda)}_{\mu\mu}(g) \quad (\forall g \in G). \tag{3.5.1}$$

Here, we use notation (3.4.19); $D^{(\lambda)}_{\mu\mu'}$ are matrices of representation $T^{(\lambda)}$, and summation runs over all *weight vectors*.[h] We emphasize that $\hat{\chi}_\lambda(g) = \hat{\chi}_\lambda(h\, g\, h^{-1})$ $(\forall h \in G)$, i.e., character $\hat{\chi}_\lambda$ is a function on conjugacy classes of group G.

For *compact and connected* Lie groups G, exponential mapping acts from Lie algebra $\mathcal{A}(G)$ *onto the whole* group G (see Ref. [32, §102]). This means that any element g of compact connected group G can be written as a single exponential $g = \exp(A)$, where $A \in \mathcal{A}(G)$.

Proposition 3.5.1. *Let G be compact connected simple Lie group, $\mathcal{A}(G)$ its Lie algebra and $\mathcal{H}$ Cartan subalgebra in $\mathcal{A}(G)$. For any element $A \in \mathcal{A}(G)$ there exists an element $h \in G$ such that*

$$\mathrm{ad}(h^{-1}) \cdot A \equiv h^{-1} \cdot A \cdot h = \theta^i \, H_i \in \mathcal{H}, \tag{3.5.2}$$

where $\{H_1, \ldots, H_r\}$ are generators of Cartan subalgebra $\mathcal{H}$, r is rank of G and θ^i are parameters.

Proof. Let $A, B \in \mathcal{A}(G)$ and $(A,\, B)_{\mathcal{A}}$ be nondegenerate, invariant scalar product[i] in $\mathcal{A}(G)$. Consider function on group G defined by

$$f(U) = \big(A,\, U \cdot y \cdot U^{-1}\big)_{\mathcal{A}},$$

where $U \in G$ and $y \in \mathcal{H}$. Since group G is compact, function $f(U)$ is bounded, and it has maximum (minimum) at $U = h \in G$. Consider element $h \cdot e^{tB}$ away from extremum point h by small right shift along an arbitrary vector $B \in \mathcal{A}(G)$ (parameter t is small). Since h is point of extremum, we have

$$\begin{aligned} 0 = \partial_t \, f(h \cdot e^{tB})|_{t=0} &= \partial_t \big(A\,,\; h \cdot e^{tB} \cdot y \cdot e^{-tB} \cdot h^{-1}\big)_{\mathcal{A}}|_{t=0} \\ &= \big(h^{-1} \cdot A \cdot h,\; [B,\, y]\big)_{\mathcal{A}} \\ &= \big([y,\, h^{-1} \cdot A \cdot h],\; B\big)_{\mathcal{A}} \;\Rightarrow\; [y,\, h^{-1} \cdot A \cdot h] = 0. \end{aligned} \tag{3.5.3}$$

Now, for simple Lie algebra $\mathcal{A}(G)$, there exists an element $y = y^i \, H_i \in \mathcal{H}$ such that the property $[y,\, x] = 0$ for $x \in \mathcal{A}(G)$ leads to $x \in \mathcal{H}$. Indeed, we choose vector $\vec{y} = (y^1, \ldots, y^r)$ from root space of $\mathcal{A}(G)$ such that $(\vec{y},\, \alpha) \equiv (y^i \alpha_i) \neq 0$ for all roots α of algebra $\mathcal{A}(G)$ (vector $\vec{y}$ is *strictly inside* some

[h] Weights μ may be degenerate, so it would be wrong to say that summation in (3.5.1) runs over all weights.

[i] For matrix algebras, this scalar product can be taken as $(A,\, B)_{\mathcal{A}} := \mathrm{Tr}(A \cdot B)$.

Weyl chamber). Then, for any root element $E_\alpha \in \mathcal{A}(G)$ we have $[y,\, E_\alpha] = (y^i\alpha_i)E_\alpha \neq 0$. We expand an element $x \in \mathcal{A}(G)$ in Cartan–Weyl basis, $x = x^i H_i + \sum_\alpha c_\alpha E_\alpha$ and find

$$[y,\, x] = 0 \;\Rightarrow\; \sum_\alpha (\vec{y}, \alpha)\, c_\alpha\, E_\alpha = 0.$$

Since $(\vec{y}, \alpha) \neq 0$ and generators E_α are linear independent, this gives $c_\alpha = 0$ $(\forall \alpha)$, i.e., $x \in \mathcal{H}$, as promised. Taking such an element $y \in \mathcal{H}$ in (3.5.3), we find that element $(h^{-1} \cdot A \cdot h)$ belongs to Cartan subalgebra $\mathcal{H}$ and can be expanded in basis $H_i \in \mathcal{H}$. □

Remark. It is shown in Section I-5.2 of the first book that every compact Lie algebra $\mathcal{A}$ is a finite sum of several Abelian subalgebras $u(1)$ and simple compact Lie algebras $\mathcal{A}_1, \mathcal{A}_2, \ldots$:

$$\mathcal{A} = u(1) + \cdots + u(1) + \mathcal{A}_1 + \mathcal{A}_2 + \cdots\,. \qquad (3.5.4)$$

In view of this fact, the result of Proposition 3.5.1 can be extended to all compact connected Lie groups.

Formula (3.5.2) shows that conjugacy classes of compact connected Lie groups G are parameterized by r parameters θ^i $(i = 1, \ldots, r)$: any element $g \in G$ can be written as $g = \exp(A)$ and, in accordance with (3.5.2), we have

$$g = h \cdot \exp(\theta^i H_i) \cdot h^{-1}, \quad (\theta^i)^* = -\theta^i, \quad h \in G. \qquad (3.5.5)$$

It is convenient to choose *Hermitian* generators $H_i \in \mathcal{H}$ in any unitary representation, then θ^i are *pure imaginary*. In this parameterization, character $\hat{\chi}_\lambda(g)$ reads

$$\hat{\chi}_\lambda(g) = \mathrm{Tr}_\lambda(T^{(\lambda)}(e^{\theta^i H_i})) = \sum_\mu \mathrm{mult}_\lambda(\mu) \cdot \exp(\theta, \mu) \equiv \chi_\lambda(\theta), \qquad (3.5.6)$$

where $(\theta, \mu) = (\theta^i \mu_i) = g^{ki}\theta_k \mu_i$ is scalar product (3.1.6), summation runs over all weights μ of representation $T^{(\lambda)}$ and $\mathrm{mult}_\lambda(\mu)$ is multiplicity of weight μ. Note that the parameters θ^i in (3.5.2) are non-uniquely defined for a given element A, so one cannot, generally speaking, parameterize conjugacy classes in group G by parameters θ^i: different sets $\{\theta^i\}$ may correspond to one and the same conjugacy class. As an example, sets θ and $\sigma_\alpha(\theta)$ are associated with one and the same conjugacy class, since, in accordance with (3.2.32), we have

$$S_\alpha \exp(\theta^i H_i)\, S_\alpha^{-1} = \exp(\sigma_\alpha(\theta), H), \qquad (3.5.7)$$

where $S_\alpha = \exp(i\pi T_2(\alpha))$ is the operator of Weyl reflection associated with root α. Since equivalent weights have the same multiplicity,

$\text{mult}_\lambda(\mu) = \text{mult}_\lambda(\sigma_\alpha(\mu))$, Eq. (3.5.7) shows that character (3.5.6) is invariant under Weyl reflections, $\chi_\lambda(\theta) = \chi_\lambda(\sigma_\alpha(\theta))$.

Formula (3.5.6) is not particularly convenient for calculating characters, since it involves (not necessarily known) multiplicities $\text{mult}_\lambda(\mu)$. Conversely, once one knows character $\chi_\lambda(\theta)$ explicitly, multiplicities $\text{mult}_\lambda(\mu)$ are obtained from the expansion of character (3.5.6) in weights μ of representation $T^{(\lambda)}$.

Our purpose here is to obtain explicit formulas for characters (3.5.1) and (3.5.6). To this end, we introduce space F of functions of the following form:

$$f(\theta) = \sum_{\mu \in \Lambda} a(\mu)\, e^{(\mu,\, \theta)}, \tag{3.5.8}$$

where summation runs over all sites μ of weight lattice Λ of Lie algebra $\mathcal{A}(G)$, vector $\theta = (\theta_1, \ldots, \theta_r)$ is pure imaginary, $i\theta$ belongs to root space of $\mathcal{A}(G)$, and $a(\mu)$ are arbitrary coefficients that depend on weight μ. We suppose that $a(\mu) \neq 0$ for finite number of weights in Λ, then the sum in (3.5.8) is finite. As all weights are expanded in fundamental weights $\lambda_{(i)}$ with integer coefficients, see (3.3.4), this sum is a polynomial in $2r$ variables

$$x_i^{\pm 1} = e^{\pm(\lambda_{(i)},\, \theta)}.$$

Thus, space F is the *space of polynomials in x_i and x_i^{-1}*. Note that characters (3.5.6) are elements of this space F.

The action of Weyl group in space F is defined by

$$S_w \cdot f(\theta) = f(w^{-1} \cdot \theta) = \sum_{\mu \in \Lambda} a(\mu) \exp(w \cdot \mu,\, \theta), \tag{3.5.9}$$

where w is an element of Weyl group W, $w \cdot \mu \equiv w(\mu)$ and we use the orthogonality property $(\mu, w^{-1} \cdot \theta) = (w \cdot \mu, \theta)$ with respect to the scalar product (3.1.6). Formula (3.5.9) defines representation of Weyl group in space F, since $S_w\, S_{w'} = S_{w \cdot w'}$, $\forall\, w, w' \in W$. For generators $w = \sigma_\alpha \in W$ given by (3.2.11), we have

$$S_{\sigma_\alpha} \cdot f(\theta) = f(\sigma_\alpha(\theta)) = \sum_{\mu \in \Lambda} a(\mu) e^{(\sigma_\alpha(\mu),\, \theta)} = \sum_{\mu \in \Lambda} a(\sigma_\alpha(\mu)) e^{(\mu,\, \theta)}, \;\; S_{\sigma_\alpha}^2 = 1, \tag{3.5.10}$$

where we use the identity $\sigma_\alpha^2 = 1$, as well as invariance of lattice Λ under Weyl reflections.

Definition 3.5.1. Function $f(\theta) \in \mathsf{F}$ is *symmetric* if $S_w \cdot f(\theta) = f(\theta)$ for all $w \in W$, and skew-symmetric if $S_w \cdot f(\theta) = \det(w)\, f(\theta)$ for all $w \in W$. Here $\det(w) = -1$ ($\det(w) = +1$) if w is a product of odd (even) number of Weyl reflections.

We introduce the notation $S_i = S_{\sigma_{\alpha^{(i)}}}$, where $\alpha^{(i)}$ are simple roots. Function $f(\theta) \in \mathsf{F}$ is symmetric if and only if $S_i \cdot f(\theta) = f(\theta)$ for all $i = 1, \ldots, r$, and skew-symmetric if and only if $S_i \cdot f(\theta) = -f(\theta)$ for all $i = 1, \ldots, r$. Indeed, Proposition 3.1.1 (see Section 3.1.3) tells that operations S_i generate the action of the whole Weyl group W in space of functions (3.5.8), so the above conditions are equivalent to those used in Definition 3.5.1.

Let α be a root, and function $f(\theta) \in \mathsf{F}$ be such that $S_{\sigma_\alpha} f(\theta) = -f(\theta)$. Then $f(\theta)$ is a linear combination of elements

$$f_\mu(\theta) = e^{(\mu,\,\theta)} - S_{\sigma_\alpha} \cdot e^{(\mu,\,\theta)} = e^{(\mu,\,\theta)} - e^{(\sigma_\alpha(\mu),\,\theta)} = e^{(\mu,\,\theta)}\big(1 - e^{-\mathsf{n}_{(\alpha,\,\mu)}(\alpha,\,\theta)}\big),$$

where $\mathsf{n}_{(\alpha,\,\mu)} = 2\frac{(\alpha,\,\mu)}{\alpha^2}$. These elements are divisible[j] by $(1 - e^{-(\alpha,\,\theta)})$, since $\mathsf{n}_{(\alpha,\,\mu)}$ are integers. So, if $f(\theta) \in \mathsf{F}$ is skew-symmetric, then it is divisible, as polynomial of variables $x_i^{\pm 1} = e^{\pm(\lambda_{(i)},\,\theta)}$, by polynomial $\big(1 - e^{-(\alpha,\,\theta)}\big)$, where α is any positive root.

> **Problem 3.5.1.** Show that for any two different positive roots $\alpha_{1,2}$, functions $\big(1 - e^{-(\alpha_{1,2},\,\theta)}\big)$, viewed as polynomials in $x_i^{\pm 1}$, are coprime. Hint: recall that root α_1 cannot be proportional to another root α_2.

Since polynomials $(1 - e^{-(\alpha,\,\theta)})$ for all positive roots α are coprime, any skew-symmetric polynomial $f(\theta) \in \mathsf{F}$ is divisible by product of factors $(1 - e^{-(\alpha,\,\theta)})$, i.e., by polynomial

$$e^{(\delta,\theta)} \prod_{\alpha>0} (1 - e^{-(\alpha,\theta)}) = \prod_{\alpha>0} (e^{(\alpha,\theta)/2} - e^{-(\alpha,\theta)/2}) \equiv D_\delta(\theta), \qquad (3.5.11)$$

where product runs over all positive roots, vector $\delta = \frac{1}{2}\sum_{\alpha>0} \alpha$ is defined in (3.4.61) and factor $e^{(\delta,\theta)} = \prod_{i=1}^r x_i$ on the left-hand side is introduced for convenience.[k] Function $D_\delta(\theta) \in \mathsf{F}$ (polynomial in $x_i^{\pm 1}$) plays a key role in what follows; it is called Weyl denominator (nomenclature has to do with

[j] By division here we mean the standard operation in algebra of polynomials. Polynomial P_1 is divisible by polynomial P_2, if there exists polynomial P_3 such that $P_1 = P_2 \cdot P_3$.
[k] Factor $e^{(\delta,\theta)}$ is irrelevant for divisibility of the polynomials, since it only multiplies the result of division by monomial $x_1^{-1} \cdots x_r^{-1}$.

the fact that function $D_\delta(\theta)$ is denominator in Weyl formula for characters, see (3.5.19) below).

Proposition 3.5.2. *The following* formula for Weyl denominator *is valid:*

$$D_\delta(\theta) \equiv e^{(\delta,\theta)} \prod_{\alpha>0} (1 - e^{-(\alpha,\theta)}) = \sum_{\sigma\in W} \epsilon(\sigma) \exp(\sigma(\delta), \theta), \tag{3.5.12}$$

where the right-hand side is a finite sum over all elements σ of Weyl group W, and $\epsilon(\sigma) \equiv \det(\sigma) = \pm 1$, depending on whether σ is a product of even or odd number of Weyl reflections.

Proof. Sum on the right-hand side of (3.5.12) is a skew-symmetric function. Indeed, consider operator $A = \sum_{w\in W} \epsilon(w) S_w$ in space F. We have

$$S_{\sigma_\alpha} \cdot A = \sum_{w\in W} \epsilon(w) S_{\sigma_\alpha \cdot w} = \sum_{w\in W} \epsilon(\sigma_\alpha^{-1} \cdot w) S_w = \det(\sigma_\alpha^{-1})\, A = -A, \tag{3.5.13}$$

where we make use of invariance of summation over finite group under left shifts, and recall that $\det(\sigma_\alpha) = -1$. Equation (3.5.13) shows that operator A (antisymmetrizer) takes any function $f(\theta) \in \mathsf{F}$ into skew-symmetric function $A \cdot f(\theta) \in \mathsf{F}$. The sum on the right-hand side of (3.5.12) can be written as

$$\sum_{\sigma\in W} \epsilon(\sigma) \exp(\sigma(\delta), \theta) = A \cdot \exp(\delta, \theta),$$

so this sum is skew-symmetric function, indeed. Thus, it is divisible by Weyl denominator (3.5.11), i.e.,

$$\sum_{\sigma\in W} \epsilon(\sigma) \exp(\sigma(\delta), \theta) = b(\theta)\, e^{(\delta,\theta)} \prod_{\alpha>0} (1 - e^{-(\alpha,\theta)}), \tag{3.5.14}$$

where factor $b(\theta) \in \mathsf{F}$ is a polynomial in variables $x_i^{\pm 1}$, which can be written as a sum (3.5.8) over finite number of weights. Recall that vector δ is a dominant weight, so exponents on the left-hand side of (3.5.14) contain weights $\mu = \sigma(\delta)$ in the interval[1] $-\delta \leq \mu \leq \delta$. The same holds for Weyl

[1]Recall that partial ordering of weights is defined as follows (see Section 3.4.1): one chooses a reference vector x in positive Weyl chamber, then for two weights one defines $\mu' \leq \mu$, if $(\mu', x) \leq (\mu, x)$.

denominator on the right-hand side of (3.5.14). Indeed, the latter can be cast in two equivalent forms involving products over positive roots:

$$e^{(\delta,\theta)}\prod_{\alpha>0}(1-e^{-(\alpha,\theta)}) = e^{-(\delta,\theta)}\prod_{\alpha'>0}(e^{(\alpha',\theta)}-1)\,.$$

Expansions of these products, e.g.,

$$e^{(\delta,\theta)}\prod_{\alpha>0}(1-e^{-(\alpha,\theta)}) = \Big(e^{(\delta,\theta)} - \sum_{\alpha>0} e^{(\delta-\alpha,\theta)} + \sum_{\alpha>0,\beta>0} e^{(\delta-\alpha-\beta,\theta)} + \dots\Big),$$

give a function of the form (3.5.8) where summation runs over weights $\mu = \delta - \sum\alpha = -\delta + \sum\alpha'$, where α, α' are positive roots. Thus, we have sums over weights in interval $-\delta \le \mu \le \delta$ in both sides of Eq. (3.5.14). By comparing coefficients of the terms $e^{(\delta,\theta)}$ in the two sides of (3.5.14), we find $b(\theta) = 1$, which gives (3.5.12). □

Remark 1. For any skew-symmetric function $f(\theta) \in \mathsf{F}$ we have

$$A\cdot f(\theta) = \sum_{w\in W}\epsilon(w)S_w\cdot f(\theta) = \left(\sum_{w\in W}\det{}^2(w)\right) f(\theta) = \mathrm{ord}(W)\, f(\theta).$$

Thus, any skew-symmetric function $f(\theta) \in \mathsf{F}$ of the form (3.5.8) can be written as

$$f(\theta) = \frac{1}{\mathrm{ord}(W)}\, A\cdot f(\theta) = \sum_{\mu\in\Lambda} b(\mu)\; A\cdot e^{(\mu,\,\theta)}, \quad b(\mu) \equiv \frac{a(\mu)}{\mathrm{ord}(W)}, \tag{3.5.15}$$

and hence it is a linear combination of linear independent elements $A\cdot e^{(\theta,\mu)} \in \mathsf{F}$, where $\mu \in \Lambda$.

Problem 3.5.2. Show that formula (3.5.15) for any skew-symmetric function $f(\theta)$ is written in the form

$$f(\theta) = \sum_{\mu\in\Lambda}\tilde{b}(\mu)\; e^{(\mu,\,\theta)}, \tag{3.5.16}$$

where coefficients $\tilde{b}(\mu) = \sum_{w\in W}\epsilon(w^{-1})b(w\cdot\mu)$ are such that $\tilde{b}(\sigma\cdot\mu) = \epsilon(\sigma)\,\tilde{b}(\mu)$ for all $\sigma \in W$.

Remark 2. The right-hand side of (3.5.12) is a skew-symmetric function. One can see independently that the function (3.5.11) is also skew-symmetric. Indeed, in accordance with Proposition 3.1.1, reflections $\sigma_{\alpha^{(i)}}$

generate the whole group W, and Proposition 3.1.2 tells that reflection $\sigma_{\alpha^{(i)}}$ permutes positive roots α, except for root $\alpha^{(i)}$ itself, which flips sign. Therefore, under reflection $\sigma_{\alpha^{(i)}}$ factors $(e^{(\alpha,\theta)/2}-e^{-(\alpha,\theta)/2})$ in (3.5.11) with $\alpha \neq \alpha^{(i)}$ get reshuffled, while the factor $(e^{(\alpha^{(i)},\theta)/2} - e^{-(\alpha^{(i)},\theta)/2})$ changes sign, so that

$$S_i \cdot \prod_{\alpha>0} \left(e^{(\alpha,\theta)/2} - e^{-(\alpha,\theta)/2}\right) = -\prod_{\alpha>0} \left(e^{(\alpha,\theta)/2} - e^{-(\alpha,\theta)/2}\right), \quad \forall\, i.$$

Remark 3. Formula (3.5.12) is remarkable in many respects. In particular, it shows that the product over positive roots on the left-hand side of (3.5.11) is determined solely by vectors $\sigma(\delta)$, which are obtained from vector δ by the action of elements σ of Weyl group W. Also, formula (3.5.12) is valid in the case of infinite-dimensional Kac–Moody algebras, which have infinite number of roots, but finite number r of simple roots, and components of weights $\lambda_{(i)}$ $(i = 1,\ldots,r)$ are discrete and real. In that case, formula (3.5.12) relates infinite product and infinite sum. For Kac–Moody algebra $\widehat{su}(2)$ this is a known Jacobi identity for elliptic θ-functions. Other Kac–Moody algebras give generalizations of this identity. The derivation of these generalizations requires appropriate regularization of infinite sum on the right-hand side of (3.5.12).

Consider now function (cf. right-hand side of (3.5.12))

$$D_{\delta+\lambda}(\theta) = \sum_{\sigma\in W} \epsilon(\sigma)\exp(\sigma(\delta+\lambda),\theta) = A\cdot \exp(\delta+\lambda,\theta) \in \mathsf{F}, \qquad (3.5.17)$$

which depends on highest weight λ of a representation $T^{(\lambda)}$ of compact simple Lie group G (and its Lie algebra). Since this function is obtained by applying antisymmetrizer A, it is skew-symmetric and hence it is divisible by the Weyl denominator (3.5.12). We show below that the result of division is equal to character (3.5.6) of representation $T^{(\lambda)}$. Namely we prove identity

$$D_\delta(\theta)\cdot \chi_\lambda(\theta) = D_{\delta+\lambda}(\theta), \qquad (3.5.18)$$

which gives the *Weyl character formula*

$$\chi_\lambda(\theta) = \frac{D_{\delta+\lambda}(\theta)}{D_\delta(\theta)} = \frac{\sum_{\sigma\in W}\epsilon(\sigma)\exp(\sigma(\delta+\lambda),\theta)}{\prod_{\alpha>0}(e^{(\alpha,\theta)/2} - e^{-(\alpha,\theta)/2})}\,. \qquad (3.5.19)$$

Note that alternative form for denominator here is a sum in (3.5.12). Note also that the identity (3.5.12) follows from the character formula (3.5.19) in the case when representation $T^{(\lambda)}$ is trivial, then $\lambda = 0$ and $\chi_\lambda(\theta)|_{\lambda=0} = 1$.

We now turn to the proof of Weyl formula (3.5.19). We begin with the following proposition.

Proposition 3.5.3. *Character (3.5.6) of representation $T^{(\lambda)}$ of simple Lie group G obeys second-order differential equation*

$$\left(\partial_{\theta^i}\,\partial_{\theta_i} + \sum_{\alpha>0} \frac{(e^{(\theta,\alpha)}+1)}{(e^{(\theta,\alpha)}-1)}\,(\alpha^i\partial_{\theta^i})\right)\chi_\lambda(\theta)$$

$$= ((\lambda+\delta,\,\lambda+\delta)-(\delta,\,\delta))\,\chi_\lambda(\theta), \tag{3.5.20}$$

where $\partial_{\theta^i}\,\partial_{\theta_i} = g^{ij}\partial_{\theta^i}\,\partial_{\theta^j} = \partial_\theta^2$, the sum runs over all positive weights α of algebra $\mathcal{A}$, and vector δ is defined in (3.4.61). Product $X_{\delta+\lambda}(\theta) \equiv \chi_\lambda(\theta)\,D_\delta(\theta)$ of Weyl denominator and character obeys

$$\partial_\theta^2\, X_{\delta+\lambda}(\theta) = (\delta+\lambda,\,\delta+\lambda)\,X_{\delta+\lambda}(\theta). \tag{3.5.21}$$

Proof. Making use of the formulas (3.4.59) and (3.4.60) for quadratic Casimir operator and definition (3.5.1), (3.5.6) of character of representation $T^{(\lambda)}$, we write the following chain of equalities:

$$\begin{aligned}
((\lambda+\delta,\,\lambda+\delta)-(\delta,\,\delta))\,\chi_\lambda(\theta) &= \mathrm{Tr}_\lambda\left(\hat{C}_2\cdot e^{\theta^i H_i}\right)\\
&= \mathrm{Tr}_\lambda\Big(\big(\sum_{i,j=1}^{r} H_i g^{ij} H_j + \sum_\alpha E_\alpha E_{-\alpha}\big)\cdot e^{\theta^i H_i}\Big)\\
&= \partial_\theta^2\,\mathrm{Tr}_\lambda\left(e^{\theta^i H_i}\right) + \mathrm{Tr}_\lambda\Big(\big(\sum_\alpha E_\alpha E_{-\alpha}\big)\cdot e^{\theta^i H_i}\Big)\\
&= \Big(\partial_\theta^2 + \sum_\alpha \tfrac{e^{(\theta,\alpha)}}{(e^{(\theta,\alpha)}-1)}\,(\alpha^i\partial_{\theta^i})\Big)\mathrm{Tr}_\lambda\left(e^{\theta^i H_i}\right),
\end{aligned} \tag{3.5.22}$$

where generators $\{H_i, E_\alpha\}$ are taken in representation $T^{(\lambda)}$. To obtain the last equality in (3.5.22), we make use of the relation

$$\mathrm{Tr}_\lambda(E_\alpha E_{-\alpha}\cdot e^{\theta^i H_i}) = \frac{e^{(\theta,\alpha)}}{(e^{(\theta,\alpha)}-1)}\,(\alpha^i\partial_{\theta^i})\,\mathrm{Tr}_\lambda(e^{\theta^i H_i}).$$

The latter is obtained in the following way:

$$\begin{aligned}
\mathrm{Tr}_\lambda(E_\alpha\cdot E_{-\alpha}\cdot e^{\theta^i H_i}) &= \mathrm{Tr}_\lambda(E_{-\alpha}\cdot e^{\theta^i H_i}\cdot E_\alpha)\\
&= \mathrm{Tr}_\lambda(E_{-\alpha}\cdot E_\alpha\cdot e^{\theta^i(H_i+\alpha_i)})\\
&= \mathrm{Tr}_\lambda(([E_{-\alpha},\,E_\alpha]+E_\alpha\cdot E_{-\alpha})\cdot e^{\theta^i(H_i+\alpha_i)})\\
&= \mathrm{Tr}_\lambda((-(\alpha^i H_i)+E_\alpha\cdot E_{-\alpha})\cdot e^{\theta^i(H_i+\alpha_i)}).
\end{aligned}$$

By writing the sum over all roots on the right-hand side of (3.5.22) as sum over positive roots, one arrives at (3.5.20).

Weyl denominator, written in factorized form, obeys

$$\frac{\partial}{\partial\theta^i}\, D_\delta(\theta) = \frac{\partial}{\partial\theta^i} \prod_{\alpha>0} (e^{(\alpha,\theta)/2} - e^{-(\alpha,\theta)/2}) = \frac{1}{2} \sum_{\alpha>0} \alpha_i \, \frac{(e^{(\alpha,\theta)}+1)}{(e^{(\alpha,\theta)}-1)}\, D_\delta(\theta)\,. \tag{3.5.23}$$

Now, we recall that elements of Weyl group $\sigma \in W$ are orthogonal transformations, $(\sigma(\alpha),\sigma(\beta)) = (\alpha,\beta)$, and derive for the right-hand side of (3.5.12) the identity

$$\partial_\theta^2\, D_\delta(\theta) = (\partial_\theta,\, \partial_\theta) \sum_{\sigma\in W} \epsilon(\sigma) \exp(\sigma(\delta),\theta) = (\delta,\,\delta)\, D_\delta(\theta)\,. \tag{3.5.24}$$

Finally, we apply operator ∂_θ^2 to product $X_{\delta+\lambda}(\theta) = (\chi_\lambda(\theta)\, D_\delta(\theta))$ and make use of (3.5.20), (3.5.23) and (3.5.24) to obtain

$$\begin{aligned}
&\partial_\theta^2\, (\chi_\lambda(\theta)\, D_\delta(\theta)) \\
&\quad = D_\delta(\theta) (\partial_\theta^2 \chi_\lambda(\theta)) + 2(\partial_{\theta_i} \chi_\lambda(\theta)) \partial_{\theta^i} D_\delta(\theta) + \chi_\lambda(\theta)\, (\partial_\theta^2 D_\delta(\theta)) \\
&\quad = D_\delta(\theta) \Big((\lambda+\delta,\, \lambda+\delta) - (\delta,\,\delta) - \sum_{\alpha>0} \frac{(e^{(\theta,\alpha)}+1)}{(e^{(\theta,\alpha)}-1)}\, (\alpha^i \partial_{\theta^i}) \Big) \chi_\lambda(\theta) \\
&\qquad + (\partial_{\theta_i} \chi_\lambda(\theta)) \sum_{\alpha>0} \alpha_i\, \frac{(e^{(\alpha,\theta)}+1)}{(e^{(\alpha,\theta)}-1)}\, D_\delta(\theta) + (\delta,\,\delta)\, \chi_\lambda(\theta)\, D_\delta(\theta) \\
&\quad = (\lambda+\delta,\, \lambda+\delta) \chi_\lambda(\theta)\, D_\delta(\theta)\,,
\end{aligned}$$

which is precisely (3.5.21). □

Corollary 3.5.4. *By employing the expansion* $e^{-(\theta,\alpha)}(1 - e^{-(\theta,\alpha)})^{-1} = \sum_{k=1}^{\infty} e^{-k(\theta,\alpha)}$, *we write the left-hand side of Eq.* (3.5.20) *as follows:*

$$\begin{aligned}
&\left(\partial_\theta^2 + \sum_{\alpha>0} \frac{1+e^{-(\theta,\alpha)}}{1-e^{-(\theta,\alpha)}}\, (\alpha,\partial_\theta) \right) \chi_\lambda(\theta) \\
&\quad = \left(\partial_\theta^2 + 2(\delta,\partial_\theta) + 2 \sum_{\alpha>0} \sum_{k=1}^{\infty} e^{-k(\theta,\alpha)} (\alpha,\partial_\theta) \right) \chi_\lambda(\theta).
\end{aligned}$$

Then Eq. (3.5.20) *becomes*

$$\left((\partial_\theta^2 + (2\delta,\partial_\theta) - c_2^{(\lambda)}) + 2 \sum_{\alpha>0} \sum_{k=1}^{\infty} e^{-k(\theta,\alpha)} (\alpha,\partial_\theta) \right) \chi_\lambda(\theta) = 0, \tag{3.5.25}$$

where $c_2^{(\lambda)} = (\lambda+\delta)^2 - (\delta)^2$ and we use shorthand notation $(\mu)^2 = (\mu, \mu)$. Now we take the character formula (3.5.6) as a sum over weights and substitute it into Eq. (3.5.25). As a result, we obtain relation

$$((\lambda+\delta)^2 - (\mu+\delta)^2)\, n_\lambda(\mu) = 2\sum_{\alpha>0}\sum_{k=1}^{\infty}(\alpha, \mu+k\alpha)\, n_\lambda(\mu+k\alpha), \quad (3.5.26)$$

where $n_\lambda(\mu) = \mathrm{mult}_\lambda(\mu)$ is the multiplicity of weight μ of representation $T^{(\lambda)}$. Relation (3.5.26) is known as *Freudenthal formula for multiplicities.*

Next we need the ordering of weights and the notion of dominant weight which have been introduced in Section 3.4.1 (see also the proof of Proposition 3.5.2). The following lemma is instrumental in what follows.

Lemma 3.5.5. (1) *Every weight μ of representation $T^{(\lambda)}$ is equivalent to dominant weight μ', and $(\lambda)^2 \geq (\mu')^2 = (\mu)^2$. (2) Let μ be weight of representation $T^{(\lambda)}$, then for any $\sigma \in W$ we have $(\lambda+\delta)^2 \geq (\mu+\sigma(\delta))^2$, where equality is reached only for $\mu = \sigma(\lambda)$.*

Proof. (1) We begin with the first part of lemma. The fact that any weight μ is equivalent to dominant weight μ' has been proven in Proposition 3.4.3. For dominant weight μ' of representation $T^{(\lambda)}$ we have, in view of (3.4.3), $\mu' = \lambda - \sum_i \mathsf{m}_i\alpha^{(i)}$, where $\mathsf{m}_i \in \mathbb{Z}_{\geq 0}$. We denote $\beta = \sum_i \mathsf{m}_i\alpha^{(i)}$ and write

$$(\lambda)^2 = (\mu')^2 + 2\sum_i \mathsf{m}_i(\mu', \alpha^{(i)}) + (\beta)^2,$$

where $(\beta)^2 \geq 0$ and $(\mu', \alpha^{(i)}) \geq 0$, since μ' is dominant weight. Therefore, $(\lambda)^2 \geq (\mu')^2 = (\mu)^2$.

(2) Let σ be an element of Weyl group W. Weight $\sigma^{-1}(\mu)$ is equivalent to weight μ, and, in accordance with (3.4.3), can be written as $\sigma^{-1}(\mu) = \lambda - \sum_i \mathsf{m}_i\alpha^{(i)}$ where $\mathsf{m}_i \in \mathbb{Z}_{\geq 0}$. Therefore,

$$\begin{aligned}(\lambda+\delta)^2 - (\mu+\sigma(\delta))^2 &= (\lambda)^2 - (\mu)^2 + 2(\delta, \lambda - \sigma^{-1}(\mu)) \\ &= (\lambda)^2 - (\mu)^2 + 2\sum_i \mathsf{m}_i(\delta, \alpha^{(i)}) \\ &= (\lambda)^2 - (\mu)^2 + \sum_i \mathsf{m}_i(\alpha^{(i)})^2 \geq 0, \qquad (3.5.27)\end{aligned}$$

where we use the relation $2\frac{(\delta,\alpha^{(i)})}{(\alpha^{(i)})^2} = 1$ that follows from (3.3.2) and (3.4.65), and also employ the first part of this lemma. The right-hand side of (3.5.27) vanishes only for $(\lambda)^2 = (\mu)^2$ and $\mathsf{m}_i = 0$, which is equivalent to $\mu = \sigma(\lambda)$. □

The function $D_{\delta+\lambda}(\theta)$ given by (3.5.17) obeys the same equation (3.5.21) as product $X_{\delta+\lambda}(\theta) \equiv (\chi_\lambda(\theta)\, D_\delta(\theta))$. Indeed, we have

$$\begin{aligned} \partial_\theta^2\, D_{\delta+\lambda}(\theta) &= \partial_{\theta^i}\, \partial_{\theta_i} \sum_{\sigma\in W} \epsilon(\sigma) \exp(\sigma(\delta+\lambda), \theta) \\ &= \sum_{\sigma\in W} \epsilon(\sigma) \big(\sigma(\delta+\lambda), \sigma(\delta+\lambda)\big) \exp\big(\sigma(\delta+\lambda), \theta\big) \\ &= (\delta+\lambda,\, \delta+\lambda)\, D_{\delta+\lambda}(\theta). \end{aligned} \tag{3.5.28}$$

Also, like the function $D_{\delta+\lambda}(\theta)$, function $X_{\delta+\lambda}(\theta) \in \mathsf{F}$ is skew-symmetric, since it is a product of skew-symmetric function $D_\delta(\theta)$ and symmetric function $\chi_\lambda(\theta)$. All these point toward the identification of $X_{\delta+\lambda}(\theta)$ and $D_{\delta+\lambda}(\theta)$. This is indeed the case.

Theorem 3.5.6. *Functions $D_\delta(\theta)$ and $D_{\delta+\lambda}(\theta)$ given by* (3.5.12) *and* (3.5.17) *obey the identity*

$$D_\delta(\theta)\, \chi_\lambda(\theta) = D_{\delta+\lambda}(\theta), \tag{3.5.29}$$

which is equivalent to Weyl character formula (3.5.19).

Proof. Since the left-hand side of (3.5.29) is skew-symmetric function in F, then, in accordance with (3.5.15) and (3.5.16), we have

$$D_\delta(\theta)\, \chi_\lambda(\theta) = \sum_{\mu'\in\Lambda} b(\mu')\, A \cdot e^{(\mu',\,\theta)} = \sum_{\mu'\in\Lambda} \tilde{b}(\mu')\, e^{(\mu',\,\theta)}, \tag{3.5.30}$$

where $\tilde{b}(\sigma\cdot\mu) = \epsilon(\sigma)\, \tilde{b}(\mu)$ for all $\sigma \in W$. Function $D_\delta(\theta)$ is a sum of exponentials $e^{(\sigma(\delta),\,\theta)}$, where $\sigma \in W$, while $\chi_\lambda(\theta)$ can be written (see (3.5.6)) as sum of exponentials $e^{(\mu,\,\theta)}$, where μ are weights of representation $T^{(\lambda)}$. Therefore, summation on the right-hand side of (3.5.30) runs over vectors $\mu' = \mu + \sigma(\delta)$.

We note that $\partial_\theta^2\, e^{(\mu',\,\theta)} = (\mu', \mu')\, e^{(\mu',\,\theta)}$ and recall Eq. (3.5.21) to observe that summation on the right-hand side of (3.5.30) runs over weights $\mu' = \mu + \sigma(\delta)$ such that

$$(\mu', \mu') = (\mu + \sigma(\delta), \mu + \sigma(\delta)) = (\delta+\lambda,\, \delta+\lambda)\,.$$

In accordance with the second part of Lemma 3.5.4, the latter equality can hold only for $\mu = \sigma(\lambda)$. Therefore, $\mu' = \sigma(\lambda+\delta)$ and summation over μ' in

(3.5.30) is actually summation over elements $\sigma \in W$. So, we rewrite (3.5.30) as follows:

$$\begin{aligned} D_\delta(\theta)\,\chi_\lambda(\theta) &= \sum_{\sigma \in W} \tilde{b}(\sigma(\lambda+\delta))\; e^{(\sigma(\lambda+\delta),\,\theta)} \\ &= \sum_{\sigma \in W} \epsilon(\sigma)\, \tilde{b}(\lambda+\delta)\; e^{(\sigma(\lambda+\delta),\,\theta)} = \tilde{b}(\lambda+\delta)\; D_{\delta+\lambda}(\theta), \end{aligned} \quad (3.5.31)$$

where we apply the property $\tilde{b}(\sigma \cdot \mu) = \epsilon(\sigma)\,\tilde{b}(\mu)$ for coefficients in (3.5.30) and make use of definition (3.5.17). By comparing coefficients of exponentials $e^{(\lambda+\delta,\,\theta)}$ in the two sides of (3.5.31), we find $\tilde{b}(\lambda+\delta) = 1$. □

3.5.2. Applications. Explicit formulas for characters and dimensions of representations of groups $SU(r+1)$, $SO(n)$ and $USp(2r)$

As the first application of Weyl formula (3.5.19), we obtain expression for dimension of representation $T^{(\lambda)}$ with highest weight λ. To this end, we note that $\dim(T^{(\lambda)}) = \chi_\lambda(0)$. Since both numerator and denominator in (3.5.19) vanish as $\theta \to 0$ (in particular, due to identity $\sum_{\sigma\in W} \epsilon(\sigma) = 0$), we have to do calculation carefully. We set $\theta = t\delta$. Making use of (3.2.34), we obtain for numerator in Weyl formula $D_{\delta+\lambda}(t\delta) = D_\delta(t(\lambda+\delta))$, and then apply factorized representation (3.5.11) to write

$$\chi_\lambda(t\delta) = \frac{D_{\delta+\lambda}(t\delta)}{D_\delta(t\delta)} = \frac{D_\delta\big(t(\lambda+\delta)\big)}{D_\delta(t\delta)} = \prod_{\alpha>0} \frac{\sinh(\frac{t}{2}(\alpha,\delta+\lambda))}{\sinh(\frac{t}{2}(\alpha,\delta))}.$$

We now take limit $t \to 0$ and obtain Weyl formula for dimension of representation $T^{(\lambda)}$ with highest weight λ:

$$\dim(T^{(\lambda)}) = \lim_{t\to 0} \chi_\lambda(t\delta) = \lim_{t\to 0} \prod_{\alpha>0} \frac{\sinh(\frac{t}{2}(\alpha,\delta+\lambda))}{\sinh(\frac{t}{2}(\alpha,\delta))} = \prod_{\alpha>0} \frac{(\lambda+\delta,\alpha)}{(\delta,\alpha)}. \quad (3.5.32)$$

Let us now use the general Weyl formula for characters, Eq. (3.5.19), to work out explicit expressions for characters of compact groups $SU(n)$, $SO(n)$ and $USp(2r)$.

Example 1. $SU(2)$. Compact Lie algebra $su(2)$ has one positive root $\alpha = 1$, in normalization where highest weights $\lambda \equiv j$ are integer or half-integer. In this case, we have $\delta = 1/2$ and Weyl formula (3.5.19) becomes (recall that $W = \mathbb{Z}_2$, see Example 1 in Section 3.2)

$$\chi_\lambda(\theta) = \frac{D_{\delta+\lambda}(\theta)}{D_\delta(\theta)} = \frac{e^{(\theta,\,\delta+\lambda)} - e^{-(\theta,\,\delta+\lambda)}}{e^{(\theta,\,\delta)} - e^{-(\theta,\,\delta)}} = \frac{e^{i\psi(1/2+\lambda)} - e^{-i\psi(1/2+\lambda)}}{e^{i\psi/2} - e^{-i\psi/2}}, \tag{3.5.33}$$

where we set $\theta = i\psi$, $\psi \in \mathbb{R}$. The result (3.5.33) coincides with formula (2.4.9) (see Section 2.4.1).

Example 2. $SU(r+1)$ and $U(r+1)$. Let us calculate character of group $SU(r+1)$ for representation labeled by Young diagram $[\lambda_1, \dots, \lambda_r] \vdash |\lambda|$, which is the highest weight representation of Lie algebra $s\ell(r+1, \mathbb{C})$ (complexification of $su(r+1)$) with highest weight $\lambda^{(\vec{n})}$ given by (3.4.36). Recall that in our formalism, the root space of $s\ell(r+1, \mathbb{C})$ is hyperplane $P_u \subset \mathbb{R}^{r+1}$ normal to vector $u = (1, \dots, 1) \in \mathbb{R}^{r+1}$. In accordance with (3.4.65), we have

$$\delta = \sum_{f=1}^{r} \lambda_{(f)} = \sum_{i=1}^{r+1} \left(\frac{r}{2} - i + 1\right) e^{(i)}, \tag{3.5.34}$$

where $\lambda_{(f)} = (\sum_{i=1}^{f} e^{(i)} - \frac{f}{r+1} u)$ are fundamental weights (3.4.22) of algebra $s\ell(r+1, \mathbb{C})$.

We recall (see item 1 in the proof of Proposition 3.1.1 in Section 3.1.3) that Weyl group W of algebra $s\ell(r+1, \mathbb{C})$ is isomorphic to symmetric group S_{r+1}, and action of an element $\sigma \in W = S_{r+1}$ can be thought of as permutation $\theta^i \to \theta^{\sigma(i)}$ of coordinates of vector $\theta = (\theta^1, \dots, \theta^{r+1})$. Then the function $\epsilon(\sigma)$ in (3.5.17) reads $\epsilon(\sigma) = \det(\sigma) = (-1)^{p(\sigma)}$, where $p(\sigma)$ is parity of permutation $\sigma \in S_{r+1} = W$. We now make use of (3.4.36) and (3.5.34) for $\lambda^{(\vec{n})}$ and δ and write the numerator (3.5.17) in Weyl formula (3.5.19) as follows:

$$\begin{aligned}
D_{\delta+\lambda^{(\vec{n})}}(\theta) &= \sum_{\sigma^{-1}\in W} \epsilon(\sigma^{-1}) \exp(\lambda^{(\vec{n})} + \delta,\, \sigma(\theta)) \\
&= \sum_{\sigma\in W} (-1)^{p(\sigma)} \exp(\lambda^{(\vec{n})} + \delta,\, \sigma(\theta)) \\
&= \sum_{\sigma\in S_{r+1}} (-1)^{p(\sigma)} \exp\left(\sum_{i=1}^{r+1} (\lambda_i^{(\vec{n})} + \delta_i)\, \theta^{\sigma(i)}\right)
\end{aligned}$$

$$= \sum_{\sigma\in S_{r+1}} (-1)^{p(\sigma)} \exp\left(\sum_{i=1}^{r+1}\left(\lambda_i - \frac{|\lambda|}{r+1} + \frac{r}{2} - i + 1\right)\theta^{\sigma(i)}\right)$$

$$= c_{|\lambda|} \sum_{\sigma\in S_{r+1}} (-1)^{p(\sigma)} \prod_{i=1}^{r+1} x_{\sigma(i)}^{\lambda_i+r-i+1} = c_{|\lambda|}\ \det(x_i^{\lambda_j+(r+1)-j}), \tag{3.5.35}$$

where $x_i = e^{\theta^i}$, $|\lambda| = \sum_i \lambda_i$ and $c_{|\lambda|}$ is normalization factor

$$c_{|\lambda|} = \exp\left(\left(-\frac{|\lambda|}{r+1} - \frac{r}{2}\right)\sum_{i=1}^{r+1}\theta^i\right). \tag{3.5.36}$$

Note that this factor equals 1 for $g \in SU(r+1)$, since the condition $\det(g) = 1$ is equivalent to $\sum_i \theta^i = (\theta, u) = 0$ (vector $(\theta^1, \ldots, \theta^{r+1})$ belongs to root space P_u of algebra $su(r+1)$). Nevertheless, at this point it is instructive to consider general case $\sum_i \theta^i \neq 0$, which enables us to generalize the character formula from group $SU(r+1)$ to group $U(r+1)$ below. In the last equality in (3.5.35) we use the definition of determinant of matrix $||A_{ij}||_{i,j=1,\ldots,r+1}$ (see Section I-1.2.2 of the first book):

$$\det(A) = \sum_{\sigma\in S_{r+1}} (-1)^{p(\sigma)} A_{\sigma(1)\,1}\cdots A_{\sigma(r+1)\,r+1}.$$

For denominator (3.5.11), we have, taking into account explicit form of positive roots (3.1.13),

$$D_\delta(\theta) = \prod_{i<j}\left(e^{(e^{(i)}-e^{(j)},\theta)/2} - e^{(e^{(j)}-e^{(i)},\theta)/2}\right)$$

$$= e^{-\sum_{i<j}(e^{(i)}+e^{(j)},\theta)/2}\prod_{i<j}\left(e^{(e^{(i)},\theta)} - e^{(e^{(j)},\theta)}\right)$$

$$= \exp\left(-\frac{r}{2}\sum_{i=1}^{r+1}\theta^i\right)\prod_{i<j}(x_i - x_j), \tag{3.5.37}$$

where we employ identity $\sum_{i<j}(e^{(i)} + e^{(j)}, \theta)/2 = (r/2)(u, \theta) = (r/2) \sum_{i=1}^{r+1}\theta^i$. The right-hand side of (3.5.37), modulo factor $c_0 = \exp(-\frac{r}{2}(\theta, u))$, is Vandermonde determinant (see Section I-1.2.2 of the first book)

$$\prod_{i<j}(x_i - x_j) = \det(x_i^{r+1-j}). \tag{3.5.38}$$

Problem 3.5.3. Prove (3.5.38) by making use of identity (3.5.12) for Weyl denominator in the case of algebra $u(r+1)$ whose root system is described in item 2.1 of Section 3.4.2.

Homogeneous polynomial $\det(x_i^{\lambda_j+(r+1)-j})$ in (3.5.35) is antisymmetric with respect to transposition of any pair of variables x_k, x_j. So, polynomial $\det(x_i^{\lambda_j+(r+1)-j})$ vanishes at $x_k = x_j$. Therefore, it is divisible by minimal polynomial (3.5.38) that has the same property. The result of the division is character (3.5.19) of representation of group $SU(r+1)$ with highest weight $\lambda^{(\vec{n})}$:

$$\chi_{\lambda^{(\vec{n})}}(\theta) = e^{-\frac{|\lambda|\sum_i \theta^i}{r+1}} \cdot \frac{\det(x_i^{\lambda_j+(r+1)-j})}{\det\left(x_i^{(r+1)-j}\right)} = \frac{\det(x_i^{\lambda_j+(r+1)-j})}{\det\left(x_i^{(r+1)-j}\right)}. \tag{3.5.39}$$

Here, we use the fact that for $SU(r+1)$, factor $e^{-\frac{|\lambda|\sum_i \theta^i}{r+1}} = c_{|\lambda|}/c_0$ is equal to 1 due to the tracelessness condition $\sum_i \theta^i = 0$, which is equivalent to

$$\prod_i x_i = 1. \tag{3.5.40}$$

Numerator and denominator in (3.5.39) are antisymmetric polynomials, so their ratio is homogeneous symmetric polynomial of variables x_k (which is determined by Young diagram $[\lambda_1, \ldots, \lambda_r] \vdash |\lambda|$). This polynomial is called *Schur symmetric polynomial.*

Now we show that the first equality in (3.5.39) can be used for calculating characters of a certain class of representations of group $U(r+1)$. Lie algebra $u(r+1)$ is central extension of $su(r+1)$ obtained by adding generator equal to unit matrix I_{r+1}. Accordingly, diagonal Cartan generators $H_i \in su(r+1)$ introduced in (3.1.8) are replaced by generators $H_i' \in u(r+1)$ given by (3.4.43):

$$H_i = e_{ii} - \frac{1}{r+1} I_{r+1} \;\Rightarrow\; H_i' = e_{ii}, \tag{3.5.41}$$

where e_{ii} are diagonal matrix units. If operator $H_i \in su(r+1)$ in defining representation has eigenvalues μ_i, then eigenvalues of $H_i' \in u(r+1)$ in defining representation are $\mu_i' = \mu_i + u_i/(r+1)$, where $u_i = 1$ are components of vector $u = (1, 1, \ldots, 1)$. Classification of all irreducible representations of algebra $u(r+1)$ is given in item 2.1 of Section 3.4.2. The class of representations we discuss consists of irreducible representations $T^{(\vec{n})}$ with highest weight $\lambda^{(\vec{n})}$ given by (3.4.45), with $n_{r+1} = 0$. These representations $T^{(\vec{n})}$ are labeled by Young diagrams $\Lambda = [\lambda_1, \ldots, \lambda_r]$, with the number of rows not exceeding r, so $T^{(\vec{n})}$ are constructed as subrepresentations in tensor product of fundamental representations $T_{(f)}$ where $f = 1, \ldots, r$, see Fig. 3.4.2 (this tensor product does not contain nontrivial one-dimensional representations $T_{(r+1)}$). In other words, we consider those representations $T^{(\vec{n})}$ of algebra $u(r+1)$ which correspond to representations of algebra $su(r+1)$. For such representations, the weights μ' of $u(r+1)$ and μ of $su(r+1)$ are related by $\mu'_i = \mu_i + |\lambda|\, u_i/(r+1)$, where $|\lambda| = \lambda_1 + \cdots + \lambda_r$. Using definition (3.5.6), one finds the relationship between characters $\chi_{\lambda^{(\vec{n})\prime}}(\theta)$ of group $U(r+1)$ and characters $\chi_{\lambda^{(\vec{n})}}(\theta)$ of group $SU(r+1)$:

$$\begin{aligned}\chi_{\lambda^{(\vec{n})\prime}}(\theta) &= \mathrm{Tr}_\lambda(T^{(\lambda)}(e^{\theta^i H'_i}))\\ &= \sum_{\mu'} \mathrm{mult}_\lambda(\mu') \cdot \exp(\theta, \mu')\\ &= \sum_{\mu} \mathrm{mult}_\lambda(\mu) \cdot \exp(\theta, \mu + |\lambda| u/(r+1))\\ &= \exp\left(\frac{|\lambda|}{r+1}(\theta, u)\right) \cdot \chi_{\lambda^{(\vec{n})}}(\theta). \end{aligned} \tag{3.5.42}$$

Here, characters $\chi_{\lambda^{(\vec{n})\prime}}(\theta)$ are labeled by highest weights (3.4.45) with $\lambda_{r+1} = 0$:

$$\lambda^{(\vec{n})\prime} = \lambda^{(\vec{n})} + \frac{|\lambda|}{r+1} u = \sum_{f=1}^{r} n_f \cdot \lambda_{(f)} = (\lambda_1, \ldots, \lambda_r, 0), \quad n_f \in \mathbb{Z}_{\geq 0}. \tag{3.5.43}$$

Here, in the second equality, we use formula (3.4.36), while fundamental weights $\lambda_{(f)}$ are given by (3.4.44). Weights (3.5.43), like weights $\lambda^{(\vec{n})}$, are associated with Young diagrams $[\lambda_1, \ldots, \lambda_r] \vdash |\lambda|$. Taking into account the relation (3.5.42) in the first equality in (3.5.39), we find that character of group $U(r+1)$ in representation with highest weight (3.5.43) is again given by Schur polynomial

$$\chi_{\lambda^{(\vec{n})\prime}}(\theta) = \frac{\det(x_i^{\lambda_j + (r+1) - j})}{\det(x_i^{(r+1)-j})}, \tag{3.5.44}$$

where $x_i = e^{\theta^i}$ are eigenvalues of element $g \in U(r+1)$ in defining representation (see (3.5.5)). Unlike in the case of group $SU(r+1)$ with a constraint (3.5.40), there are no constraints on x_i. Formula for characters (3.5.44) which defines Schur polynomials is known as *Jacobi–Trudi formula.*

In the end of this example, let us find, by making use of Weyl formula (3.5.32), explicit expressions for dimensions of irreducible representations of algebra $su(r+1)$ with highest weight $\lambda^{(\vec{n})}$, which is associated with Young diagram $[\lambda_1, \ldots, \lambda_r] \vdash |\lambda|$. For the case of algebra $su(r+1)$, expressions for vector δ, for highest weight $\lambda^{(\vec{n})}$ and for positive roots $\alpha_+^{(ij)}$ are, respectively, (3.5.34), (3.4.36) and (3.1.13), so we have

$$\begin{aligned}(\delta, \alpha_+^{(ij)}) &= \left(\sum_{k=1}^{r+1}(r/2-k+1)e^{(k)},\, e^{(i)}-e^{(j)}\right) = j-i, \\ (\lambda^{(\vec{n})}, \alpha_+^{(ij)}) &= \left(\sum_{k=1}^{r+1}(\lambda_k - \frac{|\lambda|}{r+1})e^{(k)},\, e^{(i)}-e^{(j)}\right) = \lambda_i - \lambda_j.\end{aligned} \tag{3.5.45}$$

We insert these equalities into Weyl formula (3.5.32) and obtain

$$\dim(T^{(\lambda^{(\vec{n})})}) = \prod_{1\le i<j\le r+1} \frac{\lambda_i - \lambda_j + j - i}{j-i}. \tag{3.5.46}$$

This is the desired result.

We now turn to the discussion of dimensions and characters of irreducible representations of groups $USp(2r)$ and $SO(n)$; similar approach is taken in Ref. [28].

Example 3. $\boldsymbol{USp(2r)}$. Fundamental weights $\lambda_{(f)}$ of Lie algebra $usp(2r)$ are given in (3.4.56). To find Weyl denominator (3.5.12), we calculate vector δ:

$$\delta = \sum_{i=1}^{r} \lambda_{(i)} = (r, r-1, \ldots, 2, 1) = \sum_{i=1}^{r}(r+1-i)e^{(i)}, \tag{3.5.47}$$

where we use (3.4.65). Weyl group W of $usp(2r)$ (see item 3 in the Proof of Proposition 3.1.1 in Section 3.1.3) is extension of group S_r of all permutations of coordinates θ_i ($i = 1, \ldots, r$) by sign flips of any number of coordinates θ_i. The sign flip operations can be thought of as transformations $\theta \to \epsilon \cdot \theta$ with diagonal matrix $\epsilon = \mathrm{diag}(\epsilon_1, \ldots, \epsilon_r)$, where $\epsilon_i = \pm 1$. Then, in view of (3.5.47), we have for Weyl denominator (3.5.12)

$$\begin{aligned}D_\delta(\theta) &= \sum_{w\in W} \det(w)\ \exp(\delta, w(\theta)) \\ &= \sum_{\sigma\in S_r} \det(\sigma) \sum_{\epsilon} \det(\epsilon) \exp\left(\Sigma_i (r+1-i)\epsilon_i \theta^{\sigma(i)}\right),\end{aligned} \tag{3.5.48}$$

where $\det(\epsilon) = \epsilon_1 \cdots \epsilon_r$ and $\sum_\epsilon = \sum_{\epsilon_1=\pm 1} \cdots \sum_{\epsilon_r=\pm 1}$. Then we apply the formula

$$\sum_\epsilon \det(\epsilon) \exp\left(\sum_i \epsilon_i \mu_i\right) = \prod_{i=1}^r \sum_{\epsilon_i=\pm 1} \epsilon_i \exp(\epsilon_i \mu_i) = \prod_{i=1}^r \left(e^{\mu_i} - e^{-\mu_i}\right), \tag{3.5.49}$$

and write (3.5.48) as

$$D_\delta(\theta) = \sum_{\sigma \in S_r} \det(\sigma) \prod_{i=1}^r (x_{\sigma(i)}^{r+1-i} - x_{\sigma(i)}^{-(r+1-i)}) = \det(x_j^{r+1-i} - x_j^{-(r+1-i)}), \tag{3.5.50}$$

where $x_j = e^{\theta^j}$. The numerator (3.5.17) of the Weyl formula is found in a similar way:

$$D_{\delta+\lambda}(\theta) = \det\left(x_j^{\lambda_i+r+1-i} - x_j^{-(\lambda_i+r+1-i)}\right), \tag{3.5.51}$$

where λ_i are components of the highest weight (3.4.9), (3.4.57):

$$\lambda = \sum_{i=1}^r n_i \, \lambda_{(i)} = (\lambda_1, \lambda_2, \ldots, \lambda_r),$$
$$\lambda_k = \sum_{j=k}^r n_j, \quad n_j \in \mathbb{Z}_{\geq 0}, \quad \lambda_1 \geq \lambda_2 \geq \cdots \geq \lambda_r. \tag{3.5.52}$$

Finally, Weyl formula (3.5.19) gives

$$\chi_\lambda(\theta) = \frac{\det(x_j^{\lambda_i+r-i+1} - x_j^{-(\lambda_i+r-i+1)})}{\det(x_j^{r-i+1} - x_j^{-(r-i+1)})}. \tag{3.5.53}$$

Making use of explicit expressions (3.5.47), (3.5.52) and (3.1.37), respectively, for vector δ, highest weight λ and positive roots of algebra $usp(2r)$, we find

$$\begin{aligned}(\delta,\, e^{(i)} \pm e^{(j)}) &= \left(\sum_{k=1}^r (r-k+1) e^{(k)},\, e^{(i)} \pm e^{(j)}\right) \\ &= (r-i+1) \pm (r-j+1), \\ (\lambda,\, e^{(i)} \pm e^{(j)}) &= \left(\sum_{k=1}^{r+1} \lambda_k e^{(k)},\, e^{(i)} \pm e^{(j)}\right) = \lambda_i \pm \lambda_j.\end{aligned} \tag{3.5.54}$$

Inserting these expressions into Weyl dimension formula (3.5.32), we calculate the dimension of representation $T^{(\lambda)}$:

$$\dim(T^{(\lambda)}) = \prod_{1\le i<j\le r} \frac{\lambda_i - \lambda_j + j - i}{j-i} \prod_{1\le i\le j\le r} \frac{\lambda_i + \lambda_j + 2r + 2 - i - j}{2r+2-i-j}. \tag{3.5.55}$$

Example 4. $SO(2r+1)$. Fundamental weights $\lambda_{(i)}$ of algebra $so(2r+1)$ are given in (3.4.51). Making use of (3.4.65) we obtain vector δ:

$$\delta = \sum_{i=1}^{r} \lambda_{(i)} = \left(r - \frac{1}{2}, r - \frac{3}{2}, \dots, \frac{1}{2}\right) = \sum_{i=1}^{r} \left(r + \frac{1}{2} - i\right) e^{(i)}. \tag{3.5.56}$$

Weyl group of algebra $so(2r+1)$ is isomorphic to Weyl group of $usp(2r)$ (see items 2 and 3 in the proof of Proposition 3.1.1 in Section 3.1.3). So, denominator and numerator in Weyl formula for characters are calculated in a similar way as in Example 3, and we get

$$D_\delta(\theta) = \det(x_j^{r+1/2-i} - x_j^{-(r+1/2-i)}), \tag{3.5.57}$$

$$D_{\delta+\lambda}(\theta) = \det(x_j^{\lambda_i + r + 1/2 - i} - x_j^{-(\lambda_i + r + 1/2 - i)}), \tag{3.5.58}$$

where $x_j = e^{\theta^j}$ and λ_i are components of the highest weight (3.4.9):

$$\lambda = \sum_{i=1}^{r} n_i \lambda_{(i)} = (\lambda_1, \lambda_2, \dots, \lambda_r), \ \lambda_k = \sum_{j=k}^{r-1} n_j + \frac{n_r}{2}, \quad n_j \in \mathbb{Z}_{\ge 0},$$
$$\lambda_1 \ge \lambda_2 \ge \cdots \ge \lambda_r \ge 0. \tag{3.5.59}$$

Finally, Weyl formula (3.5.19) for group $SO(2r+1)$ reads

$$\chi_\lambda(\theta) = \frac{\det(x_j^{\lambda_i + r - i + 1/2} - x_j^{-(\lambda_i + r - i + 1/2)})}{\det(x_j^{r-i+1/2} - x_j^{-(r-i+1/2)})}. \tag{3.5.60}$$

Making use of explicit expressions (3.5.56), (3.5.59) and (3.1.28), respectively, for vector δ, highest weight λ and positive roots of algebra

$so(2r+1,\mathbb{C})$, we find

$$\begin{aligned}
(\delta,\, e^{(i)} \pm e^{(j)}) &= \left(\sum_{k=1}^{r} \left(r - k + \frac{1}{2} \right) e^{(k)},\, e^{(i)} \pm e^{(j)} \right) \\
&= \left(r - i + \frac{1}{2} \right) \pm \left(r - j + \frac{1}{2} \right), \\
(\lambda,\, e^{(i)} \pm e^{(j)}) &= \left(\sum_{k=1}^{r+1} \lambda_k e^{(k)},\, e^{(i)} \pm e^{(j)} \right) = \lambda_i \pm \lambda_j, \\
(\delta,\, e^{(i)}) &= \left(r - i + \frac{1}{2} \right), \quad (\lambda,\, e^{(i)}) = \lambda_i.
\end{aligned} \tag{3.5.61}$$

By inserting the latter expressions in the Weyl formula (3.5.32), we find the dimension of representation $T^{(\lambda)}$ of group $SO(2r+1)$:

$$\begin{aligned}
\dim(T^{(\lambda)}) = &\prod_{1 \le i < j \le r} \frac{(\lambda_i - \lambda_j + j - i)}{(j-i)} \frac{(\lambda_i + \lambda_j + 2r + 1 - i - j)}{(2r+1-i-j)} \\
&\times \prod_{1 \le i \le r} \frac{(\lambda_i + r - i + \frac{1}{2})}{(r - i + \frac{1}{2})}.
\end{aligned} \tag{3.5.62}$$

Example 5. $SO(2r)$. Fundamental weights $\lambda_{(i)}$ of algebra $so(2r)$ are given in (3.4.47), while formula (3.4.65) leads to

$$\delta = \sum_{i=1}^{r} \lambda_{(i)} = (r-1, r-2, \ldots, 1, 0) = \sum_{i=1}^{r} (r-i) e^{(i)}. \tag{3.5.63}$$

Weyl group of algebra $so(2r)$ contains (see item 4 in the proof of Proposition 3.1.1 in Section 3.1.3) all permutations $\sigma \in S_r$ of coordinates θ_i of vector $\theta \in i\mathfrak{V}_r$, and operations of sign flip of any even number of coordinates θ_i. The latter can be written as $\theta \to \epsilon \cdot \theta$ with diagonal matrix $\epsilon = \mathrm{diag}(\epsilon_1, \ldots, \epsilon_r)$, where $\epsilon_i = \pm 1$ and the number of negative elements is even. Instead of formula (3.5.49) we now have

$$\sum_{\epsilon} (\epsilon_1 \cdots \epsilon_r) \exp\left(\sum_i \epsilon_i \mu_i \right) = \frac{1}{2} \prod_{i=1}^{r} (e^{\mu_i} + e^{-\mu_i}) + \frac{1}{2} \prod_{i=1}^{r} (e^{\mu_i} - e^{-\mu_i}), \tag{3.5.64}$$

which is straightforward to check by expanding the right-hand side into a sum. Making use of (3.5.12), (3.5.17), (3.5.63) and (3.5.64), we obtain for denominator and numerator in Weyl formula (3.5.19)

$$D_\delta(\theta) = \frac{1}{2}\det(x_j^{r-i} + x_j^{-(r-i)}),$$

$$D_{\delta+\lambda}(\theta) = \frac{1}{2}\det(x_j^{\lambda_i+r-i} + x_j^{-(\lambda_i+r-i)}) + \frac{1}{2}\det(x_j^{\lambda_i+r-i} - x_j^{-(\lambda_i+r-i)}), \tag{3.5.65}$$

where $x_j = e^{\theta^j}$, and λ_i are components of the highest weight (3.4.9):

$$\lambda = \sum_{i=1}^{r} n_i\,\lambda_{(i)} = (\lambda_1, \lambda_2, \ldots, \lambda_r),$$

$$\lambda_k = \sum_{j=k}^{r-2} n_j + \frac{n_{r-1}+n_r}{2}, \quad 1 \le k \le r-1,$$

$$\lambda_r = \frac{n_{r-1}-n_r}{2}, \quad n_j \in \mathbb{Z}_{\ge 0},\ \lambda_1 \ge \lambda_2 \ge \cdots \ge \lambda_r. \tag{3.5.66}$$

Finally, Weyl character formula (3.5.19) for group $SO(2r)$ reads

$$\chi_\lambda(\theta) = \frac{\det(x_j^{\lambda_i+r-i} + x_j^{-(\lambda_i+r-i)}) + \det(x_j^{\lambda_i+r-i} - x_j^{-(\lambda_i+r-i)})}{\det(x_j^{r-i} + x_j^{-(r-i)})}. \tag{3.5.67}$$

In the same way as above, we calculate

$$(\delta,\, e^{(i)} \pm e^{(j)}) = \left(\sum_{k=1}^{r}(r-k)e^{(k)},\, e^{(i)} \pm e^{(j)}\right) = (r-i) \pm (r-j),$$

$$(\lambda,\, e^{(i)} \pm e^{(j)}) = \left(\sum_{k=1}^{r+1}\lambda_k e^{(k)},\, e^{(i)} \pm e^{(j)}\right) = \lambda_i \pm \lambda_j, \tag{3.5.68}$$

and applying (3.5.32) we find the dimension of representation $T^{(\lambda)}$ of group $SO(2r)$:

$$\dim(T^{(\lambda)}) = \prod_{1\le i<j\le r} \frac{(\lambda_i - \lambda_j + j - i)}{(j-i)}\frac{(\lambda_i + \lambda_j + 2r - i - j)}{(2r-i-j)}$$

$$= \prod_{1\le i<j\le r} \frac{(\ell_i^2 - \ell_j^2)}{(m_i^2 - m_j^2)}, \tag{3.5.69}$$

where $\ell_i = \lambda_i + r - i$ and $m_i = r - i$.

Chapter 4

Finite-Dimensional Representations of Algebras $s\ell(N,\mathbb{C})$, $su(N)$ and Groups $SL(N,\mathbb{C})$ and $SU(N)$

We have already shown (see Proposition 3.4.8, Section 3.4.2) that all complex, finite-dimensional irreducible representations of Lie algebras $s\ell(N,\mathbb{C})$ and $su(N)$, and hence representations of groups $SU(N)$, can be thought of as subrepresentations in tensor products $T^{\otimes r}$ of defining representations T. These representations are conveniently studied as corresponding representations of group $SL(N,\mathbb{C})$. One should bear in mind, however, that the special class of representations of $SL(N,\mathbb{C})$ that we consider (and that is related to the representations of $SU(N)$) includes only those representations that are embedded in tensor products $T^{\otimes r}$, where T is defining representation of $SL(N,\mathbb{C})$, and it does not include representations constructed by invoking complex conjugate representations T^*.

We show in this chapter that these irreducible representations are classified in terms of irreducible representations of symmetric group (group of permutations) S_r. Irreducible representations of S_r, in their turn, are labeled by Young diagrams, which, therefore, classify complex, finite-dimensional irreducible representations of $SU(N)$, as well as representations of $SL(N,\mathbb{C})$ from our limited class. We have already pointed out in Remark 2 in Section 3.4.2 that Young diagrams appear naturally in the theory of representations of algebras $s\ell(N,\mathbb{C})$ and $su(N)$. Remarkable relationship between complex, finite-dimensional irreducible representations of group $SU(N)$ (and algebras $s\ell(N,\mathbb{C})$ and $su(N)$) and irreducible representations of symmetric groups is known as *Schur–Weyl duality*.

4.1. Preliminaries

Lie group $SL(N,\mathbb{C})$ is defined (see Sections I-1.1 and I-1.2 in the first book) as a set of complex $N \times N$ matrices with unit determinant. Its compact subgroup $SU(N) \subset SL(N,\mathbb{C})$ is a subset of matrices $U \in SL(N,\mathbb{C})$ obeying unitarity condition $U^\dagger U = I_N$. We defined Lie algebra $\mathcal{A}(SL(N,\mathbb{C}))$ of group $SL(N,\mathbb{C})$ as tangent space (at unit element) to group $SL(N,\mathbb{C})$ which is treated as real manifold. Thus, algebra $\mathcal{A}(SL(N,\mathbb{C}))$ is, by definition, real Lie algebra which is realification of complex Lie algebra $s\ell(N,\mathbb{C})$, the latter being complex vector space of all complex traceless $N \times N$ matrices.

Lie algebra $su(N)$ is real form of Lie algebra $s\ell(N,\mathbb{C})$, so there is a basis in $s\ell(N,\mathbb{C})$ which at the same time is basis in $su(N)$. Every representation of $s\ell(N,\mathbb{C})$ defines representation of basis elements of $su(N)$, and thus defines complex representation of $su(N)$. Conversely, every representation of $su(N)$ defines, through representation of its generators, a representation of $s\ell(N,\mathbb{C})$. Thus, there is one-to-one correspondence between complex, finite-dimensional irreducible representations of algebras $su(N)$ and $s\ell(N,\mathbb{C})$ (see similar discussion in the case of algebras $su(2)$ and $s\ell(2,\mathbb{C})$ in the beginning of Section 2.1). Now, it follows from Proposition I-4.6.5 that all finite-dimensional irreducible representations of $su(N)$ are contained, as subrepresentations, in regular representation of $su(N)$, and, therefore, all finite-dimensional irreducible representations of $s\ell(N,\mathbb{C})$ are contained in regular representation of $s\ell(N,\mathbb{C})$.

Remark. We emphasize again that although our main concern in this chapter are algebras $su(N)$, $s\ell(N,\mathbb{C})$ and group $SU(N)$, our analysis below is done to large extent in terms of group $SL(N,\mathbb{C})$. This should not lead to confusion. Representations T_{SL} of $SL(N,\mathbb{C})$ that we encounter belong to the class described, in the case $N = 2$, in Remark 3 in the end of Section 2.1. Let us define this class for arbitrary N. In the vicinity of a unit element of group $SL(N,\mathbb{C})$ we can write $g = I_N + t \cdot A + t^2 + \cdots$, where $g \in SL(N,\mathbb{C})$, $A \in s\ell(N,\mathbb{C})$ and t is a small parameter.[a] Then any representation T_{SL} *from this class* is related to representation $T_{s\ell}$ of algebra $s\ell(N,\mathbb{C})$ via $T_{SL}(g) = I_N + t \cdot T_{s\ell}(A) + t^2 + \cdots$. These representations of group $SL(N,\mathbb{C})$ are in one-to-one correspondence with representations of algebra

[a]Note that $g = I_N + t \cdot A + t^2 + \cdots$ is simply a relation between complex $N \times N$ matrices; it is *not* the usual relation between Lie group and Lie algebra: as we discussed above, complex Lie algebra $s\ell(N,\mathbb{C})$ is *not* (real) Lie algebra $\mathcal{A}(SL(N,\mathbb{C}))$ of Lie group $SL(N,\mathbb{C})$.

$s\ell(N,\mathbb{C})$, so we use the same notation for pertinent representations of algebra $s\ell(N,\mathbb{C})$ and group $SL(N,\mathbb{C})$. Since all complex finite-dimensional irreducible representations of Lie algebra $s\ell(N,\mathbb{C})$ are embedded in tensor products $T^{\otimes r}$ of defining representation T, the class of representations of group $SL(N,\mathbb{C})$ that we consider includes only those representations which are contained in tensor product of defining representations.[b] Working with group $SL(N,\mathbb{C})$, rather than algebra $s\ell(N,\mathbb{C})$, noticeably simplifies some statements and formulas.

We know (see Section 3.4.2 and, in particular, Proposition 3.4.8) that every finite-dimensional irreducible representation $T^{(\lambda)}$ of Lie algebra $sl(N,\mathbb{C})$ (and $su(N)$) is contained, as subrepresentation, in a tensor product $T^{\otimes r}$ of defining representations T of this algebra. Here, we show that all these irreducible representations are contained, as subrepresentations, in regular representation of group $SL(N,\mathbb{C})$. To this end, consider regular representation (2.2.21) of group $SL(N,\mathbb{C})$, associated with right shifts, and concentrate on elements close to unit element, $g = I_N + \epsilon + \cdots \in SL(N,\mathbb{C})$. We have

$$\begin{aligned}[\widetilde{T}^{(R)}(g)]\alpha(h) &= \alpha(h\cdot g) = \alpha(h + h\epsilon + \cdots) \\ &= \alpha(h) + \epsilon_{ij} h_{ki} \frac{\partial}{\partial h_{kj}} \alpha(h) + \cdots . \end{aligned} \tag{4.1.1}$$

Here $h, g \in SL(N,\mathbb{C})$, $\alpha(h)$ is a function on group, elements of traceless matrix ϵ are small, and dots denote higher order terms in ϵ_{ij}. Formula (4.1.1) defines right regular representation $\widetilde{T}^{(R)}$ of algebra $s\ell(N,\mathbb{C})$, which acts on functions $\alpha(h)$ on group $SL(N,\mathbb{C})$:

$$\widetilde{T}^{(R)} : A \to A_{ij} h_{ki} \frac{\partial}{\partial h_{kj}}, \quad A \in s\ell(N,\mathbb{C}). \tag{4.1.2}$$

From this, by taking into account the condition $\mathrm{Tr}(A) = 0$, we obtain the basis elements of $s\ell(N,\mathbb{C})$ (in regular representation $\widetilde{T}^{(R)}$) in the form

$$E_{ij} = h_{ki}\partial_{h_{kj}} - \frac{\delta_{ij}}{N} h_{k\ell}\partial_{h_{k\ell}}, \tag{4.1.3}$$

where $\partial_{h_{ki}} = \frac{\partial}{\partial h_{ki}}$ and summation over repeated indices is implicit.

---

[b]The relation $T_{SL}(g) = I_N + t\cdot T_{s\ell}(A) + t^2 + \cdots$ does not hold, e.g., for representation T^* of $SL(N,\mathbb{C})$ which is complex conjugate to defining representation T (complex Lie algebra does not have complex conjugate representations); this shows that we consider only restricted class of representations of $SL(N,\mathbb{C})$, which does not include representations embedded in $T^{\otimes p} \otimes (T^*)^{\otimes r}$ with $r > 0$.

Problem 4.1.1. Check that Eq. (4.1.3) defines representation of algebra $s\ell(N,\mathbb{C})$, i.e., check that operators E_{ij} obey defining relations of algebra $s\ell(N,\mathbb{C})$: $[E_{ij},E_{k\ell}] = \delta_{jk}E_{i\ell} - \delta_{i\ell}E_{kj}$.

Remark. To describe more general class of irreducible representations of group $SL(N,\mathbb{C})$, which includes complex conjugate representations, one introduces generalized regular representations $T_R^{\rm gen}$ defined on space of nonholomorphic functions $\alpha(h,h^*)$ (cf. (2.2.21)):

$$[T_R^{\rm gen}(g)\cdot\alpha](h,h^*) = \alpha(h\cdot g, h^*\cdot g^*) \quad g,h \in SL(N,\mathbb{C}). \tag{4.1.4}$$

We do not consider this class in what follows, and discuss only those representations of group $SL(N,\mathbb{C})$ which are embedded in tensor products of its defining representations and hence are obtained from regular representation (2.2.21) and (4.1.1).

Let $\mathcal{F}$ denote space of all functions $\alpha(h)$ on group $SL(N,\mathbb{C})$, and $\mathcal{F}^{(r)}$ be subspace in $\mathcal{F}$ consisting of homogeneous order-r polynomials in variables h_{ki}. Clearly, the representation (4.1.2) preserves the order of a monomial in h_{ki}, so subspaces $\mathcal{F}^{(r)}$ are invariant under action of algebra $s\ell(N,\mathbb{C})$ in regular representation $\widetilde{T}^{(R)}$. This means that finite-dimensional *irreducible* representations, embedded in regular representation, are realized on space of polynomials of a certain order, which is a subspace of one of the spaces $\mathcal{F}^{(r)}$.

We now note that right regular representation (2.2.21) of group $SL(N,\mathbb{C})$, considered on space $\mathcal{F}^{(1)}$ of *first-order* homogeneous polynomials made of elements of kth row $(h_{k1},h_{k2},\dots,h_{kN})$ of matrix $h \in SL(N,\mathbb{C})$, is equivalent to defining representation T of group $SL(N,\mathbb{C})$:

$$[\widetilde{T}^{(R)}(g)]h_{ki} = h_{kj}g_{ji} \quad (k=1,\dots,N), \tag{4.1.5}$$

$$T(g)\cdot\vec{e}_i = \vec{e}_j g_{ji}, \tag{4.1.6}$$

where $(\vec{e}_1,\vec{e}_2,\dots,\vec{e}_N)$ are basis vectors in space $\mathcal{V}$ of defining representation. In other words, defining representation of $SL(N,\mathbb{C})$ is a subrepresentation of regular representation, and the space of this subrepresentation is any of the spaces $\mathcal{F}_k^{(1)}$ of homogeneous first-order polynomials in variables h_{kj} (with one and the same k); we have $\mathcal{F}^{(1)} = \sum_k \mathcal{F}_k^{(1)}$. Likewise, space $\mathcal{F}^{(r)}$ of order-r homogeneous polynomials is a direct sum of subspaces $\mathcal{F}^{(r)}_{k_1,k_2,\dots,k_r}$ $(k_1 \le k_2 \le \cdots \le k_r)$, whose basis elements are monomials

$$h_{k_1i_1}h_{k_2i_2}\cdots h_{k_ri_r} \quad (k_1 \le k_2 \le \cdots \le k_r),$$

where $k_1, k_2, \ldots, k_r$ are fixed. These monomials are transformed under $SL(N,\mathbb{C})$ in the same way as basis elements

$$(\vec{e}_{i_1} \otimes \vec{e}_{i_2} \otimes \cdots \otimes \vec{e}_{i_r}), \tag{4.1.7}$$

in the space $\mathcal{V}^{\otimes r}$ of tensor product $T^{\otimes r}$ of r defining representations,

$$\begin{aligned}
&[\widetilde{T}^{(R)}(g)](h_{k_1 i_1} \cdots h_{k_r i_r}) = (h_{k_1 j_1} \cdots h_{k_r j_r}) g_{j_1 i_1} \cdots g_{j_r i_r} \\
&\quad \Rightarrow T^{\otimes r}(g) \cdot (\vec{e}_{i_1} \otimes \cdots \otimes \vec{e}_{i_r}) = (\vec{e}_{j_1} \otimes \cdots \otimes \vec{e}_{j_r}) g_{j_1 i_1} \cdots g_{j_r i_r}.
\end{aligned} \tag{4.1.8}$$

So, subrepresentation of regular representation acting in space $\mathcal{F}^{(r)}_{k_1,k_2,\ldots,k_r}$ is equivalent to $T^{\otimes r}$, and therefore all tensor products $T^{\otimes r}$ of defining representations T are contained in the regular representation. According to the consideration of Section 3.4.2 (see Proposition 3.4.8), every finite-dimensional irreducible representation $T^{(\lambda)}$ of Lie algebra $s\ell(N,\mathbb{C})$ (and group $SU(N)$) is a subrepresentation in some tensor product $T^{\otimes r}$. Thus, every finite-dimensional irreducible representation $T^{(\lambda)}$ of algebra $s\ell(N,\mathbb{C})$ (and group $SU(N)$) is a subrepresentation in right regular representation (4.1.2) and realized in subspaces $\mathcal{V}^{(\lambda)}$ of spaces of homogeneous polynomials $\mathcal{F}^{(r)}_{k_1,k_2,\ldots,k_r} \subset \mathcal{F}^{(r)} \subset \mathcal{F}$.

We now face a problem of extracting irreducible representations $T^{(\lambda)}$ from (generally speaking, reducible) representations $T^{\otimes r}$. One of the methods here is based on construction of highest weight representations, as discussed in Chapter 3. In this chapter, we follow another route, namely, we show that irreducible finite-dimensional representations of algebra $s\ell(N,\mathbb{C})$ (and group $SU(N)$) are extracted from representations $T^{\otimes r}$ by making use of certain projectors which are associated with irreducible representations of symmetric group (group of permutations) S_r.

4.2. Action of Group S_r in Tensor Product of Defining Representations

Let T be defining representation of group $SL(N,\mathbb{C})$ in N-dimensional complex space $\mathcal{V}$. With the basis $\{\vec{e}_i\}$ $(i = 1, \ldots, N)$ in $\mathcal{V}$, the action of element $g = ||g^i_j||$ of group $SL(N,\mathbb{C})$ in defining representation is written in the standard way,

$$T(g) \cdot \vec{e}_i = \vec{e}_k g^k{}_i. \tag{4.2.1}$$

Hereafter, we write "incoming" indices (corresponding to rows) and "outgoing" indices (corresponding to columns) as upper and lower indices, respectively. In terms of coordinates of vector $\vec{\psi} = \psi^i \vec{e}_i \in \mathcal{V}$, the action (4.2.1) reads

$$\psi^i \to \psi'^i = [T(g) \cdot \vec{\psi}]^i = g^i{}_j \psi^j. \tag{4.2.2}$$

Consider now tensor product $T^{\otimes r} := T \otimes \cdots \otimes T$ of r defining representations of group $SL(N, \mathbb{C})$. In accordance with general definition, it maps an element $g = ||g^i_j|| \in SL(N, \mathbb{C})$ to operator

$$T^{\otimes r}(g) := \underbrace{T(g) \otimes \cdots \otimes T(g)}_{r}, \tag{4.2.3}$$

which acts in tensor product $\mathcal{V}^{\otimes r}$ of N-dimensional spaces as follows:

$$\begin{aligned} &(T^{\otimes r}(g)) \cdot (\vec{e}_{i_1} \otimes \cdots \otimes \vec{e}_{i_r}) = (\vec{e}_{j_1} \otimes \cdots \otimes \vec{e}_{j_r}) g^{j_1}{}_{i_1} \cdots g^{j_r}{}_{i_r}, \\ &\qquad \psi^{i_1 \ldots i_r} \to \psi'^{i_1 \ldots i_r} = [(T^{\otimes r}(g)) \cdot \Psi]^{i_1 \ldots i_r} = g^{i_1}{}_{j_1} \cdots g^{i_r}{}_{j_r} \cdot \psi^{j_1 \ldots j_r}, \end{aligned} \tag{4.2.4}$$

where $(\vec{e}_{j_1} \otimes \cdots \otimes \vec{e}_{j_r})$ are basis elements in $\mathcal{V}^{\otimes r}$, $\Psi = \psi^{j_1 \ldots j_r} (\vec{e}_{j_1} \otimes \cdots \otimes \vec{e}_{j_r}) \in \mathcal{V}^{\otimes r}$ is rank-r tensor, and $\psi^{j_1 \ldots j_r}$ are its components. Unit operator $I_N^{\otimes r}$ in space $\mathcal{V}^{\otimes r}$ reads

$$\begin{aligned} &(I_N^{\otimes r}) \cdot (\vec{e}_{i_1} \otimes \cdots \otimes \vec{e}_{i_r}) = (\vec{e}_{j_1} \otimes \cdots \otimes \vec{e}_{j_r}) \delta^{j_1}_{i_1} \cdots \delta^{j_r}_{i_r}, \\ &\qquad \psi^{i_1 \ldots i_r} \to \psi'^{i_1 \ldots i_r} = [(I_N^{\otimes r}) \cdot \Psi]^{i_1 \ldots i_r} = \delta^{i_1}_{j_1} \cdots \delta^{i_r}_{j_r} \psi^{j_1 \ldots j_r}. \end{aligned} \tag{4.2.5}$$

We now introduce a set of operators $\mathsf{s}(\sigma)$ that permute factors $\vec{v}_a \in \mathcal{V}$ in element $(\vec{v}_1 \otimes \vec{v}_2 \otimes \cdots \otimes \vec{v}_r) \in \mathcal{V}^{\otimes r}$:

$$\mathsf{s}(\sigma) \cdot (\vec{v}_1 \otimes \vec{v}_2 \otimes \cdots \otimes \vec{v}_r) = (\vec{v}_{\sigma^{-1}(1)} \otimes \vec{v}_{\sigma^{-1}(2)} \otimes \cdots \otimes \vec{v}_{\sigma^{-1}(r)}), \tag{4.2.6}$$

where σ is permutation of r elements $(1, 2, \ldots, r)$:

$$\sigma = \begin{pmatrix} 1 & 2 & \cdots & r \\ \sigma(1) & \sigma(2) & \cdots & \sigma(r) \end{pmatrix} \in S_r, \tag{4.2.7}$$

and σ^{-1} is permutation, inverse to σ. Group of permutations (symmetric group) is introduced and discussed in Section I-1.1.1 of the first book. We recall the multiplication rule of two permutations $\mu, \sigma \in S_r$:

$$\mu \cdot \sigma = \begin{pmatrix} 1 & \cdots & r \\ \mu(1) & \cdots & \mu(r) \end{pmatrix} \cdot \begin{pmatrix} 1 & \cdots & r \\ \sigma(1) & \cdots & \sigma(r) \end{pmatrix}$$
$$= \begin{pmatrix} \sigma(1) & \cdots & \sigma(r) \\ \mu(\sigma(1)) & \cdots & \mu(\sigma(r)) \end{pmatrix} \cdot \begin{pmatrix} 1 & \cdots & r \\ \sigma(1) & \cdots & \sigma(r) \end{pmatrix}$$
$$= \begin{pmatrix} 1 & \cdots & r \\ \mu(\sigma(1)) & \cdots & \mu(\sigma(r)) \end{pmatrix}. \tag{4.2.8}$$

In accordance with definition (4.2.6), matrix representation of operators $\mathsf{s}(\sigma)$ is given by (cf. (4.2.5))

$$\mathsf{s}(\sigma) \cdot (\vec{e}_{i_1} \otimes \vec{e}_{i_2} \otimes \cdots \otimes \vec{e}_{i_r}) = (\vec{e}_{i_{\sigma^{-1}(1)}} \otimes \vec{e}_{i_{\sigma^{-1}(2)}} \otimes \cdots \otimes \vec{e}_{i_{\sigma^{-1}(r)}})$$
$$= (\vec{e}_{j_1} \otimes \vec{e}_{j_2} \otimes \cdots \otimes \vec{e}_{j_r}) \delta_{i_1}^{j_{\sigma(1)}} \cdot \delta_{i_2}^{j_{\sigma(2)}} \cdots \delta_{i_r}^{j_{\sigma(r)}}$$
$$\equiv (\vec{e}_{j_1} \otimes \cdots \otimes \vec{e}_{j_r}) [\mathsf{s}(\sigma)]^{j_1 \cdots j_r}_{i_1 \cdots i_r}, \tag{4.2.9}$$

$$\psi^{i_1 \cdots i_r} \to \psi'^{i_1 \cdots i_r} = [\mathsf{s}(\sigma) \cdot \Psi]^{i_1 \cdots i_r} = \delta_{j_1}^{i_{\sigma(1)}} \cdots \delta_{j_r}^{i_{\sigma(r)}} \psi^{j_1 \ldots j_r}$$
$$= \psi^{i_{\sigma(1)} \cdots i_{\sigma(r)}}. \tag{4.2.10}$$

Proposition 4.2.1. *Formula* (4.2.6) *defines mapping* s *from symmetry group* S_r *to set of linear operators in* $\mathcal{V}^{\otimes r}$. *Mapping* s *is homomorphism, and, therefore, it defines representation of group* S_r *in space* $\mathcal{V}^{\otimes r}$.

Proof. We have to show that $\mathsf{s}(\mu) \cdot \mathsf{s}(\sigma) = \mathsf{s}(\mu \cdot \sigma)$. This follows from the chain of equalities

$$\mathsf{s}(\mu) \cdot \mathsf{s}(\sigma) \cdot (\vec{v}_1 \otimes \cdots \otimes \vec{v}_r) = \mathsf{s}(\mu) \cdot (\vec{v}'_1 \otimes \cdots \otimes \vec{v}'_r)$$
$$= (\vec{v}'_{\mu^{-1}(1)} \otimes \cdots \otimes \vec{v}'_{\mu^{-1}(r)})$$
$$= (\vec{v}_{\sigma^{-1}(\mu^{-1}(1))} \otimes \cdots \otimes \vec{v}_{\sigma^{-1}(\mu^{-1}(r))})$$
$$= \mathsf{s}(\mu \cdot \sigma) \cdot (\vec{v}_1 \otimes \cdots \otimes \vec{v}_r),$$

where we use definition (4.2.6) and introduce notation $\vec{v}'_i = \vec{v}_{\sigma^{-1}(i)}$. □

Problem 4.2.1. Check that $\mathsf{s}(\mu) \cdot \mathsf{s}(\sigma) = \mathsf{s}(\mu \cdot \sigma)$ by making use of matrix representation (4.2.9), (4.2.10). In other words, show that $[\mathsf{s}(\mu) \cdot [\mathsf{s}(\sigma) \cdot \Psi]]^{i_1 \ldots i_r} = \Psi^{i_{\mu(\sigma(1))} \cdots i_{\mu(\sigma(r))}}$.

Example 1. As an illustration to Proposition 4.2.1 and Problem 4.2.1, let us check the property $\mathsf{s}(\mu)\cdot\mathsf{s}(\sigma)=\mathsf{s}(\mu\cdot\sigma)$ for two transpositions $\sigma=(13)$ and $\mu=(12)$. We consecutively apply (4.2.10), first for $\mu=(12)$ and then for $\sigma=(13)$:

$$[\mathsf{s}(\mu)\cdot[\mathsf{s}(\sigma)\cdot\Psi]]^{i_1i_2i_3}=[\mathsf{s}(\sigma)\cdot\Psi]]^{i_2i_1i_3}=\psi^{i_3i_1i_2}. \tag{4.2.11}$$

On the other hand, the definition (4.2.8) gives $\mu\cdot\sigma=\begin{pmatrix}1&2&3\\3&1&2\end{pmatrix}$, so that $[\mathsf{s}(\mu\cdot\sigma)\cdot\Psi]^{i_1i_2i_3}=\psi^{i_3i_1i_2}$, which agrees with (4.2.11). We note that in general, the consecutive action of two permutations $\sigma,\mu\in S_r$ on rank-r tensor can be fairly simply derived by using graphic representation of permutations discussed in Section I-1.1.4 of the first book.

Thus, representation s maps an element $\sigma\in S_r$ to a linear operator $\mathsf{s}(\sigma)$ acting, in accordance with (4.2.6)–(4.2.10), in space $\mathcal{V}^{\otimes r}$ of rank-r tensors. Formula (4.2.10) tells that the operator $\mathsf{s}(\sigma)$ permutes indices of tensor $\Psi\in\mathcal{V}^{\otimes r}$.

An example of operator of the type (4.2.6), (4.2.9) with $r=2$ is given by transposition operator P such that $P(\vec{v}_1\otimes\vec{v}_2)=\vec{v}_2\otimes\vec{v}_1$:

$$P(\vec{e}_{i_1}\otimes\vec{e}_{i_2})=(\vec{e}_{j_1}\otimes\vec{e}_{j_2})P^{j_1j_2}_{i_1i_2},\quad P^{j_1j_2}_{i_1i_2}=\delta^{j_1}_{i_2}\delta^{j_2}_{i_1}. \tag{4.2.12}$$

The generalization of this operator for $r>2$ is an operator $\mathsf{s}(\sigma_a)$, whose matrix in the standard basis (4.1.7) is

$$(\mathsf{s}(\sigma_a))^{j_1\dots j_r}_{k_1\dots k_r}=\delta^{j_1}_{k_1}\cdot\delta^{j_2}_{k_2}\cdots\delta^{j_{a-1}}_{k_{a-1}}\cdot\delta^{j_a}_{k_{a+1}}\delta^{j_{a+1}}_{k_a}\cdot\delta^{j_{a+2}}_{k_{a+2}}\cdots\delta^{j_r}_{k_r}, \tag{4.2.13}$$

where σ_a is neighboring transposition that permutes only vectors $\vec{v}_a$ and $\vec{v}_{a+1}$ in (4.2.6), leaving other vectors in their places. Clearly, matrix (4.2.13) is associated with the following operator in $\mathcal{V}^{\otimes r}$:

$$\mathsf{s}(\sigma_a)=\underbrace{I_N\otimes\cdot\otimes I_N}_{a-1}\otimes P\otimes\underbrace{I_N\otimes\cdot\otimes I_N}_{r-a-1}, \tag{4.2.14}$$

where P is operation of permutation of two neighboring vectors (see (4.2.12)).

Again, let T be defining representation of group $SL(N,\mathbb{C})$ in space $\mathcal{V}$. We say that X is $SL(N,\mathbb{C})$-invariant operator in $\mathcal{V}^{\otimes r}$ if and only if it commutes with the action of group $SL(N,\mathbb{C})$ in representation $T^{\otimes r}$:

$$X\cdot(T^{\otimes r}(g))=(T^{\otimes r}(g))\cdot X,\quad \forall g\in SL(N,\mathbb{C}). \tag{4.2.15}$$

Proposition 4.2.2. *Any $SL(N,\mathbb{C})$-invariant operator X, acting in space $\mathcal{V}^{\otimes r}$, is a linear combination of operators* (4.2.9), (4.2.10) :

$$X = \sum_{\sigma \in S_r} x_\sigma \mathsf{s}(\sigma), \tag{4.2.16}$$

where x_σ are complex coefficients, and summation runs over $r!$ permutations $\sigma \in S_r$. Any operator (4.2.16) *is $SL(N,\mathbb{C})$-invariant, i.e., it obeys* (4.2.15).

Proof. Invariance of operator (4.2.16) is a consequence of invariance of each of the terms in the sum, i.e., invariance of operators $\mathsf{s}(\sigma)$ for all σ. To prove the latter property, consider a set of operators A_i $(i = 1, \ldots, r)$ in $\mathcal{V}$ and construct operator $A_1 \otimes \cdots \otimes A_r$ acting in $\mathcal{V}^{\otimes r}$. Then the definition (4.2.6) yields the commutation rule

$$(A_1 \otimes A_2 \otimes \cdots \otimes A_r) \cdot \mathsf{s}(\sigma) = \mathsf{s}(\sigma) \cdot (A_{\sigma(1)} \otimes A_{\sigma(2)} \otimes \cdots \otimes A_{\sigma(r)}). \tag{4.2.17}$$

Now invariance of operators $\mathsf{s}(\sigma)$ $(\forall \sigma \in S_r)$ is straightforward:

$$\mathsf{s}(\sigma) \cdot (T(g) \otimes T(g) \otimes \cdots \otimes T(g)) = (T(g) \otimes T(g) \otimes \cdots \otimes T(g)) \cdot \mathsf{s}(\sigma). \tag{4.2.18}$$

We now show that any operator X obeying (4.2.15) has the form (4.2.16). Without loss of generality, when considering (4.2.15), we can replace $g \in SL(N,\mathbb{C})$ by $g \in GL(N,\mathbb{C})$ (since any matrix $g \in GL(N,\mathbb{C})$ is obtained from matrix $g \in SL(N,\mathbb{C})$ by multiplication by a number). To begin with, we choose, as an element g in (4.2.15), the diagonal matrix $g = \text{diag}(a_1, \ldots, a_N)$, where a_i are complex parameters. Then relation (4.2.15) takes the form

$$(a_{i_1} \cdot a_{i_2} \cdots a_{i_r} - a_{k_1} \cdot a_{k_2} \cdots a_{k_r}) X^{i_1 \ldots i_r}_{k_1 \ldots k_r} = 0 \tag{4.2.19}$$

(no summation over repeated indices). Since a_i are arbitrary, general solution to Eq. (4.2.19) for coefficients of operator X is a sum over permutations $\sigma \in S_r$ of particular solutions:

$$X^{i_1 \ldots i_r}_{k_1 \ldots k_r} = x^{k_1 \ldots k_r}_\sigma \delta^{i_{\sigma(1)}}_{k_1} \cdots \delta^{i_{\sigma(r)}}_{k_r}, \tag{4.2.20}$$

where $x^{k_1 \ldots k_r}_\sigma \in \mathbb{C}$. We insert (4.2.20) in (4.2.15) and no longer consider g^i_k as diagonal matrices. Then, we obtain the following equation:

$$(x^{k_1 \ldots k_r}_\sigma - x^{i_{\sigma(1)} \ldots i_{\sigma(r)}}_\sigma) g^{i_{\sigma(1)}}_{k_1} \cdots g^{i_{\sigma(r)}}_{k_r} = 0,$$

which has to be satisfied for all values of indices i_s and k_p. Since g^i_k are arbitrary, we have $x_\sigma^{k_1\dots k_r} = x_\sigma^{i_{\sigma(1)}\dots i_{\sigma(r)}}$. We set all indices on the right-hand side $i_\ell = 1$ and find that coefficients $x_\sigma^{k_1\dots k_r}$ are independent of indices $(k_1\dots k_r)$ and equal to a constant $x_\sigma = x_\sigma^{1\dots 1}$. Then expression (4.2.20) becomes $x_\sigma \cdot \mathsf{s}(\sigma)$. Summing up over σ all these particular solutions, we come to (4.2.16). □

Problem 4.2.2. Prove (4.2.18) by making use of matrix representations (4.2.4) and (4.2.9).

Proposition 4.2.2 shows that there are matrices (4.2.16), which are not proportional to unit matrix and which commute with all matrices of representation $T^{\otimes r}$ of group $SL(N,\mathbb{C})$. Therefore, in accordance with Schur's lemma, representation $T^{\otimes r}$ is reducible, and it is possible to extract irreducible representations from $T^{\otimes r}$.

Example 2. Consider tensor product of two defining representations $T \otimes T$ of group $SL(N,\mathbb{C})$, i.e., consider the case $r = 2$. Proposition 4.2.2 tells that there is a nontrivial operator X in $\mathcal{V} \otimes \mathcal{V}$ such that

$$X \cdot (T(g) \otimes T(g)) = (T(g) \otimes T(g)) \cdot X, \quad \forall g \in SL(N,\mathbb{C}), \tag{4.2.21}$$

and that this operator has the following form:

$$X = x_0 I_N \otimes I_N + x_1 P, \tag{4.2.22}$$

where $x_0, x_1 \in \mathbb{C}$ are arbitrary numbers, $I_N \otimes I_N$ is unit operator, and P is operator of transposition in $\mathcal{V}^{\otimes 2} = \mathcal{V} \otimes \mathcal{V}$. We construct two projectors

$$P^+ = \frac{1}{2}(I_N \otimes I_N + P), P^- = \frac{1}{2}(I_N \otimes I_N - P), \tag{4.2.23}$$

which have the form (4.2.22) and hence commute with the action of $SL(N,\mathbb{C})$. Operators P^+ and P^- form complete system of orthogonal projectors,

$$P^+P^- = P^-P^+ = 0, \quad (P^\pm)^2 = P^\pm, \quad P^+ + P^- = I_N \otimes I_N. \tag{4.2.24}$$

Since they are $SL(N,\mathbb{C})$-invariant, they divide the space $\mathcal{V}^{\otimes 2}$ into two invariant subspaces

$$\mathcal{V}^{(+)} = P^+\mathcal{V}^{\otimes 2}, \quad \mathcal{V}^{(-)} = P^-\mathcal{V}^{\otimes 2}.$$

Thus, representation $T^{\otimes 2}$ is reducible. Invariant subspaces $\mathcal{V}^{(+)}$ and $\mathcal{V}^{(-)}$ are spaces of symmetric and antisymmetric tensors in $\mathcal{V}^{\otimes 2}$, since, in accordance with (4.2.10), (4.2.12) and (4.2.23) we have

$$[P^{\pm} \cdot \Psi]^{i_1 i_2} = \frac{1}{2}(\psi^{i_1 i_2} \pm \psi^{i_2 i_1}). \tag{4.2.25}$$

We denote the pertinent representations by $T^{(+)}$ and $T^{(-)}$.

Proposition 4.2.3. *Representations $T^{(+)}$ and $T^{(-)}$ are irreducible.*

Proof. We give a proof *ad absurdum*. Let representation $T^{(+)}$ be reducible (the case of $T^{(-)}$ is treated in the same way). Then there is a nontrivial invariant subspace $\mathcal{V}^{(+)'}$ in the space $\mathcal{V}^{(+)}$ of representation $T^{(+)}$. We define operator $P^{+'}$ which acts in space $\mathcal{V}^{\otimes 2}$ and projects to subspace $\mathcal{V}^{(+)'} \subset \mathcal{V}^{(+)}$. Since $\mathcal{V}^{(+)'}$ is invariant, operator $P^{+'}$ commutes with the action of the group in representation $T^{\otimes 2}$, i.e., obeys (4.2.21), and at the same time acts nontrivially in subspace $\mathcal{V}^{(+)}$ to project on nontrivial subspace $\mathcal{V}^{(+)'} \subset \mathcal{V}^{(+)}$. In view of Proposition 4.2.2, operator $P^{+'}$ is a linear combination of unit operator and transposition operator, but these operators act in space $\mathcal{V}^{(+)}$ trivially (identically), so they cannot make operator $P^{+'}$. □

The latter example of extracting irreducible representations $T^{(+)}$ and $T^{(-)}$ from $T^{\otimes 2}$ suggests a general method of extracting irreducible representations from tensor product $T^{\otimes r}$ of defining representations of $SL(N,\mathbb{C})$ for any r, namely, among operators of the form (4.2.16), which commute with $T^{\otimes r}(g)$ ($\forall g \in SL(N,\mathbb{C})$), one finds projectors P_α (where index α labels these projectors). Each of these projectors must be *primitive*, meaning that it cannot be decomposed into a sum of other projectors of the form (4.2.16). Then each of these projectors is used to extract an irreducible representation $T^{(\alpha)}$ from $T^{\otimes r}$; the space of this representation is defined as follows:

$$\mathcal{V}^{(\alpha)} = P_\alpha \cdot \mathcal{V}^{\otimes r}. \tag{4.2.26}$$

Since projector P_α commutes with $T^{\otimes r}(g)$ for all $g \in SL(N,\mathbb{C})$, invariance of subspace $\mathcal{V}^{(\alpha)}$ follows from the chain of equalities

$$T^{\otimes r}(g) \cdot \mathcal{V}^{(\alpha)} = T^{\otimes r}(g) \cdot P_\alpha \cdot \mathcal{V}^{\otimes r} = P_\alpha \cdot T^{\otimes r}(g) \cdot \mathcal{V}^{\otimes r} \subset \mathcal{V}^{(\alpha)}, \tag{4.2.27}$$

and the representation $T^{(\alpha)}$ (homomorphic mapping from group $SL(N,\mathbb{C})$ to group of linear operators in $\mathcal{V}^{\otimes r}$) is defined by $T^{(\alpha)}(g) = P_\alpha \cdot T^{\otimes r}(g) \cdot P_\alpha$. Proof of the fact that $T^{(\alpha)}$ is irreducible is similar to the proof given in Proposition 4.2.3; its main ingredient is the requirement that projector P_α is primitive and has the form (4.2.16).

Thus, a primitive projector P_α determines an irreducible representation $T^{(\alpha)}$ embedded in $T^{\otimes r}$. Somewhat more ambitious though related problem is to expand the tensor product $T^{\otimes r}$ into a direct sum of irreducible representations. A general approach toward its solution is to construct complete set of primitive orthogonal projectors $\{P_\alpha\}$ such that

$$P_\alpha P_\beta = P_\alpha \delta_{\alpha\beta}, \quad \sum_\alpha P_\alpha = I_N^{\otimes r}. \tag{4.2.28}$$

Then the space $\mathcal{V}^{\otimes r}$ is expanded into direct sum of invariant subspaces,

$$\mathcal{V}^{\otimes r} = \sum_\alpha \mathcal{V}^{(\alpha)},$$

as desired.

In what follows, we study the second, more complicated problem of expanding $T^{\otimes r}$ into a direct sum of irreducible representations, having in mind that construction of primitive projectors (not necessarily forming complete orthogonal system) comes out as an intermediate result. In view of (4.2.16), objects to consider are operators $\mathsf{s}(\sigma)$. In accordance with Proposition 4.2.1, the map s from group S_r to the set of operators $\mathsf{s}(\sigma)$ defined by (4.2.6) determines representation s of symmetric group S_r in space $\mathcal{V}^{\otimes r}$. It is worth noting that operators (4.2.16) are *linear combinations* of matrices $\mathsf{s}(\sigma)$. Therefore, strictly speaking, operators (4.2.16) are representations of elements $\sum_\sigma x_\sigma \cdot \sigma$ of group algebra $\mathbb{C}[S_r]$ associated with group S_r. We recall (see Definition I-4.5.1 in Section I-4.6.2 of the first book) that group algebra $\mathbb{C}[G]$ associated with finite group G is defined as vector space of formal linear combinations

$$\alpha = \sum_{g\in G} \alpha_g g, \quad \alpha_g \in \mathbb{C},$$

with multiplication of elements $\alpha, \beta \in \mathbb{C}[G]$ dictated by group operation in group G:

$$\alpha \cdot \beta = \sum_{g,h\in G} \alpha_g \beta_h (g \cdot h) = \sum_{g\in G} \gamma_g g,$$

where $\gamma_g = \sum_{h\in G} \alpha_{g\cdot h^{-1}} \beta_h \in \mathbb{C}$. Any representation T of group G can be extended to group algebra via

$$T(\alpha) = \sum_{g\in G} \alpha_g T(g).$$

So, any representation of group G defines representation of algebra $\mathbb{C}[G]$, and, conversely, any representation of group algebra $\mathbb{C}[G]$ defines representation of group G.

Problem 4.2.3. Prove the latter statement. Prove that representation s defined by (4.2.9) is exact representation of group algebra $\mathbb{C}[S_r]$.

In accordance with (4.2.16), projectors P_α are images $P_\alpha = \mathsf{s}(e_\alpha)$ of some elements e_α of group algebra $\mathbb{C}[S_r]$. Since s is exact representation (see above Problem 4.2.3), the properties (4.2.28) imply

$$e_\alpha e_\beta = \delta_{\alpha\beta} e_\alpha, \quad e = \sum_{\alpha=1}^{s} e_\alpha, \tag{4.2.29}$$

where e is unit element in $\mathbb{C}[S_r]$.

Definition 4.2.1. Let $\mathcal{A}$ be an associative algebra with unit element e. Element $e_\alpha \in \mathcal{A}$ which obeys $e_\alpha^2 = e_\alpha$ is called *idempotent.*[c] Idempotent e_α is *primitive* if it cannot be decomposed into a sum of other idempotents in $\mathcal{A}$. Idempotents obeying the first of Eq. (4.2.29) are called *orthogonal*, and if idempotents obey also the second relation then they form *complete system of orthogonal idempotents.*

We show in Section 4.5.1 that if finite-dimensional associative algebra $\mathcal{A}$ with unit element possesses complete orthogonal system of primitive idempotents, then this system of idempotents determines the structure of algebra $\mathcal{A}$, and the idempotents e_α are the key objects for constructing irreducible representations of this algebra.

Thus, elements $e_\alpha \in \mathbb{C}[S_r]$, associated to primitive projectors $P_\alpha = \mathsf{s}(e_\alpha)$, which satisfy (4.2.28) form complete orthogonal system of primitive idempotents in algebra $\mathbb{C}[S_r]$ and hence completely determine the structure of algebra $\mathbb{C}[S_r]$ and its irreducible representations. On the other hand, representations $P_\alpha = \mathsf{s}(e_\alpha)$ of these idempotents enable one to expand representation $T^{\otimes r}$ of $SL(N,\mathbb{C})$ and $SU(N)$ in irreducible representations. All of the above is summarized in the following proposition.

Proposition 4.2.4. *Let element e_α be a primitive idempotent in algebra $\mathbb{C}[S_r]$. Let $\mathcal{V}$ be space of defining representation of group $SL(N,\mathbb{C})$ (and $SU(N)$). Then projector $P_\alpha = \mathsf{s}(e_\alpha)$ extracts from tensor product $\mathcal{V}^{\otimes r}$, an invariant subspace $P_\alpha \mathcal{V}^{\otimes r}$ of an irreducible representation. If elements e_α form complete orthogonal system of primitive idempotents in algebra $\mathbb{C}[S_r]$, then the set of projectors P_α defines an expansion of representation $T^{\otimes r}$ into direct sum of irreducible representations.*

[c]Term "idempotent" derives from Latin idem ("same") and potens ("power, ability"). This term was suggested by Benjamin Peirce in 1870s.

To construct elements $e_\alpha \in \mathbb{C}[S_r]$ and projectors $P_\alpha = \mathsf{s}(e_\alpha)$ explicitly, we need more facts about symmetric group S_r and its representations (the initial information about group S_r was given in Section I-1.1.4 of the first book).

4.3. Representations of Symmetric Group I. Young Symmetrizers

We present in this section the standard scheme of construction of primitive idempotents e_α in algebra $\mathbb{C}[S_r]$, which is due to Young and Frobenius. We show (see also Section 4.5.1) that besides their role in construction of projectors $P_\alpha = \mathsf{s}(e_\alpha)$ used in the theory of representations of group $SU(N)$, primitive idempotents e_α enable one to extract irreducible representations of group S_r from its regular representation.

4.3.1. Young tableaux and Young symmetrizers

We begin with reminding (see Section I-1.1.4 and, in particular, Proposition I-1.1.3 in the first book) that conjugacy classes in group S_r are associated with partitions $\lambda \vdash r$ of number r, which are graphically represented by Young diagrams with r boxes. The number of conjugacy classes coincides with the number of inequivalent irreducible representations of S_r (see Proposition I-4.6.4). So, one naturally expects that there is one-to-one correspondence between irreducible representations of group S_r and conjugacy classes, or Young diagrams with r boxes. We show here that this is indeed the case, namely, every Young diagram λ corresponds to a certain set of primitive idempotents $e_\alpha \in \mathbb{C}[S_r]$ which, as we will show, generate *equivalent* irreducible representations of group S_r. Conversely, every primitive idempotent $e_\alpha \in \mathbb{C}[S_r]$ is associated with a certain Young diagram λ, and idempotents associated with different Young diagrams generate *inequivalent* representations of $\mathbb{C}[S_r]$.

Consider Young diagram $\lambda = [\lambda_1, \ldots, \lambda_m] \vdash r$ with m rows, and assign numbers $(1, 2, \ldots, r)$ to each of its r boxes, as shown here:

$$\begin{array}{|c|c|c|c|c|c|}
\hline
a_1 & a_2 & a_3 & \cdots & \cdots & a_{k_1} \\
\hline
a_{k_1+1} & a_{k_1+2} & \cdots & \cdots & a_{k_2} \\
\cline{1-5}
\cdots & \cdots & \cdots & \cdots \\
\cline{1-4}
a_{k_{m-1}+1} & \cdots & a_{k_m} \\
\cline{1-3}
\end{array} \equiv \mathsf{T}_\lambda, \qquad (4.3.1)$$

where $k_1 = \lambda_1$, $k_2 = \lambda_1 + \lambda_2, \ldots$, $k_m = \lambda_1 + \cdots + \lambda_m = r$, and $(a_1, \ldots, a_r)$ is some permutation of numbers $(1, 2, \ldots, r)$. Young diagram with such an assignment of numbers in boxes is called *Young tableau* and is denoted by T_λ. We say that Young tableau T_λ, obtained by assignment of numbers to boxes in diagram λ, has shape of diagram λ.

Definition 4.3.1. Young tableau $\widetilde{\mathsf{T}}_{\tilde{\lambda}}$ is dual to Young tableau T_λ, if rows of tableau T_λ are columns of $\widetilde{\mathsf{T}}_{\tilde{\lambda}}$, and columns of T_λ are rows of $\widetilde{\mathsf{T}}_{\tilde{\lambda}}$. Young diagram $\tilde{\lambda}$ (shape of tableau $\widetilde{\mathsf{T}}_{\tilde{\lambda}}$) is dual to diagram λ (shape of tableau T_λ).

As an example, two tableaux T_λ and $\widetilde{\mathsf{T}}_{\tilde{\lambda}}$ are dual to each other:

$$\mathsf{T}_\lambda = \begin{array}{|c|c|c|c|c|} \hline 5 & 8 & 10 & 7 & 4 \\ \hline 2 & 9 & 3 & 6 & \\ \cline{1-4} 11 & 1 & & & \\ \cline{1-2} \end{array}\,, \qquad \widetilde{\mathsf{T}}_{\tilde{\lambda}} = \begin{array}{|c|c|c|} \hline 5 & 2 & 11 \\ \hline 8 & 9 & 1 \\ \hline 10 & 3 & \\ \cline{1-2} 7 & 6 & \\ \cline{1-2} 4 & & \\ \cline{1-1} \end{array}\,.$$

Their Young diagrams $\lambda = [5,4,2]$ and $\tilde{\lambda} = [3,3,2,2,1]$ are also dual to each other.

We treat Young tableaux T_λ mainly as a convenient tool for labeling special linear combinations of permutations (i.e., special elements of group algebra $\mathbb{C}[S_r]$; we call these elements later as Young symmetrizers), where Young diagrams with r boxes correspond to the case of group S_r. To characterize these linear combinations, we give additional definitions.

Let T_λ be Young tableau, where $\lambda = [\lambda_1,\ldots,\lambda_m]$ is Young diagram with m rows. Let $H(\mathsf{T}_\lambda)$ be a set of horizontal permutations $p \in S_r$ which permute numbers in each of the rows of tableau T_λ (and do not permute numbers from different rows). Clearly, $H(\mathsf{T}_\lambda)$ is a subgroup of group S_r. Since permutations of numbers in a row of length λ_k make group S_{λ_k}, we have

$$H(\mathsf{T}_\lambda) = S_{\lambda_1} \times S_{\lambda_2} \times \cdots \times S_{\lambda_m} \subset S_r, \quad \mathrm{ord}(H(\mathsf{T}_\lambda)) = \lambda_1! \cdots \lambda_m!. \tag{4.3.2}$$

Similarly, one defines a subgroup $V(\mathsf{T}_\lambda) \subset S_r$ of all vertical permutations, i.e., operations $q \in S_r$ that permute numbers in each of the column of tableau T_λ (and do not permute numbers from different columns). Thus,

$$V(\mathsf{T}_\lambda) = S_{\lambda'_1} \times S_{\lambda'_2} \times \cdots \times S_{\lambda'_{m'}} \subset S_r, \quad \mathrm{ord}(V(\mathsf{T}_\lambda)) = \lambda'_1! \cdots \lambda'_{m'}!, \tag{4.3.3}$$

where $[\lambda'_1,\ldots,\lambda'_{m'}]$ are heights of all m' columns of diagram λ.

We now use elements p of subgroup $H(\mathsf{T}_\lambda)$ and elements q of subgroup $V(\mathsf{T}_\lambda)$ to define two operators

$$P = \sum_{p \in H(\mathsf{T}_\lambda)} p \in \mathbb{C}[H(\mathsf{T}_\lambda)], \tag{4.3.4}$$

$$Q = \sum_{q \in V(\mathsf{T}_\lambda)} \delta_q q \in \mathbb{C}[V(\mathsf{T}_\lambda)], \tag{4.3.5}$$

where δ_q is parity of permutation q ($\delta_q = +1$ for even permutation q and $\delta_q = -1$ for odd permutation q). The summation in (4.3.4) runs over all horizontal permutations from $H(\mathsf{T}_\lambda)$, and summation in (4.3.5) runs over all vertical permutations from $V(\mathsf{T}_\lambda)$. Operators P and Q are elements of group algebra $\mathbb{C}[S_r]$ and are called horizontal symmetrizer and vertical antisymmetrizer, respectively, associated with tableau T_λ. Clearly, horizontal symmetrizer P and vertical antisymmetrizer Q are products (see (4.3.2) and (4.3.3))

$$P = P_1 \cdot P_2 \cdots P_m, \quad Q = Q_1 \cdot Q_2 \cdots Q_{m'},$$

where $P_i = \sum_{p \in S_{\lambda_i}} p$ is symmetrizer of ith row of T_λ, while $Q_k = \sum_{q \in S_{\lambda'_k}} \delta_q q$ is antisymmetrizer of kth column of T_λ.

An example is given by operators P and Q for Young tableau

$$\mathsf{T}_{[2,2]} = \begin{array}{|c|c|}\hline 1 & 2 \\ \hline 3 & 4 \\ \hline \end{array}\,. \tag{4.3.6}$$

Symmetrizers P_1 and P_2 are given by (e is unit element in S_4)

$$P_1 = \begin{pmatrix} 1 & 2 & 3 & 4 \\ 1 & 2 & 3 & 4 \end{pmatrix} + \begin{pmatrix} 1 & 2 & 3 & 4 \\ 2 & 1 & 3 & 4 \end{pmatrix} = e + (12),$$

$$P_2 = \begin{pmatrix} 1 & 2 & 3 & 4 \\ 1 & 2 & 3 & 4 \end{pmatrix} + \begin{pmatrix} 1 & 2 & 3 & 4 \\ 1 & 2 & 4 & 3 \end{pmatrix} = e + (34),$$

and operator P is their product

$$P(\mathsf{T}_{[2,2]}) = (e + (12)) \cdot (e + (34)) = e + (12) + (34) + (12) \cdot (34). \tag{4.3.7}$$

Likewise,

$$Q(\mathsf{T}_{[2,2]}) = (e - (13)) \cdot (e - (24)) = e - (13) - (24) + (13) \cdot (24). \tag{4.3.8}$$

It is worth emphasizing that products here involve only permutations p and q determined by the original Young tableau T_λ (we comment further on this point after Eq. (4.3.11)).

It follows from definitions (4.3.4) and (4.3.5) that operators $P \in \mathbb{C}[H(\mathsf{T}_\lambda)]$ and $Q \in \mathbb{C}[V(\mathsf{T}_\lambda)]$ obey

$$\begin{aligned} &p \cdot P = P \cdot p = P, \quad &q \cdot Q = Q \cdot q = \delta_q Q, \\ &\forall p \in H(\mathsf{T}_\lambda), &\forall q \in V(\mathsf{T}_\lambda). \end{aligned} \tag{4.3.9}$$

These properties mean that operators P coincide for all tableaux T_λ which are obtained from each other by permutations of numbers in rows, while operators Q coincide for tableaux related by vertical permutations q (in the latter case, we have $Q \to q \cdot Q \cdot q^{-1} = \delta_q^2 Q = Q$; see formula (4.3.19) below).

Problem 4.3.1. Making use of (4.3.9) show that modulo a numerical factor, operators P and Q are idempotents,

$$P \cdot P = \alpha P, \quad Q \cdot Q = \beta Q \Rightarrow E_P^2 = E_P, \quad E_Q^2 = E_Q, \tag{4.3.10}$$

where $E_P = \alpha^{-1} P$, $E_Q = \beta^{-1} Q$, $\alpha = \mathrm{ord}(H(\mathsf{T}_\lambda))$ and $\beta = \mathrm{ord}(V(\mathsf{T}_\lambda))$.

Consider now a product of operators (4.3.4) and (4.3.5) constructed for a given tableau T_λ,

$$Y(\mathsf{T}_\lambda) = Q(\mathsf{T}_\lambda) \cdot P(\mathsf{T}_\lambda) = \sum_{q \in V(\mathsf{T}_\lambda)} \sum_{p \in H(\mathsf{T}_\lambda)} \delta_q q \cdot p. \tag{4.3.11}$$

Operator $Y(\mathsf{T}_\lambda) \in \mathbb{C}[S_r]$ is called *Young symmetrizer* of tableau T_λ. We emphasize again that the right-hand side of (4.3.11) involves permutations q and p constructed for the initial Young tableau T_λ.[d] As an example, Young symmetrizer of tableau (4.3.6) is a product of operators (4.3.7) and (4.3.8). An important property which we establish below is that Young symmetrizers are, modulo numerical factors, primitive idempotents in $\mathbb{C}[S_r]$. Note that formulas (4.3.9) imply

$$\begin{aligned} q \cdot Y(\mathsf{T}_\lambda) = Y(\mathsf{T}_\lambda)\delta_q, \quad Y(\mathsf{T}_\lambda) \cdot p = Y(\mathsf{T}_\lambda), \\ \forall q \in V(\mathsf{T}_\lambda), \quad \forall p \in H(\mathsf{T}_\lambda). \end{aligned} \tag{4.3.12}$$

We use the latter relations below.

[d]This remark has to do with the fact that permutation p in product $q \cdot p$ in (4.3.11) can be thought of as an operation on tableau T_λ, see below. Then one is tempted to consider q as vertical permutation associated with tableau $p \cdot \mathsf{T}_\lambda$ rather than T_λ. This would be incorrect: vertical permutations associated with tableau $p \cdot \mathsf{T}_\lambda$ are not involved in our construction.

Example. Let us express, in terms of Coxeter generators $\sigma_i = (i, i+1) \in S_r$ (neighboring transpositions), Young symmetrizers $Y(\mathsf{T}_\lambda)$ defined by the following Young tableaux:

$$\mathsf{T}_{[2]} = \boxed{1\,|\,2}, \quad \mathsf{T}_{[1^2]} = \begin{array}{|c|}\hline 1 \\ \hline 2 \\ \hline\end{array}, \quad (r=2),$$

$$\mathsf{T}_{[3]} = \boxed{1\,|\,2\,|\,3}, \quad \mathsf{T}_{[1^3]} = \begin{array}{|c|}\hline 1 \\ \hline 2 \\ \hline 3 \\ \hline\end{array}, \quad \mathsf{T}_{[2,1]_1} = \begin{array}{|c|c|}\hline 2 & 1 \\ \hline 3 \\ \cline{1-1}\end{array}, \quad \mathsf{T}_{[2,1]_2} = \begin{array}{|c|c|}\hline 2 & 3 \\ \hline 1 \\ \cline{1-1}\end{array}, \quad (r=3).$$

Recall that the defining relations for group S_r, written in terms of Coxeter generators, are (see (3.1.62) in Section 3.1.3):

$$\begin{gathered}\sigma_i \cdot \sigma_{i+1} \cdot \sigma_i = \sigma_{i+1} \cdot \sigma_i \cdot \sigma_{i+1}, \quad (i = 1, \dots, r-2), \\ \sigma_i \cdot \sigma_j = \sigma_j \cdot \sigma_i, \quad (|i-j| > 1), \quad \sigma_i^2 = e, \quad (i = 1, \dots, r-1),\end{gathered} \tag{4.3.13}$$

where e is identical permutation. Making use of (4.3.11), we obtain

$$Y(\mathsf{T}_{[2]}) = (1 + \sigma_1), \quad Y(\mathsf{T}_{[1^2]}) = (1 - \sigma_1), \tag{4.3.14}$$

$$\begin{gathered}Y(\mathsf{T}_{[3]}) = (1 + \sigma_2 + \sigma_1\sigma_2)(1 + \sigma_1), \quad Y(\mathsf{T}_{[1^3]}) = (1 - \sigma_2 + \sigma_1\sigma_2)(1 - \sigma_1), \\ Y(\mathsf{T}_{[2,1]_1}) = (1 - \sigma_2)(1 + \sigma_1), \qquad Y(\mathsf{T}_{[2,1]_2}) = (1 - \sigma_1)(1 + \sigma_2).\end{gathered} \tag{4.3.15}$$

Hereafter we write 1 for unit element e. To obtain the first line in (4.3.15), we make use of remarkable factorization formula

$$\begin{aligned}E_{P^{[r]}} &= \frac{1}{r!}(1 + \sigma_{r-1} + \sigma_{r-2}\sigma_{r-1} + \sigma_{r-3}\sigma_{r-2}\sigma_{r-1} + \cdots + \sigma_1 \cdots \sigma_{r-1}) \\ &\quad \times (1 + \sigma_{r-2} + \sigma_{r-3}\sigma_{r-2} + \cdots + \sigma_1 \dots \sigma_{r-2}) \cdots (1 + \sigma_1) \\ &= \frac{1}{r!} Y(\mathsf{T}_{[r]}),\end{aligned} \tag{4.3.16}$$

valid for total symmetrizer $E_{P^{[r]}}$, which obeys $E_{P^{[r]}}^2 = E_{P^{[r]}}$ and is associated with a Young tableau with single row of length r,

$$\mathsf{T}_{[r]} = \boxed{1\,|\,2\,|\dots|\,r}.$$

Formula for $Y(\mathsf{T}_{[1^3]})$ in (4.3.15) follows from formula for $Y(\mathsf{T}_{[3]})$, since the total antisymmetrizer $E_{Q^{[1^r]}} = \frac{1}{r!} Y(\mathsf{T}_{[1^r]})$ for Young tableaux with one column of height r is obtained from total symmetrizer (4.3.16) by automorphism $\sigma_i \to -\sigma_i$ of algebra $\mathbb{C}[S_r]$. Note that upon opening parentheses, the element $Y(\mathsf{T}_{[r]}) = E_{P^{[r]}}$ in (4.3.16) equals the sum of all $r!$ permutations of numbers $(1,2,\ldots,r)$ in the row $\mathsf{T}_{[r]}$, i.e., the sum of all elements of S_r.

Problem 4.3.2. * Prove identity (4.3.16) for total symmetrizer.

Problem 4.3.3. * Prove that total antisymmetrizer $E_{Q^{[1^r]}}$ is obtained from total symmetrizer $E_{P^{[r]}}$ by replacement $\sigma_i \to -\sigma_i$. Check that $E(\mathsf{T}_{[r]}) = E_{P^{[r]}}$ in (4.3.16) is a sum of all $r!$ elements of symmetric group S_r.

Note that one can define another set of Young symmetrizers by constructing products $\widetilde{Y}(\mathsf{T}_\lambda) = P \cdot Q$ (cf. (4.3.11)). This construction would lead, generally speaking, to another basis of primitive idempotents in $\mathbb{C}[S_r]$. However, our analysis of idempotents (4.3.11) would be repeated almost literally for the set $\widetilde{Y}(\mathsf{T}_\lambda)$. This becomes clear from the fact that symmetrizers for dual Young tableaux $\widetilde{\mathsf{T}}_{\tilde{\lambda}}$ are related to symmetrizers (4.3.11) by automorphism ϕ in algebra $\mathbb{C}[S_r]$, which takes permutation $\sigma \in S_r$ to the same permutation multiplied by its parity, $\sigma \to \sigma\delta_\sigma$. This automorphism is written in terms of Coxeter generators (neighboring transpositions) as $\phi(\sigma_i) = -\sigma_i$, and it preserves defining relations (4.3.13) of group S_r. The automorphism ϕ takes symmetrizer P for tableau T_λ to antisymmetrizer $\widetilde{Q}$ for dual tableau $\widetilde{\mathsf{T}}_{\tilde{\lambda}}$, and antisymmetrizer Q for T_λ to symmetrizer $\widetilde{P}$ for $\widetilde{\mathsf{T}}_{\tilde{\lambda}}$. Accordingly, we obtain for Young symmetrizers

$$\phi(P \cdot Q) = \widetilde{Q} \cdot \widetilde{P} \Rightarrow \phi(\widetilde{Y}(\mathsf{T}_\lambda)) = Y(\widetilde{\mathsf{T}}_{\tilde{\lambda}}).$$

So, the analysis of the set of Young symmetrizers $\widetilde{Y}(\mathsf{T}_\lambda)$ repeats the analysis of our set of symmetrizers (4.3.11). In what follows, we consider symmetrizers (4.3.11) only.

It is convenient to introduce the action of symmetric group on Young tableaux. Consider the set $\mathcal{M}_\lambda$ of all Young tableaux T_λ with one and the same shape of Young diagram $\lambda = [\lambda_1,\ldots,\lambda_m] \vdash r$ (the number of these tableaux is $r!$, since there are $r!$ ways to assign r numbers $(1,2,\ldots,r)$ to boxes of λ), and define the action of symmetry group S_r on this set as follows. Permutation $\sigma \in S_r$ acts on tableau T_λ by assigning to a box with original number $i \in \{1,\ldots,r\}$ a new number $\sigma(i)$, see Fig. 4.3.1. In this way, tableau T_λ is transformed into new tableau $\sigma \cdot \mathsf{T}_\lambda = \mathsf{T}'_\lambda \in \mathcal{M}_\lambda$.

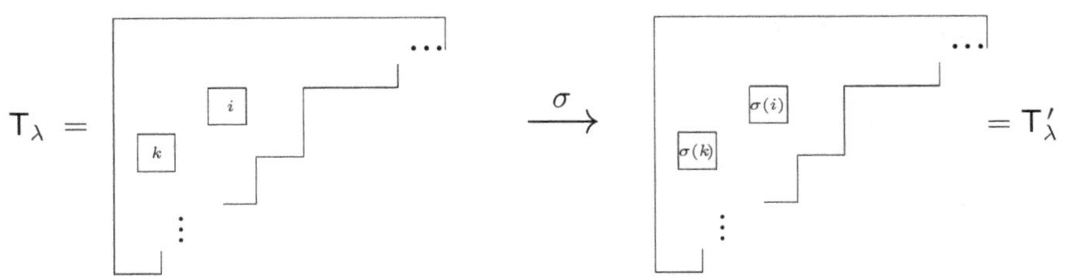

Fig. 4.3.1. Action of permutation σ on Young tableau: $\mathsf{T}_\lambda \to \sigma \cdot \mathsf{T}_\lambda = \mathsf{T}'_\lambda$.

All axioms for group action are satisfied in this case, e.g., if original tableau T_λ has number i in a certain box, then tableau $\sigma_1 \cdot \sigma_2 \cdot \mathsf{T}_\lambda$ has number $\sigma_1(\sigma_2(i))$ in this box. It is worth noting that operations $\sigma \in S_r$ permute numbers in Young tableau rather than its boxes.

As an illustration to the latter remark, let us find the action of product $q \cdot p$ on tableau $\mathsf{T}_{[2,1]_2} = \begin{array}{|c|c|}\hline 2 & 3 \\ \hline 1 \\ \cline{1-1}\end{array}$. Here $p = \sigma_2 = (2,3)$ and $q = \sigma_1 = (1,2)$ are horizontal and vertical transpositions of tableau $\mathsf{T}_{[2,1]_2}$. We have

$$p \cdot \begin{array}{|c|c|}\hline 2 & 3 \\ \hline 1 \\ \cline{1-1}\end{array} = \begin{array}{|c|c|}\hline 3 & 2 \\ \hline 1 \\ \cline{1-1}\end{array} \Longrightarrow q \cdot p \cdot \begin{array}{|c|c|}\hline 2 & 3 \\ \hline 1 \\ \cline{1-1}\end{array} = \sigma_1 \cdot \begin{array}{|c|c|}\hline 3 & 2 \\ \hline 1 \\ \cline{1-1}\end{array} = \begin{array}{|c|c|}\hline 3 & 1 \\ \hline 2 \\ \cline{1-1}\end{array}. \tag{4.3.17}$$

Note that vertical transposition q is not the transposition (13), which is the vertical transposition associated with tableau that emerges on the right-hand side of the first equality on (4.3.17).

Clearly, horizontal permutations $p \in H(\mathsf{T}_\lambda)$ constructed for Young tableau T_λ act by permuting numbers in rows of this tableau, whereas vertical permutations $q \in V(\mathsf{T}_\lambda)$ permute numbers in columns of T_λ. If tableau T'_λ is obtained from tableau T_λ by permutation $\sigma \in S_r$: $\mathsf{T}'_\lambda = \sigma \cdot \mathsf{T}_\lambda$, then elements $p \in H(\mathsf{T}_\lambda)$ and $q \in V(\mathsf{T}_\lambda)$ are related to pertinent elements $p' \in H(\mathsf{T}'_\lambda)$ and $q' \in V(\mathsf{T}'_\lambda)$ by

$$\begin{aligned} p = \sigma^{-1} \cdot p' \cdot \sigma &\Rightarrow p' = \sigma \cdot p \cdot \sigma^{-1} \\ \Rightarrow H(\mathsf{T}'_\lambda) = H(\sigma \cdot \mathsf{T}_\lambda) &= \sigma \cdot H(\mathsf{T}_\lambda) \cdot \sigma^{-1}, \end{aligned} \tag{4.3.18}$$

$$\begin{aligned} q = \sigma^{-1} \cdot q' \cdot \sigma &\Rightarrow q' = \sigma \cdot q \cdot \sigma^{-1} \\ \Rightarrow V(\mathsf{T}'_\lambda) = V(\sigma \cdot \mathsf{T}_\lambda) &= \sigma \cdot V(\mathsf{T}_\lambda) \cdot \sigma^{-1}. \end{aligned} \tag{4.3.19}$$

Let us derive the relation (4.3.18) for subgroup $H(\mathsf{T}_\lambda)$. Take, say, kth row in tableau T_λ, which contains numbers $(i_1, i_2, \ldots, i_{\lambda_k})$. Upon permutation $\sigma \in S_r$ we obtain tableau $\mathsf{T}'_\lambda = \sigma \cdot \mathsf{T}_\lambda$ with numbers $(\sigma(i_1), \ldots, \sigma(i_{\lambda_k}))$ in kth row. Permutation $p \in H(\mathsf{T}_\lambda)$ acts on kth row of tableau T_λ as follows:

$$(i_1, \ldots, i_{\lambda_k}) \to (i_{p(1)}, \ldots, i_{p(\lambda_k)}), \quad p = \begin{pmatrix} i_1 & i_2 & \cdots & i_{\lambda_k} \\ i_{p(1)} & i_{p(2)} & \cdots & i_{p(\lambda_k)} \end{pmatrix}. \tag{4.3.20}$$

The pertinent element $p' \in H(\mathsf{T}'_\lambda)$ permutes numbers $(\sigma(i_1), \ldots, \sigma(i_{\lambda_k}))$ and has the form

$$p' = \begin{pmatrix} \sigma(i_1) & \sigma(i_2) & \cdots & \sigma(i_{\lambda_k}) \\ \sigma(i_{p(1)}) & \sigma(i_{p(2)}) & \cdots & \sigma(i_{p(\lambda_k)}) \end{pmatrix}. \tag{4.3.21}$$

Making use of (4.3.20), (4.3.21) and recalling the multiplication rule (4.2.8) we find

$$\sigma^{-1}\cdot p'\cdot\sigma = \begin{pmatrix}\sigma(i_1) & \cdots & \sigma(i_{\lambda_k})\\ i_1 & \cdots & i_{\lambda_k}\end{pmatrix}\cdot\begin{pmatrix}\sigma(i_1) & \cdots & \sigma(i_{\lambda_k})\\ \sigma(i_{p(1)}) & \cdots & \sigma(i_{p(\lambda_k)})\end{pmatrix}$$
$$\times\begin{pmatrix}i_1 & \cdots & i_{\lambda_k}\\ \sigma(i_1) & \cdots & \sigma(i_{\lambda_k})\end{pmatrix} = \begin{pmatrix}i_1 & \cdots & i_{\lambda_k}\\ i_{p(1)} & \cdots & i_{p(\lambda_k)}\end{pmatrix} = p.$$

This is valid for any row, while permutations in different rows are independent, and hence we arrive at (4.3.18).

Problem 4.3.4. Find explicitly operators P and Q for Young tableau $\mathsf{T}'_{[2,2]} = \begin{array}{|c|c|}\hline 1 & 3\\\hline 2 & 4\\\hline\end{array} = (23)\cdot\mathsf{T}_{[2,2]}$, where $\mathsf{T}_{[2,2]}$ is given in (4.3.6). Check that (see (4.3.18), (4.3.19) and (4.3.7), (4.3.8))

$$P(\mathsf{T}'_{[2,2]}) = (23)\cdot P(\mathsf{T}_{[2,2]})\cdot(23)^{-1},\quad Q(\mathsf{T}'_{[2,2]}) = (23)\cdot Q(\mathsf{T}_{[2,2]})\cdot(23)^{-1}.$$

Since different tableaux T_λ and T'_λ of one and the same shape λ are related by some permutation σ of numbers in boxes, $\mathsf{T}'_\lambda = \sigma\cdot\mathsf{T}_\lambda$, the relations (4.3.18), (4.3.19) ensure that their Young symmetrizers are related by equivalence transformation

$$Y(\mathsf{T}'_\lambda) = \sigma\cdot Y(\mathsf{T}_\lambda)\cdot\sigma^{-1}. \tag{4.3.22}$$

Problem 4.3.5. Making use of definition (4.3.11), show that two tableaux

$$\mathsf{T}_{[2,1]} = \begin{array}{|c|c|}\hline 1 & 2\\\hline 3 \\\cline{1-1}\end{array} = \sigma_1\cdot\mathsf{T}_{[2,1]_1},\quad \mathsf{T}'_{[2,1]} = \begin{array}{|c|c|}\hline 1 & 3\\\hline 2 \\\cline{1-1}\end{array} = \sigma_1\cdot\mathsf{T}_{[2,1]_2},$$

are associated with Young symmetrizers $Y(\mathsf{T}_{[2,1]})$ and $Y(\mathsf{T}'_{[2,1]})$ which are related in the following way to $Y(\mathsf{T}_{[2,1]_1})$ and $Y(\mathsf{T}_{[2,1]_2})$ given by (4.3.15) (cf. (4.3.22)):

$$\begin{aligned}Y(\mathsf{T}_{[2,1]}) &= \sigma_1\cdot Y(\mathsf{T}_{[2,1]_1})\cdot\sigma_1 = (1-(13))(1+(12)),\\ Y(\mathsf{T}'_{[2,1]}) &= \sigma_1\cdot Y(\mathsf{T}_{[2,1]_2})\cdot\sigma_1 = (1-(12))(1+(13)).\end{aligned} \tag{4.3.23}$$

Here, we use the standard notations (ij) for transpositions $\sigma_1 = (12)$ and $\sigma_1\sigma_2\sigma_1 = (13)$.

Proposition 4.3.1. *Let T_λ and $\mathsf{T}'_{\lambda'}$ be two Young tableaux with shapes (Young diagrams) $\lambda\vdash r$ and $\lambda'\vdash r$. Let there exist two numbers from the* *set* $(1, 2, \ldots, r)$ *which are in one row of* $\mathsf{T}'_{\lambda'}$ *and one column of* T_λ. *We write, in accordance with* (4.3.11),

$$Y(\mathsf{T}_\lambda) = Q \cdot P, \quad Y(\mathsf{T}'_{\lambda'}) = Q' \cdot P'. \tag{4.3.24}$$

Then

$$P' \cdot Q = 0, \quad Q \cdot P' = 0, \tag{4.3.25}$$

and

$$Y(\mathsf{T}'_{\lambda'}) \cdot Y(\mathsf{T}_\lambda) = 0. \tag{4.3.26}$$

Proof. Tableaux $\mathsf{T}'_{\lambda'}$ and T_λ have two different numbers s and k in one row of $\mathsf{T}'_{\lambda'}$ and one column in T_λ. Therefore, transposition (s, k) belongs to both $H(\mathsf{T}'_{\lambda'})$ and $V(\mathsf{T}_\lambda)$. We recall the identity $(s, k)^2 = e$ and relations (4.3.9), as well as parity $\delta_{(s,k)} = -1$ and obtain

$$P' \cdot Q = P' \cdot (s, k)^2 \cdot Q = [P' \cdot (s, k)][(s, k) \cdot Q] = -P' \cdot Q, \tag{4.3.27}$$

which gives the first equality in (4.3.25). The property (4.3.26) then follows from (4.3.25) in view of representation (4.3.24). The second equality in (4.3.25) is proven by the chain of equalities similar to (4.3.27). □

Problem 4.3.6. Consider Young symmetrizers for tableaux

$$\mathsf{T}_{[2,1]} = \begin{array}{|c|c|}\hline 1 & 2 \\ \hline 3 & \\ \cline{1-1} \end{array}, \quad \mathsf{T}_{[2,1]_2} = \begin{array}{|c|c|}\hline 2 & 3 \\ \hline 1 & \\ \cline{1-1} \end{array}.$$

Check directly that

$$Y(\mathsf{T}_{[2,1]}) \cdot Y(\mathsf{T}_{[2,1]_2}) = 0, \quad Y(\mathsf{T}_{[2,1]_2}) \cdot Y(\mathsf{T}_{[2,1]}) \neq 0.$$

Thus, in the situation described in Proposition 4.3.1 we have, generally speaking, $Y(\mathsf{T}_\lambda) \cdot Y(\mathsf{T}'_{\lambda'}) \neq 0$.

Let $\lambda = [\lambda_1, \ldots, \lambda_m]$ and $\mu = [\mu_1, \ldots, \mu_k]$ be two Young diagrams, $\lambda \vdash r$, $\mu \vdash r$. Consider sequence of numbers $(\mu_1 - \lambda_1, \mu_2 - \lambda_2, \ldots)$. We write $\mu > \lambda$ if the first nonzero value of $\mu_h - \lambda_h$ is positive. We also say that diagram $\lambda \vdash r$ is included into diagram $\mu \vdash r'$ $(r' \geq r)$ and write $\lambda \subset \mu$, if λ is obtained from μ by throwing away some boxes. Note that $\mu > \lambda$ does not imply that diagram λ is included into diagram μ. Indeed, for $\mu = (3, 1)$ and $\lambda = (2, 2)$ one has $\mu > \lambda$, but $\lambda \not\subset \mu$.

Let us now turn to main combinatorial lemma.

Lemma 4.3.2. *Let T_λ and $\mathsf{T}'_{\lambda'}$ be two Young tableaux whose shapes are*

$$\lambda = [\lambda_1, \ldots, \lambda_m] \vdash r, \quad \lambda' = [\lambda'_1, \ldots, \lambda'_k] \vdash r,$$

with $\lambda' > \lambda$ or $\lambda' = \lambda$ (we write $\lambda' \geq \lambda$). Then one of the following two statements is valid:

(1) *There exist two different numbers that are contained in one row of $\mathsf{T}'_{\lambda'}$ and one column of T_λ.*
(2) *$\lambda = \lambda'$ and there exist elements $p \in H(\mathsf{T}_\lambda)$ and $q \in V(\mathsf{T}_\lambda)$ such that*[e]

$$\mathsf{T}'_{\lambda'} \equiv \mathsf{T}'_\lambda = (q \cdot p) \cdot \mathsf{T}_\lambda. \tag{4.3.28}$$

Proof. Suppose that the statement (1) is not valid. Then all numbers in the first row of tableau $\mathsf{T}'_{\lambda'}$ are placed in different columns of tableau T_λ. Therefore, there exists element $q_1 \in V(\mathsf{T}_\lambda)$ such that these numbers are contained in the first row of tableau $q_1 \cdot \mathsf{T}_\lambda$, and hence $\lambda_1 \geq \lambda'_1$. Since $\lambda' \geq \lambda$ (i.e., $\lambda'_1 \geq \lambda_1$), we have $\lambda'_1 = \lambda_1$. Numbers in the second row of tableau $\mathsf{T}'_{\lambda'}$ are again placed in different columns of tableau T_λ, and hence in different columns of tableau $q_1 \cdot \mathsf{T}_\lambda$. So, there exists element $q_2 \in V(q_1 \cdot \mathsf{T}_\lambda) = V(\mathsf{T}_\lambda)$, which does not touch elements in the first row of $q_1 \cdot \mathsf{T}_\lambda$ and puts all numbers from the second row of $\mathsf{T}'_{\lambda'}$ into the second row of tableau $q_2 \cdot q_1 \cdot \mathsf{T}_\lambda$. Thus, $\lambda_2 \geq \lambda'_2$, and since $\lambda' \geq \lambda$ (and $\lambda'_1 = \lambda_1$) we have $\lambda'_2 = \lambda_2$. We continue this procedure to the last row of λ' and find element $q_k \cdots q_2 \cdot q_1$ such that the numbers of the last, kth row of tableau $\mathsf{T}'_{\lambda'}$ are placed in the kth row of tableau $q_k \cdots q_1 \cdot \mathsf{T}_\lambda$, and $\lambda'_k = \lambda_k$. This is the last row of T_λ, otherwise $\lambda > \lambda'$ in contradiction with $\lambda' \geq \lambda$. So, we have $m = k$ and $\lambda = \lambda'$ which brings us to case (2). With $q = q_k \cdots q_2 \cdot q_1$, tableaux $q \cdot \mathsf{T}_\lambda$ and $\mathsf{T}'_{\lambda'} \equiv \mathsf{T}'_\lambda$ have the same numbers in each row, so there is an element $p' \in H(\mathsf{T}'_\lambda)$ such that $p' \cdot \mathsf{T}'_\lambda = q \cdot \mathsf{T}_\lambda$. We now consider tableau $\mathsf{T}'_\lambda = (p')^{-1} \cdot (q \cdot \mathsf{T}_\lambda)$ and make use of formula (4.3.18) with $\sigma = (p')^{-1} \cdot q$. In accordance with (4.3.18), there exists element $p^{-1} \in H(\mathsf{T}_\lambda)$ such that

$$p' = ((p')^{-1} \cdot q) \cdot p^{-1}((p')^{-1} \cdot q)^{-1} \Rightarrow p' = q \cdot p^{-1} \cdot q^{-1}.$$

Thus, we find $\mathsf{T}'_\lambda = (p')^{-1} \cdot q \cdot \mathsf{T}_\lambda = q \cdot p \cdot \mathsf{T}_\lambda$. □

Corollary 4.3.3. *If $\lambda' \vdash r$, $\lambda \vdash r$ are two Young diagrams, and $\lambda' \neq \lambda$, then*

$$Y(\mathsf{T}'_{\lambda'}) \cdot Y(\mathsf{T}_\lambda) = 0. \tag{4.3.29}$$

[e] We emphasize again that vertical permutation q belongs to $V(\mathsf{T}_\lambda)$, i.e., it is associated with tableaux T_λ — and has nothing to do with permutations which are determined by tableau $p \cdot \mathsf{T}_\lambda$.

To prove this statement, we begin with the case $\lambda' > \lambda$. Then there exist two numbers contained in one row of tableau $\mathsf{T}'_{\lambda'}$ and one column in tableau T_λ. In accordance with Proposition 4.3.1, the relation (4.3.29) holds. Let us now show that in the case $\lambda' > \lambda$, inverse identity $Y(\mathsf{T}_\lambda) \cdot Y(\mathsf{T}'_{\lambda'}) = 0$ also holds (this would mean that identity (4.3.29) holds for $\lambda > \lambda'$). Indeed, we have the second identity in (4.3.25), which is valid for any tableau $\mathsf{T}'_{\lambda'}$ with shape $\lambda' > \lambda$, including $\sigma \cdot \mathsf{T}'_{\lambda'}$ $(\forall \sigma \in S_r)$. Because of (4.3.18), horizontal symmetrizer for $\sigma \cdot \mathsf{T}'_{\lambda'}$ equals $\sigma \cdot P' \cdot \sigma^{-1}$, so one has

$$Q \cdot \sigma \cdot P' \cdot \sigma^{-1} = 0 \Longrightarrow Q \cdot \sigma \cdot P' = 0 \quad (\forall \sigma \in S_r), \tag{4.3.30}$$

where Q and P' are defined in (4.3.24). We multiply both parts of the second equality in (4.3.30) by an arbitrary function $a(\sigma) \in \mathbb{C}$ and sum up over all $\sigma \in S_r$ to obtain

$$Q \cdot a \cdot P' = 0, \quad \forall a = \sum_\sigma a(\sigma)\sigma \in \mathbb{C}[S_r]. \tag{4.3.31}$$

This, in particular, yields the identity $Y(\mathsf{T}_\lambda) \cdot Y(\mathsf{T}'_{\lambda'}) = Q \cdot (P \cdot Q') \cdot P' = 0$ (when $\lambda' > \lambda$), which is equivalent to (4.3.29) for $\lambda > \lambda'$.

Problem 4.3.7. Generalize the identity (4.3.29), i.e., show that for $\lambda' \neq \lambda$ and $\forall a \in \mathbb{C}[S_r]$ one has

$$Y(\mathsf{T}'_{\lambda'}) \cdot a \cdot Y(\mathsf{T}_\lambda) = 0. \tag{4.3.32}$$

Hint: write identity (4.3.29) in the form $Y(\mathsf{T}'_{\lambda'}) \cdot \sigma \cdot Y(\mathsf{T}_\lambda) \cdot \sigma^{-1} = 0$, where σ is an arbitrary element of S_r.

Corollary 4.3.4. *Let T_λ and T'_λ be two different tableaux with the same shape λ that are related by a permutation $g \in S_r$: $\mathsf{T}'_\lambda = g \cdot \mathsf{T}_\lambda$. Then, in accordance to Lemma 4.3.2 either element $g \in S_r$ has the form $g = q \cdot p$, where $q \in V(\mathsf{T}_\lambda)$ and $p \in H(\mathsf{T}_\lambda)$, or there are two numbers in one row of T'_λ and one column of T_λ. In the latter case, Proposition 4.3.1 tells that $Y(\mathsf{T}'_\lambda) \cdot Y(\mathsf{T}_\lambda) = 0$. Similarly, by interchanging rows and columns, one can show that either g can be written as $g = p \cdot q$ or there exist two numbers in one column of T'_λ and one row of T_λ, and in the latter case we have $Y(\mathsf{T}_\lambda) \cdot Y(\mathsf{T}'_\lambda) = 0$.*

Problem 4.3.8. Show directly that if tableau T'_λ is obtained from tableau T_λ by permutation $q \cdot p = p' \cdot q$, as written in (4.3.28), then two numbers from one column of T_λ are never placed in the same row of T'_λ.

4.3.2. Young symmetrizers and idempotents. Irreducible representations of group S_r and their dimensions

Before continuing study of Young symmetrizers in algebra $\mathbb{C}[S_r]$, we make a general comment (further discussion is postponed to Section 4.5.1). Any associative algebra $\mathcal{A}$ acts on itself (as on a vector space) by left and right multiplications. The pertinent representations of algebra $\mathcal{A}$ are *left and right regular representations.* We are interested in finite-dimensional algebras $\mathcal{A}$ with unit element e. Let such an algebra admit complete orthogonal system of idempotents $\{e_\alpha\}$, see Definition 4.2.1. Making use of completeness property (see second relation in (4.2.29)), one can expand algebra $\mathcal{A}$ in sum of left subspaces $\mathcal{L}_\alpha$ (or right subspaces $\mathcal{R}_\alpha$):

$$\mathcal{A} = \bigoplus_{\alpha=1}^{s} \mathcal{A}\cdot e_\alpha = \bigoplus_{\alpha=1}^{s} \mathcal{L}_\alpha, \quad \mathcal{A} = \bigoplus_{\alpha=1}^{s} e_\alpha \cdot \mathcal{A} = \bigoplus_{\alpha=1}^{s} \mathcal{R}_\alpha. \tag{4.3.33}$$

Each of the subspaces $\mathcal{L}_\alpha = \mathcal{A}\cdot e_\alpha$ (or $\mathcal{R}_\alpha = e_\alpha\cdot\mathcal{A}$) is invariant under left (or right) action of algebra $\mathcal{A}$. Indeed, for any $v\in\mathcal{L}_\alpha$ and any $v'\in\mathcal{R}_\alpha$ and for all $a\in\mathcal{A}$ we have

$$a\cdot v\in\mathcal{L}_\alpha, \quad v'\cdot a\in\mathcal{R}_\alpha. \tag{4.3.34}$$

Furthermore, subspaces $\mathcal{L}_\alpha$ and $\mathcal{L}_\beta$ ($\mathcal{R}_\alpha$ and $\mathcal{R}_\beta$) do not intersect, due to orthogonality of idempotents e_α and e_β ($\alpha\neq\beta$) (relation $a\cdot e_\alpha = b\cdot e_\beta$ is impossible for $\alpha\neq\beta$, since $a\cdot e_\alpha\cdot e_\alpha = a\cdot e_\alpha$ and $b\cdot e_\beta\cdot e_\alpha = 0$). Invariant subspaces $\mathcal{L}_\alpha$ ($\mathcal{R}_\alpha$) are subalgebras in $\mathcal{A}$, and, moreover, they are left (right) ideals in algebra $\mathcal{A}$ (the latter property follows from (4.3.34)). The ideals $\mathcal{L}_\alpha$ ($\mathcal{R}_\alpha$) are actually spaces of left (right) subrepresentations in regular representations of algebra $\mathcal{A}$, and formulas (4.3.33) show that algebra $\mathcal{A}$, as vector space, is expanded in direct sum of its left and right regular subrepresentations.

Proposition 4.3.5. *Young symmetrizer* (4.3.11) *associated with tableau* T_λ *equals modulo numerical factor to idempotent in algebra* $\mathbb{C}[S_r]$. *In other words, symmetrizer* $Y(\mathsf{T}_\lambda)\in\mathbb{C}[S_r]$ *obeys*

$$Y(\mathsf{T}_\lambda)\cdot Y(\mathsf{T}_\lambda) = \alpha_\lambda Y(\mathsf{T}_\lambda), \tag{4.3.35}$$

where α_λ *is a constant, equal to*

$$\alpha_\lambda = \frac{r!}{f_{(\lambda)}}, \tag{4.3.36}$$

and $f_{(\lambda)}$ *is dimension of left ideal generated by element* $Y(\mathsf{T}_\lambda)$.

Proof. We point out that Eq. (4.3.12) implies $q \cdot Y(\mathsf{T}_\lambda) \cdot p = Y(\mathsf{T}_\lambda)\delta_q$ for all $p \in H(\mathsf{T}_\lambda)$ and $q \in V(\mathsf{T}_\lambda)$. Conversely, let there exist an element $a \in \mathbb{C}[S_r]$ such that

$$q \cdot a \cdot p\delta_q = a \tag{4.3.37}$$

for all $p \in H(\mathsf{T}_\lambda)$ and $q \in V(\mathsf{T}_\lambda)$. Let us show that $a = \alpha Y(\mathsf{T}_\lambda)$, where α is a constant. Any element $a \in \mathbb{C}[S_r]$ can be written as

$$a = \sum_{g\in S_r} a(g)g, \quad a(g) \in \mathbb{C}, \tag{4.3.38}$$

and if it obeys (4.3.37), then

$$a = \sum_{g\in S_r} \delta_q a(g) q \cdot g \cdot p = \sum_{g\in S_r} \delta_q a(q^{-1} \cdot g \cdot p^{-1}) g. \tag{4.3.39}$$

We compare the coefficients of g on the right-hand sides of (4.3.38) and (4.3.39), and get

$$a(g) = a(q^{-1} \cdot g \cdot p^{-1})\delta_q \tag{4.3.40}$$

for all $p \in H(\mathsf{T}_\lambda)$, $q \in V(\mathsf{T}_\lambda)$ and $g \in S_r$. By setting $g = q \cdot p$, we find

$$a(q \cdot p) = a(e)\delta_q. \tag{4.3.41}$$

Hence, all coefficients $a(g)$ in the sum (4.3.38) coincide, modulo sign δ_q, for $g = q \cdot p$. Let us assume that $g \neq q \cdot p$ and show that in this case $a(g) = 0$. Consider tableaux T_λ and $g \cdot \mathsf{T}_\lambda$. Since $g \neq q \cdot p$, Corollary 4.3.4 ensures that there are two numbers j and k which belong to one row in $g \cdot \mathsf{T}_\lambda$ and one column in T_λ. So, for transposition $t = (j, k)$ we have $t \in H(g \cdot \mathsf{T}_\lambda)$ and $t \in V(\mathsf{T}_\lambda)$, while relations (4.3.18) give $g^{-1} \cdot t \cdot g \in H(\mathsf{T}_\lambda)$. Let us set $q_0 = t$, $p_0^{-1} = g^{-1} \cdot t \cdot g$, then $q_0^{-1} \cdot g \cdot p_0^{-1} = g$, and since $\delta_{q_0} = \delta_t = -1$ in (4.3.40), we have

$$a(g) = a(q_0^{-1} \cdot g \cdot p_0^{-1})\, \delta_{q_0} = -a(g).$$

This shows that, indeed, if $g \neq q \cdot p$, then $a(g) = 0$. Using this fact and relation (4.3.41), we rewrite (4.3.38) as follows:

$$a = \sum_{g\in S_r} a(g)g = \sum_{q\in V(\mathsf{T}_\lambda)} \sum_{p\in H(\mathsf{T}_\lambda)} \delta_q a(e) q \cdot p = \alpha Y(\mathsf{T}_\lambda), \tag{4.3.42}$$

where $\alpha = a(e)$.

Now, for an element $x \in \mathbb{C}[S_r]$ consider a product $a = Y(\mathsf{T}_\lambda)\cdot x\cdot Y(\mathsf{T}_\lambda)$. In view of (4.3.12), this product obeys (4.3.37), so a is proportional to $Y(\mathsf{T}_\lambda)$, and we have

$$Y(\mathsf{T}_\lambda)\cdot x\cdot Y(\mathsf{T}_\lambda) = \alpha_\lambda(x)Y(\mathsf{T}_\lambda), \tag{4.3.43}$$

where $\alpha_\lambda(x)$ is a numerical function on $\mathbb{C}[S_r]$ that depends on diagram $\lambda \vdash r$. In particular, for $x = e$ we obtain (4.3.35), where $\alpha_\lambda = \alpha_\lambda(e)$, and hence element $E(\mathsf{T}_\lambda) = \alpha_\lambda^{-1}Y(\mathsf{T}_\lambda)$ is idempotent, $E(\mathsf{T}_\lambda)^2 = E(\mathsf{T}_\lambda)$.

It remains to be shown that constant α_λ in (4.3.35) is given by (4.3.36). Left ideal $\mathcal{L}_{\mathsf{T}_\lambda}$ of algebra $\mathbb{C}[S_r]$ generated by idempotent $E(\mathsf{T}_\lambda)$ (or, in other words, by operator $Y(\mathsf{T}_\lambda)$) consists of all elements of $\mathbb{C}[S_r]$ that have the following form:

$$x\cdot Y(\mathsf{T}_\lambda), \quad \forall x \in \mathbb{C}[S_r]. \tag{4.3.44}$$

We denote dimension of $\mathcal{L}_{\mathsf{T}_\lambda}$ by $f_{(\lambda)}$. Let us choose complete set of basis vectors ϵ_ℓ ($\ell = 1,\ldots,r!$) in $\mathbb{C}[S_r]$ in such a way that the first $f_{(\lambda)}$ vectors make basis in $\mathcal{L}_{\mathsf{T}_\lambda}$, while other vectors ϵ_m ($m > f_{(\lambda)}$) do not belong to $\mathcal{L}_{\mathsf{T}_\lambda}$. All vectors in $\mathcal{L}_{\mathsf{T}_\lambda}$, including basis vectors ϵ_a ($a \le f_{(\lambda)}$) must have the form of product (4.3.44), so there exist elements $x_a \in \mathbb{C}[S_r]$, ($a \le f_{(\lambda)}$), such that

$$\epsilon_a = x_a\cdot Y(\mathsf{T}_\lambda), \quad 1 \le a \le f_{(\lambda)}. \tag{4.3.45}$$

We now write the product $\epsilon_\ell\cdot Y(\mathsf{T}_\lambda)$ in two ways. On the one hand, we consider product $\epsilon_\ell\cdot Y(\mathsf{T}_\lambda)$ as action of element $Y(\mathsf{T}_\lambda)$ on the right on basis element $\epsilon_\ell \in \mathbb{C}[S_r]$:

$$\epsilon_\ell\cdot Y(\mathsf{T}_\lambda) = \sum_{\ell'=1}^{r!} D_{\ell\ell'}(Y(\mathsf{T}_\lambda))\,\epsilon_{\ell'}, \tag{4.3.46}$$

where $||D_{\ell\ell'}(Y(\mathsf{T}_\lambda))||$ is a matrix of element $Y(\mathsf{T}_\lambda)$ in the right regular representation of $\mathbb{C}[S_r]$. Note that Young symmetrizer (4.3.11) is expanded as follows:

$$Y(\mathsf{T}_\lambda) = \sum_{g\in S_r} a(g)g = e + \sum_{g\neq e} a(g)g,$$

i.e., coefficient $a(e)$ of unit element e equals 1 (this follows from definition (4.3.11) of symmetrizer $Y(\mathsf{T}_\lambda)$). We then recall (see Section I-4.6.2 of the first book) the formulas for character of regular representation of finite group G, namely $\chi_R(e) = \text{ord}(G)$ and $\chi_R(g) = 0$ ($\forall g \in G,\ g \neq e$), and obtain

$$\text{Tr}(D(Y(\mathsf{T}_\lambda))) = \text{Tr}(D(e)) = r!, \tag{4.3.47}$$

where trace is evaluated over $r!$-dimensional space of regular representation. On the other hand, since $\epsilon_\ell \cdot Y(\mathsf{T}_\lambda) \in \mathcal{L}_{\mathsf{T}_\lambda}$, summation on the right-hand side of (4.3.46) involves only terms with $\ell' \leq f_{(\lambda)}$, and therefore

$$D_{\ell\ell'}(Y(\mathsf{T}_\lambda)) = 0, \quad \forall \ell' > f_{(\lambda)},\ \forall \ell. \tag{4.3.48}$$

Making use of (4.3.45), we have for $\ell \leq f_{(\lambda)}$ that

$$\epsilon_\ell \cdot Y(\mathsf{T}_\lambda) = x_\ell \cdot Y(\mathsf{T}_\lambda) \cdot Y(\mathsf{T}_\lambda) = \alpha_\lambda x_\ell \cdot Y(\mathsf{T}_\lambda) = \alpha_\lambda \epsilon_\ell. \tag{4.3.49}$$

Thus, comparison of (4.3.46) and (4.3.49) gives $D_{\ell\ell'}(Y(\mathsf{T}_\lambda)) = \alpha_\lambda\, \delta_{\ell\ell'}$ for all $\ell \leq f_{(\lambda)}$, which, in view of (4.3.48), leads to

$$\text{Tr}(D(Y(\mathsf{T}_\lambda))) = \alpha_\lambda f_{(\lambda)}. \tag{4.3.50}$$

By comparing (4.3.47) and (4.3.50), we obtain (4.3.36). □

Corollary 4.3.6. *Idempotent*

$$E(\mathsf{T}_\lambda) = \alpha_\lambda^{-1} Y(\mathsf{T}_\lambda) \tag{4.3.51}$$

is primitive.

Let us prove this *ad absurdum*. Let idempotent $E(\mathsf{T}_\lambda)$ be not primitive. Then there exist idempotents $e_1, e_2 \in \mathbb{C}[S_r]$ such that $E(\mathsf{T}_\lambda) = e_1 + e_2$ and

$$e_1^2 = e_1, \quad e_2^2 = e_2, \quad e_1 \cdot e_2 = 0 = e_2 \cdot e_1.$$

This gives the following equalities:

$$E(\mathsf{T}_\lambda) \cdot e_1 = e_1 \cdot E(\mathsf{T}_\lambda) = e_1 \Rightarrow E(\mathsf{T}_\lambda) \cdot e_1 \cdot E(\mathsf{T}_\lambda) = e_1.$$

On the other hand, in view of (4.3.43) we have $E(\mathsf{T}_\lambda) \cdot e_1 \cdot E(\mathsf{T}_\lambda) = \beta E(\mathsf{T}_\lambda)$, where β is some constant. Therefore, $e_1 = \beta E(\mathsf{T}_\lambda)$ and, taking square of this relation, we find $\beta = \beta^2$. Thus, either $\beta = 1$ or $\beta = 0$, i.e., either $e_2 = 0$ or $e_1 = 0$. This proves the corollary.

Problem 4.3.9. Show by direct calculation that Young symmetrizers constructed in (4.3.14) for $\mathbb{C}[S_2]$ and in (4.3.15) for $\mathbb{C}[S_3]$ are proportional to idempotents and are orthogonal. Hint: make use of solutions to Problems 4.3.2 and 4.3.3.

Proposition 4.3.7. *Let Young symmetrizers $Y(\mathsf{T}_\lambda)$, $Y(\mathsf{T}'_\lambda) \in \mathbb{C}[S_r]$ be associated with different tableaux T_λ and T'_λ which have the same shape $\lambda \vdash r$. Then, left regular representations of algebra $\mathbb{C}[S_r]$ in spaces of its left ideals generated by $Y(\mathsf{T}_\lambda)$ and $Y(\mathsf{T}'_\lambda)$ are equivalent. Similar statement holds for right regular representations.*

Proof. Right regular representations are treated in the same way as left regular representations, so we concentrate on the latter. Consider space of left ideal $\mathcal{L}_{\mathsf{T}_\lambda} = \mathbb{C}[S_r] \cdot Y(\mathsf{T}_\lambda)$, and let elements ϵ_a $(a = 1,\ldots,f_{(\lambda)})$ make basis in space $\mathcal{L}_{\mathsf{T}_\lambda}$. Since tableaux T_λ and T'_λ have the same shape λ, there exists permutation $\sigma \in S_r$ such that $\mathsf{T}'_\lambda = \sigma \cdot \mathsf{T}_\lambda$, and the pertinent Young symmetrizers are related by (4.3.22). Then any element in space of left ideal generated by operator $Y(\mathsf{T}'_\lambda)$,

$$\mathcal{L}_{\mathsf{T}'_\lambda} = \mathbb{C}[S_r] \cdot Y(\mathsf{T}'_\lambda) = \mathbb{C}[S_r] \cdot Y(\mathsf{T}_\lambda) \cdot \sigma^{-1},$$

is a linear combination of elements $\epsilon'_a = \epsilon_a \cdot \sigma^{-1}$, which, therefore, make basis in space $\mathcal{L}_{\mathsf{T}'_\lambda}$. Linear independence of vectors $\{\epsilon'_a\}$ is a consequence of linear independence of vectors $\{\epsilon_a\}$. Consider now left regular representation ρ in space $\mathcal{L}_{\mathsf{T}_\lambda}$. Matrix of element $x \in \mathbb{C}[S_r]$ in representation ρ is defined, in accordance with general rule (2.1.7), as $x \cdot \epsilon_a = \epsilon_b \rho_{ba}(x)$. Similarly, matrix of element $x \in \mathbb{C}[S_r]$ in left regular representation ρ' in space $\mathcal{L}_{\mathsf{T}'_\lambda}$ is defined by $x \cdot \epsilon'_a = \epsilon'_b \rho'_{ba}(x)$. On the other hand, we have

$$x \cdot \epsilon'_a = x \cdot \epsilon_a \cdot \sigma^{-1} = (\epsilon_b \rho_{ba}(x)) \cdot \sigma^{-1} = \epsilon'_b \rho_{ba}(x) \Rightarrow \rho'_{ba}(x) = \rho_{ba}(x).$$

Thus, there is basis in space $\mathcal{L}_{\mathsf{T}'_\lambda}$ (given by $\epsilon'_a = \epsilon_a \cdot \sigma^{-1}$) such that matrix $\rho'(x)$ of representation of any element $x \in \mathbb{C}[S_r]$ coincides with matrix $\rho(x)$ in representation ρ. This completes the proof. □

This proposition tells that irreducible subrepresentations in regular representation of group S_r associated with Young tableaux T_λ are equivalent for all tableaux with one and the same shape $\lambda \vdash r$. Inequivalent representations of S_r can be constructed by making use of tableaux T_λ which necessarily have different shapes (Young diagrams) $\lambda \vdash r$.

Proposition 4.3.8. *Let Young symmetrizers $Y(\mathsf{T}_\lambda)$ and $Y(\mathsf{T}'_{\lambda'})$ be associated with tableaux T_λ and $\mathsf{T}'_{\lambda'}$ which have different shapes $\lambda \neq \lambda'$ $(\lambda, \lambda' \vdash r)$. Then, left regular representations of algebra $\mathbb{C}[S_r]$ in spaces generated by symmetrizers $Y(\mathsf{T}_\lambda)$ and $Y(\mathsf{T}'_{\lambda'})$ are inequivalent. Similar statement holds for right regular representations.*

Proof. We denote by $\rho^{(\lambda)}$ and $\rho^{(\lambda')}$ left regular representations of algebra $\mathbb{C}[S_r]$, generated by Young symmetrizers $Y(\mathsf{T}_\lambda)$ and $Y(\mathsf{T}'_{\lambda'})$ and acting in spaces $\mathcal{L}_{\mathsf{T}_\lambda}$ and $\mathcal{L}_{\mathsf{T}'_{\lambda'}}$, respectively. Let us prove the proposition *ad absurdum*. Let $\lambda \neq \lambda'$ and representations $\rho^{(\lambda)}$ and $\rho^{(\lambda')}$ be equivalent. Then there exists invertible operator U such that $U \cdot \mathcal{L}_{\mathsf{T}_\lambda} = \mathcal{L}_{\mathsf{T}'_{\lambda'}}$ and we have for all $a \in \mathbb{C}[S_r]$ that

$$U^{-1} \cdot \rho^{(\lambda')}(a) \cdot U = \rho^{(\lambda)}(a).$$

Consider action of operator $\rho^{(\lambda)}(Y(\mathsf{T}_\lambda))$ in space $\mathcal{L}_{\mathsf{T}_\lambda}$ and recall (4.3.43):

$$\rho^{(\lambda)}(Y(\mathsf{T}_\lambda)) \cdot \mathcal{L}_{\mathsf{T}_\lambda} = Y(\mathsf{T}_\lambda) \cdot \mathcal{L}_{\mathsf{T}_\lambda} = \mathbb{C} \cdot Y(\mathsf{T}_\lambda), \tag{4.3.52}$$

where parameter $\mathbb{C}$ depends on the choice of element in $\mathcal{L}_{\mathsf{T}_\lambda}$ and, generally speaking, does not vanish. On the other hand, the left-hand side of (4.3.52) can be written as

$$\begin{aligned} U^{-1} \cdot \rho^{(\lambda')}(Y(\mathsf{T}_\lambda)) \cdot U \cdot \mathcal{L}_{\mathsf{T}_\lambda} &= U^{-1} \cdot \rho^{(\lambda')}(Y(\mathsf{T}_\lambda)) \cdot \mathcal{L}_{\mathsf{T}'_{\lambda'}} \\ &= U^{-1} \cdot Y(\mathsf{T}_\lambda) \cdot \mathcal{L}_{\mathsf{T}'_{\lambda'}} = 0, \end{aligned} \tag{4.3.53}$$

where we make use of (4.3.32). Equations (4.3.52) and (4.3.53) are in contradiction, which is the desired result. The proof for right regular representations is literally the same. □

We now define special type of Young tableaux which is particularly relevant in theory of representations of symmetric group S_r.

Definition 4.3.2. Young tableau in which numbers in boxes increase from left to right in each row and from top to bottom in each column is *standard* tableau.

As an example, consider three Young tableaux with one and the same shape $[6,5,4,3] \vdash 18$:

$$\begin{array}{|c|c|c|c|c|c|}\hline 5&17&18&13&15&10\\ \hline 6&16&2&12&8& \\ \cline{1-5} 14&3&1&7&& \\ \cline{1-4} 11&9&4&&& \\ \cline{1-3}\end{array}\;,\quad \begin{array}{|c|c|c|c|c|c|}\hline 5&10&13&15&17&18\\ \hline 2&6&8&12&16& \\ \cline{1-5} 1&3&7&14&& \\ \cline{1-4} 4&9&11&&& \\ \cline{1-3}\end{array}\;,\quad \begin{array}{|c|c|c|c|c|c|}\hline 1&3&7&12&16&18\\ \hline 2&6&8&14&17& \\ \cline{1-5} 4&9&11&15&& \\ \cline{1-4} 5&10&13&&& \\ \cline{1-3}\end{array}\;. \qquad (4.3.54)$$

The first two of these are not standard, while the third one is standard.

We recall that the number of inequivalent irreducible representations of symmetric group S_r is equal to the number of conjugacy classes in S_r (see Proposition I-4.6.4 in Section I-4.6.2 of the first book), and hence to the number of different Young diagrams $\lambda \vdash r$. Therefore, inequivalent irreducible representations of $\mathbb{C}[S_r]$ contained in left (right) regular representation of $\mathbb{C}[S_r]$ and generated by Young symmetrizers $Y(\mathsf{T}_\lambda)$ associated with different Young diagrams $\lambda \vdash r$ exhaust all inequivalent irreducible representations of $\mathbb{C}[S_r]$. Thus, left (right) regular representation of algebra $\mathbb{C}[S_r]$ is expanded in inequivalent representations $\rho^{(\lambda)}$ associated with different Young diagrams $\lambda \vdash r$. As we show in Section I-4.6.2 of the first book, multiplicity of each inequivalent irreducible representation $\rho^{(\lambda)}$ in the regular representation equals dimension $f_{(\lambda)}$ of this representation. On the other hand, assuming completeness of the system of Young symmetrizers (this is indeed the case, see Section 4.5.2) and recalling the expansion (4.3.33), we deduce that every inequivalent representation associated with Young diagram $\lambda \vdash r$ enters the regular representation with multiplicity equal to the number of tableaux T_λ whose symmetrizers $Y(\mathsf{T}_\lambda)$ generate *different* ideals in $\mathbb{C}[S_r]$. As we discuss in Section 4.5.2, only *standard* tableaux $\mathsf{T}_{a,\lambda}$ $(a = 1,\dots,f_{(\lambda)})$, out of $r!$ Young tableaux of shape $\lambda \vdash r$, define such symmetrizers, and the number of these tableaux equals the dimension $f_{(\lambda)}$ of the representation. Thus, left and right expansions (4.3.33) of algebra $\mathcal{A} = \mathbb{C}[S_r]$ constructed by using idempotents proportional to Young symmetrizers have the following forms:

$$\mathcal{A} = \sum_{\lambda\vdash r}\sum_{a=1}^{f_{(\lambda)}} \mathcal{A}\cdot Y(\mathsf{T}_{a,\lambda}) \equiv \sum_{\lambda,a} \mathcal{L}_{\mathsf{T}_{a,\lambda}},\quad \mathcal{A} = \sum_{\lambda\vdash r}\sum_{a=1}^{f_{(\lambda)}} Y(\mathsf{T}_{a,\lambda})\cdot\mathcal{A} \equiv \sum_{\lambda,a} \mathcal{R}_{\mathsf{T}_{a,\lambda}}, \tag{4.3.55}$$

where $\sum_{\lambda\vdash r}$ denotes the sum over all partitions of number r (all Young diagrams with r boxes), and $\sum_{a=1}^{f_{(\lambda)}}$ is the sum over all standard tableaux.

The validity of expansion (4.3.55) reflects, of course, completeness of the system of symmetrizers constructed with standard Young tableaux. We discuss standard Young tableaux, completeness of the system of the pertinent Young symmetrizers and properties of expansions (4.3.55) in more detail in Section 4.5.2.

To conclude this section, we give two formulas for dimension $f_{(\lambda)}$ of irreducible representation $\rho^{(\lambda)}$ of symmetric group S_r that is associated with Young diagram $\lambda \vdash r$ and realized on left ideal generated by Young symmetrizer $Y(\mathsf{T}_\lambda)$.

The first formula has been discovered relatively recently, in 1954 (Frame, Robinson and Thrall [42]); it is known as hook formula. Hook associated with box (i, j) in diagram λ (where i and j label row and column, respectively) is the set of boxes in λ placed either to the right of box (i, j) in row i or below box (i, j) in column j, including the box (i, j) itself. The hook formula reads

$$f_{(\lambda)} = \frac{r!}{\prod_{(i,j)\in\lambda} h_{(i,j)}}. \tag{4.3.56}$$

The product in denominator runs over all boxes (i, j) of diagram λ and $h_{(i,j)}$ is the length of the hook associated with box (i, j), i.e., the number of boxes in the hook. The second formula is due to Frobenius,

$$f_{(\lambda)} = \frac{r!}{(h_1! \dots h_m!)} \prod_{i<j} (h_i - h_j), \tag{4.3.57}$$

where m is the number of rows in diagram λ, $h_i = h_{(i,1)}$ $(i = 1, \dots, m)$ are lengths of hooks for boxes in the first column and $r = (\sum_i h_i - \frac{(m-1)m}{2})$ is the number of boxes in diagram λ. As an example, consider diagram $\mu = [6, 5, 3, 3] \vdash 17$ and write the lengths of corresponding hooks in each box (instead of usual numbers $(1, 2, \dots, 17)$). As a result, we obtain tableaux $\widetilde{\mu}$:

$$\mu = \begin{array}{|c|c|c|c|c|c|}\hline \; & \; & \; & \; & \; & \; \\ \hline & 6 & \bullet & \bullet & \bullet \\ \cline{1-5} & \bullet & \\ \cline{1-3} & \bullet & \\ \cline{1-3}\end{array} \;\Rightarrow\; \widetilde{\mu} = \begin{array}{|c|c|c|c|c|c|}\hline 9 & 8 & 7 & 4 & 3 & 1 \\ \hline 7 & 6 & 5 & 2 & 1 \\ \cline{1-5} 4 & 3 & 2 \\ \cline{1-3} 3 & 2 & 1 \\ \cline{1-3}\end{array}\,. \tag{4.3.58}$$

Proposition 4.3.9. *Hook formula* (4.3.56) *and Frobenius formula* (4.3.57) *are equivalent.*

Proof. Let us calculate the product of lengths of hooks in the first row of diagram $\lambda = [\lambda_1, \dots, \lambda_m] \vdash r$. We find

$$h_{1,1} \cdot h_{1,2} \cdots h_{1,\lambda_1}$$
$$= \frac{(\lambda_1 + m - 1)!}{(\lambda_1 - \lambda_m + m - 1)(\lambda_1 - \lambda_{m-1} + m - 2) \cdots (\lambda_1 - \lambda_2 + 1)}. \quad (4.3.59)$$

Indeed, the hook length for the first box equals $(\lambda_1 + m - 1)$, for the second box it is $(\lambda_1 + m - 2)$, etc., up to box λ_m whose hook has length $(\lambda_1 - \lambda_m + m)$. The next box has hook length $(\lambda_1 - \lambda_m + m - 2)$, if $\lambda_m < \lambda_{m-1}$; this leads to the factor $(\lambda_1 - \lambda_m + m - 1)$ in denominator in (4.3.59). If $\lambda_m = \lambda_{m-1} = \cdots = \lambda_{m-k} < \lambda_{m-k-1}$, then the hook length of the next box is $(\lambda_1 - \lambda_m + m - k - 2)$, which leads to extra factors in denominator of (4.3.59) equal to

$$(\lambda_1 - \lambda_m + m - 1) \cdot (\lambda_1 - \lambda_{m-1} + m - 2) \cdots (\lambda_1 - \lambda_{m-k} + m - k - 1).$$

This procedure continues along the first row, and factors in denominator (4.3.59) appear when the height of a column with the next box decreases. In this way, one obtains (4.3.59). The products of hook lengths of boxes in other rows is calculated in a similar way. For product of all hook lengths, we get finally

$$\prod_{(k,i)\in\lambda} h_{k,i} = \prod_{k=1}^{m} (\lambda_k + m - k)! \left(\prod_{i=1}^{m} \prod_{j=i+1}^{m} \frac{1}{(\lambda_i - \lambda_j + i - j)} \right). \quad (4.3.60)$$

Note that $(\lambda_k + m - k) = h_k$ are hook lengths of boxes $(k, 1)$ in the first column in diagram λ. Then $(\lambda_i - \lambda_j + i - j) = h_i - h_j$, and formula (4.3.60) is written as

$$\prod_{(k,i)\in\lambda} h_{(k,i)} = \prod_{i=1}^{m} h_i! \prod_{i<j} \frac{1}{(h_i - h_j)}, \quad (4.3.61)$$

which proves the equivalence of (4.3.56) and (4.3.57). □

In view of Proposition 4.3.9, it suffices to prove one of the formulas (4.3.56) or (4.3.57). Let us prove Frobenius formula (4.3.57) (proofs of formulas (4.3.56) and (4.3.57) can be found also in books [41, 46] and review [44]).

Proposition 4.3.10. *The number of standard Young tableaux whose shape is Young diagram λ with r boxes is given by Frobenius formula* (4.3.57).

Proof. We give the proof by induction. It is straightforward to check that formula (4.3.57) is valid for all Young tableaux with three and four boxes. Now, the hook lengths h_i for the first column of diagram λ with m rows, and the number of boxes λ_i in ith row are related by $h_i = \lambda_i + m - i$, so the Young diagram is uniquely defined by the hook lengths $\{h_1, \dots, h_m\}$. Let us denote the number $f_{(\lambda)}$ of standard tableaux of shape λ by $F(h_1, \dots, h_m)$. Consider a standard tableau with r boxes having shape $\lambda = [\lambda_1, \dots, \lambda_m]$. Let the box containing the last number r in this tableau sit in the ith row (this box is the last in this row). If one removes the box with number r from this tableau, one obtains a *standard* tableau with $(r-1)$ boxes and shape $\lambda' = [\lambda_1, \dots, \lambda_i - 1, \dots, \lambda_m]$. The number of *all* standard tableaux $\{\mathsf{T}_{\lambda'}\}$ having the shape λ' is $F(h_1, \dots, h_i - 1, \dots, h_m)$. Then the number $F(h_1, \dots, h_m)$ of standard tableaux $\{\mathsf{T}_\lambda\}$ having the shape λ is equal to the sum over rows i of numbers of standard tableaux $\{\mathsf{T}_{\lambda'}\}$. Therefore, in the case $\lambda_1 > \cdots > \lambda_m$ (all rows have different lengths), the function $F(h_1, \dots, h_m)$ obeys

$$F(h_1, \dots, h_m) = \sum_{i=1}^{m} F(h_1, \dots, h_i - 1, \dots, h_m), \qquad (4.3.62)$$

where summation runs over all rows of diagram λ. Let the function on the right-hand side of (4.3.57) correctly give the number of standard tableaux for all Young diagrams with $(r-1)$ boxes. Then Eq. (4.3.62) is valid not only for $\lambda_1 > \cdots > \lambda_m$, but also for diagrams λ with r boxes, which have rows of equal lengths. Indeed, let $\lambda_{i+1} = \lambda_i$. In this case removal of a box in ith row gives "wrong" Young diagram, λ' (since $\lambda'_{i+1} > \lambda'_i$) and the term with number i must vanish in the sum (4.3.62). This is indeed the case: we have $h_{i+1} = h_i - 1$, therefore the function $F(h_1, \dots, h_i - 1, h_{i+1}, \dots, h_m)$, which is given by (4.3.57) by the assumption of induction, vanishes, and does not contribute into the right-hand side of (4.3.62). So, it suffices to show that the right-hand side of (4.3.57) obeys (4.3.62), then the Frobenius formula (4.3.57) will be proven by induction. The check of the identity (4.3.62) for functions (4.3.57) is left to the problem. □

‖ **Problem 4.3.10.** * Prove the identity (4.3.62) for functions (4.3.57).

We prove in Section I-4.6.2 of the first book a remarkable identity valid for any finite group G (which is sometimes called *Burnside identity*),

$$N = \sum_{\nu=1}^{h} N_\nu^2, \tag{4.3.63}$$

where $N = \text{ord}(G)$ and N_ν are dimensions of all inequivalent irreducible representations $\rho^{(\nu)}$ $(\nu = 1, \dots, h)$ of group G. Relation (4.3.63) is valid also for any associative *simple* or *semisimple* finite-dimensional algebra $\mathcal{A}$ (see Section 4.5.1 below). In that case, N on the left-hand side of (4.3.63) denotes dimension of algebra $\mathcal{A}$. In our case of group S_r and group algebra $\mathbb{C}[S_r]$, Eq. (4.3.63) is a nontrivial combinatorial identity

$$r! = \sum_{\lambda \vdash r} (f_{(\lambda)})^2 \tag{4.3.64}$$

(summation runs over all diagrams λ with r boxes). We do not check here the identity (4.3.64) in the general case (see Problem 4.3.14).

Problem 4.3.11. * Let $p(r)$ be the number of different Young diagrams $\lambda \vdash r$. Prove that the generating function for numbers $p(r)$ is given by Euler formula (see, e.g., [41])

$$\sum_{r=0}^{\infty} p(r) x^r = \prod_{k=1}^{\infty} \frac{1}{1 - x^k}.$$

Problem 4.3.12. By making use of (4.3.56) and (4.3.57) calculate the dimension $f_{(\mu)}$ (number of standard tableaux of shape μ) of the representation S_{17} associated with diagram μ shown in (4.3.58).

Problem 4.3.13. Without using (4.3.56) and (4.3.57) derive the formula $f_{(\mu)} = \frac{(\lambda_1 - \lambda_2 + 1)}{(\lambda_1 + 1)} \cdot C^{\lambda_2}_{\lambda_1 + \lambda_2}$, for the number of standard Young tableaux (dimension of representation of group S_r), whose shape is two-row Young diagram $\mu = [\lambda_1, \lambda_2] \vdash r$. Check this formula by using (4.3.56) and (4.3.57).

Problem 4.3.14. Making use of (4.3.56) and (4.3.57), check the identity (4.3.64) for $r = 4, 5$.

4.4. Finite-Dimensional Irreducible Representations of Groups *SU* and *SL*

We recall, in the first place, that finite-dimensional representations of group $SU(N)$ and algebras $s\ell(N,\mathbb{C})$ and $su(N)$ are in one-to-one correspondence with representations of group $SL(N,\mathbb{C})$ from the class specified in Remark 3 in Section 2.1 and in the Remark in Section 4.1. So, the procedure of constructing finite-dimensional irreducible representations of $SL(N,\mathbb{C})$ from this class, which we began to develop in Section 4.2 and continue to study here is valid, word by word, for complex irreducible representations of compact group $SU(N)$ and algebras $s\ell(N,\mathbb{C})$ and $su(N)$. So, in what follows we talk about group $SL(N,\mathbb{C})$ for definiteness. Peculiarities of $SU(N)$, if any, are pointed out explicitly.

In the remainder of this chapter, we mostly consider two related problems. One is to construct all irreducible finite-dimensional representations of group $SL(N,\mathbb{C})$ (from the class specified above), i.e., complex irreducible representations of $SU(N)$, as subrepresentations in tensor product of r defining representations $T^{\otimes r}$. Another problem is to find explicitly the expansion of the tensor product $T^{\otimes r}$ in direct sum of irreducible representations. The first problem has fairly simple solution which is summarized in the end of Section 4.4.1. The recipe for the solution of the second problem is also rather simple, but the justification of this recipe needs rather cumbersome constructions which we discuss in Sections 4.5 and 4.6.

4.4.1. Finite-dimensional irreducible representations of $SL(N,\mathbb{C})$ and $SU(N)$ in spaces of symmetrized tensors

Let $\mathcal{V} = \mathbb{C}^N$ be space of defining representation T of group $SL(N,\mathbb{C})$. The space $\mathcal{V}^{\otimes r}$ is a space of representation $T^{\otimes r}$ of $SL(N,\mathbb{C})$ and at the same time the space of representation s of group algebra $\mathbb{C}[S_r]$. The latter representation is defined in (4.2.9), (4.2.10). Now, let $\lambda \vdash r$ be Young diagram with r boxes (partition of r). It determines the set of Young tableaux T_λ whose shape is λ. We have seen in Section 4.3 that to each Young tableau there corresponds Young symmetrizer $Y(\mathsf{T}_\lambda) \in \mathbb{C}[S_r]$, which generates left (right) ideal in algebra $\mathbb{C}[S_r]$, or, in other words, defines subspace of irreducible representation $\rho_{\mathsf{T}_\lambda}$ of algebra $\mathbb{C}[S_r]$ in the space of its left (right) regular representation. All representations $\rho_{\mathsf{T}_\lambda}$ associated with different Young tableaux T_λ with one and the same shape λ are equivalent, see Proposition 4.3.7, so we use notation $\rho_\lambda \equiv \rho^{(\lambda)}$ for all representations $\rho_{\mathsf{T}_\lambda}$, $\forall \mathsf{T}_\lambda$. In accordance with Proposition 4.3.5, normalized Young symmetrizers in representation s,

$$\mathsf{s}(E(\mathsf{T}_\lambda)) = \frac{1}{\alpha_\lambda}\mathsf{s}(Y(\mathsf{T}_\lambda)), \tag{4.4.1}$$

where $\alpha_\lambda = r!/f_{(\lambda)}$, are projectors in space $\mathcal{V}^{\otimes r}$ of representation $T^{\otimes r}$. These projectors extract, from the space $\mathcal{V}^{\otimes r}$, invariant subspaces

$$\mathcal{V}^{(\lambda)} = \mathsf{s}(E(\mathsf{T}_\lambda)) \cdot \mathcal{V}^{\otimes r} \sim \mathsf{s}(Y(\mathsf{T}_\lambda)) \cdot \mathcal{V}^{\otimes r} \subset \mathcal{V}^{\otimes r} \tag{4.4.2}$$

of irreducible representations $T^{(\mathsf{T}_\lambda)}$ of group $SL(N,\mathbb{C})$. Representations $T^{(\mathsf{T}_\lambda)}$ are irreducible, since idempotents $E(\mathsf{T}_\lambda)$ are primitive (see Corollary 4.3.6). Elements $g \in SL(N,\mathbb{C})$ in irreducible representation $T^{(\mathsf{T}_\lambda)}$ are written as (cf. (4.2.27))

$$T^{(\mathsf{T}_\lambda)}(g) = \alpha_\lambda^{-1}\mathsf{s}(Y(\mathsf{T}_\lambda)) \cdot T^{\otimes r}(g) = \alpha_\lambda^{-1} T^{\otimes r}(g) \cdot \mathsf{s}(Y(\mathsf{T}_\lambda)), \tag{4.4.3}$$

where the last equality follows from commutativity of operators $T^{\otimes r}(g)$ with $\mathsf{s}(a)$ for all $g \in SL(N,\mathbb{C})$ and all $a \in \mathbb{C}[S_r]$ (see Proposition 4.2.2). Operators $\mathsf{s}(Y(\mathsf{T}_\lambda))$ in $\mathcal{V}^{\otimes r}(g)$ are also called *Young symmetrizers.*

Problem 4.4.1. Making use of the fact that operator $\mathsf{s}(Y(\mathsf{T}_\lambda))$ commutes with elements $g \in SL(N,\mathbb{C})$ in representation $T^{\otimes r}$, check that mapping (4.4.3) is homomorphism.

We now prove an important proposition.

Proposition 4.4.1.

(1) *Representations $T^{(\lambda)}$ and $T^{(\lambda)'}$ associated with tableaux T_λ and T'_λ of one and the same shape, $\lambda = \lambda' \vdash r$, are equivalent.*
(2) *Representations $T^{(\lambda)}$ and $T^{(\lambda')}$ associated with tableaux T_λ and $\mathsf{T}_{\lambda'}$, which have different shapes $\lambda \vdash r$ and $\lambda' \vdash r$ (but one and the same number r of boxes), $\lambda \neq \lambda'$, are inequivalent.*

Proof. (1) For $\lambda = \lambda'$, tableau T_λ can be transformed into tableau T'_λ by a permutation $\sigma \in S_r$: $\mathsf{T}'_\lambda = \sigma \cdot \mathsf{T}_\lambda$. Then, in accordance with (4.3.22), we have $Y(\mathsf{T}'_\lambda) = \sigma \cdot Y(\mathsf{T}_\lambda) \cdot \sigma^{-1}$. Using the definition (4.4.3), we get

$$T^{(\lambda)'}(g) = A \cdot T^{(\lambda)}(g) \cdot A^{-1}, \quad \forall g \in SL(N,\mathbb{C}), \tag{4.4.4}$$

where $A = \mathsf{s}(\sigma)$ is invertible operator.

(2) Proof of the second part is similar to the proof of proposition 4.3.8. In accordance with Schur's lemma, representations $T^{(\lambda)}$ and $T^{(\lambda')}$ are inequivalent, if any intertwining operator A: $\mathcal{V}^{(\lambda')} \to \mathcal{V}^{(\lambda)}$, obeying

$$(T^{\otimes r}(g) \cdot \mathsf{s}(Y(\mathsf{T}_\lambda))) \cdot A = A \cdot (T^{\otimes r}(g) \cdot \mathsf{s}(Y(\mathsf{T}_{\lambda'}))), \quad \forall g \in SL(N, \mathbb{C}), \tag{4.4.5}$$

vanishes, $A = 0$. We make use of the fact that Young symmetrizers commute with $T^{\otimes r}(g)$ and are proportional to projectors (see (4.3.35)), and write (4.4.5) as

$$T^{\otimes r}(g) \cdot \tilde{A} = \tilde{A} \cdot T^{\otimes r}(g), \tag{4.4.6}$$

where $\tilde{A} \equiv \mathsf{s}(Y(\mathsf{T}_\lambda)) \cdot A \cdot \mathsf{s}(Y(\mathsf{T}_{\lambda'}))$ is a general form of operator from space $\mathcal{V}^{(\lambda')}$ to space $\mathcal{V}^{(\lambda)}$. So, we have to show that $\tilde{A} = 0$. In accordance with Proposition 4.2.2, any operator $\tilde{A}$ obeying (4.4.6) can be written as $\mathsf{s}(a)$, where $a \in \mathbb{C}[S_r]$. Therefore, we have

$$\tilde{A} = \frac{1}{\alpha_\lambda^2} \mathsf{s}(Y(\mathsf{T}_\lambda)) \cdot \tilde{A} \cdot \mathsf{s}(Y(\mathsf{T}_{\lambda'})) = \frac{1}{\alpha_\lambda^2} \mathsf{s}(Y(\mathsf{T}_\lambda) \cdot a \cdot Y(\mathsf{T}_{\lambda'})) = 0, \tag{4.4.7}$$

where we again use (4.3.35) in the first equality and employed (4.3.32) to derive the last equality. □

This proposition shows that inequivalent irreducible representations of group $SL(N, \mathbb{C})$ (from the class we consider), like inequivalent irreducible representations of group S_r (see Section 4.3.2), are labeled by different Young diagrams, rather than Young tableaux. We discuss this result later on (see Proposition 4.4.2).

We now look at irreducible representations $T^{(\lambda)}$ embedded in $T^{\otimes r}$, from somewhat different viewpoint. Representation $T^{\otimes r}$ of group $SL(N, \mathbb{C})$ acts in space of rank-r tensors

$$\psi = \psi^{i_1 i_2 \ldots i_r} \vec{e}_{i_1} \otimes \vec{e}_{i_2} \otimes \cdots \otimes \vec{e}_{i_r} \in \mathcal{V}^{\otimes r}. \tag{4.4.8}$$

Every Young tableau T_λ with shape $\lambda \vdash r$ determines, in accordance with (4.4.2), subspace $\mathsf{s}(Y(\mathsf{T}_\lambda))\mathcal{V}^{\otimes r}$ of rank-r tensors $\psi_{\mathsf{T}_\lambda}$ which are obtained from arbitrary tensors (4.4.8) by applying Young symmetrizers $Y(\mathsf{T}_\lambda)$:

$$
\begin{aligned}
\psi_{\mathsf{T}_\lambda}^{i_1 i_2 \dots i_r} &\equiv [\mathsf{s}(Y(\mathsf{T}_\lambda)) \cdot \psi]^{i_1 i_2 \dots i_r} = [\mathsf{s}(Q) \cdot \mathsf{s}(P) \cdot \psi]^{i_1 i_2 \dots i_r} \\
&= \left(\sum_{q \in V(\mathsf{T}_\lambda)} \delta_q \mathsf{s}(q) \right) \sum_{p \in H(\mathsf{T}_\lambda)} \psi^{p(i_1 i_2 \dots i_r)} \\
&= \sum_{q \in V(\mathsf{T}_\lambda)} \sum_{p \in H(\mathsf{T}_\lambda)} \delta_q \psi^{q \cdot p(i_1 i_2 \dots i_r)}.
\end{aligned} \tag{4.4.9}
$$

Here we use formula (4.3.11) for Young symmetrizer $Y(\mathsf{T}_\lambda) \in \mathbb{C}[S_r]$, and $p(i_1 \dots i_r)$ and $q \cdot p(i_1 \dots i_r)$ denote a new order of indices $(i_1, i_2, \dots, i_r)$ which is obtained by permutations $p \in S_r$ and $q \cdot p \in S_r$:

$$p(i_1 \dots i_r) = i_{p(1)} \dots i_{p(r)}, \quad q \cdot p(i_1 \dots i_r) = i_{q \cdot p(1)} \dots i_{q \cdot p(r)}.$$

Here q and p are vertical and horizontal permutations associated with tableau T_λ (see their definition in Section 4.3.1).

As discussed above (see Proposition 4.2.4), tensors with symmetrized components (4.4.9) form invariant subspace in $\mathcal{V}^{\otimes r}$ of irreducible representation $T^{(\lambda)}$ of group $SL(N,\mathbb{C})$. Different Young tableaux T_λ and T'_λ of the same shape λ define different Young symmetrizers, which in their turn define different tensors $\psi_{\mathsf{T}_\lambda}$ and $\psi_{\mathsf{T}'_\lambda}$. However, Proposition 4.4.1 ensures that for $\lambda = \lambda'$, the pertinent representations of group $SL(N,\mathbb{C})$ are equivalent. Thus, *inequivalent* irreducible subrepresentations $T^{(\lambda)}$ in representation $T^{\otimes r}$ are labeled by different Young diagrams $\lambda \vdash r$, rather than different tableaux T_λ. Note that multiplicity of each representation (associated with a given Young diagram) in the expansion of tensor product $T^{\otimes r}$ in direct sum of irreducible representations is, generally speaking, larger than 1.

It is convenient to associate tensors with components (4.4.9) with Young tableau T_λ whose box with number k contains tensor index i_k. As an example, a tensor of rank $r = 11$ with components $\psi_{\mathsf{T}_\lambda}^{i_1 \dots i_{11}}$ symmetrized in a special way and the corresponding tableau T_λ are given by

$$
\psi_{\mathsf{T}_\lambda}^{i_1 \dots i_{11}} = \begin{array}{|c|c|c|c|c|} \hline i_1 & i_3 & i_4 & i_5 & i_{10} \\ \hline i_2 & i_6 & i_8 & i_9 \\ \cline{1-4} i_7 & i_{11} \\ \cline{1-2} \end{array} \quad \leftrightarrow \quad \mathsf{T}_\lambda = \begin{array}{|c|c|c|c|c|} \hline 1 & 3 & 4 & 5 & 10 \\ \hline 2 & 6 & 8 & 9 \\ \cline{1-4} 7 & 11 \\ \cline{1-2} \end{array} . \tag{4.4.10}
$$

In what follows, we interchangeably use the two notations in (4.4.10). We now discuss this way of presentation by considering examples of tensors $\psi_{\mathsf{T}_\lambda}$ of rank $r = 1, 2, 3$, associated with tableaux T_λ with one, two and three boxes. We use these examples later on.

Example 1. Tableau with one box $\boxed{i_1}$ is associated with vector $\vec{\psi} = \psi^{i_1}\,\vec{e}_{i_1}$ in space $\mathcal{V} = \mathbb{C}^N$ of defining representation of group $SL(N,\mathbb{C})$. Recall that the group acts in this representation as follows (see (4.2.2)):

$$\psi^i \to \psi'^i = g^i{}_j \psi^j, \quad g \in SL(N,\mathbb{C}). \tag{4.4.11}$$

Example 2. Two Young tableaux with two boxes $\mathsf{T}_{[2]} = \begin{array}{|c|c|}\hline i_1 & i_2 \\ \hline\end{array}$ and $\mathsf{T}_{[1^2]} = \begin{array}{|c|}\hline i_1 \\ \hline i_2 \\ \hline\end{array}$, whose shapes are Young diagrams $[2]$ and $[1^2]$, are associated, respectively, with symmetric second rank tensor $\psi_s^{i_1 i_2}$ and antisymmetric tensor $\psi_a^{i_1 i_2}$. Indeed, Young tableaux $\mathsf{T}_{[2]}$ and $\mathsf{T}_{[1^2]}$ determine Young symmetrizers (4.3.14), which, in the representation s, act in space $\mathcal{V} \otimes \mathcal{V}$ and are written in the form (4.2.23):

$$\mathsf{s}(Y(\mathsf{T}_{[2]})) = \frac{1}{2}(I_N \otimes I_N + P), \quad \mathsf{s}(Y(\mathsf{T}_{[1^2]})) = \frac{1}{2}(I_N \otimes I_N - P). \tag{4.4.12}$$

By inserting these symmetrizers into (4.4.9), we obtain symmetric and antisymmetric second rank tensors with components (4.2.25):

$$\psi_s^{ij} = \frac{1}{2}(\psi^{ij} + \psi^{ji}), \quad \psi_a^{ij} = \frac{1}{2}(\psi^{ij} - \psi^{ji}). \tag{4.4.13}$$

The action of group $SL(N,\mathbb{C})$ on these tensors is determined by the rule (4.2.4) and reads

$$\psi_s^{ij} \to g^i_k\, g^j_m \psi_s^{km}, \quad \psi_a^{ij} \to g^i_k\, g^j_m \psi_a^{km}, \quad \forall g \in SL(N,\mathbb{C}).$$

Example 3. Two different Young tableaux

$$\mathsf{T}_1 \equiv \mathsf{T}_{[2,1]} = \begin{array}{|c|c|}\hline i_1 & i_2 \\ \hline i_3 & \multicolumn{1}{c}{} \\ \cline{1-1}\end{array}, \quad \mathsf{T}_2 \equiv \mathsf{T}'_{[2,1]} = \begin{array}{|c|c|}\hline i_1 & i_3 \\ \hline i_2 & \multicolumn{1}{c}{} \\ \cline{1-1}\end{array}, \tag{4.4.14}$$

which have the same shape $\lambda = [2,1]$, are associated, in accordance with (4.3.23), with Young symmetrizers

$$\mathsf{s}(Y(\mathsf{T}_1)) = (1 - \mathsf{s}(1,3))(1 + \mathsf{s}(1,2)), \quad \mathsf{s}(Y(\mathsf{T}_2)) = (1 - \mathsf{s}(1,2))(1 + \mathsf{s}(1,3)), \tag{4.4.15}$$

where (j,k) denotes transposition from symmetric group S_r, i.e., $[\mathsf{s}(j,k)\cdot\psi]^{\cdots i_j\cdots i_k\cdots} = \psi^{\cdots i_k\cdots i_j\cdots}$. By inserting symmetrizers (4.4.15) into (4.4.9), we obtain symmetrized third rank tensors with components

$$\psi_{\mathsf{T}_1}^{i_1i_2i_3} = [\mathsf{s}(Y(\mathsf{T}_1))\cdot\psi]^{i_1i_2i_3} = \psi^{i_1i_2i_3} - \psi^{i_3i_2i_1} + \psi^{i_2i_1i_3} - \psi^{i_2i_3i_1},$$
$$\psi_{\mathsf{T}_2}^{i_1i_2i_3} = [\mathsf{s}(Y(\mathsf{T}_2))\cdot\psi]^{i_1i_2i_3} = \psi^{i_1i_2i_3} - \psi^{i_2i_1i_3} + \psi^{i_3i_2i_1} - \psi^{i_3i_1i_2}. \tag{4.4.16}$$

‖ **Problem 4.4.2.** Check the validity of (4.4.16).

Spaces of tensors ψ_{T_1} and ψ_{T_2} are invariant (under the action of $SL(N,\mathbb{C})$) subspaces in $\mathcal{V}^{\otimes 3}$, and, in accordance with Proposition 4.2.4 and (4.4.3), these are subspaces of irreducible representations $T^{(\mathsf{T}_1)}$ and $T^{(\mathsf{T}_2)}$. Tableaux T_1 and T_2 have the same shape $\lambda = [2,1]$, so, due to Proposition 4.4.1, the representations $T^{(\mathsf{T}_1)}$ and $T^{(\mathsf{T}_2)}$ are equivalent. It is straightforward to construct explicitly the operator A that establishes one-to-one correspondence between spaces of tensors ψ_{T_1} and ψ_{T_2}. Indeed, tableaux T_1 and T_2, having the same shape, are related by $\mathsf{T}_2 = \sigma\cdot\mathsf{T}_1$, where in our case σ is transposition $(2,3)$. Then Young symmetrizers are related by (4.3.22), which is written in representation s as

$$\mathsf{s}(Y(\mathsf{T}_2)) = \mathsf{s}(2,3)\cdot\mathsf{s}(Y(\mathsf{T}_1))\cdot\mathsf{s}(2,3).$$

Making use of this equality, we find that the two subspaces $\mathcal{V}^{(\mathsf{T}_i)} = \mathsf{s}(Y(\mathsf{T}_2))\cdot\mathcal{V}^{\otimes 3}$ are related by $\mathcal{V}^{(\mathsf{T}_2)} = \mathsf{s}(2,3)\cdot\mathcal{V}^{(\mathsf{T}_1)}$, i.e.,

$$\begin{aligned}\psi_{\mathsf{T}_2}^{i_1i_2i_3} &= [\mathsf{s}(Y(\mathsf{T}_2))\cdot\psi]^{i_1i_2i_3} = [\mathsf{s}(2,3)\cdot\mathsf{s}(Y(\mathsf{T}_1))\cdot\mathsf{s}(2,3)\cdot\psi]^{i_1i_2i_3}\\ &= [\mathsf{s}(2,3)\cdot\mathsf{s}(Y(\mathsf{T}_1))\cdot\tilde{\psi}]^{i_1i_2i_3} = [\mathsf{s}(2,3)\cdot\tilde{\psi}_{\mathsf{T}_1}]^{i_1i_2i_3} = \tilde{\psi}_{\mathsf{T}_1}^{i_1i_3i_2},\end{aligned} \tag{4.4.17}$$

where we set $\tilde{\psi} = (\mathsf{s}(2,3)\cdot\psi)$, i.e., $\tilde{\psi}^{j_1j_2j_3} = \psi^{j_1j_3j_2}$. Due to the first formula in (4.4.16) we have

$$\begin{aligned}\tilde{\psi}_{\mathsf{T}_1}^{j_1j_2j_3} &= \tilde{\psi}^{j_1j_2j_3} - \tilde{\psi}^{j_3j_2j_1} + \tilde{\psi}^{j_2j_1j_3} - \tilde{\psi}^{j_2j_3j_1}\\ &= \psi^{j_1j_3j_2} - \psi^{j_3j_1j_2} + \psi^{j_2j_3j_1} - \psi^{j_2j_1j_3}.\end{aligned} \tag{4.4.18}$$

We see that the relevant tensors are related by invertible transformation A:

$$\psi_{\mathsf{T}_2}^{i_1i_2i_3} = (A\cdot\psi_{\mathsf{T}_1})^{i_1i_2i_3} = \tilde{\psi}_{\mathsf{T}_1}^{i_1i_3i_2}, \tag{4.4.19}$$

which is the desired result. The above argument as well as formula (4.4.19) are straightforwardly generalized to more complicated Young tableaux.

Problem 4.4.3. For two Young tableaux

$$\mathsf{T} = \begin{array}{|c|c|}\hline 1 & 2 \\ \hline 3 & 4 \\ \hline \end{array}\,, \quad \mathsf{T}' = \begin{array}{|c|c|}\hline 1 & 3 \\ \hline 2 & 4 \\ \hline \end{array}\,,$$

express the components of tensor $\psi_{\mathsf{T}_\lambda}^{i_1 i_2 i_3 i_4}$ through components of general tensor $\psi^{i_1 i_2 i_3 i_4}$. Find invertible operator A such that $\psi_{\mathsf{T}'}^{i_1 i_2 i_3 i_4} = (A \cdot \psi_{\mathsf{T}})^{i_1 i_2 i_3 i_4}$.

Remark 1. Consider Young tableau T_λ whose boxes contain indices i_a of components of rank-r tensor $\psi_{\mathsf{T}_\lambda}$ (analog of left tableau in (4.4.10))

$$\begin{array}{|c|c|c|c|c|c|}\hline i_{a_1} & i_{a_{m+1}} & \cdots & \cdots & \cdots & i_{a_r} \\ \hline i_{a_2} & i_{a_{m+2}} & \cdots & \cdots & i_{a_{r-1}} & \\ \cline{1-5} \vdots & \vdots & \vdots & & & \\ \cline{1-3} i_{a_m} & \cdots & & & & \\ \cline{1-2} \end{array} \quad \leftrightarrow \quad \psi_{\mathsf{T}_\lambda}^{i_{a_1} \dots i_{a_r}}. \tag{4.4.20}$$

Let us concentrate on a column of height k with indices $i_{b_1}, \dots, i_{b_k}$ in its boxes. Tensor $\psi_{\mathsf{T}_\lambda}$ associated with tableau T_λ is antisymmetric in these indices, as follows from definition (4.3.11) of Young symmetrizer $Y(\mathsf{T}_\lambda)$. This shows that in the case of group $SL(N, \mathbb{C})$ (or $SU(N)$), tableau T_λ with a column of height $k > N$ defines zero tensor, since indices i_{b_j} in this column take values $1, \dots, N$, and for $k > N$ some indices i_{b_j} in every component of tensor $\psi_{\mathsf{T}_\lambda}$ take one and the same value more than once. Thus, the height of columns, or, in other words, the number of rows in tableau (4.4.20) cannot exceed N, i.e., $m \leq N$. Now, if the first column has height N, then this column can be deleted from (4.4.20), since for components of the pertinent tensor we have

$$\psi^{i_{a_1} \dots i_{a_m} i_{a_{m+1}} \dots i_{a_r}} = \varepsilon^{i_{a_1} \dots i_{a_N}} \phi^{i_{a_{m+1}} \dots i_{a_r}}$$

(here $\varepsilon^{i_{a_1} \dots i_{a_N}}$ is totally antisymmetric tensor of rank N), and rank-r tensor ψ is equivalent to rank-$(r - N)$ tensor ϕ which corresponds to tableau (4.4.20) with the first column deleted. Continuing this procedure with all columns of height N we end up with understanding that inequivalent irreducible representations of $SL(N, \mathbb{C})$ (or $SU(N)$) are associated with Young tableaux which have columns of heights smaller than N, or the number of rows m is $m \leq (N - 1)$. Hereafter, we consider only tableaux having the latter property.

Proposition 4.4.2. *Representations $T^{(\lambda)}$ and $T^{(\lambda')}$ of group $SL(N,\mathbb{C})$, associated with two tableaux T_λ and $\mathsf{T}'_{\lambda'}$, whose shapes are different Young diagrams $\lambda \neq \lambda'$ with numbers of rows not exceeding $(N-1)$, are inequivalent.*

Proof. If the numbers of boxes in diagrams $\lambda \vdash r$ and $\lambda' \vdash r'$ coincide, $r = r'$, then the proposition is a consequence of the second part of Proposition 4.4.1. Inequivalence for $r \neq r'$ follows from classification of irreducible representations of Lie algebra $s\ell(N,\mathbb{C})$ which is given in item 2 of Section 3.4.2 (we recall that there is one-to-one correspondence between representations of Lie algebra $s\ell(N,\mathbb{C})$ and representations of Lie group $SL(N,\mathbb{C})$ from the class we consider). In accordance with Remarks 1 and 2 after Proposition 3.4.8, every irreducible representation of algebra $s\ell(N,\mathbb{C})$ (which is always highest weight representation) is associated with its Young diagram, and different Young diagrams correspond to different highest weights. In view of Theorem 3.4.2, representations with different highest weights are inequivalent. □

Remark 2. The argument used in the proof of Proposition 4.4.2 for $r \neq r'$ is, in fact, valid also in the case $r = r'$. It is worth noting that symmetrization of tensors determined by Young diagram λ, which is defined in Section 3.4.2, does not coincide with the symmetrization (4.4.9) (for the same diagram λ) which we utilize in this section. In the former case, one first antisymmetrizes indices in columns and then symmetrizes indices in rows of diagram λ. In the latter case, instead, one first symmetrizes indices in rows and then antisymmetrizes indices in columns. One can show, however, that the two irreducible representations obtained in these two ways are equivalent. So, this qualification does not spoil the proof of Proposition 4.4.2.

We now recall the fact that any irreducible representation of algebras $s\ell(N,\mathbb{C})$, $su(N)$ and group $SU(N)$ (i.e., any representation of group $SL(N,\mathbb{C})$ from the class we consider) is contained in some tensor product $T^{\otimes r}$ (see Proposition 3.4.8 and Section 4.1), and summarize Propositions 4.4.1 and 4.4.2 as follows.

Proposition 4.4.3. *Every complex irreducible representation of algebras $s\ell(N,\mathbb{C})$, $su(N)$ and group $SU(N)$ is associated with one and only one Young diagram λ with number of rows not exceeding $(N-1)$. Representation that is associated with a given Young diagram λ with r boxes ($\lambda \vdash r$) can be thought of as subrepresentation in $T^{\otimes r}$; it is constructed by means of Young* *symmetrizer* $Y(\mathsf{T}_\lambda)$ *corresponding to a Young tableaux* T_λ *(no matter which one) with shape* λ*. In other words, this representation is realized in space of tensors whose symmetry properties are determined by tableau* T_λ*.*

This solves the first problem stated in the beginning of Section 4.4: all complex, finite-dimensional irreducible representations of algebras $s\ell(N,\mathbb{C})$, $su(N)$ and group $SU(N)$ are listed and, moreover, a method of their explicit construction is given.

In Sections 4.4.2 and 4.4.3, we discuss, in some detail, the properties of complex irreducible representations of algebras $s\ell(N,\mathbb{C})$, $su(N)$ and group $SU(N)$ (which at the same time are representations of $SL(N,\mathbb{C})$ from the class considered). We end up this section by giving a brief and incomplete presentation of the solution to the problem of expanding the tensor product $T^{\otimes r}$ in direct sum of irreducible representations. The representation $T^{\otimes r}$ contains all irreducible representations of $SU(N)$ (and $SL(N,\mathbb{C})$) associated with Young diagrams λ with r boxes (and with number of rows not exceeding $(N-1)$), and their multiplicity equals the number $f_{(\lambda)}$ of *standard Young tableaux* of the shape λ. Note that $f_{(\lambda)}$ is equal also to the dimension of representation of symmetric group S_r associated with Young diagram λ, see Section 4.5 below. The latter is given by (4.3.56) and (4.3.57). In addition, a reducible representation $T^{\otimes r}$ with $r \geq N$ contains irreducible representations associated with Young diagrams $\lambda' \vdash r'$, where $r' < r$ and the number of rows not exceeding $(N-1)$, obtained (by making use of the procedure described in Remark 1 above) from diagrams $\lambda \vdash r$ with N rows. The method of explicit expansion of space $\mathcal{V}^{\otimes r}$ into the sum of invariant subspaces $\mathcal{V}^{(\lambda)}$, each entering $f_{(\lambda)}$ times, is studied in Sections 4.5 and 4.6.

4.4.2. Dimensions of irreducible representations of groups $SL(N,\mathbb{C})$ and $SU(N)$

Let $T^{(\lambda)}$ be finite-dimensional irreducible representation of group $SL(N,\mathbb{C})$ (and $SU(N)$) associated with diagram $\lambda \vdash r$ and acting in space $\mathcal{V}^{(\lambda)}$. It follows from the above results that dimension of this representation $\dim(\mathcal{V}^{(\lambda)})$, which we also denote by $\dim(T^{(\lambda)})$, is equal to the number of independent components of rank-r tensor $\psi_{\mathsf{T}_\lambda}$ which is symmetrized in appropriate way (see (4.4.9)). According to Proposition 4.4.1, this number is determined by Young diagram λ, rather than particular tableau T_λ. A useful formula for this dimension is

$$\dim(T^{(\lambda)}) \equiv \dim(\mathcal{V}^{(\lambda)}) = \prod_{(k,i)\in\lambda} \frac{(N+i-k)}{h_{(k,i)}}, \tag{4.4.21}$$

where the product runs over all boxes in the diagram λ, whose coordinates are (k,i) (where k and i label row and column, respectively), and $h_{(k,i)}$ is the length of the hook of the box (k,i) (hook length is defined in Section 4.3.2). Formula (4.4.21) is known as *hook formula* for dimensions of representations. It reiterates, in particular, that group $SL(N,\mathbb{C})$ does not have representations associated with diagrams whose number of rows exceeds N, see Remark in Section 4.4.1.

Proposition 4.4.4. *Formula* (4.4.21) *is equivalent to Weyl formula* (3.5.46).

Proof. To derive the formula (4.4.21) from (3.5.46), we consider Young diagram $\lambda = [\lambda_1,\dots,\lambda_m]$ with m rows (i.e., $\lambda_j = 0$, $\forall j > m$). Then Eq. (3.5.46) is written as

$$\begin{aligned}\dim(T^{(\lambda^{(\vec{n})})}) &= \prod_{1\le i<j\le N} \frac{\lambda_i-\lambda_j+j-i}{j-i} = \prod_{i=1}^{m}\prod_{j=i+1}^{N} \frac{\lambda_i-\lambda_j+j-i}{j-i} \\ &= \prod_{i=1}^{m}\left(\prod_{j=i+1}^{m} \frac{\lambda_i-\lambda_j+j-i}{j-i} \prod_{j=m+1}^{N} \frac{\lambda_i+j-i}{j-i}\right) \\ &= \left(\prod_{i=1}^{m}\prod_{j=i+1}^{m}(\lambda_i-\lambda_j+j-i)\right)\prod_{i=1}^{m}\frac{(\lambda_i+N-i)!}{(\lambda_i+m-i)!(N-i)!}.\end{aligned} \tag{4.4.22}$$

We now use formula (4.3.60) for the product of lengths of all hooks of diagram λ. We insert this formula to (4.4.22) and obtain

$$\dim(T^{(\lambda^{(\vec{n})})}) = \left(\prod_{(k,i)\in\lambda}\frac{1}{h_{(k,i)}}\right)\prod_{i=1}^{m}\frac{(\lambda_i+N-i)!}{(N-i)!}. \tag{4.4.23}$$

It remains to check the identity

$$\prod_{i=1}^{m}\frac{(\lambda_i+N-i)!}{(N-i)!} = \prod_{(k,i)\in\lambda}(N+i-k).$$

This is done straightforwardly by writing the right-hand side as a product over rows of diagram λ. □

Problem 4.4.4. (1) Making use of the formula (4.4.21), check that there are no representations of the group $SU(N)$ (and $SL(N, \mathbb{C})$) that correspond to diagrams with the number of rows exceeding N (i.e., for such diagrams, one has $\dim(\mathcal{V}^{(\lambda)}) = 0$).
(2) Let the first column in Young diagram λ have height N. Show that formula (4.4.21) is consistent with the relation (see Remark 1 in Section 4.4.1) $\dim(\mathcal{V}^{(\lambda)}) = \dim(\mathcal{V}^{(\lambda')})$, where diagram λ' is obtained from λ by deleting the first column.

Problem 4.4.5. Using hook formula (4.4.21), derive the formula for dimensions of irreducible representations of group $SU(3)$, associated with Young diagrams $\lambda = [\lambda_1, \lambda_2]$ $(\lambda_1 \geq \lambda_2 \geq 0)$ (the number of rows in λ does not exceed 2):

$$\dim(\mathcal{V}^{([\lambda_1,\lambda_2])}) = \frac{1}{2}(\lambda_2 + 1)(\lambda_1 - \lambda_2 + 1)(\lambda_1 + 2).$$

Dimensions of some particular representations of group $SL(N, \mathbb{C})$ can be computed without using hook formula. For example, consider totally antisymmetric representation associated with Young diagram

$$\lambda = [1^r] = \left.\begin{array}{|c|}\hline \\ \hline \vdots \\ \hline \\ \hline \end{array}\right\} r. \qquad (4.4.24)$$

This representation acts in the space of rank-r totally antisymmetric tensors with components $\psi_a^{i_1 \dots i_r}$. Dimension of this representation equals the number of independent components of tensor $\psi_a^{i_1 \dots i_r}$. Clearly, indices $(i_1, \dots, i_r)$ are all different, and they can be ordered, $i_1 < i_2 < \cdots < i_r$. Starting from this observation, one calculates the dimension:

$$\dim\left(\left.\begin{array}{|c|}\hline \\ \hline \vdots \\ \hline \\ \hline \end{array}\right\} r\right) = \frac{N(N-1)\cdots(N-r+1)}{r!} = \frac{N!}{r!(N-r)!} = C_N^r. \quad (4.4.25)$$

Note that formula (4.4.25) gives for $r = N - 1$ the result $\dim([1^{N-1}]) = N$, i.e., dimension of representation $T^{([1^{N-1}])}$ is same as dimension of defining representation $T^{([1])}$. This is not a coincidence. Indeed, we show in Section 4.4.3 below that $T^{([1^{N-1}])}$ is representation contragredient to defining representation $T^{([1])}$.

Similar calculation is done for totally symmetric representation of $SL(N, \mathbb{C})$, which acts in space of rank-r totally symmetric tensors associated with the tableau

$$\psi_s^{i_1\dots i_r} = \begin{array}{|c|c|c|c|}\hline i_1 & i_2 & \cdots & i_r \\ \hline\end{array}\,. \tag{4.4.26}$$

Because of symmetry, each choice $i_1 \le i_2 \le \cdots \le i_r$ corresponds to one independent component. Then

$$i_1 < i_2 + 1 < i_3 + 2 < \cdots < i_r + r - 1 \;\Rightarrow\; a_1 < a_2 < a_3 < \cdots < a_r, \tag{4.4.27}$$

where $a_k = i_k + k - 1$. In other words, we have to order numbers a_k belonging to the set $(1, 2, \dots, N + r - 1)$. This gives

$$\dim\left(\begin{array}{|c|c|c|}\hline i_1 & \cdots & i_r \\ \hline\end{array}\right) = C^r_{N+r-1} = \frac{(N+r-1)\cdots(N+1)N}{r!}. \tag{4.4.28}$$

Problem 4.4.6. Using hook formula (4.4.21), calculate dimensions of representations associated with diagrams $[1^r]$ and $[r]$, shown in (4.4.24) and (4.4.26). Compare the result with (4.4.25) and (4.4.28).

There are two other methods for calculating dimensions of representations of group $SL(N,\mathbb{C})$ that are associated with more general Young diagrams.

First method has to do with embeddings of representations of group $SL(N-1,\mathbb{C})$ in representations of group $SL(N,\mathbb{C})$. Consider representation of $SL(N,\mathbb{C})$ in the space of tensors whose symmetry properties are determined by tableau T_λ of shape λ:

$$\psi^{i_1\dots i_r} \iff \mathsf{T}_\lambda = \begin{array}{|c|c|c|c|c|c|}\hline i_1 & i_{m+1} & \cdots & \cdots & \cdots & i_r \\ \hline i_2 & i_{m+2} & \cdots & \cdots & i_{r-1} & \\ \cline{1-5} \vdots & \vdots & \vdots & & & \\ \cline{1-3} i_m & \cdots & & & & \\ \cline{1-2}\end{array}\,. \tag{4.4.29}$$

Here, tensor ψ with components $\psi^{i_1\dots i_r}$ which obeys $\mathsf{s}(Y(\mathsf{T}_\lambda)) \cdot \psi = \alpha_\lambda\, \psi$ (where coefficient α_λ is defined in (4.3.36)) and has the number of independent components equal to $\dim(T^{(\lambda)})$. Consider subgroup $SL(N-1,\mathbb{C}) \subset SL(N,\mathbb{C})$, embedded in $SL(N,\mathbb{C})$ as follows:

$$\left(\begin{array}{c|c} U & \begin{matrix} 0 \\ \vdots \\ 0 \end{matrix} \\ \hline 0 \cdots 0 & 1 \end{array}\right), \quad U \in SL(N-1,\mathbb{C}).$$

The action of this subgroup on tensor (4.4.29) does not touch indices i_k of components $\psi^{i_1\dots i_r}$, whose values are equal to N. As an example, if $i_r = N$ then components $\psi^{a_1\dots a_{r-1}N}$ with $a_i = 1,\dots,N-1$ ($\forall i < r$), make rank-$(r-1)$ tensor under group $SL(N-1,\mathbb{C})$, components $\psi^{a_1\dots a_{r-2},N,N}$, make tensor of rank $(r-2)$, etc. Needless to say, indices taking value N may occupy any place among indices $i_1,\dots,i_r$ (not necessarily in the end). The above argument tells that irreducible representation $T^{(\lambda)}_{SL(N,\mathbb{C})}$ of group $SL(N,\mathbb{C})$ associated with diagram $\lambda \vdash r$, is at the same time reducible representation of group $SL(N-1,\mathbb{C})$ which is expanded in direct sum of its irreducible representations $T^{(\lambda_a)}_{SL(N-1,\mathbb{C})}$:

$$T^{(\lambda)}_{SL(N,\mathbb{C})} = \bigoplus_b T^{(\lambda_b)}_{SL(N-1,\mathbb{C})}. \tag{4.4.30}$$

Tensors $\tilde{\psi}^{a_1\dots a_p}$ from the space of representation $T^{(\lambda_b)}_{SL(N-1,\mathbb{C})}$ differ from tensor $\psi^{i_1\dots i_r}$ by the property that $(r-p)$ indices in $\psi^{i_1\dots i_r}$ take value N, whereas others take values $1,\dots,N-1$. The assignment of indices that take value N must account for symmetry properties of tensor ψ. As an example, in the case of symmetric tensor $\psi_s^{ij} = \psi_s^{ji}$ components ψ_s^{aN} and ψ_s^{Na} are equal and define one and the same vector in the space of defining representation of $SL(N-1,\mathbb{C})$, so the defining representation of $SL(N-1,\mathbb{C})$ enters the symmetric representation of $SL(N,\mathbb{C})$ only once.

Thus, to figure out which representations enter the right-hand side of (4.4.30), we have to describe all inequivalent ways of assigning the value N to one or more indices in $\psi^{i_1\dots i_r}$. This is conveniently done in terms of Young tableaux. In this language, assigning value N to index i_k means writing symbol N in the box in tableau T_λ which originally contained index i_k. In view of antisymmetry in columns, symbol N can be present in each column not more than once. If it is present in a given column, one can place it in the lowest box of that column, again because of antisymmetry in columns. Now, if there are several columns of equal height, the symbol N can be moved to the utmost right position in the lowest row of these columns. This gives an algorithm for enumerating all inequivalent ways of assigning the value N to indices in $\psi^{i_1\dots i_r}$ and hence counting all irreducible representations of group $SL(N-1,\mathbb{C})$ emerging on the right-hand side of (4.4.30). Let us give an example of inequivalent assignments of symbol N to boxes of the following Young diagram:

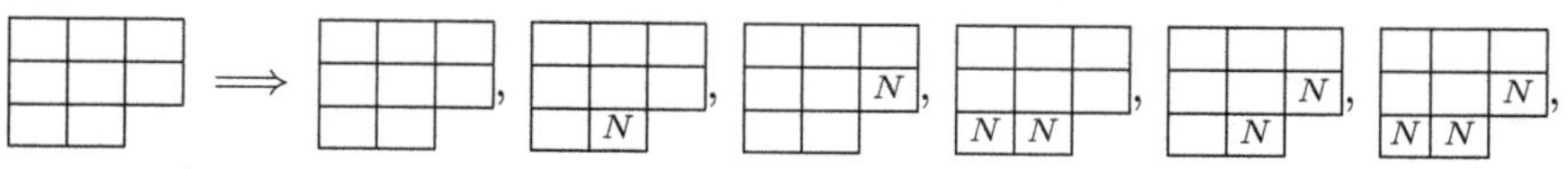

where empty boxes on the left-hand side are actually filled with indices that take all values $1,\dots,N$ and empty boxes on the right-hand side are filled with indices that take values $1,\dots,(N-1)$.

In accordance with all above, representations of group $SL(N-1,\mathbb{C})$ on the right-hand side of (4.4.30) are obtained by deleting boxes with symbol N. Thus, in our example, the expansion (4.4.30) reads

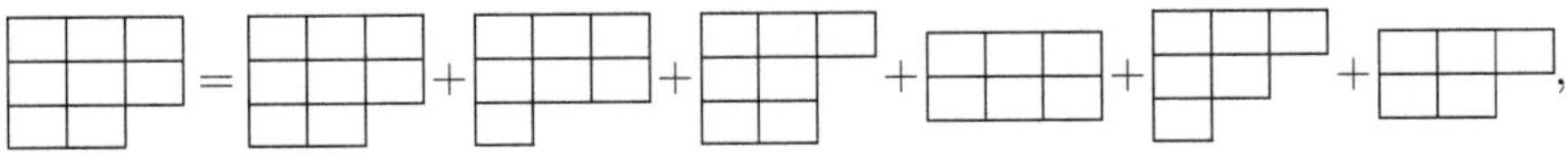

where the diagrams on the left- and right-hand sides describe representations of groups $SL(N,\mathbb{C})$ and $SL(N-1,\mathbb{C})$, respectively.

This approach gives a straightforward way to expand irreducible representations of group $SL(N,\mathbb{C})$ in irreducible representations of group $SL(N-1,\mathbb{C})$. It connects, in particular, dimensions of these representations. Indeed, in accordance with (4.4.30) we get

$$\dim(T^{(\lambda)}_{SL(N,\mathbb{C})}) = \bigoplus_b \dim(T^{(\lambda_b)}_{SL(N-1,\mathbb{C})}). \tag{4.4.31}$$

So, given the dimensions of representations $T^{(\lambda_b)}_{SL(N-1,\mathbb{C})}$ of group $SL(N-1,\mathbb{C})$ for all λ_b in (4.4.31), one calculates the dimension of representation $T^{(\lambda)}_{SL(N,\mathbb{C})}$ of group $SL(N,\mathbb{C})$.

As an example, let us consider totally symmetric representation (corresponding to tableau (4.4.26)) of group $SL(N,\mathbb{C})$. Formula (4.4.31) gives in this case

$$\begin{aligned}\dim_{SL(N,\mathbb{C})}&\left(\boxed{i_1}\boxed{i_2}\boxed{\;\dots\;}\boxed{i_r}\right)\\ &= 1+\sum_{k=1}^{r}\dim_{SL(N-1,\mathbb{C})}\left(\boxed{i_1}\boxed{\;\dots\;}\boxed{i_k}\right),\end{aligned} \tag{4.4.32}$$

where tableaux on the right-hand side denote representations of $SL(N-1,\mathbb{C})$. The term 1 on the right-hand side is dimension of trivial representation of $SL(N-1,\mathbb{C})$.

Problem 4.4.7. Check formula (4.4.32) by making use of dimensions (4.4.28). Show that identity (4.4.31) gives for totally antisymmetric representation (4.4.24)

$$\dim_{SL(N,\mathbb{C})}([1^r]) = \dim_{SL(N-1,\mathbb{C})}([1^{(r-1)}]) + \dim_{SL(N-1,\mathbb{C})}([1^r]).$$

Check the latter formula by using dimensions (4.4.25). Write the identity (4.4.31) for diagram $\lambda=[2,1^{(r-2)}]$.

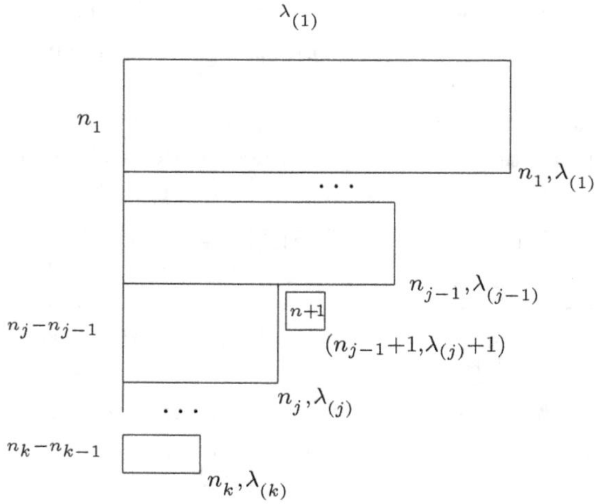

Fig. 4.4.1. Young diagram $\lambda_n = [\lambda_{(1)}^{n_1}, \lambda_{(2)}^{n_2-n_1}, \ldots, \lambda_{(k)}^{n_k-n_{k-1}}]$ with added $(n+1)$th box, whose coordinates are $(n_{j-1}+1, \lambda_{(j)}+1)$.

Second method makes use of the branching rule for representations, namely, consider tensor product of a representation associated with diagram $\lambda_n \vdash n$ shown in Fig. 4.4.1, and defining representation associated with diagram $[1] = \square$. In accordance with representation branching rule (we discuss this rule below in Section 4.6.3; see Fig. 4.6.5), we obtain expansion of this tensor product[f]

$$\lambda_n \otimes [1] = \bigoplus_{j=1}^{k+1} \lambda_{n+1}(j), \tag{4.4.33}$$

where Young diagrams λ denote for brevity the representations $T^{(\lambda)}$, and $\lambda_{n+1}(j)$ denotes diagram obtained from λ_n by adding one box with coordinates $(n_{j-1}+1, \lambda_{(j)}+1)$ in external corner of diagram λ_n, see Fig. 4.4.1. As an example,

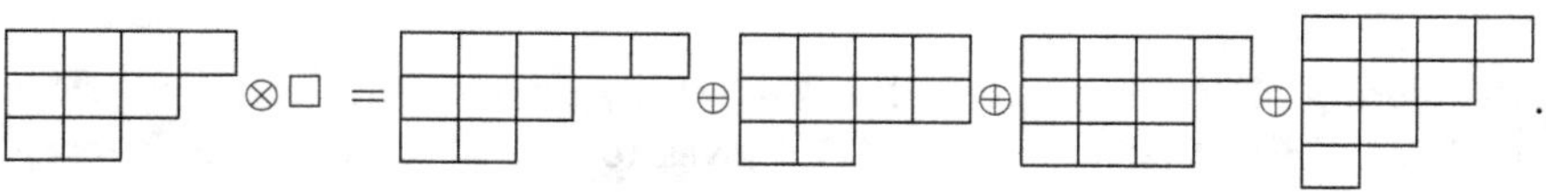

[f]Here we assume that the number n_k of rows in diagram λ_n does not exceed $(N-1)$.

Thus, we have for dimensions

$$N \cdot \dim(\lambda_n) = \sum_{j=1}^{k+1} \dim(\lambda_{n+1}(j)). \tag{4.4.34}$$

As an example, let us find dimension of representation of $SL(N,\mathbb{C})$ associated with Young diagram $[2, 1^{(r-2)}]$. We make use of (4.4.33) and get

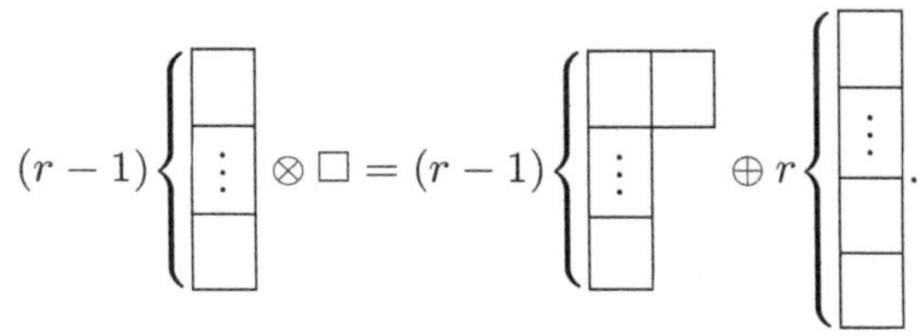

(4.4.35)

This gives for dimensions

$$N \cdot \dim([1^{r-1}]) = \dim([2, 1^{r-2}]) + \dim([1^r]).$$

We know dimensions (4.4.25) of antisymmetric representations, so we obtain

$$\begin{aligned} \dim([2, 1^{r-2}]) &= N\dim([1^{r-1}]) - \dim([1^r]) = N\, C_N^{r-1} - C_N^r \\ &= \frac{(N+1)!(r-1)}{(N-r+1)!r!}. \end{aligned} \tag{4.4.36}$$

As a cross check, for $r = N$ we obtain dimension $\dim([2, 1^{N-2}]) = N^2 - 1$ of adjoint representation of $SU(N)$ (dimension of Lie algebra $su(N)$). We show in due course that the space of adjoint representation of group $SU(N)$ indeed has tensors whose symmetry corresponds to Young diagram $[2, 1^{N-2}]$.

‖ **Problem 4.4.8.** Check the result (4.4.36) by using hook formula (4.4.21).

4.4.3. Co-defining and adjoint representations of groups $SL(N,\mathbb{C})$ and $SU(N)$

We still denote the N-dimensional complex space of defining representation of groups $SL(N,\mathbb{C})$ and $SU(N)$ by $\mathcal{V}$. Consider rank-$(N-1)$ tensor $\psi \in \mathcal{V}^{\otimes(N-1)}$, which corresponds to irreducible representation of group

$SL(N,\mathbb{C})$ (and $SU(N)$) associated with diagram

$$\lambda = [1^{(N-1)}] = \left.\begin{array}{|c|}\hline \\ \hline \vdots \\ \hline \\ \hline\end{array}\right\}(N-1).$$

In accordance with (4.4.9), this tensor is totally antisymmetric and its components $\psi^{i_1 i_2 \dots i_{N-1}}$ are in one-to-one correspondence with components

$$\bar{\psi}_i = \varepsilon_{i\, i_1 i_2 \dots i_{N-1}} \psi^{i_1 i_2 \dots i_{N-1}}, \tag{4.4.37}$$

where $\varepsilon_{i_1 i_2 \dots i_N}$ is totally antisymmetric tensor. Action of group $SL(N,\mathbb{C})$ on $\bar{\psi}$ is defined by (4.2.4) and reads

$$\begin{aligned} \bar{\psi}_i \to \bar{\psi}'_i &= \varepsilon_{i\, i_1 i_2 \dots i_{N-1}} g^{i_1}_{\ \ k_1} \cdots g^{i_{N-1}}_{\ \ k_{N-1}} \psi^{k_1 k_2 \dots k_{N-1}} \\ &= \varepsilon_{k\, i_1 i_2 \dots i_{N-1}} \psi^{i_1 i_2 \dots i_{N-1}} (g^{-1})^k_{\ \ i} = \bar{\psi}_k (g^{-1})^k_{\ \ i}, \end{aligned} \tag{4.4.38}$$

where we use the definition of determinant (see Section I-1.2.2 of the first book)

$$\varepsilon_{i_1 i_2 \dots i_n} g^{i_1}_{\ \ j_1} g^{i_2}_{\ \ j_2} \cdots g^{i_n}_{\ \ j_n} = \det(g) \varepsilon_{j_1 j_2 \dots j_n}, \tag{4.4.39}$$

and recall that $\det(g) = 1$ for $g \in SL(N,\mathbb{C})$. Thus, $\bar{\psi}_i$ transform as components of vectors from space $\bar{\mathcal{V}}$ of representation, contragredient to defining representation (see item 3 in Section I-4.1.1 of the first book),

$$\bar{\psi} \to \bar{\psi}' = (g^{-1})^{\mathsf{T}} \cdot \bar{\psi}. \tag{4.4.40}$$

The action of algebra $s\ell(N,\mathbb{C})$ on $\bar{\psi}$ is obtained from the right-hand side of (4.4.38) by inserting elements $g \in SL(N,\mathbb{C})$, close to unit matrix, $g(t) = I_N + tA + \cdots$, or $g(t)^{-1} = I_N - tA + \cdots$, where $A \in s\ell(N,\mathbb{C})$ and parameter t is small. As a result, we find

$$[A \cdot \bar{\psi}]_i = -\bar{\psi}_k A^k_{\ \ i}. \tag{4.4.41}$$

One also obtains this result by using the following chain of relations:

$$\begin{aligned} [A \cdot \bar{\psi}]_i &= \varepsilon_{i\, i_1 i_2 \dots i_{N-1}} (T^{\otimes(N-1)}_{(1)}(A))^{i_1 \dots i_{N-1}}_{\ \ k_1 \dots k_{N-1}} \psi^{k_1 k_2 \dots k_{N-1}} \\ &= -\varepsilon_{k\, i_1 i_2 \dots i_{N-1}} \psi^{i_1 i_2 \dots i_{N-1}} A^k_{\ \ i} = -\bar{\psi}_k A^k_{\ \ i}. \end{aligned}$$

Here we make use of the formula (3.4.24) for tensor product of defining representations $T^{\otimes(N-1)}_{(1)}$ and infinitesimal analog of the identity (4.4.39):

$$\varepsilon_{i_1 j_2 \dots j_n} A^{i_1}_{\ \ j_1} + \varepsilon_{j_1 i_2 \dots j_n} A^{i_2}_{\ \ j_2} + \cdots + \varepsilon_{j_1 \dots j_{n-1} i_n} A^{i_n}_{\ \ j_n} = \mathrm{Tr}(A) \varepsilon_{j_1 j_2 \dots j_n}. \tag{4.4.42}$$

Recall that $\mathrm{Tr}(A) = 0$ for all $A \in s\ell(N,\mathbb{C})$.

Thus, antisymmetric tensor $\psi \in \mathcal{V}^{\otimes(N-1)}$ associated with diagram $[1^{N-1}]$ is mapped by formula (4.4.37) to co-vector $\bar{\psi} \in \bar{\mathcal{V}}$. This mapping is invertible, and hence one-to-one,

$$\frac{1}{(N-1)!}\bar{\psi}_k \varepsilon^{k\,i_1 i_2 \dots i_{N-1}} = \psi^{i_1 i_2 \dots i_{N-1}}. \tag{4.4.43}$$

So, representation $[1^{N-1}]$ of group $SL(N,\mathbb{C})$ (and $SU(N)$) is equivalent to co-defining representation (representation contragredient to defining one).

Problem 4.4.9. Show that for $N > 2$, defining and co-defining representations T and $\bar{T}$ of $SL(N,\mathbb{C})$ and $SU(N)$ are inequivalent. Hint: compare characters $\chi_T(g) = \mathrm{Tr}(g)$ and $\chi_{\bar{T}}(g) = \mathrm{Tr}((g^{-1})^{\mathsf{T}})$.

Elements $g \in SU(N) \subset SL(N,\mathbb{C})$ are unitary, $g^{-1} = g^\dagger$, so contragredient transformation (4.4.40) can be written as

$$\bar{\psi} \to \bar{\psi}' = (g^{-1})^{\mathsf{T}} \cdot \bar{\psi} = g^* \cdot \bar{\psi}.$$

This means that co-defining representation $\bar{T}$ and complex conjugate T^* to defining representation of group $SU(N)$ are equivalent. So, representation T^* and representation $[1^{N-1}]$ (which is embedded into $T^{\otimes(N-1)}$) are also equivalent for $SU(N)$. Thus, any irreducible representation embedded into $T^{\otimes r} \otimes (T^*)^{\otimes s}$ is equivalent to subrepresentation in $T^{\otimes(r+(N-1)s)}$, i.e., any irreducible representation of $SU(N)$ is a subrepresentation in a tensor product of defining representations. This is not the case for group $SL(N,\mathbb{C})$.

Let ϕ^i and $\bar{\psi}_i$ be components of vectors from spaces $\mathcal{V}$ and $\bar{\mathcal{V}}$ of defining and co-defining representations of group $SL(N,\mathbb{C})$, respectively. Then, in accordance with (4.4.11) and (4.4.40), convolution $(\bar{\psi}_i\, \phi^i)$ is invariant under $SL(N,\mathbb{C})$, so it is an element in the space of trivial representation of $SL(N,\mathbb{C})$. The same holds for $SU(N)$. Consider now tensor product of co-defining and defining representations of $SL(N,\mathbb{C})$ (or $SU(N)$) associated with diagrams $[1^{(N-1)}]$ and $[1]$, respectively. This product acts in the space of tensors ψ:

$$\psi = \psi_k^j \vec{e}_j \otimes \vec{e}^{\,k}, \quad \vec{e}^{\,k} := \frac{1}{(N-1)!}\varepsilon^{k\,i_1 i_2 \dots i_{N-1}} \vec{e}_{i_1} \otimes \vec{e}_{i_2} \otimes \cdots \otimes \vec{e}_{i_{N-1}}, \tag{4.4.44}$$

where $\vec{e}_j$ are basis vectors in the space $\mathcal{V}$ of defining representation, while $\vec{e}^{\,k}$ are, in accordance with (4.4.43), basis vectors in the space of co-defining representation $[1^{(N-1)}]$. The product $[1^{(N-1)}] \otimes [1]$ is reducible, and its expansion in irreducible representations has the form (4.4.35) with $r = N$.

So, $[1^{(N-1)}] \otimes [1]$ is direct sum of irreducible representations $[1^N]$ and $[2, 1^{(N-2)}]$. In terms of components of tensors (4.4.44), this expansion reads

$$\psi^j_k = \frac{1}{N}\delta^j_k(\psi^i_i) + \left(\psi^j_k - \frac{1}{N}\delta^j_k(\psi^i_i)\right), \tag{4.4.45}$$

where invariant contraction (ψ^i_i) is an element of trivial representation $[1^N] = \emptyset$, while traceless second rank tensor $X^j_k = \left(\psi^j_k - \frac{1}{N}\delta^j_k(\psi^i_i)\right)$ is an element of the space of irreducible adjoint representation $[2, 1^{(N-2)}]$. This correspondence between expansions (4.4.35) (for $r = N$) and (4.4.45) follow from the rule of action of $SL(N, \mathbb{C})$ (or $SU(N)$) on contraction (ψ^i_i) and on matrix $||X^j_k||$:

$$(\psi^i_i) \to (\psi^i_i), \quad X \to g \cdot X \cdot g^{-1}.$$

It is these relations that tell that invariant vector (ψ^i_i) is associated with trivial representation $[1^N] = \emptyset$, while tensor $||X^j_k||$ is associated with diagram $[2, 1^{(N-2)}]$.

Representation $[2, 1^{(N-2)}]$ is, with qualifications given in Remark below, adjoint representation of $SU(N)$.

Remark. Representation $[2, 1^{(N-2)}]$, embedded in $T^{\otimes N}$, where T is defining representation of $SU(N)$ (and $SL(N, \mathbb{C})$), is by definition complex. On the other hand, we define Lie algebra $\mathcal{A}(G)$ of any Lie group G as real space, so adjoint representation of Lie group G that acts in its Lie algebra $\mathcal{A}(G)$ is real representation (see Section I-4.1.1 of the first book). So, representation $[2, 1^{(N-2)}]$ of group $SL(N, \mathbb{C})$ (which is sometimes also called adjoint at the expense of possible confusion) is to be understood as representation in complex algebra $s\ell(N, \mathbb{C})$, which (according to our convention) is *not* Lie algebra of Lie group $SL(N, \mathbb{C})$ considered as real manifold. In the context of group $SU(N)$, representation $[2, 1^{(N-2)}]$ is complexified adjoint representation, i.e., representation in complexification $s\ell(N, \mathbb{C})$ of real algebra $su(N) = \mathcal{A}(SU(N))$.

4.4.4. Quarks, $SU(3)$-symmetry and its breaking

Classical exposition of the material of this section is given in lectures by N.N. Bogoliubov [11]; see also book [12].

Quark charges and other quantum numbers

Let T_f be defining representation of group $SU(3)$ that acts in three-dimensional complex space $\mathcal{V}_f$ with basis $\vec{e}_i$ ($i = 1, 2, 3$). In particle physics, when studying unitary symmetry $SU(3)$, components $\psi^i = (\vec{\psi})^i$ $(i = 1, 2, 3)$ of vector $\vec{\psi} = \psi^i \, \vec{e}_i \in \mathcal{V}_f$ are associated with three quarks u, d and s:

$$(\psi^1, \, \psi^2, \, \psi^3) = (u, \, d, \, s) \tag{4.4.46}$$

and transformed as coordinates of vector[g] in defining representation of group $SU(3)$. In physics literature, this representation has notation 3 and called triplet. Three quark degrees of freedom u, d and s are called flavors, and group $SU(3)$ that acts on these degrees of freedom is called flavor group and is denoted by $SU(3)_f$ (here f refers to flavor). Quarks u, d and s have electric charges $+2/3$, $-1/3$ and $-1/3$, respectively. In other words, there is the electric charge operator Q acting in space $\mathcal{V}_f$:

$$Q \cdot \vec{e}_k = \vec{e}_i Q^i{}_k, \quad (Q \cdot \vec{\psi})^i = Q^i{}_k \psi^k, \tag{4.4.47}$$

and in flavor basis (4.4.46) the matrix of operator Q is diagonal,

$$||Q^i{}_k|| = \frac{1}{3} \begin{pmatrix} 2 & 0 & 0 \\ 0 & -1 & 0 \\ 0 & 0 & -1 \end{pmatrix} = \frac{1}{2} \left(\lambda_3 + \frac{1}{\sqrt{3}} \lambda_8 \right). \tag{4.4.48}$$

Here λ_3 and λ_8 are Cartan generators of Lie algebra $su(3)_f$ of group $SU(3)_f$. Eight generators of Lie algebra $su(3)_f$ are (see item 5 in Section I-3.3.1 of the first book) as follows:

$$\lambda_3 = \begin{pmatrix} 1 & 0 & 0 \\ 0 & -1 & 0 \\ 0 & 0 & 0 \end{pmatrix}, \quad \lambda_8 = \frac{1}{\sqrt{3}} \begin{pmatrix} 1 & 0 & 0 \\ 0 & 1 & 0 \\ 0 & 0 & -2 \end{pmatrix}, \tag{4.4.49}$$

$$\lambda_1 = \begin{pmatrix} 0 & 1 & 0 \\ 1 & 0 & 0 \\ 0 & 0 & 0 \end{pmatrix}, \quad \lambda_4 = \begin{pmatrix} 0 & 0 & 1 \\ 0 & 0 & 0 \\ 1 & 0 & 0 \end{pmatrix}, \quad \lambda_6 = \begin{pmatrix} 0 & 0 & 0 \\ 0 & 0 & 1 \\ 0 & 1 & 0 \end{pmatrix},$$
$$\lambda_2 = \begin{pmatrix} 0 & -i & 0 \\ i & 0 & 0 \\ 0 & 0 & 0 \end{pmatrix}, \quad \lambda_5 = \begin{pmatrix} 0 & 0 & -i \\ 0 & 0 & 0 \\ i & 0 & 0 \end{pmatrix}, \quad \lambda_7 = \begin{pmatrix} 0 & 0 & 0 \\ 0 & 0 & -i \\ 0 & i & 0 \end{pmatrix}, \tag{4.4.50}$$

[g]Vector ψ with components ψ^i is nothing but the quark state vector.

and in physics literature they are called *Gell-Mann matrices.* We write operator Q as linear combination

$$Q = \mathsf{I}_3 + Y/2, \tag{4.4.51}$$

where I_3 is the third component of isospin operator $\vec{\mathsf{I}} = (\mathsf{I}_1, \mathsf{I}_2, \mathsf{I}_3)$ and Y is hypercharge:

$$(\mathsf{I}_1,\, \mathsf{I}_2,\, \mathsf{I}_3) = \frac{1}{2}(\lambda_1, \lambda_2, \lambda_3), \quad Y = \frac{1}{\sqrt{3}}\lambda_8. \tag{4.4.52}$$

Thus, matrices of operators I_3 and Y in defining representation are

$$||(\mathsf{I}_3)^i{}_k|| = \frac{1}{2}\lambda_3 = \frac{1}{2}\begin{pmatrix} 1 & 0 & 0 \\ 0 & -1 & 0 \\ 0 & 0 & 0 \end{pmatrix}, \quad ||Y^i{}_k|| = \frac{1}{\sqrt{3}}\lambda_8 = \frac{1}{3}\begin{pmatrix} 1 & 0 & 0 \\ 0 & 1 & 0 \\ 0 & 0 & -2 \end{pmatrix}. \tag{4.4.53}$$

Matrix Q given by (4.4.48) is an element of Lie algebra $su(3)_f$ of group $SU(3)_f$, so its action in representation $(T_f)^{\otimes r}$ on coordinates of rank-r tensor $\psi \in \mathcal{V}_f^{\otimes r}$ is determined by the standard rule (cf. (3.4.24)):

$$\begin{aligned}(T_f^{\otimes r}(Q)\psi)^{i_1 i_2 \dots i_r} &= \left(\sum_{m=1}^{r} \delta^{i_1}_{k_1} \cdots \delta^{i_{m-1}}_{k_{m-1}} Q^{i_m}{}_{k_m} \delta^{i_{m+1}}_{k_{m+1}} \cdots \delta^{i_r}_{k_r}\right) \psi^{k_1 k_2 \dots k_r} \\ &\Leftrightarrow T_f^{\otimes r}(Q)\psi = (Q \otimes I^{\otimes(r-1)} + I \otimes Q \otimes I^{\otimes(r-2)} + \cdots \\ &\quad + I^{\otimes(r-1)} \otimes Q)\psi, \end{aligned} \tag{4.4.54}$$

where $||Q^i_k||$ is matrix (4.4.48) and I is unit operator in $\mathcal{V}_f$. Operators I_3 and Y are also elements of Lie algebra $su(3)_f$, so their action on $\psi \in \mathcal{V}_f^{\otimes k}$ is determined by the same rule (4.4.54) as the action of operator Q. It follows from (4.4.54) that charge Q, hypercharge Y and third projection of isospin I_3 are additive characteristics.

Antiquarks $(\bar{u}, \bar{d}, \bar{s}) = (\bar{\psi}_1, \bar{\psi}_2, \bar{\psi}_3)$ are associated with coordinates of conjugate vector $\bar{\psi}_j$ which belongs to the space of complex conjugate representation $\bar{\mathbf{3}}$ (equivalent to contragredient representation). In accordance with the general rule (4.4.41), we have

$$(Q \cdot \bar{\psi})_k = -\bar{\psi}_j Q^j{}_k, \tag{4.4.55}$$

i.e., charges of antiquarks $(\bar{u},\, \bar{d},\, \bar{s})$ are opposite to charges of quarks u, d, s, as one anticipates. Likewise, hypercharges Y and isospin components I_3 of antiquarks are opposite to those of quarks. More generally, additive charges of composite particles from a given multiplet[h] T are opposite to the charges of (anti)particles from antimultiplet. This follows from the fact that for any representation T its complex conjugate representation T^* is equivalent to contragredient representation $(T^{\mathsf{T}})^{-1}$.

Note that hypercharge Y can be written as a sum of two additive quantum numbers, baryon number $\mathcal{B}$ and strangeness S,

$$Y = \mathcal{B} + \mathsf{S}. \tag{4.4.56}$$

Matrices of operators $\mathcal{B}$ and S in defining representation and in basis (4.4.46) are

$$||\mathcal{B}^i{}_k|| = \frac{1}{3}\begin{pmatrix} 1 & 0 & 0 \\ 0 & 1 & 0 \\ 0 & 0 & 1 \end{pmatrix}, \quad ||\mathsf{S}^i{}_k|| = \begin{pmatrix} 0 & 0 & 0 \\ 0 & 0 & 0 \\ 0 & 0 & -1 \end{pmatrix}. \tag{4.4.57}$$

So, quarks u, d have strangeness $\mathsf{S} = 0$, while strangeness of quark s is $\mathsf{S} = -1$; baryon number of each quark equals $1/3$. So, the formula (4.4.51) for charge takes the following form:

$$Q = \mathsf{I}_3 + (\mathcal{B} + \mathsf{S})/2; \tag{4.4.58}$$

this is known as *Gell-Mann–Nishijima formula.* Note that operators $\mathcal{B}$ and S are not elements of Lie algebra $su(3)_f$ of group $SU(3)_f$. Thus, generally speaking, formula (4.4.54) (which is valid for elements of Lie algebras) is not directly applicable to charges $\mathcal{B}$ and S and their additivity is not obvious. However, one extends $SU(3)_f$ to group $U(3)_f = SU(3)_f \otimes U(1)_B$, where Abelian generator of subgroup $U(1)_{\mathcal{B}}$ is baryon number $\mathcal{B}$. Then formula (4.4.54) can be applied to operators $\mathcal{B}$ and S, and this explains, from group theory viewpoint, why baryon number and strangeness are additive quantum numbers.

Remark 1. Formulas (4.4.51), (4.4.58) are valid for quarks u, d and s, and by additivity they are valid for all particles composed of these quarks (light hadrons). It can be generalized[i] to include heavy quarks c, b and t with electric charges $(Q(c), Q(b), Q(t)) = \left(\frac{2}{3}, -\frac{1}{3}, \frac{2}{3}\right)$. There are also variants of formulas (4.4.51) and (4.4.58) in electroweak theory.

---

[h] In physics literature, multiplet is a vector in a space of representation T of group $SU(3)$. As an example, vector in space of defining representation is triplet, while vector in space of adjoint representation is octet.

[i] To avoid confusion, there is no symmetry $SU(6)_f$, even approximate, in real world with heavy quarks; see discussion of this point in the end of this section.

Remark 2. Every quark flavor u, d and s (as well as $\bar{u}$, $\bar{d}$ and $\bar{s}$) has also color and spin degrees of freedom. We discuss this point later on, when considering baryons.

Free quarks do not exist in Nature. Particles which participate in strong interactions — *hadrons* — are composite. In nonrelativistic quark model that we consider, hadrons are composed of either quark and antiquark (mesons), or three quarks (baryons). These properties are due to *confinement of color*, which we discuss further when studying baryons. We note that one of recent discoveries is the observation of hadrons with more complex structure, such as pentaquarks (composite states of four quarks and one antiquark).

Pseudoscalar mesons

In what follows, we discuss hadrons made of light quarks u, d, s and their antiquarks $\bar{u}$, $\bar{d}$, $\bar{s}$. Mesons are composite particles consisting of one quark and one antiquark. Depending on spin J and spatial parity P, one talks about scalar (spin 0, positive parity), pseudoscalar (spin 0, negative parity), vector and pseudovector mesons (spin 1, negative and positive parity, respectively), tensor mesons (spin 2), etc. The notation used in this regard is $J^P = 0^\pm, 1^\pm, 2^\pm$. In accordance with their quark content, mesons transform under irreducible representations of $SU(3)_f$ which are subrepresentations of tensor product of quark and antiquark representations (cf. general formula (4.4.45) valid for group $SU(N)$):

$$\psi^k \bar{\psi}_j = \frac{1}{3}\delta^k_j(\psi^i\bar{\psi}_i) + (\psi^k\bar{\psi}_j - \frac{1}{3}\delta^k_j(\psi^i\bar{\psi}_i)). \qquad (4.4.59)$$

Dimension of the space of tensors on the left-hand side of (4.4.59) is $3^2 = 9$. In terms of Young diagram, expansion (4.4.59) reads

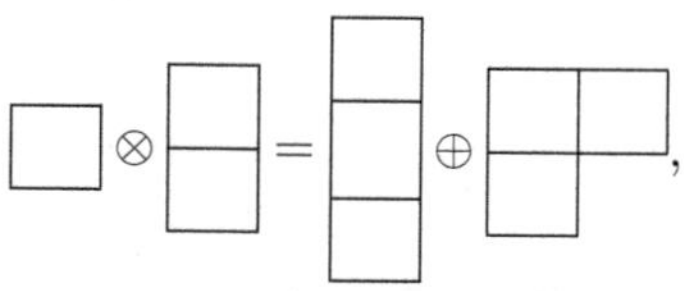

where the first diagram on the right-hand side corresponds to totally antisymmetric rank-3 tensor, which is a meson singlet $(\psi_i\bar{\psi}^i)$ (dimension of this representation is 1), while the second diagram corresponds to traceless tensor

$$M^k{}_j = (\psi^k\bar{\psi}_j - \frac{1}{3}\delta^k_j(\psi^i\bar{\psi}_i)), \qquad (4.4.60)$$

which is meson octet (as we have seen in Section 4.4.3, this is adjoint representation of group $SU(3)_f$ whose dimension is $3^2-1=8$).

The lightest hadrons are pseudoscalar mesons $(\pi^0,\pi^\pm,K^\pm,K^0,\bar{K}^0,\eta)$. They make octet under $SU(3)_f$. Unlike in the cases of other spin and parity, the ninth pseudoscalar meson η', which is singlet under $SU(3)_f$, is much heavier than mesons from the octet, and it can be discarded to the zeroth approximation. In particular, we do not discuss fairly delicate issues related to η–η'-mixing.

Traceless matrix (4.4.60) is associated with octet of pseudoscalar mesons in the following way:

$$M=\begin{pmatrix} \frac{\pi^0}{\sqrt{2}}+\frac{\eta}{\sqrt{6}} & \pi^+ & K^+ \\ \pi^- & -\frac{\pi^0}{\sqrt{2}}+\frac{\eta}{\sqrt{6}} & K^0 \\ K^- & \bar{K}^0 & -\frac{2\eta}{\sqrt{6}} \end{pmatrix},\quad \mathrm{Tr}(M)=0. \tag{4.4.61}$$

This corresponds to quark content

$$\begin{gathered}\pi^+=u\,\bar{d},\ \ \pi^-=d\,\bar{u},\ \ K^+=u\,\bar{s},\ \ K^-=s\,\bar{u},\ \ K^0=d\,\bar{s},\ \ \bar{K}^0=s\,\bar{d},\\ \pi^0=\frac{1}{\sqrt{2}}(u\,\bar{u}-d\,\bar{d}),\ \ \eta=\frac{1}{\sqrt{6}}(u\,\bar{u}+d\,\bar{d}-2s\,\bar{s}).\end{gathered} \tag{4.4.62}$$

There is also the ninth meson which we mentioned above: $\eta'=\frac{1}{\sqrt{3}}(u\,\bar{u}+d\,\bar{d}+s\,\bar{s})$; the latter corresponds to invariant vector $(\psi_i\bar{\psi}^i)$, i.e., trivial (one-dimensional) representation of $SU(3)_f$. In field-theoretic description of mesons, components $M^k{}_j$ of matrix (4.4.61) are fields that depend on space-time coordinates $x^\mu=(x^0,x^1,x^2,x^3)$. In field theory, we have relations $\pi^-=(\pi^+)^*$, $K^-=(K^+)^*$, $\bar{K^0}=(K^0)^*$, which are consistent with (4.4.62), and fields in (4.4.61) are normalized in such a way that $SU(3)_f$-invariant kinetic term $\frac{1}{2}\mathrm{Tr}\,(\partial_\mu M\cdot\partial^\mu M)$, where[j] $\partial_\mu=\partial/\partial x^\mu$, is reduced to standard (canonical) kinetic terms for meson fields,

$$\frac{1}{2}\mathrm{Tr}(\partial_\mu M\cdot\partial^\mu M)=\partial_\mu\pi^-\partial^\mu\pi^++\frac{1}{2}\partial_\mu\pi^0\partial^\mu\pi^0+\cdots+\frac{1}{2}\partial_\mu\eta\partial^\mu\eta.$$

Note that kinetic term for real (Hermitian) field π^0 (or η) traditionally has coefficient 1/2, while this factor is absent for complex field. Similar

---

[j] Our metric convention is $||\eta_{\mu\nu}||=\mathrm{diag}(+1,-1,-1,-1)$, see Section 6.4.3.

arguments determine normalization coefficients in expansions (of the type (4.4.61)) of other $SU(3)_f$-multiplets in terms of hadron fields.

In the limit of exact $SU(3)_f$ symmetry, masses of hadrons from the same $SU(3)_f$ multiplet are the same. This is not quite the case in Nature. Differences in masses of light hadrons with one and the same spin and parity are in most cases[k] due to the fact that s-quark is much heavier than u- and d-quarks:

$$m_u \simeq 2.2\,\text{MeV}, \quad m_d \simeq 4.7\,\text{MeV}, \quad m_s \simeq 96\,\text{MeV}. \tag{4.4.63}$$

Nevertheless, all these masses are small compared to typical hadronic scale which is traditionally associated with ρ-meson mass, $m_\rho = 770\,\text{MeV}$. These small values of quark masses make the approximation of massless quarks (chiral limit) reasonable. In the chiral limit, the $SU(3)_f$-symmetry is exact, which explains that it is a good approximation in Nature.

To obtain the relationship between masses of hadrons in the same $SU(3)_f$ multiplet, one considers the $SU(3)_f$-breaking part of the mass term in effective hadron Lagrangian. It emerges due to quark mass term

$$L_q = -m^i_j \bar{\psi}_i \psi^j, \tag{4.4.64}$$

where $||m^i_j|| = \text{diag}(m_u, m_d, m_s)$. It is convenient to treat m^i_j as an auxiliary field ("spurion") that transforms under representation $\mathbf{3} \times \bar{\mathbf{3}}$ (recall that we denote defining representation of $SU(3)_f$ and its complex conjugate by $\mathbf{3}$ and $\bar{\mathbf{3}}$, respectively). Then $\text{diag}(m_u, m_d, m_s)$ is the vacuum expectation value of this auxiliary field. With this interpretation, mass term (4.4.64) is invariant under $SU(3)_f$, so mass terms in the effective hadron Lagrangian must be invariant as well.

Unlike other light hadrons, mesons in pseudoscalar octet have nonzero masses exclusively due to nonzero quark masses, so the mass term in effective meson Lagrangian is proportional to m^i_j.

To understand this peculiar property, one notices that the Lagrangian of Quantum Chromodynamics, the theory of strong interactions, has enhanced symmetry in the limit of zero quark masses, $m_u, m_d, m_s \to 0$ (chiral limit), namely, the symmetry of QCD in chiral limit is $SU(3)_L \times SU(3)_R$, where $SU(3)_L$ acts on left components of quark fields and leaves right components intact, while $SU(3)_R$ acts on right components only.[l] Symmetry

[k]An important exception is precisely pseudoscalar η'-meson.

$SU(3)_L \times SU(3)_R$ *is spontaneously broken* in real world to "diagonal" $SU(3)_f$ (precisely the symmetry that we discuss in this section), that acts on left and right components of quark field on the same footing. Due to Goldstone theorem, this spontaneous symmetry breaking is accompanied by the existence of 8 *massless Nambu–Goldstone bosons*, whose number is the number of broken generators, $\dim(SU(3)_L \times SU(3)_R) - \dim(SU(3)_f) = 8$. These bosons are pseudoscalars, and they form octet under $SU(3)_f$ which is introduced in (4.4.61). In real world, quarks have nonzero masses, but they are small, and QCD is close to chiral limit. The quark mass term (4.4.64) breaks *explicitly* the symmetry $SU(3)_L \times SU(3)_R$, so pseudo-scalar mesons have nonzero masses (in this context they are pseudo-Nambu–Goldstone bosons). To the leading order in quark masses, the masses of pseudo-Nambu–Goldstone bosons are proportional to quark masses.

Naively, QCD Lagrangian is invariant also under $U(1)_B \times U(1)_A$, where $U(1)_B$ is generated by baryon number $\mathcal{B}$ (see (4.4.57)) and makes simultaneous phase rotations of all quark fields $\psi^j \to e^{i\varphi/3}\psi^j$, while $U(1)_A$ rotates left and right quark fields with opposite phases, $\psi^j_L \to e^{i\varphi}\psi^j_L$, $\psi^j_R \to e^{-i\varphi}\psi^j_R$. The "symmetry" $U(1)_A$ is also spontaneously broken, so one would naively expect the existence of the ninth light pseudo-Nambu–Goldstone boson. In quantum theory, however, there is no symmetry $U(1)_A$ at all due to axial anomaly. This shows up as large mass of the ninth pseudoscalar boson η'. Detailed treatment of these issues is given in book [43].

The mass term for pseudoscalar mesons, proportional to quark mass matrix $m^i{}_j$, must be invariant under $SU(3)_f$ once one treats $m^i{}_j$ as spurion. The only invariant, which is linear in $m^i{}_j$ and quadratic in meson fields $M^i{}_j$, is

$$L_M = -M^i{}_j \mu^j{}_k M^k{}_i = -\mathrm{Tr}(M\mu M), \tag{4.4.65}$$

where

$$\mu^j{}_k = B \cdot m^j{}_k,$$

[1]QCD Lagrangian in chiral limit reads

$$L_{\mathrm{QCD}m_i=0} = \frac{1}{2}\mathrm{Tr}(G_{\mu\nu}G^{\mu\nu}) + i\sum_j \bar{\psi}_j \gamma^\mu D_\mu \psi^j,$$

where we write explicitly only flavor index j; $G_{\mu\nu}$ is gluon field strength; D_μ is covariant derivative that involves gluon field; other notations are standard. Dirac matrices γ^μ, left and right spinors ψ^j are defined in Eqs. (6.1.3) and (6.4.39), see Chapter 6. In realistic case of massive quarks, the mass term (4.4.64) is added.

and B is a constant independent of quark flavor.[m] We insert here the matrix (4.4.61) and recall that the spurion vacuum expectation value is $\mu = \mathrm{diag}(\mu_1, \mu_2, \mu_3) \equiv \mathrm{diag}(\mu_u, \mu_d, \mu_s) = B \cdot \mathrm{diag}(m_u, m_d, m_s)$. As a result, we obtain the mass term in the following form:

$$-L_M = \sum_{i<j} (\mu_i + \mu_j) M^i{}_j M^j{}_i - L_M^{(\pi^0\eta)}, \tag{4.4.66}$$

where $L_M^{(\pi^0\eta)} = -\sum_i \mu_i (M^i{}_i)^2$ and therefore

$$-L_M^{(\pi^0\eta)} = \frac{1}{2}[(\mu_u + \mu_d)(\pi^0)^2 + \frac{1}{3}(\mu_u + \mu_d + 4\mu_s)\eta^2 + \frac{2}{\sqrt{3}}(\mu_u - \mu_d)\pi^0\eta]. \tag{4.4.67}$$

The first term in (4.4.66) reads

$$\sum_{i<j} (\mu_i + \mu_j) M^i{}_j M^j{}_i = (\mu_u + \mu_d)\pi^-\pi^+ + (\mu_u + \mu_s)K^-K^+ + (\mu_d + \mu_s)\bar{K}^0 K^0,$$

so masses squared of kaons and charged pions are

$$m_{\pi^\pm}^2 = \mu_u + \mu_d, \quad m_{K^\pm}^2 = \mu_u + \mu_s, \quad m_{K^0}^2 = \mu_d + \mu_s. \tag{4.4.68}$$

The term (4.4.67) describes masses of π^0 and η. The last term in (4.4.67) gives rise to $(\pi^0 - \eta)$-mixing: states with definite masses are linear combinations of π^0 and η, rather than π^0 and η themselves. In the approximation

$$\mu_s \gg \mu_u, \mu_d \tag{4.4.69}$$

(which is valid to reasonable precision, see (4.4.63)), $(\mu_u - \mu_d)$ is small, $(\pi^0 - \eta)$-mixing is weak and we get

$$m_{\pi^0}^2 = \mu_u + \mu_d, \quad m_\eta^2 = \frac{4}{3}\mu_s. \tag{4.4.70}$$

In this approximation, we have $m_{K^\pm}^2 = m_{K^0}^2 = \mu_s$, so there is mass formula

$$m_\eta^2 = \frac{4}{3} m_K^2. \tag{4.4.71}$$

It is valid with reasonable precision (m_π = 135–140 MeV, m_K = 494–498 MeV, m_η = 547.5 MeV; see [13]).

---

[m]In fact, this constant is expressed through parameters related to spontaneous breaking of $SU(3)_L \times SU(3)_R$-symmetry.

Problem 4.4.10. Find correction to the first formula in (4.4.70) which is due to $(\pi^0 - \eta)$-mixing. Estimate this correction making use of numerical values (4.4.63).

Problem 4.4.11. Find correction proportional to m_π^2 to mass formula (4.4.71); do not forget about $(\pi^0 - \eta)$-mixing. Estimate the relative contribution of this correction.

Isotopic symmetry

Masses of u- and d-quarks are very small in hadronic mass scale of order $m_\rho = 770\,\text{MeV}$. Therefore, it is legitimate to neglect m_u and m_d when studying the properties of hadrons (except for π-mesons). In this approximation, $SU(3)_f$-symmetry is explicitly broken by s-quark mass only, and this breaking is incomplete: there remains unbroken isospin subgroup $SU(2)_I$ embedded in $SU(3)_f$ as follows:

$$\left(\begin{array}{c|c} g & 0 \\ \hline 0 & 1 \end{array}\right) \in SU(3)_f, \quad g \in SU(2)_I. \tag{4.4.72}$$

In this approximation, masses of hadrons belonging to one isospin multiplet are equal; there are also relations for scattering amplitudes. All these relations are valid with high precision, which is much better than precision of relations based on $SU(3)_f$-symmetry.

Note that masses of neutral and charged pions are equal in chiral theory, see first formulas in (4.4.68), (4.4.70). Although this equality is rather accidental, it shows that pion masses do not break isotopic symmetry $SU(2)_I$. Note also that isotopic symmetry is explicitly broken not only because quarks u and d have different masses, but also because these quarks have different electric charges (i.e., isotopic symmetry is explicitly broken by electromagnetic interactions). For this reason, study of effects due to isotopic symmetry breaking is quite demanding, and we consider this symmetry exact in what follows. In particular, we approximate the quark mass matrix by

$$m^i{}_j = \text{diag}(0, 0, m_s).$$

We note that isotopic symmetry is exact, and $(\pi^0 - \eta)$-mixing is absent (see (4.4.67)) in a more general case $m_u = m_d = m$, so that $m^i{}_j = \text{diag}(m, m, m_s)$.

Vector mesons

Let us come back to quark–antiquark bound states $V^k_{\ j} = \psi^k \bar{\psi}_j$, and now specify to vector, rather than pseudoscalar, mesons. In this case, it is convenient to set $\mathrm{Tr}(V) = V^k_{\ k} \neq 0$. Then tensors (4.4.59) make reducible representation of $SU(3)_f$, and matrix $V^k_{\ j}$ describes nonet of vector mesons[n]

$$V^k_{\ j} = \begin{pmatrix} \dfrac{\rho^0}{\sqrt{2}} + \dfrac{\omega^0}{\sqrt{2}} & \rho^+ & K^{*+} \\ \rho^- & -\dfrac{\rho^0}{\sqrt{2}} + \dfrac{\omega^0}{\sqrt{2}} & K^{*0} \\ K^{*-} & \bar{K}^{*0} & \phi^0 \end{pmatrix}, \tag{4.4.73}$$

with masses (in MeV)

$$\begin{gathered} m_{\rho^\pm} = 766 \pm 2, \quad m_{\rho^0} = 770 \pm 1, \quad m_{\omega^0} = 782,7 \pm 0,3, \\ m_{K^{*\pm}} = 892,2 \pm 0,5, \quad m_{K^{*0}} = 896,1 \pm 0,3, \quad m_{\phi^0} = 1019,7 \pm 0,3, \end{gathered} \tag{4.4.74}$$

and quark content

$$\begin{gathered} \rho^+ = u\,\bar{d}, \quad \rho^- = d\,\bar{u}, \quad \rho^0 = \frac{1}{\sqrt{2}}(u\,\bar{u} - d\,\bar{d}), \quad \omega^0 = \frac{1}{\sqrt{2}}(u\,\bar{u} + d\,\bar{d}), \\ K^{*+} = u\,\bar{s}, \quad K^{*-} = s\,\bar{u}, \quad K^{*0} = d\,\bar{s}, \quad \bar{K}^{*0} = s\,\bar{d}, \quad \phi^0 = s\,\bar{s}. \end{gathered} \tag{4.4.75}$$

The difference between vector and pseudoscalar mesons is that the mass of ω^0-meson (analog of η-meson in (4.4.61)) is almost the same as mass of ρ-meson (analog of π-meson in (4.4.61)). Since the mass differences of hadrons (except for pseudoscalar meson η') is due to the heavier s-quark, equality $m_{\omega^0} \approx m_\rho$ suggests that s-quark is absent in ω^0-meson, like in ρ-meson. This property is used in (4.4.73) and (4.4.75).

---

[n] In terms of decomposition $3 \times \bar{3} = 8 + 1$, see (4.4.59), this situation corresponds to the case where octet and singlet, in the limit of unbroken $SU(3)_f$, have equal or very close masses. Then breaking of $SU(3)_f$ leads to mixing between singlet and eighth component of octet, and for maximal mixing we get (4.4.73).

We now choose the mass term in the vector meson Lagrangian as follows[o]:

$$
\begin{aligned}
-L_V &= \frac{m_v^2}{2} V^j{}_k V^k{}_j + \Delta V^k{}_3 V^3{}_k \\
&= m_v^2 \rho^+ \rho^- + \frac{m_v^2}{2}((\rho^0)^2 + \omega^2) + (m_v^2 + \Delta)(K^{*+}K^{*-} + K^{*0}\bar{K}^{*0}) \\
&\quad + \left(\frac{m_v^2}{2} + \Delta\right)\phi^2. \qquad (4.4.76)
\end{aligned}
$$

Here m_v^2 is the mass parameter that does not break $SU(3)_f$, Δ is $SU(3)_f$-breaking parameter; we make use of the fact that the only invariant mass term (with quark mass matrix treated as spurion) in effective vector meson Lagrangian that breaks $SU(3)_f$ is const $\cdot$ Tr(VmV). In our approximation

$$m^i{}_j = \mathrm{diag}(0, 0, m_s) \propto \mathrm{diag}(0, 0, \Delta), \qquad (4.4.77)$$

which gives (4.4.76). In this way, we obtain mass formulas for vector mesons:

$$
\begin{aligned}
&m_{\rho^0}^2 = m_{\rho^\pm}^2 = m_v^2, \quad m_{\omega^0}^2 = m_v^2, \\
&m_{K^{*0}}^2 = m_{K^{*\pm}}^2 = m_v^2 + \Delta, \quad m_\phi^2 = m_v^2 + 2\Delta,
\end{aligned}
$$

which lead to the relation

$$\Delta = m_{K^*}^2 - m_\rho^2 = m_\phi^2 - m_{K^*}^2.$$

Recalling (4.4.74), we see that this relation is indeed satisfied with reasonable precision.

Baryons

Besides mesons, quarks make baryons. In nonrelativistic quark model, baryons are bound states of three quarks, and their wave functions are products of wave functions of three quarks. Thus, baryon fields transform under irreducible representations of $SU(3)_f$ which are contained in tensor

---

[o]We assume here that in the limit of unbroken $SU(3)_f$, when $\Delta \to 0$, singlet and octet masses are equal. This assumption is justified by our results for vector meson masses, and also by other results based on quark content (4.4.75).

product of three quark representations (*three* defining representations T_f of $SU(3)_f$). We have for this tensor product

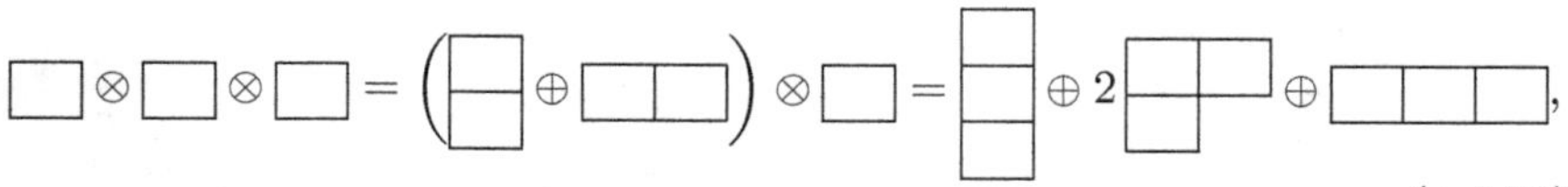

(4.4.78)

where the first diagram on the right-hand side corresponds to a singlet, the second diagram describes two octets (we already know that dimension of each of this representation equals 8), and the last diagram is associated with decuplet (in accordance with (4.4.28) with $N = 3$ and $r = 3$, dimension of this representation equals 10; this is consistent with the fact that dimension of representation on the left-hand side of (4.4.78) is $3^3 = 27$, whereas the dimension on the right hand side is $1 + 2 \times 8 + 10$).

Problem 4.4.12. Find irreducible representations of $SU(3)_f$ which would describe pentaquarks, bound states of four quarks and one antiquark

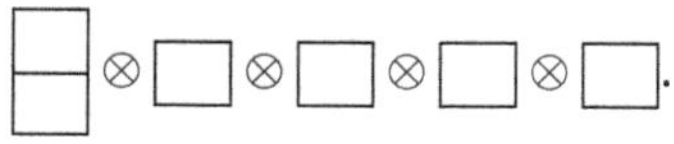

Using hook formula (4.4.21), calculate dimensions of these representations.

In fact, not all of the possibilities listed in (4.4.78) are actually realized for baryons. For instance, there are no light singlet baryon. This has to do with Fermi statistics of quarks and their color and spin.

Let us discuss this issue in some detail. We begin with the absence of singlet baryon. In *nonrelativistic quark model*, quark spin and color are accounted for by the fact that quarks (4.4.46) actually transform as a representation of wider symmetry group $SU(3)_f \otimes SU(3)_c \otimes SU(2)$, where $SU(2)$ is spin group, and $SU(3)_c$ is color group (c refers to color; this notation is used to distinguish this group from flavor group $SU(3)_f$). Let $\mathcal{V}_f$, $\mathcal{V}_c$ and $\mathcal{V}_2$ be spaces of defining representations T_f, T_c and $T^{(\frac{1}{2})}$ of groups $SU(3)_f$, $SU(3)_c$ and $SU(2)$ with bases $\vec{e}_i$ ($i = 1, 2, 3$), $\vec{e}_A$ ($A = 1, 2, 3$) and $\vec{e}^{\alpha}$ ($\alpha = 1, 2$) in $\mathcal{V}_f$, $\mathcal{V}_c$ and $\mathcal{V}_2$, respectively. Quark fields take values in space $\mathcal{V}_f \times \mathcal{V}_c \times \mathcal{V}_2$:

$$\psi = \psi^{iA}_{\alpha} \vec{e}_i \otimes \vec{e}_A \otimes \vec{e}^{\alpha} \in \mathcal{V}_f \times \mathcal{V}_c \times \mathcal{V}_2, \quad (\psi^{1A}_{\alpha}, \psi^{2A}_{\alpha}, \psi^{3A}_{\alpha}) = (u^A_{\alpha}, d^A_{\alpha}, s^A_{\alpha}). \tag{4.4.79}$$

So, every quark flavor u, d and s has color and spin degrees of freedom $A = 1,2,3$ and $\alpha = 1,2$, respectively, i.e., it transforms under defining representation of color group $SU(3)_c$ and defining representation of spin group $SU(2)$. The latter means that quarks have spin $1/2$, and hence they are fermions.

An important physical property is that observed particles are singlets (trivial representations) under color group $SU(3)_c$. This *color confinement* property still lacks solid theoretical explanation, but it is certainly valid in Nature. Antiquarks transform under representation of $SU(3)_c$ conjugate to defining one, so quark–antiquark pair can be color singlet $\psi^{iA}\bar{\psi}_{jA}$. This is precisely the color structure of a meson. The fact that baryons are made of three quarks (not two or four) is directly related to the number of colors: color group $SU(N)_c$ has $N = 3$, and tensor product of *three* defining representations of $SU(3)_c$ contains color singlet, totally antisymmetric rank-3 tensor. In other words, color structure of a baryon is $\psi^{iA}\cdot\psi^{jB}\cdot\psi^{kC} \sim \varepsilon^{ABC}(\varepsilon_{A'B'C'}\psi^{iA'}\cdot\psi^{jB'}\cdot\psi^{kC'})$.

Furthermore, Fermi-statistics of quarks means that quark fields ψ^{iA}_α anticommute with each other, $\psi^{iA}_\alpha\psi^{kB}_\beta = -\psi^{kB}_\beta\psi^{iA}_\alpha$. Therefore, fields of baryons (bound states of three quarks) are linear combinations of products $\psi^{iA}_\alpha\cdot\psi^{jB}_\beta\cdot\psi^{kC}_\gamma$, which must be antisymmetric under permutations of triples of indices (i,A,α), (j,B,β) and (k,C,γ). We do not invoke here the possible dependence of quark fields ψ^{iA}_α on spatial coordinates $\vec{r} = (x^1,x^2,x^3)$, since quarks in lowest mass baryons have zero angular momentum (we consider precisely this case), and baryon wave function $\psi^{iA}_\alpha(\vec{r}_1)\cdot\psi^{jB}_\beta(\vec{r}_2)\cdot\psi^{kC}_\gamma(\vec{r}_3)$ is totally symmetric under permutations of spatial arguments $\vec{r}_1$, $\vec{r}_2$ and $\vec{r}_3$. All this means that the field of $SU(3)_f$-singlet baryon in (4.4.78) would be written in terms of quark fields ψ^{iA}_α as follows:

$$S_{\alpha\beta\gamma} = \varepsilon_{ijk}\varepsilon_{ABC}\psi^{iA}_\alpha\psi^{jB}_\beta\psi^{kC}_\gamma .$$

Here the contraction with totally antisymmetric tensor ε_{ABC} is required to ensure that $S_{\alpha\beta\gamma}$ are color singlets. Likewise, convolution with ε_{ijk} ensures that $S_{\alpha\beta\gamma}$ is singlet under group $SU(3)_f$ as well. Since the fields ψ^{iA}_α anticommute with each other (Fermi-statistics), tensor $S_{\alpha\beta\gamma}$ must be antisymmetric under all permutations of indices α,β and γ. This is impossible, since these indices take only two values $1,2$, hence $S_{\alpha\beta\gamma} = 0$. This precisely means that singlet light baryons must be absent, which is indeed the case in Nature.

Consider now baryon (flavor) decuplet in (4.4.78), which is associated with Young tableau $\boxed{i_1 \mid i_2 \mid i_3}$. In accordance with general formula (4.4.9), field of $SU(3)_f$-decuplet is written in terms of quarks ψ^{iA}_α as follows:

$$D^{(i_1 i_2 i_3)}_{\alpha\beta\gamma} = \frac{1}{3!}\varepsilon_{ABC}\sum_{\sigma\in S_3}\psi^{i_{\sigma(1)}A}_\alpha \psi^{i_{\sigma(2)}B}_\beta \psi^{i_{\sigma(3)}C}_\gamma, \tag{4.4.80}$$

where $\sum_{\sigma\in S_3}$ denotes the sum over all permutations σ of three indices (i_1, i_2, i_3). We recall Fermi-statistics and see that tensor (4.4.80) is totally symmetric in indices (α, β, γ). Such a tensor is transformed under totally symmetric representation $T^{(3/2)}$ of spin group $SU(2)$, (see Section 2.3.2), and $T^{(3/2)}$ is embedded in tensor product $(T^{(\frac{1}{2})})^{\otimes 3}$. The representation $T^{(3/2)}$ of group $SU(2)$ is associated with Young tableau $\boxed{\alpha \mid \beta \mid \gamma}$ and has spin $j = 3/2$. We conclude that *light baryons of spin* 3/2 *make* $SU(3)_f$-*decuplet and, conversely, baryons of* $SU(3)_f$-*decuplet necessarily have spin* 3/2.

This is indeed the case in Nature. Components of baryon decuplet (4.4.80) are:

(1) Non-strange baryons, $s = 0$ (no s-quark and, thus, isospin $\mathsf{I} = 3/2$), correspond to the case in which all indices i, j, k in (4.4.80) take values $1, 2$ (isospin subgroup $SU(2)_I$ is embedded in $SU(3)_f$ as given in (4.4.72)). Omitting color and spin indices, as well as normalization factors, we have

$$\Delta^{++} = D^{111} = uuu, \quad \Delta^- = D^{222} = ddd, \quad \Delta^+ = D^{(112)} = uud,$$
$$\Delta^0 = D^{(122)} = udd, \tag{4.4.81}$$

where we use standard notations for particles $\Delta^+, \Delta^-, \Delta^0, \Delta^{++}$ with upper indices showing electric charges (recall that charges of u-, d- and s-quarks are $+2/3$, $-1/3$ and $-1/3$, respectively).

(2) Strange $s = (-1)$ baryons (one s-quark, see (4.4.57)) and isospin $\mathsf{I} = 1$ correspond to the case in which one of indices i, j, k in (4.4.80) equals 3, while others take values $1, 2$:

$$\Sigma^{*+} = D^{(113)} = uus, \quad \Sigma^{*-} = D^{(223)} = dds, \quad \Sigma^{*0} = D^{(123)} = uds. \tag{4.4.82}$$

(3) Baryons with strangeness $s = -2$ (two s-quarks) and isospin $\mathsf{I} = 1/2$ are

$$\Xi^{*0} = D^{(133)} = uss, \Xi^{*-} = D^{(233)} = dss. \tag{4.4.83}$$

(4) Baryon with strangeness $s = -3$ (three s-quarks) and isospin $\mathsf{I} = 0$ is

$$\Omega^- = D^{333} = sss. \tag{4.4.84}$$

Note that baryon Ω^- whose quark content in given in (4.4.84) would be forbidden if quarks would not have color. Indeed, in the absence of color, wave function $s_\alpha \cdot s_\beta \cdot s_\gamma$ for fermions would be antisymmetric in spin indices α, β, γ, which is impossible, since spin indices take values $(1, 2)$ only. It is the existence of light baryons with wave functions symmetric in flavor (these are Δ^{++}, Δ^- and Ω^-) that lead Bogoliubov, Struminsky and Tavkhelidze and independently Han and Nambu to introduce an extra quantum number of quarks, color.

We now study two baryon octets in (4.4.78). These are associated with two equivalent representations of group $SU(3)_f$, 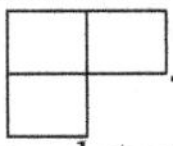. They correspond to rank-3 tensors symmetrized in special ways determined by Young tableaux T_1 and T_2 shown in (4.4.14). Explicit forms of these tensors have been found in (4.4.16). In terms of quarks, components of tensor ψ_{T_1} given by (4.4.16) are (we consider color singlet states and use Fermi-statistics for quarks)

$$\begin{aligned} B^{i_1 i_2 i_3}_{\alpha\beta\gamma} &= \varepsilon_{ABC} (\psi^{i_1 i_2 i_3}_{\mathsf{T}_1})^{ABC}_{\alpha\beta\gamma} \\ &= \varepsilon_{ABC} (\psi^{i_1 i_2 i_3} - \psi^{i_3 i_2 i_1} + \psi^{i_2 i_1 i_3} - \psi^{i_2 i_3 i_1})^{ABC}_{\alpha\beta\gamma} \\ &= \varepsilon_{ABC} (\psi^{i_1 A}_\alpha \psi^{i_2 B}_\beta \psi^{i_3 C}_\gamma - \psi^{i_3 A}_\alpha \psi^{i_2 B}_\beta \psi^{i_1 C}_\gamma + \psi^{i_2 A}_\alpha \psi^{i_1 B}_\beta \psi^{i_3 C}_\gamma \\ &\quad - \psi^{i_2 A}_\alpha \psi^{i_3 B}_\beta \psi^{i_1 C}_\gamma) \\ &= 2\varepsilon_{ABC} (\psi^{i_2 A}_\beta \psi^{i_1 B}_{[\alpha} \psi^{i_3 C}_{\gamma]} + \psi^{i_2 A}_\alpha \psi^{i_1 B}_{[\beta} \psi^{i_3 C}_{\gamma]}) = -B^{i_3 i_2 i_1}_{\alpha\beta\gamma}, \end{aligned} \tag{4.4.85}$$

where square brackets denote antisymmetrization in spin indices. Since antisymmetric in spin indices functions $\psi_{[\alpha\beta]}$ correspond to trivial representation of spin group $SU(2)$ (spin $j = 0$), each baryon component $B^{i_1 i_2 i_3}$ in (4.4.85) is in defining representation of spin group $SU(2)$ with spin $j = 1/2$. Thus, *baryon octet has spin* $1/2$, *and, conversely, light baryons with spin* $1/2$ *necessarily make* $SU(3)_f$*-octet.* Let us write quark content of the baryon octet (4.4.85) (we list here only independent components and do not show components that are related by $B^{i_1 i_2 i_3}_{\alpha\beta\gamma} = -B^{i_3 i_2 i_1}_{\alpha\beta\gamma}$)

$$\begin{aligned} p = B^{112}_{\alpha\beta\gamma} &= \varepsilon_{ABC} (u^A_\beta u^B_{[\alpha} d^C_{\gamma]} + u^A_\alpha u^B_{[\beta} d^C_{\gamma]}), \\ n = B^{122}_{\alpha\beta\gamma} &= \varepsilon_{ABC} (d^A_\beta u^B_{[\alpha} d^C_{\gamma]} + d^A_\alpha u^B_{[\beta} d^C_{\gamma]}), \end{aligned}$$

$$
\begin{aligned}
\Sigma^+ &= B^{113}_{\alpha\beta\gamma} = \varepsilon_{ABC}(u^A_\beta u^B_{[\alpha} s^C_{\gamma]} + u^A_\alpha u^B_{[\beta} s^C_{\gamma]}),\\
\Sigma^- &= B^{223}_{\alpha\beta\gamma} = \varepsilon_{ABC}(d^A_\beta d^B_{[\alpha} s^C_{\gamma]} + d^A_\alpha d^B_{[\beta} s^C_{\gamma]}),\\
\Sigma^0 &= B^{123}_{\alpha\beta\gamma} + B^{213}_{\alpha\beta\gamma} = \varepsilon_{ABC}(d^A_\beta u^B_{[\alpha} s^C_{\gamma]} + d^A_\alpha u^B_{[\beta} s^C_{\gamma]}) + (u \leftrightarrow d),\\
\Lambda^0 &= B^{132}_{\alpha\beta\gamma} = (B^{123}_{\alpha\beta\gamma} - B^{213}_{\alpha\beta\gamma}) = \varepsilon_{ABC}(s^A_\beta u^B_{[\alpha} d^C_{\gamma]} + s^A_\alpha u^B_{[\beta} d^C_{\gamma]}),\\
\Xi^0 &= B^{133}_{\alpha\beta\gamma} = \varepsilon_{ABC}(s^A_\beta u^B_{[\alpha} s^C_{\gamma]} + s^A_\alpha u^B_{[\beta} s^C_{\gamma]}),\\
\Xi^- &= B^{233}_{\alpha\beta\gamma} = \varepsilon_{ABC}(s^A_\beta d^B_{[\alpha} s^C_{\gamma]} + s^A_\alpha d^B_{[\beta} s^C_{\gamma]}),
\end{aligned} \tag{4.4.86}
$$

where we omit normalization and sign factors. Note that component Λ^0 is antisymmetric under permutation $u \leftrightarrow d$, so it is singlet under isospin subgroup $SU(2)_I \subset SU(3)_f$, while component Σ^0 (symmetric under permutation $u \leftrightarrow d$) together with Σ^+ and Σ^- make isotriplet.

Finally, we show that the second baryon octet in (4.4.78) is actually identical to the octet (4.4.85). Indeed, in terms of quarks, components of the second octet are

$$
\tilde{B}^{i_1 i_2 i_3}_{\alpha\beta\gamma} = \varepsilon_{ABC}(\psi^{i_1 i_2 i_3}_{\mathsf{T}_2})^{ABC}_{\alpha\beta\gamma} = 2\varepsilon_{ABC}(\psi^{i_1 A}_{[\alpha} \psi^{i_2 B}_{\beta]} \psi^{i_3 C}_{\gamma} + \psi^{i_3 A}_{\alpha} \psi^{i_2 B}_{[\beta} \psi^{i_1 C}_{\gamma]}). \tag{4.4.87}
$$

We compare components (4.4.85), (4.4.87) and see that $B^{i_1 i_2 i_3}_{\alpha\beta\gamma} = \tilde{B}^{i_1 i_3 i_2}_{\alpha\gamma\beta}$, so that the total wave functions are related to each other by renaming of indices. Thus, the two baryon octets on the right-hand side of (4.4.78), which are associated with standard Young tableaux

$$
\mathsf{T}_1 = \begin{array}{|c|c|}\hline i_1 & i_2 \\ \hline i_3 & \\ \cline{1-1}\end{array}, \quad \mathsf{T}_2 = \begin{array}{|c|c|}\hline i_1 & i_3 \\ \hline i_2 & \\ \cline{1-1}\end{array},
$$

are in fact indistinguishable in their quark content and spin. Indeed, there is just one baryon octet (4.4.86) in Nature.

We now turn to mass formulas for baryons composed of light quarks u, d and s, which are similar in spirit to meson mass formulas. Considering octet, we make use of antisymmetry of tensor (4.4.85) in indices i_1, i_3 and write equivalently

$$
B^k{}_j = \frac{1}{2}\varepsilon_{jlm} B^{lkm}.
$$

In terms of baryon states, this matrix is

$$B^k{}_j = \begin{pmatrix} \dfrac{\Sigma^0}{\sqrt{2}} + \dfrac{\Lambda^0}{\sqrt{6}} & \Sigma^+ & p \\ \Sigma^- & -\dfrac{\Sigma^0}{\sqrt{2}} + \dfrac{\Lambda^0}{\sqrt{6}} & n \\ \Xi^- & \Xi^0 & -\dfrac{2\Lambda^0}{\sqrt{6}} \end{pmatrix}, \quad B^k{}_k = 0. \tag{4.4.88}$$

Unlike Hermitian meson matrix (4.4.61), the latter matrix has Hermitian conjugate which describes antibaryon octet

$$\bar{B}^k{}_j = \begin{pmatrix} \dfrac{\bar{\Sigma}^0}{\sqrt{2}} + \dfrac{\bar{\Lambda}^0}{\sqrt{6}} & \bar{\Sigma}^- & \bar{\Xi}^+ \\ \bar{\Sigma}^+ & -\dfrac{\bar{\Sigma}^0}{\sqrt{2}} + \dfrac{\bar{\Lambda}^0}{\sqrt{6}} & \bar{\Xi}^0 \\ \bar{p} & \bar{n} & -\dfrac{2\bar{\Lambda}^0}{\sqrt{6}} \end{pmatrix}, \quad \bar{B}^k{}_k = 0. \tag{4.4.89}$$

Problem 4.4.13. Derive the representation (4.4.88). Hint: take notice that kinetic term for baryon octet $\bar{B}^j{}_k i\gamma^\mu\partial_\mu B^k{}_j$ must be equal to the sum of kinetic terms for each of the baryons, $\bar{p}i\gamma^\mu\partial_\mu p + \cdots + \bar{\Xi}^0 i\gamma^\mu\partial_\mu \Xi^0$.

Since baryons are fermions, mass term in baryon Lagrangian is linear in masses,

$$-L_{B,8} = m_{B,8}\bar{B}^k{}_j B^j{}_k + \alpha\bar{B}^k{}_3 B^3{}_k + \beta\bar{B}^3{}_k B^k{}_3. \tag{4.4.90}$$

The first term here is $SU(3)_f$-invariant, while the second and third terms come from independent invariants written in terms of quark mass matrix $C_1\mathrm{Tr}(\bar{B}mB)$ and $C_2\mathrm{Tr}(\bar{B}Bm)$, respectively. In view of (4.4.77), these invariants break $SU(3)_f$ down to isotopic $SU(2)_I$. We insert (4.4.88) and (4.4.89) into mass term (4.4.90) and obtain

$$\begin{gathered} m_p = m_n = m_{B,8} + \beta, \\ m_{\Sigma^\pm} = m_{\Sigma^0} = m_{B,8}, \quad m_\Lambda = m_{B,8} + \frac{2}{3}(\alpha+\beta), \\ m_{\Xi^-} = m_{\Xi^0} = m_{B,8} + \alpha, \end{gathered} \tag{4.4.91}$$

which leads to mass relation for baryons

$$3m_\Lambda + m_\Sigma = 2m_N + 2m_\Xi \quad (N = p, n), \tag{4.4.92}$$

which is known as *Gell-Mann–Okubo mass formula*. It is valid with reasonable precision, given that [13]:

$$m_N \approx 939\,\text{MeV}, \;\; m_\Lambda \approx 1116\,\text{MeV}, \;\; m_\Sigma \approx 1193\,\text{MeV}, \;\; m_\Xi \approx 1319\,\text{MeV} \tag{4.4.93}$$

(masses of Σ^0, $\Sigma^\pm$ are in interval 1189.3–1197.5 MeV, while masses of Ξ^-, Ξ^0 equal 1314.9–1321.7 MeV).

To obtain mass formulas for baryon decuplet, we write the symmetric tensor $D^{(ijk)}$ in terms of baryon variables (color and spin indices are omitted):

$$\begin{aligned}
&D^{(111)} = \Delta^{++}, \quad D^{(112)} = \frac{1}{\sqrt{3}}\Delta^+, \quad D^{(122)} = \frac{1}{\sqrt{3}}\Delta^0, \quad D^{(222)} = \Delta^-,\\
&D^{(113)} = \frac{1}{\sqrt{3}}\Sigma^{*+}, \quad D^{(223)} = \frac{1}{\sqrt{3}}\Sigma^{*-}, \quad D^{(123)} = \frac{1}{\sqrt{6}}\Sigma^{*+},\\
&D^{(133)} = \frac{1}{\sqrt{3}}\Xi^{*0}, \quad D^{(233)} = \frac{1}{\sqrt{3}}\Xi^{*-}, \quad D^{(333)} = \Omega^-.
\end{aligned} \tag{4.4.94}$$

As before, numerical factors here are chosen in such a way that invariant kinetic term $i\bar{D}_{(ijk)}\gamma^\mu\partial_\mu D^{(ijk)}$ yields the standard (canonical) kinetic terms for each of the baryons. In view of total symmetry of $D^{(ijk)}$, the only invariant mass term that breaks $SU(3)_f$ is const $\cdot\, (m^i_j\bar{D}_{(jkl)}D^{(ikl)})$. Therefore, mass term in decuplet Lagrangian reads

$$\begin{aligned}
-L_{B,10} &= m_{B,10}\bar{D}_{(ijk)}D^{(ijk)} + h_0 m^i_j \bar{D}_{(jkl)}D^{(ikl)}\\
&= m_{B,10}\bar{D}_{(ijk)}D^{(ijk)} + \delta_{10}\bar{D}_{(3kl)}D^{(3kl)},
\end{aligned}$$

where the first term on the right-hand side is invariant under $SU(3)_f$, and the second term breaks $SU(3)_f$ down to isotopic $SU(2)_I$. Together with (4.4.94), this gives mass formulas

$$m_{\Sigma^*} - m_\Delta = m_{\Xi^*} - m_{\Sigma^*} = m_\Omega - m_{\Xi^*}, \tag{4.4.95}$$

which are valid with precision of about 1% for experimental values [13]

$$\begin{aligned}
m_\Delta &\approx 1232\,\text{MeV}, \quad m_{\Sigma^*} \approx 1385\,\text{MeV},\\
m_{\Xi^*} &\approx 1530\,\text{MeV}, \quad m_\Omega \approx 1672\,\text{MeV}.
\end{aligned} \tag{4.4.96}$$

We emphasize that both the very existence of baryon Ω^- and its mass were predicted before its experimental discovery. That was a great success of quark model.

Problem 4.4.14. Check that, for baryon octet, isospins I and hypercharges $Y = B + S$ are

$$\begin{array}{c} p,n : \mathsf{I} = 1/2, \quad Y = 1; \quad \Lambda : \mathsf{I} = 0, \quad Y = 0; \\ \Sigma : \mathsf{I} = 1, \quad Y = 0; \quad \Xi : \mathsf{I} = 1/2, \quad Y = -1; \end{array} \tag{4.4.97}$$

whereas for baryon decuplet one has

$$\begin{array}{c} \Delta : \mathsf{I} = 3/2, \quad Y = 1; \quad \Sigma^* : \mathsf{I} = 1, \quad Y = 0; \\ \Xi^* : \mathsf{I} = 1/2, \quad Y = -1; \quad \Omega : \mathsf{I} = 0, \quad Y = -2. \end{array} \tag{4.4.98}$$

Check that baryon masses (4.4.93) and (4.4.96) obey, with good precision, Okubo's formula

$$m = m_\lambda + h_1 Y + h_2(\mathsf{I}(\mathsf{I}+1) - \frac{1}{4}Y^2), \tag{4.4.99}$$

where $h_1 \approx -196.3\,\mathrm{MeV}$, $h_2 \approx 38.5\,\mathrm{MeV}$, and parameter m_λ depends on representation: for octet $m_\lambda = m_8$ and decuplet $m_\lambda = m_{10}$ one has

$$m_8 \approx 1116\,\mathrm{MeV}, \quad m_{10} \approx 1308\,\mathrm{MeV}. \tag{4.4.100}$$

Problem 4.4.15. Consider the weight lattice of algebra $s\ell(3,\mathbb{C})$ (and $su(3)$), in which the fundamental weights $\lambda_{(1)}$, $\lambda_{(2)}$ and root vectors are indicated (see Figs. 3.3.2 and 3.3.3 in Section 3.3). Show that weight diagrams for baryon octet (4.4.88) and decuplet(4.4.94) have the form shown in Fig. 4.4.2. Show that in the planar weight lattice of algebra $su(3)$, one can choose isotopic spin I_3 as the horizontal axis and hypercharge Y as the vertical axis. Hint: make use of the fact that highest weights $\lambda_{(1)}$ and $\lambda_{(2)}$ correspond respectively to the triplet of quarks and to the antisymmetric combination of two quark triplets (which is the same as antitriplet); the highest weight for octet is $(\lambda_{(1)} + \lambda_{(2)})$ (adjoint representation of $su(3)$), while the highest weight for decuplet is $3 \cdot \lambda_{(1)}$.

To conclude this section, we note that attempts to include heavy quarks c, b, t into the above scheme have failed. This is due to the fact that masses of c, b, t exceed hadronic mass scale of order $m_\rho = 770\,\mathrm{MeV}$. Nevertheless, the notion that quark–antiquark pairs of any sort make mesons, and three quarks make baryons, is out of question. The only reservation is that hadrons with the heaviest t-quark do not exist: their formation time is longer than t-quark lifetime, so these hadrons have simply not enough time to get formed.

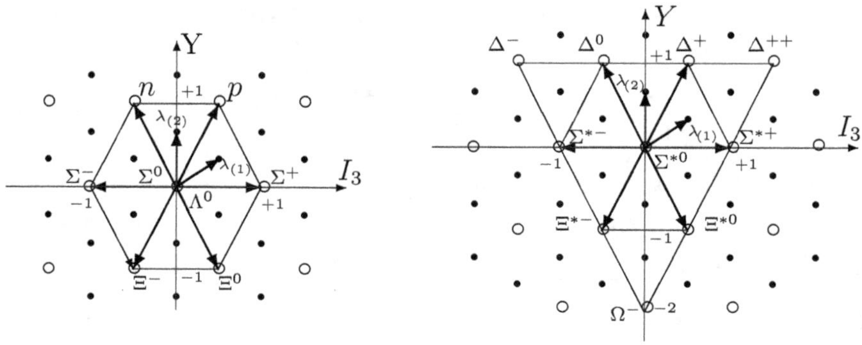

Fig. 4.4.2. Weight diagrams for octet and decuplet of baryons.

4.5. Representations of Symmetric Group II. Young–Frobenius Theory

As we already pointed out, in this chapter we study two problems: (1) construction of all complex finite-dimensional irreducible representations $T^{(\alpha)}$ of group $SU(N)$ and algebras $s\ell(N, \mathbb{C})$ and $su(N)$ (and all representations of group $SL(N, \mathbb{C})$ from our special subclass), as subrepresentations in tensor product $T^{\otimes r}$ of defining representations; (2) expansion of tensor product $T^{\otimes r}$ in irreducible representations $T^{(\alpha)}$. The first problem is already solved: all complex finite-dimensional irreducible representations of $SU(N)$ are constructed by employing Young symmetrizers, corresponding to *primitive* idempotents $e_\alpha \in \mathbb{C}[S_r]$, and, in fact, with some qualifications given in Section 4.4.1, are in one-to-one correspondence with Young diagrams. The latter give also a method of explicit construction of the irreducible representations in spaces of tensors with definite symmetry properties, see Section 4.4.1 (an alternative method based on highest weight representations is given in Section 3.4). The second problem (which is a fairly narrow case of Clebsch–Gordan expansion for representations of groups $SU(N)$) reduces to extraction of a *complete* set of *orthogonal* primitive idempotents from the entire set of primitive idempotents $e_\alpha \in \mathbb{C}[S_r]$ (Young symmetrizers associated with all Young tableaux with r boxes).

So, the above discussion (see Sections 4.2 and 4.4.1) leads us to the problem of constructing the set of idempotents e_α ($\alpha = 1, \ldots, s$) in group algebra $\mathbb{C}[S_r]$, whose properties (4.2.29) we repeat here:

$$e_\alpha e_\beta = \delta_{\alpha\beta} e_\alpha, \quad \text{idempotence and orthogonality,} \tag{4.5.1a}$$

$$e = \sum_{\alpha=1}^{s} e_\alpha, \quad \text{completeness,} \tag{4.5.1b}$$

where e is unit element in $\mathbb{C}[S_r]$. Since we are interested in *irreducible* representations of group $SL(N,\mathbb{C})$, the idempotents $e_\alpha \in \mathbb{C}[S_r]$ must be *primitive*, i.e., not decomposable into a sum of idempotents (see Definition 4.2.1). Once a set of primitive idempotents $\{e_\alpha\} \in \mathbb{C}[S_r]$ obeying (4.5.1) is known, the representation s of algebra $\mathbb{C}[S_r]$ in space $\mathcal{V}^{\otimes r}$, given by (4.2.6), enables one to construct a set of orthogonal projectors $\mathsf{s}(e_\alpha)$ in $\mathcal{V}^{\otimes r}$. This set is used for expanding the space $\mathcal{V}^{\otimes r}$ in direct sum

$$\mathcal{V}^{\otimes r} = \bigoplus_\alpha \mathcal{V}_\alpha,$$

where $\mathcal{V}_\alpha = \mathsf{s}(e_\alpha) \cdot \mathcal{V}^{\otimes r}$ are invariant (under the action of $T^{\otimes r}$) subspaces of irreducible representations $T^{(\alpha)}$ embedded in $T^{\otimes r}$. This reiterates that the problem of expanding the tensor product $T^{\otimes r}$ of defining representations in irreducible representations $T^{(\alpha)}$ (some of these irreducible representations may be equivalent) is reduced to the problem of constructing a system of primitive idempotents in group algebra $\mathbb{C}[S_r]$ obeying (4.5.1).

We point out that the existence of a complete system of primitive orthogonal idempotents is not specific to group algebra $\mathbb{C}[S_r]$. We discuss in the following subsection a fairly large class of associative finite-dimensional algebras $\mathcal{A}$ with unit element, which admit such a system of idempotents. We also show that a system of idempotents with the above properties gives a method of constructing irreducible representations of algebra $\mathcal{A}$.

4.5.1. Idempotents and irreducible representations of associative algebras. Peirce decomposition

Part of the material of this section is presented in a different way in Ref. [26].

Let $\mathcal{A}$ be an associative algebra over field $\mathbb{C}$ with unit element e. We show here that complete system of orthogonal idempotents can be constructed for fairly large class of these algebras. The choice of primitive idempotents in $\mathcal{A}$ is not unique (see Remark 1 below), but for each choice the number of primitive idempotents is finite for finite-dimensional algebra $\mathcal{A}$.

Any associative algebra $\mathcal{A}$ acts on itself (as on vector space) by left (right) multiplication. The pertinent representation of $\mathcal{A}$ is *left* (*right*) *regular representation*. We consider finite-dimensional algebras and assume that their left (right) regular representations are *completely reducible*, i.e., they can be expanded in direct sum of irreducible representations $T^{(\lambda)}$ (index λ labels these representations). An example of this situation is given by a group algebra $\mathbb{C}[G]$ of any finite group G. This algebra is associative, finite-dimensional and has unit element. All inequivalent irreducible representations $T^{(\lambda)}$ of algebra $\mathbb{C}[G]$ are contained in its regular representation, and multiplicity of representation $T^{(\lambda)}$ in regular representation is equal to dimension of $T^{(\lambda)}$ (see Section I-4.6.2 in the first book; see also formula (4.3.63)).

We consider left regular representation of algebra $\mathcal{A}$ whose properties are listed above; right regular representation is studied in the same way. Since regular representation is completely reducible, algebra $\mathcal{A}$, as vector space, is expanded in direct sum of its *simple left ideals*,

$$\mathcal{A} = \bigoplus_{\alpha=1}^{s} \mathcal{L}_\alpha, \tag{4.5.2}$$

i.e., direct sum of invariant subspaces such that $\mathcal{A} \cdot \mathcal{L}_\alpha \subset \mathcal{L}_\alpha$. Left ideals are simple if they do not contain left invariant subspaces (left ideals). In our case, the latter property holds because $\mathcal{L}_\alpha$ are spaces of irreducible representations of $\mathcal{A}$. It follows from (4.5.2) that any element $a \in \mathcal{A}$ can be written as a sum $a = \sum_\alpha a_\alpha$, where $a_\alpha \in \mathcal{L}_\alpha$. In particular, for unit element $e \in \mathcal{A}$ we have expansion (4.5.1b):

$$e = \sum_{\alpha=1}^{s} e_\alpha, \quad e_\alpha \in \mathcal{L}_\alpha. \tag{4.5.3}$$

Since $e_\beta \in \mathcal{L}_\beta$, and $\mathcal{L}_\beta$ is left invariant, the product $e_\alpha \cdot e_\beta$ belongs to $\mathcal{L}_\beta$, while the relation

$$e_\alpha = e_\alpha \cdot e = \sum_\beta e_\alpha \cdot e_\beta \in \mathcal{L}_\alpha$$

shows that $e_\alpha \cdot e_\beta = 0$ (for all $\alpha \neq \beta$) and $e_\alpha \cdot e_\alpha = e_\alpha$, so we have the properties (4.5.1a):

$$e_\alpha \cdot e_\beta = \delta_{\alpha\beta} e_\alpha. \tag{4.5.4}$$

Thus, elements e_α in (4.5.3) are orthogonal idempotents, and they are primitive, in view of the fact that $\mathcal{L}_\alpha$ are *simple* left ideals. In accordance with (4.5.3), they form complete system. Note that any element from $\mathcal{L}_\alpha$ can be written as $a \cdot e_\alpha$, where $a \in \mathcal{A}$, i.e., the whole simple left ideal $\mathcal{L}_\alpha$ is generated by an element e_α: $\mathcal{L}_\alpha = \mathcal{A} \cdot e_\alpha$. This shows that algebra $\mathcal{A}$ is expanded in sum of its left ideals $\mathcal{L}_\alpha = \mathcal{A} \cdot e_\alpha$ generated by a complete system of orthogonal primitive idempotents e_α. Conversely, any complete system of orthogonal idempotents $e_\alpha \in \mathcal{A}$ obeying (4.5.3) and (4.5.4) can be used to expand algebra $\mathcal{A}$ in sum of its left ideals $\mathcal{L}_\alpha$ (or right ideals $\mathcal{R}_\alpha$),

$$\mathcal{A} = \bigoplus_{\alpha=1}^{s} \mathcal{A} \cdot e_\alpha = \bigoplus_{\alpha=1}^{s} \mathcal{L}_\alpha, \quad \mathcal{A} = \bigoplus_{\alpha=1}^{s} e_\alpha \cdot \mathcal{A} = \bigoplus_{\alpha=1}^{s} \mathcal{R}_\alpha. \tag{4.5.5}$$

To summarize, any expansion of algebra $\mathcal{A}$ in direct sum of its simple ideals is associated with complete system of orthogonal primitive idempotents. We emphasize again that, by construction, every subspace $\mathcal{L}_\alpha = \mathcal{A} \cdot e_\alpha$ (or $\mathcal{R}_\alpha = e_\alpha \cdot \mathcal{A}$) is invariant subspace in $\mathcal{A}$ with respect to left (or right) action of algebra $\mathcal{A}$. This means that for all $v \in \mathcal{L}_\alpha$ and $v' \in \mathcal{R}_\alpha$ we have

$$a \cdot v \in \mathcal{L}_\alpha, \quad v' \cdot a \in \mathcal{R}_\alpha, \quad \forall a \in \mathcal{A}. \tag{4.5.6}$$

Left and right ideals $\mathcal{L}_\alpha$ and $\mathcal{R}_\alpha$ are subalgebras in $\mathcal{A}$; this follows from (4.5.6).

Problem 4.5.1. Consider associative algebra $\bar{\mathcal{B}}_+$ of upper triangle matrices whose basis is made of matrix units $\{e_{ij}\}$, $(1 \leq i < j \leq n)$, plus unit matrix I_n. Show that its regular representation is reducible but not completely reducible. Find the structure of left ideals in $\bar{\mathcal{B}}_+$. Show that this algebra does not have idempotents other than trivial one I_n.

The first and second expansions in (4.5.5) are called *left and right Peirce decomposition*, respectively. Making use of left and right Peirce decompositions simultaneously, we have *two-sided Peirce decomposition*

$$\mathcal{A} = \bigoplus_{\alpha,\beta=1}^{s} e_\alpha \mathcal{A} e_\beta = \bigoplus_{\alpha,\beta=1}^{s} \mathcal{A}_{\alpha,\beta}, \quad \mathcal{A}_{\alpha,\beta} = e_\alpha \mathcal{A} e_\beta. \tag{4.5.7}$$

Here the linear spaces $\mathcal{A}_{\alpha,\beta}$ are, generally speaking, neither left nor right ideals in $\mathcal{A}$. Instead, products of elements of $\mathcal{A}_{\alpha,\beta}$ obey the following relations, which resemble relations for matrix units:

$$\mathcal{A}_{\alpha,\beta} \cdot \mathcal{A}_{\gamma,\kappa} = \delta_{\beta,\gamma} \mathcal{A}_{\alpha,\kappa}. \tag{4.5.8}$$

We now consider the left action of an element $a \in \mathcal{A}$ on idempotent e_β and use right Peirce decomposition for element $a \cdot e_\beta$. We get

$$a \cdot e_\beta = \sum_\alpha e_\alpha a_{\alpha\beta}, \tag{4.5.9}$$

where $a_{\alpha\beta} = e_\alpha \cdot a \cdot e_\beta \in \mathcal{A}_{\alpha,\beta} \subset \mathcal{A}$ are coefficients of two-sided Peirce decomposition (4.5.7) of element a. Formula (4.5.9) can be interpreted as commutation rule of element a with idempotent e_β. In view of associativity of $\mathcal{A}$, Eq. (4.5.9) gives the multiplication rule for coefficients of two-sided Peirce decomposition:

$$(a \cdot b)_{\alpha\beta} = \sum_\gamma a_{\alpha\gamma} b_{\gamma\beta}, \quad \forall a, b \in \mathcal{A}. \tag{4.5.10}$$

This property, together with linearity $(a+b)_{\alpha\beta} = a_{\alpha\beta} + b_{\alpha\beta}$, shows that the coefficients of two-sided Peirce decomposition behave like elements of matrix algebra.

Formulas (4.5.8)–(4.5.10) can be used for constructing representations of associative finite-dimensional algebras of particular class (see Remark 3 below). We consider finite-dimensional algebras $\mathcal{A}$ that admit complete system of primitive orthogonal idempotents e_α ($\alpha = 1, \ldots, s$) such that subspaces $\mathcal{A}_{\alpha,\beta}$ of two-sided Peirce decomposition are *one dimensional.*[p] We choose normalization of basis vectors $e_{\alpha,\beta}$ in $\mathcal{A}_{\alpha,\beta}$ in such a way that (see (4.5.8))

$$e_{\alpha,\beta} \cdot e_{\gamma,\kappa} = \delta_{\beta\gamma} e_{\alpha,\kappa}. \tag{4.5.11}$$

Any element a of algebra $\mathcal{A}$ can be written as $a = \sum a_{\alpha,\beta} e_{\alpha,\beta}$, where $a_{\alpha,\beta}$ are now *numbers* (rather than elements of the algebra like in (4.5.9)), which obey (4.5.10). These numbers are conveniently written in Dirac notations, $a_{\alpha,\beta} = \langle\alpha|a|\beta\rangle$. To explain this formula, we construct special representation T of algebra $\mathcal{A}$, which acts in complex space $\mathcal{V}$ with basis

[p]This is actually fairly large class of algebras which includes, as we see below, group algebras $\mathbb{C}[S_r]$ and Clifford algebras.

vectors $|\alpha\rangle$ $(\alpha = 1,\dots,s)$. The latter obey the relations

$$e_{\beta,\gamma}|\alpha\rangle = |\beta\rangle\delta_{\gamma\alpha}, \quad \langle\alpha|e_{\gamma,\beta} = \delta_{\gamma\alpha}\langle\beta|,$$

where $\langle\beta|$ and $|\alpha\rangle$ are orthonormalized bra- and ket-vectors $\langle\beta|\alpha\rangle = \delta_{\alpha\beta}$. These are consistent with (4.5.11). Then basis elements $e_{\alpha,\beta} \in \mathcal{A}_{\alpha,\beta}$ and idempotents e_α are written in representation T in Dirac notations as follows (see Eq. (4.5.26) in Example 1 below):

$$e_{\alpha,\beta} = |\alpha\rangle\langle\beta|, \quad e_\alpha = e_{\alpha,\alpha} = |\alpha\rangle\langle\alpha|. \tag{4.5.12}$$

The completeness condition (4.5.1b) reads

$$\sum_\alpha |\alpha\rangle\langle\alpha| = e.$$

Then coefficients $a_{\alpha\beta} = e_\alpha \cdot a \cdot e_\beta$ of two-sided Peirce decomposition of element a in (4.5.9) and relation (4.5.10) have the following forms:

$$\begin{aligned} e_\alpha \cdot a \cdot e_\beta &= |\alpha\rangle\langle\alpha|a|\beta\rangle\langle\beta| = \langle\alpha|a|\beta\rangle e_{\alpha,\beta}, \\ \langle\alpha|a\cdot b|\beta\rangle &= \sum_\gamma \langle\alpha|a|\gamma\rangle\langle\gamma|b|\beta\rangle. \end{aligned} \tag{4.5.13}$$

This shows that factors $\langle\alpha|\, a\,|\beta\rangle \in \mathbb{C}$ in coefficients (4.5.13) of two-sided Peirce decomposition define matrices of representation T of elements $a \in \mathcal{A}$:

$$e_\beta \cdot a \cdot e_\alpha = \langle\beta|a|\alpha\rangle e_{\beta,\alpha} \;\Rightarrow\; T_{\beta\alpha}(a) := \langle\beta|a|\alpha\rangle. \tag{4.5.14}$$

In general, this representation T is reducible. We discuss this fact, as well as a way of extracting irreducible representations from T, below.

Two-sided Peirce decomposition, as we see, is used in representation theory. Besides that, it is employed for obtaining detailed information on algebra $\mathcal{A}$, e.g., matrix structure of its regular representation (see Remark 2 below). Concrete cases of two-sided Peirce decomposition are considered below in Examples 1–3.

Definition 4.5.1. Ideal, which is both left and right, is called two-sided ideal, or simply ideal. Ideal is trivial if it contains only zero element or coincides with the whole algebra. Algebra $\mathcal{A}$ is called *simple* if it does not have nontrivial ideals, and *semisimple* if it is isomorphic to direct sum of simple ideals.

Obviously, ideal is a subalgebra in $\mathcal{A}$.

Algebra $\mathcal{A}$ may have two important sets of idempotents, obeying (4.5.3) ad (4.5.4): (1) primitive idempotents $\{e_\alpha\}$ discussed above; (2) primitive *central* idempotents $\{e_A\}$, which belong to center of algebra $\mathcal{A}$ and cannot be decomposed into sum of *central* idempotents. Idempotents from set (2) obey

$$e_A \mathcal{A} e_B = \delta_{A,B} e_A \mathcal{A} = \delta_{A,B} \mathcal{A} e_A.$$

So, left, right and two-sided Peirce decompositions coincide for central idempotents, and we have

$$\mathcal{A} = \bigoplus_{A=1}^{s'} e_A \mathcal{A} = \bigoplus_{A=1}^{s'} \mathcal{A} e_A = \bigoplus_{A=1}^{s'} \mathcal{A}_A, \quad \mathcal{A}_A \equiv e_A \mathcal{A} e_A, \tag{4.5.15}$$

where $\mathcal{A}_A$ are (two-sided) ideals in $\mathcal{A}$, and s' is the number of central idempotents in $\mathcal{A}$. Thus, if algebra $\mathcal{A}$ has nontrivial central idempotents, then formula (4.5.15) gives expansion of algebra $\mathcal{A}$ in direct sum of its ideals, so this algebra is not simple. Note that semisimple algebra $\mathcal{A}$ always possesses complete system of central primitive idempotents. This follows from expansion of this algebra in direct sum of its simple subalgebras and arguments similar to those used for obtaining Eqs. (4.5.2)–(4.5.4). Expansion (4.5.15) is analogous to expansion of semisimple Lie algebra in direct sum of simple Lie algebras considered in Section I-5.2 of the first book.

We concentrate in what follows on semisimple algebras. Then for central idempotents $\{e_A\}$ we can use left (or right) Peirce decomposition (4.5.5) and expand e_A in primitive idempotents $\{e_\alpha\}$:

$$e_A = \sum_{\alpha \in A} e_\alpha, \tag{4.5.16}$$

where A is a subset of indices from the set $\{1, 2, \ldots, s\}$; in other words, central orthogonal idempotents $\{e_A\}$ making complete system are conveniently labeled by nonintersecting subsets $A \subset \{1, 2, \ldots, s\}$, which cover the entire set of indices $\{1, 2, \ldots, s\}$.

Problem 4.5.2. Show that left (and right) Peirce decomposition (4.5.5) for central idempotents $\{e_A\}$ has the form (4.5.16). Hint: substitute general expansion $e_A = \sum_{\alpha \in A} C_{A\alpha} e_\alpha$ into $e_A e_B = \delta_{AB} e_A$.

Let A_i ($i = 1, \ldots, s'$) be nonintersecting subsets in $\{1, 2, \ldots, s\}$ (which cover the entire set), defining central idempotents e_{A_i}. Let $\alpha \in A_i$, $\beta \in A_j$

and $i \neq j$, then, in view of orthogonality $e_{A_i} \cdot e_{A_j} = 0$, for any element $a \in \mathcal{A}$ coefficient $a_{\alpha,\beta} = e_\alpha \cdot a \cdot e_\beta$ of two-sided Peirce decomposition vanishes,

$$e_\alpha \cdot a \cdot e_\beta = e_\alpha \cdot e_{A_i} \cdot a \cdot e_{A_j} \cdot e_\beta = e_\alpha \cdot a \cdot e_{A_i} \cdot e_{A_j} \cdot e_\beta = 0.$$

This is consistent with (4.5.15) and tells that the two-sided Peirs decomposition (4.5.7) of semisimple algebra $\mathcal{A}$ does not contain terms $\mathcal{A}_{\alpha,\beta}$, if $\alpha \in A_i$, $\beta \in A_j$ and $i \neq j$, so that we have

$$\mathcal{A} = \bigoplus_{i=1}^{s'} \bigoplus_{\alpha,\beta \in A_i} e_\alpha \mathcal{A} e_\beta = \bigoplus_{i=1}^{s'} \bigoplus_{\alpha,\beta \in A_i} \mathcal{A}_{\alpha,\beta}, \tag{4.5.17}$$

where, again, s' is the number of primitive central idempotents e_{A_i}. So, matrices $T(a) = ||T_{\alpha\beta}(a)||$ of elements $a \in \mathcal{A}$ in representation T determined by coefficients of two-sided Peirce decomposition in (4.5.14) have block-diagonal form

$$T(a) = \bigoplus_{i=1}^{s'} T^{(A_i)}(a), \tag{4.5.18}$$

where matrices $T^{(A_i)}(a) = ||T^{(A_i)}_{\alpha\beta}(a)||$ are determined by coefficients of two-sided Peirce decomposition for ideal $\mathcal{A}_{A_i} \subset \mathcal{A}$ (cf. (4.5.14)):

$$e_\alpha \cdot a \cdot e_\beta = T^{(A_i)}_{\alpha\beta}(a) e_{\alpha,\beta}, \quad \alpha, \beta \in A_i. \tag{4.5.19}$$

This shows that representation T is reducible, and it is decomposed into direct sum of representations $T^{(A_i)}$, each constructed by using idempotent e_{A_i}. Dimension n_i of representation $T^{(A_i)}$ is equal to number $\mathrm{card}(A_i)$ of elements in subset $A_i \subset \{1, \ldots, s\}$, i.e., it is equal to the number of noncentral primitive idempotents e_α in the sum (4.5.16) with $A = A_i$.

Remark 1. The system of primitive noncentral idempotents (4.5.3), (4.5.4) is not uniquely defined. As an example, we can make transformation

$$e_\alpha \to e'_\alpha = S \cdot e_\alpha \cdot S^{-1}, \quad \forall \alpha, \tag{4.5.20}$$

where S is any invertible element in $\mathcal{A}$. This yields a new system of primitive idempotents e'_α. Note that the transformation (4.5.20) does not modify central idempotents.

Example 1. Consider finite-dimensional matrix algebra $\mathrm{Mat}_n(\mathbb{C})$. Basis in $\mathrm{Mat}_n(\mathbb{C})$ is made of matrix units e_{ij} $(i,j=1,\ldots,n)$ obeying

$$e_{ij} \cdot e_{km} = \delta_{jk} e_{im}. \tag{4.5.21}$$

In particular, e_{ii} are diagonal matrices which have 1 in ith place on diagonal and zero otherwise. Note that algebra $\mathrm{Mat}_n(\mathbb{C})$ can be viewed as an abstract algebra with basis elements obeying (4.5.21). Then matrix units e_{ij} are basis elements in defining representation of algebra $\mathrm{Mat}_n(\mathbb{C})$.

A natural choice of primitive idempotents in algebra $\mathrm{Mat}_n(\mathbb{C})$ are elements $e_i := e_{ii}$ $(i = 1,\ldots,n)$. Indeed, in view of (4.5.21), elements e_i obey (4.5.3) and (4.5.4):

$$\sum_{i=1}^{n} e_i = e, \quad e_i \cdot e_k = \delta_{ik} e_i, \tag{4.5.22}$$

where e is unit element in $\mathrm{Mat}_n(\mathbb{C})$, i.e., matrix $e = I_n$. With this system of primitive idempotents, left Peirce decomposition of an element $a = \sum_{i,j=1}^{n} a_{ij} e_{ij} \in \mathrm{Mat}_n(\mathbb{C})$, with $a_{ij} \in \mathbb{C}$, reads

$$a = a \cdot \sum_{i=1}^{n} e_i = \sum_{i=1}^{n} a_i. \tag{4.5.23}$$

Here $a_i = \sum_{k=1}^{n}(a_{ki} e_{ki})$ is an element of ith left ideal $\mathcal{L}_i$. The latter representation for $a_i \in \mathcal{L}_i$ shows that basis in $\mathcal{L}_i$ is made of vectors $\epsilon_k^{(i)} \equiv e_{ki}$ $(k = 1,\ldots,n)$, element a_i in coordinate form corresponds to ith column of matrix $||a_{kj}||$, and spaces $\mathcal{L}_i$ for all $i \in (1,\ldots,n)$ are isomorphic to $\mathbb{C}^n$. In view of (4.5.21), left action of an element $a = \sum_{j,\ell} a_{j\ell} e_{j\ell} \in \mathrm{Mat}_n(\mathbb{C})$ on $\mathcal{L}_i$ is given by

$$\epsilon_k^{(i)} \to a \cdot \epsilon_k^{(i)} = \sum_{j,\ell} a_{j\ell} e_{j\ell} \cdot \epsilon_k^{(i)} = \sum_{j} \epsilon_j^{(i)} a_{jk} \in \mathcal{L}_i. \tag{4.5.24}$$

This shows that representations of algebra $\mathrm{Mat}_n(\mathbb{C})$ in all spaces $\mathcal{L}_i = \mathbb{C}^n$ $(i = 1,2,\ldots,n)$ are equivalent to defining representation of $\mathrm{Mat}_n(\mathbb{C})$, so they are all equivalent. This means that left regular representation of $\mathrm{Mat}_n(\mathbb{C})$ is expanded in direct sum of n equivalent defining representations of dimension n. We have for any $a = \sum_{i,j} a_{ij} e_{ij} \in \mathrm{Mat}_n(\mathbb{C})$

$$e_i \cdot a \cdot e_j = a_{ij} e_{ij} \tag{4.5.25}$$

(no summation over i, j on the right-hand side), so components of two-sided Peirce decomposition (4.5.15) are one-dimensional spaces with basis vectors e_{ij}. By comparing (4.5.25) with (4.5.14), we find that the representation T (defined in (4.5.14) through two-sided Peirce decomposition) for $\mathrm{Mat}_n(\mathbb{C})$ is equivalent to defining representation of $\mathrm{Mat}_n(\mathbb{C})$, and for any element $a \in \mathrm{Mat}_n(\mathbb{C})$ it is given by matrix $T_{ij}(a) = a_{ij}$. Note that elements $e_{ij} \in \mathrm{Mat}_n(\mathbb{C})$ in defining representation can be written as given in (4.5.12):

$$e_{ij} = |i\rangle\langle j| = \vec{\epsilon}_i \otimes \vec{\epsilon}_j^{\,\mathsf{T}}, \tag{4.5.26}$$

where $\vec{\epsilon}_i$ are basis vectors with coordinates $(\vec{\epsilon}_i)_k = \delta_{ik}$. Center in algebra $\mathrm{Mat}_n(\mathbb{C})$ is trivial (consists of elements proportional to e), so this algebra does not have nontrivial (two-sided) ideals. Algebra $\mathrm{Mat}_n(\mathbb{C})$ is therefore simple.

Problem 4.5.3. Extract complete set of primitive idempotents in algebra $\mathcal{B}_+$ of upper triangle matrices with basis $\{e_{ij}\}$, where $i \leq j$. Construct pertinent left and right ideals. Show that algebra $\mathcal{B}_+$ is neither simple nor semisimple.

Remark 2. All of the above hints toward the following proposition which is valid for semisimple associative finite-dimensional algebras and is known as *Wedderburn theorem*.

Theorem 4.5.1. *Semisimple associative finite-dimensional algebra $\mathcal{A}$ over field $\mathbb{C}$ is isomorphic to direct sum of matrix algebras $\mathrm{Mat}_{n_1}(\mathbb{C}) \oplus \mathrm{Mat}_{n_2}(\mathbb{C}) \oplus \cdots \oplus \mathrm{Mat}_{n_{s'}}(\mathbb{C})$, where s' is the number of primitive central idempotents $e_{A_i} \in \mathcal{A}$, and $n_i = \mathrm{card}(A_i)$ is the number of primitive idempotents $e_\alpha \in \mathcal{A}$ entering the expansion (4.5.16) of central idempotent e_{A_i}.*

For matrix algebra $\mathrm{Mat}_{n_1}(\mathbb{C}) \oplus \cdots \oplus \mathrm{Mat}_{n_{s'}}(\mathbb{C})$ diagonal matrix units $e^{(i)}_{jj}|_{j=1,n_i} \in \mathrm{Mat}_{n_i}(\mathbb{C})$ and unit elements $e^{(i)} = \sum_{j=1}^{n_i} e^{(i)}_{jj} \in \mathrm{Mat}_{n_i}(\mathbb{C})$ are, respectively, primitive idempotents and central idempotents (see Example 1 above). Furthermore, for matrix algebra $\mathrm{Mat}_{n_1}(\mathbb{C}) \oplus \cdots \oplus \mathrm{Mat}_{n_{s'}}(\mathbb{C})$, all terms $e_\alpha \cdot \mathcal{A} \cdot e_\beta \equiv \mathcal{A}_{\alpha,\beta}$ in two-sided Peirce decomposition (4.5.7) are one dimensional. So, in view of isomorphism given in Theorem 4.5.1, all semisimple finite-dimensional algebras $\mathcal{A}$ belong to the class of algebras whose basis elements $e_{\alpha,\beta} \in \mathcal{A}_{\alpha,\beta}$ and primitive idempotents e_α have the form (4.5.12). Theorem 4.5.1 also leads to the following result (cf. Section I-4.6.2).

Proposition 4.5.2. *For semisimple finite-dimensional algebra $\mathcal{A}$, every N_ν-dimensional irreducible representation $T^{(\nu)}$ enters its regular representation $T^{(R)}$ precisely N_ν times.*

Theorem 4.5.1 and Proposition 4.5.2 also ensure that

$$\dim(\mathcal{A}) = \sum_{i=1}^{s'} n_i^2 = \sum_{i=1}^{s'} n_i \dim(T^{(A_i)}) \tag{4.5.27}$$

(cf. the first equality in (4.5.27) with (4.3.63)), where $T^{(A_i)}$ are representations given by (4.5.19).

Remark 3. Let us explain the meaning of Proposition 4.5.2 in some detail. Every left ideal $\mathcal{L}_\alpha = \mathcal{A} \cdot e_\alpha$ (and similarly every right ideal $\mathcal{R}_\alpha = e_\alpha \cdot \mathcal{A}$), generated by primitive idempotent e_α, is associated with irreducible representation of algebra $\mathcal{A}$. Indeed, one chooses basis elements ϵ_ℓ (where $\ell = 1, 2, \ldots, \dim(\mathcal{L}_\alpha)$) in linear space $\mathcal{L}_\alpha$, then left action of an element $a \in \mathcal{A}$ on $\mathcal{L}_\alpha$ is written as

$$a \cdot \epsilon_\ell = \epsilon_{\ell'} T^{(\alpha)}_{\ell'\ell}(a), \tag{4.5.28}$$

where mapping $T^{(\alpha)}$ from algebra $\mathcal{A}$ to matrices $||T^{(\alpha)}_{\ell'\ell}(a)||$ defines representation of $\mathcal{A}$. This representation is irreducible, since the left ideal $\mathcal{L}_\alpha$ is simple. Subspaces $\mathcal{A}_{\alpha,\beta}$ are one dimensional for semisimple algebra $\mathcal{A}$, so representations $T^{(\alpha)}$ for $\alpha \in A_i$ (where A_i is subset in $\{1, \ldots, s\}$ corresponding to central idempotent e_{A_i}) are equivalent to representation $T^{(A_i)}$ defined in (4.5.19). Indeed, in the case of semisimple algebra $\mathcal{A}$, one can choose basis in $\mathcal{L}_\alpha$, where $\alpha \in A_i$, consisting of n_i elements $\epsilon_\beta \equiv e_{\beta,\alpha} \in \mathcal{A}_{\beta,\alpha}$ (α is fixed, and β runs the whole subspace A_i). Then $\dim(\mathcal{L}_\alpha) = n_i$, and in analogy with (4.5.24), one has

$$\begin{aligned} a \cdot \epsilon_\beta &= \left(\sum_{i=1}^{s'} \sum_{\gamma,\delta \in A_i} e_\gamma \cdot a \cdot e_\delta \right) \cdot e_{\beta,\alpha} = \sum_{\gamma,\delta \in A_i} \langle \gamma | a | \delta \rangle e_{\gamma,\delta} \cdot e_{\beta,\alpha} \\ &= \sum_{\gamma \in A_i} \langle \gamma | a | \beta \rangle e_{\gamma,\alpha} = \sum_{\gamma \in A_i} \epsilon_\gamma T^{(A_i)}_{\gamma\beta}(a). \end{aligned} \tag{4.5.29}$$

By comparing these relations with (4.5.28), one finds that for all $\alpha \in A_i$, representations $T^{(\alpha)}$, whose number is $n_i = \mathrm{card}(A_i) = \dim(\mathcal{L}_\alpha)$, are equivalent to representation $T^{(A_i)}$ (labeled by central idempotent e_{A_i}). We emphasize the dual role of idempotent e_α in this construction. On the one hand, idempotent e_α generates irreducible representation $T^{(\alpha)}$, and, on the other hand, it is associated with basis element ϵ_α in each of the spaces $T^{(\beta)}$, where β belongs to the same set A_i as α. So, there is one-to-one correspondence between basis elements ϵ_α of the space of representation $T^{(\beta)}$ and spaces of equivalent representations $T^{(\alpha)}$. This agrees with Proposition 4.5.2.

Remark 4. Let the system of elements $e'_\alpha \in \mathcal{A}$ $(\alpha = 1, \ldots, s)$ be complete, in the sense that it generates nonintersecting left ideals $\mathcal{L}'_\alpha = \mathcal{A} \cdot e'_\alpha \subset \mathcal{A}$, and it defines an analog of left Peirce decomposition:

$$\mathcal{A} = \sum_{\alpha=1}^{s} \mathcal{A} \cdot e'_\alpha = \sum_{\alpha=1}^{s} \mathcal{L}'_\alpha. \tag{4.5.30}$$

Then unit element $e \in \mathcal{A}$ is decomposed as follows (cf. (4.5.3)):

$$e = \sum_{\alpha=1}^{s} x_\alpha \cdot e'_\alpha, \quad x_\alpha \in \mathcal{A}. \tag{4.5.31}$$

We multiply both sides of equality (4.5.31) on the left by element e'_β and obtain (see in this regard derivation of (4.5.4) from (4.5.3)) generalized orthogonality and idempotent conditions (cf. (4.5.4)):

$$e'_\beta \cdot x_\alpha \cdot e'_\alpha = \delta_{\alpha\beta} e'_\beta. \tag{4.5.32}$$

In other words, the standard orthogonality relation $e'_\alpha \cdot e'_\beta = 0$ $(\alpha \neq \beta)$ does not hold for e'_α. However, by comparing (4.5.31), (4.5.32) with formulas (4.5.3), (4.5.4) and recalling that different ideals $\mathcal{L}'_\alpha$ do not intersect, one obtains, starting from elements e'_α, complete system of orthogonal idempotents (4.5.3) and (4.5.4) by setting $e_\alpha = x_\alpha \cdot e'_\alpha$, $\forall \alpha$. Similar complete system of idempotents is constructed for right Peirce decomposition. Complete system of generalized idempotents e'_α which generates decompositions (4.5.30), (4.5.31) and obeys (4.5.32), is of importance for our purposes, since, as we find in Section 4.5.2, Young symmetrizers define just this kind of system.[q]

Example 2. Consider symmetric group S_2 that consists of two elements $\{e, \sigma_1\}$, where e is unit element and σ_1 is neighboring transposition

[q]See step 4 in recipe for constructing primitive orthogonal idempotents in $\mathbb{C}[S_r]$, Section 4.5.2.

$\sigma_1 = (1,2)$. Orthogonal primitive idempotents in $\mathbb{C}[S_2]$ are

$$\begin{gathered} e_{[2]} = \frac{1}{2}(e+\sigma_1), \quad e_{[1^2]} = \frac{1}{2}(e-\sigma_1), \\ e_{[2]} \cdot e_{[1^2]} = 0, \quad (e_{[2]})^2 = e_{[2]}, \quad (e_{[1^2]})^2 = e_{[1^2]}, \quad e_{[2]} + e_{[1^2]} = e. \end{gathered} \tag{4.5.33}$$

Idempotents $e_{[2]}$ and $e_{[1^2]}$ are central, since they commute with the only generator σ_1. Transposition $\sigma_1 = (1,2)$ acts on idempotents $e_{[2]}$ and $e_{[1^2]}$ as follows:

$$\sigma_1 \cdot e_{[2]} = e_{[2]}, \quad \sigma_1 \cdot e_{[1^2]} = -e_{[1^2]}, \tag{4.5.34}$$

i.e., it leaves $e_{[2]}$ intact and changes sign of $e_{[1^2]}$. Therefore, in accordance with definitions (4.3.4) and (4.3.5), $e_{[2]}$ and $e_{[1^2]}$ are *symmetrizer* and *antisymmetrizer* in $\mathbb{C}[S_2]$, respectively (cf. (4.3.16)). In accordance with (4.5.33), algebra $\mathbb{C}[S_2]$ is expanded in direct sum of its one-dimensional two-sided ideals $\mathcal{L}_{[2]}$ and $\mathcal{L}_{[1^2]}$, so regular representation of $\mathbb{C}[S_2]$ is decomposed into just two irreducible one-dimensional representations defined by formulas (4.5.34):

$$(1)\ T^{[2]}(e) = 1, \quad T^{[2]}(\sigma_1) = 1; \quad (2)\ T^{[1^2]}(e) = 1, \quad T^{[1^2]}(\sigma_1) = -1.$$

Idempotents $e_{[2]}$ and $e_{[1^2]}$ that generate representations $T^{[2]}$ and $T^{[1^2]}$, modulo factor 1/2 coincide with two Young symmetrizers (4.3.14) of $\mathbb{C}[S_2]$, which are associated with Young tableaux

$$T^{[2]} \leftrightarrow \mathsf{T}_{[2]} = \boxed{1\,|\,2}, \quad T^{[1^2]} \leftrightarrow \mathsf{T}_{[1^2]} = \begin{array}{|c|}\hline 1 \\ \hline 2 \\ \hline \end{array}. \tag{4.5.35}$$

Note that idempotents (4.5.33), in matrix representation s, give, in accordance with (4.2.9) and (4.2.10), projectors (4.2.23).

Example 3. Consider symmetric group S_3 and its presentation (4.3.13) in terms of two Coxeter generators σ_1, σ_2. Six elements of group S_3 in this presentation are as follows:

$$e, \ \sigma_1 = (1,2), \quad \sigma_2 = (2,3), \ \sigma_1\sigma_2, \ \sigma_2\sigma_1, \quad \sigma_1\sigma_2\sigma_1 = \sigma_2\sigma_1\sigma_2. \tag{4.5.36}$$

One chooses the following elements as complete set of primitive idempotents:

$$\begin{aligned} e_{[3]} &= \frac{1}{6}(1+\sigma_2+\sigma_1\sigma_2)(1+\sigma_1), \\ e_{[1^3]} &= \frac{1}{6}(1-\sigma_2+\sigma_1\sigma_2)(1-\sigma_1), \\ e_{[2,1]_1} &= \frac{1}{3}(1-\sigma_2+\varepsilon(1-\sigma_1)\sigma_2)(1+\sigma_1), \\ e_{[2,1]_2} &= \frac{1}{3}(1-\sigma_1)(1+\sigma_2-\varepsilon\sigma_2(1+\sigma_1)), \end{aligned} \tag{4.5.37}$$

where ε is an arbitrary parameter, and we write the unit element e as 1 for brevity. Generators $\sigma_i = (\sigma_1, \sigma_2)$ act on the left and on the right on idempotents $e_{[3]}$ and $e_{[1^3]}$ as follows:

$$\sigma_i \cdot e_{[3]} = e_{[3]} \cdot \sigma_i = e_{[3]}, \quad \sigma_i \cdot e_{[1^3]} = e_{[1^3]} \cdot \sigma_i = -e_{[1^3]}, \tag{4.5.38}$$

so idempotents $e_{[3]}$ and $e_{[1^3]}$ are central and are *symmetrizer* and *antisymmetrizer* in $\mathbb{C}[S_3]$ (cf. the first two formulas in (4.5.37) with definition (4.3.16); see also Problems 4.3.2, 4.3.3). Idempotents $e_{[2,1]_1}$ and $e_{[2,1]_2}$ are not central, and their dependence on arbitrary parameter ε shows that the choice of noncentral idempotents is not unique (see Remark **1** above).

Problem 4.5.4. Making use of defining relations (4.3.13), check that elements (4.5.37) make complete system of orthogonal idempotents in $\mathbb{C}[S_3]$, i.e., check equalities (4.5.3) and (4.5.4). Check formulas (4.5.38). Hint: make use of projector properties of operators $P_i^{\mp} = (1 \pm \sigma_i)$, namely, $P_i^+ P_i^- = 0$, $P_i^{\pm}\sigma_i = \pm P_i^{\pm}$.

Central idempotents $e_{[3]}$ and $e_{[1^3]}$ define one-dimensional (two-sided) ideals in $\mathbb{C}[S_3]$. Idempotent $e_{[3]}$ defines trivial one-dimensional representation $T^{([3])}(\sigma) = +1$ ($\forall \sigma \in S_3$), while idempotent $e_{[1^3]}$ defines one-dimensional representation $T^{([1^3])}(\sigma) = +1$ for even $\sigma \in S_3$ and $T^{([1^3])}(\sigma) = -1$ for odd $\sigma \in S_3$. Partition of unity

$$e = e_{[3]} + e_{[1^3]} + e_{[2,1]_1} + e_{[2,1]_2},$$

shows that the sum (cf. (4.5.16))

$$e_{[2,1]_1} + e_{[2,1]_2} = \frac{1}{3}(2 - \sigma_1\sigma_2 - \sigma_2\sigma_1) =: e_{[2,1]} \tag{4.5.39}$$

is also central idempotent, so that the identity $e = e_{[3]} + e_{[1^3]} + e_{[2,1]}$ is central partition of unity. Note that central idempotent $e_{[2,1]}$ is independent of free parameter ε, as anticipated: central idempotents are uniquely defined.

Consider now, as an example, left ideal $\mathcal{L}_{[2,1]_1}$ generated by noncentral idempotent $e_{[2,1]_1}$. We write this idempotent as

$$e_{[2,1]_1} = (\epsilon_1 + \varepsilon\epsilon_2), \tag{4.5.40}$$

where two elements

$$\epsilon_1 = \frac{1}{3}(1-\sigma_2)(1+\sigma_1), \quad \epsilon_2 = \frac{1}{3}(1-\sigma_1)\sigma_2(1+\sigma_1) \tag{4.5.41}$$

belong to $\mathcal{L}_{[2,1]_1}$ and can be chosen as basis in $\mathcal{L}_{[2,1]_1}$. Indeed, elements (4.5.41) obey the relations

$$\epsilon_1 \cdot \epsilon_1 = \epsilon_1, \quad \epsilon_2 \cdot \epsilon_2 = 0, \quad \epsilon_1 \cdot \epsilon_2 = 0, \quad \epsilon_2 \cdot \epsilon_1 = \epsilon_2, \tag{4.5.42}$$

which can be obtained by the left action of generators $\sigma_1, \sigma_2 \in S_3$ on elements (4.5.41) (cf. (4.5.28)):

$$\sigma_1 \cdot \epsilon_1 = \epsilon_1 + \epsilon_2, \quad \sigma_1 \cdot \epsilon_2 = -\epsilon_2, \quad \sigma_2 \cdot \epsilon_1 = -\epsilon_1, \quad \sigma_2 \cdot \epsilon_2 = \epsilon_1 + \epsilon_2. \tag{4.5.43}$$

This shows that any element $x \in \mathcal{L}_{[2,1]_1} = \mathbb{C}[S_3] \cdot e_{[2,1]_1}$ is a linear combination $x = x_1\epsilon_1 + x_2\epsilon_2$, $(x_1, x_2 \in \mathbb{C})$. Thus, left ideal $\mathcal{L}_{[2,1]_1}$ is two dimensional.

Relations (4.5.43) yield (in accordance with the rule (4.5.28)) two dimensional irreducible representation of group S_3:

$$T^{([2,1])}(\sigma_1) = \begin{pmatrix} 1 & 0 \\ 1 & -1 \end{pmatrix}, \quad T^{([2,1])}(\sigma_2) = \begin{pmatrix} -1 & 1 \\ 0 & 1 \end{pmatrix}. \tag{4.5.44}$$

Problem 4.5.5. Check the relations (4.3.13) for representation (4.5.44) of group S_3.

Problem 4.5.6. Derive the relations (4.5.42) for generators ϵ_1, ϵ_2 of subalgebra (left ideal) $\mathcal{L}_{[2,1]_1}$. Hint: make use of the formulas (4.5.43) and identity $(e_{[2,1]_1})^2 = e_{[2,1]_1}$, where element $e_{[2,1]_1}$ is defined by (4.5.40) with free parameter ε.

Consider now the second left ideal $\mathcal{L}_{[2,1]_2} \subset \mathbb{C}[S_3]$, generated by idempotent $e_{[2,1]_2}$ given by (4.5.37). We set $\varepsilon = 0$ for definiteness. This left ideal is also two dimensional, and one can chose its basis elements as

$$\epsilon_3 = \frac{1}{3}(1-\sigma_1)(1+\sigma_2), \quad \epsilon_4 = \frac{1}{3}(1-\sigma_2)\sigma_1(1+\sigma_2). \tag{4.5.45}$$

These obey

$$\sigma_1 \cdot \epsilon_3 = -\epsilon_3, \quad \sigma_1 \cdot \epsilon_4 = \epsilon_3 + \epsilon_4, \quad \sigma_2 \cdot \epsilon_3 = \epsilon_3 + \epsilon_4, \quad \sigma_2 \cdot \epsilon_4 = -\epsilon_4, \tag{4.5.46}$$

$$\epsilon_3 \cdot \epsilon_3 = \epsilon_3, \quad \epsilon_4 \cdot \epsilon_4 = 0, \quad \epsilon_3 \cdot \epsilon_4 = 0, \quad \epsilon_4 \cdot \epsilon_3 = \epsilon_4. \tag{4.5.47}$$

Then the two-sided ideal $\mathcal{L}_{[2,1]}$ generated by central idempotent (4.5.39) is four dimensional, and its basis elements $\epsilon_i \in \mathcal{L}_{[2,1]}$ $(i = 1, \ldots, 4)$ obey the relations (4.5.42), (4.5.47) of two subalgebras $\mathcal{L}_{[2,1]_1}$, $\mathcal{L}_{[2,1]_2}$ as well as crossing formulas

$$\begin{aligned} &\epsilon_1 \cdot \epsilon_3 = \epsilon_3 \cdot \epsilon_1 = 0, \quad \epsilon_2 \cdot \epsilon_3 = \epsilon_4 \cdot \epsilon_1 = 0, \\ &\epsilon_3 \cdot \epsilon_2 = \epsilon_2, \epsilon_1 \cdot \epsilon_4 = \epsilon_4, \quad \epsilon_4 \cdot \epsilon_2 = \epsilon_1, \quad \epsilon_2 \cdot \epsilon_4 = \epsilon_3. \end{aligned} \tag{4.5.48}$$

Formulas (4.5.42), (4.5.47) and (4.5.48) can be written in unified way, analogous to relation for matrix units

$$e_{ij} \cdot e_{k\ell} = \delta_{jk} e_{i\ell}, \tag{4.5.49}$$

where we set $\epsilon_1 \equiv e_{11}$, $\epsilon_2 \equiv e_{21}$, $\epsilon_3 \equiv e_{22}$ and $\epsilon_4 \equiv e_{12}$. It follows from (4.5.49) that two-sided ideal $\mathcal{L}_{[2,1]} \subset \mathbb{C}[S_3]$ is isomorphic to algebra $\mathrm{Mat}_2(\mathbb{C})$; it does not have nontrivial two-sided ideals and hence is a simple subalgebra.

Problem 4.5.7. * Prove that left ideal $\mathcal{L}_{[2,1]_2}$ is two dimensional and (for $\varepsilon = 0$) its basis elements can be chosen as in (4.5.45). Show that these elements obey (4.5.46), (4.5.47) and (4.5.48). Prove equivalence of representations determined by left ideals $\mathcal{L}_{[2,1]_1}$ and $\mathcal{L}_{[2,1]_2}$.

We recall that $\epsilon_1 = e_{[2,1]_1}|_{\varepsilon=0}$ and $\epsilon_3 = e_{[2,1]_2}|_{\varepsilon=0}$ are also primitive idempotents in $\mathbb{C}[S_3]$. Matrix coefficients (4.5.13) of two-sided Peirce decomposition (4.5.7) of elements σ_1, σ_2 are

$$\begin{pmatrix} \epsilon_1\sigma_1\epsilon_1 & \epsilon_1\sigma_1\epsilon_3 \\ \epsilon_3\sigma_1\epsilon_1 & \epsilon_3\sigma_1\epsilon_3 \end{pmatrix} = \begin{pmatrix} \epsilon_1 & 0 \\ \epsilon_2 & -\epsilon_3 \end{pmatrix} = \begin{pmatrix} e_{11} & 0 \\ e_{21} & -e_{22} \end{pmatrix} \Rightarrow T(\sigma_1) = \begin{pmatrix} 1 & 0 \\ 1 & -1 \end{pmatrix},$$

$$\begin{pmatrix} \epsilon_1\sigma_2\epsilon_1 & \epsilon_1\sigma_2\epsilon_3 \\ \epsilon_3\sigma_2\epsilon_1 & \epsilon_3\sigma_2\epsilon_3 \end{pmatrix} = \begin{pmatrix} -\epsilon_1 & \epsilon_4 \\ 0 & \epsilon_3 \end{pmatrix} = \begin{pmatrix} -e_{11} & e_{12} \\ 0 & e_{22} \end{pmatrix} \Rightarrow T(\sigma_2) = \begin{pmatrix} -1 & 1 \\ 0 & 1 \end{pmatrix}.$$

where irreducible representation T of group S_3 is obtained by applying the rule (4.5.19) and coincides (see Remark 3 above) with representation (4.5.44).

All that we have discussed in this Example 3 leads us to the conclusion that six-dimensional algebra $\mathbb{C}[S_3]$ with basis $\{e_{[3]}, e_{[1^3]}, \epsilon_1, \epsilon_2, \epsilon_3, \epsilon_4\}$ is decomposed in direct sum of three simple ideals $\mathcal{L}_{[3]}$, $\mathcal{L}_{[1^3]}$ and $\mathcal{L}_{[2,1]}$, which are isomorphic to matrix algebras $\mathrm{Mat}_1(\mathbb{C})$, $\mathrm{Mat}_1(\mathbb{C})$ and $\mathrm{Mat}_2(\mathbb{C})$. This means that algebra $\mathbb{C}[S_3]$ is semisimple. The decomposition

$$\mathbb{C}[S_3] = \mathrm{Mat}_1(\mathbb{C}) \oplus \mathrm{Mat}_1(\mathbb{C}) \oplus \mathrm{Mat}_2(\mathbb{C}) \tag{4.5.50}$$

is in accordance with Wedderburn theorem (Theorem 4.5.1).

Remark 5. We have shown in Section 4.3 that one can extract irreducible representations $T^{[3]}$, $T^{[1^3]}$ and $T^{[2,1]_1} \sim T^{[2,1]_2}$ of algebra $\mathbb{C}[S_3]$ from regular representation by Young symmetrizers associated with tableaux that have the shapes of Young diagrams

$[3] =$ 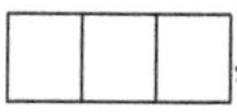, $[1^3] =$ 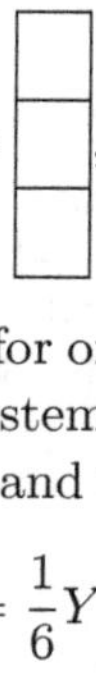, $[2,1] =$ 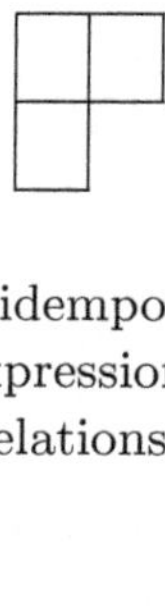.

We compare explicit formulas (4.5.37) for orthogonal idempotents $e_{[3]}$, $e_{[1^3]}$, $e_{[2,1]_1}$, $e_{[2,1]_2}$, that make complete system, with expressions (4.3.15) for Young symmetrizers of algebra $\mathbb{C}[S_3]$, and find the relations

$$\begin{gathered} e_{[3]} = \frac{1}{6} Y(\mathsf{T}_{[3]}), \quad e_{[1^3]} = \frac{1}{6} Y(\mathsf{T}_{[1^3]}), \\ e_{[2,1]_1}|_{\varepsilon=0} = \frac{1}{3} Y(\mathsf{T}_{[2,1]_1}), \quad e_{[2,1]_2}|_{\varepsilon=0} = \frac{1}{3} Y(\mathsf{T}_{[2,1]_2}). \end{gathered} \tag{4.5.51}$$

Remark 6. As a matter of fact, we have studied in the latter example the same problem as in item 2 of Section I-4.6.1 of the first book: we decomposed the regular representation of group S_3 in irreducible representations. The method we used here is different from direct method employed in the first book. The method explained in Example 3 for group S_3, based on extraction of primitive orthogonal idempotents, is universal and, as we see in Section 4.6, is applicable to groups S_r for all r.

Remark 7. Isomorphism given in (4.5.50) is a special case of general statement which we formulate without proof, but which in fact follows from the discussions in Sections 4.3.2 and 4.5.1.

Proposition 4.5.3. *Group algebra $\mathbb{C}[S_r]$ (for all $r \geq 2$) is semisimple and is isomorphic to matrix algebra*

$$\mathbb{C}[S_r] = \bigoplus_{\lambda \vdash r} \mathrm{Mat}_{f_{(\lambda)}}(\mathbb{C}),$$

where $f_{(\lambda)}$ are dimensions of irreducible representations of $\mathbb{C}[S_r]$, which are associated with Young diagrams $\lambda \vdash r$ (see formulas (4.3.56) *and* (4.3.57)).

We give further arguments in favor of this proposition in the following subsection.

4.5.2. Orthogonality and completeness of Young symmetrizers

We now make use of the results of Section 4.3 to study one of the methods of constructing, in terms of Young symmetrizers, complete orthogonal system of primitive idempotents in algebra $\mathbb{C}[S_r]$. This method is largely based on works by A. Young and F. G. Frobenius, and its main ingredient is the notion of standard Young tableaux. We first give the pertinent recipe, and then consider details of construction. In our analysis, we make use of the general theory developed in Section 4.5.1.

We recall (see Definition 4.3.2 in Section 4.3.2) that Young tableau T_λ whose numbers increase from left to right in each row and from top to bottom in each column is *standard tableau*. Examples of nonstandard and standard tableaux are given in (4.3.54). As we see below, the number of standard Young tableaux of a given shape λ is equal to dimension $f_{(\lambda)}$ of irreducible representation of symmetric group S_r, which is associated with Young diagram λ. This dimension is given by hook formula (4.3.56) or Frobenius formula (4.3.57); see Section 4.3.2.

We now give the recipe of construction of complete orthogonal system of primitive idempotents for algebra $\mathbb{C}[S_r]$:

(1) *Find all partitions* $\lambda \vdash r$ *of number* r, *or, in other words, all Young diagrams with* r *boxes, and then construct all standard Young tableaux* $\mathsf{T}_{a,\lambda}$ $(a = 1,\ldots,f_{(\lambda)})$ *for each diagram* λ.

(2) *For each standard tableau* $\mathsf{T}_{a,\lambda}$ *calculate horizontal symmetrizer* P, *vertical antisymmetrizer* Q, *and then Young symmetrizer* $Y(\mathsf{T}_{a,\lambda}) = Q \cdot P$ *(see Section* 4.3.1*).*

(3) *Using these Young symmetrizers* $Y(\mathsf{T}_{a,\lambda})$, *construct the system of Young idempotents* $E(\mathsf{T}_{a,\lambda}) = \alpha_\lambda^{-1} Y(\mathsf{T}_{a,\lambda})$, *where* $\alpha_\lambda = r!/f_{(\lambda)}$, *see* (4.3.36)*. This system is complete but not orthogonal yet.*

(4) *Orthogonalize this system by constructing new idempotents of the form*

$$E'(\mathsf{T}_{a,\lambda}) = X_{a,\lambda} \cdot E(\mathsf{T}_{a,\lambda}), \quad E''(\mathsf{T}_{a,\lambda}) = E(\mathsf{T}_{a,\lambda}) \cdot \tilde{X}_{a,\lambda}, \tag{4.5.52}$$

where $X_{a,\lambda}$ *and* $\tilde{X}_{a,\lambda}$ *are appropriate elements of algebra* $\mathbb{C}[S_r]$; *see Remark* 4 *in Section* 4.5.1*. The procedure of evaluating the factors* $X_{a,\lambda}$ *and* $\tilde{X}_{a,\lambda}$ *is described below in this Section (see Eqs.* (4.5.70) – (4.5.71)*) and is based on ordering (dominance) in the set of standard Young tableaux. The idempotent systems* (4.5.52) *are complete and orthogonal,*

$$\begin{aligned} &E'(\mathsf{T}_{a,\lambda}) \cdot E'(\mathsf{T}_{b,\mu}) = E'(\mathsf{T}_{b,\lambda})\delta_{ab}\delta_{\lambda\mu}, \\ &E''(\mathsf{T}_{a,\lambda}) \cdot E''(\mathsf{T}_{b,\mu}) = E''(\mathsf{T}_{b,\lambda})\delta_{ab}\delta_{\lambda\mu}, \\ &\lambda \vdash r, \quad \mu \vdash r, \quad a = 1,\ldots,f_{(\lambda)}, \quad b = 1,\ldots,f_{(\mu)}. \end{aligned} \tag{4.5.53}$$

This is the final step in the solution of the whole problem.

A drawback of this method is the necessity to evaluate factors $X_{a,\lambda}, \tilde{X}_{a,\lambda} \in \mathbb{C}[S_r]$ to define the modified Young idempotents (4.5.52). In Section 4.6 below, we consider alternative method of constructing complete and orthogonal system of primitive idempotents $e(\mathsf{T}_{a,\lambda}) \in \mathbb{C}[S_r]$ (these are, generally speaking, different from (4.5.52)), which does not require evaluation of modification factors $X_{a,\lambda}, \tilde{X}_{a,\lambda}$. Completeness and orthogonality properties for idempotents $e(\mathsf{T}_{a,\lambda})$ are automatic in that approach.

Nevertheless, in this section we continue with the construction started in Section 4.3, and describe the way to extract, from the entire set of Young symmetrizers $Y(\mathsf{T}_\lambda)$, the subset that forms complete system of idempotents in algebra $\mathbb{C}[S_r]$. We also consider the problem of orthogonality of the latter idempotents.

To proceed, we point out that left (right) ideals generated by Young symmetrizers $Y(\mathsf{T}_\lambda)$ and $Y(\mathsf{T}'_\lambda)$ associated with different tableaux T_λ and T'_λ of the same shape λ may coincide. As an example, choosing in (4.3.22) horizontal permutation $\sigma = p \in H(\mathsf{T}_\lambda)$ or vertical permutation

$\sigma = q \in V(\mathsf{T}_\lambda)$, we obtain

$$\begin{aligned} Y(p\cdot \mathsf{T}_\lambda) &= p\cdot Y(\mathsf{T}_\lambda)\cdot p^{-1} = p\cdot Y(\mathsf{T}_\lambda), \\ Y(q\cdot \mathsf{T}_\lambda) &= q\cdot Y(\mathsf{T}_\lambda)\cdot q^{-1} = \delta_q Y(\mathsf{T}_\lambda)\cdot q^{-1}. \end{aligned} \tag{4.5.54}$$

This follows from (4.3.12). In view of (4.5.54), left ideals $\mathcal{L}_{p\cdot\mathsf{T}_\lambda}$ generated by idempotents $Y(p\cdot\mathsf{T}_\lambda)$ coincide, for all $p\in H(\mathsf{T}_\lambda)$, with left ideal $\mathcal{L}_{\mathsf{T}_\lambda}$ generated by idempotent $Y(\mathsf{T}_\lambda)$, and right ideals $\mathcal{R}_{q\cdot\mathsf{T}_\lambda}$ generated by idempotents $Y(q\cdot\mathsf{T}_\lambda)$ coincide, for all $q\in V(\mathsf{T}_\lambda)$, with right ideal $\mathcal{R}_{\mathsf{T}_\lambda}$ generated by $Y(\mathsf{T}_\lambda)$,

$$\mathbb{C}[S_r]\cdot Y(p\cdot\mathsf{T}_\lambda) = \mathbb{C}[S_r]\cdot Y(\mathsf{T}_\lambda), \quad Y(q\cdot\mathsf{T}_\lambda)\cdot\mathbb{C}[S_r] = Y(\mathsf{T}_\lambda)\cdot\mathbb{C}[S_r]. \tag{4.5.55}$$

Therefore, Young symmetrizers $Y(\mathsf{T}_\lambda)$, $Y(p\cdot\mathsf{T}_\lambda)$ and $Y(q\cdot\mathsf{T}_\lambda)$ should not be included together in complete system of independent idempotents used to construct expansions (4.3.33), (4.5.5). As we see below, one way of removing these redundant idempotents is to choose only those Young symmetrizers that are associated with standard Young tableaux.

We emphasize that so far we define the completeness of the system of idempotents $\alpha_\lambda^{-1}Y(\mathsf{T}_\lambda)$ as completeness of the system of ideals generated by $Y(\mathsf{T}_\lambda)$ (see Remark 4 in Section 4.5.1). In other words, we do not require yet the orthogonality property (4.5.4). It is sufficient at this point that the idempotents generate nonintersecting ideals in $\mathbb{C}[S_r]$ which cover the whole algebra $\mathbb{C}[S_r]$, i.e., the idempotents define expansion with properties (4.5.30) and (4.5.31).

Clearly, there exist Young tableaux which cannot be transformed into standard ones by only horizontal, or only vertical permutation. As an example we take tableaux (4.3.54) and consider the left tableau T'_μ below

$$\mathsf{T}'_\mu = \begin{array}{|c|c|c|c|c|c|} \hline 5 & 17 & 18 & 13 & 15 & 10 \\ \hline 6 & 16 & 2 & 12 & 8 \\ \cline{1-5} 14 & 3 & 1 & 7 \\ \cline{1-4} 11 & 9 & 4 \\ \cline{1-3} \end{array} \xrightarrow{p'} \begin{array}{|c|c|c|c|c|c|} \hline 5 & 10 & 13 & 15 & 17 & 18 \\ \hline 2 & 6 & 8 & 12 & 16 \\ \cline{1-5} 1 & 3 & 7 & 14 \\ \cline{1-4} 4 & 9 & 11 \\ \cline{1-3} \end{array} \xrightarrow{q} \begin{array}{|c|c|c|c|c|c|} \hline 1 & 3 & 7 & 12 & 16 & 18 \\ \hline 2 & 6 & 8 & 14 & 17 \\ \cline{1-5} 4 & 9 & 11 & 15 \\ \cline{1-4} 5 & 10 & 13 \\ \cline{1-3} \end{array} = \mathsf{T}_\mu. \tag{4.5.56}$$

We need composition of two permutations $q\cdot p' = p'\cdot q'$ to transform T'_μ into standard tableau T_μ:

$$\mathsf{T}_\mu = q\cdot p'\cdot\mathsf{T}'_\mu = p'\cdot q'\cdot\mathsf{T}'_\mu,$$

where p' is horizontal permutation that orders numbers in rows of the tableau T'_μ from left to right, while q is vertical permutation that orders numbers in columns of the tableau $p' \cdot \mathsf{T}'_\mu$ from top to bottom. In accordance with (4.3.19), we use the notation $q' = p'^{-1} \cdot q \cdot p'$ for vertical permutation that permutes numbers in columns of tableau T'_μ.

Problem 4.5.8. Show that vertical permutation $q' = p'^{-1} \cdot q \cdot p' \in V(\mathsf{T}'_\mu)$, where p' and q are given in (4.5.56), reads

$$q' = \begin{pmatrix} 1 & 2 & 3 & 4 & 5 & 6 & 7 & 8 & 9 & 10 & 11 & 12 & 13 & 14 & 15 & 16 & 17 & 18 \\ 4 & 2 & 9 & 18 & 14 & 6 & 13 & 15 & 17 & 10 & 5 & 7 & 12 & 11 & 8 & 16 & 3 & 1 \end{pmatrix}.$$

The example (4.5.56), in fact, is of general character.

Proposition 4.5.4. *Any tableau T'_λ can be transformed into standard tableau T_λ of the same shape λ by two consecutive permutations p' and q, where p' is horizontal permutation that arranges numbers in rows of tableau T'_λ in ascending order from left to right, while q is vertical permutation that orders numbers in columns of tableau $p' \cdot \mathsf{T}'_\lambda$ in ascending order from top to bottom. Thus,*

$$\mathsf{T}_\lambda = q \cdot p' \cdot \mathsf{T}'_\lambda \Rightarrow \mathsf{T}'_\lambda = (p')^{-1} \cdot q^{-1} \cdot \mathsf{T}_\lambda = q^{-1} \cdot p^{-1} \cdot \mathsf{T}_\lambda, \tag{4.5.57}$$

where T_λ is standard tableau, and $p = q \cdot p' \cdot q^{-1}$ is horizontal permutation of tableau T_λ.

Proof. We prove this proposition *ad absurdum*. We denote the number in box (i,j) of tableau T_λ by $a_{(i,j)}$, where i and j label row and column, respectively. Let the tableau T_λ obtained after the second, vertical permutation q that orders numbers in columns of tableau $p' \cdot \mathsf{T}'_\lambda$, be nonstandard. This means that there is a row i in T_λ where larger number $a_{(i,j)}$ is on the left of smaller number $a_{(i,k)}$, i.e., $k > j$, while $a_{(i,k)} < a_{(i,j)}$. Now, as a result of permutation q, all numbers $\{a_{(1,k)}, \ldots, a_{(i,k)}\}$ (the amount of which is equal to i) in kth column of T_λ are smaller than numbers $\{a_{(i,j)}, a_{(i+1,j)}, \ldots\}$ in jth column:

$$a_{(1,k)} < a_{(2,k)} < \cdots < a_{(i,k)} < a_{(i,j)} < a_{(i+1,j)} < \cdots. \tag{4.5.58}$$

Since the first permutation p' ordered the numbers in rows of tableau T'_λ from left to right, jth column of tableau T_λ contains at least i numbers, which are smaller than numbers $\{a_{(1,k)}, \ldots, a_{(i,k)}\}$ in kth column of T_λ. However, in view of (4.5.58), numbers in jth column of tableau T_λ, which are smaller than $\{a_{(1,k)}, \ldots, a_{(i,k)}\}$, can occupy only the places above the box (i,j), and their amount is at most $(i-1)$. So, we come to contradiction, i.e., the tableau T_λ is standard. The fact that permutation $p = q \cdot p' \cdot q^{-1}$ belongs to $H(\mathsf{T}_\lambda)$ is a consequence of (4.3.18). □

Corollary 4.5.5. *The whole set of tableaux of shape $\lambda \vdash r$ is divided into classes $\mathcal{M}(\mathsf{T}_\lambda)$, each containing tableaux T'_λ, which, upon two permutations p' and q in (4.5.57), is transformed into standard tableau T_λ. Young symmetrizers associated with standard tableau T_λ and tableau $\mathsf{T}'_\lambda \in \mathcal{M}(\mathsf{T}_\lambda)$ are related by*

$$Y(\mathsf{T}'_\lambda) = q^{-1} \cdot p^{-1} \cdot Y(\mathsf{T}_\lambda) \cdot p \cdot q = q^{-1} \cdot p^{-1} \cdot Y(\mathsf{T}_\lambda) \cdot q \tag{4.5.59}$$

(*here $p \in H(\mathsf{T}_\lambda)$ and $q \in V(\mathsf{T}_\lambda)$). These symmetrizers are not orthogonal:*

$$Y(\mathsf{T}'_\lambda) \cdot Y(\mathsf{T}_\lambda) = q^{-1} \cdot p^{-1} \cdot Y(\mathsf{T}_\lambda) \cdot q \cdot Y(\mathsf{T}_\lambda) = \delta_q \alpha_\lambda q^{-1} \cdot p^{-1} \cdot Y(\mathsf{T}_\lambda) \neq 0. \tag{4.5.60}$$

Remark 1. Classes $\mathcal{M}(\mathsf{T}_\lambda)$ are *not* equivalence classes. In particular, for some Young diagrams, $\lambda \vdash r$ there exist two *standard* tableaux T_λ and T'_λ, such that $\mathsf{T}'_\lambda \in \mathcal{M}(\mathsf{T}_\lambda)$, but $\mathsf{T}_\lambda \notin \mathcal{M}(\mathsf{T}'_\lambda)$. Consider an example of two standard tableaux of shape $\lambda = [3,2]$:

$$\mathsf{T}_\lambda = \begin{array}{|c|c|c|}\hline 1 & 2 & 3 \\ \hline 4 & 5 & \\ \cline{1-2}\end{array}\,, \quad \mathsf{T}'_\lambda = \begin{array}{|c|c|c|}\hline 1 & 3 & 5 \\ \hline 2 & 4 & \\ \cline{1-2}\end{array}\,. \tag{4.5.61}$$

These are related by (4.5.57) with $q = (2,5)$, $p' = (2,4)\cdot(3,5)$ and $p = q \cdot p' \cdot q^{-1} = (2,3)\cdot(4,5)$, so that $\mathsf{T}'_\lambda \in \mathcal{M}(\mathsf{T}_\lambda)$. However, $\mathsf{T}_\lambda \notin \mathcal{M}(\mathsf{T}'_\lambda)$, since there are two numbers 1 and 2 in one row of T_λ and in one column of T'_λ. This implies that there are no consecutive permutations $p \in H(\mathsf{T}_\lambda)$ and $q' \in V(\mathsf{T}'_\lambda)$ such that $\mathsf{T}'_\lambda = q' \cdot p \cdot \mathsf{T}_\lambda$. Note that, in accordance with Corollary 4.3.4, we have $Y(\mathsf{T}_\lambda) \cdot Y(\mathsf{T}'_\lambda) = 0$. However, Young symmetrizers associated with standard tableaux (4.5.61) are not orthogonal, since their product in reverse order, again due to Corollary 4.3.4 obeys (4.5.57), (4.5.59) and (4.5.60), which give

$$\begin{aligned} Y(\mathsf{T}'_\lambda) \cdot Y(\mathsf{T}_\lambda) &= -\alpha_{[3,2]}(2,5)\cdot(4,5)\cdot(2,3)\cdot Y(\mathsf{T}_\lambda) \\ &= \alpha_{[3,2]}(2,4)\cdot(3,5)\cdot Y(\mathsf{T}_\lambda) \neq 0, \end{aligned} \tag{4.5.62}$$

where constant $\alpha_{[3,2]} = 5!/5 = 24$ is obtained from (4.3.36). In the last equality in (4.5.62), we move transposition $(2,5)$ to the right and use the first identity in (4.3.12), which in our case reads $(2,5)\cdot Y(\mathsf{T}_\lambda) = -Y(\mathsf{T}_\lambda)$.

Problem 4.5.9. Check the relation (4.5.62) by inserting explicit formulas for symmetrizers $Y(\mathsf{T}_\lambda)$ and $Y(\mathsf{T}'_\lambda)$, associated with tableaux (4.5.61).

So, the system of Young symmetrizers associated with *standard Young tableaux*, pretends to form complete system of idempotents. However, in view of Remark 1, this system is not, generally speaking, orthogonal.

Remark 2. We pointed out in Section 4.3.2 that for any Young symmetrizer $Y(\mathsf{T}_\lambda)$ and any $x \in \mathbb{C}[S_r]$, the following relation holds (see (4.3.43)) $Y(\mathsf{T}_\lambda) \cdot x \cdot Y(\mathsf{T}_\lambda) = \alpha_\lambda(x) Y(\mathsf{T}_\lambda)$, where $\alpha_\lambda(x)$ is a complex number. Let $\alpha_\lambda(x) \neq 0$ for an element x. Then, we multiply the latter relation by x on the right (or left), and we find that elements

$$Y'(\mathsf{T}_\lambda) = x \cdot Y(\mathsf{T}_\lambda), \quad Y''(\mathsf{T}_\lambda) = Y(\mathsf{T}_\lambda) \cdot x, \tag{4.5.63}$$

are again proportional to idempotents. With this modification, left ideals generated by symmetrizers $Y(\mathsf{T}_\lambda)$ and $Y'(\mathsf{T}_\lambda)$ (and right ideals generated by symmetrizers $Y(\mathsf{T}_\lambda)$ and $Y''(\mathsf{T}_\lambda)$) coincide, so the first expansion in (4.5.5) (in left ideals) does not change, if we use idempotent $\alpha_\lambda(x)^{-1} Y'(\mathsf{T}_\lambda)$ instead of $E(Y(\mathsf{T}_\lambda)) = \alpha_\lambda^{-1} Y(\mathsf{T}_\lambda)$. Likewise, the second expansion (in right ideals) in (4.5.5) does not change if we replace idempotent $E(Y(\mathsf{T}_\lambda))$ with $\alpha_\lambda(x)^{-1} Y''(\mathsf{T}_\lambda)$. This modification is used in what follows to make relevant Young symmetrizers orthogonal.

Let us introduce ordering of standard tableaux. Let T_λ and T'_λ be two standard tableaux of the same shape λ. We still assign to each box its coordinates (i, j) labeling its row and column, respectively. The number 1 in both tableaux is always placed in box $(1, 1)$.

Definition 4.5.2. Compare the numbers in tableaux T_λ and T'_λ in each row from left to right. Let the first unequal number appear in ith row in box (i, j), and it is equal to k in tableau T_λ and k' in tableau T'_λ. If $k' > k$, then the standard tableau T'_λ *dominates* over standard tableau T_λ.

All standard tableaux T_λ of one and the same shape λ are ordered in this manner. Clearly, if a standard tableau T'_λ dominates over all other tableaux T_λ, the numbers in tableau T'_λ are ordered in columns: the first column of height m_1 contains numbers $(1, 2, \ldots, m_1)$, numbers in the second column of height m_2 are $(m_1 + 1, m_1 + 2, \ldots, m_1 + m_2)$, etc. Likewise, if a standard tableau T_λ is dominated by all other tableaux T'_λ (i.e., T_λ does not dominate over any other tableau), the numbers in T_λ are ordered in rows. It is worth

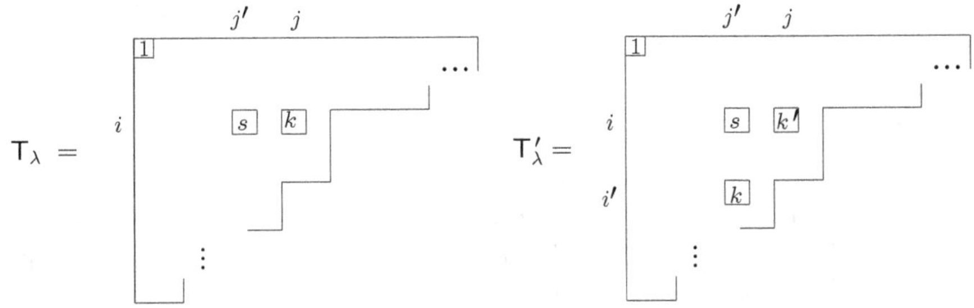

Fig. 4.5.1. Boxes with numbers k, k' and s in standard Young tableaux T_λ and T'_λ.

emphasizing that this ordering applies to standard tableaux of one and the same shape only.

Proposition 4.5.6. *Let two standard tableaux T_λ and T'_λ have the same shape λ, and let tableau T'_λ dominate over T_λ. Then there are two numbers which belong to one and the same row of tableau T_λ and one and the same column of T'_λ.*

Proof. Consider the situation where the first unequal number appears in tableaux T_λ and T'_λ in ith row, in box (i, j), and this number equals k in tableau T_λ and k' in tableau T'_λ. Then we have $k' > k$. Since all numbers in previous boxes $((i'', j'')$ with $i'' < i$ and $i'' = i$, $j'' < j)$ coincide, the number k in standard tableau T'_λ cannot be placed in or above ith row, but is placed in one of the lower rows with $i' > i$, say, in box (i', j'). Then we have $j' < j$ (see Fig. 4.5.1). To prove the latter inequality, we note that if $j' \geq j$ (i.e., column j' is either to the right of column j or coincident with column j), then for *standard* tableau T'_λ we would have $k' \leq s < k$, where s is the number in the box (i, j'). This would contradict the relation $k' > k$. Now, all numbers in the boxes before (i, j) coincide in tableaux T_λ and T'_λ, so both tableaux have the same number s $(1 \leq s < k)$ in box (i, j'), see Fig. 4.5.1. Thus, the two numbers s and k are in one row of tableau T_λ and one column of tableau T'_λ. □

Corollary 4.5.7. *If standard tableau T'_λ dominates over standard tableau T_λ, then, in accordance with Propositions* 4.3.1 *and* 4.5.7, *we have*

$$Y(\mathsf{T}_\lambda) \cdot Y(\mathsf{T}'_\lambda) = 0. \tag{4.5.64}$$

Remark 3. Let standard tableau T'_λ dominate over standard tableau T_λ. In view of Corollary 4.3.4, either (1) there are two numbers in one and the same row of tableau T'_λ and one and the same column of T_λ; then, in accordance with Proposition 4.3.1, we have both (4.5.64) and

$$Y(\mathsf{T}'_\lambda) \cdot Y(\mathsf{T}_\lambda) = 0, \tag{4.5.65}$$

i.e., symmetrizers T'_λ and T_λ are orthogonal; or (2) tableau T'_λ can be transformed into standard tableau T_λ by two permutations $p' \in H(\mathsf{T}'_\lambda)$ and $q \in V(\mathsf{T}_\lambda)$ (see (4.5.57)); in that case, in accordance with the rule (4.3.22), Young symmetrizers constructed with standard tableaux T_λ and T'_λ are related by (4.5.59) and obey equality (4.5.60) which shows that the orthogonality relation does not hold for $Y(\mathsf{T}'_\lambda)$ and $Y(\mathsf{T}_\lambda)$.

We now come back to completeness of Young symmetrizers constructed with standard tableaux.

Proposition 4.5.8. *Let $\{\mathsf{T}_{a,\lambda}\}$ $(a = 1, \ldots, f_{(\lambda)})$ be the set of all standard tableaux whose shape is given by Young diagram $\lambda \vdash r$. Elements $X'_a \equiv X_a \cdot Y(\mathsf{T}_{a,\lambda})$ $(a = 1, \ldots, f_{(\lambda)})$ are linear independent for all $X_a \in \mathbb{C}[S_r]$, i.e., left ideals $\mathcal{L}_{\mathsf{T}_{a,\lambda}}$ and $\mathcal{L}_{\mathsf{T}_{b,\lambda}}$ do not intersect for $a \neq b$.*

Proof. We give the proof *ad absurdum*. We order the set $\mathsf{T}_{a,\lambda}$ in such a way that $\mathsf{T}_{b,\lambda}$ dominates over $\mathsf{T}_{a,\lambda}$ for $b > a$. Let there exist constants α_a, not all equal to zero, such that

$$\alpha_1 X'_1 + \alpha_2 X'_2 + \cdots + \alpha_f X'_f = 0, \tag{4.5.66}$$

where we write $f_{(\lambda)} = f$ for brevity. We act on the right on the two sides of (4.5.66) by dominating symmetrizer $Y(\mathsf{T}_{f,\lambda})$ and make use of (4.3.35) and (4.5.64). We get $\alpha_f X'_f = 0$, i.e., $\alpha_f = 0$ and there is no term $\alpha_f X'_f$ in (4.5.66). We then act on the two sides of (4.5.66) by symmetrizer $Y(\mathsf{T}_{f-1,\lambda})$ and make use of (4.5.64) and the fact that $\alpha_f = 0$. We find $\alpha_{f-1} = 0$. We continue this procedure and find that all coefficients α_a in (4.5.66) vanish, in contradiction with the original assumption. In particular, for any $X'_a \in \mathcal{L}_{\mathsf{T}_{a,\lambda}}$, we have $X'_a \neq X'_b$ if $a \neq b$. □

It follows from Proposition 4.5.8 that left ideals $\mathcal{L}_{\mathsf{T}_{a,\lambda}}$ and $\mathcal{L}_{\mathsf{T}_{b,\lambda}}$ do not intersect for $a \neq b$. Moreover, the orthogonality (4.3.29) of Young symmetrizers constructed with tableaux T_λ and $\mathsf{T}_{\lambda'}$ of different shapes $\lambda \neq \lambda'$ shows that $\mathcal{L}_{\mathsf{T}_{a,\lambda}} \cap \mathcal{L}_{\mathsf{T}_{b,\lambda'}} = \emptyset$ for all a, b. We make use of the fact that the number $f_{(\lambda)}$ of standard tableaux $\mathsf{T}_{a,\lambda}$ of shape $\lambda \vdash r$ coincides with the dimension of left ideal $\mathcal{L}_{\mathsf{T}_{a,\lambda}}$ (this follows from comparison of the identities (4.3.63) and (4.3.64)), and conclude that Young idempotents

$$E(\mathsf{T}_{a,\lambda}) = \alpha_\lambda^{-1} Y(\mathsf{T}_{a,\lambda}), \tag{4.5.67}$$

constructed with all standard tableaux $\mathsf{T}_{a,\lambda}$, where λ runs over all Young drams $\lambda \vdash r$ and $a = 1, \dots, f_{(\lambda)}$, make complete system of idempotents (here completeness is understood in the sense discussed in Remark 4. in Section 4.5.1). Thus, the first expansion (in left ideals) in (4.3.55) is valid. The correctness of the second expansion (in right ideals) in (4.3.55) is shown in the same way. We discuss in what follows the expansion in left ideals only; the case of right ideals is treated in perfect analogy.

Completeness of idempotent system (4.5.67) can be written, using (4.3.55), as the partition of unit element e in elements of left ideals generated by $E(\mathsf{T}_{a,\lambda})$ (cf. (4.5.31)) :

$$e = \sum_{\lambda \vdash r} \sum_{a=1}^{f_{(\lambda)}} X_{a,\lambda} \cdot E(\mathsf{T}_{a,\lambda}), \quad X_{a,\lambda} \in \mathbb{C}[S_r]. \tag{4.5.68}$$

As we pointed out in Remarks 1 and 3, idempotents $E(\mathsf{T}_{a,\lambda})$ are not, generally speaking, orthogonal. However, in accordance with (4.5.68) (see in this regard the derivation of (4.5.4) from expansion (4.5.3)) there exists modification (4.5.63)

$$E(\mathsf{T}_{a,\lambda}) \to E'(\mathsf{T}_{a,\lambda}) = X_{a,\lambda} \cdot E(\mathsf{T}_{a,\lambda}), \tag{4.5.69}$$

which gives orthogonal system of idempotents. Let us explain how to evaluate the coefficients $X_{a,\lambda}$ explicitly.

For a given Young diagram $\lambda \vdash r$ we take the dominant standard tableau $\mathsf{T}_{f,\lambda}$ (here $f = f_{(\lambda)}$) and act on both sides of (4.5.68) by idempotent $E(\mathsf{T}_{f,\lambda})$. This idempotent is orthogonal to all idempotents $E(\mathsf{T}_{a,\lambda'})$ with $\lambda' \neq \lambda$, and obeys the relation $E(\mathsf{T}_{a,\lambda}) \cdot E(\mathsf{T}_{f,\lambda}) = 0$ ($\forall a < f$), which follows from the dominance of $\mathsf{T}_{f,\lambda}$. Therefore, we obtain from (4.5.68) the identity $E(\mathsf{T}_{f,\lambda}) = X_{f,\lambda} \cdot E(\mathsf{T}_{f,\lambda})$, so we have to set $X_{f_{(\lambda)},\lambda} = e$ in expansion (4.5.68) and definition (4.5.69) of new idempotents $E'(\mathsf{T}_{f_{(\lambda)},\lambda})$. We now extract in expansion (4.5.68) the terms that correspond to Young diagram $\lambda \vdash r$:

$$e = \dots + \sum_{a=1}^{f-1} X_{a,\lambda} \cdot E(\mathsf{T}_{a,\lambda}) + E(\mathsf{T}_{f,\lambda}) + \cdots, \quad X_{a,\lambda} \in \mathbb{C}[S_r], \tag{4.5.70}$$

where $f = f_{(\lambda)}$. We act on both sides of (4.5.68), (4.5.70) by the next-to-dominant idempotent $E(\mathsf{T}_{f-1,\lambda})$. All terms denoted by dots in (4.5.70), and also the terms with $X_{a,\lambda} \cdot E(\mathsf{T}_{a,\lambda})$ and $a < (f-1)$ vanish because Young idempotents are orthogonal for different Young diagrams, and because the standard tableau $\mathsf{T}_{f-1,\lambda}$ dominates over tableaux $\mathsf{T}_{a,\lambda}$ with $a < (f-1)$. There are two options (see Remark 3 above) : either $E(\mathsf{T}_{f,\lambda}) \cdot E(\mathsf{T}_{f-1,\lambda}) = 0$ and then $X_{f-1,\lambda} = e$, or there exist two permutations $p \in H(\mathsf{T}_{f-1,\lambda})$ and $q \in V(\mathsf{T}_{f-1,\lambda})$ such that $\mathsf{T}_{f,\lambda} = p \cdot q \cdot \mathsf{T}_{f-1,\lambda}$, and then, in accordance with (4.5.60), we have

$$E(\mathsf{T}_{f,\lambda}) \cdot E(\mathsf{T}_{f-1,\lambda}) = \delta_q q^{-1} \cdot p^{-1} \cdot E(\mathsf{T}_{f-1,\lambda}). \tag{4.5.71}$$

In the latter case, we get an equation

$$E(\mathsf{T}_{f-1,\lambda}) = X_{f-1,\lambda} \cdot E(\mathsf{T}_{f-1,\lambda}) + \delta_q q^{-1} \cdot p^{-1} \cdot E(\mathsf{T}_{f-1,\lambda}), \tag{4.5.72}$$

which determines the coefficient $X_{f-1,\lambda} = e - \delta_q q^{-1} \cdot p^{-1}$. We then act on (4.5.70) on the right by idempotent $E(\mathsf{T}_{f-2,\lambda})$ and make use of the same argument as above to find the next coefficient $X_{f-2,\lambda}$. This procedure continues and gives constructive method of evaluation of the coefficients $X_{a,\lambda}$ and hence gives operators (4.5.69), which form the complete system of orthogonal idempotents.

Example. We consider as an example the case of algebra $\mathbb{C}[S_5]$, and study Young symmetrizers associated with Young diagram $\nu = [3,2]$. There are five standard tableaux of this shape:

$$\begin{gathered}
\mathsf{T}_{1,\nu} = \begin{array}{|c|c|c|}\hline 1 & 2 & 3 \\ \hline 4 & 5 \\ \cline{1-2}\end{array}\,, \quad \mathsf{T}_{2,\nu} = \begin{array}{|c|c|c|}\hline 1 & 2 & 4 \\ \hline 3 & 5 \\ \cline{1-2}\end{array}\,, \quad \mathsf{T}_{3,\nu} = \begin{array}{|c|c|c|}\hline 1 & 2 & 5 \\ \hline 3 & 4 \\ \cline{1-2}\end{array}\,, \\
\mathsf{T}_{4,\nu} = \begin{array}{|c|c|c|}\hline 1 & 3 & 4 \\ \hline 2 & 5 \\ \cline{1-2}\end{array}\,, \quad \mathsf{T}_{5,\nu} = \begin{array}{|c|c|c|}\hline 1 & 3 & 5 \\ \hline 2 & 4 \\ \cline{1-2}\end{array}\,.
\end{gathered} \tag{4.5.73}$$

Here tableaux on the right dominate over tableaux on the left. The result of Proposition 4.5.6 is explicit in (4.5.73). As an example, numbers 1 and 2 are in one row of tableau $\mathsf{T}_{1,\nu}$ and one column in tableaux $\mathsf{T}_{4,\nu}$ and $\mathsf{T}_{5,\nu}$, which dominate over $\mathsf{T}_{1,\nu}$. Note that almost all pairs of tableaux show the opposite pattern: if $\mathsf{T}_{b,\nu}$ dominates over $\mathsf{T}_{a,\nu}$ $(b > a)$, then there exists a pair of numbers that are placed in one row of $\mathsf{T}_{b,\nu}$ and one column of $\mathsf{T}_{a,\nu}$.

The only exception (see Remark 1 to Proposition 4.5.4 above) is a pair of tableaux $\mathsf{T}_{1,\nu}$ and $\mathsf{T}_{5,\nu}$ in (4.5.73). Therefore, the orthogonality identities

$$Y(\mathsf{T}_{a,\nu}) \cdot Y(\mathsf{T}_{b,\nu}) = \alpha_\nu Y(\mathsf{T}_{b,\nu})\delta_{ab} \tag{4.5.74}$$

are valid for all $a, b = 1, \ldots, 5$ except for $(a = 5, b = 1)$. For $(a = 5, b = 1)$ we have equality (4.5.62),

$$Y(\mathsf{T}_{5,\nu}) \cdot Y(\mathsf{T}_{1,\nu}) = \alpha_\nu (2,4) \cdot (3,5) \cdot Y(\mathsf{T}_{1,\nu}) \neq 0, \tag{4.5.75}$$

where $\alpha_\nu = \alpha_{[3,2]} = 24$.

We now consider partition of unit element $e \in S_5$ given by (4.5.68), and write explicitly only the contributions coming from tableaux (4.5.73):

$$e = \cdots + \sum_{a=1}^{5} X_{a,\nu} \cdot E(\mathsf{T}_{a,\nu}) + \cdots, \quad X_{a,\lambda} \in \mathbb{C}[S_5], \tag{4.5.76}$$

where $E(\mathsf{T}_{a,\nu}) = \alpha_\nu^{-1} Y(\mathsf{T}_{a,\nu})$ and dots denote contributions related to diagrams $\lambda \neq \nu$. We make use of the above general procedure for evaluating the coefficients $X_{a,\lambda}$ and act on the right on the two sides of (4.5.76) by idempotents $E(\mathsf{T}_{b,\nu})$ with $b = 5, 4, 3, 2$. We use formula (4.5.74) and find the relations $E(\mathsf{T}_{b,\nu}) = X_{b,\nu} \cdot E(\mathsf{T}_{b,\nu})$ $(b = 5, 4, 3, 2)$, which give

$$X_{b,\nu} = e, \quad b = 5, 4, 3, 2. \tag{4.5.77}$$

Finally, we multiply the two sides of (4.5.76) on the right by idempotent $E(\mathsf{T}_{1,\nu})$ and make use of formula (4.5.74) for $a = b = 1$ and also relation (4.5.75). We obtain equation $E(\mathsf{T}_{1,\nu}) = X_{1,\nu} \cdot E(\mathsf{T}_{1,\nu}) + (2,4) \cdot (3,5) \cdot E(\mathsf{T}_{1,\nu})$ whose solution is

$$X_{1,\nu} = (e - (2,4) \cdot (3,5)) \in \mathbb{C}[S_5]. \tag{4.5.78}$$

So, the system of orthogonal idempotents constructed with standard tableaux (4.5.73) is

$$\begin{gathered} E'(\mathsf{T}_{1,\nu}) = (e - (2,4) \cdot (3,5)) \cdot E(\mathsf{T}_{1,\nu}) = \alpha_\nu^{-1}(e - (2,4) \cdot (3,5)) \cdot Y(\mathsf{T}_{1,\nu}), \\ E'(\mathsf{T}_{b,\nu}) = E(\mathsf{T}_{b,\nu}) = \alpha_\nu^{-1} Y(\mathsf{T}_{b,\nu}), \quad b = 2, \ldots, 5. \end{gathered} \tag{4.5.79}$$

These idempotents define the (part of) the left Peirce decomposition.

Problem 4.5.10. Construct operators (4.5.79) explicitly and check by direct calculation that they obey $E'(\mathsf{T}_{a,\nu}) \cdot E'(\mathsf{T}_{b,\nu}) = \delta_{ab} E'(\mathsf{T}_{a,\nu})$ $(a, b = 1, \ldots, 5)$.

With this example we complete our study of the method (summarized in the beginning of this section) for constructing the complete orthogonal system of primitive idempotents for $\mathbb{C}[S_r]$. Its key element is left (and right) Peirce decomposition for unit element $e \in S_r$ in terms of Young symmetrizers associated with standard Young tableaux $\mathsf{T}_{a,\lambda}$:

$$e = \sum_{\lambda \vdash r} \sum_{a=1}^{f_{(\lambda)}} X_{a,\lambda} \cdot E(\mathsf{T}_{a,\lambda}) = \sum_{\lambda \vdash r} \sum_{a=1}^{f_{(\lambda)}} E(\mathsf{T}_{a,\lambda}) \cdot \tilde{X}_{a,\lambda}, \tag{4.5.80}$$

whose integral part is the evaluation of factors $X_{a,\lambda}, \tilde{X}_{a,\lambda} \in \mathbb{C}[S_r]$. As we pointed out, the calculation of these factors is the drawback of this method. This drawback is absent in the alternative method studied in Section 4.6.

4.5.3. Schur–Weyl duality

We recall (see Section 4.4.1) that representation s of algebra $\mathbb{C}[S_r]$ is defined in (4.2.9), (4.2.10) and acts in space $\mathcal{V}^{\otimes r}$, where $\mathcal{V}$ is complex N-dimensional vector space of defining representation T of group $SL(N, \mathbb{C})$. Let $\lambda \vdash r$ be partition of number r (Young diagram with r boxes). Every Young diagram λ defines a set of standard Young tableaux $\mathsf{T}_{a,\lambda}$, $(a = 1, \ldots, f_{(\lambda)})$, where $f_{(\lambda)}$ is given by (4.3.56), (4.3.57). Every standard tableau defines, in its turn, Young symmetrizer $Y(\mathsf{T}_{a,\lambda}) \in \mathbb{C}[S_r]$, which generates the left (right) ideal $\mathcal{L}_{\mathsf{T}_{a,\lambda}}$ $(\mathcal{R}_{\mathsf{T}_{a,\lambda}})$. This ideal is the subspace of irreducible representation $\rho_a^{(\lambda)}$ of algebra $\mathbb{C}[S_r]$ in its regular representation. All representations $\rho_a^{(\lambda)}$ $(a = 1, \ldots, f_{(\lambda)})$ are equivalent (see Proposition 4.3.7), so we use the notation $\rho^{(\lambda)}$. We also know (see Section 4.2) that operators

$$\mathsf{s}(E(\mathsf{T}_{a,\lambda})) = \frac{1}{\alpha_\lambda} \mathsf{s}(Y(\mathsf{T}_{a,\lambda}))$$

are projectors in space $\mathcal{V}^{\otimes r}$; they commute with action of group $SL(N, \mathbb{C})$ in representation $T^{\otimes r}$ and define in $\mathcal{V}^{\otimes r}$ invariant subspaces $\mathcal{V}^{(\lambda)}$ of irreducible representations $T_a^{(\lambda)}$ of group $SL(N, \mathbb{C})$. Representations $T_a^{(\lambda)}$ are irreducible because idempotents $E(\mathsf{T}_{a,\lambda})$ are primitive (see Corollary 4.3.6). Moreover, due to Proposition 4.4.1, representations $T_a^{(\lambda)}$ are equivalent for all a, $T_a^{(\lambda)} \equiv T^{(\lambda)}$. Finally, representations $T^{(\lambda)}$ and $T^{(\lambda')}$ constructed with two tableaux T_λ and $\mathsf{T}'_{\lambda'}$ of different shapes $\lambda \neq \lambda'$, where the number of rows in λ and λ' do not exceed $(N - 1)$, are inequivalent (see Proposition 4.4.2). In addition, we have to remember that projectors $\mathsf{s}(E(\mathsf{T}_{a,\lambda}))$, in the case of diagrams λ with number of rows exceeding N, are trivial and map the space $\mathcal{V}^{\otimes r}$ into zero vector (see Sections 4.4.1 and 4.4.2).

So, there is correspondence between irreducible representations $\rho^{(\lambda)}$ of symmetric group S_r (algebra $\mathbb{C}[S_r]$) and irreducible representations $T^{(\lambda)}$ of group $SL(N,\mathbb{C})$ (and $SU(N)$), which are extracted from tensor product $T^{\otimes r}$ of its defining representations. This correspondence is known as *Schur–Weyl duality*. For $r \leq N$, it is expressed as the expansion

$$\mathcal{V}^{\otimes r} = \mathsf{s}(e)\mathcal{V}^{\otimes r} = \sum_{\lambda \vdash r}\sum_{a=1}^{f_{(\lambda)}} \mathsf{s}(E(\mathsf{T}_{a,\lambda}) \cdot \tilde{X}_{a,\lambda})\mathcal{V}^{\otimes r} = \sum_{\lambda \vdash r}\sum_{a=1}^{f_{(\lambda)}} \mathcal{V}^{(a,\lambda)}, \quad (4.5.81)$$

where $\mathcal{V}^{(a,\lambda)} = \mathsf{s}(E(\mathsf{T}_{a,\lambda}) \cdot \tilde{X}_{a,\lambda})\mathcal{V}^{\otimes r}$ are invariant subspaces of irreducible representations $T^{(\lambda)}$ of group $SL(N,\mathbb{C})$, and we use the right Peirce decomposition for unit element $e \in S_r$ in (4.5.80). Expansion (4.5.81) can be written as a remarkable formula

$$\mathcal{V}^{\otimes r} = \sum_{\lambda \vdash r} \mathcal{R}_\lambda \otimes \mathcal{V}^{(\lambda)}, \quad (4.5.82)$$

where summation runs over all Young diagrams λ with r boxes, $\mathcal{V}^{(\lambda)}$ is the space of irreducible representation $T^{(\lambda)}$ of group $SL(N,\mathbb{C})$, which corresponds to Young diagram λ, and $\mathcal{R}_\lambda$ is the space of irreducible representation $\rho^{(\lambda)}$ of group S_r that corresponds to the same diagram λ. To cast (4.5.81) into the form (4.5.82), we make use of isomorphism of spaces $\mathcal{V}^{(a,\lambda)} \equiv \mathcal{V}^{(\lambda)}$ for all $a = 1, \dots, f_{(\lambda)}$, which we write (to distinguish $\mathcal{V}^{(a,\lambda)}$ with different a) as follows: $\mathcal{V}^{(a,\lambda)} = e_{a,\lambda} \otimes \mathcal{V}^{(\lambda)}$, where $e_{a,\lambda}$ $(a = 1, \dots, f_{(\lambda)})$ are basis vectors in space $\mathcal{R}_\lambda$ of representation $\rho^{(\lambda)}$. We emphasize (see Remark 3 in Section 4.5.1) that basis vectors $e_{a,\lambda}$ in $\mathcal{R}_\lambda$ are labeled by idempotents $E(\mathsf{T}_{a,\lambda})$ or standard tableaux $\mathsf{T}_{a,\lambda}$ of shape λ. Thus, in accordance with decomposition (4.5.82), for $r \leq N$ one can choose the basis in space $\mathcal{V}^{\otimes r}$ as tensor product of basis vectors $e_{a,\lambda}$ of space $\mathcal{R}_\lambda$ and basis vectors in space $\mathcal{V}^{(\lambda)}$ of irreducible representation $T^{(\lambda)}$ of group $SL(N,\mathbb{C})$. Recall that the action of group S_r in space $\mathcal{V}^{\otimes r}$ in representation (4.2.9) commutes with action of group $SL(N,\mathbb{C})$ in representation $T^{\otimes r}$. In this situation, one says that group S_r *centralizes* group $SL(N,\mathbb{C})$ in space $\mathcal{V}^{\otimes r}$ and, conversely, group $SL(N,\mathbb{C})$ *centralizes* group S_r.

Remark 1. One can use formula (4.4.33) to illustrate expansions (4.5.81) and (4.5.82) and the relevance of standard Young tableaux in these expansions. One applies consecutively the formula (4.4.33) and obtains, recalling the relationship (4.4.29), that

$$\begin{array}{|c|}\hline i_1\\\hline\end{array}\otimes\begin{array}{|c|}\hline i_2\\\hline\end{array}\otimes\cdots\otimes\begin{array}{|c|}\hline i_r\\\hline\end{array}=\left(\begin{array}{|c|}\hline 1\\\hline 2\\\hline\end{array}\oplus\begin{array}{|c|c|}\hline 1&2\\\hline\end{array}\right)\otimes\begin{array}{|c|}\hline 3\\\hline\end{array}\otimes\cdots\otimes\begin{array}{|c|}\hline r\\\hline\end{array}$$

$$=\left(\begin{array}{|c|}\hline 1\\\hline 2\\\hline 3\\\hline\end{array}\oplus\begin{array}{|c|c|}\hline 1&3\\\hline 2\\\cline{1-1}\end{array}\oplus\begin{array}{|c|c|}\hline 1&2\\\hline 3\\\cline{1-1}\end{array}\oplus\begin{array}{|c|c|c|}\hline 1&2&3\\\hline\end{array}\right)\otimes\begin{array}{|c|}\hline 4\\\hline\end{array}\otimes\cdots$$

$$=\left(\begin{array}{|c|}\hline 1\\\hline 2\\\hline 3\\\hline 4\\\hline\end{array}\oplus\begin{array}{|c|c|}\hline 1&4\\\hline 2\\\cline{1-1} 3\\\cline{1-1}\end{array}\oplus\begin{array}{|c|c|}\hline 1&3\\\hline 2\\\cline{1-1} 4\\\cline{1-1}\end{array}\oplus\begin{array}{|c|c|}\hline 1&2\\\hline 3\\\cline{1-1} 4\\\cline{1-1}\end{array}\oplus\begin{array}{|c|c|c|}\hline 1&3&4\\\hline 2\\\cline{1-1}\end{array}\oplus\begin{array}{|c|c|c|}\hline 1&2&4\\\hline 3\\\cline{1-1}\end{array}\right.$$

$$\left.\oplus\begin{array}{|c|c|c|}\hline 1&2&3\\\hline 4\\\cline{1-1}\end{array}\oplus\begin{array}{|c|c|}\hline 1&3\\\hline 2&4\\\hline\end{array}\oplus\begin{array}{|c|c|}\hline 1&2\\\hline 3&4\\\hline\end{array}\oplus\begin{array}{|c|c|c|c|}\hline 1&2&3&4\\\hline\end{array}\right)\otimes\cdots \tag{4.5.83}$$

$$=\cdots=\sum_{\lambda\vdash r}\sum_{\mathsf{T}_{a,\lambda}}\psi_{\mathsf{T}_{a,\lambda}}, \tag{4.5.84}$$

where we use shorthand notations for the content of boxes $\begin{array}{|c|}\hline k\\\hline\end{array}\equiv\begin{array}{|c|}\hline i_k\\\hline\end{array}$, and write in the last equality the sum $\sum_{\lambda\vdash r}$ over all Young diagrams λ with r boxes and the sum $\sum_{\mathsf{T}_{a,\lambda}}$ over all standard tableaux $\mathsf{T}_{a,\lambda}$ of the shape λ; $\psi_{\mathsf{T}_{a,\lambda}}$ are components of tensor symmetrized in accordance with standard tableau $\mathsf{T}_{a,\lambda}$ with appropriate orthogonalization factor $\tilde{X}_{a,\lambda}$. Note that formula (4.5.83) as well as formulas in (4.5.84) indicated by dots are valid until diagrams with the number of rows exceeding N appear. This explains our condition $r \leq N$ in relations (4.5.81) and (4.5.82). Note also that here and below, we encounter only standard Young tableaux. Since all representations of $SL(N,\mathbb{C})$ corresponding to one and the same Young diagram λ are equivalent (see Proposition 4.4.1), formula (4.5.83) can be written as

$$T^{\otimes r}=\sum_{\lambda\vdash r}f_{(\lambda)}T^{(\lambda)}. \tag{4.5.85}$$

Note also that basis elements of irreducible representation λ of group S_r are labeled by standard tableaux $\mathsf{T}_{a,\lambda}$, so the sum in (4.5.84) is the result of expansion of the right-hand sides of (4.5.81) and (4.5.82) in basis vectors. Finally, we note that elements $\tilde{X}_{a,\lambda}$ are not explicit in (4.5.84), since these elements are hidden in definition of tensors $\psi_{\mathsf{T}_{a,\lambda}} \in \mathcal{V}^{(a,\lambda)}$ (see in this regard the definition of space $\mathcal{V}^{(a,\lambda)}$ in (4.5.81)).

Remark 2. In the case $r \geq N$, the correspondence between irreducible representations of the symmetric group S_r (of the algebra $\mathbb{C}[S_r]$) and irreducible subrepresentations of the group $SL(N,\mathbb{C})$ ($SU(N)$) in tensor product $T^{\otimes r}$ is not one-to-one. Indeed, there are no representations of the group $SL(N,\mathbb{C})$ ($SU(N)$), which would correspond to nontrivial representations of algebra $\mathbb{C}[S_r]$ with Young diagrams $\lambda \vdash r$ having the number of rows exceeding N. The one-to-one correspondence between irreducible representations of the group $SL(N,\mathbb{C})$ ($SU(N)$), embedded in $T^{\otimes r}$, is established not with representations of the algebra $\mathbb{C}[S_r]$, but with a set of representations of special factor algebra, which is obtained from $\mathbb{C}[S_r]$ by imposing on its generators additional constraint $A_{N+1} = 0$, where the central element $A_{N+1} \equiv E_{Q[1^{N+1}]}$ is the antisymmetrizer of rank $(N+1)$ in $\mathbb{C}[S_r]$ (definition of $E_{Q[1^r]}$ is given in Problem 4.3.3).

4.6. Representations of Symmetric Group III. Okounkov–Vershik Approach

In this section, we study another method of constructing complete system of primitive orthogonal idempotents in algebra $\mathbb{C}[S_r]$ and hence all irreducible representations of symmetric group S_r. This method was originated by A. Jucys and G. Murphy and its final version is given by A. Okounkov and A. Vershik [47]. As we alluded to above, this method is free of orthogonalization drawback of Young–Frobenius theory. Our presentation is somewhat simplified as compared with the original version [47] due to the use of intertwining operators U_k. This simplification is introduced in Ref. [45] in more general context.

4.6.1. Jucys–Murphy elements and intertwining operators in algebra $\mathbb{C}[S_n]$

We make use of Coxeter presentation of symmetric group S_n, in which the generators are neighboring transpositions $\sigma_k = (k, k+1)$ ($k = 1, 2, \ldots, n-1$). They obey the relations (4.3.13) which we write again for references:

$$\begin{aligned} \sigma_i\sigma_{i+1}\sigma_i = \sigma_{i+1}\sigma_i\sigma_{i+1}, \quad i = 1, 2, \ldots, n-2, \ \text{braid relations}; \\ [\sigma_i, \sigma_k] = 0, \quad |i-k| > 1, \ \text{locality relations}; \\ \sigma_i^2 = e. \end{aligned} \tag{4.6.1}$$

One defines a set of elements $\{y_m\}$ $(m = 1, \ldots, n)$ in group algebra $\mathbb{C}[S_n]$:

$$\begin{aligned} y_1 = 0, \quad y_m &= \sum_{k=1}^{m-1} \sigma_k \cdots \sigma_{m-2}\sigma_{m-1}\sigma_{m-2} \cdots \sigma_k \\ &= \sum_{k=1}^{m-1} \sigma_{m-1} \cdots \sigma_{k+1}\sigma_k\sigma_{k+1} \cdots \sigma_{m-1} \\ &= (1, m) + (2, m) + \cdots + (m-1, m), \ \ m = 2, \ldots, n. \end{aligned} \tag{4.6.2}$$

The trivial element $y_1 = 0$ is included here for convenience. Objects y_m are called *Jucys–Murphy elements*; they play key role in the construction below. Note that Jucys–Murphy elements can be defined via induction procedure

$$y_1 = 0, \quad y_{m+1} = \sigma_m + \sigma_m \cdot y_m \cdot \sigma_m. \tag{4.6.3}$$

The first few elements, in terms of generators σ_i, are

$$y_2 = \sigma_1, \quad y_3 = \sigma_2 + \sigma_2\sigma_1\sigma_2, \quad y_4 = \sigma_3 + \sigma_3\sigma_2\sigma_3 + \sigma_3\sigma_2\sigma_1\sigma_2\sigma_3, \ldots. \tag{4.6.4}$$

It follows from definitions (4.6.2), (4.6.3) and relations (4.6.1) that elements y_m obey

$$y_{m+1} \cdot \sigma_m = \sigma_m \cdot y_m + 1, \quad \sigma_m \cdot y_{m+1} = y_m \cdot \sigma_m + 1, \tag{4.6.5}$$

$$[y_m, \sigma_i] = 0, \quad m \neq i, i+1. \tag{4.6.6}$$

Hereafter, we denote the identical permutation $e \in S_n$ by 1, and an element $c \cdot e$, where c is a constant, is written simply as c.

Definition 4.6.1. Algebra $\mathcal{A}$ over field $\mathbb{K}$ is generated by elements $\{g_1, g_2, \ldots, g_n\}$ if and only if any element from $\mathcal{A}$ can be written as a polynomial in elements $\{g_1, g_2, \ldots, g_n\}$ with coefficients from $\mathbb{K}$.

Remark 1. It follows from (4.6.1) that algebra $\mathbb{C}[S_k]$ $(\forall k = 2, \ldots, n)$ is generated by the last generator σ_{k-1} together with previous generators $\{\sigma_1, \ldots, \sigma_{k-2}\}$. The latter, in turn, generate subalgebra $\mathbb{C}[S_{k-1}] \subset \mathbb{C}[S_k]$.

Let Z_k be central subalgebra in $\mathbb{C}[S_k]$. Then there are natural embeddings of group algebras $\mathbb{C}[S_k] \subset \mathbb{C}[S_n]$, and the pertinent central subalgebras Z_k are embedded in $\mathbb{C}[S_k]$:

$$\begin{array}{ccccc} \mathbb{C}[S_2] & \subset \mathbb{C}[S_3] & \subset \cdots & \subset \mathbb{C}[S_{n-1}] & \subset \mathbb{C}[S_n] \\ \cup & \cup & \cdots & \cup & \cup \\ Z_2 & Z_3 & \cdots & Z_{n-1} & Z_n \end{array} \tag{4.6.7}$$

We define subalgebra $GZ_n \subset \mathbb{C}[S_n]$ generated by elements of all central subalgebras $Z_1, Z_2, \ldots, Z_n$, where Z_1 has just one unit element $e \in \mathbb{C}[S_n]$. Clearly, GZ_n is commutative; it is called *Gelfand–Tsetlin subalgebra* (by analogy to representations of linear and orthogonal Lie algebras; see Section 4.7 below). In fact, this is the maximal commutative (Abelian) subalgebra in $\mathbb{C}[S_n]$, i.e., every element z that commutes with all elements of GZ_n belongs to GZ_n. Another remarkable property is that $GZ_1 \subset \cdots \subset GZ_{n-1} \subset GZ_n$. From the definition (4.6.2) of Jucys–Murphy elements, it follows that

$$y_m = z_m - z_{m-1}, \tag{4.6.8}$$

where each element $z_m = \sum_{1\le i<j\le m}(i,j)$ is equal to the sum of all transpositions in $\mathbb{C}[S_m]$ and thus belongs to the central subalgebra $Z_m \subset \mathbb{C}[S_m]$. Therefore, Jucys–Murphy elements $y_2, \ldots, y_n$ are commutative and belong to Gelfand–Tsetlin subalgebra GZ_n. Moreover, it can be shown (see [47]) that Abelian subalgebra Y_n, which is generated by Jucys–Murphy elements $\{y_2, \ldots, y_n\}$, coincides with Gelfand–Tsetlin subalgebra GZ_n. In this way, we arrive at the following proposition.

Proposition 4.6.1. *Jucys–Murphy elements $\{y_2, \ldots, y_n\}$ form complete system of commuting elements in $\mathbb{C}[S_n]$:*

$$[y_i, y_j] = 0, \quad \forall i, j. \tag{4.6.9}$$

Completeness here means that any element $y \in \mathbb{C}[S_n]$ that commutes with all elements y_i is a polynomial in y_i.

Problem 4.6.1. Show explicitly that $[y_i, y_j] = 0$ $(\forall i, j)$. Hint: make use of the last representation in (4.6.2).

Remark 2. There are numerous polynomial relations between Jucys–Murphy elements $\{y_1, \ldots, y_n\}$, so that these elements are not independent in algebraic sense.

Problem 4.6.2. Making use of definition (4.6.2), prove the identities

$$(y_2+1)(y_3+1)(y_3-2)=0, \quad (y_2+1)(y_3+1)(y_4+1)(y_4-3)=0. \tag{4.6.10}$$

We now define *intertwining operators* U_k in algebra $\mathbb{C}[S_n]$:

$$\begin{aligned} U_k &= \sigma_k \cdot y_k - y_k \cdot \sigma_k = y_{k+1} \cdot \sigma_k - \sigma_k \cdot y_{k+1} \\ &= (y_{k+1}-y_k)\sigma_k - 1 = \sigma_k(y_k - y_{k+1}) + 1, \quad k=1,\dots,n-1. \end{aligned} \tag{4.6.11}$$

Proposition 4.6.2. *Commutation relations between intertwining operator U_k with generators σ_m and Jucys–Murphy elements y_m are*

$$U_k \cdot \sigma_k = -\sigma_k \cdot U_k, \quad U_k \cdot \sigma_m = \sigma_m \cdot U_k \quad (|k-m|>1). \tag{4.6.12}$$

$$\begin{aligned} U_k \cdot y_m &= y_m \cdot U_k \quad (m \neq k, k+1), \\ U_k \cdot y_k &= y_{k+1} \cdot U_k, \quad U_k \cdot y_{k+1} = y_k \cdot U_k. \end{aligned} \tag{4.6.13}$$

Relations between U_k and $\sigma_{k\pm1}$ are irrelevant for our purposes, so we do not write them down. Also, intertwining operators U_k obey braid group relations

$$U_k \cdot U_{k+1} \cdot U_k = U_{k+1} \cdot U_k \cdot U_{k+1}, \tag{4.6.14}$$

while the operator U_k squared is expressed through y_k and y_{k+1} :

$$U_k^2 = (1+y_{k+1}-y_k)(1-y_{k+1}+y_k) = 1-(y_{k+1}-y_k)^2. \tag{4.6.15}$$

So, U_k^2 belongs to Abelian subalgebra Y_n.

Proof. The first relation in (4.6.12) is obtained by using $\sigma_k^2 = 1$ (see (4.6.1)):

$$U_k \cdot \sigma_k = (\sigma_k \cdot y_k - y_k \cdot \sigma_k)\sigma_k = -\sigma_k(\sigma_k \cdot y_k - y_k \cdot \sigma_k) = -\sigma_k \cdot U_k.$$

The second relation in (4.6.12) and the first relation in (4.6.13) follow straightforwardly from (4.6.6) and locality relations (4.6.1). To prove the second relation in (4.6.13) (the third relation in (4.6.13) is proven in the same way), we use the definitions (4.6.3), (4.6.11) and identities (4.6.5), (4.6.9):

$$\begin{aligned} y_{k+1} \cdot U_k &= y_{k+1} \cdot (\sigma_k(y_k - y_{k+1}) + 1) = (\sigma_k \cdot y_k + 1)(y_k - y_{k+1}) + y_{k+1} \\ &= \sigma_k(y_k - y_{k+1})y_k + y_k = U_k \cdot y_k. \end{aligned}$$

The relation (4.6.15) is obtained as follows:

$$\begin{aligned}U_k \cdot U_k &= (\sigma_k \cdot (y_k - y_{k+1}) + 1) \cdot U_k = \sigma_k \cdot U_k \cdot (y_{k+1} - y_k) + U_k \\ &= (y_k - y_{k+1}) \cdot (y_{k+1} - y_k) + 1 = (1 + y_{k+1} - y_k) \cdot (1 - y_{k+1} + y_k),\end{aligned}$$

where we make use of equalities (4.6.13) and the last representation in (4.6.11).

Let us now turn to braid relation (4.6.14). We first rewrite the right-hand side of (4.6.14) by trying to move Jucys–Murphy elements y_m to the right by using (4.6.11)–(4.6.13):

$$\begin{aligned}&U_{k+1} \cdot U_k \cdot U_{k+1} \\ &\quad = (\sigma_{k+1} \cdot (y_{k+1} - y_{k+2}) + 1) \cdot U_k \cdot U_{k+1} \\ &\quad = \sigma_{k+1} \cdot U_k \cdot U_{k+1} \cdot (y_k - y_{k+1}) + U_k \cdot U_{k+1} \\ &\quad = \sigma_{k+1} \cdot \sigma_k \cdot \sigma_{k+1} \cdot (y_{k+1} - y_{k+2}) \cdot (y_k - y_{k+2}) \cdot (y_k - y_{k+1}) \\ &\qquad + \sigma_{k+1} \cdot \sigma_k \cdot (y_k - y_{k+2}) \cdot (y_k - y_{k+1}) \\ &\qquad + (y_{k+1} - y_{k+2}) \cdot (y_k - y_{k+1}) \\ &\qquad + \sigma_{k+1} \cdot (y_k - y_{k+2}) + \sigma_k \sigma_{k+1} \cdot (y_{k+1} - y_{k+2}) \cdot (y_k - y_{k+2}) \\ &\qquad + \sigma_k \cdot (y_k - y_{k+2}) + 1,\end{aligned} \tag{4.6.16}$$

where we use the expression

$$\begin{aligned}U_k \cdot U_{k+1} &= (\sigma_k \cdot (y_k - y_{k+1}) + 1) \cdot U_{k+1} \\ &= \sigma_k \cdot U_{k+1} \cdot (y_k - y_{k+2}) + U_{k+1} \\ &= \sigma_k \cdot \sigma_{k+1} \cdot (y_{k+1} - y_{k+2}) \cdot (y_k - y_{k+2}) + \sigma_k \cdot (y_k - y_{k+2}) \\ &\quad + \sigma_{k+1} \cdot (y_{k+1} - y_{k+2}) + 1.\end{aligned}$$

Making use of the same formula, as well as relations (4.6.13) and the last representation in (4.6.11), we find for the left-hand side of (4.6.14):

$$\begin{aligned}(U_k \cdot U_{k+1}) \cdot U_k &= \sigma_k \cdot \sigma_{k+1} \cdot U_k \cdot (y_k - y_{k+2}) \cdot (y_{k+1} - y_{k+2}) \\ &\quad + \sigma_k \cdot U_k \cdot (y_{k+1} - y_{k+2}) + \sigma_{k+1} \cdot U_k \cdot (y_k - y_{k+2}) + U_k \\ &= \sigma_k \cdot \sigma_{k+1} \cdot \sigma_k \cdot (y_k - y_{k+1}) \cdot (y_k - y_{k+2}) \cdot (y_{k+1} - y_{k+2}) \\ &\quad + \sigma_k \cdot \sigma_{k+1} \cdot (y_k - y_{k+2}) \cdot (y_{k+1} - y_{k+2}) \\ &\quad + (y_k - y_{k+1}) \cdot (y_{k+1} - y_{k+2}) \\ &\quad + \sigma_k \cdot (y_k - y_{k+2}) + \sigma_{k+1} \cdot \sigma_k \cdot (y_k - y_{k+1}) \cdot (y_k - y_{k+2}) \\ &\quad + \sigma_{k+1} \cdot (y_k - y_{k+2}) + 1.\end{aligned} \tag{4.6.17}$$

By comparing the right-hand sides in (4.6.16) and (4.6.17), we obtain (4.6.14). □

Corollary 4.6.3. *In accordance with* (4.6.13), *commutation of intertwining operator* U_k *with any function* $f(y_1,\ldots,y_n)$ *is reduced to transposition of two Jucys–Murphy elements* y_k *and* y_{k+1} :

$$U_k \cdot f(y_1,\ldots,y_k,y_{k+1},\ldots,y_n) = f(y_1,\ldots,y_{k+1},y_k,\ldots,y_n) \cdot U_k. \quad (4.6.18)$$

This gives a necessary condition for the function $f(y_1,\ldots,y_n)$ *to belong to the center* Z_n *of algebra* $\mathbb{C}[S_n]$, *namely, function* $f(y_1,\ldots,y_n)$ *must be symmetric in all its arguments.*

This requirement is, in fact, also sufficient.

Proposition 4.6.4. *Any symmetric function* f_s *of variables* $\{y_1,\ldots,y_n\}$ *belongs to the center of group algebra* $\mathbb{C}[S_n]$:

$$[\sigma_k, f_s(y_1,\ldots,y_n)] = 0, \quad \forall k. \quad (4.6.19)$$

Proof. To prove that symmetric function $f_s(y_1,\ldots,y_n)$ is central, we use the symmetry in two variables y_k and y_{k+1} and write

$$\begin{aligned} &f_s(y_1,\ldots,y_k,y_{k+1},\ldots,y_n) \\ &\quad = \tilde{f}(y_1,\ldots,y_{k-1},(y_k \cdot y_{k+1}),(y_k + y_{k+1}),\, y_{k+2},\ldots,y_n). \end{aligned}$$

We then use locality relations (4.6.6) and equalities

$$[\sigma_k, y_k \cdot y_{k+1}] = 0, \quad [\sigma_k, y_k + y_{k+1}] = 0,$$

which can be checked by employing induction formula (4.6.3) and relations (4.6.1). □

Remark 3. Central elements $z_n \in \mathbb{C}[S_n]$ occurring in (4.6.8) and equal to the sums of all transpositions can be written as the simplest symmetric functions

$$z_n = y_1 + y_2 + \cdots + y_n.$$

4.6.2. Idempotents and spectrum of Jucys–Murphy operators

We now turn to construction of primitive orthogonal idempotents $e_\alpha \in \mathbb{C}[S_n]$ obeying (4.5.3), (4.5.4). Their definition is not unique, so we additionally require that idempotents e_α are eigenvectors of Jucys–Murphy operators in left regular representation:

$$y_k \cdot e_\alpha = a_k^{(\alpha)} e_\alpha \quad (a_k^{(\alpha)} \in \mathbb{C}), \quad \forall k = 1, \dots, n. \tag{4.6.20}$$

This choice of idempotents is always possible in algebra $\mathbb{C}[S_n]$. As an example, idempotents (4.5.33) in algebra $\mathbb{C}[S_2]$ obey the relations (4.5.34), which, in view of $\sigma_1 = y_2$, coincide with (4.6.20).

Problem 4.6.3. Show that in algebra $\mathbb{C}[S_3]$, idempotents (4.5.37) with specific choice $\varepsilon = 1/2$ have the property (4.6.20):

$$\begin{gathered}(y_1, y_2, y_3) \cdot e_{[3]} = (0, 1, 2) e_{[3]}, \quad (y_1, y_2, y_3) \cdot e_{[1^3]} = (0, -1, -2) e_{[1^3]}, \\ (y_1, y_2, y_3) \cdot e_{[2,1]_1} = (0, 1, -1) e_{[2,1]_1}, \quad (y_1, y_2, y_3) \cdot e_{[2,1]_2} = (0, -1, 1) e_{[2,1]_2}.\end{gathered} \tag{4.6.21}$$

It follows from (4.6.20) that idempotents e_α commute with Jucys–Murphy elements $\{y_k\}$. Indeed, we multiply the two sides of (4.6.20) on the left by e_β:

$$e_\beta \cdot y_k \cdot e_\alpha = \delta_{\alpha\beta} a_k^{(\alpha)} e_\alpha,$$

and sum over all α. As a result, taking into account completeness condition (4.5.3), we obtain

$$e_\beta \cdot y_k = a_k^{(\beta)} e_\beta \quad \forall \beta. \tag{4.6.22}$$

By comparing this relation for $\beta = \alpha$ with condition (4.6.20), we find

$$[y_k, e_\alpha] = 0, \quad \forall k. \tag{4.6.23}$$

Since Jucys–Murphy elements $\{y_k\}$ make complete set of commuting elements in $\mathbb{C}[S_n]$ (see Proposition 4.6.1), idempotents e_α are some functions of $\{y_m\}$.

Thus, in accordance with (4.6.20), every primitive idempotent e_α, which we search for, is characterized by a set of eigenvalues

$$\Lambda^{(\alpha)} = (a_1^{(\alpha)}, a_2^{(\alpha)}, \dots, a_n^{(\alpha)}) \in \mathrm{Spec}(y_1, y_2, \dots, y_n). \tag{4.6.24}$$

Here $\mathrm{Spec}(y_1, y_2, \dots, y_n)$ denotes the set of spectra $(a_1^{(\alpha)}, a_2^{(\alpha)}, \dots, a_n^{(\alpha)})$, characteristic of all primitive orthogonal idempotents e_α that make a complete system. Every spectrum $\Lambda^{(\alpha)}$ corresponds to unique primitive idempotent $e_\alpha \in \mathbb{C}[S_n]$, otherwise the set of commuting elements $\{y_2, \dots, y_n\}$ would be incomplete (see Proposition 4.6.1). Note also that since $(a_1^{(\alpha)}, \dots, a_n^{(\alpha)}) \neq (a_1^{(\beta)}, \dots, a_n^{(\beta)})$, the properties (4.6.20) and (4.6.22) ensure orthogonality of different idempotents, $e_\alpha \cdot e_\beta = 0$. This is also consistent with our original proposal.

> **Problem 4.6.4.** Derive orthogonality of elements e_α and e_β for $\Lambda^{(\alpha)} \neq \Lambda^{(\beta)}$, using Eqs. (4.6.20) and (4.6.22).

Let us now consider the special arrangement of integers in the left upper corner of infinite-dimensional matrix

0	1	2	3	...
−1	0	1	2	...
−2	−1	0	1	...
−3	−2	−1	0	...
⋮	⋮	⋮	⋮	...

Namely, the main diagonal has zeroes, the upper next-to-main diagonal has $+1$, the lower next-to-main diagonal has -1, etc. We call the number assigned in this way to a box as *content* of this box. A box with coordinates (k, m), where k and m label row and column, respectively, has content $(m - k)$.

Now we put into this corner a standard Young tableau T_{λ_n} which has the shape of Young diagram λ_n with n boxes. As a result, we obtain the standard Young tableau in which the box with coordinates (k, m) has both the original number i and content $(m-k)$. This box is represented as $\boxed{{}^{i}m{-}k}$. Then, by setting $a_i = (m-k)$, we assign to every tableau T_{λ_n} (whose boxes have both numbers and contents) a sequence of n integers $(a_1, a_2, \dots, a_n)$. We call this sequence a *content vector* of length n. The set of all content vectors of length n obtained from all standard tableaux T_λ of a given shape $\lambda \vdash n$ is denoted by $\mathrm{Cont}(\lambda)$. In other words, we associate the set $\mathrm{Cont}(\lambda)$ to each Young diagram $\lambda \vdash n$.

As an example, consider standard tableau

$$\mathsf{T}_{\lambda_8} = \begin{array}{|c|c|c|c|}\hline 1 & 2 & 4 & 6 \\ \hline 3 & 5 & 8 \\ \cline{1-3} 7 \\ \cline{1-1}\end{array}$$

($\lambda_8 = [4, 3, 1]$) and put the contents in its boxes. We arrive at the tableau

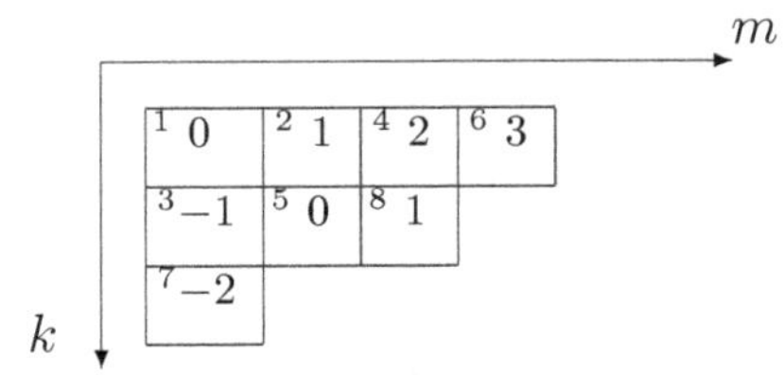

(4.6.25)

which uniquely defines the sequence $(a_1, a_2, \ldots, a_8) = (0, 1, -1, 2, 0, 3, -2, 1) \in \mathrm{Cont}(\lambda_8)$.

We show below (and this is the main result of this section) that every primitive idempotent $e_\alpha \in \mathbb{C}[S_n]$ obeying (4.6.20) determines, in one-to-one way, a standard tableau $\mathsf{T}_{\lambda_n} \leftrightarrow e_\alpha$ in such a way that the sequence of eigenvalues (4.6.24) of Jucys–Murphy elements on vector e_α coincides with the content vector for boxes in that standard tableau T_{λ_n}:

$$(a_1^{(\alpha)}, a_2^{(\alpha)}, \ldots, a_n^{(\alpha)}) = (a_1, a_2, \ldots, a_n). \tag{4.6.26}$$

Content a_1 of the first box in standard tableau T_{λ_n} is always equal to zero, which is consistent with the eigenvalue $a_1^{(\alpha)} = 0$ of element y_1 defined in (4.6.2).

We note that idempotents obeying (4.6.20) do not, generally speaking, coincide with normalized Young symmetrizers (4.3.51) constructed with standard tableaux T_{λ_n}. As an example, idempotents (4.5.37) obey (4.6.21) for $\varepsilon = \frac{1}{2}$, but in this case they are not equal to Young symmetrizers determined by (4.5.37) with $\varepsilon = 0$.

So, on the one hand, primitive idempotents $e_\alpha \in \mathbb{C}[S_n]$ obeying (4.6.20) are associated with standard Young tableaux with n boxes and content sequence $(a_1, \ldots, a_n)$, and on the other hand, each of these idempotents generates, upon left (right) action on it by the whole algebra $\mathbb{C}[S_n]$, irreducible representation of this algebra (see Section 4.5.1). Thus, in this approach too, there is a correspondence between irreducible representations of algebra $\mathbb{C}[S_n]$ (group S_n) and Young tableaux. We describe below (see Section 4.6.4) a method of explicit construction of idempotents obeying (4.6.20) with eigenvalues $(a_1^{(\alpha)}, \ldots, a_n^{(\alpha)})$ which are given by contents of the standard Young tableaux. Notably, this method automatically gives a complete system of orthogonal primitive idempotents.

We point out that the right-hand side of (4.6.26) is a sequence of integers. Therefore, our first step in establishing the relationship (4.6.26) is the following proposition.

Proposition 4.6.5. *Spectrum of Jucys–Murphy operators y_j for algebra $\mathbb{C}[S_n]$ is such that*

$$\mathrm{Spec}(y_j) \subset [1-j, j-1] \quad (\forall j = 1, 2, \ldots, n), \tag{4.6.27}$$

where $[1-j, j-1]$ denotes the set of integers

$$[1-j, j-1] = \{1-j, \ldots, -2, -1, 0, 1, 2, \ldots, j-1\}. \tag{4.6.28}$$

We emphasize that formula (4.6.27) does not imply that the spectrum of operator y_j contains *all* integers from interval $[1-j, j-1]$. Say, the spectrum of operator y_2 does not contain eigenvalue 0, see (4.6.21) and formula (4.6.29) below.

Proof. We prove (4.6.27) by induction. The last relation in (4.6.1) shows that element $y_2 = \sigma_1$ obeys characteristic equation

$$(y_2 - 1)(y_2 + 1) = 0. \tag{4.6.29}$$

So, we have $\mathrm{Spec}(y_2) = \{-1, +1\}$, and the spectrum of operator y_2 obeys (4.6.27). Let us assume that the spectrum of y_{j-1} obeys (4.6.27) for some $j \geq 3$. We would like to show that the spectrum of y_j also obeys (4.6.27). Consider characteristic equation for element y_{j-1} (cf. (4.6.29)):

$$\prod_\alpha (y_{j-1} - a^{(\alpha)}_{j-1}) = 0, \tag{4.6.30}$$

where $a^{(\alpha)}_{j-1} \in \mathrm{Spec}(y_{j-1})$. We introduce a notation for the left-hand side of (4.6.30):

$$f(y_{j-1}) \equiv \prod_\alpha (y_{j-1} - a^{(\alpha)}_{j-1}).$$

Making use of intertwining operators U_{j-1} with properties (4.6.13), (4.6.15), we find that Eq. (4.6.30) yields the identity

$$0 = U_{j-1} \cdot f(y_{j-1}) \cdot U_{j-1} = f(y_j) \cdot U^2_{j-1} = f(y_j) \cdot (1 - (y_{j-1} - y_j)^2), \tag{4.6.31}$$

which means that

$$\mathrm{Spec}(y_j) \subset \mathrm{Spec}(y_{j-1}) \cup (\mathrm{Spec}(y_{j-1}) + 1) \cup (\mathrm{Spec}(y_{j-1}) - 1),$$

and hence proves (4.6.27). □

Let $\Lambda = (a_1, \ldots, a_i, a_{i+1}, \ldots, a_n)$ be a set of eigenvalues of Jucys–Murphy elements, which occurs when these elements act on primitive idempotent $e_\Lambda \in \mathbb{C}[S_n]$ (see (4.6.20)). We say that idempotent e_Λ has spectral set $\Lambda \in \mathrm{Spec}(y_1,\dots,y_n)$.

Proposition 4.6.6. *Eigenvalues of neighboring Jucys–Murphy operators y_i and y_{i+1} on one and the same idempotent e_Λ do not coincide. In other words, the set Λ is such that $a_i \neq a_{i+1}$ for all $i = 1,\dots,n-1$.*

Proof. We give the proof *ad absurdum*. Let there be a primitive idempotent $e_\Lambda \neq 0$ such that $a_i = a_{i+1}$, i.e.,

$$y_i \cdot e_\Lambda = a_i e_\Lambda = y_{i+1} \cdot e_\Lambda. \tag{4.6.32}$$

Then we use the intertwining operator (4.6.11) to obtain

$$e_\Lambda \cdot U_i \cdot e_\Lambda = e_\Lambda \cdot (\sigma_i \cdot (y_i - y_{i+1}) + 1) \cdot e_\Lambda = e_\Lambda \cdot e_\Lambda = e_\Lambda. \tag{4.6.33}$$

On the other hand, in view of (4.6.20) and (4.6.22) we have

$$e_\Lambda \cdot U_i \cdot e_\Lambda = e_\Lambda \cdot (\sigma_i \cdot y_i - y_i \cdot \sigma_i) \cdot e_\Lambda = e_\Lambda \cdot (\sigma_i \cdot a_i - a_i \cdot \sigma_i) \cdot e_\Lambda = 0,$$

which contradicts (4.6.33). □

Let $e_{\Lambda'} \in \mathbb{C}[S_n]$ be idempotent with spectral set $\Lambda' = (a'_1,\dots,a'_n)$. We make use of (4.6.13) and get ($k = 2,\dots,n$):

$$a_k e_{\Lambda'} \cdot U_i \cdot e_\Lambda = e_{\Lambda'} \cdot U_i \cdot y_k \cdot e_\Lambda = e_{\Lambda'} \cdot y_{\sigma_i(k)} \cdot U_i \cdot e_\Lambda = a'_{\sigma_i(k)} e_{\Lambda'} \cdot U_i \cdot e_\Lambda, \tag{4.6.34}$$

where $\sigma_i = (i, i+1)$ is neighboring transposition. Thus, if $e_{\Lambda'} \cdot U_i \cdot e_\Lambda \neq 0$, then $a'_{\sigma_i(k)} = a_k$, which is equivalent to equality

$$\Lambda' \equiv s_i \cdot \Lambda = (a_1,\dots,a_{i+1},a_i,\dots,a_n) \in \mathrm{Spec}(y_1,\dots,y_n). \tag{4.6.35}$$

Here operator s_i transposes in the set Λ the eigenvalues a_i and a_{i+1}, and leave other eigenvalues in their places. On the other hand, Eq. (4.6.34) gives

$$e_{\Lambda''} \cdot U_i \cdot e_\Lambda = 0, \quad \forall \Lambda'' \neq s_i \cdot \Lambda. \tag{4.6.36}$$

It is worth recalling that every spectral set Λ corresponds to at most one primitive idempotent $e_\Lambda \in \mathbb{C}[S_n]$, which follows from completeness of the set of commuting elements $\{y_2,\dots,y_n\}$ (see Proposition 4.6.1).

Proposition 4.6.7. *Let idempotent $e_\Lambda \in \mathbb{C}[S_n]$ have spectral set*

$$\Lambda = (a_1,\dots,a_i,a_{i+1},\dots,a_n) \in \mathrm{Spec}(y_1,\dots,y_n). \tag{4.6.37}$$

(1) *If* $|a_{i+1} - a_i| \neq 1$, *then there exists idempotent* $e_{\Lambda'} \in \mathbb{C}[S_n]$ *whose spectral set is given by* (4.6.35), *and the following identities are satisfied:*

$$U_i \cdot e_\Lambda = e_{\Lambda'} \cdot U_i, \quad e_\Lambda \cdot U_i = U_i \cdot e_{\Lambda'}. \tag{4.6.38}$$

(2) *If* $\Lambda' = s_i \Lambda \notin \mathrm{Spec}(y_1, \dots, y_n)$, *then idempotent* $e_\Lambda \in \mathbb{C}[S_n]$ *obeys*

$$U_i \cdot e_\Lambda = 0, \quad e_\Lambda \cdot U_i = 0, \tag{4.6.39}$$

and $|a_{i+1} - a_i| = 1$.

Proof. (1) Consider the case $|a_{i+1} - a_i| \neq 1$. Let us prove that there exists idempotent $e_{\Lambda'}$ whose spectral set is given by (4.6.35). Consider product

$$U_i \cdot e_\Lambda = (\sigma_i(y_i - y_{i+1}) + 1) \cdot e_\Lambda = (\sigma_i(a_i - a_{i+1}) + 1) \cdot e_\Lambda. \tag{4.6.40}$$

Since the operator $(\sigma_i(a_i - a_{i+1}) + 1)$ is invertible for $|a_{i+1} - a_i| \neq 1$, we have $U_i \cdot e_\Lambda \neq 0$. Making use of completeness property (4.5.3) and recalling (4.6.36), we find

$$U_i \cdot e_\Lambda = \sum_\alpha e_\alpha \cdot U_i \cdot e_\Lambda = e_{\Lambda'} \cdot U_i \cdot e_\Lambda, \tag{4.6.41}$$

where the right-hand side would vanish if there were no idempotent $e_{\Lambda'} \neq 0$ with spectral set (4.6.35). Since $U_i \cdot e_\Lambda \neq 0$, such an idempotent does exist. Now, we again use completeness relation (4.5.3) and property (4.6.36) to obtain, in addition to (4.6.41), that

$$e_{\Lambda'} \cdot U_i = e_{\Lambda'} \cdot U_i \cdot \sum_\alpha e_\alpha = e_{\Lambda'} \cdot U_i \cdot e_\Lambda. \tag{4.6.42}$$

By comparing (4.6.41) and (4.6.42), we arrive at the first equality in (4.6.38). The second equality in (4.6.38) is either derived in a similar way, or obtained from the first equality:

$$\begin{aligned} U_i \cdot e_\Lambda = e_{\Lambda'} \cdot U_i &\Rightarrow U_i^2 \cdot e_\Lambda \cdot U_i = U_i \cdot e_{\Lambda'} \cdot U_i^2 \\ &\Rightarrow (1 - (a_{i+1} - a_i)^2) \cdot e_\Lambda \cdot U_i = U_i \cdot e_{\Lambda'} \cdot (1 - (a_{i+1} - a_i)^2) \\ &\Rightarrow e_\Lambda \cdot U_i = U_i \cdot e_{\Lambda'}, \end{aligned}$$

where we use the identity (4.6.15) and the property $(a_{i+1} - a_i)^2 \neq 1$.

(2) If $\Lambda' = s_i \cdot \Lambda$ does not belong to $\mathrm{Spec}(y_1, \dots, y_n)$, i.e., there is no idempotent with spectral set Λ' in the complete system of idempotents, then we sum up equalities (4.6.36) over all spectral sets Λ'' and use completeness property (4.5.3) to obtain the first identity in (4.6.39). The second identity

in (4.6.39) is proven in the same way. Let us show that in this case one has $|a_i - a_{i+1}| = 1$. Let us write the first relation in (4.6.39) explicitly:

$$0 = U_i \cdot e_\Lambda = (\sigma_i(y_i - y_{i+1}) + 1) \cdot e_\Lambda = (\sigma_i(a_i - a_{i+1}) + 1) \cdot e_\Lambda. \quad (4.6.43)$$

If $|a_i - a_{i+1}| \neq 1$, then the operator $\Big(\sigma_i(a_i - a_{i+1}) + 1\Big)$ is invertible, which contradicts (4.6.43). □

Corollary 4.6.8. *We multiply both sides of the first equality in* (4.6.38) *on the right by* U_i *and make use of* (4.6.15) *to obtain*

$$U_i \cdot e_\Lambda \cdot U_i = (1 - (a_{i+1} - a_i)^2)e_{\Lambda'}, \quad (4.6.44)$$

where we recall the eigenvalues (4.6.35) *of operators* y_i, y_{i+1} *on idempotent* $e_{\Lambda'}$. *Then, if* $|a_{i+1} - a_i| \neq 1$, *the relation* (4.6.44) *gives explicit formula that relates idempotents* $e_{\Lambda'}$ *and* e_Λ *whose spectral sets obey* (4.6.35):

$$e_{\Lambda'} = \frac{1}{(1 - (a_{i+1} - a_i)^2)} U_i \cdot e_\Lambda \cdot U_i \in \mathbb{C}[S_n]. \quad (4.6.45)$$

Problem 4.6.5. Show that the element $e_{\Lambda'}$ given by (4.6.45) has the property (4.5.4), once this property holds for the element e_Λ.

Corollary 4.6.9. *Recall that in view of* (4.6.23), *idempotents* e_Λ *are elements of commutative Gelfand–Tsetlin subalgebra* Y_n, *i.e., they are polynomials in generators* $(y_2, \dots, y_n) \in Y_n$. *Since* e_Λ *are eigenvectors of operators* $(y_2, \dots, y_n)$, *these polynomials obey the identities*

$$e_\Lambda \cdot (y_2 - a_2) = e_\Lambda \cdot (y_3 - a_3) = \cdots = e_\Lambda \cdot (y_n - a_n) = 0,$$

which are analogous to identities (4.6.10).

Corollary 4.6.10. *Using the explicit expressions* (4.6.11) *for* U_i *and spectral sets* (4.6.35) *and* (4.6.37) *for* $e_{\Lambda'}$ *and* e_Λ, *we write Eq.* (4.6.38) *as follows*:

$$\begin{aligned} (\sigma_i(a_i - a_{i+1}) + 1) \cdot e_\Lambda &= e_{\Lambda'} \cdot ((a_i - a_{i+1})\sigma_i - 1), \\ (\sigma_i(a_{i+1} - a_i) + 1) \cdot e_{\Lambda'} &= e_\Lambda \cdot ((a_{i+1} - a_i)\sigma_i - 1). \end{aligned} \quad (4.6.46)$$

Also, if $\Lambda' \notin \mathrm{Spec}(y_1, \dots, y_n)$ and hence $a_{i+1} = a_i \pm 1$, then $e_{\Lambda'} = 0$, and the relations (4.6.46) take the form

$$(\sigma_i \mp 1) \cdot e_\Lambda = 0, \quad e_\Lambda \cdot (\sigma_i \mp 1) = 0 \quad (4.6.47)$$

(upper and lower signs are correlated with the signs in $a_{i+1} = a_i \pm 1$).

In fact, equality $a_{i+1} = a_i \pm 1$ is not only necessary but also sufficient for the validity of (4.6.47) and the property $\Lambda' \notin \mathrm{Spec}(y_1, \dots, y_n)$.

Proposition 4.6.11. *If the spectral set Λ obeys $a_{i+1} = a_i \pm 1$ for some $i \le n-1$, then idempotent e_Λ obeys* (4.6.47) *and $\Lambda' \notin \mathrm{Spec}(y_1, \dots, y_n)$.*

Proof of this proposition is quite nontrivial, see [47] and also remark 1 in Section 4.6.3.

We summarize the above results, including Propositions 4.6.5–4.6.7 and 4.6.11, as follows.

Theorem 4.6.12. *Let a spectral set of idempotent e_Λ be equal to*

$$\Lambda = (a_1, \dots, a_i, a_{i+1}, \dots, a_n) \in \mathrm{Spec}(y_1, \dots, y_n). \tag{4.6.48}$$

Then, in accordance with (4.6.27) *we have $a_i \in [1-j, j-1]$ and this set Λ obeys the following conditions*:

(1) *$a_i \neq a_{i+1}$ for all $i \le n-1$ (see Proposition* 4.6.6);
(2) *if $a_{i+1} \neq a_i \pm 1$, then the transposition of eigenvalues a_i and a_{i+1} in (4.6.48) gives again a spectral set*

$$\Lambda' = (a_1, \dots, a_{i+1}, a_i, \dots, a_n) \in \mathrm{Spec}(y_1, \dots, y_n); \tag{4.6.49}$$

the pertinent idempotent $e_{\Lambda'}$ is given by (4.6.45);
(3) *if $a_{i+1} = a_i \pm 1$, then $e_{\Lambda'} = 0$, and idempotent e_Λ forms one-dimensional representation of element σ_i (see* (4.6.47)) :

$$\sigma_i \cdot e_\Lambda = \pm e_\Lambda, \quad e_\Lambda \cdot \sigma_i = \pm e_\Lambda. \tag{4.6.50}$$

Corollary 4.6.13. *Sequences $(\dots, a_i, a_{i+1}, a_{i+2}, \dots) = (\dots, a, a \pm 1, a, \dots)$ cannot be present in spectral set (4.6.48), since in view of part* (3) *of Theorem* 4.6.12, *idempotent e_Λ would generate one-dimensional representations* (4.6.50) *of elements σ_i and σ_{i+1}, namely $\sigma_i = \pm 1$ and $\sigma_{i+1} = \mp 1$, which would contradict braid relation* (4.6.1).

Definition 4.6.2. A set of sequences $(a_1, a_2, \dots, a_n)$ (of length n) of integers $a_i \in [1-i, i-1]$ obeying

$$\begin{aligned}
&(1) \quad a_1 = 0;\\
&(2) \quad a_r = z \Rightarrow \{z+1, z-1\} \cap \{a_1, \dots, a_{r-1}\} \neq \emptyset \ \forall r > 1, \ z \neq 0;\\
&(3) \quad a_i = a_j = z (i < j) \Rightarrow \{z+1, \ z-1\} \subset \{a_{i+1}, \dots, a_{j-1}\}
\end{aligned} \tag{4.6.51}$$

is called *set of content vectors* of length n and denoted by Cont_n.

We show below that Cont_n is a union of all sets $\mathrm{Cont}(\lambda)$ for all Young diagrams $\lambda \vdash n$. But before this we need a number of results.

Theorem 4.6.14. *Spectral set* $\Lambda = (a_1, \ldots, a_n) \in \mathrm{Spec}(y_1, y_2, \ldots, y_n)$, *has the properties* (4.6.51), *i.e.*,

$$\mathrm{Spec}(y_1, y_2, \ldots, y_n) \subset \mathrm{Cont}_n.$$

Proof. In accordance with Proposition 4.6.5, eigenvalues a_i are integers in the interval $[1-i, i-1]$. Let us prove (1), (2) and (3) in (4.6.51) for all spectral sets $\Lambda \in \mathrm{Spec}(y_1, \ldots, y_n)$.

(1) Property (1) follows from definition $y_1 = 0$, see (4.6.2).

(2) In accordance with part (2) of Theorem 4.6.12, there is the following transformation of the spectral sets. If, in the set (4.6.48), we have $a_r \neq a_{r-1} \pm 1$, then we can transpose the elements a_r and a_{r-1} and obtain another spectral set

$$\Lambda' = (a_1, \ldots, a_{r-2}, a_r, a_{r-1}, \ldots, a_n) \in \mathrm{Spec}(y_1, \ldots, y_n).$$

Now, if $a_r \neq a_{r-2} \pm 1$, then we can transpose a_r and a_{r-2} in Λ' to obtain yet another spectral set $\Lambda'' = (\ldots, a_{r-3}, a_r, a_{r-2}, a_{r-1}, \ldots) \in \mathrm{Spec}(y_1, \ldots, y_n)$. Thus, if there were no elements in sequence $\{a_1, \ldots, a_{r-1}\}$ equal to $(a_r + 1)$ or $(a_r - 1)$, then these transformations would enable us to move the value $a_r = z \neq 0$ to the first position on the left, still remaining with some spectral set, which would contradict the property (1) in (4.6.51).

(3) Without loss of generality, we assume that none of the eigenvalues $\{a_{i+1}, \ldots, a_{j-1}\}$ equals z. The transformation discussed in the context of property (2) shows that one of the eigenvalues $\{a_{i+1}, \ldots, a_{j-1}\}$ is necessarily equal to either $(z+1)$ or $(z-1)$, otherwise one could move $a_j = z$ to the $(i+1)$th place neighboring $a_i = z$, which would contradict the property (1) in Theorem 4.6.12. Let $(z-1) \in \{a_{i+1}, \ldots, a_{j-1}\}$ (the case $(z+1) \in \{a_{i+1}, \ldots, a_{j-1}\}$ is completely analogous). We have to show that the sequence $\{a_{i+1}, \ldots, a_{j-1}\}$ contains $(z+1)$ as well. The idea of the proof is that in the opposite case, allowed (in the sense of Theorem 4.6.12) permutations of $\{z, a_{i+1}, \ldots, (z-1), \ldots, a_{j-1}, z\}$ can be used to put the eigenvalues $z, (z-1), z$ one near another in this order, which would contradict Corollary 4.6.13. The latter property can

be proven by induction, using parts (1) and (3) of Theorem 4.6.12 and Corollary 4.6.13. We leave the completion of this simple but lengthy proof to the reader (see [47, proof of Theorem 5.1]). □

Consider a set Tab_n of all standard Young tableaux T_{λ_n} with the shapes of all possible diagrams λ_n having n boxes.

Proposition 4.6.15. *There is one-to-one correspondence between elements of the set* Cont_n *and elements of the set* Tab_n.

Proof. Recall that we assign contents a_i to boxes of standard Young tableaux T_{λ_n}, namely, for a box with coordinates (k, m) and number i in it we assign content $a_i = (m - k)$. So, every standard tableau T_{λ_n} defines content vector $(a_1, \ldots, a_n)$, which, as we see momentarily, have properties listed in Definition 4.6.2. This means that every element in Tab_n defines an element of Cont_n (we also have to show the inverse, see below). Indeed, the properties of content vectors Λ, given in Definition 4.6.2, have simple geometric interpretation (see Fig. 4.6.1) from the viewpoint of the correspondence of numbers $a_i \in \Lambda \in \mathrm{Cont}_n$ to the contents of the boxes of the standard Young tableaux T_{λ_n}. A box with number r and content $z \neq 0$ can enter a standard tableau T_{λ_n} if either (1) this tableau has a box with number $s < r$ and content $(z + 1)$, which is placed above the box r, or (2) tableau T_{λ_n} has a box with number $t < r$ and content $(z-1)$, which is placed on the left of box r. This shows that the property (2) of Definition 4.6.2 holds for tableau T_{λ_n}. Also, if two boxes with number i and j $(j > i)$ in a standard tableau T_{λ_n} have the same content z, then they belong to one diagonal in T_{λ_n}, and this tableau necessarily has boxes with contents $(z-1)$

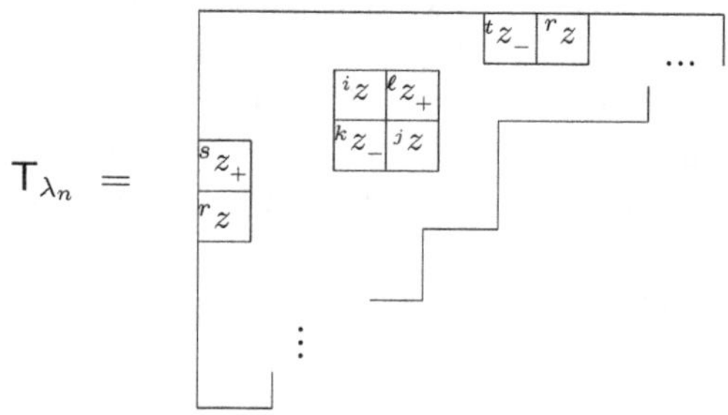

Fig. 4.6.1. Graphic representation of properties (2) and (3) in Definition 4.6.2 for components of content vectors corresponding to tableau T_{λ_n}. Notations here are $z_- = z - 1$ and $z_+ = z + 1$.

and $(z+1)$ and numbers k, ℓ such that $i < k, \ell < j$ (see Fig. 4.6.1). This proves the validity of the property (3) of Definition 4.6.2. We see that every tableau T_{λ_n} defines some content vector $(a_1, \dots, a_n) \in \mathrm{Cont}_n$.

Let us now prove the inverse, i.e., that every content vector $(a_1, \dots, a_n) \in \mathrm{Cont}_n$ defines a standard tableau T_{λ_n}. We give the proof by induction. The set Cont_1 consists of one content vector $a_1 = 0$; this corresponds to tableau with one box. In accordance with the property (2) in Definition 4.6.2, the set Cont_2 consists of two vectors $(a_1, a_2) = (0,1)$ and $(a_1, a_2) = (0,-1)$ corresponding to tableaux (4.5.35):

$$(0,1) \leftrightarrow \mathsf{T}_{[2]} = \boxed{{}^1 0}\boxed{{}^2 1}, \quad (0,-1) \leftrightarrow \mathsf{T}_{[1^2]} = \begin{array}{|c|}\hline {}^1 0 \\ \hline {}^2 \text{-}1 \\ \hline \end{array}. \tag{4.6.52}$$

Now, let a vector $(a_1, \dots, a_n) \in \mathrm{Cont}_n$ define a standard tableau $\mathsf{T}_{\lambda_n} \in \mathrm{Tab}_n$, so that the contents of boxes in the latter coincide with $(a_1, \dots, a_n)$. Consider content vector $(a_1, \dots, a_n, a_{n+1}) \in \mathrm{Cont}_{n+1}$, obtained by adding a component a_{n+1} to vector $(a_1, \dots, a_n) \in \mathrm{Cont}_n$ in such a way that properties (2) and (3) in Definition 4.6.2 hold. Let the maximal and minimal contents in boxes of tableau T_{λ_n} be equal to z_1 and z_2, respectively (see Fig. 4.6.2), i.e., $z_1 \le a_1, \dots, a_n \le z_2$. Then, in accordance with property (2) in Definition 4.6.2, we have $(z_1 - 1) \le a_{n+1} \le (z_2 + 1)$ and maximal and minimal possible values of a_{n+1} in content vector $(a_1, \dots, a_{n+1}) \in \mathrm{Cont}_{n+1}$ are equal to $a_{n+1} = z_1 - 1$ and $a_{n+1} = z_2 + 1$, respectively. If $a_{n+1} = z_1 - 1$, then content vector $(a_1, \dots, a_{n+1}) \in \mathrm{Cont}_{n+1}$ uniquely defines standard tableau $\mathsf{T}_{\lambda_{n+1}}$ with the $(n+1)$th box added below the box s with content z_1, and if $a_{n+1} = z_2 + 1$, then $(a_1, \dots, a_{n+1}) \in \mathrm{Cont}_{n+1}$ uniquely defines standard tableau $\mathsf{T}_{\lambda_{n+1}}$, with $(n+1)$th box placed to the right of box r with content z_2 (see Fig. 4.6.2). Now, in view of the property (3) in Definition 4.6.2, the component a_{n+1} in content vector $(a_1, \dots, a_{n+1}) \in \mathrm{Cont}_{n+1}$ obeys $a_{n+1} \neq z_1$ and $a_{n+1} \neq z_2$ (otherwise $(z_1 - 1)$ or $(z_2 + 1)$ belong to the set $(a_1, \dots, a_n)$). Let us consider the remaining case $z_1 < a_{n+1} < z_2$ and show that component a_{n+1} in content vector $(a_1, \dots, a_n, a_{n+1}) \in \mathrm{Cont}_{n+1}$ can take only those values z which correspond to external corners of tableau T_{λ_n} (see Fig. 4.6.2). Indeed, by adding $(n+1)$th box with content $a_{n+1} = z$, where $z_1 < z < z_2$, we can meet only the possibilities shown in Fig. 4.6.3. It is only the case (D) which has the property (3) in Definition 4.6.2. The resulting content vector $(a_1, \dots, a_n, a_{n+1}) \in \mathrm{Cont}_{n+1}$ uniquely defines standard tableau $\mathsf{T}_{\lambda_{n+1}}$, which is obtained by adding $(n+1)$th box in the external corner of standard tableau T_{λ_n}. Together with the cases $a_{n+1} = z_1 - 1$ and

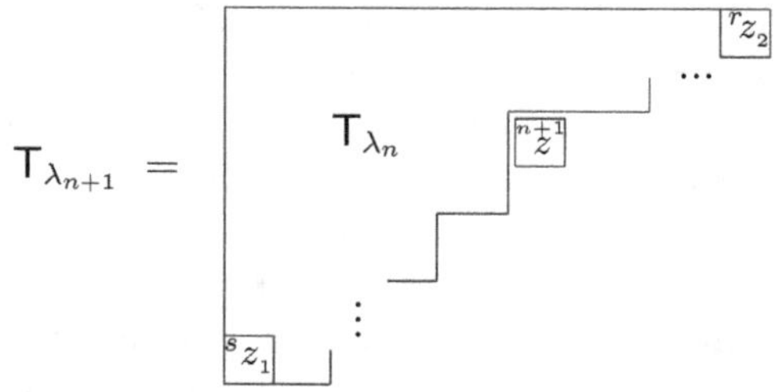

Fig. 4.6.2. Standard tableau $\mathsf{T}_{\lambda_{n+1}}$ obtained from standard tableau T_{λ_n} by adding a box with number $(n+1)$ (and with content $a_{i+1} = z$) to an external corner of diagram λ_n. Here z_1 and z_2 are minimal and maximal contents of boxes in tableau T_{λ_n}.

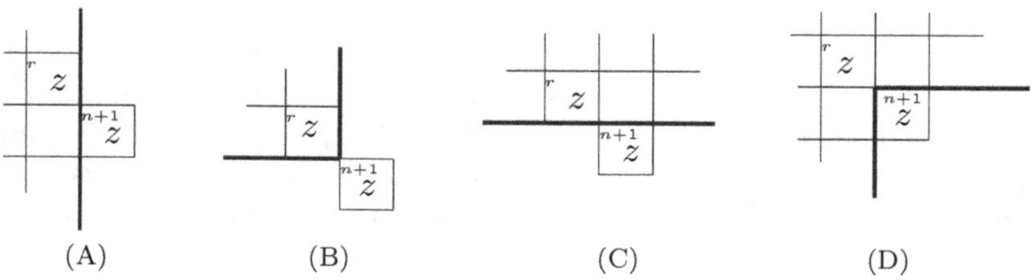

Fig. 4.6.3. The ways to add content $a_{n+1} = z$ to content vector $(a_1, \ldots, a_n) \in \mathrm{Cont}_n$, the latter corresponding to standard tableau T_{λ_n} which has a box $r \leq n$ with the same content $a_r = z$ (thick lines show the boundary of tableau T_{λ_n}). Only the case (D) has property (3) in Definition 4.6.2. The content vector $(a_1, \ldots, a_n, a_{n+1}) \in \mathrm{Cont}_{n+1}$ obtained in this way uniquely defines standard tableau $\mathsf{T}_{\lambda_{n+1}}$.

$a_{n+1} = z_2 + 1$, this proves that every content vector from the set Cont_n uniquely defines standard tableau T_{λ_n}. We thus complete the proof of one-to-one correspondence $\mathsf{T}_{\lambda_n} \leftrightarrow \mathrm{Cont}_n$. □

Remark. Note that induction procedure for obtaining content vector $(a_1, \ldots, a_n, a_{n+1}) \in \mathrm{Cont}_{n+1}$ from content vector $(a_1, \ldots, a_n) \in \mathrm{Cont}_n$ resembles the process in which a box with number $(n+1)$ and content a_{n+1} is thrown in a corner with Young tableau T_{λ_n} and rotated as shown in Fig. 4.6.4. In this process, the box with content a_{n+1} always slides down (case B in Fig. 4.6.3 is considered unstable) and occupies external corner of tableau T_{λ_n}.

Finally, our main proposition is as follows.

Theorem 4.6.16. *There is one-to-one correspondence between the set* Tab_n *of standard Young tableaux with* n *boxes, the set* $\mathrm{Spec}(y_1, \ldots, y_n)$ *and the set of primitive orthogonal idempotents* $e_\Lambda \in \mathbb{C}[S_n]$. *The latter may be labeled by the spectral sets* $\Lambda \in \mathrm{Spec}(y_1, \ldots, y_n)$.

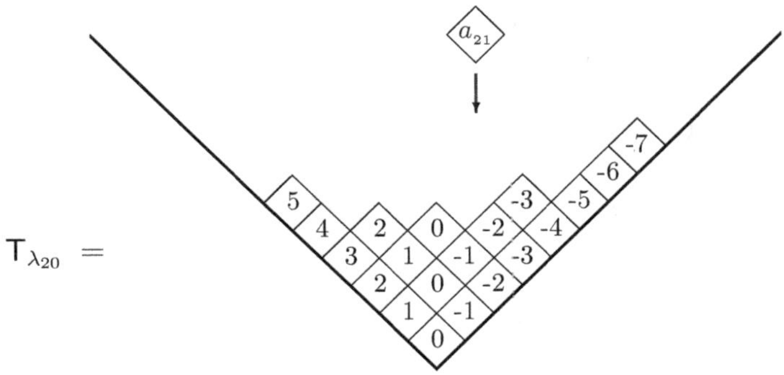

Fig. 4.6.4. Process (case (D) in Fig. 4.6.3) of adding content a_{21} to content vector $(a_1,\ldots,a_{20}) \in \mathrm{Cont}_{20}$ associated with standard tableau $\mathsf{T}_{\lambda_{20}}$. This process results in content vector $(a_1,\ldots,a_{20},a_{21}) \in \mathrm{Cont}_{21}$ obeying all requirements in Definition 4.6.2 and associated, in unique way, with the standard tableau $\mathsf{T}_{\lambda_{21}}$.

Proof. In accordance with Proposition 4.6.15, there is one-to-one correspondence $\mathrm{Tab}_n \leftrightarrow \mathrm{Cont}_n$. One can show (see Ref. [47]) that the number of elements in the set $\mathrm{Tab}_n \simeq \mathrm{Cont}_n$ is equal to the number of elements in $\mathrm{Spec}(y_1,\ldots,y_n)$. Then the result of Theorem 4.6.14 that $\mathrm{Spec}(y_1,\ldots,y_n) \subset \mathrm{Cont}_n$ means that $\mathrm{Spec}(y_1,\ldots,y_n) \leftrightarrow \mathrm{Cont}_n$, which gives finally one-to-one correspondence $\mathrm{Tab}_n \leftrightarrow \mathrm{Spec}(y_1,\ldots,y_n)$. □

We now study a concrete example that illustrates Theorem 4.6.16.

Example (Continues Example 3 in Section 4.5.1). Consider group S_3 and its algebra $\mathbb{C}[S_3]$. Three commuting Jucys–Murphy elements (see (4.6.4)), generating subalgebra $Y_3 \subset \mathbb{C}[S_3]$, are

$$y_1 = 0, \quad y_2 = \sigma_1, \quad y_3 = \sigma_2 + \sigma_2\sigma_1\sigma_2.$$

We act by elements y_1, y_2, y_3 on the right on primitive idempotents (4.5.37). We find that for $\varepsilon = 1/2$, all idempotents (4.5.37) are eigenvectors of operators y_k:

$$\begin{aligned}
(y_1,\, y_2,\, y_3)\cdot e_{[3]} &= (0,\, 1,\, 2)e_{[3]},\\
(y_1,\, y_2,\, y_3)\cdot e_{[1^3]} &= (0,\, -1,\, -2)e_{[1^3]},\\
(y_1,\, y_2,\, y_3)\cdot e_{[2,1]_1} &= (0,\, 1,\, -1)e_{[2,1]_1},\\
(y_1,\, y_2,\, y_3)\cdot e_{[2,1]_2} &= (0,\, -1,\, 1)e_{[2,1]_2}.
\end{aligned} \tag{4.6.53}$$

Here, on the right-hand sides, we give the eigenvalues of operators (y_1, y_2, y_3).

Problem 4.6.6. Check that idempotents (4.5.37) with $\varepsilon = 1/2$ are expressed through Jucys–Murphy elements:

$$
\begin{gathered}
e_{[3]} = \frac{1}{6}(1+y_2)\cdot(1+y_3), \quad e_{[1^3]} = \frac{1}{6}(1-y_2)\cdot(1-y_3), \\
e_{[2,1]_1} = \frac{1}{6}(2-y_3)\cdot(1+y_2), \quad e_{[2,1]_2} = \frac{1}{6}(1-y_2)\cdot(2+y_3).
\end{gathered}
\tag{4.6.54}
$$

On the other hand, consider three diagrams

$$
[3] = \begin{array}{|c|c|c|}\hline 0 & 1 & 2 \\ \hline \end{array}, \quad [1^3] = \begin{array}{|c|}\hline 0 \\ \hline -1 \\ \hline -2 \\ \hline \end{array}, \quad [2,1] = \begin{array}{|c|c|}\hline 0 & 1 \\ \hline -1 & \multicolumn{1}{c}{} \\ \cline{1-1} \end{array}, \tag{4.6.55}
$$

where contents are shown in boxes. We see that contents here coincide precisely with the eigenvalues of operators y_k on the pertinent idempotents, as given in (4.6.53), i.e., with spectral sets of these idempotents. However, the correspondence between Young diagrams with their contents (4.6.55) and idempotents (4.5.37) is not one-to-one, since one and the same diagram $[2,1]$ with its content corresponds to two different idempotents $e_{[2,1]_1}$, $e_{[2,1]_2}$. Clearly, this is due to the fact that one can associate with diagram $[2,1]$ (with its content) two sequences of contents, $(0,1,-1)$ and $(0,-1,1)$. To obtain one-to-one correspondence, we put numbers in boxes of diagrams (4.6.55) in such a way that they become standard Young tableaux. In this way, we obtain four standard Young tableaux with contents:

$$
[3] = \begin{array}{|c|c|c|}\hline {}^{1}0 & {}^{2}1 & {}^{3}2 \\ \hline \end{array}, \quad [1^3] = \begin{array}{|c|}\hline {}^{1}0 \\ \hline {}^{2}{-1} \\ \hline {}^{3}{-2} \\ \hline \end{array}, \quad [2,1]_1 = \begin{array}{|c|c|}\hline {}^{1}0 & {}^{2}1 \\ \hline {}^{3}{-1} & \multicolumn{1}{c}{} \\ \cline{1-1} \end{array}, \quad [2,1]_2 = \begin{array}{|c|c|}\hline {}^{1}0 & {}^{3}1 \\ \hline {}^{2}{-1} & \multicolumn{1}{c}{} \\ \cline{1-1} \end{array}. \tag{4.6.56}
$$

These tableaux are now in one-to-one correspondence with four idempotents (4.6.54) and four elements of the set $\mathrm{Spec}(y_1, y_2, y_3)$:

$$
(a_1, a_2, a_3) = (0,1,2), \quad (0,-1,-2), \quad (0,1,-1), \quad (0,-1,1),
$$

which are written here in accordance with formulas (4.6.53).

4.6.3. Colored Young graph and branching rule for representations

We now represent the results on symmetric group S_n given in Section 4.6.2 in somewhat different form, by introducing the notion of *Young graph*, see

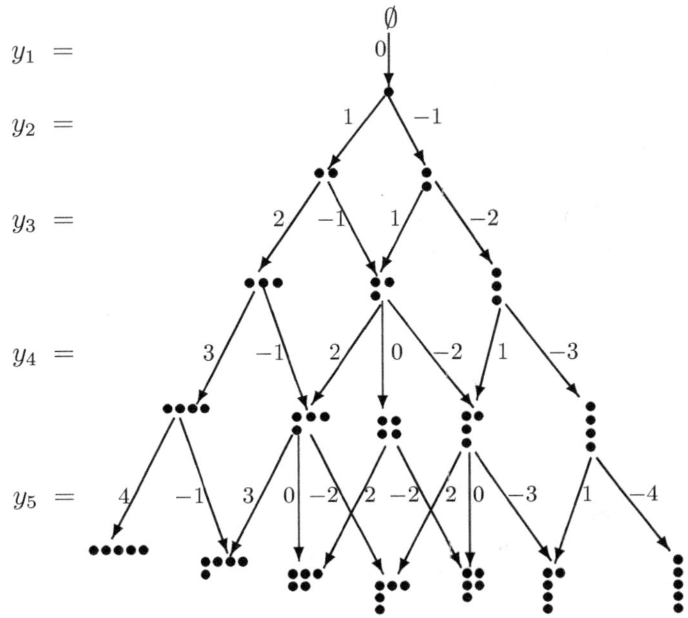

Fig. 4.6.5. Colored Young graph for S_5. Every path from the node $\emptyset$ to node with Young diagram $\lambda \vdash 5$ in the lowest level of the graph corresponds to idempotent e_Λ or standard Young tableau T_λ. Indices attached to links of this path are equal to eigenvalues of operators y_i shown on the left (see also Fig. 4.6.6), and the sequence of these indices along the path give spectral set Λ, or, equivalently, contents of boxes in the standard tableau T_λ.

Fig. 4.6.5. Nodes of Young graph of group S_n are all Young diagrams with $r \leq n$ boxes, while its links join diagrams λ_{r-1} with $(r-1)$ boxes and diagrams λ_r with r boxes. These links are drawn in accordance with *branching rule* for irreducible representations of groups S_r (recall that Young diagrams are associated with irreducible representations of S_r).

Let us explain what is branching of irreducible representations. Irreducible representation of group S_r is, generally speaking, reducible representation of its subgroup $S_{r-1} \subset S_r$. Therefore, irreducible representation $T^{(\lambda_r)}$ of group S_r labeled by Young diagram $\lambda_r \vdash r$ contains several irreducible representations $T^{(\lambda_{r-1})}$ of subgroup $S_{r-1} \subset S_r$, which are labeled by Young diagrams $\lambda_{r-1} \vdash (r-1)$. Representation $T^{(\lambda_{r-1})}$ is contained in representation $T^{(\lambda_r)}$ if diagram λ_r is obtained from diagram λ_{r-1} by adding a box to the latter, as shown in Fig. 4.6.2. The connection between irreducible representation $T^{(\lambda_{r-1})}$ and irreducible representations $T^{(\lambda_r)}$ containing $T^{(\lambda_{r-1})}$ is called *branching of representation* $T^{(\lambda_{r-1})}$.

We attach to links of Young graph that connect diagrams λ_{r-1} with diagrams λ_r the contents a_r of the last, rth box[r] which is added to diagram λ_{r-1} for obtaining diagram λ_r (we denote this box by λ_r/λ_{r-1}). Young graph with attached contents a_i is conveniently called *colored*.[s] We show colored Young graph for group S_5 in Fig. 4.6.5. There, boxes in Young diagrams are replaced by thick dots for the sake of presentation.

Every path in colored Young graph of group S_n, that starts at the upper node $\emptyset$ and ends up at a node at the lowest level that is labeled by Young diagram λ_n with n boxes, determines a set of standard tableaux embedded into each other: $\mathsf{T}_{\lambda_1} \subset \mathsf{T}_{\lambda_2} \subset \cdots \subset \mathsf{T}_{\lambda_n}$. The number i in box λ_i/λ_{i-1} of Young tableau T_{λ_n} is equal to the number of the step at which this box appears. Thus, if a box appears when diagram λ_{i-1} with $(i-1)$ boxes transforms into diagram λ_i with i boxes, then this box is equipped with number i, and its content a_i is attached to the link. So, every path in Young graph defines also a sequence of contents $(a_1, a_2, \ldots, a_n)$ attached to links of this path, and this sequence of contents is precisely the content vector of Young tableau T_{λ_n}. The latter tableau has shape of Young diagram λ_n labeling the last node of the path. Thus, every path from node $\emptyset$ to node λ_n defines standard Young tableau $\mathsf{T}_{\lambda_n} \in \mathrm{Tab}_n$ with its contents, and hence, in accordance with the results Section 4.6.2 (see the relation (4.6.26) and Theorem 4.6.16) this path defines spectral set $\Lambda = (a_1, \ldots, a_n) \in \mathrm{Spec}(y_1, \ldots, y_n)$ (set of eigenvalues of Jucys–Murphy elements) and idempotent $e_\Lambda = e(\mathsf{T}_{\lambda_n})$.

Let $X(n)$ be a set of all paths of length n in colored Young graph that start from node $\emptyset$, go downwards and end up at the Young diagram with n boxes. In other words, $X(n)$ is the set of paths with n links in colored Young graph for group S_n. Then all of the above can be summarized as follows.

---

[r]In more general case of a sequence of algebras $\mathcal{A}_n$ $(n = 1, 2, \ldots)$ such that $\mathcal{A}_n \subset \mathcal{A}_{n+1}$, an irreducible representation T_n of subalgebra $\mathcal{A}_n$ may enter irreducible representation T_{n+1} of algebra $\mathcal{A}_{n+1}$ several times. The number m of these embeddings is called multiplicity of branching of representation T_n in representation T_{n+1}. In that case one often attaches this multiplicity m to the link joining representations T_n and T_{n+1}. A graph showing branching of algebras $\mathcal{A}_1 \subset \mathcal{A}_2 \subset \cdots \subset \mathcal{A}_n$ with attached multiplicities is often called *Bratelli diagram*. In our case of nondegenerate spectrum of Jucys–Murphy operators (see Proposition 4.6.1), all multiplicities in Young graph are equal to 1.

[s]To the best of our knowledge, such a colored graph was first considered by O. Ogievetsky and one of the authors of this book [45].

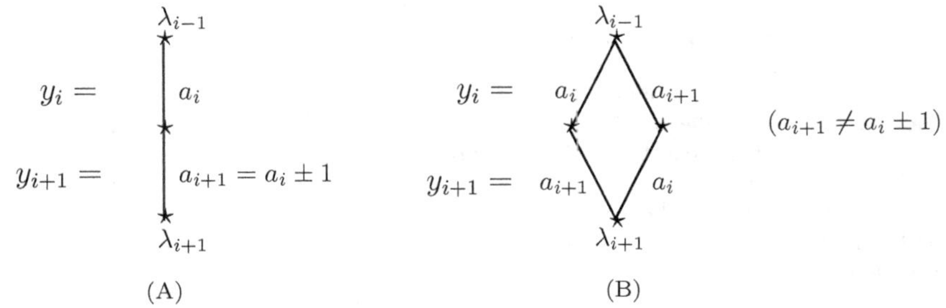

Fig. 4.6.6. Figures (A) and (B) show all possible subgraphs in Young graph of Fig. 4.6.5, which consist of subpaths of length 2 joining vertices with diagrams λ_{i-1} and λ_{i+1}. These are the segments of paths in Young graph that correspond to primitive idempotents on which operators y_i, y_{i+1} are diagonalized (indices at links of graphs (A) and (B) are eigenvalues of elements y_i, y_{i+1}). In the case of graph (A), this is idempotent e_Λ whose eigenvalues a_i, a_{i+1} (of elements y_i, y_{i+1}) in spectral set Λ obey the relation $a_{i+1} = a_i \pm 1$. In the case of graph (B), these are two idempotents e_Λ and $e_{\Lambda'}$ whose spectral sets Λ and Λ', differ by transposition of two eigenvalues a_i and a_{i+1} such that $a_{i+1} \neq a_i \pm 1$.

Theorem 4.6.17. *There is one-to-one correspondence between the set* Tab_n *of standard Young tableaux with n boxes (with their contents), the set* $\mathrm{Spec}(y_1,\ldots,y_n)$*, the complete set of primitive orthogonal idempotents* $\{e_\Lambda\}$ *in algebra* $\mathbb{C}[S_r]$ *and the set $X(n)$ of all paths of length n in the colored Young graph:*

$$\mathrm{Tab}_n \simeq \mathrm{Spec}(y_1,\ldots,y_n) \simeq \{e_\Lambda\} \simeq X(n). \tag{4.6.57}$$

Remark 1. All subpaths in colored Young graph that consist of two links and connect node λ_{i-1} at level $(i-1)$ and node λ_{i+1} at level $(i+1)$ make subgraphs A and B shown in Fig. 4.6.6 (stars show Young diagrams). It is straightforward to see that with assignment of indices in subgraphs A and B in Fig. 4.6.6, and given the "painting" (indices) on links at the first two levels of Young graph of Fig. 4.6.5, "painting" of the entire Young graph can be reconstructed step by step. Thus, subgraphs of Fig. 4.6.6 determine completely the structure of colored Young graph of group S_n.

Let us explain the fact that all subgraphs with intervals of length 2 in Young graph in Fig. 4.6.5 have the form given in Fig. 4.6.6.

For any given $i \leq n-1$, elements y_i, y_{i+1}, $\sigma_i \in \mathbb{C}[S_n]$ obey the closed structural relations (see (4.6.1), (4.6.5) and (4.6.9)):

$$\sigma_i^2 = 1, \sigma_i \cdot y_{i+1} = y_i \cdot \sigma_i + 1, \quad y_i \cdot y_{i+1} = y_{i+1} \cdot y_i, \tag{4.6.58}$$

and they generate a non-Abelian subalgebra $H_{2(i)}$ in algebra $\mathbb{C}[S_n]$. Irreducible representations of algebra $\mathbb{C}[S_n]$ are, generally speaking, reducible representations of its subalgebra $H_{2(i)}$. It is shown in [47] that these representations of $\mathbb{C}[S_n]$ are decomposed into direct sum of irreducible representations of $H_{2(i)}$. In other words, these representations are decomposable as representations of $H_{2(i)}$. Our purpose is to find all irreducible representations ρ of subalgebras $H_{2(i)} \subset \mathbb{C}[S_n]$ $(\forall i)$, which emerge in irreducible representations of algebra $\mathbb{C}[S_n]$. We note in this regard that algebra $\mathbb{C}[S_{i+1}]$ is generated by its subalgebras $\mathbb{C}[S_i]$ and $H_{2(i)}$. So, all steps from i to $(i+1)$ are reduced to the analysis of representations of subalgebras $H_{2(i)}$.

We construct irreducible representations ρ of subalgebra $H_{2(i)}$ in a way, similar to the construction of highest weight representations of Lie algebras. We begin with noticing that there is always a common eigenvector v_1 of y_i and y_{i+1} in the space $\mathcal{V}$ of representation ρ:

$$y_i \cdot v_1 = a_i v_1, \quad y_{i+1} \cdot v_1 = a_{i+1} v_1, \tag{4.6.59}$$

where a_i and a_{i+1} are eigenvalues. Let us act by element σ_i on v_1 and consider first the case in which the new vector $\sigma_i \cdot v_1$ and vector v_1 are linear independent. In this case, in view of (4.6.58), we find that the action of subalgebra $H_{2(i)}$ on vector v_1 generates two-dimensional invariant subspace $\mathcal{V}$ with basis v_1 and $v_2 := \sigma_i \cdot v_1$:

$$\begin{gathered} \sigma_i \cdot (v_1, v_2) = (v_2, v_1), \quad y_i \cdot (v_1,\, v_2) = (a_i\, v_1, a_{i+1}\, v_2 - v_1), \\ y_{i+1} \cdot (v_1,\, v_2) = (a_{i+1}\, v_1, a_i\, v_2 + v_1). \end{gathered} \tag{4.6.60}$$

Using the standard prescription (2.1.7), we obtain from these formulas two-dimensional matrix representation ρ for generators of algebra $H_{2(i)}$:

$$\rho(\sigma_i) = \begin{pmatrix} 0 & 1 \\ 1 & 0 \end{pmatrix}, \quad \rho(y_i) = \begin{pmatrix} a_i & -1 \\ 0 & a_{i+1} \end{pmatrix}, \quad \rho(y_{i+1}) = \begin{pmatrix} a_{i+1} & 1 \\ 0 & a_i \end{pmatrix}. \tag{4.6.61}$$

We are interested in the case $a_i \neq a_{i+1}$ (see Proposition 4.6.6). In this case, matrices $\rho(y_i)$, $\rho(y_{i+1})$ in (4.6.61) can be simultaneously diagonalized by equivalence transformation $\rho(y) \to \tilde{\rho}(y) = A^{-1} \cdot \rho(y) \cdot A$, where

$$A = \begin{pmatrix} 1 & b_i \\ 0 & 1 \end{pmatrix}, \quad b_i := (a_i - a_{i+1})^{-1}.$$

This gives matrix representation $\tilde{\rho}$ of algebra $H_{2(i)}$, which is equivalent to (4.6.61):

$$\tilde{\rho}(\sigma_i) = \begin{pmatrix} -b_i & (1-b_i^2) \\ 1 & b_i \end{pmatrix}, \quad \tilde{\rho}(y_i) = \begin{pmatrix} a_i & 0 \\ 0 & a_{i+1} \end{pmatrix}, \quad \tilde{\rho}(y_{i+1}) = \begin{pmatrix} a_{i+1} & 0 \\ 0 & a_i \end{pmatrix}. \tag{4.6.62}$$

It is written in new basis

$$(v_\Lambda, v_{\Lambda'}) := (v_1, v_2) \cdot A = (v_1, (a_i - a_{i+1})^{-1}\, v_1 + v_2). \tag{4.6.63}$$

We explain the notations v_Λ and $v_{\Lambda'}$ below. If $b_i \neq \mp 1 \Leftrightarrow a_{i+1} \neq a_i \pm 1$, then two-dimensional representation (4.6.62) is irreducible and can be a subrepresentation of the representation of subalgebra $H_{2(i)} \subset \mathbb{C}[S_i]$ in irreducible representation of the whole algebra $\mathbb{C}[S_i]$. For $b_i = \mp 1 \Leftrightarrow a_{i+1} = a_i \pm 1$ two-dimensional representation (4.6.62) is reducible (but not decomposable) and the space $\mathcal{V}$ of this representation has one-dimensional invariant subspace $\mathcal{V}_{\Lambda'}$ with basis vector

$$v_{\Lambda'} = (a_i - a_{i+1})^{-1} v_1 + v_2 = (\sigma_i \mp 1) v_\Lambda. \tag{4.6.64}$$

In accordance with (4.6.62), subalgebra $H_{2(i)}$ acts on this vector as follows:

$$y_i v_{\Lambda'} = a_{i+1} v_{\Lambda'}, \quad y_{i+1} v_{\Lambda'} = a_i v_{\Lambda'}, \quad \sigma_i v_{\Lambda'} = \mp v_{\Lambda'}, \quad (a_{i+1} = a_i \pm 1) \tag{4.6.65}$$

(signs are correlated). This reducible, but not decomposable representation (4.6.62) cannot be a subrepresentation (see [47]) of subalgebra $H_{2(i)} \subset \mathbb{C}[S_n]$ in irreducible representation of the whole algebra $\mathbb{C}[S_n]$. For this reason, these representations of $H_{2(i)}$ are irrelevant for our consideration.

Now, we turn to the case when vectors $\sigma_i \cdot v_1$ and $v_1 \equiv v_\Lambda$ are linear dependent, $\sigma_i \cdot v_\Lambda \sim v_\Lambda$. In this case, it follows from (4.6.58) that the vector $v_1 \equiv v_\Lambda$ is the basis vector of one-dimensional representation of subalgebra $H_{2(i)}$:

$$y_i v_\Lambda = a_i v_\Lambda, \quad y_{i+1} v_\Lambda = a_{i+1} v_\Lambda, \quad \sigma_i v_\Lambda = \pm v_\Lambda, \quad (a_{i+1} = a_i \pm 1) \tag{4.6.66}$$

(signs are correlated). Note that the representation (4.6.66) is obtained from the reducible (the case $a_{i+1} = a_i \pm 1$) representation $\tilde{\rho}$ given by (4.6.62), if the invariant subspace (4.6.65) is trivial, $v_{\Lambda'} = 0$.

Thus, all irreducible representations of subalgebra $H_{2(i)} \subset \mathbb{C}[S_n]$ that emerge in the representations of the whole algebra $\mathbb{C}[S_n]$ are either two-dimensional representation (4.6.62) or one-dimensional representation (4.6.66).

We proceed to consider irreducible representations ρ of subalgebra $H_{2(i)} \subset \mathbb{C}[S_n]$ embedded in irreducible representations $\bar{\rho}$ of algebra $\mathbb{C}[S_n]$. The space $\mathcal{V}$ of representation ρ is embedded in space $\bar{\mathcal{V}}$ of representation $\bar{\rho}$, and vector $v_1 \in \mathcal{V}$ we have considered in the beginning of our discussion can be chosen as common eigenvector of all elements $(y_1 \ldots, y_i, y_{i+1}, \ldots, y_n)$ in space $\bar{\mathcal{V}}$, which corresponds to a spectral set $\Lambda = (a_1, \ldots, a_i, a_{i+1}, \ldots, a_n)$; see (4.6.48). In accordance with (4.6.62), eigenvalues a_i and a_{i+1} of operators y_i and y_{i+1} are transposed for vector $v_{\Lambda'}$ as compared to v_Λ. This explains the notation v_Λ and $v_{\Lambda'}$, in (4.6.63). (cf. Λ and Λ' in (4.6.48) and (4.6.49)).

Graphic representation of the two irreducible representations (4.6.66) and (4.6.62) of subalgebra $H_{2(i)} \subset \mathbb{C}[S_n]$ is given in Fig. 4.6.6. Figures (A) and (B) show one-dimensional (the case $a_{i+1} = a_i \pm 1$) and two-dimensional (the case $a_{i+1} \neq a_i \pm 1$) representations of subalgebra $H_{2(i)}$, respectively. Different paths from upper vertex to lower one correspond to basis vectors on which operators y_i, y_{i+1} are diagonalized. These are the vector $v_\Lambda = e_\Lambda$ in one-dimensional case (A), and vectors v_Λ and $v_{\Lambda'}$ defined by (4.6.63) in two-dimensional case (B). Indices attached to edges are equal to eigenvalues of elements y_i, y_{i+1} (entering sets Λ, Λ').

Example. Let us consider the following path in Fig. 4.6.5:

$$\{\emptyset \overset{0}{\to} \bullet \overset{1}{\to} \bullet\bullet \overset{-1}{\to} \begin{smallmatrix}\bullet & \bullet\\ \bullet & \end{smallmatrix} \overset{0}{\to} \begin{smallmatrix}\bullet & \bullet\\ \bullet & \bullet\end{smallmatrix} \overset{2}{\to} \begin{smallmatrix}\bullet & \bullet & \bullet\\ \bullet & \bullet & \end{smallmatrix}\}. \tag{4.6.67}$$

It defines Young tableau

$$\mathsf{T}_{\lambda_5} = \begin{array}{|c|c|c|}\hline {}^{1}\,0 & {}^{2}\,1 & {}^{5}\,2 \\ \hline {}^{3}\,{-1} & {}^{4}\,0 & \\ \cline{1-2}\end{array}. \tag{4.6.68}$$

The shape of this tableau is determined by the last node of this path (Young diagram $\lambda_5 = \begin{smallmatrix}\bullet&\bullet&\bullet\\ \bullet&\bullet&\end{smallmatrix}$), while numbers in boxes of tableau T_{λ_5} show the order in which "boxes" $\bullet$ appear in nodes (Young diagrams) along this path. Contents of the boxes in (4.6.68) correspond to the sequence $(a_1, \dots, a_5) = (0, 1, -1, 0, 2)$ and are given by indices attached to links along the path. Numbers $(a_1, \dots, a_5) = (0, 1, -1, 0, 2)$ are eigenvalues of Jucys–Murphy elements $(y_1, y_2, y_3, y_4, y_5)$ on primitive idempotent $e(\mathsf{T}_{\lambda_5})$ that is associated with tableau (4.6.68).

Remark 2. In accordance with the results of Sections 4.3.2 and 4.5, dimension of irreducible representation of algebra $\mathbb{C}[S_n]$ (group S_n), that corresponds to Young diagram λ_n is equal to the number of standard tableaux T_{λ_n} of the shape λ_n, or, as we have just demonstrated, to the number of paths of length n in Young graph, that lead to this diagram from node $\emptyset$. This number $f_{(\lambda_n)}$ is given by formulas (4.3.56) and (4.3.57) in the end of Section 4.3.2.

Remark 3. Young graph shown in Fig. 4.6.5 (without indices attached to links) gives also the branching rule of representations of group $SL(N, \mathbb{C})$. This rule enables one to decompose tensor product of defining representations $T^{\otimes 5}$ into irreducible representations $T^{(\lambda)}$ where λ are Young diagrams with five boxes. Here we take $N > 5$, since there are no representations of group $SL(N, \mathbb{C})$ whose diagrams λ have N or more rows (see Fig. 4.6.8). The multiplicity of representation $T^{(\lambda)}$ in $T^{\otimes 5}$ is equal to the number of paths starting at vertex $\emptyset$ of the graph and ending at diagram $\lambda \vdash 5$. As an example, the decomposition of representation $T^{\otimes 4}$ given by (4.5.83) is determined by the upper four levels of Young graph in Fig. 4.6.5. The fact that one and the same Young graph with r levels determines both branching of representations of group S_r and branching of representations of group $SL(N, \mathbb{C})$ in the decomposition of its representation $T^{\otimes r}$ is the manifestation of Schur–Weyl duality (see Section 4.5.3).

So far, we have completely characterized colored Young graph for group S_n. It remains to make use of this graph to invent induction procedure of constructing (in terms of Jucys–Murphy elements) complete system of primitive orthogonal idempotents e_α of algebra $\mathbb{C}[S_n]$. This is done in the following section.

4.6.4. Young graph and inductive construction of idempotents

We need the following proposition.

Proposition 4.6.18. *Let y be a variable, and $\{a_1, a_2, \dots, a_n\}$ be a set of parameters such that $a_i \neq a_k$ for $i \neq k$. Then functions $P_k(y) = \prod_{\substack{i\neq k\\ i=1}}^{n} \frac{(y-a_i)}{(a_k-a_i)}$ obey the identity*

$$\sum_{k=1}^{n} P_k = 1. \tag{4.6.69}$$

Let now y be linear operator in n-dimensional vector space $\mathcal{V}$ and let y obey characteristic identity $\prod_{i=1}^{n}(y-a_i) = 0$. Then $a_1, a_2, \dots, a_n$ are eigenvalues of operator y, and operators $P_k(y) = \prod_{\substack{i\neq k\\ i=1}}^{n} \frac{(y-a_i)}{(a_k-a_i)}$ make complete system of orthogonal projectors in $\mathcal{V}$:

$$P_k \cdot P_i = \delta_{ki} P_k, \quad \sum_{k=1}^{n} P_k = 1. \tag{4.6.70}$$

Proof. To prove the identity (4.6.69), we consider a polynomial $p(y) = (\sum_{k=1}^{n} P_k - 1)$. This polynomial vanishes at n points $y = a_i$ $(i = 1, \dots, n)$ and hence

$$p(y) = c \cdot (y - a_1) \cdot (y - a_2) \cdots (y - a_n). \tag{4.6.71}$$

Due to the definition of P_k, polynomial $p(y)$ is at most of order $(n-1)$, so the factor c on the right-hand side of (4.6.71) vanishes, and we get (4.6.69).

We now come to the second part. It follows from characteristic identity $\prod_i (y-a_i) = 0$ that $y \cdot P_k = a_k P_k$. Hence one gets the first projector relation in (4.6.70). The second equality in (4.6.70) is proven in the same way as (4.6.69) above. The space $\mathcal{V}$ is decomposed in direct sum $\mathcal{V} = \sum_k \mathcal{V}_k$, where $\mathcal{V}_k = P_k \mathcal{V}$ are spaces of eigenvectors of operator y with eigenvalues a_k. □

Consider Young diagram $\lambda_n = [\lambda_{(1)}^{n_1}, \lambda_{(2)}^{n_2-n_1}, \dots, \lambda_{(k)}^{n_k-n_{k-1}}]$ with n_k rows and n boxes:

$$n = \sum_{i=1}^{k} (n_i - n_{i-1}) \lambda_{(i)},$$

where we set $n_0 = 0$. In other words, the first n_1 rows have length $\lambda_{(1)}$, the next $(n_2 - n_1)$ rows have length $\lambda_{(2)}$, etc., and, finally, the last $(n_k - n_{k-1})$ rows have length $\lambda_{(k)}$. This diagram is shown in Fig. 4.6.7, where $(n_i, \lambda_{(i)})$ are coordinates of boxes in corners of diagram λ_n. Recall that each diagram λ defines a set of standard Young tableaux $\{\mathsf{T}_{1,\lambda}, \ldots, \mathsf{T}_{f_{(\lambda)},\lambda}\}$, whose number $f_{(\lambda)}$ is given by formulas (4.3.56), (4.3.57). We are going to use the following definition.

Definition 4.6.3. Standard tableau T_{λ_r} is included in standard tableau T_{λ_n}, where $n > r$, if T_{λ_r} is obtained from T_{λ_n} by deleting the last boxes with numbers $i = r+1, \ldots, n$. In this case, we write $\mathsf{T}_{\lambda_r} \subset \mathsf{T}_{\lambda_n}$.

Consider a standard Young tableau T_{λ_n} whose shape is the diagram λ_n shown in Fig. 4.6.7. Let us assume that we already know primitive idempotent $e(\mathsf{T}_{\lambda_n}) \in \mathbb{C}[S_n] \subset \mathbb{C}[S_{n+1}]$, which is associated with the tableau T_{λ_n} and obeys

$$(y_1, y_2, \ldots, y_n) \cdot e(\mathsf{T}_{\lambda_n}) = (0, a_2, \ldots, a_n) e(\mathsf{T}_{\lambda_n}), \tag{4.6.72}$$

where $(0, a_2, \ldots, a_n)$ are contents of boxes in tableau T_{λ_n}. We recall the branching rule encoded in Young graph for S_{n+1} that tells that all diagrams $\lambda^{(r)}_{n+1}$ with $(n+1)$ boxes are obtained from the diagram $\lambda_n \subset \lambda^{(r)}_{n+1}$ by adding one box to λ_n in any of its external corners with coordinates $(n_{r-1}+1, \lambda_{(r)}+1)$, where $r = 1, 2, \ldots, k+1$; see Fig. 4.4.1 in Section 4.4.2. By definition, corners with coordinates $(1, \lambda_{(1)}+1)$ and $(n_k+1, 1)$ are also external for λ_n.

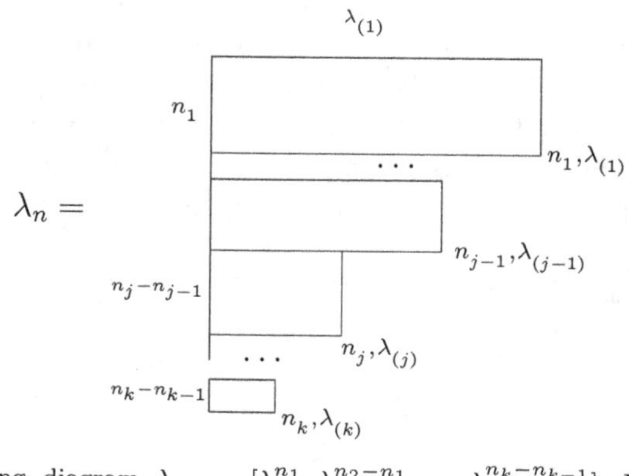

Fig. 4.6.7. Young diagram $\lambda_n = [\lambda_{(1)}^{n_1}, \lambda_{(2)}^{n_2-n_1}, \ldots, \lambda_{(k)}^{n_k-n_{k-1}}]$. Pairs of numbers $(n_j, \lambda_{(j)})$ are the coordinates of boxes in inner corners of diagram λ_n.

Proposition 4.6.19. *Idempotent $e(\mathsf{T}_{\lambda_n})$, considered as an element of algebra $\mathbb{C}[S_{n+1}]$, obeys the following identity:*

$$e(\mathsf{T}_{\lambda_n}) \cdot \prod_{r=1}^{k+1} (y_{n+1} - a_{n+1}^{(r)}) = 0, \tag{4.6.73}$$

where $a_{n+1}^{(r)} = (\lambda_{(r)} - n_{r-1})$ are contents of possible added boxes $\lambda_{n+1}^{(r)}/\lambda_n$ with coordinates $(n_{r-1}+1, \lambda_{(r)}+1)$ (here $r = 1, \dots, k+1$).

Proof. Consider colored Young graph for S_{n+1}. Links starting from diagram λ_n shown in Fig. 4.6.7 have indices $a_{n+1}^{(r)}$ equal to contents of new boxes with coordinates $(n_{r-1}+1, \lambda_{(r)}+1)$ which can be added to diagram λ_n in its external corners to obtain a new diagram $\lambda_{n+1}^{(r)}$ with $(n+1)$ boxes; see Fig. 4.4.1 in Section 4.4.2. This means that the path leading to standard tableau T_{λ_n} branches at node λ_n into paths leading to standard tableaux $\mathsf{T}_{\lambda_{n+1}^{(r)}}$ such that $\mathsf{T}_{\lambda_n} \subset \mathsf{T}_{\lambda_{n+1}^{(r)}}$, so we have for idempotents $e(\mathsf{T}_{\lambda_{n+1}^{(r)}}) \in \mathbb{C}[S_{n+1}]$ and $e(\mathsf{T}_{\lambda_n}) \in \mathbb{C}[S_n] \subset \mathbb{C}[S_{n+1}]$ the relations

$$\begin{aligned} &e(\mathsf{T}_{\lambda_n}) \cdot e(\mathsf{T}_{\lambda_{n+1}^{(r)}}) = e(\mathsf{T}_{\lambda_{n+1}^{(r)}}), \\ &\text{whereas } e(\mathsf{T}_{\lambda_n}) \cdot e(\mathsf{T}_{\lambda_{n+1}}) = 0 (\forall \mathsf{T}_{\lambda_n} \not\subset \mathsf{T}_{\lambda_{n+1}}). \end{aligned} \tag{4.6.74}$$

This shows that idempotent $e(\mathsf{T}_{\lambda_n})$ is decomposed into a sum of idempotents $e(\mathsf{T}_{\lambda_{n+1}^{(r)}})$:

$$e(\mathsf{T}_{\lambda_n}) = e(\mathsf{T}_{\lambda_n}) \cdot e = e(\mathsf{T}_{\lambda_n}) \cdot \sum_{\lambda \vdash (n+1)} \sum_{\mathsf{T}_\lambda} e(\mathsf{T}_\lambda) = \sum_{r=1}^{k+1} e(\mathsf{T}_{\lambda_{n+1}^{(r)}}), \tag{4.6.75}$$

where the third expression involves summation over all Young diagrams λ with $(n+1)$ boxes and over all standard tableaux T_λ with shape λ, and the last expression is the sum over all $(k+1)$ standard tableaux $\mathsf{T}_{\lambda_{n+1}^{(r)}}$ obtained from T_{λ_n} by adding one of the boxes $(n_{r-1}+1, \lambda_{(r)}+1)$ with content $a_{n+1}^{(r)} = (\lambda_{(r)} - n_{r-1})$. Now we recall that $e(\mathsf{T}_{\lambda_{n+1}^{(r)}})(y_{n+1} - a_{n+1}^{(r)}) = 0$, which, in view of (4.6.75), gives identity (4.6.73). □

Consider two tableaux $\mathsf{T}_{\lambda_{n+1}^{(j)}}$ and $\mathsf{T}_{\lambda_n} \subset \mathsf{T}_{\lambda_{n+1}^{(j)}}$, where Young diagram $\lambda_{n+1}^{(j)}$ is obtained by adding $(n+1)$th box with coordinates $(n_{j-1}+1, \lambda_{(j)}+1)$ to diagram λ_n, see Figs. 4.6.7 and 4.4.1. Then the following property holds.

Proposition 4.6.20. *Primitive orthogonal idempotent* $e(\mathsf{T}_{\lambda^{(j)}_{n+1}})$ *in algebra* $\mathbb{C}[S_{n+1}]$ *associated with standard tableau* $\mathsf{T}_{\lambda^{(j)}_{n+1}}$ *is determined by induction formula*

$$e(\mathsf{T}_{\lambda^{(j)}_{n+1}}) = e(\mathsf{T}_{\lambda_n}) \cdot \prod_{\substack{r=1 \\ r\neq j}}^{k+1} \frac{\left(y_{n+1} - a^{(r)}_{n+1}\right)}{\left(a^{(j)}_{n+1} - a^{(r)}_{n+1}\right)}, \tag{4.6.76}$$

where $e(\mathsf{T}_{\lambda_n}) \in \mathbb{C}[S_n]$ *is primitive orthogonal idempotent associated with standard tableau* $\mathsf{T}_{\lambda_n} \subset \mathsf{T}_{\lambda^{(j)}_{n+1}}$.

Proof. In view of identity (4.6.73), element (4.6.76) obeys

$$y_{n+1} \cdot e(\mathsf{T}_{\lambda^{(j)}_{n+1}}) = a^{(j)}_{n+1} e(\mathsf{T}_{\lambda^{(j)}_{n+1}}). \tag{4.6.77}$$

Together with (4.6.72) and commutativity of elements y_i, this gives

$$(y_1, y_2, \dots, y_n, y_{n+1}) \cdot e(\mathsf{T}_{\lambda^{(j)}_{n+1}}) = (0, a_2, \dots, a_n, a^{(j)}_{n+1}) e(\mathsf{T}_{\lambda^{(j)}_{n+1}}).$$

So, formula (4.6.76) is consistent with the fact that $(n+1)$th box in $\mathsf{T}_{\lambda^{(j)}_{n+1}}$ has coordinates $(n_{j-1}+1, \lambda_{(j)}+1)$, its content is $a^{(j)}_{n+1} = \lambda_{(j)} - n_{j-1}$, and the complete spectral set of operator $e(\mathsf{T}_{\lambda^{(j)}_{n+1}})$ is given by content vector $(0, a_2, \dots, a_n, a^{(j)}_{n+1})$, as it should. We recall the formulas (4.6.70) (see Proposition 4.6.18) and make use of the relation $[e(\mathsf{T}_{\lambda_n})]^2 = e(\mathsf{T}_{\lambda_n})$ to obtain that operators (4.6.76) define the system of orthogonal idempotents

$$e(\mathsf{T}_{\lambda^{(j)}_{n+1}}) \cdot e(\mathsf{T}_{\lambda^{(j)}_{n+1}}) = e(\mathsf{T}_{\lambda^{(j)}_{n+1}}), \quad \text{whereas } e(\mathsf{T}_{\lambda^{(j)}_{n+1}}) \cdot e(\mathsf{T}_{\lambda^{(r)}_{n+1}}) = 0 \quad (j \neq r),$$

where tableau $\mathsf{T}_{\lambda^{(r)}_{n+1}}$ is obtained from T_{λ_n} by adding a new $(n+1)$th box, but with different coordinates $(n_{r-1}+1, \lambda_{(r)}+1)$. Also, we make use of the identity (4.6.69) from Proposition 4.6.18 for idempotents (4.6.76) and obtain, in agreement with branching rule for representations, the spectral decomposition (4.6.75). The latter yields, by induction (assuming that completeness relation is valid for idempotents $e(\mathsf{T}_{\lambda_n}) \in \mathbb{C}[S_n]$), completeness relation for idempotents $e(\mathsf{T}_{\lambda^{(j)}_{n+1}}) \in \mathbb{C}[S_{n+1}]$:

$$\sum_{\lambda_{n+1}} \sum_{\mathsf{T}_{\lambda_{n+1}}} e(\mathsf{T}_{\lambda_{n+1}}) = \sum_{\lambda_n} \sum_{\mathsf{T}_{\lambda_n}} \sum_{j} e(\mathsf{T}_{\lambda^{(j)}_{n+1}}) = e, \tag{4.6.78}$$

where e is unit element in $\mathbb{C}[S_{n+1}]$ and index j enumerates all outer corners of diagram λ_n. □

Corollary 4.6.21. *Inductive procedure* (4.6.76) *for tableaux* $\mathsf{T}_{\lambda_r} \subset \mathsf{T}_{\lambda_n}$, *where* $n > r$, *immediately gives an identity* (*cf.* (4.6.74))

$$e(\mathsf{T}_{\lambda_r}) \cdot e(\mathsf{T}_{\lambda_n}) = e(\mathsf{T}_{\lambda_n}) \cdot e(\mathsf{T}_{\lambda_r}) = e(\mathsf{T}_{\lambda_n}).$$

If the standard tableau T_{λ_r} *is not included in standard tableau* T_{λ_n}, *then we have* $e(\mathsf{T}_{\lambda_r}) \cdot e(\mathsf{T}_{\lambda_n}) = 0$.

Making use of inductive formula (4.6.76), the initial condition $e\left(\boxed{1}\right) = 1$ and the procedure of construction of colored Young graph, we can now evaluate, step by step, complete systems of primitive orthogonal idempotents $e_\alpha \in \mathbb{C}[S_n]$, obeying the key property (4.6.20) (and associated with standard Young tableaux T_{λ_n} with n boxes).

Example. Let us make use of the inductive procedure to construct idempotent associated with standard Young tableau T_{λ_5} shown in (4.6.68). We go step by step along the path (4.6.67) in the Young graph of Fig. 4.6.5. The first step $\{\emptyset \stackrel{0}{\to} \bullet\}$ gives, by definition, $e\left(\boxed{1}\right) = 1$. At the second step, we add to the box $\boxed{1}$ the second box $\boxed{2}$. Let us forget for a while about the path (4.6.67) and point out that the latter step can be made in two ways, as shown by asterisks,

$$\begin{array}{cc} \boxed{1} & * \\ * & \end{array}.$$

Therefore, the formula (4.6.73) has the form (4.6.29):

$$e\left(\boxed{1}\right) \cdot (y_2 - 1) \cdot (y_2 + 1) = 0,$$

and, in accordance with (4.6.76), we have two idempotents for $\mathbb{C}[S_2]$ (cf. (4.5.33)):

$$e\left(\boxed{1}\boxed{2}\right) = \frac{(1+y_2)}{2}, \quad e\left(\begin{array}{c}\boxed{1}\\ \boxed{2}\end{array}\right) = \frac{(1-y_2)}{2}. \tag{4.6.79}$$

We now come back to the path (4.6.67). Its second box at the second step $\{\bullet \stackrel{1}{\to} \bullet\bullet\}$ is in the same row with the first box, so content of the second box equals $+1$, and the pertinent idempotent is given by $e\left(\boxed{1}\boxed{2}\right)$ in (4.6.79). At the third step, the box with number 3 can in principle appear at one of the two places shown by asterisks,

$$\begin{array}{ccc} \boxed{1} & \boxed{2} & * \\ * & & \end{array},$$

so the formula (4.6.73) reads (cf. the first identity in (4.6.10))

$$e\left(\ytableaushort{12}\right)\cdot(y_3+1)\cdot(y_3-2)=0.$$

In accordance with (4.6.76), we again have two possible idempotents (cf. $e_{[2,1]_1}$ and $e_{[3]}$ in (4.6.54)):

$$e\left(\ytableaushort{12,3}\right)=e\left(\ytableaushort{12}\right)\cdot\frac{(y_3-2)}{(-1-2)}=\frac{(1+y_2)}{2}\cdot\frac{(2-y_3)}{3}, \tag{4.6.80}$$

$$e\left(\ytableaushort{123}\right)=e\left(\ytableaushort{12}\right)\cdot\frac{(y_3+1)}{(2+1)}=\frac{(1+y_2)}{2}\cdot\frac{(1+y_3)}{3}. \tag{4.6.81}$$

The third step in the path (4.6.67) is $\{\bullet\bullet \overset{-1}{\to} \begin{smallmatrix}\bullet&\bullet\\ \bullet&\end{smallmatrix}\}$, so its next idempotent is given by (4.6.80). At the fourth step, the box with number 4 can in principle be added in one of the three places,

$$\begin{array}{|c|c|l}\hline 1 & 2 & * \\ \hline 3 & \multicolumn{1}{|c}{*} & \\ \cline{1-1} \multicolumn{1}{c}{*} & \multicolumn{1}{c}{} & \end{array},$$

so an analog of the identity (4.6.73) reads

$$e\left(\ytableaushort{12,3}\right)\cdot y_4\cdot(y_4-2)\cdot(y_4+2)=0.$$

Using this formula together with (4.6.76) and (4.6.80), we construct three idempotents

$$e\left(\ytableaushort{124,3}\right)=e\left(\ytableaushort{12,3}\right)\cdot\tfrac{y_4\cdot(y_4+2)}{2\cdot 4}=\tfrac{1}{6\cdot 8}(1+y_2)\cdot(2-y_3)\cdot y_4\cdot(y_4+2),$$

$$e\left(\ytableaushort{12,3,4}\right)=e\left(\ytableaushort{12,3}\right)\cdot\tfrac{y_4\cdot(y_4-2)}{2\cdot(-2-2)}=\tfrac{1}{6\cdot 8}(1+y_2)\cdot(2-y_3)\cdot y_4\cdot(2-y_4),$$

$$e\left(\ytableaushort{12,34}\right)=e\left(\ytableaushort{12,3}\right)\cdot\tfrac{(y_4+2)\cdot(y_4-2)}{2\cdot(-2)}$$

$$=\tfrac{1}{6\cdot 4}(1+y_2)\cdot(2-y_3)\cdot(2+y_4)\cdot(2-y_4). \tag{4.6.82}$$

The fourth box is added along the path (4.6.67) as follows, $\{\begin{smallmatrix}\bullet&\bullet\\ \bullet&\end{smallmatrix} \overset{0}{\to} \begin{smallmatrix}\bullet&\bullet\\ \bullet&\bullet\end{smallmatrix}\}$, so the fourth idempotent is given by (4.6.82).

Finally, at the last step, the box with number 5 can appear in two places, $\begin{array}{|c|c|l}\hline 1&2&*\\ \hline 3&4&\\ \cline{1-2} \multicolumn{1}{c}{*}&\multicolumn{1}{c}{}&\end{array}$, the identity (4.6.73) reads $e\left(\begin{array}{|c|c|}\hline 1&2\\ \hline 3&4\\ \hline\end{array}\right)\cdot(y_5-2)\cdot(y_5+2)=0$ and idempotent associated with tableau T_{λ_5} given by (4.6.68), is

$$
\begin{aligned}
e\left(\begin{array}{|c|c|c|}\hline 1&2&5\\ \hline 3&4&\multicolumn{1}{c}{}\\ \cline{1-2}\end{array}\right) &= e\left(\begin{array}{|c|c|}\hline 1&2\\ \hline 3&4\\ \hline\end{array}\right)\cdot\frac{(y_5+2)}{(2+2)} \\
&= \frac{(1+y_2)\cdot(2-y_3)\cdot(2+y_4)\cdot(2-y_4)\cdot(2+y_5)}{2^5\cdot 3}.
\end{aligned}
\tag{4.6.83}
$$

Problem 4.6.7. Show that total symmetrizer $e([n])$ and total antisymmetrizer $e([1^n])$ for algebra $\mathbb{C}[S_n]$ are written in terms of Jucys–Murphy elements y_i as follows:

$$
\begin{aligned}
e([n]) &= \frac{1}{n!}(1+y_2)\cdot(1+y_3)\cdots(1+y_n) \\
&= \frac{1}{n!}(1+(1,2))\cdot(1+(1,3)+(2,3))\cdots \\
&\quad\times(1+(1,n)+\cdots+(n-1,n)), \\
e([1^n]) &= \frac{1}{n!}(1-y_2)\cdot(1-y_3)\cdots(1-y_n) \\
&= \frac{1}{n!}(1-(1,2))\cdot(1-(1,3)-(2,3))\cdots \\
&\quad\times(1-(1,n)-\cdots-(n-1,n)).
\end{aligned}
\tag{4.6.84}
$$

Compare idempotents $e([r])$ with operators $E_{P[r]}$ in (4.3.16) and $e([1^r])$ with operators, obtained from $E_{P[r]}$ by replacement $\sigma_i \to -\sigma_i$.

4.6.5. Projection operators and characters of irreducible representations of $U(N)$. Symmetric polynomials

Consider N-dimensional complex space $\mathcal{V}$ of defining representation T of group $SL(N,\mathbb{C})$. Let $\{\vec{e}_1,\dots,\vec{e}_N\}$ be a basis in the space $\mathcal{V}$. Linear operator A acts in tensor product $\mathcal{V}_N^{\otimes n}=\mathcal{V}_N\otimes\cdots\otimes\mathcal{V}_N$ in standard way:

$$
A\cdot(\vec{e}_{i_1}\otimes\cdots\otimes\vec{e}_{i_n}) = (\vec{e}_{k_1}\otimes\cdots\otimes\vec{e}_{k_n})A^{k_1\dots k_n}_{i_1\dots i_n},
$$

where $A^{k_1\dots k_n}_{i_1\dots i_n} \in \mathbb{C}$ are elements of matrix of operator A. We denote the set of all linear operators in $\mathcal{V}^{\otimes n}$ by $\mathcal{L}(\mathcal{V}^{\otimes n})$. We also define partial trace $\mathrm{Tr}_{(m)}$ of an operator A over the subspace $\mathcal{V}$ which is in the mth place in tensor product $\mathcal{V}^{\otimes n}$, as mapping $\mathcal{L}(\mathcal{V}^{\otimes n}) \to \mathcal{L}(\mathcal{V}^{\otimes (n-1)})$, which maps operator $A \in \mathcal{L}(\mathcal{V}^{\otimes n})$ to the following operator $\mathrm{Tr}_{(m)}(A) \in \mathcal{L}(\mathcal{V}^{\otimes (n-1)})$:

$$\mathrm{Tr}_{(m)}(A) \cdot (\vec{e}_{i_1} \otimes \cdots \otimes \vec{e}_{i_{n-1}}) = (\vec{e}_{k_1} \otimes \cdots \otimes \vec{e}_{k_{n-1}}) B^{k_1\dots k_{n-1}}_{i_1\dots i_{n-1}},$$

where

$$B^{k_1\dots k_{m-1}\, k_{m+1}\dots k_n}_{i_1\dots i_{m-1}\, i_{m+1}\dots i_n} \equiv \sum_{i_m} A^{k_1\dots k_{m-1}\, i_m\, k_{m+1}\dots k_n}_{i_1\dots i_{m-1}\, i_m\, i_{m+1}\dots i_n} \;\Leftrightarrow\; B = \mathrm{Tr}_{(m)}(A). \tag{4.6.85}$$

Clearly, evaluating the total trace over space $\mathcal{V}^{\otimes n}$ can be understood as taking consecutively partial traces over all spaces $\mathcal{V}$ entering $\mathcal{V}^{\otimes n}$:

$$\mathbf{Tr} = \mathrm{Tr}_{(1)}\mathrm{Tr}_{(2)} \dots \mathrm{Tr}_{(n)} \;\Rightarrow\; \mathbf{Tr}(A) = \sum_{i_1,\dots,i_n} A^{i_1\dots i_n}_{i_1\dots i_n}. \tag{4.6.86}$$

We define in Section 4.2 the action of algebra $\mathbb{C}[S_n]$ in space $\mathcal{V}^{\otimes n}$. To this end, elements of algebra $\mathbb{C}[S_n]$ are written in (matrix) representation s (see (4.2.9), (4.2.10)). Let us denote the representation s of primitive idempotent $e(\mathsf{T}_\lambda) \in \mathbb{C}[S_n]$, where T_λ is a standard Young tableau of shape $\lambda \vdash n$, as follows:

$$P(\mathsf{T}_\lambda) = \mathsf{s}(e(\mathsf{T}_\lambda)). \tag{4.6.87}$$

We emphasize that hereafter we consider idempotents $e(\mathsf{T}_\lambda)$ constructed in Okounkov–Vershik approach. This set of idempotents is, generally speaking, different from the set of primitive idempotents $E(\mathsf{T}_\lambda) = \frac{1}{\alpha_\lambda} Y(\mathsf{T}_\lambda)$ constructed in terms of Young symmetrizers. Therefore, projectors (4.6.87) are, generally speaking, different from projectors (4.4.1).

Recall that $\mathcal{V}^{\otimes n}$ is the space of representation $T^{\otimes n}$ (tensor product of defining representations T) of group $SL(N, \mathbb{C})$ (and its subgroup $SU(N)$); the representation $T^{\otimes n}$ is reducible and is expanded in irreducible representations $T^{(\lambda)}$. The latter are labeled by Young diagrams $\lambda \vdash n$ and act in invariant subspaces $\mathcal{V}^{(\lambda)} \subset \mathcal{V}^{\otimes n}$, which are extracted from space $\mathcal{V}^{\otimes n}$ by applying either Young symmetrizers (4.4.1), or projectors (4.6.87). This expansion of representations $T^{\otimes n}$ in irreducible representations and extraction of invariant subspaces $\mathcal{V}^{(\lambda)} \subset \mathcal{V}^{\otimes n}$ is illustrated by Young branching graph shown in Fig. 4.6.8. This graph follows from (4.4.33), (4.5.83) and due to Schur–Weyl duality is given by Young branching graph for group S_n (see Fig. 4.6.5).

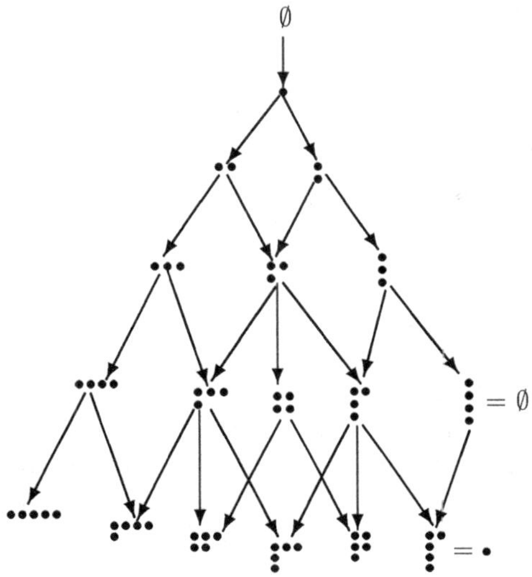

Fig. 4.6.8. Young branching graph for representations of group $SL(4,\mathbb{C})$ (and $SU(4)$) that act in spaces $\mathcal{V}^{\otimes n}$ ($n \leq 5$). This graph is given by Young graph for group S_5 (see Fig. 4.6.5). Every path from vertex $\emptyset$ at upper level to vertex with Young diagram $\lambda \vdash 5$ at the lowest level corresponds to standard Young tableau T_λ and projector $P(\mathsf{T}_\lambda)$. Therefore, the multiplicity of representation associated with Young diagram λ in tensor product of defining representations is equal to the number of paths from vertex $\emptyset$ to vertex λ. We show in this graph what happens with representations of $SL(4,\mathbb{C})$ and $SU(4)$ when one encounters Young diagrams λ with columns of height $m \geq 4$ (see Remark in Section 4.4.1). In the case $n > 5$, branching yields a new Young subgraph emanating from representation $[1^4] = \emptyset$, which emerges at the fourth level. This new subgraph will repeat the graph shown in this figure.

Proposition 4.6.22. *Total traces* (4.6.86) *of projectors* $P(\mathsf{T}_\lambda)$ *and* $P(\mathsf{T}'_\lambda)$ *which are associated with two Young tableaux* T_λ *and* T'_λ *of one and the same shape* $\lambda \vdash n$ *are equal (i.e., the total trace is independent of the assignment of numbers in boxes of diagram* λ*) and are equal to dimension of space* $\mathcal{V}^{(\lambda)}$ *of representation* $T^{(\lambda)}$:

$$\mathbf{Tr}(P(\mathsf{T}_\lambda)) = \mathbf{Tr}(P(\mathsf{T}'_\lambda)) = \dim(\mathcal{V}^{(\lambda)}). \tag{4.6.88}$$

Proof. Two different tableaux T_λ and T'_λ of one and the same shape λ correspond to two different paths (with the end node λ) in Young graph and define two spectral sets Λ and Λ', which transform into each other by several neighboring transpositions (4.6.35) of eigenvalues a_i. In other words, there exists a product $\mathsf{s}_{i_1} \cdots \mathsf{s}_{i_k}$ of transpositions defined in (4.6.35) (according to part (1) of Proposition 4.6.7, we associate to each transposition s_i an element U_i), which transforms the spectral set Λ into spectral set Λ'. Therefore, idempotents $e(\mathsf{T}_\lambda)$ and $e(\mathsf{T}'_\lambda)$ are related by several equivalence transformations (4.6.38):

$$e(\mathsf{T}'_\lambda) = (U_{i_1} \cdots U_{i_k}) \cdot e(\mathsf{T}_\lambda) \cdot (U_{i_1} \cdots U_{i_k})^{-1}, \tag{4.6.89}$$

where we make use of the fact that operators U_i are nondegenerate for $a_{i+1} \neq a_i \pm 1$. This gives

$$\mathbf{Tr}(P(\mathsf{T}'_\lambda)) = \mathbf{Tr}(\mathsf{s}(U_{i_1} \cdots U_{i_k}) \cdot P(\mathsf{T}_\lambda) \cdot \mathsf{s}(U_{i_1} \cdots U_{i_k})^{-1}) = \mathbf{Tr}(P(\mathsf{T}_\lambda)).$$

Since $e(\mathsf{T}_\lambda)$ is idempotent, the operator $P(\mathsf{T}_\lambda)$ is a projector, $P^2 = P$, so the eigenvalues of operator $P(\mathsf{T}_\lambda)$ are either 0 or +1, and eigenvectors with eigenvalue +1 belong to the image of projector $P(\mathsf{T}_\lambda)$, which is the subspace $\mathcal{V}^{(\lambda)}$. The trace $\mathbf{Tr}(P(\mathsf{T}_\lambda))$ is equal to the sum of eigenvalues of operator $P(\mathsf{T}_\lambda)$, and this sum is in turn equal to dimension of subspace $\mathcal{V}^{(\lambda)}$. □

Problem 4.6.8. Insert expressions for symmetrizer and antisymmetrizer (4.6.84) in (4.6.88) and obtain dimensions (4.4.25) and (4.4.28) of representations $T^{([1^n])}$ and $T^{([n])}$.

Expression (4.6.88) is rather difficult to employ for calculating dimensions of concrete irreducible representations of groups $SU(N)$ and $SL(N,\mathbb{C})$. More convenient hook formula (4.4.21) has been discussed in Section 4.4.2; it is obtained from the expression (3.5.46), which in turn is derived from Weyl formula for characters of group $SU(N)$.

Note that the space $\mathcal{V} = \mathbb{C}^N$ of defining representation T of group $SU(N)$ is at the same time the space of defining representation T' of group $U(N)$. Clearly, invariant subspaces $\mathcal{V}^{(\lambda)}$ in $\mathcal{V}^{\otimes n}$ for tensor products $T^{\otimes n}$ and $T'^{\otimes n}$ of defining representations of groups $SU(N)$ and $U(N)$ also coincide. Therefore, we can use the same projection operators (4.6.87) both for group $SU(N)$ and group $U(N)$. We should bear in mind, however, that invariant subspace $\mathcal{V}^{([1^N])} \subset \mathcal{V}^{\otimes n}$ obtained by using total antisymmetrizer $P(\mathsf{T}_{[1^N]})$ is the space of trivial representation in the case of $SU(N)$, while in the case of $U(N)$ it is the space of nontrivial determinant representation $T'^{([1^N])}$: $g \to \det(g)$, where $g \in U(N)$. In what follows, we consider more general case of matrices $g \in U(N)$. Since any matrix $g \in U(N)$ can be written as $g = \det^{1/N}(g) \cdot h$, where $h \in SU(N)$, the case of group $SU(N)$ is obtained by imposing $\det(g) = 1$.

Proposition 4.6.23. *Let g be an element of $U(N)$, and λ_n be Young diagram with n boxes. Total trace*

$$\chi_{\lambda_n}(g) := \mathbf{Tr}(\underbrace{(g\otimes\cdots\otimes g)}_{n}P(\mathsf{T}_{\lambda_n})) \tag{4.6.90}$$

is independent of the assignment of numbers in boxes of standard tableau T_{λ_n} (i.e., it depends on the shape of Young diagram λ_n only) and is character of group $U(N)$ in irreducible representation $T^{(\lambda_n)}$ corresponding to diagram λ_n.

We emphasize that the projector $P(\mathsf{T}_{\lambda_n})$ itself does depend on the assignment of numbers in the standard tableau T_{λ_n}, i.e., it depends on the choice of tableau T_{λ_n}.

Proof. Operators of elements $g\in U(N)$ in irreducible representation $T^{(\lambda_n)}$ are written as (cf. (4.4.3))

$$T^{(\lambda_n)}(g) = P(\mathsf{T}_{\lambda_n})\cdot T^{\otimes n}(g) = T^{\otimes n}(g)\cdot P(\mathsf{T}_{\lambda_n}), \tag{4.6.91}$$

where $T^{\otimes n}(g) = g^{\otimes n}$ (since T is defining representation). Total trace of the two sides of (4.6.91) gives formula (4.6.90), which according to definition (3.5.1) gives character of representation T_{λ_n}. The fact that the right-hand side of (4.6.90) depends only on the shape of diagram λ_n, but not on the choice of standard tableau T_{λ_n}, follows from (4.6.89) and the relation $[g^{\otimes n}, s(U_{i_1}\cdots U_{i_k})] = 0$, which is a consequence of Proposition 4.2.2. □

Matrix $g = ||g^i_j|| \in U(N)$ is diagonalizable (see Proposition 3.5.1), and can be reduced to diagonal form $X = \mathrm{diag}(x_1,\dots,x_N)$ by the equivalence transformation $g\to h^{-1}\cdot g\cdot h$. As we have shown in Section 3.5.2, character $\chi_{\lambda_n}(g) = \chi_{\lambda_n}(X)$ is symmetric order-n polynomial of N variables $\{x_i\}$, which is called Schur polynomial for diagram λ_n. General construction of Schur symmetric polynomials is given by determinant formula (3.5.39) in Section 3.5.2. That formula is not quite explicit, as it defines Schur polynomials as ratios of two antisymmetric polynomials (see, however, Proposition 4.6.27 below). Here, we make use of the formula for character (4.6.90) to find explicit expressions for Schur polynomials for some simple Young diagrams.

Examples. We calculate characters (4.6.90) for representations of group $U(N)$ that correspond to simple idempotents (4.5.33) and (4.5.37) in algebras $\mathbb{C}[S_2]$ and $\mathbb{C}[S_3]$. We use the following notations (see also (4.2.13),

(4.2.14)):

$$P_k = \mathsf{s}(\sigma_k) = I^{\otimes(k-1)} \otimes P \otimes I^{\otimes(n-k-1)}, \quad g_k = I^{\otimes(k-1)} \otimes g \otimes I^{\otimes(n-k)}, \tag{4.6.92}$$

where P is transposition matrix. We insert the expressions (4.5.33) for idempotents $\mathsf{s}(e(\mathsf{T}_{[2]}))$ and $\mathsf{s}(e(\mathsf{T}_{[1^2]}))$ in formula (4.6.90) and obtain (for $n = 2$)

$$\begin{aligned} \chi_{[2]}(g) &= \frac{1}{2}\,\mathbf{Tr}(g_1\, g_2\, (1 + P_1)) = \frac{1}{2}\,(\mathrm{Tr}\, g)^2 + \frac{1}{2}\,\mathrm{Tr}(g^2), \\ \chi_{[1^2]}(g) &= \frac{1}{2}\,\mathbf{Tr}(g_1\, g_2\, (1 - P_1)) = \frac{1}{2}\,(\mathrm{Tr}\, g)^2 - \frac{1}{2}\,\mathrm{Tr}(g^2). \end{aligned} \tag{4.6.93}$$

Insertion of idempotents (4.5.37) in formula (4.6.90) gives for $n = 3$:

$$\begin{aligned} \chi_{[3]}(g) &= \frac{1}{6}\,\mathbf{Tr}(g_1\, g_2\, g_3(1 + P_2 + P_1P_2)(1 + P_1)) \\ &= \frac{1}{6}\,p_1^3 + \frac{1}{2}\,p_1\,p_2 + \frac{1}{3}\,p_3, \end{aligned} \tag{4.6.94}$$

$$\begin{aligned} \chi_{[1^3]}(g) &= \frac{1}{6}\,\mathbf{Tr}(g_1\, g_2\, g_3\,(1 - P_2 + P_1P_2)(1 - P_1)) \\ &= \frac{1}{6}\,p_1^3 - \frac{1}{2}\,p_1\,p_2 + \frac{1}{3}\,p_3, \end{aligned} \tag{4.6.95}$$

$$\begin{aligned} \chi_{[2,1]_1}(g) &= \frac{1}{3}\,\mathbf{Tr}(g_1\, g_2\, g_3(1 - P_2 + \varepsilon(1 - P_1)\, P_2)(1 + P_1)) = \frac{1}{3}p_1^3 - \frac{1}{3}\,p_3, \\ \chi_{[2,1]_2}(g) &= \frac{1}{3}\mathbf{Tr}(g_1\, g_2\, g_3(1 - P_1)(1 + P_2 - \varepsilon\, P_2\,(1 + P_1))) = \frac{1}{3}p_1^3 - \frac{1}{3}\,p_3, \end{aligned} \tag{4.6.96}$$

where we use shorthand notation $p_k \equiv \mathrm{Tr}(g^k)$.

> **Problem 4.6.9.** * Prove the equalities in (4.6.93), (4.6.94), (4.6.95) and (4.6.96).

Let us cast the matrix g in diagonal form $g \to \mathrm{diag}(x_1, \ldots, x_N)$, then characters (4.6.93), (4.6.94), (4.6.95) and (4.6.96) are expressed in terms of symmetric functions

$$\chi_{[2]}(g) = \sum_{i\leq j} x_i x_j, \quad \chi_{[1^2]}(g) = \sum_{i<j} x_i x_j,$$
$$\chi_{[3]}(g) = \sum_{i\leq j\leq k} x_i x_j x_k, \quad \chi_{[1^3]}(g) = \sum_{i<j<k} x_i x_j x_k, \tag{4.6.97}$$

$$\chi_{[2,1]_1}(g) = 2\sum_{i<j<k} x_i x_j x_k + \sum_{i\neq j} x_i x_j^2 = \chi_{[2,1]_2}(g). \tag{4.6.98}$$

Note that characters $\chi_{[2,1]_1}$ and $\chi_{[2,1]_2}$ in (4.6.96) do not depend on arbitrary parameter ε. Moreover, as anticipated (see Proposition 4.6.23) we find $\chi_{[2,1]_1} = \chi_{[2,1]_2} \equiv \chi_{[2,1]}$, i.e., characters do not depend on assignment of numbers in Young tableau T_λ that defines the representation, and depend only on the shape of T_λ (i.e., on Young diagram λ).

Problem 4.6.10. Making use of (4.6.93), (4.6.94), (4.6.95) and (4.6.96), check identities for characters that follow from the representation product rule (4.4.33):

$$\chi_{[2]}(g)\cdot\chi_{[1]}(g) = \chi_{[3]}(g) + \chi_{[2,1]}(g), \quad \chi_{[1^2]}(g)\cdot\chi_{[1]}(g) = \chi_{[1^3]}(g) + \chi_{[2,1]}(g), \tag{4.6.99}$$

where $\chi_{[1]} = \mathrm{Tr}(g)$ is character of defining representation. For the case $N=3$, deduce formulas (4.6.97) and (4.6.98) from the Jacobi–Trudi formula (3.5.44).

Proposition 4.6.24. *Characters of totally symmetric and totally antisymmetric representations of group $U(N)$ are, respectively (cf. (4.6.93) and (4.6.94)),*

$$\chi_{[k]}(g) = \sum_{i_1\leq i_2\leq\cdots\leq i_k} x_{i_1} x_{i_2}\cdots x_{i_k}, \quad \chi_{[1^k]}(g) = \sum_{i_1<i_2<\cdots<i_k} x_{i_1} x_{i_2}\cdots x_{i_k}, \tag{4.6.100}$$

where $x_1, x_2, \ldots, x_N$ are eigenvalues of matrix $g \in U(N)$. The generating functions for characters $\chi_{[k]}(g)$ and $\chi_{[1^k]}(g)$ read

$$H(z) \equiv \sum_{k=0}^{\infty} \chi_{[k]}(g) z^k = \prod_{k=1}^{N} \frac{1}{(1 - z\,x_k)},$$
$$E(z) \equiv \sum_{k=0}^{N} \chi_{[1^k]}(g) z^k = \prod_{k=1}^{N} (1 + z\,x_k), \tag{4.6.101}$$

where we set $\chi_{[1^k]}(g) = \chi_{[k]}(g) = 1$ for $k=0$.

Proof. Representations of group $U(N)$, that correspond to Young diagrams $[1^k]$ and $[k]$, act in spaces of totally antisymmetric and totally symmetric rank-k tensors (see Section 4.4.2). Let matrix $g \in U(N)$ be cast in diagonal form $g = \text{diag}(x_1, \ldots, x_N)$. Then the action of the element g on components of rank-k tensor $\psi \in \mathcal{V}^{\otimes k}$ reads

$$\psi^{i_1,\ldots,i_k} \to [g \cdot \psi]^{i_1,\ldots,i_k} = (x_{i_1} \cdots x_{i_k})\psi^{i_1,\ldots,i_k}, \tag{4.6.102}$$

so elements $g \in U(N)$ in this representation are also diagonal matrices with diagonal elements $(x_{i_1} \cdots x_{i_k})$. Indices in the independent components $\psi_a^{i_1,\ldots,i_k}$ and $\psi_s^{i_1,\ldots,i_k}$ of totally antisymmetric tensor $\psi_a \in \mathcal{V}^{\otimes k}$ and totally symmetric tensor $\psi_s \in \mathcal{V}^{\otimes k}$ can be ordered as follows: $i_1 < i_2 < \cdots < i_k$ and $i_1 \leq i_2 \leq \cdots \leq i_k$, respectively. Therefore, by summing up all eigenvalues of operator g written in (4.6.102) we obtain expressions (4.6.100) for characters of totally symmetric and totally antisymmetric representations.

Proof of the fact that functions $H(z)$ and $E(z)$ given by (4.6.101) are indeed generating functions of characters $\chi_{[k]}$ and $\chi_{[1^k]}$ is left to the reader. □

Corollary 4.6.25. *In the case of diagonalizable matrices $g \in U(N)$ generating functions $H(z)$ and $E(z)$ given by* (4.6.101) *have determinant representations*

$$H(z) \equiv \sum_{k=0}^{N} \chi_{[k]}(g) z^k = \det{}^{-1}(I_N - z\, g), \tag{4.6.103}$$

$$E(z) \equiv \sum_{k=0}^{N} \chi_{[1^k]}(g) z^k = \det(I_N + z\, g). \tag{4.6.104}$$

These identities follow directly from equalities (4.6.101), *since the right-hand sides of the latter are products of eigenvalues of matrices $(I_N - z\,g)^{-1}$ and $(I_N + z\,g)$, i.e., their determinants.*

Problem 4.6.11. ⋆ Without assuming that a matrix g is diagonalizable, show that the generating function $E(z)$ for characters $\chi_{[1^k]}(g)$ is given by a characteristic polynomial obtained by expanding the determinant (4.6.104) in z. In the case of diagonalizable $(N \times N)$ matrix g, prove the characteristic identity (Hamilton–Cayley theorem; see the proof in the general case in book [29]):

$$\sum_{r=0}^{N} (-g)^r \, \chi_{[1^{(N-r)}]}(g) = 0. \tag{4.6.105}$$

Problem 4.6.12. Prove the identity $(\mathrm{Tr}(g))^n = \sum_{\lambda\vdash n} f_{(\lambda)}\chi_\lambda(g)$, where $f_{(\lambda)}$ is the number of standard tableaux whose shape is Young diagram λ. Hint: make use of the formula (4.5.85). Check this identity explicitly for $n=3$, making use of equalities (4.6.94) and (4.6.96).

We introduce standard notations for characters $\chi_{[k]}(g)$ and $\chi_{[1^k]}(g)$ in (4.6.100):

$$h_k(x) = \sum_{i_1\leq i_2\leq\cdots\leq i_k} x_{i_1}x_{i_2}\cdots x_{i_k}, \quad e_k(x) = \sum_{i_1<i_2<\cdots<i_k} x_{i_1}x_{i_2}\cdots x_{i_k}. \tag{4.6.106}$$

Then the generating functions (4.6.101) are written as follows:

$$H(z) = \sum_{k=0}^{\infty} h_k(x)z^k = \prod_{k=1}^{N}\frac{1}{(1-z\,x_k)}, \quad E(z) = \sum_{k=0}^{N} e_k(x)z^k = \prod_{k=1}^{N}(1+z\,x_k). \tag{4.6.107}$$

Functions $h_k(x)$ and $e_k(x)$ are dubbed the kth *complete* and *elementary* symmetric polynomials of N variables $x_1, x_2, \ldots, x_N$, respectively. One also introduces yet another set of symmetric polynomials

$$p_k(x) = x_1^k + x_2^k + \cdots + x_N^k = \mathrm{Tr}(g^k), \tag{4.6.108}$$

called kth *Newton's power sums*. We define generating function for power sums (4.6.108):

$$\tilde{P}(z) = \sum_{k=1}^{\infty}\mathrm{Tr}(g^k)\,\frac{z^k}{k}, \quad P(z) \equiv \partial_z\tilde{P} = \sum_{k=1}^{\infty} p_k(x)\,z^{k-1} = \sum_{i=1}^{N}\frac{x_i}{1-zx_i}. \tag{4.6.109}$$

Proposition 4.6.26. *Generating functions $H(z)$, $E(z)$ and $\tilde{P}(z)$, defined by* (4.6.107) *and* (4.6.109), *are related as follows:*

$$H(z) = \exp(\tilde{P}(z)), \quad E(z) = \exp(-\tilde{P}(-z)). \tag{4.6.110}$$

Sets of symmetric polynomials (4.6.106) *and* (4.6.108) *are related by*

$$h_r(x) - e_1(x)h_{r-1}(x) + e_2(x)h_{r-2}(x) - \cdots + (-1)^r e_r(x) = 0, \tag{4.6.111}$$

$$rh_r(x) - p_1(x)h_{r-1}(x) - p_2(x)h_{r-2}(x) - \cdots - p_r(x) = 0, \tag{4.6.112}$$

$$r\,e_r(x) - p_1(x)e_{r-1}(x) + p_2(x)e_{r-2}(x) - \cdots + (-1)^r p_r(x) = 0, \tag{4.6.113}$$

where $r = 1,\ldots,N$. Equalities (4.6.112) *and* (4.6.113) *are* Newton's identities.

Proof. One makes use of the identity $\det(A) = \exp(\mathrm{Tr}\ln(A))$, valid for any matrix A, as well as expansion $\ln(1-z) = -\sum_{k=1}^{\infty} z^k/k$ to cast the determinants in (4.6.103) and (4.6.104) in the forms $\exp(\tilde{P}(z))$ and $\exp(-\tilde{P}(-z))$, respectively. This leads to the formulas (4.6.110) which, as well as (4.6.107), give $H(z)E(-z) = 1$, i.e.,

$$\left(\sum_{k=0}^{\infty} h_k(x) z^k\right) \cdot \left(\sum_{k=0}^{N} e_k(x)(-z)^k\right) = 1.$$

By expanding the left-hand side here in z we obtain (4.6.111). Now, we note that the functions $H(z)$, $E(z)$ and $P(z)$ are related by

$$\partial_z H(z) = P(z)H(z), \quad \partial_z E(z) = P(-z)E(z); \tag{4.6.114}$$

these relations are obtained by differentiating (4.6.110). We insert here the expansions of $H(z)$, $E(z)$ and $P(z)$ in z, and obtain (4.6.112) and (4.6.113). □

It is known that one can choose N elementary symmetric polynomials $\{e_1(x), \ldots, e_N(x)\}$ as generators in algebra $\mathcal{P}$ of all symmetric polynomials of N variables $\{x_1, \ldots, x_N\}$ over field[t] $\mathbb{K} = \mathbb{R}, \mathbb{C}$. We expand the two sides of (4.6.110) in z and find that functions $\{h_k(x) = \chi_{[k]}(x)\}$ and $\{e_k(x) = \chi_{[1^k]}(x)\}$ can be expressed as polynomials in functions $\{p_k(x)\}$ with rational coefficients. Furthermore, Eq. (4.6.111) shows that functions $\{e_k(x) = \chi_{[1^k]}(x)\}$ are polynomials (with integer coefficients) in functions $\{h_k(x)\}$, where $k = 1, \ldots, N$. Therefore, one can choose finite sets of power sums $\{p_1(x), \ldots, p_N(x)\}$ or complete polynomials $\{h_1(x), \ldots, h_N(x)\}$ as generators in algebra $\mathcal{P}$.

Problem 4.6.13. Derive expressions (4.6.93) and (4.6.94) for characters $\chi_{[k]}, \chi_{[1^k]}$ in terms of power sums $p_r = \mathrm{Tr}(g^r)$ for $k = 2, 3$ from Eq. (4.6.110).

As an example important for applications, we show explicitly that symmetric Schur polynomials $\chi_\lambda(x) \in \mathcal{P}$, that correspond to Young diagrams $\lambda = [\lambda_1, \ldots \lambda_m]$ with $m \leq N$, are combinations of complete symmetric polynomials $h_k(x) = \chi_{[k]}(x)$ $(k = 1, \ldots, N)$.

Proposition 4.6.27. *Let $\lambda = [\lambda_1, \ldots, \lambda_N]$ be Young diagram with $m \leq N$ rows (i.e., $\lambda_k = 0$ for all $k > m$). We define $(N \times N)$ matrix $A(x)$ with*

---

[t]We consider the cases $\mathbb{K} = \mathbb{R}, \mathbb{C}$ for definiteness, although important cases here are also fields $\mathbb{Z}$ and $\mathbb{Q}$.

elements $A_{ij}(x) = h_{\lambda_i+j-i}(x)$, *where* $h_k(x) = 0$ *for* $k < 0$. *Then Schur polynomials* $\chi_\lambda(x)$ *admit determinant representation*

$$\chi_\lambda(x) = \det(A(x)) = \det(||h_{\lambda_i+j-i}(x)||). \tag{4.6.115}$$

Proof. (See also Ref. [46]). Recall that Schur polynomials can be defined using Jacobi–Trudi formula (3.5.44):

$$\chi_\lambda(x) = \frac{\det(||x_i^{\lambda_j+N-j}||)}{\det(||x_i^{N-j}||)}. \tag{4.6.116}$$

Here $||x_i^{\lambda_j+N-j}||$ and $||x_i^{N-j}||$ are $(N \times N)$ matrices, i.e., $i,j = 1,\ldots,N$ are row and column numbers. Let us show that the ratio of two polynomials in (4.6.116) is equal to symmetric polynomial on the right-hand side of (4.6.115). Let p be an integer such that $1 \le p \le N$ and let $e_r^{(p)}(x')$ be rth elementary symmetric polynomial of variables $x' = (x_1,\ldots,x_{p-1},x_{p+1},\ldots,x_N)$. Then, making use of generating functions (4.6.107), we get

$$\left(\sum_k h_k(x)z^k\right)\cdot\left(\sum_{r=0}^{N-1} e_r^{(p)}(x')(-z)^r\right) = \frac{1}{1-x_p\,z}. \tag{4.6.117}$$

The sum over k on left-hand side of (4.6.117) can be extended to the range $k \in [-\infty,\infty]$, once one recalls that $h_k = 0$ for all $k < 0$. Expanding the two sides of (4.6.117) in z we derive the formula

$$\sum_{j=1}^{N} h_{q+j-N}(x)(-1)^{N-j}e_{N-j}^{(p)}(x') = (x_p)^q.$$

For $q = \lambda_i + N - i$, it gives for the product of matrices

$$||h_{\lambda_i+j-i}(x)||\cdot||(-1)^{N-j}e_{N-j}^{(p)}(x')|| = ||(x_p)^{\lambda_i+N-i}||. \tag{4.6.118}$$

We now recall that $\det||h_{j-i}|| = 1$, which follows from $h_0(x) = 1$ and $h_k(x) = 0$ for $k < 0$, and find that formula (4.6.118) for $\lambda = [0,0,\ldots,0]$ gives

$$\det(||(-1)^{N-i}e_{N-i}^{(j)}||) = \det(||(x_j)^{N-i}||). \tag{4.6.119}$$

The result (4.6.115) then follows from calculation of determinants of the two sides of matrix relation (4.6.118) and then employing formulas (4.6.116) and (4.6.119). □

Problem 4.6.14. Making use of the determinant representation (4.6.115), derive the formula (cf. the first of Eqs. (4.6.99))

$$\chi_{[2,1]} = \det \begin{pmatrix} h_2(x) & h_3(x) \\ h_0(x) & h_1(x) \end{pmatrix} = h_2(x)\, h_1(x) - h_3(x).$$

Remark. Instead of standard Young tableaux whose shape is Young diagram λ (see Definition 4.3.2 in Section 4.3.2), consider tableaux defined in the following way. We put into boxes of diagram $\lambda \vdash k$ numbers $\{1, 2, \ldots, N\}$ in such a way that (i) they do not decrease in rows from left to right; (ii) they strictly increase in columns from top to bottom. An example is

$$\mathsf{T} = \begin{array}{|c|c|c|c|c|c|} \hline 1 & 2 & 2 & 3 & 3 & 3 \\ \hline 2 & 3 & 3 & 4 & 5 & \\ \cline{1-5} 3 & 4 & 4 & & & \\ \cline{1-3} 5 & 6 & & & & \\ \cline{1-2} \end{array} \, . \tag{4.6.120}$$

Tableaux of this type are called *semi-standard*. Let $n_{\mathsf{T}}(i)$ be the multiplicity of number i in semi-standard tableau T. We associate with the tableau T a monomial

$$x^{\mathsf{T}} = \prod_{i=1}^{N} x_i^{n_{\mathsf{T}}(i)}. \tag{4.6.121}$$

As an example, the monomial associated with semi-standard tableau (4.6.120) is $x^{\mathsf{T}} = x_1 x_2^3 x_3^6 x_4^3 x_5^2 x_6$. Schur polynomials $s_\lambda(x_1, \ldots, x_N)$ are usually defined in literature as homogeneous polynomials, which are sums of monomials (4.6.121) over all semi-standard tableaux T of shape λ that contain numbers $\{1, 2, \ldots, N\}$ in their boxes:

$$s_\lambda(x_1, \ldots, x_N) = \sum_{\mathsf{T}} x^{\mathsf{T}}. \tag{4.6.122}$$

It is straightforward to see that in the particular cases $\lambda = [k]$ and $\lambda = [1^k]$, one has $s_{[k]}(x) = \chi_{[k]}(x)$ and $s_{[1^k]}(x) = \chi_{[1^k]}(x)$, where characters $\chi_{[k]}(x)$ and $\chi_{[1^k]}(x)$ are defined in (4.6.100). The general proof of equality $s_\lambda(x) = \chi_\lambda(x)$ where $\chi_\lambda(x)$ are defined in (4.6.115) and (4.6.116) makes use of combinatorics of Young tableaux (see [41]) and is beyond the scope of this book.

We only point out a hint in favor of the equality $s_\lambda(x) = \chi_\lambda(x)$. Consider the action (4.6.102) of diagonal matrix $g = \mathrm{diag}(x_1, \ldots, x_N) \in U(N)$

in space of tensors $\psi^{i_1\dots i_k}$ whose symmetry properties are dictated by Young diagram $\lambda \vdash k$ (recall that indices $i_1,\dots,i_k$ of the symmetrized tensor are put in boxes of diagram λ). Then indices in eigenvalues $(x_{i_1}\cdot x_{i_2}\cdots x_{i_k})$ in every column of diagram λ can be ordered in such a way that they strictly increase from top to bottom, and do not decrease in every row from left to right. Therefore, every eigenvalue $(x_{i_1}\cdot x_{i_2}\cdots x_{i_k})$ is a monomial (4.6.121), and the sum of these monomials give, on the one hand, character of representation $\chi_\lambda(x)$, and, on the other hand, Schur function (4.6.122) (similar arguments are used in the proof of equalities (4.6.100) in Proposition 4.6.24).

4.7. Concluding Remarks. Gelfand–Tsetlin Basis

In conclusion, we point out that our discussion of the theory of finite-dimensional representations of linear groups and linear Lie algebras is not quite complete. We have not considered at least two important issues. One has to do with Gelfand–Tsetlin basis in the spaces of irreducible representations of algebras $s\ell(N,\mathbb{C})$, $g\ell(N,\mathbb{C})$ and their real forms. Gelfand–Tsetlin construction enables one to solve the problem of degeneracy of weights (see item B in Section 3.2) that label basis vectors in spaces of irreducible representations of Lie algebras $s\ell(N,\mathbb{C})$ and also $g\ell(N,\mathbb{C})$ and $u(N)$. Also, Gelfand–Tsetlin basis is a convenient tool for studying and classifying irreducible representations. The second topic is Clebsch–Gordan expansion of tensor product of two arbitrary irreducible representations of algebra $s\ell(N,\mathbb{C})$ for $N>2$ (the case $N=2$ is studied in Section 2.4), or, in other words, Littlewood–Richardson rule for expansion of tensor product of two irreducible representations of symmetric group S_r. Readers interested in these topics are invited to consult the books [14, 28, 31, 32].

Here we only sketch the idea of Gelfand–Tsetlin construction. We consider spaces $\mathcal{V}^{(\lambda)}$ of finite-dimensional irreducible representations $T^{(\lambda)}$ of Lie algebra $g\ell(N,\mathbb{C})$ with generators E_{ij} $(i,j=1,\dots,N)$ obeying structure relations

$$[E_{ij},E_{k\ell}]=\delta_{jk}E_{i\ell}-\delta_{i\ell}E_{kj}. \tag{4.7.1}$$

Highest weight representations of algebra $g\ell(N,\mathbb{C})$ has been considered in item 2.1 of Section 3.4.2. Weight degeneracy means that enveloping algebra $\mathcal{U}(g\ell(N,\mathbb{C}))$ $(N>2)$ contains operators which, on the one hand, are not products of Cartan elements and Casimir operators (Casimir operators of algebras $s\ell(N,\mathbb{C})$ and $g\ell(N,\mathbb{C})$ coincide) and, on the other hand, commute with all Cartan elements in $g\ell(N, \mathbb{C})$. Recall that Cartan subalgebra in $g\ell(N, \mathbb{C})$ includes unit central element and has dimension N, see item 2.1 in Section 3.4.2. We also recall (see Section 3.4.3) that the values of Casimir operators on all vectors of space $\mathcal{V}^{(\lambda)}$ can be expressed through components of the highest weight $\lambda = (\lambda_1, \ldots, \lambda_N)$ of representation $T^{(\lambda)}$. We have seen in item 2.1 of Section 3.4.2 that integer-valued components of the highest weight λ are ordered (see (3.4.46)):

$$\lambda_1 \geq \lambda_2 \geq \cdots \geq \lambda_N \geq 0, \tag{4.7.2}$$

and correspond to Young diagram $\Lambda = [\lambda_1, \ldots, \lambda_N]$. Consider subalgebra $g\ell(N-1, \mathbb{C}) \subset g\ell(N, \mathbb{C})$ embedded in $g\ell(N, \mathbb{C})$ as follows:

$$\left(\begin{array}{c|c} A & \begin{matrix} 0 \\ \vdots \\ 0 \end{matrix} \\ \hline 0 \ldots 0 & 0 \end{array}\right), \quad A \in g\ell(N-1, \mathbb{C}). \tag{4.7.3}$$

In other words, subalgebra $g\ell(N-1, \mathbb{C})$ has generators $E_{ij} \in g\ell(N, \mathbb{C})$ with $i, j = 1, \ldots, (N-1)$. Irreducible representation $T^{(\lambda)}$ of algebra $g\ell(N, \mathbb{C})$ is reducible for subalgebra $g\ell(N-1, \mathbb{C})$, and is decomposed into the sum of its irreducible representations $T^{(\lambda')}$. The latter have dimensions smaller than $\dim(T^{(\lambda)})$ and are labeled by their highest weights $\lambda' = (\lambda'_1, \ldots, \lambda'_{N-1})$. Here the components are ordered in the same way as in (4.7.2): $\lambda'_1 \geq \cdots \geq \lambda'_{N-1} \geq 0$. Weight λ' corresponds to Young diagram $\Lambda' = [\lambda'_1, \ldots, \lambda'_{N-1}]$ and defines eigenvalues (as functions of λ'_i) of Casimir operators for $g\ell(N-1, \mathbb{C})$. Casimir operators of $g\ell(N-1, \mathbb{C})$ commute with Casimir operators and Cartan elements of $g\ell(N, \mathbb{C})$. Therefore, we can use their eigenvalues, and hence components of highest weight $\lambda' = (\lambda'_1, \ldots, \lambda'_{N-1})$, to label vectors in subspace $\mathcal{V}^{(\lambda')} \subset \mathcal{V}^{(\lambda)}$, together with the components of the highest weight $\lambda = (\lambda_1, \ldots, \lambda_N)$. We now recall the arguments used in Section 4.4.2 to introduce the first method of calculating dimensions of representations of group $SL(N, \mathbb{C})$, which is based on embeddings of representations of subgroup $SL(N-1, \mathbb{C})$ in representations of group $SL(N, \mathbb{C})$. These arguments can as well be applied to embeddings of representations of subalgebra $g\ell(N-1, \mathbb{C})$ in representation of algebra $g\ell(N, \mathbb{C})$. In this way, we find that representation $T^{(\lambda')}$ of algebra $g\ell(N-1, \mathbb{C})$ that corresponds to Young diagram $\Lambda' = [\lambda'_1, \ldots, \lambda'_{N-1}]$, can be embedded in representation $T^{(\lambda)}$ of algebra $g\ell(N, \mathbb{C})$ only if the following condition holds:

$$\lambda_i \geq \lambda'_i \geq \lambda_{i+1}, \quad i = 1, 2, \ldots, N-1. \tag{4.7.4}$$

The multiplicity of each irreducible representation $T^{(\lambda')}$ in representation $T^{(\lambda)}$ is equal to 1. Consider now a chain of embeddings of algebras $g\ell(1,\mathbb{C}) \subset g\ell(2,\mathbb{C}) \subset \cdots \subset g\ell(N,\mathbb{C})$, and decompose the space $\mathcal{V}^{(\lambda)}$ of irreducible representation of algebra $g\ell(N,\mathbb{C})$ into subspaces $\mathcal{V}^{(\lambda')}$ of irreducible representations of algebra $g\ell(N-1,\mathbb{C})$, which in turn are decomposed into subspaces $\mathcal{V}^{(\lambda'')}$ of representations of algebra $g\ell(N-2,\mathbb{C})$, etc., down to spaces of irreducible representations of algebra $g\ell(1,\mathbb{C})$. Each space $\mathcal{V}^{(\lambda_{(K)})}$ of irreducible representation of subalgebra $g\ell(K,\mathbb{C})$ is labeled by its highest weight $\lambda_{(K)} = (\lambda_{1(K)},\dots,\lambda_{K(K)})$, whose components obey (see (4.7.2), (4.7.4))

$$\lambda_{1(K)} \geq \cdots \geq \lambda_{K(K)}, \quad \lambda_{i(K)} \geq \lambda_{i(K-1)} \geq \lambda_{i+1(K)}, \quad i = 1,2,\dots,K-1. \tag{4.7.5}$$

Since the spaces of irreducible representations of subalgebra $g\ell(1,\mathbb{C})$ are one-dimensional, the set of components of all highest weights $\lambda_{(K)}$ ($K = 1,\dots,N$) labels one-dimensional subspace $\mathcal{V}^{(\lambda_{(1)})} \subset \mathcal{V}^{(\lambda_{(2)})} \subset \cdots \subset \mathcal{V}^{(\lambda_{(N)})} = \mathcal{V}^{(\lambda)}$ in unique way. The sets obeying (4.7.5) are conveniently represented by *Gelfand–Tsetlin schemes*

$$\begin{array}{c} \lambda_{1(N)} \quad \lambda_{2(N)} \quad \cdots \quad \lambda_{N-1(N)} \quad \lambda_{N(N)} \\ \lambda_{1(N-1)} \quad \lambda_{2(N-1)} \quad \cdots \quad \lambda_{N-1(N-1)} \\ \cdots\cdots\cdots\cdots\cdots\cdots\cdots\cdots \\ \lambda_{1(2)} \quad \lambda_{2(2)} \\ \lambda_{1(1)} \end{array} \tag{4.7.6}$$

Unit basis vectors of subspaces $\mathcal{V}^{(\lambda_{(1)})}$ labeled by schemes (4.7.6) make basis in $\mathcal{V}^{(\lambda)}$. This is precisely *Gelfand–Tsetlin basis.* To determine the action of representation of algebra $g\ell(N,\mathbb{C})$ in Gelfand–Tsetlin basis, one has to figure out the action of generators $E_{i,j} \in g\ell(N,\mathbb{C})$ on schemes (4.7.6). We do not derive the explicit formulas here; they can be found in books [14, 28, 31, 32].

Problem 4.7.1. * Construct Gelfand–Tsetlin bases in spaces $\mathcal{V}^{([3])}$ and $\mathcal{V}^{([2,1])}$ of representations of algebra $u(3)$ labeled by Young diagrams [3] and [2, 1]. Find the correspondence between Gelfand–Tsetlin schemes (basis vectors of spaces $\mathcal{V}^{([3])}$ and $\mathcal{V}^{([2,1])}$) and members of baryon decuplet and octet (see formulas (4.4.81)–(4.4.84) and (4.4.86) in Section 4.4.4).

Chapter 5

Finite-Dimensional Representations of Groups *SO*, *Sp* and Lie Algebras *so*, *sp*

We consider in Sections 5.1–5.3 a class of complex finite-dimensional irreducible representations of groups of SO and Sp families, which are subrepresentations in tensor products of *defining representations* of these groups. The representations from this class exhaust all irreducible representations of *groups* of SO and Sp which are associated with representations of complex Lie algebras *so* and *sp*. In the case of Sp family, we do not get anything new when turning from *groups* to *algebras sp*. Indeed, as we have shown in Section 3.4.2 (item 5), all irreducible representations of Lie algebras *sp* are contained in tensor products of their defining representations and are in one-to-one correspondence with representations of groups Sp of the class we consider. So, our analysis covers *all* complex finite-dimensional irreducible representations of groups of Sp-series associated with representations of algebras of *sp*-series. This is not the case for algebras *so*: as we discussed in Section 3.4.2 (items 3 and 4), these algebras have spinor representations which cannot be embedded in tensor products of defining representations and do not have analogs among the representations of *groups* SO. Instead, the spinor representations of *so* algebras are generated by representations of groups Spin, which are locally isomorphic to groups SO. We begin the discussion of spinor representations of algebras *so* in Section 5.4 and consider groups Spin in some detail in Chapter 6.

5.1. Tensor Representations of Groups $O(N,\mathbb{C})$, $SO(N,\mathbb{C})$ and Their Subgroups $O(p,q)$, $SO(p,q)$

We begin this section with a concise summary of the facts which in one or another way have been discussed in Sections I-1.2.5, I-3.1.3 and I-3.3.1 of the first book. We think it makes sense to collect these facts in one place and consider them in some detail. All results concerning representations of groups $O(p,q)$ and their Lie algebras $so(p,q)$ are valid, of course (with $p=1$ and $q=N-1$), for multidimensional Lorentz group $O(1,N-1)$ and its Lie algebra.

5.1.1. Pseudo-orthogonal group $O(p,q)$ and Lie algebra $so(p,q)$

Consider N-dimensional *real* vector space $\mathcal{V}_N$ equipped with nondegenerate scalar product of vectors $\vec{a},\vec{b} \in \mathcal{V}_N$, namely, $(\vec{a},\vec{b}) \in \mathbb{R}$. One chooses orthogonal, with respect to this scalar product, basis $\{\vec{e}_i\}$, $(i=1,\dots,N)$, for which matrix $(\vec{e}_i,\vec{e}_j) = \eta_{ij}$ is diagonal, $\eta = \mathrm{diag}(\eta_1,\eta_2,\dots,\eta_N)$ and $\eta_i \neq 0$ for all i. Matrix η is called metric in space $\mathcal{V}_N$. By dilation of the basis vectors $\vec{e}_j \to \lambda_j\vec{e}_j$ (where $\lambda_j \in \mathbb{R}$) one can always make each η_i equal to one of the two numbers, $\eta_i = \pm 1$. Recall that if $\eta_i = +1$ for all i (or $\eta_i = -1$ for all i), then the space $\mathcal{V}_N$ is called Euclidean and is denoted by $\mathbb{R}^N$, whereas if the set $\{\eta_j\}$ contains both $+1$ and -1, i.e.,

$$\eta = I_{p,q} = \mathrm{diag}(\underbrace{+1,\dots,+1}_{p},\underbrace{-1,\dots,-1}_{q}), \tag{5.1.1}$$

where $p+q=N$, then the space $\mathcal{V}_N$ is called pseudo-Euclidean and is denoted by $\mathbb{R}^{p,q}$. Space $\mathbb{R}^{1,N-1}$ is *Minkowski space.*

Consider linear transformation of basis in space $\mathbb{R}^{p,q}$:

$$\vec{e}_j{}' = \Lambda \cdot \vec{e}_j = \vec{e}_k\, \Lambda^k{}_j, \quad ||\Lambda^k{}_j|| \in \mathrm{Mat}_N(\mathbb{R}), \tag{5.1.2}$$

where Λ is real matrix. Then metric $\eta_{ij} = (\vec{e}_i,\vec{e}_j)$ transforms as follows:

$$\eta'_{ij} = (\vec{e}_i{}',\vec{e}_j{}') = \Lambda^k{}_i\, \Lambda^n{}_j\, (\vec{e}_k,\vec{e}_n) = \Lambda^k{}_i\, \Lambda^n{}_j\, \eta_{kn}. \tag{5.1.3}$$

We distinguish not only upper and lower indices in the notation for matrix elements $\Lambda^k{}_i$ but also left (incoming) and right (outgoing) indices; these are k and i, respectively.

Definition 5.1.1. Pseudo-orthogonal group $O(p,q)$ is the set of linear operators Λ in $\mathbb{R}^{p,q}$, which transform the basis in $\mathbb{R}^{p,q}$ in such a way that metric (5.1.1) remains intact:

$$\eta'_{ij} = \Lambda^k{}_i \, \eta_{kn} \, \Lambda^n{}_j = \eta_{ij} \Leftrightarrow \Lambda^{\mathsf{T}} \, \eta \, \Lambda = \eta. \tag{5.1.4}$$

In other words, pseudo-orthogonal group $O(p,q)$ is the group of real $(p+q) \times (p+q)$ matrices $\Lambda = ||\Lambda^k{}_i||$, which obey (5.1.4); the group operation is the usual product of matrices.

Clearly, group $O(p,q)$ is the set of transformations (5.1.2) which preserve the scalar products of basis vectors, $(\vec{e}_i{}', \vec{e}_j{}') = (\vec{e}_i, \vec{e}_j)$ in $\mathbb{R}^{p,q}$, and hence scalar products of any vectors in $\mathbb{R}^{p,q}$.

We now recall the structure of the Lie group $O(p,q)$ (see item 12 in Section I-2.1.2 of the first book), including the definition of connected component of unit element $SO^{\uparrow}(p,q) \subset O(p,q)$. This information is used in Section 6, where we study groups $\mathsf{Spin}(p,q)$ which cover the connected components $SO^{\uparrow}(p,q)$.

It follows from (5.1.4) that $\det(\Lambda) = \pm 1$. Therefore, the group $O(p,q)$ consists of two disconnected subsets $O_+(p,q)$ and $O_-(p,q)$, for which $\det(\Lambda) = +1$ (proper transformations) and $\det(\Lambda) = -1$ (improper transformations), respectively. Subset of proper transformations $O_+(p,q) \equiv SO(p,q)$ is a subgroup in $O(p,q)$. Let us write Λ as block matrix

$$\Lambda = \begin{pmatrix} X & Y \\ Z & W \end{pmatrix}, \tag{5.1.5}$$

where X, Y, Z and W are real matrices of size $p \times p$, $p \times q$, $q \times p$ and $q \times q$, respectively.

Problem 5.1.1. Making use of (5.1.4), show that $|\det(X)| \geq 1$, $|\det(W)| \geq 1$.

It follows from the result of this problem (see also Section I-2.1.2, item 12 of the first book) that one can define proper "orthochronous" subgroup $SO^{\uparrow}(p,q) \subset SO(p,q)$ as a set of elements $\Lambda \in O(p,q)$ which obey $\det(\Lambda) = +1$ and whose blocks X and W in (5.1.5) satisfy $\det(X) \geq 1$, $\det(W) \geq 1$. Subgroup $SO^{\uparrow}(p,q) \subset O(p,q)$ is a connected component of unit element $I \in O(p,q)$, and the entire manifold of group $O(p,q)$

$(p \geq 1,\ q \geq 1)$ consists of four disconnected components:

$$O(p,q) = (SO^{\uparrow}(p,q) \cdot I) \cup (SO^{\uparrow}(p,q) \cdot T) \cup (SO^{\uparrow}(p,q) \cdot P) \\ \cup\, (SO^{\uparrow}(p,q) \cdot PT), \tag{5.1.6}$$

where $T = \mathrm{diag}(-1, +1, \ldots, +1)$ and $P = \mathrm{diag}(+1, \ldots, +1, -1)$. Thus, the union of components $SO^{\uparrow}(p,q)$ and $(SO^{\uparrow}(p,q) \cdot PT)$ is the subgroup of proper transformations $SO(p,q) \subset O(p,q)$, while union of components $(SO^{\uparrow}(p,q) \cdot T)$ and $(SO^{\uparrow}(p,q) \cdot P)$ is improper part $O_{-}(p,q) \subset O(p,q)$.

Note that (5.1.4) can be written as $\eta = (\Lambda^{\mathsf{T}})^{-1}\eta\Lambda^{-1}$, so one has

$$\eta^{-1} = \Lambda\, \eta^{-1}\, \Lambda^{\mathsf{T}} \;\Leftrightarrow\; \eta^{ij} = \Lambda^{i}{}_{k}\, \eta^{kn}\, \Lambda^{j}{}_{n}, \tag{5.1.7}$$

where coefficients η^{kj} (with upper indices) are elements of inverse matrix η^{-1}: $\eta_{ik}\eta^{kj} = \delta_i^j$; for the standard metric (5.1.1), we have $\eta^{-1} = \eta$. The condition (5.1.4) can also be written as one of the relations which we use in the following:

$$\eta_{ij}(\Lambda^{-1})^{j}{}_{n} = \Lambda^{k}{}_{i}\eta_{kn} \;\Leftrightarrow\; (\Lambda^{-1})_{in} = \Lambda_{ni} \;\Leftrightarrow\; (\Lambda^{-1})_{i}{}^{n} = \Lambda^{n}{}_{i}, \tag{5.1.8}$$

$$\Lambda_{i}{}^{k}\, \eta_{kn}\, \Lambda_{j}{}^{n} = \eta_{ij} \;\Leftrightarrow\; \Lambda_{k}{}^{i}\, \eta^{kn}\, \Lambda_{n}{}^{j} = \eta^{ij}. \tag{5.1.9}$$

Hereafter, indices are raised and lowered with metric (5.1.1), e.g., $\Lambda_{ni} = \eta_{nj}\Lambda^{j}{}_{i}$, $\Lambda_{n}{}^{i} = \eta^{ij}\Lambda_{nj}$, etc.

By considering elements $\Lambda \in SO(p,q)$ in the vicinity of unit element, $\Lambda = I + tA + \cdots$ with small parameter t, we obtain from (5.1.4) to the first order in t

$$A^{k}{}_{i}\, \eta_{kj} + \eta_{in}\, A^{n}{}_{j} = 0 \Leftrightarrow A^{i}{}_{k}\, \eta^{kj} + \eta^{in}\, A^{j}{}_{n} = 0. \tag{5.1.10}$$

The set of real $(N \times N)$ matrices A obeying (5.1.10) is a real vector space tangent to group manifold $O(p,q)$ at unit element. This is precisely Lie algebra $so(p,q)$; if elements A and A' belong to $so(p,q)$, then their commutator $[A, A']$ obeys (5.1.10), so $[A, A'] \in so(p,q)$.

We recall that η is symmetric matrix and write the relation (5.1.10) as $\eta_{jk}\, A^{k}{}_{i} = -\eta_{ik}\, A^{k}{}_{j}$. We see that matrix $\eta \cdot A$ is antisymmetric, so it can be expanded in the basis of antisymmetric matrices:

$$\eta \cdot A = \omega_{ij}(e_{ij} - e_{ji}),$$

where e_{ij} are matrix unities, and $\omega_{ij} = -\omega_{ji} \in \mathbb{R}$. Thus, one can choose the basis in algebra $so(p,q)$ as follows (see also item 7 in Section I-3.3.1 of the first book):

$$L_{ij} = \eta^{-1} \cdot (e_{ij} - e_{ji}), \quad 1 \le i < j \le p+q. \tag{5.1.11}$$

Making use of the rule $e_{ij} \cdot h \cdot e_{k\ell} = h_{jk}\, e_{i\ell}$, where $h = ||h_{ik}||$, we find the structure relations for generators (5.1.11) of algebra $so(p,q)$:

$$[L_{ij},\, L_{k\ell}] = \eta^{-1}_{jk} L_{i\ell} + \eta^{-1}_{j\ell} L_{ki} + \eta^{-1}_{ik} L_{\ell j} + \eta^{-1}_{i\ell} L_{jk}. \tag{5.1.12}$$

These relations are valid with arbitrary choice of the symmetric metric η.

Recall that we have defined in the first book orthogonal group of complex $N \times N$ matrices O obeying orthogonality condition

$$O^{\mathsf{T}}\, \eta\, O = \eta, \tag{5.1.13}$$

where η is an arbitrary symmetric real nondegenerate matrix. Unlike in the case $O(p,q)$, these groups are isomorphic for different η. We denoted this group as $O(N,\mathbb{C})$, and made a choice $\eta = I_N$. Subgroup $SO(N,\mathbb{C}) \subset O(N,\mathbb{C})$ is extracted from $O(N,\mathbb{C})$ by imposing the condition $\det(O) = 1$. One again considers matrices near unit element, $O = I_N + t\,A + t^2 + \cdots \in SO(N,\mathbb{C})$, where parameter t is small, and using the defining relation $O^{\mathsf{T}}\, O = I_N$, finds the relation for complex matrices A

$$A + A^{\mathsf{T}} = 0. \tag{5.1.14}$$

Thus, it is natural to define *complex* Lie algebra $so(N,\mathbb{C})$ whose elements are complex matrices A obeying (5.1.14). Similar to the case of $SL(N,\mathbb{C})$, this is not Lie algebra $\mathcal{A}(SO(N,\mathbb{C}))$ of the Lie group $SO(N,\mathbb{C})$, which is considered as real manifold. Indeed, Lie algebras (including $\mathcal{A}(SO(N,\mathbb{C}))$) of real Lie groups are *real*. Like in the case of linear groups, the study of complex Lie algebra $so(N,\mathbb{C})$ and its representations is convenient, since Lie algebras $so(p,q)$ are real forms of $so(N,\mathbb{C})$, and representations of $so(N,\mathbb{C})$ are at the same time complex representations of $so(p,q)$. Complete classification of real forms of algebra $so(N,\mathbb{C})$ is given in Section I-3.3.2 of the first book.

In what follows, we study representations of group $O(p,q)$, which are at the same time representations of its Lie algebra $so(p,q)$. As we pointed out already, these representations of $O(p,q)$ do not generate all representations of algebra $so(p,q)$, for example, spinor representations of $so(p,q)$ cannot be obtained in this way.

5.1.2. Tensor representations of groups $O(p,q)$

Particularly interesting for physics are *real* representations of groups $O(p,q)$.

Consider vector $\vec{x} = x^j \vec{e}_j \in \mathbb{R}^{p,q}$ with coordinates $x^j \in \mathbb{R}$. Under the basis transformation (5.1.2), these coordinates transform in accordance with (2.1.8):

$$x'^{\,j} = \Lambda^j{}_k \, x^k = x^k \, (\Lambda^{-1})_k{}^j , \tag{5.1.15}$$

where in the second equality we use (5.1.8). It follows from Definition 5.1.1 that these transformations with matrices Λ obeying (5.1.4), preserve scalar product

$$(\vec{y}, \vec{x}) = y^i \eta_{ij} x^j = y_j \, x^j , \quad \forall \vec{x}, \vec{y} \in \mathbb{R}^{p,q}, \tag{5.1.16}$$

where we use coordinates with lowered indices

$$x_j = (\vec{x}, \vec{e}_j) = x^k \, \eta_{kj} . \tag{5.1.17}$$

These coordinates are transformed as follows:

$$x'_j = x_k \, (\Lambda^{-1})^k{}_j = \Lambda_j{}^k \, x_k , \tag{5.1.18}$$

which follows from (5.1.15) and (5.1.17) (and ensures invariance of (5.1.16)). We call (5.1.18) *covariant* coordinate transformation, while transformation (5.1.15) is *contravariant.* Accordingly, coordinates x_j (with lower indices) and x^j are *covariant* and *contravariant*, respectively. Note that the inverse of (5.1.17) is

$$x^k = \eta^{kj} \, x_j . \tag{5.1.19}$$

Metrics η_{jk} and η^{kj} can also be used to lower and raise indices of basis vectors $\vec{e}^{\,j}$ and $\vec{e}_j$. In terms of basis vectors $\vec{e}^{\,j}$, the transformation (5.1.2) reads

$$\vec{e}'^{\,j} = \Lambda \cdot \vec{e}^{\,j} = (\Lambda^{-1})^j{}_k \, \vec{e}^{\,k} = \vec{e}^{\,k} \, \Lambda_k{}^j . \tag{5.1.20}$$

Matrices $\Lambda = ||\Lambda^k{}_j||$ obeying (5.1.4) are matrices of defining representation T of pseudo-orthogonal group $O(p,q)$. Representation T acts in space $\mathbb{R}^{p,q}$, and this action is given by (5.1.2) and (5.1.15). Representation $\widetilde{T}$, contragredient to defining representation, is defined in a standard way (see Example 3 in Section I-4.1.1 of the first book):

$$\widetilde{T}(\Lambda) = (\Lambda^{-1})^{\mathsf{T}}, \quad \forall \Lambda = ||\Lambda^{k}{}_{j}|| \in O(p,q).$$

It also acts in vector space $\mathbb{R}^{p,q}$; this action is defined by (5.1.18) and (5.1.20). Due to the second relation in (5.1.4), representations T and $\widetilde{T}$ are equivalent:

$$(\Lambda^{-1})^{\mathsf{T}} = \eta\, \Lambda\, \eta^{-1}, \tag{5.1.21}$$

whereas equalities (5.1.17) and (5.1.19) establish the correspondence between vectors in the spaces $\mathbb{R}^{p,q}$ of these representations.

Consider tensor product of several representations T and $\widetilde{T}$ of group $O(p,q)$:

$$T^{(s+t)} = \underbrace{T \otimes \cdot \otimes T}_{s} \otimes \underbrace{\widetilde{T} \otimes \cdot \otimes \widetilde{T}}_{t} = T^{\otimes s} \otimes \widetilde{T}^{\otimes t}, \quad s+t=r. \tag{5.1.22}$$

This representation acts in tensor product $(\mathbb{R}^{p,q})^{\otimes r}$. Elements in this tensor product are

$$\mathsf{t} = \mathsf{t}^{k_1 \dots k_s}{}_{k_{s+1} \dots k_r}\ \vec{e}_{k_1} \otimes \dots \otimes \vec{e}_{k_s} \otimes \vec{e}^{k_{s+1}} \otimes \dots \otimes \vec{e}^{k_r} \in (\mathbb{R}^{p,q})^{\otimes r}, \tag{5.1.23}$$

where $\mathsf{t}^{k_1 \dots k_s}{}_{k_{s+1} \dots k_r}$ are real numbers called components of element t. They have s upper indices and $t = r - s$ lower indices. In accordance with the formulas (5.1.2) and (5.1.20) for action of group $O(p,q)$ on basis vectors $\vec{e}_j, \vec{e}^k \in \mathbb{R}^{p,q}$, components (5.1.23) transform as follows:

$$\mathsf{t}^{k_1 \dots k_s}_{j_1 \cdots j_t} \to \Lambda^{k_1}_{\ n_1} \cdots \Lambda^{k_s}_{\ n_s}\ \mathsf{t}^{n_1 \dots n_s}_{\ell_1 \dots \ell_t}\ (\Lambda^{-1})^{\ell_1}_{\ j_1} \cdots (\Lambda^{-1})^{\ell_t}_{\ j_t}. \tag{5.1.24}$$

Definition 5.1.2. Rank $r = (s+t)$ *tensor* in space $\mathbb{R}^{p,q}$ is an element (5.1.23) in space $(\mathbb{R}^{p,q})^{\otimes r}$, whose components $\mathsf{t}^{k_1 \cdots k_s}_{j_1 \cdots j_t}$ transform under basis transformation (5.1.2), (5.1.20) in $\mathbb{R}^{p,q}$ in accordance with (5.1.24). Representation (5.1.22) acting in space of tensors (5.1.23) as given by (5.1.24) is called *tensor representation.*

In other words, rank $(s+t)$ tensor is a vector in the space of representation $T^{(s+t)}$. Examples of tensors are vectors $x^i \vec{e}_i$ and $x_i \vec{e}^i$ in $\mathbb{R}^{p,q}$ of rank $(1+0)$ and $(0+1)$, respectively, while, in view of (5.1.4) and (5.1.7), metrics η and η^{-1} are invariant tensors of ranks $(0+2)$ and $(2+0)$.

Suppose there are two tensors t and t' of ranks $r, r' \geq 1$. Then one can construct a new tensor t'' of a different rank by making use of contraction procedure, namely, components of tensor t'' are obtained by summation of some upper (lower) indices of components of tensor t with some lower (upper) indices of components of t'. The fact that one obtains tensor is guaranteed by the form of transformations (5.1.24). As an example, consider contraction of tensors t and t' of ranks $((k+t)+s)$ and $(g+(k+r))$ in k indices:

$$\mathsf{t}^{j_1\dots j_k n_1\dots n_t}_{k_1\dots k_s} \cdot \mathsf{t}'^{\,i_1\dots i_g}_{\ \ j_1\dots j_k \ell_1\dots\ell_r} = \mathsf{t}''^{n_1\dots n_t i_1\dots i_g}_{\ \ k_1\dots k_s \ell_1\dots \ell_r}.$$

The result is tensor of rank $((t+g)+(s+r))$.

Clearly, one can raise and lower indices of tensors (like in the case of vectors, see (5.1.19)) by contracting with metrics η^{jk} and η_{kj}:

$$\begin{aligned} &\mathsf{t}^{k_1\dots k_s}_{j_1\dots j_t} \cdot \eta^{j_1 i_1} \cdots \eta^{j_t i_t} \cdot \eta_{k_1\ell_1} \cdots \eta_{k_s \ell_s} = \mathsf{t}^{i_1\dots i_t}_{\ell_1\dots\ell_s} \\ &\quad = \mathsf{t}_{\ell_1\dots\ell_r} \eta^{\ell_{s+1} i_1} \cdots \eta^{\ell_r i_t} = \mathsf{t}^{i_1\dots i_r} \eta_{i_{t+1}\ell_1} \cdots \eta_{i_r \ell_s}, \end{aligned} \tag{5.1.25}$$

where $r = s+t$. These relations establish one-to-one correspondence between tensors of ranks $(s+t)$, $(t+s)$, $(0+r)$ and $(r+0)$, so the pertinent representations of group $O(p,q)$ are equivalent:

$$T^{(s+t)} \simeq T^{(t+s)} \simeq T^{(0+r)} \simeq T^{(r+0)} \equiv T^{\otimes r}. \tag{5.1.26}$$

Therefore, there is no point to consider these representations separately; it suffices to study one of them, e.g., $T^{(r+0)} \equiv T^{\otimes r}$. The latter acts in space of tensors

$$\mathsf{t} = \mathsf{t}^{k_1\dots k_r}\ \vec{e}_{k_1} \otimes \cdots \otimes \vec{e}_{k_r} \ \in\ (\mathbb{R}^{p,q})^{\otimes r}. \tag{5.1.27}$$

Representation $T^{\otimes r}$ is reducible for $r > 1$. As an example, tensors with components

$$\mathsf{t}^{k_1\dots k_{r-2}\ell_{r-1}\ell_r}\ \eta_{\ell_{r-1}\ell_r}\ \eta^{k_{r-1}k_r} \tag{5.1.28}$$

form an invariant subspace in $(\mathbb{R}^{p,q})^{\otimes r}$. Since tensors with components $\mathsf{t}^{k_1\dots k_{r-2}} = \mathsf{t}^{k_1\dots k_{r-2}\ell_{r-1}\ell_r}\ \eta_{\ell_{r-1}\ell_r}$ form space of representation $T^{\otimes(r-2)}$, representation $T^{\otimes r}$ contains representation $T^{\otimes(r-2)}$. Continuing with this observation, we find that representation $T^{\otimes(r-2)}$ contains representation $T^{\otimes(r-4)}$, etc., so $T^{\otimes r}$ contains all representations $T^{\otimes k}$, where $k = r-2$, $r-4, \dots$.

Remark. In what follows, we also consider complex tensors, i.e., tensors with complex components $\mathsf{t}^{k_1\dots k_r}$. These belong to spaces of complex representations of groups $SO(p,q)$. We still denote these representations by $T^{\otimes r}$; it will be clear from the context whether our discussion applies to real, or complex representation or both. Complex representations $T^{\otimes r}$ and their subrepresentations are at the same time representations of groups $SO(N, \mathbb{C})$. When talking about representations of $SO(N, \mathbb{C})$, we will have in mind only those representations which are contained in $T^{\otimes r}$, as these are in one-to-one correspondence with (complex) representations of $SO(p, q)$. Using the same arguments as in the case of linear groups $SL(N, \mathbb{C})$ (see Remark in Section 4.1), one establishes one-to-one correspondence between the class of irreducible representations of groups $SO(N, \mathbb{C})$ and $SO(p, q)$, embedded in $T^{\otimes r}$, with the class of irreducible representations of algebras $so(N, \mathbb{C})$ and $so(p, q)$, which does not include spinor representations. We also emphasize that like in the case of linear groups $SL(N, \mathbb{C})$, representations embedded in $T^{\otimes r}$ do not exhaust all representations of $SO(N, \mathbb{C})$; as an example, representation T^* complex conjugate to defining representation T does not belong to the class we consider. We do not study representations $(T^*)^{\otimes r}$ with $r > 0$ and their subrepresentations.

5.1.3. Extracting irreducible representations of groups $O(p, q)$ and $SO(p, q)$ from representation $T^{\otimes r}$

In this and the following sections, we consider two related problems. The first is the classification of all finite-dimensional irreducible representations of groups $O(p, q)$ and $SO(p, q)$ as subrepresentations in tensor product $T^{\otimes r}$ of r defining representations. The second is the decomposition of the tensor product $T^{\otimes r}$ in irreducible representations. The first problem is solved in a relatively straightforward way, as we see in this section (many of these results are summarized in Ref. [27]). The recipe for solving the second problem is given in Section 5.1.4. The justification of this recipe is based on the notion of Brauer algebra and study of its representations; this is done in Section 5.2. Our analysis to a large extent will be valid for both complex and real representations of $O(p, q)$ and $SO(p, q)$. We will explicitly point out the differences between real and complex case in appropriate places.

Let us consider the problem of extraction of irreducible representations of group $O(p, q)$ from representation $T^{\otimes r}$. We note, in the first place, that any representation of group G is at the same time a representation of its subgroup $H \subset G$. An irreducible representation of G may be a reducible representation of H. On the other hand, if representation of a subgroup $H \subset G$ obtained from representation of group G is irreducible, then it is irreducible as a representation of entire group G. Group $O(p, q)$ is a subgroup of group $GL(N, \mathbb{R})$ with $N = p + q$. We use the same notation T both for defining representation of group $O(p,q)$ and for defining representation of $GL(N,\mathbb{R})$, which act in N-dimensional space $\mathcal{V}_N$ (in the case of real representations $\mathcal{V}_N \equiv \mathbb{R}^{p,q}$, $p+q=N$). We first consider group $GL(N,\mathbb{R})$ and its representation $T^{\otimes r}$ which acts in the space of rank-r tensors $\mathcal{V}_N^{\otimes r}$. Since representation $T^{\otimes r}$ is at the same time a representation of the subgroup $SL(N,\mathbb{R}) \subset GL(N,\mathbb{R})$, irreducible representations of group $GL(N,\mathbb{R})$ are extracted from $T^{\otimes r}$ in much the same way as irreducible representations of $SL(N,\mathbb{R})$, namely, one makes use of Young symmetrizers $\mathsf{s}(Y(\mathsf{T}_\lambda))$ or projectors $P(\mathsf{T}_\lambda)$, where T_λ are standard Young tableaux whose shapes are Young diagrams $\lambda \vdash r$ with number of rows smaller than N. A particular property of group $GL(N,\mathbb{R})$ is that each of these diagrams defines several irreducible representations of $GL(N,\mathbb{R})$, see formula (5.1.29). In the context of *complex* representations of group $SL(N,\mathbb{C})$, Young symmetrizers $\mathsf{s}(Y(\mathsf{T}_\lambda))$ and projectors $P(\mathsf{T}_\lambda) = \mathsf{s}(e(\mathsf{T}_\lambda))$ are discussed in detail in Sections 4.4.1 and 4.6.5, respectively; see formulas (4.4.2), (4.4.3) and (4.6.91).

We note that in accordance with definition (4.2.9) of the matrix representation s of symmetric group algebra $\mathbb{C}[S_r]$ in space $\mathcal{V}_N^{\otimes r}$, and also due to explicit form of idempotents $Y(\mathsf{T}_\lambda)$ and $e(\mathsf{T}_\lambda)$, all projection matrices $\mathsf{s}(Y(\mathsf{T}_\lambda))$ and $P(\mathsf{T}_\lambda) = \mathsf{s}(e(\mathsf{T}_\lambda))$ are real. Therefore, they can be used for extracting both *complex* and *real* irreducible representation of real matrix group $GL(N,\mathbb{R})$ and its subgroup $SL(N,\mathbb{R})$. We emphasize that we do not pretend to translate all results valid for *complex* representations of $SL(N,\mathbb{C})$ to the case of *real* representations of $SL(N,\mathbb{R})$. It is sufficient for our purposes that Proposition 4.2.2 and the reasoning afterwards are also valid for group $SL(N,\mathbb{R})$, namely, one and the same system of primitive idempotents extracts irreducible representations from $T^{\otimes r}$ in both complex and real cases. We also make use of the fact that real irreducible representations of $SL(N,\mathbb{R})$ that are associated with different Young diagrams (with number of rows $m < N$) are inequivalent.

It is worth noting that trivial representation of group $SL(N,\mathbb{K})$ (where $\mathbb{K}$ is the field of either complex or real numbers) which is extracted from $T^{\otimes N}$ by using complete antisymmetrizer $\mathsf{s}(Y(\mathsf{T}_{[1^N]})) = P(\mathsf{T}_{[1^N]})$ becomes, in the case of group $GL(N,\mathbb{K})$, a nontrivial one-dimensional representation $T^{(0)}$, such that $T^{(0)}(g) = \det(g)$ for $g \in GL(N,\mathbb{K})$. For this reason, all irreducible representations of $GL(N,\mathbb{K})$ are labeled by Young diagrams with number of rows $m \leq N$ in contrast to the number of rows $m < N$ sufficient to label representations in the case of $SL(N,\mathbb{K})$. The Young diagram $\lambda = [\lambda_1, \ldots, \lambda_{N-1}, \lambda_N]$ with N rows describes the following irreducible representation of group $GL(N,\mathbb{K})$:

$$T^{(\lambda)} = T^{(\lambda')} \otimes (T^{(0)})^{\otimes \lambda_N}, \tag{5.1.29}$$

where $\lambda' = [\lambda_1 - \lambda_N, \dots, \lambda_{N-1} - \lambda_N]$ and $T^{(0)}$ is the determinant representation introduced above. Note that the tensor product of one-dimensional representations $T^{(0)}$ in (5.1.29) can be replaced by a usual product and also that integer exponent λ_N can be replaced by an arbitrary parameter ξ. In this way, one finds that inequivalent irreducible representations of groups $GL(N,\mathbb{K})$ are labeled by a diagram λ' and parameter ξ.

As we pointed out above, irreducible representations $T^{(\lambda)}$ of group $GL(N,\mathbb{R})$ (obtained from $T^{\otimes r}$ by making use of projectors $\mathsf{s}(Y(\mathsf{T}_\lambda))$ or $P(\mathsf{T}_\lambda)$) are at the same time representations (in general, reducible) of subgroup $O(p,q) \subset GL(N,\mathbb{R})$. One more way to see this property is to note that the action of group $O(p,q)$ (and, in fact, of any matrix subgroup of $GL(N,\mathbb{R})$) in space $\mathcal{V}_N^{\otimes r}$ commutes with action in this space of the operators $\mathbb{C}[S_r]$ in representation s (see (4.2.15)):

$$\mathsf{s}(X) \cdot (\underbrace{g \otimes \cdots \otimes g}_{r}) = (\underbrace{g \otimes \cdots \otimes g}_{r}) \cdot \mathsf{s}(X), \quad \forall g \in GL(N,\mathbb{C}), \tag{5.1.30}$$

where $X \in \mathbb{C}[S_r]$. This means that Young symmetrizers $\mathsf{s}(Y(\mathsf{T}_\lambda))$ (or projectors $P(\mathsf{T}_\lambda)$) extract subspaces $\bar{\mathcal{V}}^{(\lambda)}$ in $\mathcal{V}_N^{\otimes r}$, which are invariant under $O(p,q)$. As an example, consider the space $\mathcal{V}_N^{\otimes 2}$ of second rank tensors with components t^{ik}. Young symmetrizers (4.4.12) (or projectors $P(\mathsf{T}_\lambda) = \mathsf{s}(e(\mathsf{T}_\lambda))$ constructed with idempotents (4.5.33)) extract subspaces $\bar{\mathcal{V}}^{([2])}$ and $\bar{\mathcal{V}}^{([1^2])}$ of symmetric and antisymmetric tensors t_s and t_a with components

$$\mathsf{t}_s^{ik} \equiv \mathsf{t}^{(ik)} = \frac{1}{2}(\mathsf{t}^{ik} + \mathsf{t}^{ki}), \quad \mathsf{t}_a^{ik} \equiv \mathsf{t}^{[ik]} = \frac{1}{2}(\mathsf{t}^{ik} - \mathsf{t}^{ki}). \tag{5.1.31}$$

These tensors transform independently under $O(p,q)$, and make two invariant subspaces in $\mathcal{V}_N^{\otimes 2}$.

Now, we are interested in *irreducible* representations of group $O(p,q)$. In view of example (5.1.28), we have to check whether invariant subspaces $\bar{\mathcal{V}}^{(\lambda)} \subset \mathcal{V}_N^{\otimes r}$ consist of traceless tensors only:

$$\mathsf{t}^{k_1 k_2 \dots k_r} \eta_{k_i k_j} = 0, \quad \forall i,j \in \{1,2,\dots,r\}. \tag{5.1.32}$$

If components of tensors symmetrized by $\mathsf{s}(Y(\mathsf{T}_\lambda))$ do not obey (5.1.32) the pertinent subrepresentation in space $\bar{\mathcal{V}}^{(\lambda)}$ is reducible, and one has to extract "traceless" representation $T^{(\lambda)}$ of group $O(p,q)$ which acts in subspace $\mathcal{V}^{(\lambda)} \subset \bar{\mathcal{V}}^{(\lambda)}$ containing only traceless tensors t of rank r. In other words, representation $T^{(\lambda)}$ is obtained by imposing additional tracelessness conditions (5.1.32). We emphasize that the requirement that tensors t are traceless is necessary, but not sufficient for irreducibility of pertinent representations of group $O(p,q)$, see the following.

Decomposition into traceless subrepresentations is possible for any representation obtained by making use of the symmetrizer $\mathsf{s}(Y(\mathsf{T}_\lambda))$ and acting in space $\bar{\mathcal{V}}^{(\lambda)}$. As an example, such decompositions for tensors of the second and third ranks are

$$\mathsf{t}^{k_1k_2} \to \mathsf{t}^{k_1k_2} - \frac{1}{N}\eta^{k_1k_2}(\eta_{i_1i_2}\mathsf{t}^{i_1i_2}),$$

$$\begin{aligned}\mathsf{t}^{k_1k_2k_3} \to \mathsf{t}^{k_1k_2k_3} &- \frac{N+1}{(N-1)(N+2)}(\eta^{k_1k_2}\mathsf{t}^{**k_3} + \eta^{k_1k_3}\mathsf{t}^{*k_2*} + \eta^{k_2k_3}\mathsf{t}^{k_1**})\\ &+\frac{1}{(N-1)(N+2)}(\eta^{k_1k_2}(\mathsf{t}^{k_3**} + \mathsf{t}^{*k_3*}) + \eta^{k_1k_3}(\mathsf{t}^{k_2**} + \mathsf{t}^{**k_2})\\ &+\eta^{k_2k_3}(\mathsf{t}^{*k_1*} + \mathsf{t}^{**k_1})),\end{aligned}$$

where we use shorthand notations $\mathsf{t}^{k**} = \eta_{ij}\mathsf{t}^{kij}$, $\mathsf{t}^{*k*} = \eta_{ij}\mathsf{t}^{ikj}$ and $\mathsf{t}^{**k} = \eta_{ij}\mathsf{t}^{ijk}$. Thus, the representation $T^{\otimes r}$ contains not only traceless representations $T^{(\lambda)}$ associated with Young diagrams $\lambda \vdash r$ with r boxes but also representations of rank smaller than r which are obtained by contracting tensors from representations $\bar{\mathcal{V}}^{(\lambda)}$ with metric η. The latter tensors are again subject to the procedures of symmetrization and extraction of traceless part, and the process continues. These new subrepresentations are associated with Young diagrams where the number of boxes is less than r. We discuss the decomposition of representation $T^{\otimes r}$ into irreducible traceless representations in Section 5.1.4.

As an example, we again consider spaces $\bar{\mathcal{V}}^{([2])}$ and $\bar{\mathcal{V}}^{([1^2])}$ of symmetric and antisymmetric tensors t_s and t_a, see Eq. (5.1.31). Representation $T^{([1^2])}$ of group $O(p,q)$ acts in space $\mathcal{V}^{([1^2])} \equiv \bar{\mathcal{V}}^{([1^2])}$ of antisymmetric tensors t_a; it is traceless and irreducible. Representation in space $\bar{\mathcal{V}}^{([2])}$ of symmetric tensors t_s with components $\mathsf{t}^{(ik)}$ is, on the other hand, reducible, and it is decomposed into two irreducible representations acting in invariant subspaces $\mathcal{V}^{([2])}$ and $\mathcal{V}^{([\emptyset])}$. The former is the space of traceless symmetric tensors t'_s, while the latter contains symmetric tensor t_0, whose components are proportional to irreducible tensor of zero rank $\eta_{rs}\mathsf{t}^{(rs)}$:

$$\begin{aligned}\mathsf{t}_s'^{ik} &= \mathsf{t}^{(ik)} - \frac{1}{N}\eta^{ik}(\mathsf{t}^{(rs)}\eta_{rs}),\\ \mathsf{t}_0^{ik} &= \eta^{ik}(\eta_{rs}\mathsf{t}^{(rs)}).\end{aligned} \tag{5.1.33}$$

In accordance with (5.1.31) and (5.1.33), irreducible tensors with components $\mathsf{t}_a^{i_1 i_2}$, $\mathsf{t'}_s^{i_1 i_2}$ and $\mathsf{t}_0^{i_1 i_2}$ are extracted from a general tensor with components $\mathsf{t}^{k_1 k_2}$ by making use of normalized projectors

$$(\Pi_a)^{i_1 i_2}_{k_1 k_2} = \frac{1}{2}\left(I^{i_1 i_2}_{k_1 k_2} - P^{i_1 i_2}_{k_1 k_2}\right),$$
$$(\Pi_s)^{i_1 i_2}_{k_1 k_2} = \frac{1}{2}\left(I^{i_1 i_2}_{k_1 k_2} + P^{i_1 i_2}_{k_1 k_2}\right) - \frac{1}{N}K^{i_1 i_2}_{k_1 k_2}, \quad (\Pi_0)^{i_1 i_2}_{k_1 k_2} = \frac{1}{N}K^{i_1 i_2}_{k_1 k_2}, \tag{5.1.34}$$

where $P^{i_1 i_2}_{k_1 k_2} = \delta^{i_1}_{k_2}\delta^{i_2}_{k_1}$ is the transposition matrix, $I^{i_1 i_2}_{k_1 k_2} = \delta^{i_1}_{k_1}\delta^{i_2}_{k_2}$ is the unit matrix, and $K^{i_1 i_2}_{k_1 k_2} = \eta^{i_1 i_2}\eta_{k_1 k_2}$.

Problem 5.1.2. Show that operators Π_α, where $\alpha = a, s, 0$, obey projector relations $\Pi_\alpha \Pi_\beta = \delta_{\alpha\beta}\Pi_\alpha$ and completeness condition $\Pi_a + \Pi_s + \Pi_0 = I$.

Thus, irreducible representations of orthogonal groups $O(N, \mathbb{C})$ (from the class we consider, see Remark in Section 5.1.2; we do not make this qualification in what follows) and their real subgroups $O(p, q)$ act in spaces of traceless tensors of rank r, symmetrized in accordance with Young tableaux T_λ, where the diagram λ has r boxes. Like in the linear group case, all representations associated with tableaux T_λ of one and the same shape λ are equivalent.[a] We recall that the number of rows in pertinent Young diagrams cannot exceed N for group $GL(N, \mathbb{C})$ and $(N-1)$ for group $SL(N, \mathbb{C})$. Clearly, these limitations are also valid for matrix subgroups $O(p, q) \subset GL(N, \mathbb{C})$ and $SO(p, q) \subset SL(N, \mathbb{C})$. We now show that the existence of metric η leads to another restriction on Young diagrams which determine inequivalent representations of special orthogonal groups $SO(N, \mathbb{C})$ and their real subgroups $SO(p, q)$. This new restriction leads to a substantial difference between classifications of irreducible representations of orthogonal and linear groups (see Sections 3.4.2 and 4.4.1).

Proposition 5.1.1. *Inequivalent irreducible representations $T^{(\lambda)}$ of group $SO(N, \mathbb{C})$ ($N = 2r$ or $N = 2r+1$) and its real subgroups $SO(p, q)$, where $N = p+q$, are associated with different Young diagrams λ where the number of rows is not greater than r.*

Proof. We begin with *reducible* representations of $SO(N, \mathbb{C})$ acting in the space of all tensors whose symmetry is determined by Young tableau T_λ with the shape of diagram λ, i.e., for the beginning, we do not impose on vectors of representation spaces the tracelessness conditions (5.1.32).

---

[a]The proof is basically a repetition of the proof of Proposition 4.4.1, where all operators are understood as acting in spaces of traceless tensors.

All such representations corresponding to different tableaux T_λ (but with the fixed shape λ) are equivalent. At the expense of slight abuse of notations, in this proof, we denote these *reducible* representations also by $T^{(\lambda)}$.

We now note an important difference between the systems of such representations of orthogonal groups $SO(N,\mathbb{C})$ (and their subgroups $SO(p,q)$), on the one hand, and corresponding irreducible representations of linear groups $SL(N,\mathbb{C})$ (and their subgroups $SL(N,\mathbb{R})$ and $SU(p,q)$), on the other. In the linear group case, as we know, all representations $T^{(\lambda)}$ constructed with Young symmetrizers $\mathsf{s}(Y(\mathsf{T}_\lambda))$ are inequivalent if they have different Young diagrams where the numbers of rows are less than N. In the orthogonal group case, on the other hand, there exist different Young diagrams $\lambda \vdash r$, $\lambda' \vdash r, \ldots$ with the number of rows smaller than N, which define equivalent representations $T^{(\lambda)}, T^{(\lambda')}, \ldots$ of group $SO(N,\mathbb{C})$ and their real subgroups $SO(p,q)$. Indeed, since there exists invariant metric η, defining and co-defining representations T and $\tilde{T}$ of orthogonal groups are equivalent, so representations $T^{\otimes k}$ and $\tilde{T}^{\otimes k}$ are also equivalent for any $k > 0$. Now, in the case of special linear group $SL(N,\mathbb{C})$ and all its subgroups, such as $SO(N,\mathbb{C})$ and $SO(p,q) \subset SO(N,\mathbb{C})$, irreducible subrepresentation of $T^{\otimes k}$ acting in space of antisymmetric tensors with components $\mathsf{t}^{[r_1\cdots r_k]}$ is equivalent to irreducible representation $\tilde{T}^{\otimes(N-k)}$ acting in the space of antisymmetric tensors with components $\tilde{\mathsf{t}}_{[r_1\cdots r_{N-k}]}$. This is due to the existence of invariant totally antisymmetric tensor with components $\varepsilon_{i_1\ldots i_N}$. Indeed, there is one-to-one correspondence between elements t and $\tilde{\mathsf{t}}$ in the above spaces, consistent with the action of $SL(N,\mathbb{C})$:

$$\mathsf{t}^{[j_1\ldots j_k]}\varepsilon_{j_1\ldots j_k,r_1\ldots r_{N-k}} = \tilde{\mathsf{t}}_{[r_1\ldots r_{N-k}]}. \tag{5.1.35}$$

Since representations $\tilde{T}^{\otimes(N-k)}$ and $T^{\otimes(N-k)}$ are equivalent for group $SO(N,\mathbb{C})$ and its subgroups, the correspondence (5.1.35) shows that antisymmetric representations on tensors with components $\mathsf{t}_1^{[i_1\ldots i_k]}$ and $\mathsf{t}_2^{[i_1\ldots i_{N-k}]}$ are equivalent. Note that these two representations are associated with Young diagrams $\lambda_1 = [1^k]$ and $\lambda_2 = [1^{N-k}]$. The one-to-one correspondence between these tensors is given by (in the case of $SO(p,q)$)

$$\mathsf{t}_1^{[j_1\cdots j_k]} \leftrightarrow \mathsf{t}_2^{[i_1\cdots i_{N-k}]}\varepsilon_{i_1\cdots i_{N-k}i_{N-k+1}\cdots i_N}\eta^{i_{N-k+1}j_1}\cdots\eta^{i_N j_k}. \tag{5.1.36}$$

Thus, antisymmetric representations of group $SO(N,\mathbb{C})$ are determined by Young diagrams $[1^k]$ with $2k \le N$.

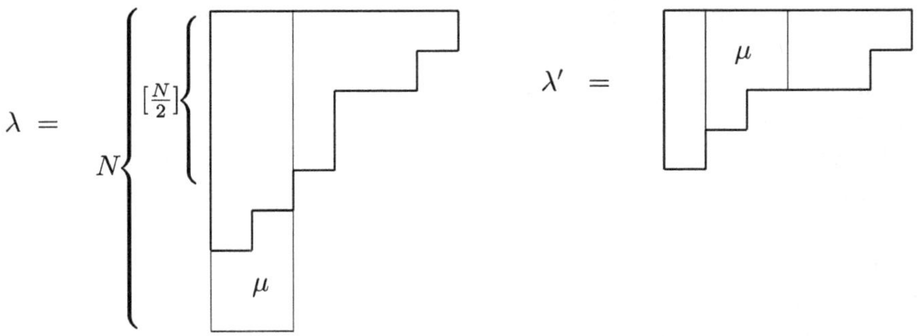

Fig. 5.1.1. Young diagram λ with two columns higher than $[N/2]$. These two columns are replaced by two columns denoted by μ which are inserted as shown in diagram λ' on the right. Representations $T^{(\lambda)}$ and $T^{(\lambda')}$ are equivalent.

The correspondence (5.1.36) is straightforwardly generalized to tensors whose components have not only indices $j_1, \ldots, j_k$ (indices $i_1, \ldots, i_{N-k}$) in which they are antisymmetric but also indices of other types. As an example, representations which act in the space of tensors with components $\mathsf{t}^{[j_1 \cdots j_k]j}$ and $\mathsf{t}^{[i_1 \cdots i_{N-k}]j}$ are equivalent. Further generalization is as follows. Recall that indices of components of tensor t of representation $T^{(\lambda)}$ are symmetrized by projector $\mathsf{s}(Y(\mathsf{T}_\lambda))$ which first symmetrizes indices in rows and then antisymmetrizes indices in columns of Young diagram $\lambda = [\lambda_1, \ldots, \lambda_m]$ (where $m < N$ like in the case of $SL(N, \mathbb{C})$). Therefore, we make use of equivalence (5.1.36) to replace any column whose height k exceeds $N/2$ by a column of smaller height $(N - k)$. By putting this column in the right place (see Fig. 5.1.1) we obtain diagram λ' that defines representation $T^{(\lambda')}$ equivalent to $T^{(\lambda)}$. We apply this procedure to all columns of height larger than $N/2$ and replace the diagram λ by the diagram λ', the number of rows in the latter diagram does not exceed $[N/2]$ where $[x]$ is integer part of number x. In other words, height of any column in diagram λ' does not exceed r for $N = 2r$ and $N = 2r + 1$. Now, since reducible representations $T^{(\lambda)}$ and $T^{(\lambda')}$ are equivalent, their decompositions into irreducible representations are also equivalent, so it is sufficient to consider only diagram λ'. Proposition 5.1.1 is proven: the number of rows in relevant Young diagrams does not exceed r for groups $SO(N, \mathbb{C})$ with $N = 2r, 2r + 1$ and their subgroups $SO(p, q)$, $p + q = N$. This result agrees with Remark 3 in Section 3.4.2 made on the basis of weight theory classification of representations of Lie algebras $so(2r, \mathbb{C})$ and $so(2r + 1, \mathbb{C})$. $\square$

Remark 1. Equivalence (5.1.35) is absent for linear groups $GL(N,\mathbb{C})$ and general orthogonal groups $O(N,\mathbb{C})$ (and their real subgroups $O(p,q)$). Representations of these groups on tensors $\tilde{\mathsf{t}}_{[j_1\cdots j_{N-k}]}$ and $\mathsf{t}^{[i_1\cdots i_k]}\varepsilon_{i_1\cdots i_{N-k}j_1\cdots j_k}$ are inequivalent, since the transformations of these tensors (see (5.1.24)) differ by a factor $\det(\Lambda)$ which is different from 1 for these groups. As an example, $\det(\Lambda) = -1$ for improper elements of $\Lambda \in O(p,q)$. So, the two sides of (5.1.35) would involve elements of spaces of different representations, which cannot be the case. This fact shows that irreducible representations of general orthogonal groups $O(N,\mathbb{C})$ (and their subgroups $O(p,q)$, $p+q = N$) do not have the property stated in Proposition 5.1.1 and valid for representations of special groups $SO(N,\mathbb{C})$ (and their subgroups $SO(p,q)$).

We discuss the latter point in more detail. It follows from Proposition 5.1.1 that the number of boxes in the first two rows in Young diagrams for group $SO(N,\mathbb{C})$ does not exceed $2r$ where $r = [N/2]$. In the case of group $O(N,\mathbb{C})$ somewhat similar property holds, as follows from the result of the following problem.

Problem 5.1.3. ⋆ Show that irreducible finite-dimensional representations of group $O(N,\mathbb{C})$ (and its real subgroups $O(p,q)$ with $p+q = N$) are associated with Young diagrams whose numbers of boxes in the first two rows do not exceed N. Hint: derive the formula for dimension of representation $T^{(\lambda')}$ of $O(N,\mathbb{C})$ associated with diagram λ' with two columns of heights a and b $(a \geq b)$:

$$\dim(T^{(\lambda')}) = \frac{(a-b+1)}{(a+1)!b!}\,\frac{N!(N+2)!(N-a-b+1)}{(N-a+1)!(N-b+2)!}. \tag{5.1.37}$$

Note that formula (5.1.37) is valid for representations of $SO(N,\mathbb{C})$ as well, and that it is invariant under transformation $a \to (N-a)$ which is consistent with the procedure on Fig. 5.1.1.

As follows from the discussion above, equivalence (5.1.36) holds for general orthogonal groups modulo multiplication by one-dimensional determinant representation $T^{(0)}(\Lambda) = \det(\Lambda)$:

$$T^{([1^k])} = T^{(0)} \cdot T^{([1^{N-k}])}, \tag{5.1.38}$$

where $T^{(0)}(\Lambda) = +1$ and $T^{(0)}(\Lambda) = -1$, respectively, for proper and improper elements $\Lambda \in O(N,\mathbb{C})$. Therefore, it follows from the result of Problem 5.1.3 and Proposition 5.1.1 that inequivalent irreducible representations $\bar{T}^{(\lambda)}$ of general orthogonal groups $O(N,\mathbb{C})$ and their subgroups

$O(p,q)$ with $p+q=N$ are associated with Young diagrams λ with number of rows $m \leq [N/2] = r$ and are given by (cf. (5.1.29))

$$\bar{T}^{(\lambda)}(\Lambda) = T^{(\lambda)}(\Lambda) \cdot \det{}^{\xi}(\Lambda), \tag{5.1.39}$$

where $T^{(\lambda)}$ is irreducible representation of $SO(N,\mathbb{C})$ (or $SO(p,q)$) defined by diagram λ, and ξ is a free parameter (if ξ is integer and even, then the determinant factor in (5.1.39) is trivial, since $\det(\Lambda) = \pm 1$).

Remark 2. The above procedure of extraction of finite-dimensional irreducible representation works for all orthogonal groups $SO(N,\mathbb{C})$ and their subgroups $SO(p,q)$. However, this procedure is incomplete in the case of groups $SO(N,\mathbb{C})$ and $SO(p,q)$ with even $N = p+q = 2r$. This is again due to the existence of invariant rank $2r$ totally antisymmetric tensor ε. Making use of this tensor, one divides representation of $SO(p,q)$, acting on antisymmetric rank r tensors, into two irreducible representations, which act in the spaces of selfdual tensors $\mathsf{t}^{(+)}$ and anti-selfdual tensors $\mathsf{t}^{(-)}$:

$$\frac{\alpha}{r!}\varepsilon_{i_1\cdots i_r k_1\cdots k_r}\mathsf{t}_{(+)}^{k_1\cdots k_r} = \mathsf{t}_{(+)\,i_1\cdots i_r}, \quad \frac{\alpha}{r!}\varepsilon_{i_1\cdots i_r k_1\cdots k_r}\mathsf{t}_{(-)}^{k_1\cdots k_r} = -\mathsf{t}_{(-)i_1\cdots i_r}, \tag{5.1.40}$$

where we choose $\alpha = e^{-i\pi(r+q)/2}$ (one of the solutions of equation $\alpha^2 = (-1)^{(r+q)}$). Selfdual and anti-selfdual tensors for $N=4$ are considered in Section I-3.3.3 of the first book.

Problem 5.1.4. Write (5.1.40) as projection equalities

$$(P_\pm)^{i_1\cdots i_r}{}_{k_1\cdots k_r}\mathsf{t}^{k_1\cdots k_r} = 0, \quad (P_\pm)^{i_1\cdots i_r}{}_{k_1\cdots k_r} \equiv A^{i_1\cdots i_r}{}_{k_1\cdots k_r} \pm \frac{\alpha}{r!}\varepsilon^{i_1\cdots i_r}{}_{k_1\cdots k_r}, \tag{5.1.41}$$

where $A^{i_1\cdots i_r}{}_{k_1\cdots k_r}$ is rank-r antisymmetrizer. Check the coefficients in definitions (5.1.40) of (anti-) selfdual tensors, i.e., show that $P_\pm^2 = P_\pm$. Hint: recall that indices of tensors in pseudo-Euclidean space $\mathbb{R}^{p,q}$ are raised and lowered with metric $\eta = I_{p,q}$; this means, in particular, that $\varepsilon^{k_1k_2\cdots k_{p+q}} = (-1)^q\varepsilon_{k_1k_2\cdots k_{p+q}}$.

In Section 6.4.1 (see Proposition 6.4.1), we show that irreducible (anti-) selfdual representations of $SO(2r,\mathbb{C})$ and $SO(p,q)$ are constructed as subrepresentations in tensor product of two fundamental spinor representations of algebras $so(2r,\mathbb{C})$ and $so(p,q)$. We emphasize that in the case of general orthogonal groups $O(2r,\mathbb{C})$ and $O(p,q)$ with $(p+q) = 2r$, the representation acting on antisymmetric tensors of rank r is irreducible, i.e., it cannot be split into selfdual and anti-selfdual representations. This is due to the fact that improper elements Λ from groups $O(2r,\mathbb{C})$ and $O(2r-q,q)$ move selfdual tensors into anti-selfdual, and vice versa (since left-hand sides of conditions (5.1.40) acquire determinant factors $\det(\Lambda)$).

Remark 3. We emphasize that (anti-)selfdual *real* tensors $\mathsf{t}_{(-)}$ and $\mathsf{t}_{(+)}$ can be defined by (5.1.40) for real α only, i.e., when $(r+q)$ is even. This holds for Euclidean orthogonal group $SO(4k)$ (in particular, for $SO(4)$), but does not hold for multidimensional Lorentz group $SO(1,2r-1)$ with $r=2k$, including physical Lorentz group $SO(1,3)$. If $(r+q)$ is odd, then selfdual and anti-selfdual representations of group $SO(p,q)$ with $p+q=2r$ are necessarily complex.

With all these qualifications, we enumerate all inequivalent complex irreducible representations of groups $SO(N,\mathbb{C})$ (from the class we consider) and $SO(p,q)$.

Proposition 5.1.2. *For odd $N=(p+q)=(2r+1)$, any complex irreducible representation of $SO(N,\mathbb{C})$ (from the class we consider) and $SO(p,q)$ is given by traceless (in all pairs of indices) tensor, whose symmetry is defined by a Young diagram with number of rows smaller than or equal to r. For even $N=p+q=2r$, any complex irreducible representation of $SO(N,\mathbb{C})$ and $SO(p,q)$ is given by either* (i) *traceless tensor, whose symmetry is defined by a Young diagram with number of rows smaller than r or* (ii) *traceless tensor, whose symmetry is defined by a Young diagram λ with number of rows equal to r; this tensor is either selfdual with respect to all columns of height r in diagram λ or anti-selfdual with respect to all these columns.*

The last statement in Proposition 5.1.2 implies that (anti-)selfdual representations are not equivalent to any other representations of $SO(N,\mathbb{C})$ and $SO(p,q)$. Proposition equivalent to this property is proven in book [27]. Importantly, the last statement in Proposition 5.1.2 also means that if a tensor corresponding to Young diagram λ is selfdual with respect to one of the columns of height $r=N/2$, then it is selfdual with respect to all other such columns; tensors defined by any Young diagram λ, which are selfdual with respect to the set of r indices in one column and anti-selfdual with respect to the set of r indices in another column, identically vanish. The latter statement is the subject of the following problem.

Problem 5.1.5. ⋆ Let $\mathsf{t}^{k_1\dots k_r;\ell_1\dots\ell_r}$ be components of traceless rank-$2r$ tensor $\mathsf{t} \in \mathcal{V}^{\otimes 2r}$, where $\mathcal{V}$ is the space of defining representation of group $SO(2r-q,q)$ (or group $SO(2r,\mathbb{C})$). Let these components be symmetrized in accordance with the tableau $\begin{array}{|c|c|}\hline k_1 & \ell_1 \\ \hline \vdots & \vdots \\ \hline k_r & \ell_r \\ \hline \end{array}$.

Prove the identity $(P_+)^{i_1\cdots i_r}{}_{k_1\cdots k_r}(P_-)^{j_1\cdots j_r}{}_{\ell_1\cdots\ell_r}\mathsf{t}^{k_1\cdots k_r;\ell_1\cdots\ell_r} = 0$, where (anti-) selfdual projectors $P_\pm$ are defined in (5.1.40), (5.1.41).

We illustrate the Proposition 5.1.2 and Problem 5.1.5 by a simple example.

Example. Consider groups $SO(2)$ and $SO(1,1)$, and tensor t of rank $k=1$ with components t^ℓ. Its (anti-)selfdual parts $\mathsf{t}_{(\pm)}$ have components $\mathsf{t}^\ell_{(\pm)}$ given by

$$SO(2):\;\; \mathsf{t}^k_{(\pm)} = \mathsf{t}^k \pm i\varepsilon^{k\ell}\mathsf{t}_\ell \Rightarrow \mathsf{t}^1_{(\pm)} = \mathsf{t}^1 \pm i\mathsf{t}^2 = \pm i\mathsf{t}^2_{(\pm)},$$

$$SO(1,1):\;\; \mathsf{t}^k_{(\pm)} = \mathsf{t}^k \mp \varepsilon^{k\ell}\mathsf{t}_\ell \Rightarrow \mathsf{t}^1_{(\pm)} = \mathsf{t}^1 \pm \mathsf{t}^2 = \pm\mathsf{t}^2_{(\pm)},$$

where $\varepsilon^{k\ell} = -\varepsilon^{\ell k}$, $\varepsilon^{12} = +1$ and $\varepsilon^{k\ell}\,\varepsilon^{\ell s} = -\delta^{ks}$. In other words, in terms of (anti-)selfdual components $\mathsf{t}^\pm$, the rank 1 tensor t reads

$$SO(2):\;\; \mathsf{t}^\ell \to \mathsf{t}^\pm = \mathsf{t}^1 \pm i\mathsf{t}^2, \quad SO(1,1):\;\; \mathsf{t}^\ell \to \mathsf{t}^\pm = \mathsf{t}^1 \pm \mathsf{t}^2. \tag{5.1.42}$$

Tensors $\mathsf{t}_{(\pm)}$ form two spaces of one-dimensional representations $T^{(\pm)}$ of groups $SO(2)$ and $SO(1,1)$ (complex in the former case and real in the latter).

Consider now complex representations $T^{([k])}$ of group $SO(2)$ or $SO(1,1)$, which act in spaces of totally symmetric traceless tensors t of rank k with components $\mathsf{t}^{i_1\cdots i_k}$, $(i_1,\dots,i_k = 1,2)$:

$$\mathsf{t}^{\cdots i\cdots j\cdots} = \mathsf{t}^{\cdots j\cdots i\cdots}, \quad \eta_{ij}\mathsf{t}^{\cdots ij\cdots} = 0, \tag{5.1.43}$$

where $\eta = \mathrm{diag}(1,1)$ for group $SO(2)$ and $\eta = \mathrm{diag}(1,-1)$ for group $SO(1,1)$. Their Young diagrams $[k]$ have one row, or, in other words, have k columns of height 1. Proposition 5.1.2 tells that representations $T^{([k])}$ are reducible, and are split into two one-dimensional representations[b] acting on tensors with components $\mathsf{t}^{++\cdots+}$ and $\mathsf{t}^{--\cdots-}$. Hence, tensor t is either

[b]The fact that the group $SO(2)$ has only two types of complex irreducible representations acting in spaces of k-rank tensors with components $\mathsf{t}^{++\cdots+}$ and $\mathsf{t}^{--\cdots-}$ is behind the property that massless particles of spin $k/2$ have at most two helicities $+k/2$ and $-k/2$.

selfdual or anti-selfdual with respect to each of the columns of unit height (each of the k indices of components $\mathrm{t}^{i_1\cdots i_k}$). Indeed, in the case of group $SO(2)$ (group $SO(1,1)$ is considered in the same way), we obtain for mixed components $\mathrm{t}^{\cdots+-\cdots}$ the identity

$$\mathrm{t}^{\cdots+-\cdots} = \mathrm{t}^{\cdots 1-\cdots} + i\mathrm{t}^{\cdots 2-\cdots} = (\mathrm{t}^{\cdots 11\cdots} - i\mathrm{t}^{\cdots 12\cdots}) + i(\mathrm{t}^{\cdots 21\cdots} - i\mathrm{t}^{\cdots 22\cdots}) = 0,$$

where we make use of (5.1.42) and (5.1.43). This shows that traceless symmetric tensor t of rank k has two nonvanishing components with indices being either all $+$ or all $-$.

5.1.4. Irreducible tensor representations of orthogonal groups. Oscillating Young tableaux

Making use of the above rules, one attempts to expand the tensor product $T^{\otimes n}$ of defining representations of orthogonal groups into irreducible representations of these groups. Like in the case of group $SU(N)$, *all finite-dimensional irreducible representations*[c] of group $SO(p,q)$ are contained in the set of reducible representations $\{T^{\otimes n}\}$, where $n = 1, 2, \ldots$. Recall that we also consider a special class of representations of group $SO(N,\mathbb{C})$, see Remark in Section 5.1.2; all representations from this class are also contained in the set $\{T^{\otimes n}\}$.

We have characterized all finite-dimensional irreducible representations of $SO(p,q)$ in Section 5.1.3; they are associated with Young diagrams λ whose numbers of rows do not exceed $[N/2] = r$ (we set, as before, $N = p+q$), see Proposition 5.1.2. However, the representation branching rules have specific properties in the orthogonal group case as compared to branching rules for linear groups studied in Section 4.6.3. As an example, consider tensor product $T^{(\lambda_n)} \otimes T$, where $T^{(\lambda_n)}$ is irreducible representation of group $SO(p,q)$ determined by Young diagram λ_n with n boxes, and T is defining representation of $SO(p,q)$. $T^{(\lambda_n)} \otimes T$ is reducible representation which splits into the sum of irreducible representations associated with Young diagrams with not only $(n+1)$ boxes, but also $(n-1)$ boxes (cf. (4.4.33)),

$$T^{(\lambda_n)} \otimes T^{([1])} \equiv \lambda_n \otimes [1] = \left(\bigoplus_{j=1}^{k+1} \lambda_{n+1}(j)\right) \bigoplus \left(\bigoplus_{j=1}^{k} \lambda_{n-1}(j)\right), \quad (5.1.44)$$

[c]Except for "spinor" representations, which are not, in fact, representations of group $SO(p,q)$ but rather representations of its Lie algebra and its covering group $\mathrm{Spin}(p,q)$. Groups $\mathrm{Spin}(p,q)$ and their representations are studies in Sections 6.2.1 and 6.4.3 of chapter 6.

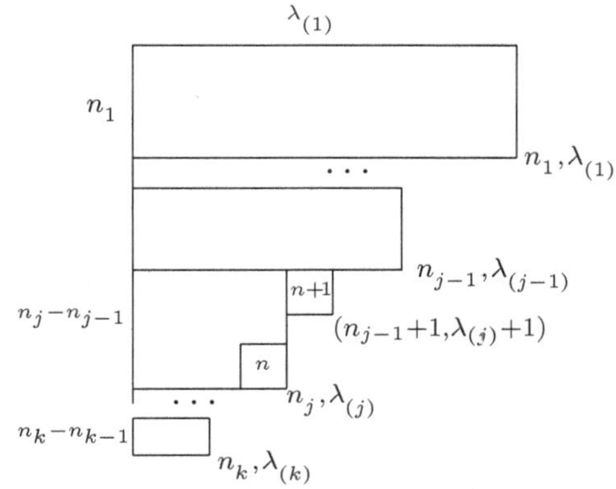

Fig. 5.1.2. Young diagram $\lambda_n = [\lambda_{(1)}^{n_1}, \lambda_{(2)}^{n_2-n_1}, \ldots, \lambda_{(k)}^{n_k-n_{k-1}}]$ with k blocks (each of length $\lambda_{(k)}$ and height n_k), with an $(n+1)$th box added in external corner with coordinates $(n_{j-1}+1, \lambda_{(j)}+1)$ and an n-th box, marked for deletion, in internal corner with coordinates $(n_j, \lambda_{(j)})$.

where we denote representations $T^{(\lambda)}$ by their Young diagrams λ; diagrams $\lambda_{n+1}(j)$ are obtained from λ_n by adding one box to external corner with coordinates $(n_{j-1}+1, \lambda_{(j)}+1)$ (see Fig. 5.1.2), while diagrams $\lambda_{n-1}(j)$ are obtained from λ_n by erasing one box in internal corner with coordinates $(n_j, \lambda_{(j)})$ (internal corners in λ_n are the boxes which are the rightmost in their rows and lowest in their columns at the same time; see Fig. 5.1.2). Number k in (5.1.44) is the number of blocks in the diagram λ_n in Fig. 5.1.2. Thus, addition of a box to the Young diagram λ_n is governed by the same rules that work for group $SL(N, \mathbb{C})$ and yields representations with tensors of rank $(n+1)$. Alternatively, a box is removed from one of the internal corners, and the resulting tensor has rank $(n-1)$. Here, we assume for simplicity that the number of rows m of diagram λ_n does not exceed $(N/2-1)$, so that adding a box to the first column does not violate the condition $m \leq N/2$ (this assumption is relaxed in Fig. 5.1.3).

Let us explain in more detail the reason for the appearance of diagrams $\lambda_{n-1}(j)$ in the expansion (5.1.44). We recall that the elements of spaces of irreducible representations of $SO(p,q)$ are traceless tensors. Representation $T^{(\lambda_n)} \otimes T$ acts in space $\mathcal{V}^{(\lambda_n)} \otimes \mathcal{V}$, whose elements are rank-$(n+1)$ tensors ψ with components $\psi^{i_1 \ldots i_n; j}$. First, n indices of these tensors are symmetrized in accordance with the diagram λ_n, and are traceless in the first n indices. However, contraction of one of the indices $(i_1, \ldots, i_n)$ with the last index, $\psi^{i_1 \ldots i_n; j} \eta_{i_k j} = \bar{\psi}^{i_1 \ldots \hat{i}_k \ldots i_n}$ (here notation $\hat{i}_k$ means that

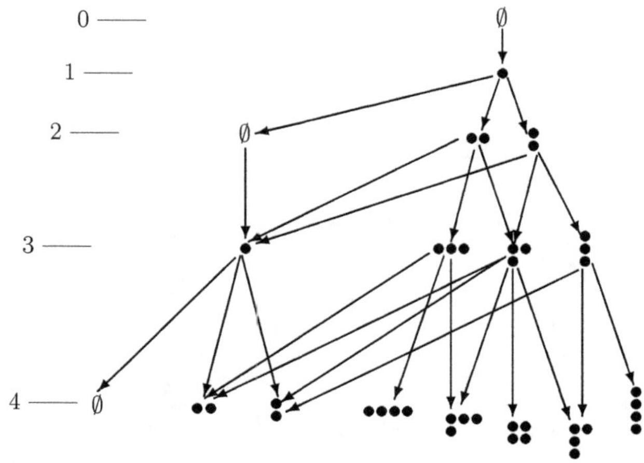

Fig. 5.1.3. Young graph showing expansion of tensor product of four defining representations of group $SO(p, q)$ into irreducible representations. The graph demonstrates branching of representations of $SO(p, q)$ (for $N = p + q \geq 8$). Continuation of this graph contains, at some level, diagrams with number of rows exceeding $[N/2]$. Once they appear, these diagrams must be immediately replaced by equivalent diagrams, in the sense of Fig. 5.1.1.

the index is removed) is not necessarily equal to zero; it defines a rank-$(n-1)$ tensor $\bar{\psi}$. Tensors $\bar{\psi}$ with components $\bar{\psi}^{i_1 \dots \hat{i}_k \dots i_n}$ are traceless by construction and inherit the symmetry of tensors $\psi \in \mathcal{V}^{(\lambda_n)}$. So, tensors $\bar{\psi}$ make an irreducible representation of group $SO(p, q)$. Symmetry of these rank-$(n-1)$ tensors corresponds to Young diagram with $(n-1)$ boxes which is subdiagram of original diagram λ_n, and all these diagrams $\lambda_{n-1}(j)$ are indeed obtained from λ_n by removing one box in one of the internal corners of λ_n.

Complete proof of the facts that expansion (5.1.44) contains all representations associated with diagrams $\lambda_{n-1}(j)$ and that multiplicity of each of these representations in (5.1.44) is equal to 1 is beyond the scope of this book.

This rule of addition or removal of one box in λ_n is illustrated for $n = 0, 1, 2, 3$ by Young branching graph shown in Fig. 5.1.3 (this is an analog of Young graph for linear groups $SL(N, \mathbb{C})$, shown in Fig. 4.6.8). The boxes of Young diagrams there are denoted by dots. Each Young diagram λ at nth level in Fig. 5.1.3 is connected by oriented links with some diagrams at $(n+1)$th level, i.e., one level below. These links show addition or removal of boxes, in accordance with representation branching rule.

A simple example of expansion of the tensor product $T^{\otimes 2}$ into the sum of irreducible representations of $SO(p,q)$ is given by formulas (5.1.31) and (5.1.33) which we write as completeness property (see Problem 5.1.2)

$$\mathsf{t}^{ik} = \mathsf{t}_a^{ik} + \bar{\mathsf{t}}_s^{ik} + \frac{1}{N}\mathsf{t}_0^{ik} \Leftrightarrow I = \Pi_1 + \Pi_2 + \Pi_3. \tag{5.1.45}$$

In terms of Young diagrams, this expansion reads

$$\ydiagram{1} \otimes \ydiagram{1} = \ydiagram{1,1} \oplus \ydiagram{2} \ominus \emptyset, \tag{5.1.46}$$

where $\emptyset$ denotes trivial representation of $SO(p,q)$. This example is illustrated by the second level in Fig. 5.1.3, which has the same Young diagrams as the right-hand side of (5.1.46). We stress that, in the case $(p+q) = 2$, the first representation in the sum in the right-hand side of (5.1.46) is equivalent to trivial representation $\emptyset$, and in the case $(p+q) = 3$, this representation is equivalent to representation $\ydiagram{1}$. Therefore, generally speaking, the notation (5.1.46) is acceptable only for the case $(p+q) \geq 4$. As a more complicated example, consider the expansion of the tensor product of three defining representations into irreducible representations of $SO(p,q)$ for $p+q \geq 6$. We have the following chain of equalities:

$$\ydiagram{1} \otimes \ydiagram{1} \otimes \ydiagram{1} = \left(\ydiagram{1,1} \oplus \ydiagram{2} \oplus \emptyset\right) \otimes \ydiagram{1} = \left(\ydiagram{1,1,1} \oplus 2\,\ydiagram{2,1} \oplus \ydiagram{3} \oplus 3\,\ydiagram{1}\right). \tag{5.1.47}$$

Here, the right-hand side is the third level in the Young graph in Fig. 5.1.3. Finally, consider tensor product of four defining representations for the case $(p+q) \geq 8$ (cf. (4.5.83)):

$$\ydiagram{1} \otimes \ydiagram{1} \otimes \ydiagram{1} \otimes \ydiagram{1} = \left(\ydiagram{1,1,1} \oplus 2\,\ydiagram{2,1} \oplus \ydiagram{3} \oplus 3\,\ydiagram{1}\right) \otimes \ydiagram{1}$$

$$= \left(\ydiagram{1,1,1,1} \oplus 3\,\ydiagram{2,1,1} \oplus 3\,\ydiagram{3,1} \oplus 2\,\ydiagram{2,2} \oplus \ydiagram{4} \oplus 6\,\ydiagram{1,1} \oplus 6\,\ydiagram{2} \oplus 3\,\emptyset\right). \tag{5.1.48}$$

This result is also illustrated by Young graph in Fig. 5.1.3, namely, the sum (5.1.48) corresponds to the fourth, lowest level in Fig. 5.1.3. Clearly, coefficients (multiplicities) of each Young diagram λ in (5.1.47) or (5.1.48) are equal, by construction, to the number of paths in Young graph of

Fig. 5.1.3 from the upper vertex $\emptyset$ to the vertex with diagram λ at the third or fourth level. It is worth noting that multiplicities of equivalent representations λ given in the right-hand side of (5.1.48) are *not* equal to the numbers of standard tableaux of shape λ (as was the case for group $SL(N,\mathbb{C})$, see Eq. (4.5.83) and Young graph in Fig. 4.6.5). This has to do with the fact that paths from the upper vertex $\emptyset$ to vertex λ at the fourth level in Fig. 5.1.3 go through both diagrams that are obtained by adding boxes (as was the case for Young graph in Fig. 4.6.5 for group $SL(N,\mathbb{C})$) and diagrams obtained by removing boxes. As an example, one of the paths in Fig. 5.1.3 from vertex $\emptyset$ at the upper level to vertex $\substack{\bullet\\ \bullet}$ at the fourth level corresponds to the following sequence of diagrams:

$$\{\,\emptyset \to \bullet \to \bullet\bullet \to \bullet \to \substack{\bullet\\ \bullet}\,\}. \tag{5.1.49}$$

These sequences are called *oscillating Young tableaux* (we discuss oscillating Young tableaux in Section 5.2.3 below). The number of oriented links (arrows) in an oscillating Young tableau is called the length of this tableau. As an example, the length of the tableau (5.1.49) equals 4.

We summarize our discussion in the following proposition.

Proposition 5.1.3. *Let $\{T^{(\lambda)}\}$ be a set of irreducible representations of group $SO(p,q)$, associated with Young diagrams λ, whose number of rows m obeys $m \leq r$ for $N = (p+q) = (2r+1)$, and $m < r$ for $N = (p+q) = 2r$. In the case $N = (p+q) = 2r$, we enlarge this set by including representations $T^{(\lambda')}$ associated with diagrams λ' with number of rows $m = r$ (such a representation is reducible and decomposes into a sum of selfdual and anti-selfdual representations). Tensor product $T^{\otimes n}$ of defining representations T of group $SO(p,q)$ is the direct sum of irreducible representations $T^{(\lambda)}$ and representations $T^{(\lambda')}$, where λ (or λ') is the last Young diagram in oscillating Young tableau of length n,*

$$\Lambda = \{\lambda_0 = \emptyset, \lambda_1, \ldots, \lambda_n = \lambda\}. \tag{5.1.50}$$

Multiplicity of representation $T^{(\lambda)}$ (or $T^{(\lambda')}$) in $T^{\otimes n}$ is equal to the number of oscillating Young tableaux (5.1.50), *i.e., to the number of paths in Young graph for $T^{\otimes n}$ that start from vertex $\emptyset$ at the zeroth level and end at diagram λ (or λ') at the lowest, nth level. If at some level of this Young graph one encounters a diagram λ'' with more than $[N/2]$ rows, this diagram must be replaced by the equivalent complementary diagram, see Fig.* 5.1.1 *and caption to Fig.* 5.1.3.

We note that decomposition (5.1.40) of representation $[1^r]$ into selfdual and anti-selfdual representations in the case of groups $SO(p,q)$ $(p+q=2r)$ and $SO(2r,\mathbb{C})$ is not encoded in Young graph for $T^{\otimes r}$ (see Fig. 5.1.3 for $T^{\otimes 4}$ and $p+q=8$). One has to bear in mind that all representations of these groups associated with diagrams λ with number of rows equal to r, that emerge in the Young graph, are reducible, but one should not decompose any of them into the two irreducible representations (selfdual and anti-selfdual) until the very end, for consistency of branching, i.e., addition or removal of a box at the next level of the graph.

Projectors $P^{(\lambda)}$ that extract irreducible[d] representations from tensor product $T^{\otimes n}$ generalize projectors (5.1.34) found for $T^{\otimes 2}$. They are constructed in terms of idempotents of Brauer algebra (an extension of algebra $\mathbb{C}[S_n]$) which act in the space $\mathcal{V}^{\otimes n}$ of representation $T^{\otimes n}$. We define Brauer algebra and describe the construction of its idempotents in the following section.

Problem 5.1.6. Prove that adjoint representation of $SO(N,\mathbb{C})$ corresponds to Young diagram 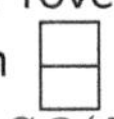 $=[1^2]$. Check that tensor product of two adjoint representations of $SO(N,\mathbb{C})$ for $N \geq 6$ has the following expansion:

$$[1^2]\otimes[1^2]=[1^4]\oplus[2,1^2]\oplus[2^2]\oplus[2]\oplus[1^2]\oplus\emptyset, \tag{5.1.51}$$

where representation $[1^4]$ in the right-hand side of (5.1.51) is equivalent to representations $[1^2]$ and $[1^3]$, respectively, for $N=6$ and $N=7$. Show that representation $[1^4]$ for $N=8$ and representation $[2,1^2]$ for $N=6$ are reducible and decomposed into selfdual and anti-selfdual parts. Modify decomposition (5.1.51) for the cases $N=3,4,5$. Hint: making use of formula (5.1.37), compare dimension of representation on the left-hand side of (5.1.51) with dimension of the sum of representations on the right-hand side of (5.1.51).

5.2. Brauer Algebra $\mathcal{B}r_n$ and Its Representations

We study in this section the representation theory for Brauer algebra $\mathcal{B}r_n$ and make use of the technique that generalizes Okounkov–Vershik approach [47] developed for the group algebra $\mathbb{C}[S_n]$. We make use of the approach of Refs. [48, 49] (see also Refs. [50, 51]).

[d]Modulo the selfduality subtlety for diagrams with r rows for groups with $p+q=2r$.

In the end of this section, we present an explicit construction of complete system of orthogonal idempotents in Brauer algebra $\mathcal{B}r_n$. Their special representation in the space $\mathcal{V}^{\otimes n}$ of rank-n tensors (here $\mathcal{V}$ is either complex space of defining representation of group $SO(N,\mathbb{C})$ or real space of defining representation of group $SO(p,q)$) gives projectors that extract from $\mathcal{V}^{\otimes n}$ invariant subspaces $\mathcal{V}^{(\lambda)}$ of irreducible representations.

5.2.1. Brauer algebra $\mathcal{B}r_n$. Jucys–Murphy elements for $\mathcal{B}r_n$

We recall the matrices P and K defined in (5.1.34)

$$P^{j_1j_2}_{i_1i_2} = \delta^{j_1}_{i_2}\delta^{j_2}_{i_1}, \quad K^{j_1j_2}_{i_1i_2} = \eta^{j_1j_2}\,\eta_{i_1i_2}, \tag{5.2.1}$$

where $\eta_{i_1i_2}$ is metric in space $\mathcal{V}$ of defining representation of subgroup $SO(p,q) \subset SO(N,\mathbb{C})$. We define the set of operators P_k and K_k ($k = 1,\ldots,n-1$) which act in the space $\mathcal{V}^{\otimes n}$ of rank-n tensors as follows (cf. (4.2.9)):

$$\begin{aligned} &P_k \cdot (\vec{e}_{i_1} \otimes \cdots \otimes \vec{e}_{i_k} \otimes \vec{e}_{i_{k+1}} \otimes \cdots \otimes \vec{e}_{i_n}) \\ &\quad = (\vec{e}_{i_1} \otimes \cdots \otimes \vec{e}_{i_{k+1}} \otimes \vec{e}_{i_k} \otimes \cdots \otimes \vec{e}_{i_n}) \\ &\quad = (\vec{e}_{j_1} \otimes \cdots \otimes \vec{e}_{j_k} \otimes \vec{e}_{j_{k+1}} \otimes \cdots \otimes \vec{e}_{j_n}) \\ &\qquad \times \delta^{j_1}_{i_1} \cdots \delta^{j_{k-1}}_{i_{k-1}} P^{j_kj_{k+1}}_{i_ki_{k+1}} \delta^{j_{k+2}}_{i_{k+2}} \cdots \delta^{j_n}_{i_n}, \end{aligned} \tag{5.2.2}$$

$$\begin{aligned} &K_k \cdot (\vec{e}_{i_1} \otimes \cdots \otimes \vec{e}_{i_k} \otimes \vec{e}_{i_{k+1}} \otimes \cdots \otimes \vec{e}_{i_n}) \\ &\quad = (\vec{e}_{j_1} \otimes \cdots \otimes \vec{e}_{j_k} \otimes \vec{e}_{j_{k+1}} \otimes \cdots \otimes \vec{e}_{j_n}) \\ &\qquad \times \delta^{j_1}_{i_1} \cdots \delta^{j_{k-1}}_{i_{k-1}} K^{j_kj_{k+1}}_{i_ki_{k+1}} \delta^{j_{k+2}}_{i_{k+2}} \cdots \delta^{j_n}_{i_n}. \end{aligned} \tag{5.2.3}$$

In notations without indices, the matrices of these operators read

$$P_k = I^{\otimes(k-1)} \otimes P \otimes I^{\otimes(n-k-1)}, \quad K_k = I^{\otimes(k-1)} \otimes K \otimes I^{\otimes(n-k-1)}. \tag{5.2.4}$$

By making use of the identities

$$\begin{aligned} &K^2 = N \cdot K, \quad PK = KP = K, \quad K_kK_{k+1}K_k = K_{k+1}, \\ &P_kK_{k+1}K_k = P_{k+1}K_k, \end{aligned}$$

etc., one finds that the operators $P_k \equiv \mathsf{s}(\sigma_k)$ and $K_k \equiv \mathsf{s}(\kappa_k)$ form representation s (which acts in space $\mathcal{V}^{\otimes n}$) of generators σ_k and κ_k of a certain associative algebra with defining relations

$$\begin{gathered}
\sigma_i^2 = e, \quad \kappa_i^2 = \omega\kappa_i, \quad \sigma_i\kappa_i = \kappa_i\sigma_i = \kappa_i, \quad i = 1,\dots,n-1, \\
\sigma_i\sigma_j = \sigma_j\sigma_i, \quad \kappa_i\kappa_j = \kappa_j\kappa_i, \quad \sigma_i\,\kappa_j = \kappa_j\,\sigma_i, \quad |i-j| > 1, \\
\sigma_i\sigma_{i+1}\sigma_i = \sigma_{i+1}\sigma_i\sigma_{i+1}, \quad \kappa_i\kappa_{i+1}\kappa_i = \kappa_i, \quad \kappa_{i+1}\kappa_i\kappa_{i+1} = \kappa_{i+1}, \\
\sigma_i\kappa_{i+1}\kappa_i = \sigma_{i+1}\kappa_i, \quad \kappa_{i+1}\kappa_i\sigma_{i+1} = \kappa_{i+1}\sigma_i, \quad i = 1,\dots,n-2,
\end{gathered} \tag{5.2.5}$$

where e denotes unity and ω is a numerical parameter that is introduced for generality. To have an agreement with (5.2.2)–(5.2.4), parameter ω of algebra (5.2.5) in the case considered is equal to N. However, we are going to employ relations valid in the general case, so we keep ω arbitrary. Note that generators σ_i are invertible while κ_i are not invertible.

Definition 5.2.1. Associative algebra $\mathcal{B}r_n(\omega)$ over the field of complex numbers with unity and generators σ_k and κ_k (where $k = 1,\dots,n-1$) obeying defining relations (5.2.5) is called *Brauer algebra.*

All basis elements of algebra $\mathcal{B}r_n(\omega)$ are products of generators σ_k and κ_k. We show (see discussion before Fig. 5.2.1) that algebra $\mathcal{B}r_n(\omega)$ is finite dimensional, and its dimension is

$$\dim(\mathcal{B}r_n) = (2n-1)!! \equiv (2n-1)\cdot(2n-3)\cdots 3\cdot 1. \tag{5.2.6}$$

We compare the relations in (5.2.5), that involve only σ_k, with relations (4.6.1) and observe that elements σ_k generate subalgebra $\mathbb{C}[S_n] \subset \mathcal{B}r_n(\omega)$, so Brauer algebra $\mathcal{B}r_n(\omega)$ can be viewed as an extension of group algebra $\mathbb{C}[S_n]$ of symmetric group S_n. Accordingly, like in (4.2.13), operators P_k, given by (5.2.2) and (5.2.4), are representations of generators σ_k (neighboring transpositions) of group S_n. Brauer algebra plays the same role in the representation theory for orthogonal groups $SO(N,\mathbb{C})$ and their real subgroups $SO(p,q)$, as group algebra $\mathbb{C}[S_n]$ does in the case of $SL(N,\mathbb{C})$ and $SU(2)$ (see Section 4.2). The basic fact behind this property is the analog of Proposition 4.2.2 (the proof of this analog is completely the same as the proof of Proposition 4.2.2).

Proposition 5.2.1. *Let T be defining representation of group $O(N,\mathbb{C})$ in space $\mathcal{V}$ with metric η: $g \in O(N,\mathbb{C})$ if and only if $g^{\mathsf{T}}\eta g = \eta$. Operator X acting in $\mathcal{V}^{\otimes n}$ is $O(N,\mathbb{C})$-invariant,*

$$X\cdot(T^{\otimes n}(g)) = (T^{\otimes n}(g))\cdot X, \quad \forall g \in O(N,\mathbb{C}), \tag{5.2.7}$$

if and only if $X = \mathsf{s}(a)$, *where* $a \in \mathcal{B}r_n(N)$ *and* s *is representation of* $\mathcal{B}r_n(N)$ *given by (5.2.2), (5.2.3). In other words, the* $O(N,\mathbb{C})$*-invariant operator* X *is a linear combination*

$$X = \sum_i x_i \mathsf{s}(a_i), \tag{5.2.8}$$

where summation runs over all basis elements a_i *of algebra* $\mathcal{B}r_n(N)$ *(index* i *runs from* 1 *to* $(2n-1)!!$*) and* x_i *are complex coefficients.*

Remark. In the case $\eta = I_{p,q}$, Proposition 5.2.1 is also valid for subgroup $O(p,q) \subset O(N,\mathbb{C})$ $(p+q=N)$. Proposition 5.2.1 is also valid almost always for special orthogonal groups $SO(N,\mathbb{C})$ and $SO(N-q,q)$. The exception for these groups occurs only in even-dimensional case $N=2r$, and for representations $T^{\otimes n}$ with $n \geq r$. In that case, there is a way to construct invariant operators in $\mathcal{V}^{\otimes n}$, which do not have the form (5.2.8). The construction involves totally antisymmetric tensor with components $\varepsilon_{i_1 \ldots i_{2r}}$. Namely, these invariant operators are projectors (5.1.41) on spaces of selfdual and anti-selfdual totally antisymmetric rank-r tensors. The (anti-)selfdual representations of groups $SO(2r,\mathbb{C})$ and $SO(2r-q,q)$ are discussed in detail in the end of Section 5.1.3.

There is a very convenient graphic representation of elements of Brauer algebra $\mathcal{B}r_n(\omega)$. It is analogous to representation of elements of symmetric group S_n introduced in Section I-1.1.4 of the first book. Generator $\sigma_i \in \mathcal{B}r_n(\omega)$ is represented by a braid

$\sigma_i =$ 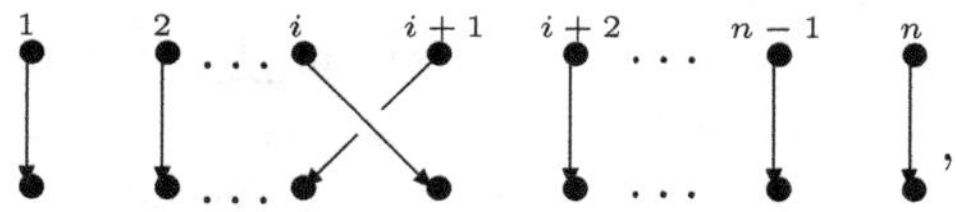 , (5.2.9)

and, since $\sigma_i^2 = 1$ (see (5.2.5)) it is irrelevant whether the line connecting i and $i+1$ is above or below the line connecting $i+1$ and i. Due to locality relation $\sigma_i\sigma_j = \sigma_j\sigma_i$ for $i+1<j<n-1$ (see (5.2.5)), we have

$\sigma_j\sigma_i =$ 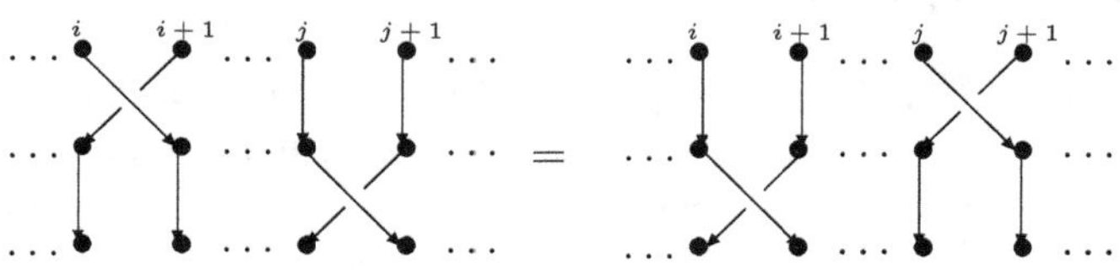 $= \sigma_i\sigma_j.$

Braid relation $\sigma_{i+1}\sigma_i\ \sigma_{i+1} = \sigma_i\sigma_{i+1}\sigma_i$ in (5.2.5) has the following graphic representation:

$\sigma_{i+1}\sigma_i\sigma_{i+1} =$

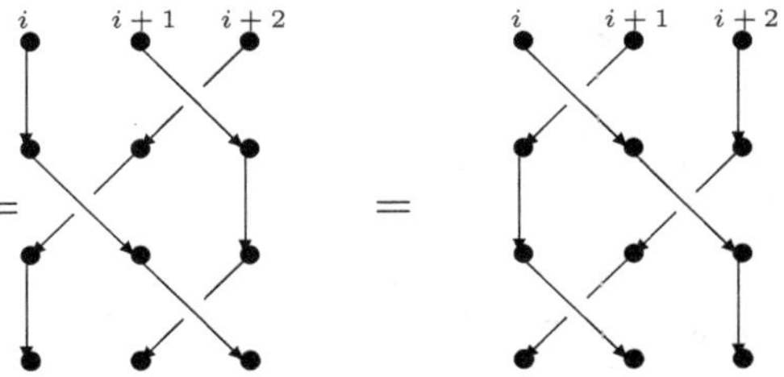

$= \sigma_i\sigma_{i+1}\sigma_i.$

Finally, graphic representation of element $\kappa_i \in \mathcal{B}r_n(\omega)$ is

$\kappa_i =$

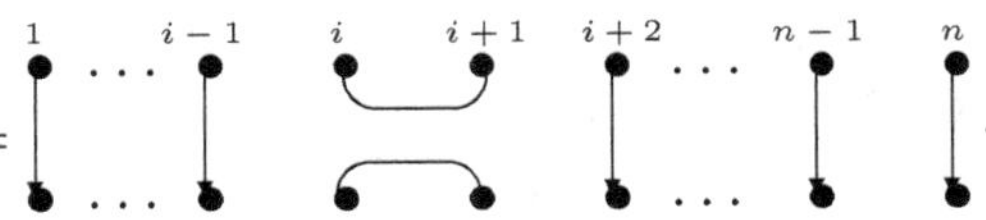

. (5.2.10)

Making use of the graphs (5.2.9), (5.2.10) for generators $\sigma_i, \kappa_i \in \mathcal{B}r_n(\omega)$, one constructs a braid representation of any product of these generators. In doing so, one encounters loops:

$= \omega.$ (5.2.11)

These must be replaced by the constant ω, as shown above.

Problem 5.2.1. Making use of representations (5.2.9)–(5.2.11) show graphically all defining relations (5.2.5).

Graphic representation (5.2.9), (5.2.10) implies that any monomial of algebra $\mathcal{B}r_n$ is represented by a graph (see Fig. 5.2.1) with n lines and $2n$ vertices, belonging either to the upper or to the lower row. Each of these $2n$ vertices is connected with one of the remaining $(2n-1)$ vertices in upper or lower rows. Due to locality and braid properties, the inner part of the graph is irrelevant. Different graphs of this type define different basis elements in algebra $\mathcal{B}r_n$, and their number (dimension of $\mathcal{B}r_n$) is calculated as follows. The first of $2n$ vertices can be connected to one of $(2n-1)$ other vertices. The next of the remaining $(2n-2)$ vertices has $(2n-3)$ ways to be connected with the remaining "free" vertices, etc. So, the product $(2n-1)\cdot(2n-3)\cdot(2n-5)\cdots$ is precisely the number of graphs we consider, i.e., the dimension (5.2.6) of algebra $\mathcal{B}r_n$.

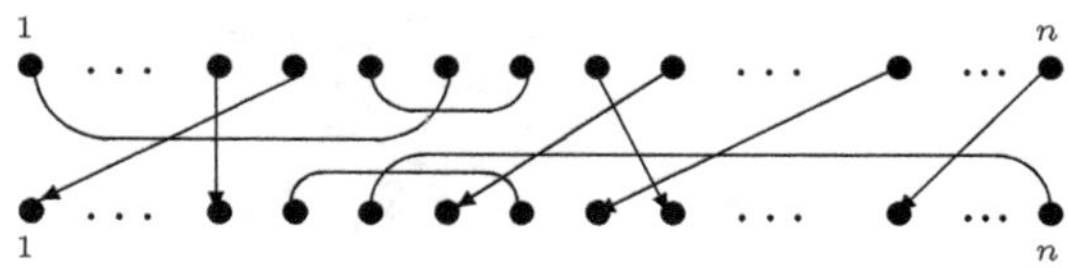

Fig. 5.2.1. Graphic representation of a monomial in algebra $\mathcal{B}r_n$, i.e., a product of generators $\sigma_i, \kappa_j \in \mathcal{B}r_n$.

In representation s of algebra $\mathcal{B}r_n(\omega)$, given by (5.2.2), (5.2.3), every link in graphs (5.2.9) and (5.2.10) that connects vertices k in upper row and j in lower row, represents delta-symbol $\delta^{i_k}_{i_j}$, while lines in graph (5.2.10) connecting two upper (lower) vertices k and j give metrics $\eta^{i_k i_j}$ $(\eta_{i_k i_j})$. Therefore, in representation s, loop (5.2.11) is a contraction of two metrics, so the parameter ω of algebra $\mathcal{B}r_n(\omega)$ has to be equal $\omega = \eta^{ik}\eta_{ik} = N$, as we know already.

In what follows, we omit parameter ω in notation of Brauer algebra. In Brauer algebra $\mathcal{B}r_n$, we define a set of elements $\{y_m\}$ $(m = 1, \dots, n)$ (cf. (4.6.2)):

$$
\begin{aligned}
y_1 = 0, \quad y_m &= \sum_{k=1}^{m-1} (\sigma_{k,m} - \kappa_{k,m}) \\
&= \sum_{k=1}^{m-1} \sigma_{m-1} \cdots \sigma_{k+1} (\sigma_k - \kappa_k) \sigma_{k+1} \cdots \sigma_{m-1}, \\
& \quad m = 2, 3, \dots, n,
\end{aligned} \tag{5.2.12}
$$

where

$$
\begin{aligned}
\sigma_{k,m} &= \sigma_{m-1} \cdots \sigma_{k+1} \cdot \sigma_k \cdot \sigma_{k+1} \cdots \sigma_{m-1} = (k, m), \\
\kappa_{k,m} &= \sigma_{m-1} \cdots \sigma_{k+1} \cdot \kappa_k \cdot \sigma_{k+1} \cdots \sigma_{m-1}.
\end{aligned} \tag{5.2.13}
$$

Problem 5.2.2. Making use of graphic representation (5.2.9), (5.2.10), draw graphs for elements (5.2.13).

Operators y_m are called *Jucys–Murphy elements* for Brauer algebra, and play similar role in the construction of primitive idempotents in algebra $\mathcal{B}r_n$, as Jucys–Murphy elements (4.6.2) in the case of algebra $\mathbb{C}[S_n]$. Note that Jucys–Murphy elements may be defined via induction procedure

$$y_1 = 0, \quad y_{m+1} = \sigma_m - \kappa_m + \sigma_m \cdot y_m \cdot \sigma_m. \tag{5.2.14}$$

It follows from definitions (5.2.12), (5.2.14) and relations (5.2.5) that elements y_m obey (we use these identities in what follows)

$$y_{m+1} \cdot \sigma_m = \sigma_m \cdot y_m + 1 - \kappa_m, \quad \sigma_m \cdot y_{m+1} = y_m \cdot \sigma_m + 1 - \kappa_m, \tag{5.2.15}$$

$$\kappa_m(y_m + y_{m+1} + \omega - 1) = 0, \quad (y_m + y_{m+1} + \omega - 1)\kappa_m = 0, \tag{5.2.16}$$

$$[y_m + y_{m+1}, \sigma_m] = 0, \quad [y_m, \sigma_i] = 0, \quad m \neq i, i+1. \tag{5.2.17}$$

Hereafter, unit element $e \in \mathcal{B}r_n$ multiplied by complex number c is written merely as c. Let us prove the first equality in (5.2.16) by induction (the second equality is proven in the same way). We use definition (5.2.14) on the left-hand side of (5.2.16) and employ (5.2.5) to get

$$\begin{aligned}
\kappa_m(y_m + y_{m+1} + \omega - 1) &= \kappa_m(y_m + (\sigma_m - \kappa_m + \sigma_m y_m \sigma_m) + \omega - 1) \\
&= \kappa_m y_m (1 + \sigma_m) \\
&= \kappa_m(\sigma_{m-1} - \kappa_{m-1} + \sigma_{m-1} y_{m-1} \sigma_{m-1})(1 + \sigma_m) \\
&= \kappa_m \sigma_{m-1} y_{m-1} \sigma_{m-1} (1 + \sigma_m).
\end{aligned}$$

So, equivalent form of the first equality in (5.2.16) reads

$$\kappa_k \sigma_{k-1} y_{k-1} \sigma_{k-1} (1 + \sigma_k) = 0. \tag{5.2.18}$$

For $k = 2$, this equality is trivially satisfied, since $y_1 = 0$. Let us show that equality (5.2.18) is valid for $k = m+1$ provided that it is valid for $k = m$. We write

$$\begin{aligned}
\kappa_{m+1} \sigma_m y_m \sigma_m (1 + \sigma_{m+1}) &= \kappa_{m+1} \sigma_m \sigma_{m+1}^2 y_m \sigma_m (1 + \sigma_{m+1}) \\
&= \kappa_{m+1} \kappa_m y_m \sigma_{m+1} \sigma_m (1 + \sigma_{m+1}) \\
&= \kappa_{m+1} \kappa_m y_m (1 + \sigma_m) \sigma_{m+1} \sigma_m \\
&= \kappa_{m+1} \kappa_m (\sigma_{m-1} - \kappa_{m-1} + \sigma_{m-1} y_{m-1} \sigma_{m-1}) \\
&\quad \times (1 + \sigma_m) \sigma_{m+1} \sigma_m \\
&= \kappa_{m+1} [\kappa_m \sigma_{m-1} y_{m-1} \sigma_{m-1} (1 + \sigma_m)] \\
&\quad \times \sigma_{m+1} \sigma_m \\
&= 0,
\end{aligned}$$

which is the desired result. Here, we have used identities $\sigma_{m+1}^2 = 1$, $\kappa_{m+1} \sigma_m \sigma_{m+1} = \kappa_{m+1} \kappa_m$, $\sigma_{m+1} y_m = y_m \sigma_{m+1}$, definition (5.2.14), etc.

Problem 5.2.3. Making use of (5.2.5), check that the element $y_2 = (\sigma_1 - \kappa_1)$ obeys third-order characteristic identity

$$(y_2 - 1)(y_2 + 1)(y_2 + \omega - 1) = 0 \Rightarrow \mathrm{Spec}(y_2) = \{+1, -1, 1 - \omega\}. \tag{5.2.19}$$

We now define a subalgebra $Y_n \subset \mathcal{B}r_n$, which is generated by all Jucys–Murphy elements $\{y_2, \dots, y_n\}$. The following proposition holds [48] (it is an analog of Proposition 4.6.1 valid in the case of algebra $\mathbb{C}[S_n]$).

Proposition 5.2.2. *Jucys–Murphy elements* $\{y_2, \dots, y_n\}$ *given by* (5.2.12) *form complete set of commuting elements in* $\mathcal{B}r_n$:

$$[y_i, y_j] = 0, \quad \forall i, j, \tag{5.2.20}$$

i.e., algebra Y_n *is the maximal commutative subalgebra in* $\mathcal{B}r_n$. *Here, completeness of a set of commuting elements means that any element* $y \in \mathcal{B}r_n$ *that commutes with all elements* y_i, *is a polynomial in* y_i.

Proof. The fact that Jucys–Murphy elements commute with each other follows from the relations ($k < \ell < r$):

$$[\sigma_{k,\ell} - \kappa_{k,\ell}, \sigma_{k,r} - \kappa_{k,r} + \sigma_{\ell,r} - \kappa_{\ell,r}] = 0,$$

which in turn follow from identities

$$\sigma_{k,\ell}\sigma_{k,r} = \sigma_{\ell,r}\sigma_{k,\ell}, \quad \sigma_{k,\ell}\kappa_{k,r} = \kappa_{\ell,r}\sigma_{k,\ell},$$
$$\kappa_{k,\ell}(\sigma_{k,r} - \kappa_{\ell,r}) = 0, \quad \kappa_{k,\ell}(\sigma_{\ell,r} - \kappa_{k,r}) = 0,$$

and their mirror counterparts. All these identities are straightforwardly proven by making use of graphic representation (5.2.9) and (5.2.10) (see Problem 5.2.2). We do not give proof of the fact that Y_n is the maximal commutative subalgebra in $\mathcal{B}r_n$. □

5.2.2. Intertwining elements and idempotents in algebra $\mathcal{B}r_n$. Spectrum of Jucys–Murphy operators

Let us introduce *intertwining* elements $U_k \in \mathcal{B}r_n$ (cf. (4.6.11))

$$\begin{aligned} U_k &= [\sigma_k, y_k]W_k = W_k[\sigma_k, y_k] \\ &= W_k((y_{k+1} - y_k)\sigma_k - 1) = W_k(\sigma_k(y_k - y_{k+1}) + 1), \quad k = 1, \dots, n-1, \end{aligned} \tag{5.2.21}$$

where we introduce the notation

$$W_k \equiv (y_k + y_{k+1} + \omega - 1),$$

and use identities (5.2.16), (5.2.17), which can be written as follows:

$$\kappa_k W_k = W_k \kappa_k = 0, \quad \sigma_k W_k = W_k \sigma_k. \tag{5.2.22}$$

The following proposition holds (cf. Proposition 4.6.2).

Proposition 5.2.3. *Elements U_k obey*

$$U_k y_k = y_{k+1} U_k, \quad U_k y_{k+1} = y_k U_k, \quad U_k y_i = y_i U_k \quad \text{for} \quad i \neq k, k+1,$$

$$U_k[\sigma_k, y_k] = (1 + y_{k+1} - y_k)(1 - y_{k+1} + y_k) W_k, \tag{5.2.23}$$

$$U_{k+1} U_k U_{k+1} = U_k U_{k+1} U_k, \quad \kappa_k U_k = U_k \kappa_k = 0. \tag{5.2.24}$$

Proof. Intertwining operator U_k in Brauer algebra $\mathcal{B}r_n$ differs from intertwining operator (4.6.11) in algebra $\mathbb{C}[S_n]$ by a factor W_k, which, due to (5.2.22) vanished upon multiplication by element κ_k. We recall that algebra $\mathbb{C}[S_n]$ is obtained from $\mathcal{B}r_n$ by setting $\kappa_k = 0$ and therefore find that the proof of formulas (5.2.23) and the first formula in (5.2.24) is the same as the proof of formulas (4.6.13)–(4.6.15) in Proposition 4.6.2. The second formula in (5.2.24) is a straightforward consequence of (5.2.22). □

Elements U_k are used to obtain information on the spectrum of Jucys–Murphy elements $\{y_j\}$ in algebra $\mathcal{B}r_n$ (cf. Proposition 4.6.5).

Proposition 5.2.4. *Spectrum of elements $y_j \in \mathcal{B}r_n$ obeys*

$$\text{Spec}(y_j) \subset \{[1 - j, j - 1], [2 - j, j - 2] + 1 - \omega\}, \tag{5.2.25}$$

where $j \geq 2$, the notation $[-z, z]$ refers to the set of integers (cf. (4.6.28))

$$[-z, z] = \{-z, \ldots, -1, 0, 1, \ldots, z\}, \quad z \in \mathbb{Z}_{\geq 0}, \tag{5.2.26}$$

and $[-z, z] + a$ denotes the set of numbers shifted by a:

$$[-z, z] + a = \{a - z, \ldots, a - 1, a, a + 1, \ldots, a + z\}.$$

Proof. We give the proof by induction. The property (5.2.25) obviously holds for y_2, see (5.2.19). Assume that the property (5.2.25) holds for y_{j-1}, where $(j-1) \geq 2$:

$$\text{Spec}(y_{j-1}) \subset \{[2-j, j-2], [3-j, j-3] + 1 - \omega\}. \qquad (5.2.27)$$

Let $f(y_{j-1}) = 0$ be characteristic identity for operator y_{j-1} (cf. (5.2.19)). Then

$$\begin{aligned} 0 &= U_{j-1} f(y_{j-1})[\sigma_{j-1}, y_{j-1}] = f(y_j) U_{j-1}[\sigma_{j-1}, y_{j-1}] \\ &= f(y_j)(1 + y_{j-1} - y_j)(1 + y_j - y_{j-1})(y_j + y_{j-1} + \omega - 1). \end{aligned}$$

We use here the last equality in (5.2.23). So, we have

$$\text{Spec}(y_j) \subset \text{Spec}(y_{j-1}) \cup \{\text{Spec}(y_{j-1}) \pm 1\} \cup \{-\text{Spec}(y_{j-1}) + 1 - \omega\},$$

which, in view of (5.2.27), proves (5.2.25) for y_j. □

Our purpose is to construct complete set of primitive orthogonal idempotents $e_\alpha \in \mathcal{B}r_n$ labeled by index α. They should obey the relations (4.5.3), (4.5.4):

$$e_\alpha \cdot e_\beta = \delta_{\alpha\beta} e_\alpha, \quad \sum_\alpha e_\alpha = e. \qquad (5.2.28)$$

Like in Okounkov–Vershik method for algebra $\mathbb{C}[S_n]$ (see Section 4.6.2), we require that idempotents $e_\alpha \in \mathcal{B}r_n$ should be eigenvectors of Jucys–Murphy elements (5.2.12) in left regular representation of algebra $\mathcal{B}r_n$:

$$y_k \cdot e_\alpha = a_k^{(\alpha)} e_\alpha \quad (a_k^{(\alpha)} \in \mathbb{R}) \quad \forall k = 1, \ldots, n. \qquad (5.2.29)$$

This choice of idempotents in algebra $\mathcal{B}r_n$ is always possible, since Jucys–Murphy elements $\{y_1, \ldots, y_n\}$ commute with each other. Like in the case of algebra $\mathbb{C}[S_n]$, one can show that idempotents e_α obeying (5.2.28) and (5.2.29) commute[e] with all Jucys–Murphy elements y_k, and hence they are functions of y_k. Note that in view of Proposition 5.2.4, all eigenvalues $a_k^{(\alpha)}$ in (5.2.29) are real numbers.

Example. Algebra $\mathcal{B}r_2$ contains, besides trivial element $y_1 \equiv 0$, one Jucys–Murphy element $y_2 = \sigma_1 - \kappa_1$ which obeys the characteristic identity

[e]One multiplies both parts of (5.2.19) on the left by e_β and sums over α.

(5.2.19). Making use of this identity, we find complete system of primitive orthogonal idempotents in $\mathcal{B}r_2$:

$$\begin{gathered} e_a = \tfrac{(1-y_2)(y_2+\omega-1)}{2(\omega-2)} = \tfrac{1}{2}(1-\sigma_1), \quad e_0 = \tfrac{(1-y_2)(1+y_2)}{\omega(2-\omega)} = \tfrac{1}{\omega}\kappa_1, \\ e_s = \tfrac{(1+y_2)(y_2+\omega-1)}{2\omega} = \tfrac{1}{2}(1+\sigma_1) - \tfrac{1}{\omega}\kappa_1. \end{gathered} \tag{5.2.30}$$

These idempotents obey the relations (5.2.28) and also (5.2.29) (see (5.2.19)):

$$y_2 \cdot e_a = (-1)e_a, \quad y_2 \cdot e_0 = (1-\omega)e_0, \quad y_2 \cdot e_s = (+1)e_s, \tag{5.2.31}$$

and, in representation s given by (5.2.2), (5.2.3) and $\omega = N$, coincide with projectors (5.1.34). The latter projectors, as we know, extract irreducible representations from tensor product of two defining representations of group $SO(p,q)$.

Every primitive idempotent $e_\alpha \in \mathcal{B}r_n$ (eigenvector of all elements y_i $(i = 1,\dots,n)$) defines a spectral set

$$\Lambda_\alpha = (a_1^{(\alpha)}, \dots, a_n^{(\alpha)}) \in \mathbb{R}^n,$$

where $a_i^{(\alpha)}$ are eigenvalues of y_i and, by definition, $a_1^{(\alpha)} = 0$. We denote the set of spectral sets by $\mathrm{Spec}(y_1,\dots,y_n)$.

Proposition 5.2.5. *Let $\Lambda_\alpha = (a_1^{(\alpha)}, \dots, a_n^{(\alpha)})$ be spectral set corresponding to primitive idempotent $e_\alpha \in \mathcal{B}r_n$.*

(1) *If $a_k^{(\alpha)} + a_{k+1}^{(\alpha)} \neq 1-\omega$, then $a_k^{(\alpha)} \neq a_{k+1}^{(\alpha)}$.*
(2) *If $a_k^{(\alpha)} + a_{k+1}^{(\alpha)} \neq 1-\omega$ and $\sigma_k \cdot e_\alpha = \pm e_\alpha$, then*

$$(a_{k+1}^{(\alpha)} - a_k^{(\alpha)}) = \pm 1, \quad \kappa_k \cdot e_\alpha = 0, \tag{5.2.32}$$

where upper and lower signs are correlated.
(3) *If $a_k^{(\alpha)} + a_{k+1}^{(\alpha)} \neq 1-\omega$ and $(a_{k+1}^{(\alpha)} - a_k^{(\alpha)}) \neq \pm 1$, then*

$$\Lambda'_\alpha = (a_1^{(\alpha)}, \dots, a_{i+1}^{(\alpha)}, a_i^{(\alpha)}, \dots, a_n^{(\alpha)}) \in \mathrm{Spec}(y_1,\dots,y_i,y_{i+1},\dots,y_n).$$

Proof. We apply operator e_α on the left and on the right to both sides of the relation $U_k = [\sigma_k, y_k]W_k$ (see (5.2.21) and make use of (5.2.29)). The result is $e_\alpha U_k e_\alpha = 0$. We then apply idempotent e_α on the left and on the right to the two sides of the identity $W_k(\sigma_k(y_k - y_{k+1}) + 1) = U_k$, which also follows from (5.2.21). We have

$$(e_\alpha \sigma_k e_\alpha (a_k^{(\alpha)} - a_{k+1}^{(\alpha)}) + e_\alpha)(a_k^{(\alpha)} + a_{k+1}^{(\alpha)} + \omega - 1) = 0.$$

We see that if $a_k^{(\alpha)} + a_{k+1}^{(\alpha)} \neq 1 - \omega$, then

$$e_\alpha \sigma_k e_\alpha (a_k^{(\alpha)} - a_{k+1}^{(\alpha)}) + e_\alpha = 0, \tag{5.2.33}$$

which implies $a_k^{(\alpha)} \neq a_{k+1}^{(\alpha)}$, otherwise $e_\alpha = 0$. This proves part (1) of the proposition. We now insert the condition[f] $\sigma_k \cdot e_\alpha = \pm e_\alpha$ in (5.2.33). We obtain $e_\alpha(1 \pm (a_k^{(\alpha)} - a_{k+1}^{(\alpha)})) = 0$, which gives the first relation in (5.2.32). We apply the two sides of the second equality in (5.2.14) to idempotent e_α on the left and make use of the relation $\sigma_k \cdot e_\alpha = \pm e_\alpha$ and the first equality in (5.2.32) to obtain $\kappa_k \cdot e_\alpha = 0$, which completes the proof of part (2) of the proposition.

The proof of part (3) is analogous to the proof of similar statement in the case of group algebra $\mathbb{C}[S_n]$ (see Proposition 4.6.7). □

Remark. The inverse of part (2) of Proposition 5.2.5 also holds. Namely, if $a_k^{(\alpha)} + a_{k+1}^{(\alpha)} \neq (1-\omega)$, then the property of the spectral set $(a_{k+1}^{(\alpha)} - a_k^{(\alpha)}) = \pm 1$ leads to two relations: $\sigma_k \cdot e_\alpha = \pm e_\alpha$ and $\kappa_k \cdot e_\alpha = 0$.

There exists the following Brauer algebra analog of Theorem 4.6.12 that describes the properties of spectral sets Λ_α. Some of the propositions of this theorem has been proven above, and we do not give the proofs of the others (see, e.g., [48]).

Theorem 5.2.6. *Let*

$$\Lambda = (a_1, \ldots, a_i, a_{i+1}, \ldots, a_n) \in \text{Spec}(y_1, \ldots, y_i, y_{i+1}, \ldots, y_n), \tag{5.2.34}$$

be a spectral set of a primitive idempotent $e_\Lambda \in \mathcal{B}r_n$. *Then* $a_1 = 0$ *and*

(1) $a_j \in \{[1-j, j-1], [2-j, j-2] + 1 - \omega\}, \forall j \geq 2;$
(2) $a_i + a_{i+1} \neq (1-\omega) \Rightarrow a_i \neq a_{i+1}, i = 1, \ldots, n-1;$
(3a) $a_i + a_{i+1} \neq (1-\omega)$ *and* $a_{i+1} = a_i \pm 1 \Rightarrow \sigma_i \cdot e_\Lambda = \pm e_\Lambda,\ \kappa_i \cdot e_\Lambda = 0;$
(3b) $a_i + a_{i+1} \neq (1-\omega)$ *and* $a_{i+1} \neq a_i \pm 1 \Rightarrow \Lambda' = (a_1, \ldots, a_{i+1}, a_i, \ldots, a_n) \in \text{Spec}(y_1, \ldots, y_i, y_{i+1}, \ldots, y_n),\ \kappa_i \cdot e_\Lambda = 0, \kappa_i \cdot e_{\Lambda'} = 0;$

[f]Since $\sigma_k^2 = 1$, elements σ_k have only two eigenvalues ± 1.

(4) $a_i + a_{i+1} = (1-\omega) \Rightarrow$ *there exists* $m \in \mathbb{Z}_{\geq 0}$ *and* $(2m+1)$ *spectral sets* $\Lambda^{(k)} = (a_1, \dots, a_{i-1}, a_i^{(k)}, a_{i+1}^{(k)}, a_{i+2}, \dots, a_n) \in \mathrm{Spec}(y_1, \dots, y_n)$ *(where* $k = 1, 2, \dots, 2m+1$*), including* $\Lambda \in \{\Lambda^{(k)}\}$*, for which* $a_i^{(k)} + a_{i+1}^{(k)} = (1-\omega)$, $\sum_{k=1}^{2m+1} a_i^{(k)} = m(1-\omega)$, $\sum_{k=1}^{2m+1} a_{i+1}^{(k)} = (m+1)(1-\omega)$.

One can show that each spectral set of length n defines one and only one primitive idempotent of algebra $\mathcal{B}r_n$. In other words, there is one-to-one correspondence between the set of spectral sets (5.2.34) and the set of primitive orthogonal idempotents in algebra $\mathcal{B}r_n$. All spectral sets have properties listed in Theorem 5.2.6.

5.2.3. Oscillating Young tableaux and their content vectors

Let (s, t) be coordinates of a box in Young diagram λ. Recall (see Sec. 4.6.2) that we assign to this box a number $(s-t)$, which is its *content*. An example is the diagram $[4, 3, 1]$ with contents in its boxes:

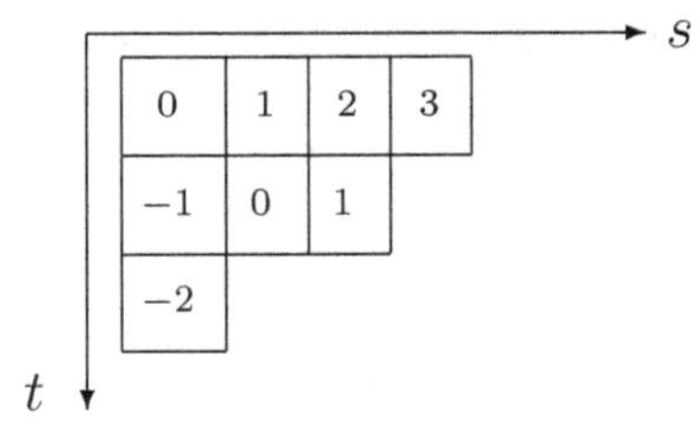

Consider now a sequence of Young diagrams $\Lambda = \{\lambda_0, \lambda_1, \dots, \lambda_n\}$ which starts from a trivial diagram $\lambda_0 = \emptyset$, and in which the $(k+1)$th diagram λ_{k+1} is obtained from the previous diagram λ_k by either adding a box in one of external corners of diagram λ_k or removing a box from one of internal corners of λ_k (an example is given in Fig. 5.2.2). Recall that such a sequence of diagrams is called oscillating Young tableau, see Section 5.1.4. An example of oscillating Young tableau is given in (5.1.49). We define content of the relevant (added/removed) box (s, t) by $(s-t)$, if the box is added, and by $1 - \omega - (s-t)$, if the box is removed. We show contents of the added/removed box in Fig. 5.2.2 as numbers at arrows. So, each oscillating Young tableau determines a sequence of transformations of

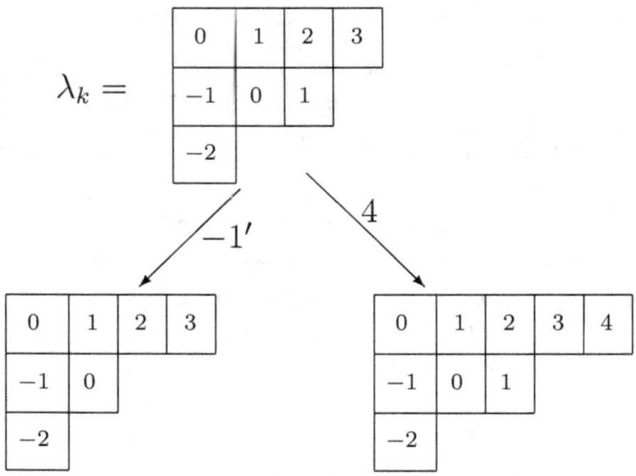

Fig. 5.2.2. Addition or removal of one box at a step $\lambda_k \to \lambda_{k+1}$ in oscillating Young tableau $\Lambda = (\dots, \lambda_k, \lambda_{k+1}, \dots)$. For given λ_k, diagram λ_{k+1} may take various shapes, two of which are shown. Numbers $(-1')$ and 4 on arrows show contents of removed or added box; we define, to simplify formulas, notation $(-n') = (1-\omega) - n$.

Young diagrams

$$\Lambda = \{\emptyset \xrightarrow{a_1=0} \lambda_1 \xrightarrow{a_2} \lambda_2 \xrightarrow{a_3} \lambda_3 \xrightarrow{a_4} \cdots \xrightarrow{a_n} \lambda_n\}, \tag{5.2.35}$$

where each transformation $\lambda_{k-1} \to \lambda_k$ defines content a_k of added or removed box. In this way, one associates with every oscillating Young tableau Λ of length n its *content vector* of length n

$$\Lambda = (a_1, a_2, \dots, a_n),$$

where we have (cf. Proposition 5.2.4)

$$a_i \in \{[1-i, i-1], [2-i, i-2] + 1 - \omega\}. \tag{5.2.36}$$

Definition 5.2.2. The set of sequences $(a_1, a_2, \dots, a_n)$ of numbers a_i obtained as contents of all oscillating Young tableaux of length n (and hence obeying (5.2.36)) is the *set of content vectors of length* n denoted by Cont_n.

The following proposition holds.

Theorem 5.2.7. *There is one-to-one correspondence between the set of spectral sets* $\mathrm{Spec}(y_1, \dots, y_n)$ *and the set of content vectors* Cont_n.

In view of the proposition after Theorem 5.2.6, this means that *there is one-to-one correspondence between the set of content vectors* Cont_n *and the set of primitive orthogonal idempotents in algebra* $\mathcal{B}r_n$.

Like in the case of algebra $\mathbb{C}[S_n]$ studied in Sections 4.6.2 and 4.6.3, Theorems 5.2.6 and 5.2.7 lead to the notion of colored Young graph for algebra $\mathcal{B}r_n$.

5.2.4. Oscillating Young graph for algebra $\mathcal{B}r_n$

The results given in Sections 5.2.1, 5.2.2 and 5.2.3 can be presented in terms of colored (in a similar sense as in Section 4.6.3) Young graph for algebra $\mathcal{B}r_n$. This graph is shown in Fig. 5.2.3 for $\mathcal{B}r_4$. It encodes all information on irreducible representations of algebras $\mathcal{B}r_k$, with $k = 1, 2, 3, 4$, including branching rules of these representations.

Each vertex of the graph is labeled by double index $\{\lambda; k\}$ where λ denotes Young diagram and k is the level of this diagram in the graph

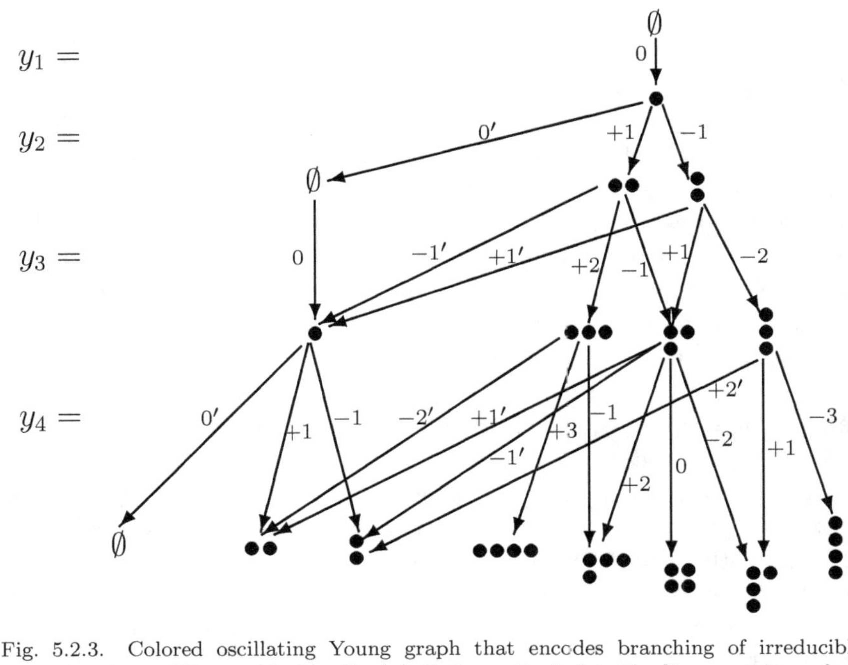

Fig. 5.2.3. Colored oscillating Young graph that encodes branching of irreducible representations of Brauer algebra $Br_4(\omega)$. Indices attached to the lines are eigenvalues of Jucys–Murphy operators y_k in algebra Br_4; to simplify notations, we define $\pm n' \equiv 1 - \omega \pm n$.

(levels are counted from the top, and counting starts from 0). Vertex $\{\lambda; k\}$ is associated with irreducible representation $\rho^{\{\lambda;k\}}$ of algebra $\mathcal{B}r_k$. As an example, each vertex $\{\lambda; 4\}$ at the lowest level of the graph in Fig. 5.2.3 is associated with the irreducible representation $\rho^{\{\lambda;4\}}$ of algebra $\mathcal{B}r_4$ (we discuss this point later). Paths staring from the upper vertex $\emptyset$ at level 0 and ending at the lowest level (these are paths of length 4 in the case considered) determine oscillating Young tableaux (5.2.35) which in turn are associated with spectral sets $\Lambda = (a_1, \ldots, a_n)$. As an example, one of the three paths from the upper vertex $\{\emptyset; 0\}$ to vertex $\{\emptyset; 4\}$ at the fourth level, defines oscillating tableau $\Lambda = \{\emptyset \overset{a_1=0}{\rightarrow} [1] \overset{a_2=-1}{\rightarrow} [1^2] \overset{a_3=+1'}{\rightarrow} [1] \overset{a_4=0'}{\rightarrow} \emptyset\}$ and spectral set $\Lambda = (0, -1, 2-\omega, 1-\omega)$. Thus, all paths in oscillating Young graph are associated with primitive idempotents $e_\Lambda \in \mathcal{B}r_4$, which are eigenvectors of Jucys–Murphy elements $(y_1, \ldots, y_4)$ with eigenvalues $\Lambda = (a_1, a_2, a_3, a_4)$. The eigenvalues (a_1, a_2, a_3, a_4) are shown on links and are equal to contents of added or removed boxes of the Young diagrams. Each primitive idempotent e_Λ generates, once the entire algebra $\mathcal{B}r_4$ acts on it on the left (right), irreducible subrepresentation, in regular representation, which acts in the space of left ideal $\mathcal{L}_\Lambda$ (right ideal $\mathcal{R}_\Lambda$) (see Section 4.5.1). Therefore, each path Λ in oscillating Young graph, that ends at vertex $\{\lambda; 4\}$, is associated with irreducible representation of Brauer algebra $\mathcal{B}r_4$ in the space $\mathcal{L}_\Lambda$. This is precisely the representation by $\rho^{\{\lambda;4\}}$. Different paths labeled by different sets Λ_i and ending at one and the same vertex $\{\lambda; 4\}$ are associated with different idempotents, and hence different, but equivalent representations $\rho^{\{\lambda;4\}}$ that act in isomorphic spaces $\mathcal{L}_{\Lambda_i}$. The latter property follows from the general analysis of semisimple associative algebras in Section 4.5.1 and, in particular, from Remark 3 there. So, irreducible representations of algebra $\mathcal{B}r_k$ are labeled by vertices $\{\lambda; k\}$, and the multiplicity of representation $\rho^{\{\lambda;k\}}$ in regular representation is equal to the number of paths leading to the vertex $\{\lambda; k\}$. On the other hand, in accordance with duality discussed in Remark 3 in Section 4.5.1, different paths ending at the same vertex $\{\lambda; 4\}$ and associated with primitive idempotents e_{Λ_i} with different eigenvalue sets Λ_i label basis vectors in the space of representation $\rho^{\{\lambda;4\}}$. Therefore, the number of different paths from upper vertex $\{\emptyset; 0\}$ to vertex $\{\lambda; 4\}$ is also equal to the dimension of representation $\rho^{\{\lambda;4\}}$. We note again that the equality between the multiplicity of equivalent irreducible representations $\rho^{\{\lambda;4\}}$ in regular representation, on the one hand, and the dimension of these representations, on the other, is a particular case of Proposition 4.5.2 in Section 4.5.1.

> **Problem 5.2.4.** Find dimensions of irreducible representations of algebras $\mathcal{B}r_3$ and $\mathcal{B}r_4$ associated with Young diagrams at the third and fourth levels of Young graph in Fig. 5.2.3, respectively. Making use of the formula (4.5.27), calculate dimensions of algebras $\mathcal{B}r_3$ and $\mathcal{B}r_4$ and compare them with (5.2.6).

Colored oscillating Young graph in Fig. 5.2.3 has subgraphs of length 2 which connect vertices $\{\lambda; i-1\}$ and $\{\lambda'; i+1\}$ at levels $(i-1)$ and $(i+1)$. All types of these subgraphs are shown in Fig. 5.2.4 (cf. Fig. 4.6.6 in Section 4.6.2). Figures 5.2.4(A) and 5.2.4(B) show subgraphs corresponding to spectral conditions (3a) and (3b) in Proposition 5.2.6. In these two subgraphs, transition from the upper diagram λ to lower one λ' involves additions of boxes only, so these subgraphs are the same as in the case of algebra $\mathbb{C}[S_4]$, see Fig. 4.6.5 in Section 4.6.3). Subgraphs C and D in Fig. 5.2.4 correspond to spectral condition (4) in Proposition 5.2.6. Note that every path from the set of $(2m+1)$ paths in subgraph D in

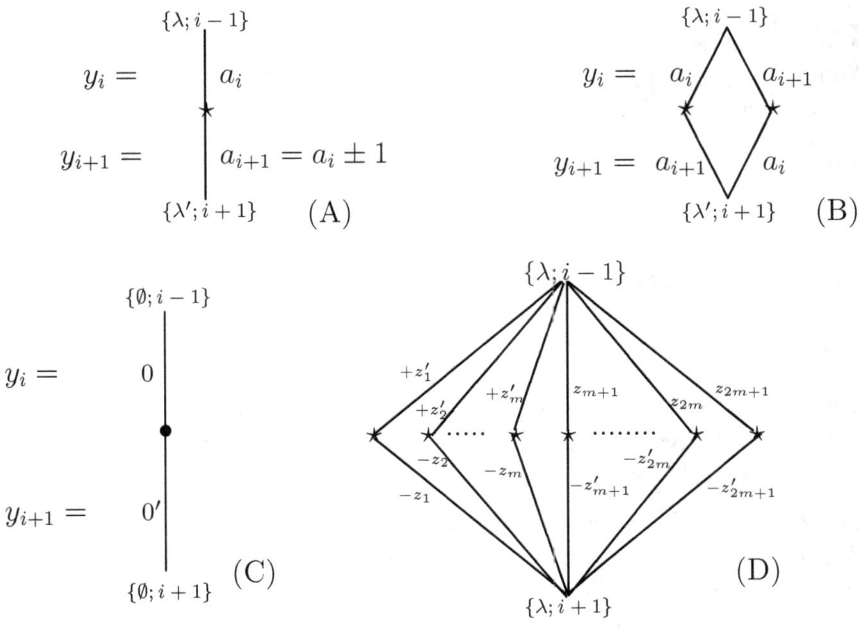

Fig. 5.2.4. Subgraphs in oscillating Young graph in Fig. 5.2.3 with paths of length 2 that connect vertices $\{\lambda; i-1\}$ and $\{\lambda'; i+1\}$. Stars show intermediate Young diagrams at level i. We use in (C) and (D) shorthand notation $\pm z' = (1-\omega) \pm z$ for eigenvalues $a_i^{(k)}$ and $a_{i+1}^{(k)}$ such that $a_i^{(k)} + a_{i+1}^{(k)} = (1-\omega)$; see the case (4) in Proposition 5.2.6.

Fig. 5.2.4 has at the first step an addition (removal) of a box, and at the second step removal (addition) of the same box, so that the diagram λ' at level $(i+1)$ coincides with the diagram λ at level $(i-1)$. Subgraph C in Fig. 5.2.4 corresponds to the case $m=0$ (one path) of subgraph in Fig. 5.2.4(D), and occurs in Fig. 5.2.3 at levels $(i-1,i,i+1)=(0,1,2)$ and $(i-1,i,i+1)=(2,3,4)$. Subgraphs of the type D in Fig. 5.2.4 with three paths and $m=1$ appear in Fig. 5.2.3 at levels $(1,2,3)$ and $(2,3,4)$. Subgraphs with $m>1$ $(5,7,\ldots$ paths) occur in oscillating Young graphs for algebras $\mathcal{B}r_n$ with $n>4$ and larger.

We again emphasize that irreducible representations of group algebra $\mathbb{C}[S_n]$ (of symmetric group S_n) are labeled by Young diagrams $\lambda \vdash n$ with n boxes, and primitive idempotents in $\mathbb{C}[S_n]$ are associated with standard Young tableaux of shape $\lambda \vdash n$. Irreducible representations of Brauer algebra $\mathcal{B}r_n$ are labeled by a pair $\{\lambda; n\}$ where λ is Young diagram and n is the last level in the oscillating Young graph for algebra $\mathcal{B}r_n$, which contains the diagram λ. Primitive idempotents in algebra $\mathcal{B}r_n$ are associated with oscillating Young tableaux $\Lambda=\{\lambda_1=\emptyset,\lambda_2,\ldots,\lambda_n=\lambda\}$ of length n, or, in other words, with paths of length n (with last node λ) in the oscillating Young graph for algebra $\mathcal{B}r_n$. The set of all such paths is denoted by $X(n)$.

The analog of Theorem 4.6.17 is also valid for algebra $\mathcal{B}r_n$.

Theorem 5.2.8. *There is one-to-one correspondence between the set* Tab_n *of oscillating Young tableaux of length* n *equipped with content vectors, the set* $\mathrm{Spec}(y_1,\ldots,y_n)$ *and the set* $X(n)$ *of paths of length* n *in colored oscillating Young graph,*

$$\mathrm{Tab}_n \simeq \mathrm{Spec}(y_1,\ldots,y_n) \simeq X(n). \tag{5.2.37}$$

5.2.5. Primitive idempotents in Brauer algebra and invariant projectors for representations of orthogonal groups

The construction of primitive orthogonal idempotents in Brauer algebra $\mathcal{B}r_n$ to large extent repeats the similar construction studied in Section 4.6.4 in the case of algebra $\mathbb{C}[S_n]$.

Colored oscillating Young graph for algebra $\mathcal{B}r_n$ (see example in Fig. 5.2.3) encodes the rules for constructing the complete set of primitive orthogonal idempotents for $\mathcal{B}r_n$. The completeness of this set follows from the fact that the set of commutative Jucys–Murphy elements is complete.

Let $\{\lambda; n\}$ be a vertex in Young graph at level n with diagram (the same diagram is considered in Section 4.6.4):

$\lambda =$

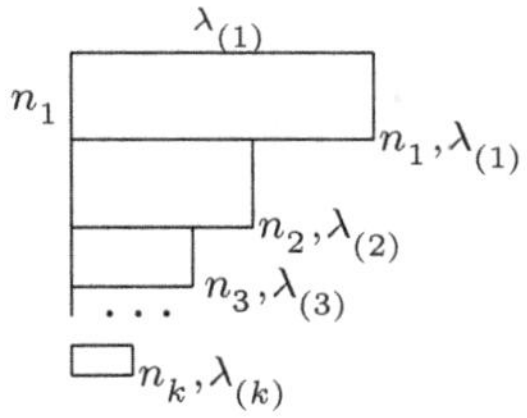

Here $(n_i, \lambda_{(i)})$ are coordinates of boxes in inner corners of diagram $\lambda = [\lambda_{(1)}^{n_1}, \lambda_{(2)}^{n_2-n_1}, \dots, \lambda_{(k)}^{n_k-n_{k-1}}]$; the number of boxes n' in $\lambda \vdash n'$ obeys $n' \le n$. Consider, in oscillating Young graph of algebra $\mathcal{B}r_{n+1}$, a path $T_{\{\lambda;n\}}$ from the upper vertex $\{\emptyset; 0\}$ to vertex $\{\lambda; n\}$, i.e., a path associated with oscillating Young tableau $\Lambda = \{\lambda_1 = \emptyset, \lambda_2, \dots, \lambda_n = \lambda\}$. Let $E_{T_{\{\lambda;n\}}} \in \mathcal{B}r_n$ be primitive idempotent associated with $T_{\{\lambda;n\}}$. We make use of branching rule dictated by oscillating Young graph, and find that for continuing the path $T_{\{\lambda;n\}}$ to level $(n+1)$ we should add, to the oscillating tableau $\Lambda = \{\lambda_1 = \emptyset, \lambda_2, \dots, \lambda_n = \lambda\}$, another diagram λ_{n+1}. The latter diagram is obtained from λ by either adding a box (with content $(\lambda_{(r)} - n_{r-1})$) in one of the external corners of diagram λ, or removing a box (with content $(\lambda_{(r)} - n_r)$) in one of its internal corners. Once the contents of added or removed boxes are known, all eigenvalues (equal to $(\lambda_{(r)} - n_{r-1})$ or $(1 - \omega + n_r - \lambda_{(r)})$) of element y_{n+1} are determined, so we have the identity

$$E_{T_{\{\lambda;n\}}} \cdot \prod_{r=1}^{k+1} (y_{n+1} - (\lambda_{(r)} - n_{r-1})) \prod_{r=1}^{k} (y_{n+1} - (1 - \omega + n_r - \lambda_{(r)})) = 0,$$

where $\lambda_{(k+1)} = n_0 = 0$. Thus, upon appropriate normalization, the primitive idempotent associated with new diagram $\lambda_{n+1} = \lambda'$ obtained by adding to diagram λ box with coordinates $(n_{j-1} + 1, \lambda_{(j)} + 1)$ is

$$E_{T_{\{\lambda';n+1\}}} = E_{T_{\{\lambda;n\}}} \cdot \prod_{\substack{r=1\\ r\neq j}}^{k+1} \frac{(y_{n+1} - (\lambda_{(r)} - n_{r-1}))}{((\lambda_{(j)} - n_{j-1}) - (\lambda_{(r)} - n_{r-1}))}$$

$$\times \prod_{r=1}^{k} \frac{(y_{n+1} - (1 - \omega + n_r - \lambda_{(r)}))}{((\lambda_{(j)} - n_{j-1}) - (1 - \omega + n_r - \lambda_{(r)}))}. \quad (5.2.38)$$

With the new diagram $\lambda_{n+1} = \lambda''$, obtained from λ by removing a box with coordinates $(n_j, \lambda_{(j)})$, one associates the primitive idempotent

$$E_{T_{\{\lambda'';n+1\}}} = E_{T_{\{\lambda;n\}}} \cdot \prod_{r=1}^{k+1} \frac{(y_{n+1} - (\lambda_{(r)} - n_{r-1}))}{((1-\omega+(n_j-\lambda_{(j)})) - (\lambda_{(r)} - n_{r-1}))}$$

$$\times \prod_{\substack{r=1 \\ r\neq j}}^{k} \frac{(y_{n+1} - (1-\omega+n_r-\lambda_{(r)}))}{((n_j-\lambda_{(j)}) - (n_r - \lambda_{(r)}))}. \qquad (5.2.39)$$

Making use of these formulas, and also "initial data" $E_{T_{\{\emptyset;0\}}} = 1$ we can find, step by step, explicit expressions for all primitive orthogonal idempotents in Brauer algebra $\mathcal{B}r_n$. Orthogonality and completeness of this system of primitive idempotents follows from their construction automatically (see Section 4.6.4, Proposition 4.6.18).

Example. Let us derive an explicit expression for symmetrizer $E_{T_{\{[n];n\}}}$ in algebra $\mathcal{B}r_n$. In the left regular representation of algebra $\mathcal{B}r_n$, idempotent $E_{T_{\{[n-1];n-1\}}}$ projects out a subspace in which element y_n has three eigenvalues. Indeed, in accordance with representation branching rule, there is one way of removing the box with content $(n-2)$ from the diagram $[n-1]$ and two ways of adding a box, the latter are shown by asterisks in the following diagram:

$$\begin{array}{|c|c|c|c|c|}\hline 0 & 1 & \cdots & n{-}3 & n{-}2 \\ \hline \end{array} \; * \qquad (5.2.40)$$
$$*$$

Thus, the above procedure of constructing idempotents tells that y_n obeys characteristic equation

$$E_{T_{\{[n-1];n-1\}}} \cdot (y_n - (n-1)) \cdot (y_n + 1) \cdot (y_n + \omega + n - 3) = 0. \qquad (5.2.41)$$

Since we obtain symmetrizer $E_{T_{\{[n];n\}}}$ from the diagram (5.2.40) by adding the box with content $(n-1)$, Eq. (5.2.41) yields recurrence relation for symmetrizer $(n \geq 2)$

$$E_{T_{\{[n];n\}}} = E_{T_{\{[n-1];n-1\}}} \cdot \frac{(y_n+1)\cdot(y_n+\omega+n-3)}{n\cdot(2n-2+\omega)}. \qquad (5.2.42)$$

By solving this recurrence relation, we find an explicit formula for symmetrizer (cf. (5.2.30) for $n = 2$)

$$E_{T_{\{[n];n\}}} = \frac{(y_2+1)\cdots(y_n+1)}{n!} \cdot \frac{(y_2+\omega-1)\cdot(y_3+\omega)\cdots(y_n+\omega+n-3)}{\omega\cdot(2+\omega)\cdots(2n-4+\omega)}. \qquad (5.2.43)$$

Problem 5.2.5. Show that explicit formula for antisymmetrizer $E_{T_{\{[1^n];n\}}} \in \mathcal{B}r_n$ reads ($n \geq 2$):

$$E_{T_{\{[1^n];n\}}} = \frac{(1-y_2)\cdots(1-y_n)}{n!} \cdot \frac{(y_2+\omega-1)\cdots(y_n+\omega-n+1)}{(\omega-2)\cdots(\omega+2-2n)}. \tag{5.2.44}$$

Remark 1. Thus, we have given induction procedure based on formulas (5.2.38) and (5.2.39), which enables one to construct explicitly the complete system of orthogonal idempotents in Brauer algebra $\mathcal{B}r_n$. These idempotents in representation s given by (5.2.2), (5.2.3) define invariant projectors $\mathsf{s}(E_{T_{\{\lambda;n\}}})$ in space of rank-n tensors $\mathcal{V}^{\otimes n}$. Here, as before, $\mathcal{V}$ is the space of defining representation of group $SO(N,\mathbb{C})$, and of its real subgroup $SO(N-q,q)$ (for real representations of $SO(N-q,q)$, the space $\mathcal{V}$ is a real: $\mathcal{V}=\mathbb{R}^{N-q,q}$). Projectors $\mathsf{s}(E_{T_{\{\lambda;n\}}})$ commute with the action of group $SO(N,\mathbb{C})$ (group $SO(N-q,q)$) in space $\mathcal{V}^{\otimes n}$ and single out invariant subspaces $\mathcal{V}^{(\lambda)} \subset \mathcal{V}^{\otimes n}$ of irreducible[g] representations $T^{(\lambda)}$ (recall that the number of boxes in λ may be smaller than n). We stress here that the eigenvalues of elements $y_n \in \mathcal{B}r_n(\omega)$ (for real ω) are real numbers (it follows from Proposition 5.2.4). Thus, in the representation s, projectors $\mathsf{s}(E_{T_{\{\lambda;n\}}})$ are real matrices and they can be used for extracting of real irreducible representations of subgroups $SO(N-q,q)$. The form of the projector $\mathsf{s}(E_{T_{\{\lambda;n\}}})$ depends on the oscillating Young tableau $\Lambda = \{\emptyset \to \lambda_1 \to \cdots \to \lambda_{n-1} \to \lambda\}$ that has the diagram λ at the last level n. Generally, there are several paths $T_{\{\lambda;n\}}$ in oscillating Young graph (several oscillating Young tableaux Λ) which have one and the same Young diagram at the last vertex. However, like in the linear group case, irreducible representations obtained with projectors $\mathsf{s}(E_{T_{\{\lambda;n\}}})$, that are associated with different oscillating Young tableaux (different paths) but have the same final diagram λ, are equivalent. In other words, the irreducible representations themselves are labeled by the Young diagrams rather than oscillating Young tableaux. Indeed, it is the Young diagram that determines the rank of nontrivial part of traceless tensor and its symmetry. Therefore, the projectors $\mathsf{s}(E_{T_{\{\lambda;n\}}})$ on these tensors (with different $T_{\{\lambda;n\}}$) should give equivalent representations. More instructive explanation of this fact is based on intertwining operators

[g] An important exception is that for $N=2r$ and diagrams λ with r rows, the pertinent representations are reducible and split into a sum of selfdual and anti-selfdual representations, see Remarks 1–3 in Section 5.1.3.

U_k given by (5.2.21); see also formula (4.6.89) and Proposition 4.4.1 valid in the case of group $SL(N, \mathbb{C})$.

Remark 2. In the case of $SO(N, \mathbb{C})$, there is yet another class of equivalence relationships, over and beyond the equivalence discussed in Remark 1. The equivalence we point out here is between representations associated with different Young diagrams λ and λ', which are related by rearrangements shown in Fig. 5.1.1. Like in the above case, primitive idempotents $\mathsf{s}(E_{T_{\{\lambda;n\}}})$ and $\mathsf{s}(E_{T_{\{\lambda';n\}}})$ in Brauer algebra, constructed for different paths $T_{\{\lambda;n\}}$ and $T_{\{\lambda';n\}}$ in oscillating Young graph for $\mathcal{B}r_n$ (in accordance with different oscillating Young tableaux), as well as their images in representation s are different. Nevertheless, irreducible representations of group $SO(N, \mathbb{C})$, embedded in tensor product $T^{\otimes n}$ and extracted with projectors $\mathsf{s}(E_{T_{\{\lambda;n\}}})$ and $\mathsf{s}(E_{T_{\{\lambda';n\}}})$, are equivalent.

Remark 3. Branching graph of irreducible representations (embedded in tensor product of defining representations $T^{\otimes n}$) of group $SO(N, \mathbb{C})$, shown for $n = 4$ in Fig. 5.1.3, and branching graph of irreducible representations of Brauer algebra $\mathcal{B}r_n(\omega)$ (similar to that shown in Fig. 5.2.3) *coincide only for* $n \leq [N/2]$. In this case, there is a one-to-one correspondence between irreducible representations (embedded in $T^{\otimes n}$) of group $SO(N, \mathbb{C})$, and representations of Brauer algebra $\mathcal{B}r_n(\omega = N)$ (in this case, one speaks about Schur–Weyl–Brauer duality; see Section 4.5.3 on the Schur–Weyl duality). In the case $n > [N/2]$, Proposition 5.1.3 implies that branching graph of representations of group $SO(N, \mathbb{C})$, unlike branching graph of representations of Brauer algebra, does not contain Young diagrams with number of rows exceeding $[N/2]$. The branching graphs for $SO(N, \mathbb{C})$ (see Fig. 5.1.3) and $\mathcal{B}r_n(\omega)$ (see Fig. 5.2.3) differ also by the interpretation of the number of paths $T_{\{\lambda;n\}}$ leading to one and the same Young diagram λ at level n. In the former, $SO(N, \mathbb{C})$ case this number is the multiplicity of the irreducible[h] representation $T^{(\lambda)}$ in tensor product $T^{\otimes n}$ (see Section 5.1.4), while in the latter, $\mathcal{B}r_n$ case it is not only multiplicity of irreducible representation $\rho^{\{\lambda,n\}}$ in regular representation of $\mathcal{B}r_n(\omega)$, but also the dimension of representation $\rho^{\{\lambda,n\}}$. We recall in this regard that representation $\rho^{\{\lambda,n\}}$ is labeled by both Young diagram λ and its level n in Young graph (see Section 5.2.4). Representations $\rho^{\{\lambda,n\}}$ and $\rho^{\{\lambda,n'\}}$ with the same Young diagram λ which, however, sit at different levels n and n' of

[h]Modulo (anti-)selfduality subtlety for even $N = 2r$.

Young graph, are representations of different Brauer algebras $\mathcal{B}r_n(\omega)$ and $\mathcal{B}r_{n'}(\omega)$.

Like in the case of the Schur–Weyl duality (which has to do with representations of the group $SL(N,\mathbb{C})$; see Remark 2 in Section 4.5.3), one-to-one correspondence is established between irreducible representations of group $SO(N,\mathbb{C})$, embedded in $T^{\otimes r}$, and special representations of a factor algebra, which is obtained from $\mathcal{B}r_n(\omega)$ by imposing on its generators an additional constraint $A_{N+1} = 0$, where the element $A_{N+1} \equiv E_{T_{\{[1^{N+1}];N+1\}}}$ is an antisymmetrizer of rank $(N+1)$ in $\mathcal{B}r_n(\omega)$ (definition of $E_{T_{\{[1^{N+1}];N+1\}}}$ is given in (5.2.44)). Note that in the algebra $\mathcal{B}r_n(\omega)$, the condition $A_{N+1} = 0$ $(N < n)$ is consistent only for a special parameter value $\omega = N$. This is exactly the case when Brauer algebra can be taken in the representations (5.2.2), (5.2.3).

Problem 5.2.6. Show that antisymmetrizer $A_n \equiv E_{T_{\{[1^n];n\}}}$ for algebra $\mathcal{B}r_n(\omega)$, given in (5.2.44), satisfies the recurrence relation

$$\begin{aligned} A_n &= \frac{1}{n}(1-\sigma_1)\left(\frac{1}{2}-\sigma_2\right)\cdots\left(\frac{1}{n-1}-\sigma_{n-1}\right)A_{n-1} \\ &= \frac{1}{n}(1-\sigma_{n-1}+\sigma_{n-2}\sigma_{n-1}-\cdots-(-1)^n\sigma_1\cdots\sigma_{n-1})A_{n-1}, \end{aligned} \tag{5.2.45}$$

and thus coincides with the antisymmetrizer in the algebra $\mathbb{C}[S_n]$ (see Problem 101). Prove that the conditions $A_{N+1} = 0$ and $A_N \neq 0$ imply $\omega = N$. Hint: use the identity $\kappa_{N+1} \cdot A_{N+1} \cdot \kappa_{N+1} = \frac{(\omega-N)}{N+1}\kappa_{N+1} \cdot A_N$.

5.3. Tensor Representation of Groups $Sp(2r,\mathbb{C})$ and Their Subgroups $Sp(2r,\mathbb{R})$, $USp(2r)$, $Sp(p,r-p)$

Let us recall the definition of group $Sp(2r,\mathbb{C})$ and its subgroups $Sp(2r,\mathbb{R})$, $USp(2r)$, $Sp(p,q)$, and also the definitions of the pertinent Lie algebras (see Sections I-1.2.5 and I-3.1.3 in the first book and item 3 in Section 3.1.2). Symplectic group $Sp(2r,\mathbb{C})$ is the set of complex $(2r\times 2r)$ matrices T obeying

$$T^{\mathsf{T}} \cdot J \cdot T = J, \tag{5.3.1}$$

where J is a nondegenerate antisymmetric matrix

$$J := \begin{pmatrix} 0 & I_r \\ -I_r & 0 \end{pmatrix}, \quad J = -J^{\mathsf{T}}, \tag{5.3.2}$$

which is called symplectic metric. In other words, $Sp(2r, \mathbb{C})$ is a group of linear operators in $\mathbb{C}^{2r}$ which preserve antisymmetric bilinear form $f(\vec{x}, \vec{y}) = x_i \, J_{ij} \, y_i$, where x_i are coordinates in $\mathbb{C}^{2r}$. Note that any other choice of symplectic metric J in (5.3.1) would lead to a group isomorphic to $Sp(2r, \mathbb{C})$.

Problem 5.3.1. Prove that for all $T \in Sp(2r, \mathbb{C})$ the condition $\det(T) = +1$ is fulfilled. Hint: calculate the Pfaffian of both sides of relation (5.3.1).

Choosing a curve in the space of complex symplectic matrices near unit matrix $T(t) = I_{2r} + t\, A + O(t^2)$ (where t is a small parameter), we get

$$A^{\mathsf{T}} \cdot J + J \cdot A = 0. \tag{5.3.3}$$

In accordance with the viewpoint adopted in this and the first book [1], all Lie groups, including those consisting of complex matrices, are real manifolds, whose tangent spaces (Lie algebras) are real vector spaces. Thus, Lie algebra $\mathcal{A}(Sp(2r, \mathbb{C}))$ is real vector space whose elements are complex matrices A obeying (5.3.3). It is then natural to define complex Lie algebra $sp(2r, \mathbb{C})$, which is a complex vector space of matrices which have the property (5.3.3). We emphasize that algebras $\mathcal{A}(Sp(2r, \mathbb{C}))$ are realifications of $sp(2r, \mathbb{C})$ and these two algebras are different objects.

Lie group $Sp(2r, \mathbb{C})$ has several important subgroups whose Lie algebras are real forms of complex Lie algebra $sp(2r, \mathbb{C})$. The first subgroup $Sp(2r, \mathbb{R})$ is a set of *real* $2r \times 2r$ matrices T obeying (5.3.1). The second subgroup $USp(2r)$ is a set of complex $2r \times 2r$ matrices T that are both symplectic (obey (5.3.1)) and unitary (obey $T^\dagger \cdot T = I_n$), so that $USp(2r) = Sp(2r, \mathbb{C}) \cap U(2r)$. Other subgroups in $Sp(2r, \mathbb{C})$ are denoted by $Sp(p, q)$ (here $p + q = r$). These subgroups are sets of complex $2r \times 2r$ matrices T, which obey both symplectic condition (5.3.1) and pseudo-unitarity condition

$$T^\dagger \cdot \tilde{I}_{2p,2q} \cdot T = \tilde{I}_{2p,2q}, \quad \tilde{I}_{2p,2q} := \begin{pmatrix} I_{p,q} & 0 \\ 0 & I_{p,q} \end{pmatrix}, \tag{5.3.4}$$

where matrix $I_{p,q}$ is defined in (5.1.1). Thus, $Sp(p, q) = Sp(2r, \mathbb{C}) \cap U(2p, 2q)$. Clearly, $Sp(r, 0) = Sp(0, r) = USp(2r)$, so group $USp(2r)$ is sometimes denoted by $Sp(r)$.

Vector space of real matrices A obeying (5.3.3) is Lie algebra $sp(2r, \mathbb{R})$. This is Lie algebra of Lie group $Sp(2r, \mathbb{R})$, i.e., $sp(2r, \mathbb{R}) = \mathcal{A}(Sp(2r, \mathbb{R}))$. Clearly, $sp(2r, \mathbb{R})$ is a real form of algebra $sp(2r, \mathbb{C})$. By considering curves in group manifold of $Sp(p,q)$, $r = (p+q)$, near unit element, we find, from (5.3.1) and (5.3.4), that Lie algebra $sp(p,q) = \mathcal{A}(Sp(p,q))$ is a real vector space of $(2r \times 2r)$ complex matrices A which obey both (5.3.3) and

$$A^{\dagger} \cdot \tilde{I}_{2p,2q} + \tilde{I}_{2p,2q} \cdot A = 0. \tag{5.3.5}$$

Likewise, Lie algebra $usp(2r) = \mathcal{A}(USp(2r))$ is real vector space of complex matrices A obeying both (5.3.3) and (5.3.5), where we take unit matrix I_{2r} instead of $\tilde{I}_{2p,2q}$ in the latter condition. Thus, $usp(2r) = sp(r,0) = sp(0,r)$. All algebras $sp(p,q)$, including compact algebra $usp(2r)$, are real forms of algebra $sp(2r,\mathbb{C})$.

Let $T_{(1)}$ be defining complex representation of Lie algebra $sp(2r,\mathbb{C})$. In Section 3.4.2, item 5, we have shown that *all* fundamental representations $T_{(f)}$ (here $f = 1,\ldots,r$) of algebra $sp(2r,\mathbb{C})$ are equivalent to wedge products $T_{(1)}^{\wedge f}$ of defining representations. Taking into account of this fact, we have shown that *all* complex finite-dimensional irreducible representations, which are highest weight representations with highest weights $\lambda = (\lambda_1,\ldots,\lambda_r)$ (where $\lambda_1 \geq \lambda_2 \geq \cdots \geq \lambda_r \geq 0$) (i) are irreducible subrepresentations in tensor products of fundamental representations $T_{(f)}$ and (ii) are associated with Young diagram $[\lambda_1,\ldots,\lambda_r]$. This result implies that all finite-dimensional complex irreducible representations of algebra $sp(2r,\mathbb{C})$ can be embedded in tensor products of defining representations, and *there are no $sp(2r,\mathbb{C})$-analogs of spinor representations (which exist in the case of algebras $so(n,\mathbb{C})$)*.

Now, let T be defining complex representation of Lie group $Sp(2r,\mathbb{C})$, which is at the same time complex defining representation of all subgroups $Sp(2r,\mathbb{R})$, $USp(2r)$, $Sp(p,q)$ in $Sp(2r,\mathbb{C})$. Representation T of group $Sp(2r,\mathbb{C})$ is naturally related to defining representation $T_{(1)}$ of Lie algebra $sp(2r,\mathbb{C})$. Indeed, near unit element, any matrix $U \in Sp(2r,\mathbb{C})$ has the form $U = \exp(tA)$, where $A \in sp(2r,\mathbb{C})$ and t is a small *complex* parameter.[i] The relationship between defining representation T of group $Sp(2r,\mathbb{C})$ and defining representation $T_{(1)}$ of algebra $sp(2r,\mathbb{C})$ is given by (cf. Remark 3 in Section 2.1)

$$T(\exp(tA)) = I_{2r} + tT_{(1)}(A) + O(t^2). \tag{5.3.6}$$

In what follows, we use the same notation T for defining representation of both group $Sp(2r,\mathbb{C})$ and algebra $sp(2r,\mathbb{C})$. The relation (5.3.6), as usual, is

---

[i] Note that this is *not* the usual relationship between a Lie group G and its Lie algebra $\mathcal{A}(G)$: in that case, the parameter t is real.

straightforwardly generalized to tensor products of defining representations of group $Sp(2r,\mathbb{C})$ and algebra $sp(2r,\mathbb{C})$. Without mentioning explicitly, we consider in what follows only those representations of group $Sp(2r,\mathbb{C})$ which are subrepresentations in tensor products of its defining representations. They correspond, in one-to-one way, to *all* representations of algebra $sp(2r,\mathbb{C})$ (since all representations of this algebra are subrepresentations in tensor products of its defining representation), as well as to *all* complex finite-dimensional representations of subgroups $Sp(2r,\mathbb{R})$, $USp(2r)$ and $Sp(p,q)$. Thus, we will study irreducible representations of group $Sp(2r,\mathbb{C})$, which are contained in one of tensor products of its defining representations, and obtain in this way all irreducible complex representations of algebra $sp(2r,\mathbb{C})$ and groups $Sp(2r,\mathbb{R})$, $USp(2r)$, $Sp(p,q)$ (but not all irreducible representations of group $Sp(2r,\mathbb{C})$ itself[j]). Furthermore, once we find a method of expanding a tensor product of defining representations of group $Sp(2r,\mathbb{C})$ in irreducible representations, we automatically find the method of doing the same for algebra $sp(2r,\mathbb{C})$ and groups $Sp(2r,\mathbb{R})$, $USp(2r)$, $Sp(p,q)$. We note that here we follow the same route as in the case of algebras $s\ell(N,\mathbb{C})$ and groups $SU(N)$ in Chapter 4 and in case of algebras $so(N,\mathbb{C})$ and groups $SO(p,q)$ in Sections 5.1 and 5.2.5, where we consider only representations of groups $SL(N,\mathbb{C})$ and $SO(N,\mathbb{C})$, which are contained in tensor products of their defining representations.

Thus, our purpose is to extract irreducible representations $T^{(\lambda)}$ in tensor product $T^{\otimes n}$ of defining representations T. Representation $T^{\otimes n}$ of group $Sp(2r,\mathbb{C})$ acts in the space $\mathcal{V}^{\otimes n}$ of rank-n tensors. Note that defining representation T and co-defining representation $\widetilde{T}$ are equivalent for symplectic groups (similar to orthogonal groups; see Section 5.1.2). Vectors from the spaces $\mathcal{V}$ and $\widetilde{\mathcal{V}}$ of representations T and $\widetilde{T}$ are transformed to each other by symplectic metric. A general tensor of mixed type

$$\mathsf{t} = \mathsf{t}^{k_1\ldots k_s}{}_{k_{s+1}\ldots k_{s+t}}\vec{e}_{k_1}\otimes\cdots\otimes\vec{e}_{k_s}\otimes\vec{e}^{\,k_{s+1}}\otimes\cdots\otimes\vec{e}^{\,k_{s+t}}\in\mathcal{V}^{\otimes s}\otimes\widetilde{\mathcal{V}}^{\otimes t}, \tag{5.3.7}$$

where $\vec{e}_k$ and $\vec{e}^{\,k_{s+t}}$ are basis vectors in spaces $\mathcal{V}$ and $\widetilde{\mathcal{V}}$, is transformed into tensor $\mathsf{t}'\in\mathcal{V}^{\otimes(s+t)}$ by raising indices in its components:

$$\mathsf{t}'^{\,k_1\ldots k_s k_{s+1}\ldots k_{s+t}} = J^{k_{s+1}\ell_{s+1}}\cdots J^{k_{s+t}\ell_{s+t}}\mathsf{t}^{k_1\ldots k_s}{}_{\ell_{s+1}\ldots\ell_{s+t}}. \tag{5.3.8}$$

[j]In particular, we do not consider irreducible representations which are embedded in the tensor product $(T^*)^{\otimes r}$, where T^* is complex conjugate to defining representation.

Therefore, from the representation theory viewpoint, it is unnecessary to study mixed representations $T^{\otimes s} \otimes \widetilde{T}^{\otimes t}$; it suffices to consider only representations $T^{\otimes(s+t)}$, and extract irreducible representations from them. This is fully consistent with the above discussion of the relationship between representations of group $Sp(2r, \mathbb{C})$ and algebra $sp(2r, \mathbb{C})$.

Invariant subspaces $\mathcal{V}^{(\lambda)} \subset \mathcal{V}^{\otimes n}$ of irreducible representations $T^{(\lambda)}$ are constructed for groups $Sp(2r, \mathbb{C})$ in a very similar way to the case of groups $SO(N, \mathbb{C})$ and $SO(p, q)$. Spaces $\mathcal{V}^{(\lambda)}$ contain tensors that have the symmetry dictated by Young diagram λ and are traceless with respect to antisymmetric symplectic metric J (cf. the condition (5.1.32)):

$$\mathsf{t}^{k_1 k_2 \ldots k_n} J_{k_i k_j} = 0, \quad \forall i, j \in \{1, 2, \ldots, n\}. \tag{5.3.9}$$

Thus, the method of constructing invariant subspaces for groups $SO(N, \mathbb{C})$ and $SO(p, q)$ should be modified by merely replacing (pseudo-)Euclidean metric η by symplectic metric J.

Example. Let us expand representation $T^{\otimes 2}$ in irreducible representations. To this end, we consider the space $\mathcal{V}^{\otimes 2}$, i.e., the space of second rank tensors t with components t^{ik}. In the first place, like in the cases of groups $SL(N, \mathbb{C})$ and $SO(N, \mathbb{C})$, we represent a tensor t as a sum of symmetric and antisymmetric parts, $\mathsf{t} = \mathsf{t}_s + \mathsf{t}_a$, see (5.1.31). Now, symmetric tensor t_s with components $\mathsf{t}^{(ik)}$ is traceless with respect to antisymmetric metric J, so symmetric tensors form the space of irreducible representation of group $Sp(2r, \mathbb{C})$. The space of antisymmetric tensors t_a splits into two invariant subspaces $\mathcal{V}^{([1^2])}$ and $\mathcal{V}^{(\emptyset)}$, which are composed of traceless antisymmetric tensors with components

$$\mathsf{t}_a^{\prime ik} = \mathsf{t}^{[ik]} + \frac{1}{2r} J^{ik} (J_{rs} \mathsf{t}^{[rs]}), \quad \mathsf{t}'_a \in \mathcal{V}^{([1^2])}, \tag{5.3.10}$$

and tensor that forms trivial representation (cf. (5.1.33)):

$$\mathsf{t}_0^{ik} = J^{ik} (J_{rs}\, \mathsf{t}^{[rs]}), \quad \mathsf{t}_0 \in \mathcal{V}^{(\emptyset)}, \tag{5.3.11}$$

where J^{ik} are components of matrix $\bar{J} = ||J^{ik}||$ inverse to the matrix of metric $J = ||J_{ik}||$, i.e., $J^{ik} J_{kj} = \delta^i_j$ and $J_{ik} J^{ik} = (-2\,r)$.

It follows from definition of tensor t_s and formulas (5.3.10) and (5.3.11) that irreducible tensors with components $\mathsf{t}_a^{\prime i_1 i_2}$, $\mathsf{t}_s^{i_1 i_2}$ and $\mathsf{t}_0^{i_1 i_2}$ are extracted from a general tensor with components $\mathsf{t}^{k_1 k_2}$ by normalized projectors (cf. (5.1.34))

$$(\Pi_s)^{i_1 i_2}_{k_1 k_2} = \frac{1}{2}\left(I^{i_1 i_2}_{k_1 k_2} + P^{i_1 i_2}_{k_1 k_2}\right),$$
$$(\Pi_a)^{i_1 i_2}_{k_1 k_2} = \frac{1}{2}\left(I^{i_1 i_2}_{k_1 k_2} - P^{i_1 i_2}_{k_1 k_2}\right) + \frac{1}{2r}\tilde{K}^{i_1 i_2}_{k_1 k_2}, \quad (\Pi_0)^{i_1 i_2}_{k_1 k_2} = -\frac{1}{2r}\tilde{K}^{i_1 i_2}_{k_1 k_2}, \tag{5.3.12}$$

where

$$P^{i_1 i_2}_{k_1 k_2} = \delta^{i_1}_{k_2}\delta^{i_2}_{k_1}, \quad I^{i_1 i_2}_{k_1 k_2} = \delta^{i_1}_{k_1}\delta^{i_2}_{k_2}, \quad \tilde{K}^{i_1 i_2}_{k_1 k_2} = J^{i_1 i_2} J_{k_1 k_2}, \tag{5.3.13}$$

$P^{i_1 i_2}_{k_1 k_2}$ is the matrix of transposition operator, $I^{i_1 i_2}_{k_1 k_2}$ is the unit matrix and $\tilde{K}^{i_1 i_2}_{k_1 k_2}$ is the matrix of rank 1.

Problem 5.3.2. Show that operators Π_α given by (5.3.12) form complete set of orthogonal projectors, i.e.,

$$\Pi_a + \Pi_s + \Pi_0 = I, \quad \Pi_\alpha \cdot \Pi_\beta = \Pi_\alpha \delta_{\alpha\beta}. \tag{5.3.14}$$

To find a solution to the general problem of decomposition of tensor product $T^{\otimes n}$ of defining representations of group $Sp(2r,\mathbb{C})$ (or defining representations of groups $Sp(2r,\mathbb{R})$, $USp(2r)$ and $Sp(p, r-p)$) into irreducible representations $T^{(\lambda)}$, we have to construct a complete system of orthogonal primitive projectors $P^{(\lambda)}$ (here index λ labels these projectors), which single out invariant subspaces $\mathcal{V}^{(\lambda)}$ in space $\mathcal{V}^{\otimes n}$ (i.e., in space of rank-n tensors). Notably, this problem is solved by the same method that is used for orthogonal groups. Indeed, by making use of matrices (5.3.13) we construct operators acting in $\mathcal{V}^{\otimes n}$, (cf. (5.2.2), (5.2.3)):

$$\mathsf{s}(\sigma_k)\cdot(\vec{e}_{i_1}\otimes\cdots\otimes\vec{e}_{i_k}\otimes\vec{e}_{i_{k+1}}\otimes\cdots\otimes\vec{e}_{i_n})$$
$$= -\left(\vec{e}_{j_1}\otimes\cdots\otimes\vec{e}_{j_k}\otimes\vec{e}_{j_{k+1}}\otimes\cdots\otimes\vec{e}_{j_n}\right)\delta^{j_1}_{i_1}\cdots\delta^{j_{k-1}}_{i_{k-1}}P^{j_k j_{k+1}}_{i_k i_{k+1}}\delta^{j_{k+2}}_{i_{k+2}}\cdots\delta^{j_n}_{i_n}, \tag{5.3.15}$$

$$\mathsf{s}(\kappa_k)\cdot(\vec{e}_{i_1}\otimes\cdots\otimes\vec{e}_{i_k}\otimes\vec{e}_{i_{k+1}}\otimes\cdots\otimes\vec{e}_{i_n})$$
$$= \left(\vec{e}_{j_1}\otimes\cdots\otimes\vec{e}_{j_k}\otimes\vec{e}_{j_{k+1}}\otimes\cdots\otimes\vec{e}_{j_n}\right)\delta^{j_1}_{i_1}\cdots\delta^{j_{k-1}}_{i_{k-1}}\tilde{K}^{j_k j_{k+1}}_{i_k i_{k+1}}\delta^{j_{k+2}}_{i_{k+2}}\cdots\delta^{j_n}_{i_n}, \tag{5.3.16}$$

where $k = 1, 2, \ldots, n-1$ (take note of minus sign on the right-hand side of (5.3.15)). In abstract notations, these operators are

$$\begin{aligned} \mathsf{s}(\sigma_k) &= -I^{\otimes(k-1)} \otimes P \otimes I^{\otimes(n-k-1)} \equiv -P_k, \\ \mathsf{s}(\kappa_k) &= I^{\otimes(k-1)} \otimes \tilde{K} \otimes I^{\otimes(n-k-1)} \equiv \tilde{K}_k. \end{aligned} \tag{5.3.17}$$

These operators define representation of Brauer algebra $\mathcal{B}r_n(\omega)$ with generators $\{\sigma_k, \kappa_k\}$ $(k = 1, \ldots, n-1)$ and defining relations (5.2.5) with parameter $\omega = -2r$. This representation acts in space $\mathcal{V}^{\otimes n}$.

Problem 5.3.3. Show that operators (5.3.15) and (5.3.16) define representation of Brauer algebra $\mathcal{B}r_r(\omega)$ with $\omega = -2r$. Hint: make use of the relations $P\tilde{K} = \tilde{K}P = -\tilde{K}$, $(\tilde{K})^2 = -2r\tilde{K}$, $\tilde{K}_k\tilde{K}_{k+1}\tilde{K}_k = \tilde{K}_k$, $P_k\,\tilde{K}_{k+1}\,\tilde{K}_k = P_{k+1}\,\tilde{K}_k$, etc.

The representation theory for Brauer algebra $\mathcal{B}r_r(\omega)$ has been studied in Section 5.2, where we have derived induction procedure (see formulas (5.2.38), (5.2.39) in Section 5.2.5) for constructing complete set of primitive orthogonal idempotents $E_{T_{\{\lambda;n\}}} \in \mathcal{B}r_r(\omega)$. These idempotents, in representation s defined by (5.3.15), (5.3.16), determine projectors $\mathsf{s}(E_{T_{\{\lambda;n\}}}) = P^{(\lambda)}$ acting in space $\mathcal{V}^{\otimes n}$ of rank-n tensors, which is the space of reducible representation $T^{\otimes n}$ of group $Sp(2r, \mathbb{C})$. Projectors $\mathsf{s}(E_{T_{\{\lambda;n\}}})$ single out invariant subspaces $\mathcal{V}^{(\lambda)}$ of symplectic groups in $\mathcal{V}^{\otimes n}$. They are labeled by Young diagram λ at the lowest level n in oscillating Young tableau $T_{\{\lambda;n\}}$. Thus, all said in Remarks 1–3 in the end of Section 5.2.5 applies to symplectic groups without any essential change.

Remark 1. In accordance with (5.2.38) (5.2.39) and in view of the fact that the eigenvalues of Jucys–Murphy elements $y_i \in \mathcal{B}r_n(\omega)$, for real ω, are real (see Proposition 5.2.4), operators $E_{T_{\{\lambda;n\}}} \in \mathcal{B}r_n(\omega)$ are elements of Brauer algebra over the field of real numbers and matrix projectors $\mathsf{s}(E_{T_{\{\lambda;n\}}})$ constructed by making use of these operators, are also real. So, they can be used for extracting real irreducible representations of group $Sp(2r, \mathbb{R}) \subset Sp(2r, \mathbb{C})$ out of tensor product $T^{\otimes n}$ of real defining representations T of group $Sp(2r, \mathbb{R})$.

Remark 2. There is equivalence between totally antisymmetric representations of groups $Sp(2r, \mathbb{K})$ and $Sp(r-q, q)$, which are associated with

Young diagrams $[1^k]$ and $[1^{N-k}]$ (cf. (5.1.36)):

$$\mathrm{t}_1^{[j_1\dots j_k]} \leftrightarrow \mathrm{t}_2^{[i_1\dots i_{N-k}]}\varepsilon_{i_1\dots i_{N-k}i_{N-k+1}\dots i_N}J^{i_{N-k+1}j_1}\dots J^{i_N j_k}, \quad N=2r. \tag{5.3.18}$$

Like in the case of orthogonal groups $SO(2r,\mathbb{K})$ and $SO(p,2r-p)$ (see Proposition 5.1.1 in Section 5.1.3), this leads to equivalence of representations associated with Young diagram λ' with number of rows $m>r$, and representation determined by Young diagram λ complementary to λ' in the sense of Fig. 5.1.1. Thus, irreducible inequivalent representations of groups $Sp(2r,\mathbb{K})$ and $Sp(r-q,q)$ are associated with Young diagrams with number of rows $m\le r$.

Remark 3. In contrast to the case of orthogonal groups $SO(2r,\mathbb{K})$ and $SO(p,2r-p)$ (see Remarks 2 and 3 in Section 5.1.3), representation $T^{\wedge r}$ of symplectic groups $Sp(2r,\mathbb{C})$ and $Sp(p,r-p)$, which acts in space $\mathcal{V}^{\wedge r}$ of antisymmetric and traceless (in the sense of Eq. (5.3.9)) rank-r tensors, is irreducible, i.e., there is no way of decomposing it into something like a sum of selfdual and anti-selfdual representations. Indeed, let $t^{k_1\dots k_r}$ be components of totally antisymmetric and traceless tensor, $t^{k_1k_2\dots k_r}J_{k_1k_2}=0$. Define dual tensor $\tilde{t}$ with components

$$\tilde{t}^{k_1\dots k_r}=(-1)^{[\frac{r}{2}]}\frac{\mathrm{Pf}(J)}{r!}J^{k_1\ell_1}\dots J^{k_r\ell_r}\varepsilon_{\ell_1\dots\ell_r i_1\dots i_r}t^{i_1\dots i_r}, \tag{5.3.19}$$

where $\mathrm{Pf}(J)$ is Pfaffian of matrix $||J_{ij}||$ and $[\frac{r}{2}]$ is integer part of $\frac{r}{2}$. It turns out that $\tilde{t}^{k_1\dots k_r}=t^{k_1\dots k_r}$, so it is impossible, in the case of groups $Sp(2r,\mathbb{C})$ and $Sp(p,r-p)$, to make up from tensors t and $\tilde{t}$ *two* independent linear combinations (selfdual and anti-selfdual tensors, cf. (5.1.40)).

Problem 5.3.4. ★ Prove that components $t^{k_1k_2\dots k_r}$ of antisymmetric and traceless tensor t coincide with components (5.3.19) of dual tensor $\tilde{t}$.

Thus, in view of this result, all representations of groups $Sp(2r,\mathbb{C})$ and $Sp(p,r-p)$ with number of rows $m=r$ are irreducible and do not split into direct sums of selfdual and anti-selfdual representations (cf. the second part of Proposition 5.1.2 given in the context of groups $SO(2r,\mathbb{C})$ and $SO(2r-q,q)$).

5.4. Spinor Representations of Lie Algebras $so(N, \mathbb{C})$

As before, $\mathfrak{V}_r$ denote r-dimensional root spaces of rank-r algebras $so(2r, \mathbb{C})$ and $so(2r+1, \mathbb{C})$. We have shown in Section 3.4.2, items 3 and 4, that there are two special irreducible representations of algebra $so(2r, \mathbb{C})$ with fundamental weights

$$\lambda_{(r-1)} = \left(\frac{1}{2}, \dots, \frac{1}{2}, -\frac{1}{2}\right) \in \mathfrak{V}_r, \quad \lambda_{(r)} = \left(\frac{1}{2}, \dots, \frac{1}{2}, \frac{1}{2}\right) \in \mathfrak{V}_r, \tag{5.4.1}$$

and one special irreducible representation of algebra $so(2r+1, \mathbb{C})$ with fundamental weight

$$\lambda_{(r)} = \left(\frac{1}{2}, \dots, \frac{1}{2}, \frac{1}{2}\right) \in \mathfrak{V}_r. \tag{5.4.2}$$

These representations cannot be embedded in tensor products $T^{\otimes n}$ of defining representations T of algebras $so(2r, \mathbb{C})$ and $so(2r+1, \mathbb{C})$. So, these representations[k] cannot descend from representations of groups $SO(N, \mathbb{C})$ and its real subgroups $SO(p, q)$. They are called *spinor representations.* We explicitly construct in this section irreducible representations of Lie algebras $so(2r, \mathbb{C})$ and $so(2r+1, \mathbb{C})$ with fundamental weights (5.4.1) and (5.4.2). To this end, we introduce the realization of Lie algebras $so(N, \mathbb{C})$ in terms of complex Clifford algebras $\mathcal{C}\ell_N$. Clifford algebras $\mathcal{C}\ell_N$ and their representations are studied in Section 1.3.

5.4.1. Spinor representations of Lie algebras $so(2r, \mathbb{C})$

Consider Clifford algebra $\mathcal{C}\ell_N$ with even number $N = 2r$ of generators:

$$(\Gamma_1, \dots, \Gamma_{2r}) = \sqrt{2}\,(z_1, \dots, z_r, \bar{z}_r, \dots, \bar{z}_1), \tag{5.4.3}$$

where free fermions $\{z_k, \bar{z}_i\}$ obey relations (1.3.19), i.e.,

$$[z_i, z_k]_+ = 0, \ [\bar{z}_i, \bar{z}_k]_+ = 0, \ [z_i, \bar{z}_k]_+ = \delta_{ik} I \quad (\forall i, k = 1, \dots, r). \tag{5.4.4}$$

In accordance with (5.4.3) and (5.4.4), defining relations for generators $\Gamma_m \in \mathcal{C}\ell_{2r}$ have the form (1.3.20), i.e.,

$$\Gamma_m \Gamma_n + \Gamma_n \Gamma_m = 2\, \eta_{mn}\, I, \quad \eta_{nm} = \delta_{n, N-m+1}, \tag{5.4.5}$$

where $m, n = 1, \dots, 2r$, and $||\eta_{mn}||$ is antidiagonal matrix.

---

[k] As we discuss in Chapter 6, these are representations of groups $\mathsf{Spin}(N, \mathbb{C})$ and $\mathsf{Spin}(p, q)$ which cover groups $SO(N, \mathbb{C})$ and $SO(p, q)$ $(p + q = N)$. In physics literature, these representations are sometimes called double-valued representations of orthogonal groups.

Standard set of basis elements of algebra $\mathcal{C}\ell_{2r}$ is given in (1.3.5). We single out second-order basis elements (cf. (1.3.34)):

$$L_{mn} \equiv \frac{1}{4}(\Gamma_m \Gamma_n - \Gamma_n \Gamma_m) = \frac{1}{2}\Gamma_{[mn]} \in \mathcal{C}\ell_{2r}. \tag{5.4.6}$$

Elements L_{mn} obey commutation relations for generators of Lie algebra $so(2r, \mathbb{C})$ (cf. (5.1.12)):

$$[L_{nm}, L_{k\ell}] = \eta_{mk}L_{n\ell} + \eta_{n\ell}L_{mk} - \eta_{nk}L_{m\ell} - \eta_{m\ell}L_{nk}, \tag{5.4.7}$$

with antidiagonal metric $\eta = \eta^{-1}$ defined in (5.4.5). Relations (5.4.7) are obtained by making use of identities

$$L_{n\ell} = \frac{1}{2}(\Gamma_n\Gamma_\ell - \eta_{n\ell}I), \quad [L_{nm}, \Gamma_k] = \eta_{mk}\Gamma_n - \eta_{nk}\Gamma_m, \tag{5.4.8}$$

which in turn follow from (5.4.5). We recall (see item 2 of Section 3.1.1) that the choice of antidiagonal metric η in (5.4.7) is convenient for constructing root systems of Lie algebras of *so*-type.

Let us choose Cartan–Weyl basis in Lie algebra $so(2r, \mathbb{C})$ with elements (5.4.6) in the following way. The generators of Cartan subalgebra are ($k = 1, \dots, r$)

$$H_k = \frac{1}{4}(\Gamma_k\Gamma_{k'} - \Gamma_{k'}\Gamma_k) = \frac{1}{2}I - \bar{z}_k\, z_k, \quad [H_m, H_k] = 0. \tag{5.4.9}$$

where $k' = 2r - k + 1$. Positive roots correspond to root generators ($i < j$)

$$E_{ij} = \frac{1}{4}(\Gamma_i\Gamma_j - \Gamma_j\Gamma_i) = z_i\, z_j, \quad E_{ij'} = \frac{1}{4}(\Gamma_i\Gamma_{j'} - \Gamma_{j'}\Gamma_i) = -\bar{z}_j\, z_i, \tag{5.4.10}$$

while root generators with negative roots are ($i < j$)

$$E_{i'j} = \frac{1}{4}(\Gamma_{i'}\Gamma_j - \Gamma_j\Gamma_{i'}) = \bar{z}_i\, z_j, \quad E_{i'j'} = \frac{1}{4}(\Gamma_{i'}\Gamma_{j'} - \Gamma_{j'}\Gamma_{i'}) = -\bar{z}_j\, \bar{z}_i. \tag{5.4.11}$$

The fact that elements (5.4.10) and (5.4.11) are root generators of algebra $so(2r, \mathbb{C})$, follows from the relations ($i < j$)

$$\begin{aligned} [H_k, E_{ij}] &= (\delta_{kj} + \delta_{ki})E_{ij}, \quad [H_k, E_{ij'}] = (\delta_{ki} - \delta_{kj})E_{ij'}, \\ [H_k, E_{i'j}] &= (\delta_{kj} - \delta_{ki})E_{i'j}, \quad [H_k, E_{i'j'}] = -(\delta_{kj} + \delta_{ki})E_{i'j'}, \end{aligned} \tag{5.4.12}$$

which are identical to the structure relations (3.1.22) and (3.1.23).

We have constructed in Section 1.3.2 2^r-dimensional irreducible representation ρ of free fermion algebra (5.4.4) which, in view of (5.4.3), defines representation of Clifford algebra $\mathcal{C}\ell_{2r}$. Recall that representation ρ acts (see (1.3.25)) in fermion Fock space $\mathcal{F}$ with basis vectors (1.3.22):

$$|j_1,\ldots,j_k\rangle = \bar{z}_{j_1}\cdots\bar{z}_{j_k}|0\rangle \quad (1 \le j_1 < \cdots < j_k \le r). \tag{5.4.13}$$

Here, vacuum vector $|0\rangle$ is defined by the relations (1.3.21), i.e.,

$$z_k|0\rangle = 0 \quad (\forall k = 1,\ldots,r). \tag{5.4.14}$$

Lie algebra $so(2r,\mathbb{C})$ with generators (5.4.6) is embedded in Clifford algebra $\mathcal{C}\ell_{2r}$, so representation ρ of free fermion algebra and algebra $\mathcal{C}\ell_{2r}$ is at the same time representation of Lie algebra $so(2r,\mathbb{C})$. Let us show that this representation of $so(2r,\mathbb{C})$ is reducible and splits into direct sum of two irreducible representations with highest weights (5.4.1).

We note that due to (5.4.14), vacuum vector $|0\rangle \in \mathcal{F}$ obeys

$$E_{ij}|0\rangle = 0 = E_{ij'}|0\rangle \quad (i<j), \quad H_k|0\rangle = \frac{1}{2}|0\rangle \quad (\forall k = 1,\ldots,r), \tag{5.4.15}$$

where E_{ij} and $E_{ij'}$ are positive root generators (5.4.10), and H_k are Cartan generators. We recall that components of weight vector are eigenvalues of Cartan generators, and it follows from (5.4.15) that vacuum $|0\rangle \in \mathcal{F}$ is the highest weight vector of irreducible representation of algebra $so(2r,\mathbb{C})$ with highest weight $\lambda_{(r)} = (\frac{1}{2},\ldots,\frac{1}{2})$, which is one of the weights in (5.4.1). By acting on the highest weight vector $|\lambda_{(r)}\rangle \equiv |0\rangle$ by negative root generators (5.4.11), we create the whole invariant space $\mathcal{V}_{(r)} \subset \mathcal{F}$ of irreducible representation with highest weight $\lambda_{(r)}$. It follows from the form of generators (5.4.11) that space $\mathcal{V}_{(r)}$ consists of linear combinations of basis vectors (5.4.13) which are even degree monomials in creation operators $\bar{z}_k$:

$$|j_1,\ldots,j_{2k}\rangle = \bar{z}_{j_1}\cdots\bar{z}_{j_{2k}}|0\rangle \quad (1 \le j_1 < \cdots < j_{2k} \le r). \tag{5.4.16}$$

Now, consider vector

$$|\lambda_{(r-1)}\rangle \equiv \bar{z}_r|0\rangle \in \mathcal{F}, \tag{5.4.17}$$

which does not belong to space $\mathcal{V}_{(r)}$. It obeys the relations

$$E_{ij}|\lambda_{(r-1)}\rangle = 0 = E_{ij'}|\lambda_{(r-1)}\rangle \quad (i<j),$$

$$H_k|\lambda_{r-1}\rangle = \frac{1}{2}|\lambda_{(r-1)}\rangle (\forall k = 1,\ldots,r-1), \quad H_r|\lambda_{(r-1)}\rangle = -\frac{1}{2}|\lambda_{(r-1)}\rangle.$$

These relations show that $|\lambda_{(r-1)}\rangle$ is the highest weight vector with highest weight $\lambda_{(r-1)} = (\frac{1}{2}, \dots, \frac{1}{2}, -\frac{1}{2})$, see (5.4.1). By acting on the highest weight vector $|\lambda_{(r-1)}\rangle$ by generators (5.4.11) we create invariant space $\mathcal{V}_{(r-1)} \subset \mathcal{F}$ of irreducible representation with highest weight $\lambda_{(r-1)}$. This space $\mathcal{V}_{(r-1)}$ consists of linear combinations of basis vectors (5.4.13) which are odd degree monomials in creation operators $\bar{z}_k$:

$$|j_1, \dots, j_{2k-1}\rangle = \bar{z}_{j_1} \cdots \bar{z}_{j_{2k-1}}|0\rangle \quad (1 \le j_1 < \cdots < j_{2k-1} \le r). \tag{5.4.18}$$

Thus, the whole Fock space $\mathcal{F}$ of dimension 2^r (see (1.3.24)) splits into direct sum of two subspaces $\mathcal{V}_{(r-1)}$ and $\mathcal{V}_{(r)}$, which are the spaces of (2^{r-1})-dimensional irreducible representations of algebra $so(2r, \mathbb{C})$ with highest weights $\lambda_{(r-1)}$ and $\lambda_{(r)}$ given by (5.4.1).

Problem 5.4.1. Show that the spaces $\mathcal{V}_{(r-1)}$ and $\mathcal{V}_{(r)}$ have dimensions $\dim(\mathcal{V}_{(r-1)}) = \dim(\mathcal{V}_{(r)}) = 2^{r-1}$.

Finally, let us construct spinor representations with highest weights (5.4.1) in the case when the metric in the definition of Lie algebra $so(2r, \mathbb{C})$ is chosen in the standard diagonal form, $\eta_{ik} = \delta_{ik}$. To this end, we make the linear change of variables (1.3.18), i.e.,

$$\begin{gathered} \Gamma_{2k-1} = z_k + \bar{z}_k = \mathsf{\Gamma}_k + \mathsf{\Gamma}_{k'}, \quad \Gamma_{2k} = i(\bar{z}_k - z_k) = i(\mathsf{\Gamma}_{k'} - \mathsf{\Gamma}_k), \\ z_k = \frac{1}{2}(\Gamma_{2k-1} + i\Gamma_{2k}), \quad \bar{z}_k = \frac{1}{2}(\Gamma_{2k-1} - i\Gamma_{2k}). \end{gathered} \tag{5.4.19}$$

The new generators Γ_k obey the standard Clifford algebra relations (1.3.1) with diagonal metric,

$$\Gamma_m \Gamma_n + \Gamma_n \Gamma_m = 2\delta_{mn}\, I. \tag{5.4.20}$$

We then construct elements (cf. (1.3.34) and (5.4.6))

$$M_{(k,j)} = \frac{1}{4}(\Gamma_k \Gamma_j - \Gamma_j \Gamma_k) = \frac{1}{2}(\Gamma_k \Gamma_j - \delta_{kj}) \in \mathcal{C}\ell_{2r}. \tag{5.4.21}$$

They obey commutation relations (5.1.12) for generators of Lie algebra $so(2r, \mathbb{C})$ with Euclidean metric $\eta = I_{2r}$, i.e.,

$$[M_{(n,m)},\, M_{(k,\ell)}] = \eta_{mk} M_{(n,\ell)} + \eta_{n\ell} M_{(m,k)} - \eta_{nk} M_{(m,\ell)} - \eta_{m\ell} M_{(n,k)}. \tag{5.4.22}$$

In terms of generators $\Gamma_m \in \mathcal{C}\ell_{2r}$, Cartan–Weyl basis elements (5.4.9)–(5.4.11) are $(i < j)$:

$$
\begin{aligned}
H_k &= \frac{1}{2}I - \bar{z}_k\, z_k = -i\ M_{(2k-1,2k)} \quad (k = 1,\dots,r),\\
E_{ij} &= z_i\ z_j = \frac{1}{4}(\Gamma_{2i-1} + i\Gamma_{2i})(\Gamma_{2j-1} + i\Gamma_{2j}) = \frac{1}{2}M^{(+1,+1)}_{(2i-1,2i;2j-1,2j)},\\
E_{ij'} &= -\bar{z}_j\ z_i = \frac{1}{4}(\Gamma_{2i-1} + i\Gamma_{2i})(\Gamma_{2j-1} - i\Gamma_{2j}) = \frac{1}{2}M^{(+1,-1)}_{(2i-1,2i;2j-1,2j)},\\
E_{i'j} &= \bar{z}_i\ z_j = \frac{1}{4}(\Gamma_{2i-1} - i\Gamma_{2i})(\Gamma_{2j-1} + i\Gamma_{2j}) = \frac{1}{2}M^{(-1,+1)}_{(2i-1,2i;2j-1,2j)},\\
E_{i'j'} &= -\bar{z}_j\ \bar{z}_i = \frac{1}{4}(\Gamma_{2i-1} - i\Gamma_{2i})(\Gamma_{2j-1} - i\Gamma_{2j}) = \frac{1}{2}M^{(-1,-1)}_{(2i-1,2i;2j-1,2j)},
\end{aligned}
\tag{5.4.23}
$$

where we make use of the second equality in (5.4.21) and the following notations for root generators:

$$
\begin{aligned}
M^{(\epsilon,\eta)}_{(2i-1,2i;2j-1,2j)} \equiv{}& (M_{(2i-1,2j-1)} + i\epsilon M_{(2i,2j-1)}) + i\,\eta(M_{(2i-1,2j)}\\
&+ i\epsilon M_{(2i,2j)}).
\end{aligned}
$$

The latter root generators have been introduced in the first book, Section I-6.2.3, for constructing root generators of Lie algebras $so(N,\mathbb{C})$ with diagonal metric $\eta = I_{2r}$. Thus, the construction of irreducible spinor representations with highest weights (5.4.1) in the case of diagonal metric is completely equivalent to their construction for antidiagonal metric.

Remark. Recall (see Section 1.3.1) that the longest element in Clifford algebra $\mathcal{C}\ell_{2r}$ is

$$
\mathsf{\Gamma}_{2r+1} = (-i)^{\nu}\ \Gamma_1 \cdot \Gamma_2 \cdots \Gamma_{2r} \in \mathcal{C}\ell_{2r}. \tag{5.4.24}
$$

It anticommutes with all generators Γ_k (or $\mathsf{\Gamma}_k$) of algebra $\mathcal{C}\ell_{2r}$, and squares to unity:

$$
[\mathsf{\Gamma}_{2r+1},\ \Gamma_k]_+ = 0 = [\mathsf{\Gamma}_{2r+1},\ \mathsf{\Gamma}_k]_+, \quad (\mathsf{\Gamma}_{2r+1})^2 = I. \tag{5.4.25}
$$

The first property implies that the element (5.4.24) commutes with all generators (5.4.6), or (5.4.21), of Lie algebra $so(2r,\mathbb{C})$:

$$
[\mathsf{\Gamma}_{2r+1}, L_{mn}] = 0 = [\mathsf{\Gamma}_{2r+1}, M_{(m,n)}].
$$

In accordance with Schur's lemma, element $\mathsf{\Gamma}_{2r+1}$ is proportional to unit operator in irreducible complex representations of algebra $so(2r,\mathbb{C})$ that are based on realizations (5.4.6) and (5.4.21). In view of (5.4.25), the proportionality coefficient is either +1 or −1. In free fermion Fock space, element (5.4.24) reads

$$\Gamma_{2r+1} = (z_1\bar{z}_1 - \bar{z}_1 z_1)\cdot(z_2\bar{z}_2 - \bar{z}_2 z_2)\cdots(z_r\bar{z}_r - \bar{z}_r z_r) = 2^r\, H_1 \cdot H_2 \cdots H_r. \tag{5.4.26}$$

It acts in spaces of odd and even representations $\mathcal{V}_{(r-1)}$ and $\mathcal{V}_{(r)}$ as follows:

$$\Gamma_{2r+1}\cdot \mathcal{V}_{(r-1)} = -I\cdot \mathcal{V}_{(r-1)}, \quad \Gamma_{2r+1}\cdot \mathcal{V}_{(r)} = +I\cdot \mathcal{V}_{(r)},$$

so that the spaces $\mathcal{V}_{(r-1)}$ and $\mathcal{V}_{(r)}$ are extracted out from the Fock space $\mathcal{F}$ by projector operators

$$P_- = \frac{1}{2}(I - \Gamma_{2r+1}), \quad P_+ = \frac{1}{2}(I + \Gamma_{2r+1}). \tag{5.4.27}$$

Note that the sign in the definition (5.4.24) can be chosen arbitrarily, so signs in (5.4.27) depend on the definition of element Γ_{2r+1}.

Complex Clifford algebra $\mathcal{C}\ell_{2r}$ and Lie algebra $so(2r,\mathbb{C})$ act in 2^r-dimensional complex Fock space $\mathcal{F}$ in accordance with representations (5.4.19) and (5.4.23).

Definition 5.4.1. Vectors in 2^r-dimensional space $\mathcal{F}$ with basis (5.4.13) are called *Dirac spinors* in 2r-dimensional space. Vectors in 2^{r-1}-dimensional complex subspaces $\mathcal{V}_{(r-1)}, \mathcal{V}_{(r)} \subset \mathcal{F}$ invariant under $so(2r,\mathbb{C})$-action are called *Weyl spinors* in $2r$-dimensional space–time.

5.4.2. Spinor representations of Lie algebras $so(2r+1,\mathbb{C})$

To construct spinor representations of algebra $so(2r+1,\mathbb{C})$ with highest fundamental weight (5.4.2), we make use of Clifford algebra $\mathcal{C}\ell_N$ with odd number $N = 2r+1$ of generators obeying defining relations (5.4.20). We have seen in Section 1.3.1 that the relevant representations of Clifford algebra $\mathcal{C}\ell_{2r+1}$ can be obtained from irreducible representation ρ of Clifford algebra $\mathcal{C}\ell_{2r}$ by employing homomorphism $\mathcal{C}\ell_{2r+1} \to \mathcal{C}\ell_{2r}$. The latter homomorphism maps the first $2r$ generators of Clifford algebra $\mathcal{C}\ell_{2r+1}$ to $2r$ generators Γ_k $(k = 1,\ldots,2r)$ of its subalgebra $\mathcal{C}\ell_{2r}$, while the last generator

$\Gamma_{2r+1} \in \mathcal{C}\ell_{2r+1}$ is mapped into the longest element of $\mathcal{C}\ell_{2r}$,

$$\Gamma_{2r+1} \to \pm(-i)^{\nu}\, \Gamma_1 \cdot \Gamma_2 \cdots \Gamma_{2r} \in \mathcal{C}\ell_{2r}. \tag{5.4.28}$$

Two signs in (5.4.28) lead to two inequivalent irreducible representations of algebra $\mathcal{C}\ell_{2r+1}$ which differ by the eigenvalues ± 1 of the central element

$$\Gamma \equiv (-i)^{\nu}\, \Gamma_1 \cdots \Gamma_{2r} \cdot \Gamma_{2r+1} \in \mathcal{C}\ell_{2r+1}. \tag{5.4.29}$$

We realize the algebra $\mathcal{C}\ell_{2r}$ in terms of free fermions (see formula (5.4.19)), then there are two representations of the last generator of algebra $\mathcal{C}\ell_{2r+1}$:

$$\rho^+ :\ \Gamma_{2r+1} = +\mathsf{\Gamma}_{2r+1}, \qquad \rho^- :\ \Gamma_{2r+1} = -\mathsf{\Gamma}_{2r+1}, \tag{5.4.30}$$

$$(\Gamma_{2r+1})^2 = I, \quad [\Gamma_{2r+1}, z_k]_+ = 0 = [\Gamma_{2r+1}, \bar{z}_k]_+ \ \ (k = 1, \dots, r),$$

where element $\mathsf{\Gamma}_{2r+1} \in \mathcal{C}\ell_{2r}$ is given by (5.4.24), (5.4.26).

Let us consider representation ρ^+. We again define Clifford algebra $\mathcal{C}\ell_{2r+1}$ with antidiagonal metric, which is done by choosing the generators $\mathsf{\Gamma}'_k \in \mathcal{C}\ell_{2r+1}$ as follows (cf. (5.4.3)):

$$\begin{aligned}(\mathsf{\Gamma}'_1, \dots, \mathsf{\Gamma}'_r, \mathsf{\Gamma}'_{r+1}, \mathsf{\Gamma}'_{r+2}, \dots, \mathsf{\Gamma}'_{2r+1}) &= \sqrt{2}(z_1, \dots, z_r, \frac{1}{\sqrt{2}}\mathsf{\Gamma}_{2r+1}, \bar{z}_r, \dots, \bar{z}_1) \\ &= (\mathsf{\Gamma}_1, \dots, \mathsf{\Gamma}_r, \mathsf{\Gamma}_{2r+1}, \mathsf{\Gamma}_{r+1}, \dots, \mathsf{\Gamma}_{2r}).\end{aligned} \tag{5.4.31}$$

To construct the complete basis of algebra $so(2r+1, \mathbb{C})$, we add to the basis elements (5.4.6), (5.4.9)–(5.4.11) of algebra $so(2r, \mathbb{C})$ extra root generators

$$\begin{aligned} L_{i,2r+1} &= \frac{1}{4}(\mathsf{\Gamma}_i\, \mathsf{\Gamma}_{2r+1} - \mathsf{\Gamma}_{2r+1}\, \mathsf{\Gamma}_i) = \frac{1}{2} z_i\, \mathsf{\Gamma}_{2r+1}, \\ L_{i',2r+1} &= \frac{1}{4}(\mathsf{\Gamma}_{i'}\, \mathsf{\Gamma}_{2r+1} - \mathsf{\Gamma}_{2r+1}\, \mathsf{\Gamma}_{i'}) = \frac{1}{2} \bar{z}_i\, \mathsf{\Gamma}_{2r+1}, \\ i &= 1, \dots, r, \quad i' = 2r - i + 1, \end{aligned} \tag{5.4.32}$$

where $L_{i,2r+1}$ and $L_{i',2r+1}$ are positive and negative root generators, respectively (cf. (3.1.24)):

$$[H_k, L_{i,2r+1}] = \delta_{ki}\, L_{i,2r+1}, \quad [H_k, L_{i',2r+1}] = -\delta_{ki}\, L_{i',2r+1}.$$

Note that the choice in the above construction of another representation ρ^- (instead of ρ^+) of Clifford algebra $\mathcal{C}\ell_{2r+1}$, defined in (5.4.30), would result merely in the change of sign in additional generators (5.4.32). The signs of additional generators in (5.4.32) can be flipped by equivalence transformation

$$\Gamma_{2r+1} L_{i,2r+1} \Gamma_{2r+1} = -L_{i,2r+1}, \quad \Gamma_{2r+1} L_{i',2r+1} \Gamma_{2r+1} = -L_{i',2r+1},$$

so inequivalent representations ρ^+ and ρ^- of Clifford algebra $\mathcal{C}\ell_{2r+1}$ define equivalent representations of algebra $so(2r+1, \mathbb{C})$.

Problem 5.4.2. Show that the generators (5.4.32), together with the generators (5.4.6), (5.4.9)–(5.4.11) obey defining commutation relations of algebra $so(2r+1, \mathbb{C})$.

The presence of additional generators (5.4.32) in algebra $so(2r+1, \mathbb{C})$ leads to the property that two irreducible representations of algebra $so(2r, \mathbb{C}) \subset so(2r+1, \mathbb{C})$, acting in spaces $\mathcal{V}_{(r-1)} \subset \mathcal{F}$ and $\mathcal{V}_{(r)} \subset \mathcal{F}$, are unified into a single irreducible representation of algebra $so(2r+1, \mathbb{C})$ with highest weight (5.4.2). The latter representation acts in the whole Fock space $\mathcal{F}$. Indeed, vectors in space $\mathcal{V}_{(r-1)}$ transform into vectors from space $\mathcal{V}_{(r)}$, and vice versa, by additional generators (5.4.32). As an example, vacuum vector $|0\rangle$ is obtained from vector (5.4.17) by positive root generator $L_{r,2r+1}$:

$$L_{r,2r+1}\, \bar{z}_r\, |0\rangle = -\frac{1}{2}\, |0\rangle. \tag{5.4.33}$$

Thus, highest weight vector (5.4.17) of subalgebra $so(2r, \mathbb{C}) \subset so(2r+1, \mathbb{C})$ is no longer highest weight vector of the whole algebra $so(2r+1, \mathbb{C})$. The Fock space $\mathcal{F}$ has a single highest weight vector $|0\rangle$ which is annihilated by all positive root generators of algebra $so(2r+1, \mathbb{C})$. Indeed, the vacuum $|0\rangle$ is the only vector that obeys both the relations (5.4.15) and the relations

$$L_{k,2r+1}\, |0\rangle = 0, \quad k = 1, \ldots, r,$$

where $L_{k,2r+1}$ are additional positive root generators. The vacuum vector has weight $\lambda_{(r)} = (\frac{1}{2}, \ldots, \frac{1}{2})$. So, algebra $so(2r+1, \mathbb{C})$ has only one irreducible spinor representation with highest fundamental weight (5.4.2); this representation emanates from highest weight vector $|0\rangle$ and acts in the entire Fock space $\mathcal{F}$. Dimension of this representations is equal to $\dim(\mathcal{F}) = 2^r$.

Let us note that for the choice of generators $\Gamma_k \in \mathcal{C}\ell_{2r+1}$ obeying defining relations (5.4.20) with diagonal metric $\eta_{kn} = \delta_{kn}$, additional to

(5.4.21) basis elements are

$$M_{(k,2r+1)} = \frac{1}{2}\Gamma_k \Gamma_{2r+1} \in so(2r+1,\mathbb{C}) \quad (k=1,\ldots,2r), \tag{5.4.34}$$

where generators Γ_k are related to free fermions via (5.4.19). In terms of elements (5.4.34), root generators (5.4.32) read $(i = 1,\ldots,r)$

$$M^{(+)}_{(i)} = M_{(2i-1,2r+1)} + iM_{(2i,2r+1)} = z_i \Gamma_{2r+1},$$

$$M^{(-)}_{(i)} = M_{(2i-1,2r+1)} - iM_{(2i,2r+1)} = \bar{z}_i \Gamma_{2r+1}.$$

Root generators $M^{(+)}_{(i)}$ and $M^{(-)}_{(i)}$ have been used in the first book, Section I-6.2.3, for constructing root systems of algebras $so(2r+1,\mathbb{C})$ defined with diagonal metric η.

Chapter 6

Groups $\mathsf{Spin}(p, q)$ and Their Finite-Dimensional Representations

Complex and real representations (including spinor) of real Lie algebras $so(p, q)$ are important in physics applications. These representations are at the same time representations of groups $\mathsf{Spin}(p, q)$ which are locally isomorphic to groups $SO(p, q)$ and cover connected component $SO^{\uparrow}(p, q)$ of group $SO(p, q)$.

We define groups $\mathsf{Spin}(p, q)$ and study their representations in this chapter. To this end, we introduce real Clifford algebras $\mathcal{C}\ell_{(p,q)}$ associated with spaces $\mathbb{R}^{p,q}$, and construct representations of these algebras.

6.1. Clifford Algebras and Their Representations

We have introduced in Section 1.3 complex Clifford algebra $\mathcal{C}\ell_N$ with N generators (see Definition 1.3.1). For even $N = 2\nu$, we have shown that algebra $\mathcal{C}\ell_{2\nu}$ is isomorphic to free fermion algebra $\mathcal{A}_\nu$ (fermion oscillator algebra) and constructed 2^ν-dimensional irreducible complex representation ρ of algebra $\mathcal{C}\ell_{2\nu}$ that acts in Fock space $\mathcal{F}$ of fermion oscillators. We have shown also that ρ is exact representation of algebra $\mathcal{C}\ell_{2\nu}$, which has minimum possible dimension 2^ν, and that other irreducible complex representations of algebra $\mathcal{C}\ell_{2\nu}$ are 2^ν-dimensional and equivalent to ρ. We study in this section real forms $\mathcal{C}\ell_{(p,q)}$ $(p + q = N)$ of complex Clifford algebra $\mathcal{C}\ell_N$ for arbitrary (not necessarily even) N, and both complex and real representations of $\mathcal{C}\ell_{(p,q)}$.

6.1.1. Real Clifford algebras $\mathcal{C}\ell_{(p,q)}$

To begin with, we give a general definition of Clifford algebra. Let $\mathcal{V}$ be vector space over field $\mathbb{K}'$ equipped with bilinear scalar product $(\cdot, \cdot)$.

We map every vector $v \in \mathcal{V}$ into a generator $\hat{v}$ of some associative algebra (over field $\mathbb{K}$ such that $\mathbb{K}' \subseteq \mathbb{K}$) with unity I and defining relations[a]

$$\hat{v} \cdot \hat{u} + \hat{u} \cdot \hat{v} = 2(v, u)I. \tag{6.1.1}$$

Algebra $\mathcal{C}\ell(\mathcal{V}, (\cdot,\cdot))$ with this property is called Clifford algebra associated with vector space $\mathcal{V}$.

Let us choose $\mathcal{V}$ as pseudo-Euclidean space $\mathbb{R}^{p,q}$ $(p + q = N)$ with metric (5.1.1), $\eta = \text{diag}(+1, \ldots, +1, -1, \ldots, -1)$, and basis vectors $\vec{e}_m$ $(m = 1, \ldots, N)$ such that $(\vec{e}_m, \vec{e}_k) = \eta_{mk}$, where $(\cdot,\cdot)$ is scalar product in $\mathbb{R}^{p,q}$. In accordance with above construction, every basis vector $\vec{e}_m \in \mathbb{R}^{p,q}$ is mapped into generator Γ_m of associative algebra with unity I and defining relations

$$\Gamma_m \Gamma_n + \Gamma_n \Gamma_m = 2\eta_{mn} I. \tag{6.1.2}$$

Definition 6.1.1. Associative algebra $\mathcal{C}\ell^{\mathbb{R}}_{(p,q)}$, over field $\mathbb{R}$, with unit element I and generators Γ_m $(m = 1, 2, \ldots, N)$ obeying (6.1.2) is called *Clifford algebra associated with space* $\mathbb{R}^{p,q}$.

In what follows, we also consider complexification of algebra $\mathcal{C}\ell^{\mathbb{R}}_{(p,q)}$ which we denote by $\mathcal{C}\ell^{\mathbb{C}}_{(p,q)}$.

Example. We define 4×4 matrices γ_m with block structure

$$\gamma_m = \begin{pmatrix} \mathbf{0} & \sigma_m \\ \tilde{\sigma}_m & \mathbf{0} \end{pmatrix}, \quad m = 0, 1, 2, 3. \tag{6.1.3}$$

Here we follow tradition and start numbering m from zero. We denote zero 2×2 block by $\mathbf{0}$ and 2×2 blocks σ_m and $\tilde{\sigma}_m$ are components of matrix 4-vectors

$$\sigma = (\sigma_0, \sigma_1, \sigma_2, \sigma_3), \quad \tilde{\sigma} = (\sigma_0, -\sigma_1, -\sigma_2, -\sigma_3), \quad \sigma_0 \equiv I_2, \tag{6.1.4}$$

where Pauli matrices $\sigma_1, \sigma_2, \sigma_3$ are given by (2.1.16). The latter obey identities

$$\sigma_i \sigma_k = i\varepsilon_{ikj}\sigma_j + \delta_{ik} I_2 \;\Rightarrow\; \sigma_i\sigma_k + \sigma_k\sigma_i = 2\delta_{ik} I_2 \quad (i, k, j = 1, 2, 3). \tag{6.1.5}$$

Due to these identities, matrices (6.1.3) obey

$$\gamma_m \gamma_n + \gamma_n \gamma_m = 2\eta_{mn} I_4, \; ||\eta_{mn}|| = \text{diag}(+1, -1, -1, -1), \tag{6.1.6}$$

[a]Mapping $v \to \hat{v}$ must be consistent with all operations in $\mathcal{V}$. As an example, vector $(\alpha v + \beta u)$ must be mapped into element $(\alpha\hat{v} + \beta\hat{u})$ for all $v, u \in \mathcal{V}$ and $\alpha, \beta \in \mathbb{K}'$.

i.e., they generate four-dimensional representation of Clifford algebra $\mathcal{C}\ell^{\mathbb{K}}_{(1,3)}$. Matrices (6.1.3) are called *Dirac matrices* in chiral representation. In general, any set of four 4×4 matrices obeying (6.1.6) is called a set of Dirac matrices. As we point out below (see Remark 2 in Section 6.1.2), all these sets $\{\gamma_n\}$, $\{\tilde{\gamma}_n\}$ are equivalent: $\tilde{\gamma}_n = U\gamma_n U^{-1}$.

Remark 1. Clifford algebra obtained from $\mathcal{C}\ell^{\mathbb{K}}_{(p,q)}$ $(p + q = N)$ by permutation $\sigma \in S_N$ of generators,

$$(\Gamma_1, \ldots, \Gamma_N) \;\to\; (\Gamma_{\sigma(1)}, \ldots, \Gamma_{\sigma(N)}), \tag{6.1.7}$$

(with components of diagonal metric η permuted accordingly), is isomorphic to the original algebra $\mathcal{C}\ell^{\mathbb{K}}_{(p,q)}$, so we do not distinguish these algebras in what follows.

Remark 2. In the case $\mathbb{K} = \mathbb{R}$, we call Clifford algebra *real* and use shorthand notation $\mathcal{C}\ell_{(p,q)} = \mathcal{C}\ell^{\mathbb{R}}_{(p,q)}$. For $\mathbb{K} = \mathbb{C}$, Clifford algebra is *complex*. In the latter case, the complex transformation of generators

$$\Gamma_m \;\to\; \Gamma_m \; (m = 1, \ldots, p), \quad \Gamma_m \to i\,\Gamma_m \; (m = p+1, \ldots, p+q), \tag{6.1.8}$$

yields defining relations of algebra $\mathcal{C}\ell^{\mathbb{C}}_{(N,0)}$ with Euclidean metric $\eta_{mn} = \delta_{mn}$:

$$\Gamma_m \Gamma_n + \Gamma_n \Gamma_m = 2\delta_{mn}\, I. \tag{6.1.9}$$

Therefore, all complex Clifford algebras $\mathcal{C}\ell^{\mathbb{C}}_{(p,q)}$ (for given $p + q = N$) are isomorphic to algebra $\mathcal{C}\ell^{\mathbb{C}}_{(N,0)}$. Like in Section 1.3.1, we use the notation $\mathcal{C}\ell_N$ for complex Clifford algebra $\mathcal{C}\ell^{\mathbb{C}}_{(N,0)}$.

Since the transformation (6.1.8) that changes metric signature in (6.1.2) is complex, one would suspect that there are no isomorphisms between real Clifford algebras $\mathcal{C}\ell_{(p,q)}$ for different p and q. This is not the case, however.

Proposition 6.1.1. *There are the following isomorphisms between real Clifford algebras* $\mathcal{C}\ell_{(p,q)}$:

$$\mathcal{C}\ell_{(p,q)} = \mathcal{C}\ell_{(q+1,p-1)} \quad (p \geq 1), \tag{6.1.10}$$

$$\mathcal{C}\ell_{(p,q)} = \mathcal{C}\ell_{(p-4,q+4)} \quad (p \geq 4). \tag{6.1.11}$$

Proof. Let Γ_i be generators of algebra $\mathcal{C}\ell_{(p,q)}$ obeying the relations (6.1.2) with metric $\eta = I_{p,q}$. Let us make a change of generators $\Gamma_i \to \Gamma'_i$:

$$\Gamma'_i = \begin{cases} \Gamma_1, & i = 1, \\ \Gamma_{N-i+2}\Gamma_1, & i = 2, \dots, N. \end{cases} \tag{6.1.12}$$

It is straightforward to see that for $p \geq 1$, elements Γ'_i obey the relations (6.1.2) with metric $\eta = I_{q+1,p-1}$, so they generate algebra $\mathcal{C}\ell_{(q+1,p-1)}$. Since the transformation (6.1.12) is invertible, it establishes isomorphism (6.1.10).

Consider now the case $p \geq 4$ and make another transformation

$$\Gamma'_i = \begin{cases} \Gamma_i\Gamma_1\Gamma_2\Gamma_3\Gamma_4, & i = 1, 2, 3, 4, \\ \Gamma_i, & i = 5, \dots, N. \end{cases} \tag{6.1.13}$$

Upon permutation of indices $(1, \dots, p+q) \to (5, \dots, p+q, 1, 2, 3, 4)$, elements Γ'_i in (6.1.13), obey the relations (6.1.2) with metric $\eta = I_{p-4,q+4}$. This establishes isomorphism (6.1.11). □

Let us note that it is sometimes convenient to label real Clifford algebras $\mathcal{C}\ell_{(p,q)}$ by the dimension $N = p + q$ of space $\mathbb{R}^{p,q}$ and signature $s = (p - q)$, instead of parameters $p, q \in \mathbb{Z}_{\geq 0}$. Then the second isomorphism (6.1.11) relates those real algebras $\mathcal{C}\ell_{(p,q)}$ whose signatures coincide modulo 8. This periodicity is important when studying real algebras $\mathcal{C}\ell_{(p,q)}$ (it shows up, in particular, in the properties of their real representations, see Sections 6.3.1 and 6.4.2), and also when discussing homotopy groups of orthogonal and symplectic Lie groups; it is known as *Cartan–Bott periodicity.*

Proposition 6.1.2. *Real Clifford algebra $\mathcal{C}\ell_{(p,q)}$ with finite $N = (p+q)$ is finite dimensional, and its dimension is*

$$\dim(\mathcal{C}\ell_{(p,q)}) = 2^N. \tag{6.1.14}$$

Proof. The Proof is literally the same as for Proposition 1.3.1 (see Section 1.3.1) valid in the case of complex Clifford algebra $\mathcal{C}\ell_N$. It makes use of the fact that basis of real algebra $\mathcal{C}\ell_{(p,q)}$, like basis of complex algebra $\mathcal{C}\ell_N$, is made of monomials

$$\begin{gathered} \Gamma_A = \{I, \Gamma_{m_1}, \Gamma_{m_1m_2}, \Gamma_{m_1m_2m_3}, \dots, \Gamma_{12\dots N}\}, \\ \Gamma_{m_1m_2\dots m_k} \equiv \Gamma_{m_1}\Gamma_{m_2}\cdots\Gamma_{m_k}, \end{gathered} \tag{6.1.15}$$

where the product in each monomial $\Gamma_{m_1m_2\dots m_k}$ is ordered: $1 \leq m_1 < m_2 < m_3 < \dots \leq N = p + q$. We leave the completion of the proof to the reader. □

Definition 6.1.2. We call a monomial $\Gamma_{m_1 m_2 \dots m_k}$ *even (odd)*, if k is even (odd). Linear combinations of even (odd) monomials are called even (odd) elements.

This definition induces $\mathbb{Z}_2$-grading in Clifford algebra.

Consider the last, longest basis element in (6.1.15):

$$\Gamma_{1\dots N} = \Gamma_1 \cdot \Gamma_2 \cdots \Gamma_N = \frac{(-1)^q}{N!} \varepsilon^{i_1 \dots i_N} \Gamma_{i_1} \cdots \Gamma_{i_N}, \tag{6.1.16}$$

where we write it both as the ordered product and as the antisymmetric product of all generators $\Gamma_m \in \mathcal{C\ell}_{(p,q)}$. Here ε is a totally antisymmetric tensor in space $\mathbb{R}^{p,q}$:

$$\begin{aligned} \varepsilon_{\dots i_k \dots i_\ell \dots} &= -\varepsilon_{\dots i_\ell \dots i_k \dots}, \quad \varepsilon_{12\dots N} = +1, \\ \varepsilon^{j_1 \dots j_k}{}_{i_{k+1} \dots i_N} &= \varepsilon_{i_1 i_2 \dots i_N} \eta^{i_1 j_1} \dots \eta^{i_k j_k} \\ \Rightarrow \; \varepsilon_{i_1 i_2 \dots i_N} &= \det(||\eta_{ij}||) \varepsilon^{i_1 i_2 \dots i_N} = (-1)^q \varepsilon^{i_1 i_2 \dots i_N}. \end{aligned} \tag{6.1.17}$$

Making use of (6.1.2) we obtain an identity for the element (6.1.16):

$$(\Gamma_{1\dots N})^2 = (-1)^{N(N-1)/2} (\Gamma_N \cdots \Gamma_2 \cdot \Gamma_1)(\Gamma_1 \cdot \Gamma_2 \cdots \Gamma_N) = \alpha_{(p,q)} I, \tag{6.1.18}$$

$$\alpha_{(p,q)} = (-1)^{\frac{N(N-1)}{2} + q} = (-1)^{\nu + q} = e^{i\pi(\nu+q)},$$

where $\nu = \left[\frac{N}{2}\right]$ is an integer part of $\frac{N}{2}$. Formula (6.1.18) generalizes relation (1.3.9) to the case $q \neq 0$.

> **Problem 6.1.1.** Show that the coefficient $\alpha_{(p,q)}$ equals $(-1)^{\frac{s}{2}}$ for even N and $(-1)^{\frac{s-1}{2}}$ for odd N, where $s = p - q$ is signature.

It is convenient to define normalized element

$$\Gamma_{N+1} = e^{-i\frac{\pi}{2}(\nu+q)} \Gamma_{1\dots N} = e^{-i\frac{\pi}{2}(\nu+q)} \Gamma_1 \cdot \Gamma_2 \cdots \Gamma_N, \tag{6.1.19}$$

such that

$$\Gamma_{N+1}^2 = I. \tag{6.1.20}$$

Element (6.1.19) belongs to complex algebra $\mathcal{C\ell}^{\mathbb{C}}_{(p,q)}$ for odd $(\nu + q)$ and real algebra $\mathcal{C\ell}_{(p,q)}$ for even $(\nu + q)$. Coefficient in (6.1.19) is defined by (6.1.20) modulo sign, $\Gamma_{N+1} \to -\Gamma_{N+1}$.

Let us show that the analysis of real algebra $\mathcal{C\ell}_{(p,q)}$ with odd number of generators $p + q = 2\nu + 1$ can be reduced to that of algebra $\mathcal{C\ell}_{(p-1,q)}$ or

$\mathcal{C}\ell_{(p,q-1)}$ with even number of generators (see similar argument for complex algebras $\mathcal{C}\ell_N$ in Section 1.3.1). For odd $N = p + q = 2\nu + 1$, the element (6.1.19) reads

$$\Gamma_{N+1} = \Gamma_{2\nu+2} = e^{-i\frac{\pi}{2}(\nu+q)}\Gamma_1 \cdot \Gamma_2 \cdots \Gamma_{2\nu+1}. \tag{6.1.21}$$

We recall the definition (6.1.16), and make use of the relations (6.1.2) to obtain

$$\begin{gathered}\Gamma_{1\dots N} \cdot \Gamma_k = \Gamma_k \cdot \Gamma_{1\dots N} \;\Leftrightarrow\; \Gamma_{N+1} \cdot \Gamma_k = \Gamma_k \cdot \Gamma_{N+1}, \\ \forall k = 1, \dots, N = 2\nu + 1.\end{gathered} \tag{6.1.22}$$

Therefore, for odd N, elements (6.1.16) and (6.1.21) are central for algebra $\mathcal{C}\ell_{(p,q)}$, as they commute with all generators $\Gamma_k \in \mathcal{C}\ell_{(p,q)}$, and hence with all elements of algebra $\mathcal{C}\ell_{(p,q)}$. Recall that $\Gamma_{1\dots N} \in \mathcal{C}\ell_{(p,q)}$, while Γ_{N+1} belongs to real algebra $\mathcal{C}\ell_{(p,q)}$ only for even $(\nu+q)$. It is now convenient to consider algebra $\mathcal{C}\ell_{(p,q)}$ with arbitrary arrangement of its generators (6.1.7), i.e., arbitrary order of diagonal elements $\eta_{ii} = +1$ and $\eta_{ii} = -1$ (whose number is p and q, respectively) in metric $\eta = \mathrm{diag}(\eta_1, \dots, \eta_{NN})$. Let us take one of the generators of algebra $\mathcal{C}\ell_{(p,q)}$, say, the last one $\Gamma_N = \Gamma_{2\nu+1}$. Since $\Gamma_N^2 = \eta_{NN} I$, any element $a \in \mathcal{C}\ell_{(p,q)}$ can be written as $a = a_1 + \Gamma_N \cdot a_2$, where a_1, a_2 are elements of algebra $\mathcal{C}\ell_{(p',q')}$ with generators $\{\Gamma_1, \dots, \Gamma_{2\nu}\}$, and $p' = p - 1, q' = q$ for $\eta_{NN} = +1$, while $p' = p, q' = q - 1$ for $\eta_{NN} = -1$. So, there is a decomposition

$$\mathcal{C}\ell_{(p,q)} = \mathcal{C}\ell_{(p',q')} + \Gamma_{2\nu+1} \cdot \mathcal{C}\ell_{(p',q')}. \tag{6.1.23}$$

Let us replace here the last generator $\Gamma_N = \Gamma_{2\nu+1}$ by the central element $\Gamma_{1\dots N} \in \mathcal{C}\ell_{(p,q)}$, i.e., substitute (this follows from (6.1.16))

$$\Gamma_{2\nu+1} = (-1)^{q'}\Gamma_{1\dots N} \cdot (\Gamma_{2\nu} \cdots \Gamma_1). \tag{6.1.24}$$

Then the decomposition (6.1.23) becomes

$$\mathcal{C}\ell_{(p,q)} = \mathcal{C}\ell_{(p',q')} + \Gamma_{1\dots N} \cdot \mathcal{C}\ell_{(p',q')}, \tag{6.1.25}$$

so algebra $\mathcal{C}\ell_{(p,q)}$ can be considered as central extension of its subalgebra $\mathcal{C}\ell_{(p',q')}$. The right-hand side of (6.1.25) is generated by $p' + q' = 2\nu$ generators $\{\Gamma_1, \dots, \Gamma_{2\nu}\}$ of Clifford algebra $\mathcal{C}\ell_{(p',q')}$ and an additional central element $\Gamma_{1\dots N}$. In any irreducible representation of algebra $\mathcal{C}\ell_{(p,q)}$, the central element $\Gamma_{1\dots N}$ is proportional to unity, so the generator (6.1.24) is proportional to product $\Gamma_{2\nu} \cdots \Gamma_1 \in \mathcal{C}\ell_{(p',q')}$.

Remark 3. In a particular case of even $(\nu+q)$ we have $\Gamma_{1\dots N}^2 = I$, so any element $a \in \mathcal{C}\ell_{(p,q)}$ can be written as $a = a_1 \cdot P_+ + a_2 \cdot P_-$, where $P_\pm = (I \pm \Gamma_{1\dots N})/2$ and $a_1, a_2 \in \mathcal{C}\ell_{(p',q')}$. In other words, instead of (6.1.25) we can use the decomposition

$$\mathcal{C}\ell_{(p,q)} = \mathcal{C}\ell_{(p',q')} \cdot P_+ + \mathcal{C}\ell_{(p',q')} \cdot P_- = \mathcal{C}\ell^{(+)}_{(p',q')} \oplus \mathcal{C}\ell^{(-)}_{(p',q')}, \tag{6.1.26}$$

where $\mathcal{C}\ell^{(-)}_{(p',q')}$ and $\mathcal{C}\ell^{(+)}_{(p',q')}$ are two subalgebras $\mathcal{C}\ell_{(p,q)}$ isomorphic to $\mathcal{C}\ell_{(p',q')}$. Furthermore, if $b_1 \in \mathcal{C}\ell^{(-)}_{(p',q')}$ and $b_2 \in \mathcal{C}\ell^{(+)}_{(p',q')}$, then $b_1 \cdot b_2 = 0 = b_2 \cdot b_1$. Therefore, for even $(\nu+q)$, algebra $\mathcal{C}\ell_{(p,q)}$ is isomorphic to direct sum of its two subalgebras $\mathcal{C}\ell_{(p',q')}$. If $(\nu+q)$ is odd, then (6.1.18) gives $\Gamma_{1\dots N}^2 = -I$, hence eigenvalues of element $\Gamma_{1\dots N}$ are equal to $(\pm i)$, and analogs of projectors $P_\pm$ are complex, $P_\pm = (I \pm i\Gamma_{1\dots N})/2$. Therefore, one cannot construct in this case the decomposition (6.1.26) of real algebra $\mathcal{C}\ell_{(p,q)}$ into direct sum of two subalgebras $\mathcal{C}\ell_{(p',q')}$.

6.1.2. Matrix representations of complex Clifford algebras $\mathcal{C}\ell_N$ and their real forms $\mathcal{C}\ell_{(p,q)}$

As before, we set $p, q \in \mathbb{Z}_{\geq 0}$ and $(p+q) = N$. We have shown in Section 6.1.1 that complex Clifford algebra $\mathcal{C}\ell_N$ associated with Euclidean metric $\eta = I_N$ is isomorphic to any of complex algebras $\mathcal{C}\ell^{\mathbb{C}}_{(p,q)}$, associated with metric $\eta = I_{p,q}$. This isomorphism is defined by the transformation (6.1.8) of generators of $\mathcal{C}\ell^{\mathbb{C}}_{(p,q)}$ into generators of $\mathcal{C}\ell_N$. The inverse transformation expresses generators $\Gamma_m \in \mathcal{C}\ell^{\mathbb{C}}_{(p,q)}$ through generators of complex algebra $\mathcal{C}\ell_N$. The elements $\Gamma_m \in \mathcal{C}\ell^{\mathbb{C}}_{(p,q)}$ obtained in this way can be considered also as generators of real Clifford algebra $\mathcal{C}\ell_{(p,q)}$. This means that complex irreducible representations of complex algebra $\mathcal{C}\ell_N$ coincide with complex irreducible representations of any real algebra $\mathcal{C}\ell_{(p,q)}$. We have also shown in Section 6.1.1 that complex irreducible representations of algebra $\mathcal{C}\ell_{(p,q)}$ with $p+q = 2\nu+1$ can be induced from the representations of algebra $\mathcal{C}\ell_{(p',q')}$, where $p' = p, q' = q-1$, or $p' = p-1, q' = q$ (and $p'+q' = 2\nu$).

All of the above means that the study of complex representations of real Clifford algebras $\mathcal{C}\ell_{(p,q)}$, and also of complex Clifford algebras $\mathcal{C}\ell_N$ with even $(p+q) = N = 2\nu$ and odd $(p+q) = N = 2\nu+1$ is reduced to the analysis of complex representations of complex Clifford algebra $\mathcal{C}\ell_{2\nu}$ *with even number of generators*. We note in passing that real Clifford algebras $\mathcal{C}\ell_{(p,q)}$ are not isomorphic, generally speaking, for different p and q and fixed $(p+q) = N$ (with the exception of isomorphisms in Proposition 6.1.1 in previous Section), so their *real* representations (which we study in

Section 6.4) may be different for different p and q; their properties have to be studied on case-by-case basis.

Exact 2^ν-dimensional complex irreducible representation ρ of algebra $\mathcal{C}\ell_{2\nu}$ with even number of generators has been constructed in Section 1.3.2 (see also Section 5.4.1). There we have used isomorphism of algebra $\mathcal{C}\ell_{2\nu}$ and free fermion algebra $\mathcal{A}_\nu$. Recall that isomorphism $\mathcal{C}\ell_{2\nu} = \mathcal{A}_\nu$ is established by the following formulas (see (1.3.17), (1.3.18), (5.4.19)):

$$\begin{gathered} z_\alpha = \frac{1}{2}(\Gamma_{2\alpha-1} + i\Gamma_{2\alpha}), \quad \bar{z}_\alpha = \frac{1}{2}(\Gamma_{2\alpha-1} - i\,\Gamma_{2\alpha}), \\ \Gamma_{2\alpha-1} = z_\alpha + \bar{z}_\alpha, \quad \Gamma_{2\alpha} = i(\bar{z}_\alpha - z_\alpha), \end{gathered} \tag{6.1.27}$$

where $\Gamma_k \in \mathcal{C}\ell_{2\nu}$, and fermionic generators $z_\alpha, \bar{z}_\beta \in \mathcal{A}_\nu$ obey (see (1.3.19), (5.4.4)):

$$[z_\alpha, z_\beta]_+ = 0, \ [\bar{z}_\alpha, \bar{z}_\beta]_+ = 0, \ [z_\alpha, \bar{z}_\beta]_+ = \delta_{\alpha\beta} \quad (\forall \alpha, \beta = 1, \dots, \nu). \tag{6.1.28}$$

Now, in Section 1.3.2 (see also Section 5.4.1) we have constructed irreducible Fock representation ρ of free fermion algebra (6.1.28) which acts in Fock space $\mathcal{F}$. Fock space $\mathcal{F}$ consists of linear combinations of basis vectors obtained by the action of creation operators $\bar{z}_\alpha$ on vacuum vector $|0\rangle \in \mathcal{F}$ ($k = 0, 1, \dots, \nu$):

$$|\alpha_1, \dots, \alpha_k\rangle = \bar{z}_{\alpha_1} \cdots \bar{z}_{\alpha_k} |0\rangle \ (1 \le \alpha_1 < \alpha_2 < \cdots < \alpha_k \le \nu). \tag{6.1.29}$$

The vacuum vector is defined by relations

$$z_\alpha |0\rangle = 0 \quad (\forall \alpha = 1, \dots, \nu), \tag{6.1.30}$$

and elements z_α are annihilation operators. Dimension of space $\mathcal{F}$ equals the number of independent vectors (6.1.29) (see (1.3.24)):

$$\dim(\mathcal{F}) = 2^\nu = 2^{N/2}.$$

Let us write again for references the action of fermionic operators in Fock space (see (1.3.25)):

$$\begin{gathered} z_\beta |\alpha_1, \dots, \alpha_k\rangle = \sum_{i=1}^{k} (-1)^{i-1} \, \delta_{\beta\alpha_i} \, |\alpha_1, \dots, \alpha_{i-1}, \alpha_{i+1}, \dots, \alpha_k\rangle, \\ \bar{z}_\beta |\alpha_1, \dots, \alpha_k\rangle = (-1)^i |\alpha_1, \dots, \alpha_i, \beta, \alpha_{i+1}, \dots, \alpha_k\rangle \quad (\alpha_i < \beta < \alpha_{i+1}), \\ \bar{z}_\beta |\alpha_1, \dots, \alpha_k\rangle = 0 \quad (\beta = \alpha_i). \end{gathered} \tag{6.1.31}$$

Relations (6.1.27) and (6.1.31) define then the irreducible representation ρ of Clifford algebra $\mathcal{C\ell}_{2\nu}$ in Fock space. To cast it in matrix form, we enumerate the basis vectors (6.1.29) in an arbitrary way, $\{|\alpha_1, \ldots, \alpha_k\rangle\} \to \{v_i\}$, where $i = 1, \ldots, 2^\nu$; explicit example is given below. Then operators z_β and $\bar{z}_\beta$, acting as in (6.1.31), define 2^ν-dimensional matrices $\rho(z_\beta)$ and $\rho(\bar{z}_\beta)$, such that the standard relations hold:

$$z_\beta \cdot v_i = v_j \rho_{ji}(z_\beta), \quad \bar{z}_\beta \cdot v_i = v_j \rho_{ji}(\bar{z}_\beta). \tag{6.1.32}$$

The definition (1.3.30) of Hermitian conjugation in $\mathcal{F}$ yields $\bar{z}_\beta = z_\beta^\dagger$ so matrices $\rho(\bar{z}_\beta)$ are Hermitian conjugate to matrices $\rho(z_\beta)$, $\rho_{ij}(\bar{z}_\beta) = \rho_{ji}(z_\beta)^*$. It follows from (6.1.27) that matrices of generators $\Gamma_m \in \mathcal{C\ell}_{2\nu}$, are Hermitian, $\rho_{ij}(\Gamma_m) = \rho_{ji}(\Gamma_m)^*$ (see explicit formulas (6.1.50), (6.1.51) below).

Problem 6.1.2. Prove that matrices $\rho(\bar{z}_\beta)$ are Hermitian conjugate (with respect to Hermitian scalar product (1.3.31)) to matrices $\rho(z_\beta)$.

Remark 1. Complex algebra $\mathcal{C\ell}_{2\nu}$ of dimension $2^{2\nu}$ has exact representation ρ which is realized by all complex matrices $2^\nu \times 2^\nu$. So, there is isomorphism

$$\mathcal{C\ell}_{2\nu} = \mathrm{Mat}_{2^\nu}(\mathbb{C}). \tag{6.1.33}$$

By arguments of Section 4.5.1 (see Example 1), algebra $\mathcal{C\ell}_{2\nu}$ is simple. We discuss isomorphism (6.1.33) below in more detail, see Remark 4 in the end of this subsection.

Remark 2. We recall (see Proposition 1.3.2 in Section 1.3.2) that complex 2^ν-dimensional representation ρ of Clifford algebra $\mathcal{C\ell}_{2\nu}$ given by (6.1.27) and (6.1.31) is irreducible, and that all other complex irreducible representations of Clifford algebra $\mathcal{C\ell}_{2\nu}$ have dimension 2^ν and are equivalent to representation ρ. The identification (6.1.8) ensures that equivalence of all complex representations holds also for real Clifford algebra $\mathcal{C\ell}_{(p,q)}$ with $(p+q) = 2\nu$. Proposition that all complex irreducible representations of real Clifford algebra $\mathcal{C\ell}_{(p,q)}$ (or complex algebra $\mathcal{C\ell}_{2\nu}$) are equivalent and have dimension 2^ν is known as Pauli theorem.

Let us present an alternative proof of the Pauli theorem. We recall the remarkable identity (4.3.63) (see also (4.5.27)) valid for all finite groups and associative semisimple algebras, and apply it to (simple) Clifford algebra $\mathcal{C}\ell_{2\nu}$. The left-hand side of (4.3.63) is the dimension $2^{2\nu}$ of algebra $\mathcal{C}\ell_{2\nu}$. We know that there exists 2^ν-dimensional representation ρ of algebra $\mathcal{C}\ell_{2\nu}$, so the right-hand side of (4.3.63), which is the sum over all irreducible inequivalent representations of algebra $\mathcal{C}\ell_{2\nu}$, is saturated by only one term $(N_\rho)^2 = (2^\nu)^2$. This shows that ρ is the only inequivalent representation of Clifford algebra $\mathcal{C}\ell_{2\nu}$, i.e., that all irreducible representations are equivalent to ρ.

Remark 3. Let ρ be a 2^ν-dimensional matrix representation of algebra $\mathcal{C}\ell_{2\nu}$. Then permutations $\sigma \in S_{2\nu}$ of its matrices

$$\rho(\Gamma_1), \dots, \rho(\Gamma_{2\nu}) \;\to\; \rho(\Gamma_{\sigma(1)}), \dots, \rho(\Gamma_{\sigma(2\nu)}) \tag{6.1.34}$$

preserve the relations (6.1.9) and hence define other representations of $\mathcal{C}\ell_{2\nu}$. Pauli theorem, however, ensures that all these representations are equivalent.

Examples of matrix representations of algebras $\mathcal{A}_\nu$ and $\mathcal{C}\ell_{2\nu}$

1. Representation ρ of fermion algebra $\mathcal{A}_\nu$ with defining relations (6.1.28) in the case $\nu = 1$ is two dimensional, and basis (6.1.29) in Fock space has two vectors

$$v_1 = |0\rangle, \quad v_2 = \bar{z}_1|0\rangle. \tag{6.1.35}$$

Now, in accordance with (6.1.31) we obtain

$$z_1 \cdot v_1 = 0,\; z_1 \cdot v_2 = v_1,\; \bar{z}_1 \cdot v_1 = v_2,\; \bar{z}_1 \cdot v_2 = 0.$$

So, matrix representation ρ in basis (v_1, v_2) reads

$$\begin{gathered} z_1 \cdot v_i = v_k \rho_{ki}(z_1),\; \bar{z}_1 \cdot v_i = v_k\, \rho_{ki}(\bar{z}_1) \\ \Rightarrow\; \rho(z_1) = \begin{pmatrix} 0 & 1 \\ 0 & 0 \end{pmatrix} \equiv e_+, \rho(\bar{z}_1) = \begin{pmatrix} 0 & 0 \\ 1 & 0 \end{pmatrix} \equiv e_-. \end{gathered} \tag{6.1.36}$$

In what follows, we omit for brevity the notation ρ when this does not lead to confusion, and instead of $\rho(z)$ write z, etc. Using representation (6.1.36) in formulas (6.1.27), we obtain irreducible representation of generators (Γ_1, Γ_2) of Clifford algebra $\mathcal{C}\ell_2$:

$$\Gamma_1 = \sigma_1,\; \Gamma_2 = \sigma_2 \Rightarrow \Gamma_3 = (-i)\Gamma_1\Gamma_2 = \sigma_3, \tag{6.1.37}$$

where σ_k are Pauli matrices, and $\Gamma_3 = \sigma_3$ is the matrix Γ_{N+1} defined in (6.1.19) (with $N = 2$, i.e., $\nu = 1$, $q = 0$).

We add to matrices (Γ_1, Γ_2) the third matrix $\Gamma_3 = \sigma_3$ and get the set of generators which, in view of (6.1.5), obey relations (6.1.9) for Clifford algebra $\mathcal{C}\ell_3$ and hence define its two-dimensional representation.

2. For $\nu = 2$, representation ρ of fermion algebra is four dimensional, and basis elements (6.1.29) in Fock space are

$$v_1 = |0\rangle, \; v_2 = \bar{z}_2\,|0\rangle, \; v_3 = \bar{z}_1|0\rangle, \; v_4 = \bar{z}_1\,\bar{z}_2\,|0\rangle \tag{6.1.38}$$

(note the assignment of indices in definitions of v_2 and v_3). The action of operators z_α, $\bar{z}_\alpha$ is given by (6.1.31), and, in accordance with (6.1.32), we obtain in matrix form

$$z_1 = \begin{pmatrix} 0 & 0 & 1 & 0 \\ 0 & 0 & 0 & 1 \\ 0 & 0 & 0 & 0 \\ 0 & 0 & 0 & 0 \end{pmatrix} = e_+ \otimes I_2, \; \bar{z}_1 = \begin{pmatrix} 0 & 0 & 0 & 0 \\ 0 & 0 & 0 & 0 \\ 1 & 0 & 0 & 0 \\ 0 & 1 & 0 & 0 \end{pmatrix} = e_- \otimes I_2,$$

$$z_2 = \begin{pmatrix} 0 & 1 & 0 & 0 \\ 0 & 0 & 0 & 0 \\ 0 & 0 & 0 & -1 \\ 0 & 0 & 0 & 0 \end{pmatrix} = \sigma_3 \otimes e_+, \; \bar{z}_2 = \begin{pmatrix} 0 & 0 & 0 & 0 \\ 1 & 0 & 0 & 0 \\ 0 & 0 & 0 & 0 \\ 0 & 0 & -1 & 0 \end{pmatrix} = \sigma_3 \otimes e_-. \tag{6.1.39}$$

Here the tensor product $\otimes$ of matrices (in this particular case of size 2×2 each) is defined as follows (see Sections I-1.2.2 and I-4.3.1 in the first book). Let A and W be 2×2 and $n \times n$ matrices, respectively. Then matrix $A \otimes W$ has size $2n \times 2n$ and reads

$$A \otimes W = \begin{pmatrix} A_{11}W & A_{12}W \\ A_{21}W & A_{22}W \end{pmatrix}, \tag{6.1.40}$$

where A_{ij} are matrix elements of A, and $A_{ij}W$ are $n \times n$ blocks. In the notation (6.1.40), the usual multiplication of $2n \times 2n$ matrices obeys (see Section I-1.2.2 in the first book)

$$(A \otimes W) \cdot (B \otimes V) = (A \cdot B) \otimes (W \cdot V). \tag{6.1.41}$$

One also has

$$(A + B) \otimes W = A \otimes W + B \otimes W, \quad A \otimes (W + V) = A \otimes W + A \otimes V. \tag{6.1.42}$$

We note that arbitrary 4×4 matrix can be written in block 2×2 form

$$\begin{pmatrix} W & V \\ U & T \end{pmatrix} = \begin{pmatrix} 1 & 0 \\ 0 & 0 \end{pmatrix} \otimes W + \begin{pmatrix} 0 & 1 \\ 0 & 0 \end{pmatrix} \otimes V + \begin{pmatrix} 0 & 0 \\ 1 & 0 \end{pmatrix} \otimes U + \begin{pmatrix} 0 & 0 \\ 0 & 1 \end{pmatrix} \otimes T$$
$$= \frac{1}{2} I_2 \otimes (W+T) + \frac{1}{2}\sigma_3 \otimes (W-T) + e_+ \otimes V + e_- \otimes U,$$

where blocks W, V, U, T are (2×2) matrices. This is a particular case of the following general fact.

Problem 6.1.3. Show that any complex $(n \cdot m) \times (n \cdot m)$ matrix A can be written as $A = (e_{ij} \otimes E_{\alpha\beta}) A_{i\alpha, j\beta}$, where $A_{i\alpha,j\beta} \in \mathbb{C}$, and e_{ij}, $E_{\alpha\beta}$ are $(n \times n)$ and $(m \times m)$ matrix units, respectively.

Note that the relations (6.1.41) and (6.1.42) become evident, once one recalls that tensor product $(A \otimes B)$ of $(n \times n)$ and $(m \times m)$ matrices A and B acts in tensor product $\mathcal{V}_n \otimes \mathcal{V}_m$ of two vector spaces $\mathcal{V}_n$ and $\mathcal{V}_m$ as $(A \otimes B)(\mathcal{V}_n \otimes \mathcal{V}_m) = (A \cdot \mathcal{V}_n) \otimes (B \cdot \mathcal{V}_m)$ (see Section I-4.3.1 of the first book).

Making use of (6.1.39) and (6.1.27), we obtain irreducible representation for algebra $\mathcal{C}\ell_4$:

$$\Gamma_1 = \sigma_1 \otimes I_2, \; \Gamma_2 = \sigma_2 \otimes I_2, \; \Gamma_3 = \sigma_3 \otimes \sigma_1, \; \Gamma_4 = \sigma_3 \otimes \sigma_2. \tag{6.1.43}$$

Matrix (6.1.19) of the longest element in this case ($N = 2\nu = 4$, $q = 0$) is

$$\Gamma_5 = -\Gamma_1 \cdots \Gamma_4 = \sigma_3 \otimes \sigma_3. \tag{6.1.44}$$

Together with (6.1.43), this gives representation of algebra $\mathcal{C}\ell_5$. Note that the representation (6.1.3) of algebra $\mathcal{C}\ell_{(1,3)}$, upon the change $\Gamma_1 = \gamma_0$, $\Gamma_{k+1} = -i\gamma_k$ $(k = 1, 2, 3)$ defines representation of algebra $\mathcal{C}\ell_4$

$$\Gamma_1 = \sigma_1 \otimes I_2, \; \Gamma_{k+1} = \sigma_2 \otimes \sigma_k, \; \Gamma_5 = -\Gamma_1 \cdots \Gamma_4 = \sigma_3 \otimes I_2 \tag{6.1.45}$$

(here Γ_5 is the longest element). The latter representation is different from (6.1.43), (6.1.44), but equivalent to (6.1.43), (6.1.44) due to Pauli theorem.

Matrix representations of algebras $\mathcal{C}\ell_{2\nu}$ are constructed for arbitrary ν in a similar way. Construction (6.1.40) is straightforwardly generalized to $2^\nu \times 2^\nu$ matrices, namely, matrix $A^{(1)} \otimes A^{(2)} \otimes \cdots \otimes A^{(\nu)}$, where $A^{(i)}$ are

2×2 matrices, is

$$\begin{aligned} & A^{(1)} \otimes A^{(2)} \otimes \ldots \otimes A^{(\nu)} \\ & \quad = \begin{pmatrix} A^{(1)}_{11} \cdot (A^{(2)} \otimes \cdots \otimes A^{(\nu)}) & A^{(1)}_{12} \cdot (A^{(2)} \otimes \cdots \otimes A^{(\nu)}) \\ A^{(1)}_{21} \cdot (A^{(2)} \otimes \cdots \otimes A^{(\nu)}) & A^{(1)}_{22} \cdot (A^{(2)} \otimes \cdots \otimes A^{(\nu)}) \end{pmatrix}. \end{aligned}$$

Matrix $(A^{(2)} \otimes \cdots \otimes A^{(\nu)})$ has dimension $2^{\nu-1} \times 2^{\nu-1}$ and is given by

$$A^{(2)} \otimes \cdots \otimes A^{(\nu)} = \begin{pmatrix} A^{(2)}_{11} \cdot (A^{(3)} \otimes \cdots \otimes A^{(\nu)}) & A^{(2)}_{12} \cdot (A^{(3)} \otimes \cdots \otimes A^{(\nu)}) \\ A^{(2)}_{21} \cdot (A^{(3)} \otimes \cdots \otimes A^{(\nu)}) & A^{(2)}_{22} \cdot (A^{(3)} \otimes \cdots \otimes A^{(\nu)}) \end{pmatrix},$$

etc. The generalization of (6.1.41) reads

$$\begin{aligned} & (A^{(1)} \otimes A^{(2)} \otimes \cdots \otimes A^{(\nu)}) \cdot (B^{(1)} \otimes B^{(2)} \otimes \cdots \otimes B^{(\nu)}) \\ & \quad = (A^{(1)} \cdot B^{(1)}) \otimes (A^{(2)} \cdot B^{(2)}) \otimes \cdots \otimes (A^{(\nu)} \cdot B^{(\nu)}). \end{aligned}$$

We also have analog of (6.1.42):

$$\begin{aligned} & A^{(1)} \otimes \cdots \otimes A^{(k-1)} \otimes (A^{(k)} + B^{(k)}) \otimes A^{(k+1)} \otimes \ldots \otimes A^{(\nu)} \\ & \quad = A^{(1)} \otimes \cdots \otimes A^{(k-1)} \otimes A^{(k)} \otimes A^{(k+1)} \otimes \cdots \otimes A^{(\nu)} \\ & \qquad + A^{(1)} \otimes \cdots \otimes A^{(k-1)} \otimes B^{(k)} \otimes A^{(k+1)} \otimes \cdots \otimes A^{(\nu)}. \end{aligned}$$

With these formulas, representations (6.1.36) and (6.1.39) are generalized as follows:

$$z_\alpha = \underbrace{\sigma_3 \otimes \sigma_3 \otimes \cdots \sigma_3}_{\alpha-1} \otimes e_+ \otimes \underbrace{I_2 \cdots \otimes I_2}_{\nu-\alpha}, \tag{6.1.46}$$

$$\bar{z}_\alpha = \underbrace{\sigma_3 \otimes \sigma_3 \otimes \cdots \sigma_3}_{\alpha-1} \otimes e_- \otimes \underbrace{I_2 \cdots \otimes I_2}_{\nu-\alpha}. \tag{6.1.47}$$

This representation is 2^ν-dimensional, as anticipated. The assignment of basis vectors $(v_1, \ldots, v_{2^\nu})$ in Fock space, which gives the matrix representation (6.1.46), (6.1.47), is constructed by induction, namely, let $(v'_1, \ldots, v'_{2^{\nu-1}})$ be assignment of basis Fock vectors (6.1.29) that results in representation (6.1.46), (6.1.47) for $(\nu-1)$-dimensional free fermion algebra $\{z_\alpha, \bar{z}_\alpha\}$ $(\alpha = 1, \ldots, \nu-1)$. We add one more fermion $(z_\nu, \bar{z}_\nu)$ to this algebra, and redefine, in expressions (6.1.29) for vectors $(v'_1, \ldots, v'_{2^{\nu-1}})$, creation operators: $\bar{z}_\alpha \to \bar{z}_{\alpha+1}$. In notations (6.1.29), this is a redefinition

$$|\alpha_1,\ldots,\alpha_k\rangle \to |\alpha_1+1,\ldots,\alpha_k+1\rangle \;\Rightarrow\; v'_k \to v_k \quad (k=1,\ldots,2^{\nu-1}).$$

In this way, we define a new set of Fock vectors $(v_1,\ldots,v_{2^{\nu-1}})$. Let us construct another set of $2^{\nu-1}$ Fock vectors: $v_{2^{\nu-1}+k} = \bar{z}_1 \cdot v_k$ where $k = 1,\ldots,2^{\nu-1}$. Then the set of 2^ν vectors

$$(v_1, v_2, \ldots, v_{2^\nu}) \equiv (v_1, v_2, \ldots, v_{2^{\nu-1}},\; \bar{z}_1\cdot v_1,\; \bar{z}_1\cdot v_2, \ldots, \bar{z}_1\cdot v_{2^{\nu-1}})$$

is the desired assignment.

Problem 6.1.4. Show that the above assignment of vectors $(v_1, v_2, \ldots, v_{2^\nu})$ gives, via (6.1.31) and (6.1.32), representation (6.1.46), (6.1.47) of ν-dimensional free fermion algebra. Obtain basis (6.1.38) from basis (6.1.35) in this way.

Representation (6.1.46), (6.1.47) can be obtained also by another induction procedure, namely, let matrices (6.1.46) and (6.1.47) be representation of algebra $\mathcal{A}_\nu$ of free fermions $\{z_\alpha, \bar{z}_\beta\}$ $(\alpha,\beta = 1,\ldots,\nu)$. Then $2^{(\nu+1)}$-dimensional representation ρ' of algebra $\mathcal{A}_{\nu+1}$ is constructed as follows:

$$\begin{aligned} &\rho'(z_\alpha) = z_\alpha \otimes I_2,\; \rho'(\bar{z}_\alpha) = \bar{z}_\alpha \otimes I_2,\; (\alpha = 1,\ldots,\nu),\\ &\rho'(z_{\nu+1}) = \Gamma_{2\nu+1}\otimes e_+,\; \rho'(\bar{z}_{\nu+1}) = \Gamma_{2\nu+1}\otimes e_-, \end{aligned} \tag{6.1.48}$$

where $\Gamma_{2\nu+1}$ is the longest element (6.1.19) for $q=0$, $N=2\nu$, which is obtained by using (6.1.27) and in representation (6.1.46), (6.1.47) reads (see also (5.4.26))

$$\begin{aligned} \Gamma_{2\nu+1} &= (-i)^\nu \Gamma_1\cdots\Gamma_N = (-i\Gamma_1\Gamma_2)\cdot(-i\Gamma_3\Gamma_4)\cdots(-i\Gamma_{N-1}\Gamma_N)\\ &= (z_1\bar{z}_1 - \bar{z}_1 z_1)\cdots(z_\nu\bar{z}_\nu - \bar{z}_\nu z_\nu) = \underbrace{\sigma_3\otimes\sigma_3\otimes\cdots\sigma_3}_{\nu} \end{aligned} \tag{6.1.49}$$

(cf. Γ_3 in (6.1.37) and Γ_5 in (6.1.44) for $\nu=1$ and $\nu=2$).

Problem 6.1.5. Show that if $\{z_\alpha, \bar{z}_\beta\}$ are generators of free fermion algebra $\mathcal{A}_\nu$, then operators (6.1.48) generate algebra $\mathcal{A}_{\nu+1}$. Making use of induction procedure (6.1.48), construct representation (6.1.46), (6.1.47) for $\mathcal{C}\ell_{2\nu}$, starting from representation (6.1.36) of $\mathcal{C}\ell_2$. Show explicitly that operators (6.1.46), (6.1.47) obey relations (6.1.28).

We now plug matrices (6.1.46), (6.1.47) in (6.1.27) and obtain matrix representation of generators Γ_k:

$$\Gamma_{2\alpha-1} = \underbrace{\sigma_3 \otimes \sigma_3 \otimes \cdots \sigma_3}_{\alpha-1} \otimes \sigma_1 \otimes \underbrace{I_2 \cdots \otimes I_2}_{\nu-\alpha}, \tag{6.1.50}$$

$$\Gamma_{2\alpha} = \underbrace{\sigma_3 \otimes \sigma_3 \otimes \cdots \sigma_3}_{\alpha-1} \otimes \sigma_2 \otimes \underbrace{I_2 \cdots \otimes I_2}_{\nu-\alpha}. \tag{6.1.51}$$

Proposition 1.3.2 ensures that this 2^ν-dimensional representation of algebra $\mathcal{C}\ell_{2\nu}$ is irreducible, and all other irreducible representations are related to it by equivalence transformations. Representation of Clifford algebra $\mathcal{C}\ell_{2\nu}$ given by (6.1.50), (6.1.51) was constructed by Brauer and Weyl in Ref. [19].

Problem 6.1.6. Show explicitly that in even case $N = 2\nu$ the longest element $\Gamma_{N+1} \in \mathcal{C}\ell_N$ in representation (6.1.50), (6.1.51), i.e., matrix (6.1.49), obeys

$$\Gamma^2_{N+1} = I, \; [\Gamma_{N+1}, \Gamma_m]_+ = 0 \quad (\forall m = 1, \ldots, N). \tag{6.1.52}$$

Problem 6.1.7. Construct representation of Clifford algebra $\mathcal{C}\ell_{2\nu}$ by defining free fermions in an alternative way (cf. (6.1.27))

$$z_\alpha = \frac{1}{2}(\Gamma_\alpha + i\Gamma_{\nu+\alpha}), \; \bar{z}_\alpha = \frac{1}{2}(\Gamma_\alpha - i\Gamma_{\nu+\alpha}) \quad (\alpha = 1, \ldots, \nu),$$

(generators of algebra $\mathcal{C}\ell_{2\nu}$ are divided into two groups in a different manner). Does this representation reduce to (6.1.50), (6.1.51) by permutation of generators Γ_k?

Gamma-matrices in representation (6.1.50), (6.1.51) obey $\Gamma^*_{2\alpha-1} = \Gamma^{\mathsf{T}}_{2\alpha-1} = \Gamma_{2\alpha-1}$ and $\Gamma^*_{2\alpha} = \Gamma^{\mathsf{T}}_{2\alpha} = -\Gamma_{2\alpha}$, which follow from the explicit expressions for Pauli matrices and which can be written in unified way:

$$\Gamma^*_m = \Gamma^{\mathsf{T}}_m = (-1)^{m+1}\Gamma_m. \tag{6.1.53}$$

We also have $(\Gamma_k)^2 = I_{2^\nu}$. Furthermore, all Γ-matrices (6.1.50), (6.1.51) are Hermitian and unitary:

$$\Gamma^\dagger_k = \Gamma_k, \quad \Gamma^\dagger_k \Gamma_k = I_{2^\nu}. \tag{6.1.54}$$

Remark 4. To prove isomorphism $\mathcal{C}\ell_{2\nu} = \mathrm{Mat}_{2^\nu}(\mathbb{C})$ (see (6.1.33)), we construct explicitly matrix units in algebra $\mathrm{Mat}_{2^\nu}(\mathbb{C})$ in terms of elements of algebra $\mathcal{C}\ell_{2\nu}$. Let e_{ij} $(i,j=1,2)$ be matrix units in $\mathrm{Mat}_2(\mathbb{C})$, then matrix units in $\mathrm{Mat}_{2^\nu}(\mathbb{C})$ can be written as follows:

$$e_{i_1j_1}\otimes e_{i_2j_2}\otimes\cdots\otimes e_{i_\nu j_\nu}=\prod_{\alpha=1}^{\nu}e^{(\alpha)}_{i_\alpha j_\alpha}, \tag{6.1.55}$$

$$e^{(\alpha)}_{i_\alpha j_\alpha}:=I_2^{\otimes(\alpha-1)}\otimes e_{i_\alpha j_\alpha}\otimes I_2^{\otimes(\nu-\alpha)}. \tag{6.1.56}$$

Matrices (6.1.56), and hence (6.1.55), can be expressed through free fermions (6.1.46), (6.1.47):

$$e^{(\alpha)}_{11}=z_\alpha\bar{z}_\alpha,\ e^{(\alpha)}_{22}=\bar{z}_\alpha z_\alpha,\ e^{(\alpha)}_{12}=\prod_{\beta=1}^{\alpha-1}[z_\beta,\bar{z}_\beta]\cdot z_\alpha,\ e^{(\alpha)}_{21}=\prod_{\beta=1}^{\alpha-1}[z_\beta,\bar{z}_\beta]\cdot\bar{z}_\alpha. \tag{6.1.57}$$

Free fermions are in turn expressed through generators $\Gamma_k\in\mathcal{C}\ell_{2\nu}$ via (6.1.27). So, the relations (6.1.55), (6.1.57) establish the explicit relationship between elements of algebra $\mathcal{C}\ell_{2\nu}$ and matrix units in algebra $\mathrm{Mat}_{2^\nu}(\mathbb{C})$. Note that diagonal matrices in (6.1.55), $\prod_\alpha e^{(\alpha)}_{i_\alpha i_\alpha}$, expressed through free fermions (or generators of algebra $\mathcal{C}\ell_{2\nu}$) in accordance with the first two formulas in (6.1.57), form complete system of primitive idempotents (see Section 4.5.1) in Clifford algebra $\mathcal{C}\ell_{2\nu}$.

Consider now complex Clifford algebra $\mathcal{C}\ell_N$ with odd number of generators $N=2\nu+1$. In that case, the longest element (6.1.21) for $q=0$ reads

$$\Gamma_{N+1}=(-i)^\nu\Gamma_1\cdots\Gamma_{2\nu}\cdot\Gamma_{2\nu+1}\ \in\ \mathcal{C}\ell_{2\nu+1}. \tag{6.1.58}$$

It is central in algebra $\mathcal{C}\ell_{2\nu+1}$, and the relation $\Gamma^2_{N+1}=I$ (see (1.3.11)) shows that it can have only two eigenvalues ±1. Schur's lemma ensures that in complex irreducible representation of algebra $\mathcal{C}\ell_{2\nu+1}$ element (6.1.58) is proportional to unity; the proportionality coefficient is either $+1$ or -1. So, algebra $\mathcal{C}\ell_{2\nu+1}$ has at least two (exactly two, as we show below) inequivalent irreducible representations ρ_+ and ρ_- with

$$\rho_+(\Gamma_{N+1})=+I_+,\ \rho_-(\Gamma_{N+1})=-I_-, \tag{6.1.59}$$

where $I_\pm$ are unit operators in spaces of representations $\rho_\pm$. Making use of (6.1.59) and definition (6.1.58) of Γ_{N+1}, we express one of the generators

$\Gamma_m \in \mathcal{C}\ell_N$, say $\Gamma_N = \Gamma_{2\nu+1}$ (in irreducible representation ρ_+ or ρ_-) through the other generators Γ_m $(m = 1, \ldots, N-1 = 2\nu)$:

$$\begin{aligned} \rho_+(\Gamma_{2\nu+1}) &= +(-i)^\nu \rho_+(\Gamma_1 \cdots \Gamma_{2\nu}), \\ \rho_-(\Gamma_{2\nu+1}) &= -(-i)^\nu \rho_-(\Gamma_1 \cdots \Gamma_{2\nu}) \end{aligned} \tag{6.1.60}$$

(cf. isomorphism (1.3.16) and formula (6.1.19) with $q = 0$ and $N = 2\nu$). Representations ρ_+ and ρ_- of subalgebra $\mathcal{C}\ell_{2\nu} \subset \mathcal{C}\ell_{2\nu+1}$ with generators $\{\Gamma_1, \ldots, \Gamma_{2\nu}\}$ coincide (modulo equivalence transformation), $\rho_+ = \rho_- = \rho$. Therefore, inequivalent representations ρ_+ and ρ_- of the whole algebra $\mathcal{C}\ell_{2\nu+1}$ differ only by the sign of the last generator $\Gamma_{2\nu+1}$ (see (6.1.60)):

$$\rho_+(\Gamma_{2\nu+1}) = +(-i)^\nu \rho(\Gamma_1 \cdots \Gamma_{2\nu}),\ \rho_-(\Gamma_{2\nu+1}) = -(-i)^\nu \rho(\Gamma_1 \cdots \Gamma_{2\nu}). \tag{6.1.61}$$

Thus, the two 2^ν-dimensional inequivalent irreducible representations ρ_+ and ρ_- of algebra $\mathcal{C}\ell_{2\nu+1}$ are constructed in terms of 2^ν-dimensional irreducible representation ρ of algebra $\mathcal{C}\ell_{2\nu}$ by (6.1.61) and differ only by the value $(+1)$ or (-1) of central element Γ_{N+1}.

Now, isomorphism (1.3.15) of algebra $\mathcal{C}\ell_{2\nu+1}$ and isomorphism (6.1.33) give

$$\mathcal{C}\ell_{2\nu+1} = \mathrm{Mat}_{2^\nu}(\mathbb{C}) \oplus \mathrm{Mat}_{2^\nu}(\mathbb{C}). \tag{6.1.62}$$

Then Wedderburn theorem (Theorem 4.5.1) tells that algebra $\mathcal{C}\ell_{2\nu+1}$ is semisimple. Therefore, identity (4.3.63) does apply to algebra $\mathcal{C}\ell_{2\nu+1}$ and now has the form

$$2^{2\nu+1} = (2^\nu)^2 + (2^\nu)^2, \tag{6.1.63}$$

where the left-hand side is the dimension of algebra $\mathcal{C}\ell_{2\nu+1}$, while the right-hand side is the sum of dimensions squared of irreducible representations $\rho_\pm$ of algebra $\mathcal{C}\ell_{2\nu+1}$. It follows from (6.1.63) that algebra $\mathcal{C}\ell_{2\nu+1}$ does not have other inequivalent irreducible representations besides ρ_+ and ρ_-. It is clear from isomorphism (6.1.62) and formula (6.1.63) that irreducible 2^ν-dimensional representations ρ_+ and ρ_- of algebra $\mathcal{C}\ell_{2\nu+1}$ are not exact.

Problem 6.1.8. Construct Fock spaces $\mathcal{V}_+$ and $\mathcal{V}_-$ of inequivalent representations ρ_+ and ρ_- of Clifford algebra $\mathcal{C}\ell_{2\nu+1}$ in terms of free fermions.

We summarize all of the above, together with Proposition 1.3.2, as follows.

Proposition 6.1.3. *Complex algebra*

$$\mathcal{C}\ell_{2\nu} = \mathrm{Mat}_{2^\nu}(\mathbb{C})$$

has unique (modulo equivalence transformations) irreducible representation ρ *of dimension* 2^ν*, e.g., representation* (6.1.50), (6.1.51). *All other irreducible representations have the same dimension* 2^ν *and are equivalent to* ρ.

Complex algebra

$$\mathcal{C}\ell_{2\nu+1} = \mathrm{Mat}_{2^\nu}(\mathbb{C}) \oplus \mathrm{Mat}_{2^\nu}(\mathbb{C})$$

has exactly two inequivalent irreducible representations $\rho_\pm$ *of dimension* 2^ν*, which differ by the sign of central element* (6.1.58): $\rho_+(\Gamma_{N+1}) = -\rho_-(\Gamma_{N+1}) = I$*, where* $N = 2\nu + 1$*. Representations* $\rho_\pm$ *are constructed in terms of irreducible representation* ρ *of subalgebra* $\mathcal{C}\ell_{2\nu} \subset \mathcal{C}\ell_{2\nu+1}$ *with generators* $\Gamma_1, \Gamma_2, \ldots, \Gamma_{2\nu}$*; additional generator* $\Gamma_{2\nu+1}$ *has one of the two forms* (6.1.61).

We now discuss complex irreducible representations of real algebras $\mathcal{C}\ell_{(p,q)}$ with odd number $(p+q) = N = 2\nu+1$ of generators Γ_m $(m = 1, \ldots, 2\nu+1)$, obeying defining relations (6.1.2). Complex representations of real algebras $\mathcal{C}\ell_{(p,q)}$ are the same as representations of complex algebras $\mathcal{C}\ell_{2\nu+1}$, so our discussion to large extent parallels the above analysis of representations of complex algebras $\mathcal{C}\ell_{2\nu+1}$. As we have already pointed out, the relation (6.1.20) shows that central element Γ_{N+1} has only two eigenvalues ± 1. Then real algebra $\mathcal{C}\ell_{(p,q)}$ with odd $(p+q) = 2\nu+1$, like complex algebra $\mathcal{C}\ell_{2\nu+1}$, has two inequivalent irreducible representations ρ_+ and ρ_- with different eigenvalues (6.1.59) of central element Γ_{N+1}. We can again express one of the generators of $\mathcal{C}\ell_{(p,q)}$, say, the last one $\Gamma_N = \Gamma_{2\nu+1}$ (in irreducible representation ρ_+ or ρ_-) through the other generators (cf. (6.1.60)):

$$\begin{aligned} \rho_+(\Gamma_{2\nu+1}) &= +(-i)^{\nu+q}\eta_{NN}\,\rho_+(\Gamma_1\cdots\Gamma_{2\nu}), \\ \rho_-(\Gamma_{2\nu+1}) &= -(-i)^{\nu+q}\eta_{NN}\,\rho_-(\Gamma_1\cdots\Gamma_{2\nu}). \end{aligned} \tag{6.1.64}$$

For metric η with $\eta_{NN} = -1$, generators $(\Gamma_1, \ldots, \Gamma_{2\nu})$ form subalgebra $\mathcal{C}\ell_{(p,q-1)} \subset \mathcal{C}\ell_{(p,q)}$, while for $\eta_{NN} = +1$ they generate subalgebra $\mathcal{C}\ell_{(p-1,q)} \subset \mathcal{C}\ell_{(p,q)}$. Both subalgebras $\mathcal{C}\ell_{(p,q-1)}$ and $\mathcal{C}\ell_{(p-1,q)}$ have even number of generators $(p+q-1) = 2\nu$.

Problem 6.1.9. Making use of the definitions (6.1.64), prove that $(\rho_\pm\,(\Gamma_{2\nu+1}))^2 = \eta_{NN}$, where $N = 2\nu+1$. Hint: make use of the relation $(-1)^q = \eta_{11}\cdots\,\eta_{2\nu,2\nu}\eta_{NN}$.

Now, take $\eta_{NN} = -1$ (or $\eta_{NN} = +1$), and let ρ be irreducible 2^ν-dimensional representation of subalgebra $\mathcal{C}\ell_{(p,q-1)} \subset \mathcal{C}\ell_{(p,q)}$ (or subalgebra $\mathcal{C}\ell_{(p-1,q)}$) with even number of generators $\{\Gamma_1, \ldots, \Gamma_{2\nu}\}$. We define two 2^ν-dimensional representations $\rho_\pm$ of the whole algebra $\mathcal{C}\ell_{(p,q)}$ as follows:

$$\rho_+(\Gamma_m) = \rho_-(\Gamma_m) = \rho(\Gamma_m) \quad (\forall m = 1, \ldots, 2\nu), \tag{6.1.65}$$

$$\rho_\pm(\Gamma_{2\nu+1}) = \pm(-i)^{\nu+q}\, \eta_{NN}\, \rho(\Gamma_1 \cdots \Gamma_{2\nu}), \tag{6.1.66}$$

i.e., the representations of the first 2ν generators coincide, $\rho_+ = \rho_- = \rho$, while representations of the last generator $\rho_\pm(\Gamma_{2\nu+1})$ differ by sign, in accordance with (6.1.64). These representations of algebra $\mathcal{C}\ell_{(p,q)}$ with odd $p + q = 2\nu + 1$ are inequivalent, irreducible and determined completely by representation ρ of subalgebra $\mathcal{C}\ell_{(p,q-1)}$ (or $\mathcal{C}\ell_{(p-1,q)}$). The fact that there are no other inequivalent representations is guaranteed by the relation (6.1.63) valid also for real algebra $\mathcal{C}\ell_{(p,q)}$.

Remark 5. Analogs of isomorphisms (6.1.33), (6.1.62) are more subtle for real Clifford algebras $\mathcal{C}\ell_{(p,q)}$. This has to do with the fact that linear relations like (6.1.27) between generators $\Gamma_k \in \mathcal{C}\ell_{(p,q)}$ and free fermions are complex. So, relations between generators Γ_k and matrix units are complex too. Also, in the odd case $N = (p + q) = 2\nu + 1$ analog of decomposition (1.3.14) of real algebra $\mathcal{C}\ell_{(p,q)}$ into the direct sum of two real subalgebras is valid only if central element Γ_{N+1} belongs to real algebra $\mathcal{C}\ell_{(p,q)}$, which is the case (see (6.1.19)) only for even $(\nu + q)$. Isomorphisms between real Clifford algebras $\mathcal{C}\ell_{(p,q)}$ and real matrix algebras are studied in detail, e.g., in review [52].

6.1.3. Weyl representations of Clifford algebras $\mathcal{C}\ell_N$ and $\mathcal{C}\ell_{(p,q)}$

Here, we introduce yet another induction method of constructing irreducible representations of Clifford algebra $Cl_{2\nu}$ with defining relations (6.1.9). It leads to representations different from those given by (6.1.50), (6.1.51) (but, of course, equivalent to them due to Pauli theorem). This method is based on the following fact.

Proposition 6.1.4. *Let* $\{\Gamma'_1, \ldots, \Gamma'_{2\nu-1}\}$ *be generators of Clifford algebra* $Cl_{2\nu-1}$ *obeying* $\Gamma'_m\Gamma'_k + \Gamma'_k\Gamma'_m = 2\delta_{km}$ $(k, m = 1, \ldots, 2\nu - 1)$. *Then generators* $\{\Gamma_1, \ldots, \Gamma_{2\nu}\}$ *of Clifford algebra* $Cl_{2\nu}$ *can be chosen as follows:*

$$\Gamma_1 = \sigma_1 \otimes I_{2^{\nu-1}}, \ \Gamma_{m+1} = \sigma_2 \otimes \Gamma'_m \quad (m = 1, \ldots, 2\nu - 1). \tag{6.1.67}$$

Proof. The Proof is a straightforward check that elements (6.1.67) obey defining relations (6.1.9) for algebra $\mathcal{C}l_{2\nu}$. □

Formula (6.1.67) enables one to construct representation of generators $\{\Gamma_1, \ldots, \Gamma_{2\nu}\}$ of algebra $\mathcal{C}l_{2\nu}$, as 2^ν-dimensional matrices, in terms of $2^{\nu-1}$-dimensional representation ρ' of algebra $\mathcal{C}l_{2\nu-1}$. The $2^{\nu-1}$-dimensional representation ρ' of algebra $\mathcal{C}l_{2\nu-1}$ can in turn be constructed in terms of $2^{\nu-1}$-dimensional representation ρ' of algebra $\mathcal{C}l_{2\nu-2}$; the last generator in algebra $\mathcal{C}l_{2\nu-1}$ is the longest element defined in accordance with (6.1.49):

$$\Gamma'_{2\nu-1} = (-i)^{\nu-1}\Gamma'_1 \cdots \Gamma'_{2\nu-2}. \tag{6.1.68}$$

Thus, we have induction procedure: at its step, representation of algebra $\mathcal{C}l_{2\nu}$ is constructed in terms of representation of algebra $\mathcal{C}l_{2\nu-2}$.

Problem 6.1.10. Show explicitly that two inequivalent representations of algebra $\mathcal{C}l_{2\nu-1}$, which differ by the sign of the last generator, $\Gamma'_{2\nu-1} \to -\Gamma'_{2\nu-1}$, define through (6.1.67) two equivalent representation of algebra $\mathcal{C}l_{2\nu}$. Hint: consider equivalence transformation with matrix $(\sigma_1 \otimes \Gamma'_{2\nu-1})$.

Note that if representation ρ' of algebra $\mathcal{C}\ell_{2\nu-2}$ is Hermitian, $(\Gamma'_k)^\dagger = \Gamma'_k$, then, due to (6.1.68) we have $\Gamma'^\dagger_{2\nu-1} = \Gamma'_{2\nu-1}$, so representation (6.1.67) of algebra $\mathcal{C}\ell_{2\nu}$ is also Hermitian.[b] Therefore, by starting the induction procedure (6.1.67) from Hermitian representation ρ' of Clifford algebra $\mathcal{C}\ell_2$ in terms of Pauli matrices (6.1.37) we obtain Hermitian representations of algebras $\mathcal{C}\ell_{2\nu+1}$ for all ν:

$$\begin{gathered}\Gamma_n = \sigma_2^{\otimes(n-1)} \otimes \sigma_1 \otimes I_2^{\otimes(\nu-n)} \;(n = 1, \ldots, \nu),\; \Gamma_{\nu+1} = \sigma_2^{\otimes\nu},\\ \Gamma_{\nu+1+k} = \sigma_2^{\otimes(\nu-k)} \otimes \sigma_3 \otimes I_2^{\otimes(k-1)} \;(k = 1, \ldots, \nu),\; (\Gamma_m)^\dagger = \Gamma_m \;(\forall m).\end{gathered} \tag{6.1.69}$$

Upon removing the last generator $\Gamma_{2\nu+1} = \sigma_3 \otimes I_2^{\otimes(\nu-1)}$, from this set (this generator obeys (6.1.19), see Problem 6.1.11 below) we obtain Hermitian representation of generators of subalgebra $\mathcal{C}\ell_{2\nu} \subset \mathcal{C}\ell_{2\nu+1}$.

We make use of explicit form of Pauli matrices σ_1 and σ_2 and definition (6.1.40) of direct product of matrices and see that representation (6.1.67)

[b]We call matrix representation of Clifford algebra Hermitian if in this representation all its generators Γ_k are Hermitian matrices.

of $\mathcal{C}\ell_{2\nu}$ has block form

$$\Gamma_m = \begin{pmatrix} \mathbf{0} & \mathbf{s}_m \\ \overline{\mathbf{s}}_m & \mathbf{0} \end{pmatrix}, \quad m = 1, \ldots, 2\nu, \tag{6.1.70}$$

where $2^{\nu-1}$-dimensional matrices $\mathbf{s}_m$ and $\overline{\mathbf{s}}_m$ are

$$\mathbf{s}_1 = I_{2^{\nu-1}}, \quad \mathbf{s}_{k+1} = -i\Gamma'_k,$$
$$\overline{\mathbf{s}}_1 = I_{2^{\nu-1}}, \quad \overline{\mathbf{s}}_{k+1} = i\Gamma'_k, \quad k = 1, \ldots, 2\nu - 1.$$

Due to (6.1.9), they obey $(m, n = 1, \ldots, 2\nu)$

$$\mathbf{s}_m\overline{\mathbf{s}}_n + \mathbf{s}_n\overline{\mathbf{s}}_m = 2\delta_{mn}I_{2^{\nu-1}}, \quad \overline{\mathbf{s}}_m\mathbf{s}_n + \overline{\mathbf{s}}_n\mathbf{s}_m = 2\delta_{mn}I_{2^{\nu-1}}. \tag{6.1.71}$$

Matrix of the longest element (6.1.19) of algebra $\mathcal{C}\ell_{2\nu}$ (here $q = 0$) in representation (6.1.67), (6.1.69), (6.1.70) reads

$$\Gamma_{2\nu+1} = (-i)^\nu\Gamma_1\cdots\Gamma_{2\nu} = \sigma_3 \otimes I_{2^{\nu-1}} = \begin{pmatrix} I_{2^{\nu-1}} & \mathbf{0} \\ \mathbf{0} & -I_{2^{\nu-1}} \end{pmatrix}, \tag{6.1.72}$$

where $I_{2^{\nu-1}}$ is unit matrix. The reader is invited to compare matrix (6.1.72) of representation (6.1.67) with matrix (6.1.49) of representation (6.1.50), (6.1.51).

Problem 6.1.11. Check the second equality in (6.1.72) by plugging in explicit representation of Γ_k (6.1.67) and using (6.1.68).

Representations of complex algebra $\mathcal{C}\ell_{2\nu}$ in which gamma-matrices Γ_m $(m = 1, \ldots, 2\nu)$ have block structure (6.1.70) and matrix $\Gamma_{2\nu+1}$ is given by (6.1.72) are called *Weyl representations.* Note that element Γ_1 has the standard form in Weyl representation (6.1.67), (6.1.70):

$$\Gamma_1 = \begin{pmatrix} \mathbf{0} & I_{2^{\nu-1}} \\ I_{2^{\nu-1}} & \mathbf{0} \end{pmatrix} = \sigma_1 \otimes I_{2^{\nu-1}}, \tag{6.1.73}$$

while matrices $\mathbf{s}_k$ and $\overline{\mathbf{s}}_k$ are related as follows: $\overline{\mathbf{s}}_k = \mathbf{s}_k^\dagger = -\mathbf{s}_k$ $(k \neq 1)$. They form 2ν-dimensional vectors (cf. (6.1.4))

$$\mathbf{s} = (I_{2^{\nu-1}}, -i\Gamma'_1, \ldots, -i\Gamma'_{2\nu-1}), \quad \overline{\mathbf{s}} = (I_{2^{\nu-1}}, i\Gamma'_1, \ldots, i\Gamma'_{2\nu-1}), \tag{6.1.74}$$

where $\{\Gamma'_1, \ldots, \Gamma'_{2\nu-1}\}$ is representation of Clifford algebra $\mathcal{C}\ell_{2\nu-1}$ with odd number of generators.

Example. At the first step of the induction procedure outlined here (or, equivalently, making use of (6.1.69), with $\nu = 2$), one obtains four-dimensional representation (6.1.45) of algebra $\mathcal{C}\ell_4$:

$$\Gamma_1 = \sigma_1 \otimes I_2, \; \Gamma_2 = \sigma_2 \otimes \sigma_1, \; \Gamma_3 = \sigma_2 \otimes \sigma_2, \; \Gamma_4 = \sigma_2 \otimes \sigma_3. \tag{6.1.75}$$

Clearly, representation (6.1.75) is different from representation (6.1.43); however, these representations are equivalent by Pauli theorem.

Problem 6.1.12. Find explicit form of equivalence transformation relating representations (6.1.43) and (6.1.75).

Finally, we multiply $\Gamma_2, \Gamma_3, \Gamma_4$ by i in (6.1.75) (see (6.1.8)) and, upon shifting indices $m \to (m-1)$, get back to representation of Clifford algebra $\mathcal{C}\ell_{1,3}$ written in (6.1.3).

Now we turn to general real Clifford algebras $\mathcal{C}\ell_{(p,q)}$. Let $\{\Gamma_1, \ldots, \Gamma_{2\nu}\}$ be matrices of an irreducible representation of generators of algebra $\mathcal{C}\ell_{2\nu}$ obeying (6.1.9). Then irreducible representation of real algebra $\mathcal{C}\ell_{(p,q)}$ with $p + q = 2\nu$ is obtained by introducing (see (6.1.8))

$$\gamma_1 = \Gamma_1, \ldots, \; \gamma_p = \Gamma_p, \; \gamma_{p+1} = i\Gamma_{p+1}, \ldots, \; \gamma_{p+q} = i\Gamma_{p+q}, \tag{6.1.76}$$

where matrices $\{\gamma_1, \ldots, \gamma_{p+q}\}$ obey (see (6.1.2))

$$\gamma_m \gamma_n + \gamma_n \gamma_m = 2\eta_{mn} I_{2^\nu}, \quad ||\eta_{mn}|| = I_{p,q}. \tag{6.1.77}$$

Representations of all basis elements (6.1.15) of algebra $\mathcal{C}\ell_{(p,q)}$ is defined in a standard way ($k = 0, \ldots, p+q$)

$$\gamma_{m_1 m_2 \cdots m_k} = \gamma_{m_1} \gamma_{m_2} \cdots \gamma_{m_k}, \quad 1 < m_1 < m_2 < \cdots < m_k, \tag{6.1.78}$$

i.e., representation (6.1.76) of generators is extended to the representation of all elements as a homomorphism.

We choose Weyl representation (6.1.76) of $\Gamma_k \in \mathcal{C}\ell_{2\nu}$ in (6.1.76) and obtain

$$\gamma_1 = \sigma_1 \otimes I_{2^{\nu-1}}, \; \gamma_{m+1} = \sigma_2 \otimes \gamma'_m \; (m = 1, \ldots, 2\nu - 1), \tag{6.1.79}$$

where

$$\begin{gathered} \gamma'_1 = \Gamma'_1, \ldots, \gamma'_{p-1} = \Gamma'_{p-1}, \; \gamma'_p = i\Gamma'_p, \ldots, \gamma'_{2\nu-2} = i\Gamma'_{2\nu-2}, \\ \gamma'_{2\nu-1} = i\Gamma'_{2\nu-1} = (-i)^{\nu-2}\Gamma'_1 \cdots \Gamma'_{2\nu-2} = (-i)^{\nu+q-1}\gamma'_1 \cdots \gamma'_{2\nu-2} \end{gathered} \tag{6.1.80}$$

is matrix representation of generators of the algebra $\mathcal{C}\ell_{p-1,q}$. Matrices (6.1.70), (6.1.73) have block structure, and so do matrices of representation

$\mathcal{C}\ell_{(p,q)}$:

$$\gamma_m = \begin{pmatrix} \mathbf{0} & \mathbf{s}_m \\ \bar{\mathbf{s}}_m & \mathbf{0} \end{pmatrix} \quad (m = 1, \ldots, 2\nu), \tag{6.1.81}$$

where $2^{\nu-1}$-dimensional blocks[c] $\mathbf{s}_m$ and $\bar{\mathbf{s}}_m$ obey (cf. (6.1.71))

$$\mathbf{s}_m \, \bar{\mathbf{s}}_n + \mathbf{s}_n \, \bar{\mathbf{s}}_m = 2\eta_{mn} \, I_{2^{\nu-1}}, \quad \bar{\mathbf{s}}_m \, \mathbf{s}_n + \bar{\mathbf{s}}_n \, \mathbf{s}_m = 2\eta_{mn} \, I_{2^{\nu-1}}, \tag{6.1.82}$$

$$\bar{\mathbf{s}}_1 = \mathbf{s}_1 = I_{2^{\nu-1}}, \; \bar{\mathbf{s}}_m = -\mathbf{s}_m \; (m = 2, \ldots, 2\nu). \tag{6.1.83}$$

Blocks $\mathbf{s}_m$ and $\bar{\mathbf{s}}_m$ can again be treated as components of matrix vectors (cf. (6.1.4), (6.1.74))

$$\begin{aligned} \mathbf{s} &= (\mathbf{s}_1, \ldots, \mathbf{s}_{2\nu}) = (I_{2^{\nu-1}}, -i\Gamma'_1, \ldots, -i\Gamma'_{p-1}, \Gamma'_p \ldots, \Gamma'_{2\nu-1}), \\ \bar{\mathbf{s}} &= (\bar{\mathbf{s}}_1, \ldots, \bar{\mathbf{s}}_{2\nu}) = (I_{2^{\nu-1}}, i\Gamma'_1, \ldots, i\Gamma'_{p-1}, -\Gamma'_p \ldots, -\Gamma'_{2\nu-1}). \end{aligned} \tag{6.1.84}$$

Matrices $\mathbf{s}_m$ and $\bar{\mathbf{s}}_m$ in (6.1.84) generalize matrices σ_m and $\tilde{\sigma}_m$ that we have encountered in four-dimensional case (6.1.4). Formula (6.1.84) involves matrices $\{\Gamma'_1, \ldots, \Gamma'_{2\nu-2}\}$ which are generators of algebra $\mathcal{C}\ell_{2\nu-2}$ in arbitrary representation, while matrix $\Gamma'_{2\nu-1}$ is defined in (6.1.68) and obeys

$$(-i)^{\nu-1}\Gamma'_1 \cdots \Gamma'_{2\nu-2} \cdot \Gamma'_{2\nu-1} = 1.$$

We recall (6.1.16) and write the longest element (6.1.19) of algebra $\mathcal{C}\ell_{(p,q)}$ with $(p + q) = 2\nu$ as follows:

$$\gamma_{2\nu+1} = (-i)^{\nu+q}\gamma_1 \cdots \gamma_{2\nu} = \frac{(-i)^{\nu}(i)^q}{(2\nu)!}\varepsilon^{i_1 \ldots i_{2\nu}}\gamma_{i_1} \cdots \gamma_{i_{2\nu}}. \tag{6.1.85}$$

It has the same form in representation (6.1.81) as the matrix (6.1.72) of element $\Gamma_{2\nu+1} \in \mathcal{C}\ell_{2\nu}$:

$$\gamma_{2\nu+1} = \sigma_3 \otimes I_{2^{\nu-1}} = \begin{pmatrix} I_{2^{\nu-1}} & \mathbf{0} \\ \mathbf{0} & -I_{2^{\nu-1}} \end{pmatrix}. \tag{6.1.86}$$

Definition 6.1.3. Representation of real algebra $\mathcal{C}\ell_{(p,q)}$ with even number of generators $(p + q) = 2\nu$, in which generators $\{\gamma_1, \ldots, \gamma_{2\nu}\}$ have block structure (6.1.81) and the longest element has the form (6.1.86) is called *Weyl representation.*

---

[c]We use the same notation $\mathbf{s}_m$, $\bar{\mathbf{s}}_m$ for (generally, different) blocks in (6.1.81) and in (6.1.74); this will not lead to confusion.

One may also say that representation (6.1.70), (6.1.72) of complex algebra $\mathcal{C}\ell_{2\nu}$ and representation (6.1.81), (6.1.86) of real algebra $\mathcal{C}\ell_{(p,q)}$ belong to the class of Weyl representations.

6.2. Spinor Groups $\mathsf{Pin}(p,q)$ and $\mathsf{Spin}(p,q)$

In what follows, we treat $N = p + q$ as an arbitrary positive integer, unless it is specified that N is even (or odd).

6.2.1. Definitions of spinor groups $\mathsf{Pin}(p,q)$ and $\mathsf{Spin}(p,q)$

Let γ_k be generators of real algebra $\mathcal{C}\ell_{(p,q)}$. Consider an irreducible representation of γ_k without specifying explicit form of this representation. As usual, we denote the element of the algebra and its representation by the same symbol. Objects of interest are first-order elements

$$X = x^m \gamma_m \in \mathcal{C}\ell_{(p,q)}, \tag{6.2.1}$$

where $\vec{x} = (x^1, \ldots, x^N)$ are vectors in $\mathbb{R}^{p,q}$, and summation over repeated indices is assumed hereafter. Making use of (6.1.77), we get

$$X^2 = \vec{x}^2 I, \quad \vec{x}^2 \equiv \eta_{mn} x^m x^n. \tag{6.2.2}$$

Let U be an invertible element of Clifford algebra $\mathcal{C}\ell_{(p,q)}$ such that the transformations

$$X \to X' = U \cdot X \cdot U^{-1}, \quad X \to X' = -U \cdot X \cdot U^{-1}, \tag{6.2.3}$$

take any first-order element X into first-order element $X' = x'^m \gamma_m \in \mathcal{C}\ell_{(p,q)}$, where $\vec{x}' = (x'^1, \ldots, x'^N) \in \mathbb{R}^{p,q}$ (note that the first and second transformations in (6.2.3) are related by reflection $X' \to -X'$). In other words, transformations (6.2.3) are linear mappings $\mathbb{R}^{p,q} \to \mathbb{R}^{p,q}$. Due to (6.2.2), these transformations preserve squares of vectors:

$$(X')^2 = U \cdot X^2 \cdot U^{-1} = X^2 \Rightarrow \vec{x}^{\,2} = \vec{x}'^2.$$

So, mappings (6.2.3) define some rotations in $\mathbb{R}^{p,q}$ that belong to group $O(p,q)$:

$$x'^k = \Lambda^k{}_m x^m, \quad ||\Lambda^k{}_m|| \in O(p,q). \tag{6.2.4}$$

In view of (6.2.4), transformations (6.2.3) can be written as

$$\gamma_m \to U \cdot \gamma_m \cdot U^{-1} = \gamma_k \Lambda^k{}_m, \tag{6.2.5}$$

$$\gamma_m \to -U \cdot \gamma_m \cdot U^{-1} = \gamma_k \Lambda^k{}_m. \tag{6.2.6}$$

In even case, $N = p+q = 2\nu$, we can make the replacement $U \to U \cdot \gamma_{2\nu+1}$ and reduce the transformation (6.2.5) (for given Λ) to (6.2.6). Therefore, in the even case it is sufficient to consider only one of these transformations. In the odd case, $N = p+q = 2\nu+1$, transformations (6.2.5) and (6.2.6) cannot be related in a similar way. To see this, consider central element (6.1.19) (odd analog of the longest element (6.1.85))

$$\gamma_{N+1} = (-i)^{\nu+q}\gamma_{1\dots N} = \frac{(-i)^{\nu}(i)^{q}}{(2\nu+1)!}\varepsilon^{i_1\dots i_{2\nu+1}}\gamma_{i_1}\cdots\gamma_{i_{2\nu+1}} \in \mathcal{C}\ell^{\mathbb{C}}_{2\nu+1}.$$

$$\gamma_{1\dots N} \equiv \gamma_1\cdots\gamma_{2\nu+1}, \tag{6.2.7}$$

Transformation (6.2.5) does not change sign of this central element, while transformation (6.2.6) does:

$$\gamma_{1\dots N} \to (\pm 1)^{2\nu+1} U \cdot \gamma_{1\dots N} \cdot U^{-1} = \pm\gamma_{1\dots N},$$

where upper and lower signs refer to transformations (6.2.5) and (6.2.6), respectively. Note that right-hand side of transformations (6.2.5), (6.2.6) also give

$$\gamma_{1\dots N} \to \frac{(-1)^q}{(2\nu+1)!}\varepsilon^{i_1\dots i_{2\nu+1}}\gamma_{k_1}\cdots\gamma_{k_{2\nu+1}}\Lambda^{k_1}_{i_1}\cdots\Lambda^{k_{2\nu+1}}_{i_{2\nu+1}} = \det(\Lambda)\gamma_{1\dots N}.$$

So, formulas (6.2.5) and (6.2.6) are self-consistent, if and only if $\det(\Lambda) = +1$ and $\det(\Lambda) = -1$, respectively. We take this fact into account by writing, in odd case ($p+q = 2\nu+1$), transformations (6.2.5) and (6.2.6) as follows:

$$\gamma_m \to U \cdot \gamma_m \cdot U^{-1} = \gamma_k \Lambda^k{}_m, \quad \det(\Lambda) = +1, \tag{6.2.8}$$

$$\gamma_m \to -U' \cdot \gamma_m \cdot U'^{-1} = \gamma_k \Lambda'^k{}_m, \quad \det(\Lambda') = -1, \tag{6.2.9}$$

where $U, U' \in \mathcal{C}\ell_{(p,q)}$. The two transformations (6.2.8) and (6.2.9) are needed to describe proper and improper rotations from $O(p,q)$. Note now, that in odd case, $\Lambda \in SO(p,q)$ is proper rotation, whereas $\Lambda' = -\Lambda \in O_-(p,q)$ is improper rotation. Thus, transformations (6.2.8) and (6.2.9) are related by

$$U' \leftrightarrow U, \quad \Lambda' \leftrightarrow -\Lambda\,. \tag{6.2.10}$$

If element $U \in \mathcal{C}\ell_{(p,q)}$ in (6.2.8) defines proper rotation $\Lambda \in SO(p,q)$, then the same element $U' = U$, being used in (6.2.9), defines improper rotation $-\Lambda \in O_-(p,q)$, and transformation (6.2.9) can be viewed as composition of proper rotation (6.2.8) and reflection $\gamma_m \to -\gamma_m$. This establishes for odd case one-to-one correspondence between the set of all proper rotations (6.2.8) and set of all improper rotations (6.2.9).

We have defined transformations (6.2.3) for elements U of real algebra $\mathcal{C}\ell_{(p,q)}$. In principle, one could consider more general transformations with elements U belonging to complex Clifford algebra $\mathcal{C}\ell^{\mathbb{C}}_{(p,q)} = \mathcal{C}\ell_N$. However, this does not give anything new. To see this, we note that element U in (6.2.3), (6.2.5) and (6.2.6) can be redefined,

$$U \to a \cdot U, \tag{6.2.11}$$

where a is an arbitrary constant or central element of algebra $\mathcal{C}\ell^{\mathbb{C}}_{(p,q)}$ for even or odd case.

Proposition 6.2.1. *Making use of* (6.2.11), *one can always redefine element* $U \in \mathcal{C}\ell^{\mathbb{C}}_{(p,q)}$ *in* (6.2.5) *and* (6.2.6) *in such a way that the resulting element belongs to real Clifford algebra* $\mathcal{C}\ell_{(p,q)}$.

Proof. We treat the relations (6.2.5) and (6.2.6) as equalities between elements of algebra $\mathcal{C}\ell^{\mathbb{C}}_{(p,q)}$:

$$U \cdot \gamma_m = +\gamma_k \cdot U\Lambda^k{}_m, \quad U \cdot \gamma_m = -\gamma_k \cdot U\Lambda^k{}_m,$$

and make complex conjugation of these equalities in accordance with the rule (see Section I-1.2.1 in the first book):

$$\left(\psi_0 I + \sum_{k=1}^{N} \psi_k^{[m_1\ldots m_k]} \gamma_{m_1} \cdots \gamma_{m_k}\right)^*$$
$$= \psi_0^* I + \sum_{k=1}^{N} (\psi_k^{[m_1\ldots m_k]})^* \gamma_{m_1} \cdots \gamma_{m_k} \in \mathcal{C}\ell^{\mathbb{C}}_{(p,q)}. \tag{6.2.12}$$

We also recall that matrix $||\Lambda^k{}_m|| \in O(p,q)$ is real, and obtain

$$U^* \cdot \gamma_m = \pm\gamma_k \cdot U^*\Lambda^k{}_m \;\Rightarrow\; U^* \cdot \gamma_m \cdot U^{*-1} = \pm\gamma_k\Lambda^k{}_m, \tag{6.2.13}$$

where element U^* is complex conjugate to element $U \in \mathcal{C}\ell^{\mathbb{C}}_{(p,q)}$. It follows from (6.2.5), (6.2.6) and (6.2.13) that $U^{-1} \cdot U^* \cdot \gamma_m = \gamma_m \cdot U^{-1} \cdot U^*$ for all generators γ_m. So, element $U^{-1} \cdot U^*$ belongs to center of algebra $\mathcal{C}\ell^{\mathbb{C}}_{(p,q)}$.

In even case $p + q = 2\nu$, the center is trivial, so we have $U^{-1} \cdot U^* = a \cdot I$, i.e., $U^* = a \cdot U$, where a is a complex parameter such that $aa^* = 1$. This gives $(a^{1/2} \cdot U)^* = (a^{1/2} \cdot U) \in \mathcal{C}\ell_{(p,q)}$, so the desired redefinition is

$U \to a^{1/2} \cdot U$. In odd case, the center of algebra $\mathcal{C}\ell^{\mathbb{C}}_{(p,q)}$ contains two basis elements $\{I, \gamma_{1\ldots N}\}$, and we obtain

$$U^{-1} \cdot U^{*} = (a_1 \cdot I + a_2 \cdot \gamma_{1\ldots N}) \equiv a, \tag{6.2.14}$$

where a_1, a_2 are complex parameters, such that $a \cdot a^{*} = I$. One can always define square root of a, such that $a = a^{1/2} \cdot a^{1/2}$, and central element $a^{1/2}$ obeys $a^{1/2} \cdot (a^{1/2})^{*} = I$. Thus, in odd case too, replacement $U \to a^{1/2} \cdot U$ reduces the element U of complex algebra $\mathcal{C}\ell^{\mathbb{C}}_{(p,q)}$ to element of real algebra $\mathcal{C}\ell_{(p,q)}$. □

Definition 6.2.1. All invertible elements U of real algebra $\mathcal{C}\ell_{(p,q)}$ that make transformations (6.2.5) (in both even and odd cases) form a group with respect to multiplication in algebra $\mathcal{C}\ell_{(p,q)}$. This group is denoted by $CL(p, q)$ and called *Clifford group.*[d] The set of all transformations (6.2.5), (6.2.6) is also group, which we denote by $\overline{CL}(p, q)$; transformations (6.2.5), (6.2.6) with different U are treated as different elements of $\overline{CL}(p, q)$.

In the even case, $(p + q) = 2\nu$, transformations (6.2.5) can be reduced to (6.2.6) by replacement $\Lambda' \to \Lambda$, $U' \to U \cdot \gamma_{2\nu+1}$, so we have $\overline{CL}(p, q) = CL(p, q)$. In the odd case, transformations (6.2.8) and (6.2.9) are proper and improper rotations from $O(p, q)$. Two consecutive transformations from the class (6.2.9) give transformation of the form (6.2.8), whereas two consecutive transformations, one from the class (6.2.9) and another from the class (6.2.8), give transformation of the form (6.2.9). So, in odd case, group $\overline{CL}(p, q)$ can be viewed as a group of pairs (U, z) where $U \in CL(p, q)$ and $z = \pm 1$ with $z = +1$ and $z = -1$ corresponding to proper and improper rotations, respectively. Multiplication rule in this $\overline{CL}(p, q)$ is $(U_1, z_1) \cdot (U_2, z_2) = (U_1 \cdot U_2, z_1 \cdot z_2)$. Thus, in odd case $(p + q) = 2\nu + 1$ we have $\overline{CL}(p, q) = CL(p, q) \times \mathbb{Z}_2$ and group $(CL(p, q), +1) = CL(p, q)$ is a subgroup in $\overline{CL}(p, q)$.

Problem 6.2.1. Prove that in odd-dimensional case $(p + q) = 2\nu + 1$, there is isomorphism $O(p, q) = SO(p, q) \times \mathbb{Z}_2$.

Group $CL(p, q)$ has a subgroup $CL^{(0)}(p, q)$ with even (see Definition 6.1.2 of even and odd elements) elements $U \in \mathcal{C}\ell_{(p,q)}$. Let $CL^{(1)}(p, q)$ be a subset of $CL(p, q)$ with odd elements. Clearly, if $g \in CL^{(0)}(p, q)$, then

[d]The term "Clifford group" is sometimes used for completely different object, the set of *all* invertible elements in $\mathcal{C}\ell_{(p,q)}$.

$\gamma \cdot g \in CL^{(1)}(p,q)$, where $\gamma \in CL^{(1)}(p,q)$ is an arbitrary odd element. Conversely, if $g \in CL^{(1)}(p,q)$, then $\gamma^{-1} \cdot g \in CL^{(0)}(p,q)$. This establishes one-to-one correspondence between sets $CL^{(0)}(p,q)$ and $CL^{(1)}(p,q)$.

Proposition 6.2.2. *Consider the even case $(p+q) = 2\nu$, and take an element $U = a + b \in CL(p,q)$, where a and b are even and odd elements of algebra $\mathcal{C}\ell_{(p,q)}$, respectively. Then either $a = 0$ or $b = 0$.*

Proof. We prove this proposition *ad absurdum*. Assume that $U = a + b \in CL(p,q)$, where both even element a and odd element b of algebra $\mathcal{C}\ell_{(p,q)}$ are nonzero. It follows from (6.2.5) that $(a+b)\gamma_m = \gamma_k \Lambda^k{}_m (a+b)$. Since $b \cdot \gamma_m$ and $\gamma_k \cdot b$ are even, whereas $a \cdot \gamma_m$ and $\gamma_k \cdot a$ are odd, we get

$$a \cdot \gamma_m = \gamma_k \Lambda^k{}_m \cdot a, \quad b \cdot \gamma_m = \gamma_k \Lambda^k{}_m \cdot b. \tag{6.2.15}$$

Then element $\gamma_{2\nu+1}$ given by (6.1.85) obeys

$$a \cdot \gamma_{2\nu+1} = \det(\Lambda)\gamma_{2\nu+1} \cdot a, \quad b \cdot \gamma_{2\nu+1} = \det(\Lambda)\gamma_{2\nu+1} \cdot b. \tag{6.2.16}$$

On the other hand, in view of relations $\gamma_m \gamma_{2\nu+1} = -\gamma_{2\nu+1}\gamma_m$ ($\forall m = 1, \ldots, 2\nu$), even and odd elements satisfy

$$a \cdot \gamma_{2\nu+1} = +\gamma_{2\nu+1} \cdot a, \quad b \cdot \gamma_{2\nu+1} = -\gamma_{2\nu+1} \cdot b. \tag{6.2.17}$$

We compare these equalities with (6.2.16) and see that in the case $\det(\Lambda) = +1$ we have $b = 0$, while for $\det(\Lambda) = -1$ we have $a = 0$. □

Remark 1. Consider even case, $p + q = 2\nu$. Then formula (6.2.5) (or, equivalently, (6.2.6)) defines homomorphic mapping

$$T : CL(p,q) \to O(p,q). \tag{6.2.18}$$

By comparing (6.2.16) with (6.2.17) we see that even elements $U \in CL^{(0)}(p,q)$ are mapped by T to elements Λ from proper group $SO(p,q)$, since in this case $\det(\Lambda) = +1$. Likewise, odd elements $U \in CL^{(1)}(p,q)$ are mapped to improper elements, $T(U) = \Lambda \in O_-(p,q)$, for which $\det(\Lambda) = -1$. Therefore, we have

$$T : CL^{(0)}(p,q) \to SO(p,q), \quad T : CL^{(1)}(p,q) \to O_-(p,q). \tag{6.2.19}$$

In other words, $CL(p,q) = CL^{(0)}(p,q) \cup CL^{(1)}(p,q)$.

Remark 2. Consider now odd case, $N = p + q = 2\nu + 1$. Making use of odd central element $\gamma_{1\ldots N} \equiv i^{\nu+q}\gamma_{N+1} \in \mathcal{C}\ell_{(p,q)}$, given by (6.2.7), we establish one-to-one correspondence $b = a \cdot \gamma_{1\dots N}$ between sets of odd and even elements $a \in \mathcal{C}\ell_{(p,q)}$ and $b \in \mathcal{C}\ell_{(p,q)}$, and also between even and odd elements $U \in CL^{(0)}(p,q)$ and $U \cdot \gamma_{1\dots N} \in CL^{(1)}(p,q)$, respectively; these elements make identical transformations (6.2.8).

Proposition 6.2.3. *In odd case, every element $U \in CL(p,q)$ can be written as $U = z \cdot a$, where z is central element of algebra $\mathcal{C}\ell_{(p,q)}$ and $a \in CL^{(0)}(p,q)$.*

‖ **Problem 6.2.2.** ⋆ Prove Proposition 6.2.3.

Now, in odd-dimensional case, central element (6.2.7) changes under transformation (6.2.8) as follows:

$$\gamma_{N+1} \to \gamma_{N+1} = \det(\Lambda)\gamma_{N+1} \Rightarrow \det(\Lambda) = +1, \tag{6.2.20}$$

while under transformation (6.2.9) we get

$$\gamma_{N+1} \to -\gamma_{N+1} = \det(\Lambda')\gamma_{N+1} \Rightarrow \det(\Lambda') = -1. \tag{6.2.21}$$

So, as we pointed out above, transformations (6.2.8) and (6.2.9) define proper and improper rotations from $O(p,q)$ and therefore there are homomorphisms (cf. (6.2.18) and (6.2.19)):

$$\begin{gathered} T : \overline{CL}(p,q) = CL(p,q) \times \mathbb{Z}_2 \to O(p,q), \\ T : (CL(p,q), +1) \to SO(p,q), \quad T : (CL(p,q), -1) \to O_-(p,q). \end{gathered} \tag{6.2.22}$$

Elements $(U, +1) \in (CL^{(0)}(p,q), +1)$ and $(U \cdot \gamma_{1\dots N}, +1) \in (CL^{(1)}(p,q), +1)$ are mapped to one and the same element of $SO(p,q)$. Therefore, homomorphic mapping of subgroup of even elements

$$T : (CL^{(0)}(p,q), +1) = CL^{(0)}(p,q) \to SO(p,q) \tag{6.2.23}$$

is such that its image coincides with the image of the mapping T of entire group $(CL(p,q), +1) = CL(p,q)$.

Let $N = (p+q)$ be arbitrary. Define $\nu = [\frac{N}{2}]$ and consider irreducible 2^ν-dimensional representation ρ of Clifford algebra $\mathcal{C}\ell_{(p,q)}$, and hence the representation ρ of group $CL(p,q)$. The set of matrices $\rho(U)$ with $U \in CL(p,q)$ forms matrix group $CL_\rho(p,q)$. Accordingly, the set of pairs $(\rho(U), \pm 1)$, defining transformations (6.2.8) and (6.2.9), forms group $\overline{CL}_\rho(p,q)$.

In even case $p+q=2\nu$, all irreducible representations ρ are equivalent and have dimension 2^ν, so all matrix groups $CL_\rho(p,q)$ are isomorphic. In odd case $N=p+q=2\nu+1$, all irreducible representations ρ also have dimension 2^ν and they are equivalent either to representation ρ_+ or to representation ρ_-; these representations differ by the sign of central element (see (6.1.59)):

$$\rho_+(\gamma_{N+1}) = +I, \quad \rho_-(\gamma_{N+1}) = -I. \tag{6.2.24}$$

So, we have two representations of Clifford group $\overline{CL}_{\rho_+}(p,q)$ and $\overline{CL}_{\rho_-}(p,q)$.

Proposition 6.2.4. (1) *In even case* $(p+q=2\nu)$, *homomorphism* (6.2.18) *maps* $CL(p,q)$ *onto the whole group* $O(p,q)$. *This means that for any matrix* $\Lambda \in O(p,q)$, *there exists element* $U \in CL(p,q)$ *such that the relation* (6.2.5) *holds. Moreover, for given* Λ, *the element* $U \in CL(p,q)$ *is unique modulo multiplication by a numerical constant* a. (2) *In odd case* $(p+q=2\nu+1)$, *homomorphism* (6.2.22) *maps group* $\overline{CL}(p,q)$ *onto the whole group* $O(p,q)$. *In more detail, for any matrix* $\Lambda \in SO(p,q)$, *there exists element* $U=U(\Lambda) \in CL(p,q)$ *such that relation* (6.2.8) *holds, and for any matrix* $\Lambda' \in O_-(p,q)$, *there exists element* $U'=U'(\Lambda') \in CL(p,q)$ *such that* (6.2.9) *holds. Moreover, elements* U, U' *are unique modulo multiplication by invertible central elements* $(a_1 I + a_2\gamma_{1\ldots N}) \in \mathcal{C}\ell_{(p,q)}$, *where* $a_1, a_2 \in \mathbb{R}$.

Proof. (1) Consider the case $(p+q)=2\nu$, and let ρ be exact irreducible representation of algebra $\mathcal{C}\ell_{(p,q)}$. Elements $\rho^{(\Lambda)}(\gamma_m) = \rho(\gamma_k)\Lambda^k{}_m$ obey the same defining relations (6.1.77) as $\rho(\gamma_k)$, and hence mapping $\rho^{(\Lambda)}$ defines a new irreducible representation of algebra $\mathcal{C}\ell_{(p,q)}$. All irreducible representations of $\mathcal{C}\ell_{(p,q)}$ are equivalent, so there exists matrix $A \in \mathrm{Mat}_{2^\nu}(\mathbb{C})$ such that $A \cdot \rho(\gamma_m) \cdot A^{-1} = \rho(\gamma_k)\Lambda^k{}_m$. Since ρ is isomorphism of algebras $\mathcal{C}\ell^{\mathbb{C}}_{(p,q)}$ and $\mathrm{Mat}_{2^\nu}(\mathbb{C})$ (see (6.1.33)), there exists element U of complex algebra $\mathcal{C}\ell^{\mathbb{C}}_{(p,q)}$ such that $A=\rho(U)$. In view of Proposition 6.2.1, element $U \in \mathcal{C}\ell^{\mathbb{C}}_{(p,q)}$ can always be moved into algebra $\mathcal{C}\ell_{(p,q)}$. This proves the existence, for any matrix $\Lambda \in O(p,q)$, of element $U \in \mathcal{C}\ell_{(p,q)}$ obeying (6.2.5). Now, let U and U' be two elements obeying (6.2.5) for one and the same matrix $\Lambda \in O(p,q)$. Then, for all generators γ_m of algebra $\mathcal{C}\ell_{(p,q)}$, we have $U^{-1}\cdot U' \cdot \gamma_m = \gamma_m \cdot U^{-1} \cdot U'$, so element $U^{-1}\cdot U'$ is central in $\mathcal{C}\ell_{(p,q)}$. Since center of $\mathcal{C}\ell_{(p,q)}$ is trivial, we obtain $U^{-1}\cdot U' = a \cdot I$, i.e., $U' = a \cdot U$, where a is a numerical constant.

(2) Let us turn to odd case, $N = (p+q) = 2\nu + 1$. Let ρ be irreducible 2^ν-dimensional representation of algebra $\mathcal{C}\ell_{(p,q)}$. We choose for definiteness $\rho = \rho_+$, i.e., $\rho(\gamma_{N+1}) = +I$ (the case $\rho = \rho_-$ is treated in the same way). Elements $\rho_+^{(\Lambda)}(\gamma_m) = \rho(\gamma_k)\Lambda^k{}_m$ and $\rho_-^{(\Lambda)}(\gamma_m) = -\rho(\gamma_k)\Lambda^k{}_m$ give new irreducible representations $\rho_+^{(\Lambda)}$ and $\rho_-^{(\Lambda)}$ of algebra $\mathcal{C}\ell_{(p,q)}$ for a given matrix $\Lambda \in O(p,q)$. The representations $\rho_+^{(\Lambda)}$ and $\rho_-^{(\Lambda)}$ are not equivalent to each other, since they have different signs of central element (see (6.2.24)): $\rho_+^{(\Lambda)}(\gamma_{N+1}) = \det(\Lambda)\rho(\gamma_{N+1}) = -\rho_-^{(\Lambda)}(\gamma_{N+1})$. For $\det(\Lambda) = +1$ and $\det(\Lambda) = -1$, representation ρ is equivalent to $\rho_+^{(\Lambda)}$ and $\rho_-^{(\Lambda)}$, respectively, so there exist matrices $A_\pm(\Lambda) \in \mathrm{Mat}_{2^\nu}(\mathbb{C})$ such that

$$\begin{aligned} A_+ \cdot \rho(\gamma_m) \cdot A_+^{-1} &= \rho(\gamma_k)\Lambda^k{}_m \quad (\det(\Lambda) = +1), \\ A_- \cdot \rho(\gamma_m) \cdot A_-^{-1} &= -\rho(\gamma_k)\Lambda^k{}_m \quad (\det(\Lambda) = -1). \end{aligned} \tag{6.2.25}$$

Now, if $\Lambda \in SO(p,q)$, then $-\Lambda \in O_-(p,q)$. Therefore, we have from (6.2.25) that $A_-(-\Lambda) = A_+(\Lambda)$ so instead of two matrix functions $A_+(\Lambda)$ and $A_-(\Lambda')$ of elements $\Lambda \in SO(p,q)$ and $\Lambda' \in O_-(p,q)$ we can consider only one function $A_+(\Lambda)$. Irreducible representation ρ gives for central element $\rho(\gamma_{N+1}) = +I$ and in this representation we have (see (6.1.33))

$$\rho(\mathcal{C}\ell^{\mathbb{C}}_{(p,q)}) = \mathrm{Mat}_{2^\nu}(\mathbb{C}).$$

It follows from this isomorphism that there exist elements $U_+(\Lambda) = U_-(-\Lambda) \in \mathcal{C}\ell^{\mathbb{C}}_{(p,q)}$ such that $\rho(U_\pm) = A_\pm$, while Proposition 6.2.1 ensures that $U_\pm$ can be moved, by using (6.2.11), to elements of real algebra $U_+(\Lambda) = U_-(-\Lambda) \in \mathcal{C}\ell_{(p,q)}$. So, for any proper matrix $\Lambda \in SO(p,q)$, there exists element $U = U_+ \in \mathcal{C}\ell_{(p,q)}$ that makes transformation (6.2.5), and for any improper matrix $\Lambda' = -\Lambda \in O_-(p,q)$ there exists element $U = U_- \in \mathcal{C}\ell_{(p,q)}$ giving (6.2.6). Proof of uniqueness of $U_\pm$ (modulo multiplication by central elements) is the same as in part (1) of this proof. It is worth noting that in any representation ρ_+ or ρ_- of algebra $\mathcal{C}\ell_{(p,q)}$, central elements are numerical constants. □

There is arbitrariness in relating the matrices $\Lambda \in O(p,q)$ to elements $U \in CL(p,q)$, which is due to the possibility to multiply U by a number (in even case) or central element (in odd case) (Proposition 6.2.4). To remove (partially) this arbitrariness, it is natural to consider, instead of Clifford groups $\overline{CL}(p,q)$ (in odd case) and $CL(p,q)$ (in even case), their subgroups whose elements are normalized.

Definition 6.2.2. We define subgroups $\overline{\mathsf{Pin}}(p,q) \subset \overline{CL}(p,q)$ and $\mathsf{Pin}(p,q) \subset CL(p,q)$, whose elements $(U,\pm 1)$ and U are normalized by additional relations

$$\det(\rho(U)) = 1, \tag{6.2.26}$$

or $\det(\rho(U)) = -1$, where ρ is irreducible representation of Clifford algebra $\mathcal{C}\ell_{(p,q)}$.

In the end of this subsection, we show that the case $\det(\rho(U)) = -1$ occurs only in two- and three-dimensional cases $p+q=2$ and $p+q=3$, and only for odd elements U. We emphasize that the condition $\det(\rho(U)) = \pm 1$ is valid for any irreducible representation of algebra $\mathcal{C}\ell_{(p,q)}$ equivalent to representation ρ (equivalence transformations do not change determinant). In odd case $(p+q) = 2\nu+1$, there are two inequivalent irreducible 2^ν-dimensional representations ρ_+ and ρ_- which are related by replacement $\gamma_k \to -\gamma_k$. Due to Proposition 6.2.3, we have for any $U \in CL(p,q)$ that $\rho_+(U) = z_U \cdot \rho_-(U)$ where z_U is a constant which, in view of $\det(\rho(U)) = \pm 1$, can be chosen as $z_U = 1$. Thus, in odd case, two inequivalent representations $\rho_\pm$ of algebra $\mathcal{C}\ell_{(p,q)}$ define two equivalent representations of group $\overline{\mathsf{Pin}}(p,q)$ or group $\mathsf{Pin}(p,q)$.

Note also that irreducible representations ρ of Clifford algebras have even dimension 2^ν, so that $\det(\rho(U)) = \det(\rho(-U))$, and subgroups $\overline{\mathsf{Pin}}(p,q)$ and $\mathsf{Pin}(p,q)$ include both elements U and elements $-U$. In accordance with (6.2.5) and (6.2.6) these elements make the same transformation $\Lambda \in O(p,q)$. This means that $\overline{\mathsf{Pin}}(p,q)$ and $\mathsf{Pin}(p,q)$ cover $O(p,q)$ at least twice.

In what follows, we consider, instead of group $CL^{(0)}(p,q)$, its subgroup $\mathsf{Pin}^{(0)}(p,q)$ whose elements U are also normalized by (6.2.26). Clearly, $\mathsf{Pin}^{(0)}(p,q)$ is a subgroup in group $\mathsf{Pin}(p,q)$, and homomorphisms (6.2.19), (6.2.23) map group $\mathsf{Pin}^{(0)}(p,q)$ to group $SO(p,q)$.

Consider a set of second order basis elements

$$\Sigma_{mn} = \frac{1}{4}(\gamma_m\gamma_n - \gamma_n\gamma_m) = \frac{1}{2}\gamma_{[mn]} \in \mathcal{C}\ell_{(p,q)}. \tag{6.2.27}$$

The elements Σ_{nm} obey commutation relations for generators L_{mn} of Lie algebra of group $SO(p,q)$ (cf. (5.1.12) and (5.4.7))

$$[\Sigma_{nm}, \Sigma_{k\ell}] = \eta_{mk}\Sigma_{n\ell} + \eta_{n\ell}\Sigma_{mk} - \eta_{nk}\Sigma_{m\ell} - \eta_{m\ell}\Sigma_{nk}. \tag{6.2.28}$$

These relations are derived by making use of defining relations (6.1.77), which lead to the identities (cf. (5.4.8))

$$[\Sigma_{nm}, \gamma_k] = \eta_{mk}\gamma_n - \eta_{nk}\gamma_m. \tag{6.2.29}$$

Definition 6.2.3. Subgroup Spin(p, q) in group Pin$^{(0)}(p, q)$ is a set of all elements in Pin$^{(0)}(p, q)$, which belong to connected component of unit element.

Proposition 6.2.5. *All elements $U \in$ Spin(p, q) in a neighborhood of unit element are represented, in unique way, as*

$$U = \exp\Big(\frac{1}{2}\omega^{mn}\Sigma_{mn}\Big) \in \mathcal{C}\ell_{(p,q)}, \tag{6.2.30}$$

where $\omega^{mn} = -\omega^{nm}$ are real parameters. Conversely, any element of the form (6.2.30) *belongs to* Spin(p, q).

Proof. We show, in the first place, that elements(6.2.30) belong to group Pin$^{(0)}(p, q)$. Relations (6.2.29) guarantee that the transformation $\gamma_m \to \gamma'_m = U \cdot \gamma_m \cdot U^{-1}$, where U is given by (6.2.30), results in new generator

$$\gamma'_m = \gamma_j \Lambda^j{}_m, \quad ||\Lambda^j{}_m|| = \exp||\omega^j{}_m||, \quad \omega^j{}_m = \omega^{jk}\eta_{km}, \tag{6.2.31}$$

which agrees with (6.2.5). Hence, we have for even element U that $U \in CL^{(0)}(p, q) \subset CL(p, q)$. Now, in any representation ρ of algebra $\mathcal{C}\ell_{(p,q)}$ we get $\mathrm{Tr}(\rho(\Sigma_{mn})) = 0$, so, using the general formula $\det(e^A) = e^{\mathrm{Tr}A}$ valid for any matrix A, we obtain $\det(\rho(U)) = 1$. Thus, $U \in$ Pin$^{(0)}(p, q)$.

> **Problem 6.2.3.** Prove relation (6.2.31). Hint: make use of the identity
>
> $$e^A \cdot B \cdot e^{-A} = 1 + \sum_{k=1}^{\infty} \frac{1}{k!} \underbrace{[A, [A, \ldots, [A}_{k}, B] \ldots]].$$

Now, consider an element $U \in$ Pin$^{(0)}(p, q) \subset CL^{(0)}(p, q)$ that obeys (6.2.5), where matrix $\Lambda^k{}_m \in SO(p, q)$ is close to unit matrix, $\Lambda^k_m = (\delta^k_m + \omega^k{}_m + \cdots)$. In this case, Eq. (6.2.5) with $U = \exp(T) = I + T + \cdots$ gives

$$[T, \gamma_m] = \gamma_k \omega^k{}_m. \tag{6.2.32}$$

Let T and T' be two elements with the latter property, then $[\gamma_m, T - T'] = 0$, so $(T - T')$ commutes with $\mathcal{C}\ell_{(p,q)}$. Due to Schur's lemma, in any irreducible representation ρ we have $\rho(T - T') = \mathrm{const} \cdot I$, so matrix $\rho(T)$ is unique modulo an addition of a matrix proportional to unit matrix. The latter freedom is removed for $U \in \mathsf{Pin}^{(0)}(p,q)$ by requirement that $\det(\rho(U)) = 1$, i.e., $\mathrm{Tr}(\rho(T)) = 0$. It follows from (6.2.29) and (6.2.32) that T can be chosen as $T = \frac{1}{2}\omega^{mn}\Sigma_{mn}$ (note that $\mathrm{Tr}(\rho(\Sigma_{mn})) = 0$); this choice is unique. So, in the vicinity of unit element we have for any $U \in \mathsf{Pin}^{(0)}(p,q)$

$$\rho(U) = I + \frac{1}{2}\omega^{mn}\rho(\Sigma_{mn}) + \cdots = \exp\Big(\frac{1}{2}\omega^{mn}\rho(\Sigma_{mn})\Big) \Rightarrow U \in \mathsf{Pin}^{(0)}(p,q).$$

Since the vicinity of unit element belongs to connected component of $\mathsf{Pin}^{(0)}(p,q)$, we have $U \in \mathsf{Spin}(p,q) \subset \mathsf{Pin}^{(0)}(p,q)$. □

Proposition 6.2.5 means that elements Σ_{mn} make basis in Lie algebra $\mathsf{spin}(p,q)$ of group $\mathsf{Spin}(p,q)$. Since the defining relations (6.2.28) are the same as for algebra $so(p,q)$, we have isomorphism $\mathsf{spin}(p,q) = so(p,q)$, and group $\mathsf{Spin}(p,q)$ is locally isomorphic to group $SO(p,q)$.

Remark 3. As we will see below, the group $\mathsf{Spin}(N) \equiv \mathsf{Spin}(N,0)$ is compact and hence (see [32, §102]) exponential mapping covers the whole group $\mathsf{Spin}(N)$. In other words, any element $U \in \mathsf{Spin}(N)$ is represented in the exponential form (6.2.30). In the case $N = 2\nu$, which corresponds to Euclidean space $\mathbb{R}^{2\nu}$, one can show that any element of even subgroup $\mathsf{Pin}^{(0)}(2\nu)$ is represented in the exponential form (6.2.30). In other words, the group $\mathsf{Pin}^{(0)}(2\nu)$ is isomorphic to the group $\mathsf{Spin}(2\nu)$.

Problem 6.2.4. * Show that if $U \in \mathsf{Spin}(p,q)$, then $(-U)$ also belongs to group $\mathsf{Spin}(p,q)$. Show that any element $\gamma_{i_1}\cdots\gamma_{i_{2k}} \in \mathsf{Pin}^{(0)}(2\nu)$ of even order can be represented in the exponential form (6.2.30).

Homomorphisms (6.2.19) and (6.2.23): $CL^{(0)}(p,q) \to SO(p,q)$, defined by (6.2.5) and (6.2.8), define mappings $\mathsf{Pin}^{(0)}(p,q) \to SO(p,q)$ and $\mathsf{Spin}(p,q) \to SO^{\uparrow}(p,q)$ (the latter property follows from that fact that $\mathsf{Spin}(p,q)$ is a connected component of unit element in $\mathsf{Pin}^{(0)}(p,q)$). The following proposition holds (see, e.g., [55]).

Proposition 6.2.6. *Group* $\mathsf{Pin}^{(0)}(p,q)$ *covers group* $SO(p,q)$ *twice, and group* $\mathsf{Spin}(p,q)$ *covers group* $SO^{\uparrow}(p,q)$ *twice.*

Proof. Let two elements $U, U' \in \mathsf{Pin}^{(0)}(p,q)$ make transformation (6.2.5) with one and the same $\Lambda \in SO(p,q)$. Then $[U^{-1}\cdot U', \gamma_m] = 0$ for all generators $\gamma_m \in \mathcal{C}\ell_{(p,q)}$, hence $U^{-1}\cdot U' = z$, i.e., $U' = z\cdot U$, where z is central element in $\mathcal{C}\ell_{(p,q)}$. In an irreducible representation ρ we have

$\rho(z) = \alpha I_{2^\nu}$, where $(p+q) = 2\nu, 2\nu+1$, and since U, U' are elements of real algebra $\mathcal{C}\ell_{(p,q)}$, numerical parameter α is real. It follows from normalization (6.2.26) that $\det(\rho(z)) = 1$, therefore $\alpha^{2^\nu} = 1$, and $\alpha = \pm 1$. Thus, there are precisely two elements $\pm U$ in group $\mathsf{Pin}^{(0)}(p,q)$ that are mapped by (6.2.5) to one and the same element $\Lambda \in SO(p,q)$. Hence, group $\mathsf{Pin}^{(0)}(p,q)$ covers group $SO(p,q)$ precisely twice. Clearly, connected component of unit element in $\mathsf{Pin}^{(0)}(p,q)$, which is group $\mathsf{Spin}(p,q)$, covers twice (connected) subgroup $SO^{\uparrow}(p,q) \subset SO(p,q)$. □

In accordance with this proposition, there are isomorphisms

$$\mathsf{Spin}(p,q)/\mathbb{Z}_2 = SO^{\uparrow}(p,q), \quad \mathsf{Pin}^{(0)}(p,q)/\mathbb{Z}_2 = SO(p,q). \tag{6.2.33}$$

Note that in the case $p \neq 0$, $q \neq 0$, the group $\mathsf{Pin}^{(0)}(p,q)$ has disconnected components, since the same holds for group $SO(p,q)$. Groups $\mathsf{Spin}(p,q)$, $\mathsf{Pin}^{(0)}(p,q)$ and $\mathsf{Pin}(p,q)$ are called spinor groups.[e]

Let us point out that group $SO^{\uparrow}(p,q)$ has precisely two types of nonhomotopic closed curves only for $p \leq 1, q > 2$ (or $q \leq 1, p > 2$). In these cases, groups $\mathsf{Spin}(N)$ and $\mathsf{Spin}(1, N-1)$ are universal covering groups of $SO(N)$ and $SO^{\uparrow}(1, N-1)$, respectively. For $p \geq 2$ and $q \geq 2$, group $\mathsf{Spin}(p,q)$ is not universal covering of group $SO^{\uparrow}(p,q)$. We do not prove this fact here[f] (see, e.g., [54]).

In the rest of this chapter, we follow tradition of physics literature, and call some coordinates in $\mathbb{R}^{p,q}$ timelike, and others spacelike.

Definition 6.2.4. Let metric η in space $\mathbb{R}^{p,q}$ have diagonal form. Basis vector $\vec{e}_k \in \mathbb{R}^{p,q}$ with positive (negative) metric component $\eta_{kk} = (\vec{e}_k, \vec{e}_k)$ is called *timelike* (*spacelike*). Accordingly, coordinate x^k of vector $\vec{x} = x^i \vec{e}_i \in \mathbb{R}^{p,q}$ is timelike (spacelike), if basis vector $\vec{e}_k \in \mathbb{R}^{p,q}$ is timelike (spacelike).

---

[e]Some authors use another nomenclature concerning spinor groups in pseudo-Euclidean spaces, namely, the group that we denote by $\mathsf{Pin}^{(0)}(p,q)$ is sometimes denoted by $\mathsf{Spin}(p,q)$, while the connected component of unity, which we denote by $\mathsf{Spin}(p,q)$, is denoted by $\mathsf{Spin}^{\uparrow}(p,q)$.

[f]Universal covering group covers group $SO^{\uparrow}(p,q)$ as many times, as there are elements in fundamental group $\pi_1(SO^{\uparrow}(p,q))$ (group π_1 is defined, e.g., in books [23, 59]). Since the maximal compact subgroup in $SO^{\uparrow}(p,q)$ is $SO(p) \times SO(q)$, one shows that $\pi_1(SO^{\uparrow}(p,q)) = \pi_1(SO(p)) \times \pi_1(SO(q))$, where $\pi_1(SO(1)) = 1$, $\pi_1(SO(2)) = Z$ and $\pi_1(SO(p)) = Z_2$ for $p > 2$. So, group $\mathsf{Spin}(p,q)$, which covers $SO^{\uparrow}(p,q)$ twice, can be universal covering only for $p < 2$ or $q < 2$. The exceptional cases are $p = 1, q = 2$ and $p = 2, q = 1$, when $\pi_1(SO^{\uparrow}(1,2)) = \pi_1(SO^{\uparrow}(2,1)) = Z$; note also that $\mathsf{Spin}(1,2) = SL(2,\mathbb{R})$, and $\pi_1(SL(2,\mathbb{R})) = Z$, which explicitly shows that group $\mathsf{Spin}(1,2)$ is not simply connected.

It is worth noting that some books (e.g., [22]) use the opposite convention,[g] with timelike basis vectors having negative metric components, $\eta_{kk} < 0$. The latter convention is equivalent to ours, but needs a few technical qualifications, e.g., in definition of charge conjugation and description of Majorana representations of group $\mathsf{Spin}(p,q)$ (see Section 6.4.2 below).

Consider even case $p+q = 2\nu$ and recall the definition $\gamma^k = \eta^{ki}\gamma_i$. Besides transformations from group $\mathsf{Spin}(p,q)$, it is of interest, from the viewpoint of spatial and temporal reflections, to study the full group $\mathsf{Pin}(p,q)$, which is mapped by (6.2.5) to full group $O(p,q)$. The spatial and temporal reflections are associated with basis elements (6.1.15) of algebra $\mathcal{C}\ell_{(p,q)}$. Indeed, the first-order basis element $U_k = \gamma_k \in \mathcal{C}\ell_{(p,q)}$ generates, through (6.2.3), reflection of all coordinates x^m except for x^k: $x^m \to (-1)^{1-\delta_{km}} x^m$. This follows from the relations

$$U_k \cdot \gamma_m \cdot U_k^{-1} = \gamma_k \cdot \gamma_m \cdot \gamma^k = (-1)^{1-\delta_{km}} \gamma_m \tag{6.2.34}$$

(no summation over k). Similarly, second-order basis element $U_{m_1,m_2} = \gamma_{m_1}\gamma_{m_2}$ $(m_1 < m_2)$ generates the reflection of two coordinates x^{m_1} and x^{m_2}.

Problem 6.2.5. Show that in even case, $p+q = 2\nu$, reflection of all coordinates, $x_m \to -x_m$ is generated by the longest element $U = \gamma_{1\ldots 2\nu} \in CL^{(0)}(p,q)$. Show that operators

$$T_t = \gamma_t \cdot \gamma_{1\ldots 2\nu},\ P_s = \gamma_s \cdot \gamma_{1\ldots 2\nu}\ (t = 1,\ldots,p;\ s = p+1,\ldots,2\nu),$$

generate reflection of one timelike coordinate x^t $(t = 1,\ldots p)$ and one spacelike coordinate x^s $(s = p+1,\ldots,\nu)$, respectively.

All of the above shows that arbitrary element of the full spinor group $\mathsf{Pin}(p,q)$ can be written as

$$U = U' \cdot \gamma_{m_1\ldots m_k}, \tag{6.2.35}$$

where $\gamma_{m_1\ldots m_k}$ are basis elements of Clifford algebra (see (6.1.15)), while U' is an element of group $\mathsf{Spin}(p,q)$, which in turn is a product of elements (6.2.30).

[g]In our convention, physical Minkowski space has metric $\mathrm{diag}(+1,-1,-1,-1)$, while some other authors choose $\mathrm{diag}(-1,+1,+1,+1)$.

Problem 6.2.6. Prove the latter statement. Hint: make use of the fact that any element $\Lambda \in O(p,q)$ can be written as $\Lambda = \Lambda' \cdot R$, where $\Lambda' \in SO^{\uparrow}(p,q)$ and R is some reflection, and then recall Proposition 6.2.4 and Definition 6.2.2.

In the even case $(p+q) = 2\nu$, element (6.2.35) with even k belongs to subgroup $\mathsf{Pin}^{(0)}(p,q)$, which is mapped by (6.2.5) to proper group $SO(p,q)$, while for odd k element (6.2.35) belongs to the component of $\mathsf{Pin}(p,q)$ which is mapped to improper component $O_-(p,q)$. In the odd case $(p+q) = 2\nu+1$, any element of $\mathsf{Pin}(p,q)$ can also be written in the form (6.2.35); proper and improper transformations $\Lambda \in SO(p,q)$ are given by mappings (6.2.5) and (6.2.6), respectively.

Problem 6.2.7. Show that in odd case, $(p+q) = 2\nu + 1$, transformations (6.2.5) with $U = \gamma_{m_1 \ldots m_k} \in \mathcal{C}\ell_{(p,q)}$ are reflections of even number of coordinates (for any k), so that Λ belongs to proper subgroup $SO(p,q)$.

Remark 4. Let ρ be irreducible representation of Clifford algebra. Then for $(p+q) > 3$ we have $\det(\rho(\gamma_k)) = +1$ for all generators γ_k.

Problem 6.2.8. Prove the last statement. Hint: make use of the fact that either $\gamma_k^2 = +1$, or $\gamma_k^2 = -1$, so eigenvalues of any generator are equal to either ± 1, or $\pm i$. For $N = p+q \geq 4$, one has $\dim(\rho) = 4 \cdot 2^{[(N-4)/2]}$, while the even number of eigenvalues $+1(+i)$ equals, as follows from the explicit form (6.1.50), (6.1.51), to even number of eigenvalues $-1(-i)$.

Therefore, for $(p+q) > 3$, element (6.2.35) always has $\det(\rho(U)) = +1$, since we have, by definition, $\det(\rho(U')) = +1$.

For $(p+q) \leq 3$ one can have the situation in which element (6.2.35) is normalized by $\det(\rho(U)) = -1$. Indeed, in that case $\rho(\gamma_k)$ are 2×2 matrices proportional to Pauli matrices (or equivalent to them), and for $\eta_{kk} = +1$, one has $\rho(\gamma_k)^2 = +1$, i.e., $\rho(\gamma_k) = \sigma_k$ and $\det(\rho(\gamma_k)) = -1$.

6.2.2. Representations of algebras spin(p, q) and groups Spin(p, q)

Let ρ be representation of algebra $\mathcal{C}\ell_{(p,q)}$. Then ρ is also representation of Lie algebra $\mathsf{spin}(p,q)$ whose generators (6.2.27) are proportional to second-order basis elements of algebra $\mathcal{C}\ell_{(p,q)}$, and also representation of group $\mathsf{Spin}(p,q)$: $U \to \rho(U)$. As we see below, in even case $(p+q) = 2\nu$

irreducible 2^ν-dimensional representation ρ of algebra $\mathcal{C}\ell_{(p,q)}$ generates reducible representation of Lie group $\mathsf{Spin}(p,q)$ and its Lie algebra $\mathsf{spin}(p,q)$, which is decomposed into direct sum of two inequivalent irreducible spinor representations Δ_+ and Δ_- of dimension $2^{\nu-1}$ (equivalents of representations $T_{(r)}$ and $T_{(r-1)}$, see Section 5.4.1). This agrees with general analysis in Section 3.4.2, item 3. On the other hand, in odd case $(p+q) = 2\nu+1$, we have two inequivalent 2^ν-dimensional representations ρ_+ and ρ_- of algebra $\mathcal{C}\ell^{\mathbb{C}}_{(p,q)}$. Representations ρ_+ and ρ_- on generators (6.2.27) define two equivalent representations of algebra $\mathsf{spin}(p,q) = so(p,q)$ (see Section 5.4.2), so in odd case there is only one fundamental spinor 2^ν-dimensional representation of algebra $so(p,q)$. This is also in full agreement with the general analysis in Section 3.4.2, item 4.

We begin with proving the above statement in even case (see Remark 2 below for odd case). To this end, we consider representation of algebra $\mathsf{spin}(p,q) = so(p,q)$ which is generated by Weyl representation (6.1.81) of Clifford algebra. In this case, basis elements $\Sigma_{nm} \in \mathsf{spin}(p,q)$ given by (6.2.27) have block-diagonal form

$$\rho(\Sigma_{mn}) = \frac{1}{4}(\gamma_m\gamma_n - \gamma_n\gamma_m) = \begin{pmatrix} \frac{1}{4}(\mathbf{s}_m\overline{\mathbf{s}}_n - \mathbf{s}_n\overline{\mathbf{s}}_m) & \mathbf{0} \\ \mathbf{0} & \frac{1}{4}(\overline{\mathbf{s}}_m\mathbf{s}_n - \overline{\mathbf{s}}_n\mathbf{s}_m) \end{pmatrix}$$
$$= \begin{pmatrix} \mathbf{s}_{mn} & \mathbf{0} \\ \mathbf{0} & \overline{\mathbf{s}}_{mn} \end{pmatrix}. \tag{6.2.36}$$

Therefore, this representation of Lie algebra $\mathsf{spin}(p,q)$ is completely reducible, and so is the representation of group $\mathsf{Spin}(p,q)$:

$$U = \begin{pmatrix} \exp(\frac{1}{2}\omega^{mn}\mathbf{s}_{mn}) & \mathbf{0} \\ \mathbf{0} & \exp(\frac{1}{2}\omega^{mn}\overline{\mathbf{s}}_{mn}) \end{pmatrix} = \begin{pmatrix} A & \mathbf{0} \\ \mathbf{0} & \overline{A} \end{pmatrix} \tag{6.2.37}$$

(we have in particular that $\det(A) = \det(\overline{A}) = +1$). As we now show, it decomposes into direct sum of two irreducible representations Δ_+ and Δ_- that are inequivalent to each other.

Proposition 6.2.7. *For even $(p+q) = 2\nu$, diagonal blocks $\mathbf{s}_{mn}$ and $\overline{\mathbf{s}}_{mn}$ in (6.2.36) define two irreducible and inequivalent representations Δ_+ and Δ_- of Lie algebra $\mathsf{spin}(p,q) = so(p,q)$. Accordingly, diagonal blocks A and $\overline{A}$ in (6.2.37) define two irreducible and inequivalent $2^{\nu-1}$-dimensional representations Δ_+ and Δ_- of group $\mathsf{Spin}(p,q)$.*

Proof. We give the proof *ad absurdum*, and begin with inequivalence. Let representations of algebra $\mathsf{spin}(p,q)$ defined by matrices $\mathbf{s}_{mn}$ and $\bar{\mathbf{s}}_{mn}$ be equivalent. Then there exists nondegenerate $2^{\nu-1}$-dimensional matrix S such that $\bar{\mathbf{s}}_{mn} = S\mathbf{s}_{mn}S^{-1}$. We set $m = 1$ or $n = 1$ here, and recall (6.1.83). As a result, we have $\bar{\mathbf{s}}_n = S\mathbf{s}_n S^{-1}$, $\mathbf{s}_n = S\bar{\mathbf{s}}_n S^{-1}$. Therefore, we can define matrix $\gamma = \begin{pmatrix} \mathbf{0} & S \\ -S & \mathbf{0} \end{pmatrix}$ which anticommutes with all generators of algebra $\mathcal{C}\ell_{(p,q)}$ in representation (6.1.81), and hence commutes with matrix (6.1.85): $\gamma\gamma_{2\nu+1} = \gamma_{2\nu+1}\gamma$. This contradicts identity $\gamma\gamma_{2\nu+1} = -\gamma_{2\nu+1}\gamma$ that follows from the explicit form (6.1.86) of matrix $\gamma_{2\nu+1}$.

Let us now turn to irreducibility. Let complex representation of algebra $\mathsf{spin}(p,q)$ with matrices $\mathbf{s}_{mn}$ be reducible (the same argument works for $\bar{\mathbf{s}}_{mn}$). Then there exists nonzero matrix $S' \neq \lambda I_{2^{\nu-1}}$ such that $S' \cdot \mathbf{s}_{mn} = \mathbf{s}_{mn} \cdot S'$. We set $m = 1$ (or $n = 1$) here, and find that $S' \cdot \mathbf{s}_n = \mathbf{s}_n \cdot S'$. This shows that matrix $\gamma' = \begin{pmatrix} S' & \mathbf{0} \\ \mathbf{0} & S' \end{pmatrix} \neq \lambda I_{2^\nu}$ commutes with all γ_m, which contradicts irreducibility of the representation of algebra $\mathcal{C}\ell_{(p,q)}$. □

In even case, invariant subspaces $\mathcal{V}_+$ and $\mathcal{V}_-$ of irreducible representations Δ_+ and Δ_- of Lie algebra $\mathsf{spin}(p,q) = so(p,q)$ are extracted from the space $\mathcal{F}$ of irreducible representation (6.1.81) of algebra $\mathcal{C}\ell_{(p,q)}$ by projection operators

$$P_+ = \frac{1}{2}(I_{2^\nu} + \gamma_{2\nu+1}) = \begin{pmatrix} I_{2^{\nu-1}} & \mathbf{0} \\ \mathbf{0} & \mathbf{0} \end{pmatrix}, \; P_- = \frac{1}{2}(I_{2^\nu} - \gamma_{2\nu+1}) = \begin{pmatrix} \mathbf{0} & \mathbf{0} \\ \mathbf{0} & I_{2^{\nu-1}} \end{pmatrix}, \tag{6.2.38}$$

which are written here in Weyl representation. These projectors correspond (with $\nu = r$) to projectors (5.4.27) used in constructing two irreducible spinor representations $T_{(r-1)}$ and $T_{(r)}$ with highest weights $\lambda_{(r-1)}$ and $\lambda_{(r)}$ (see Section 5.4.1). So, the two irreducible representations Δ_+ and Δ_- of Lie algebra $\mathsf{spin}(p,q) = so(p,q)$ are equivalent (with $\nu = r$) to two fundamental spinor representations $T_{(r-1)}$ and $T_{(r)}$, announced in Section 3.4.2 (item 3) and explicitly constructed in Section 5.4.1. Following Cartan [56], representations $T_{(r-1)} = \Delta_+$ and $T_{(r)} = \Delta_-$ are called semispinor in mathematical literature, while they are called Weyl representations in physics nomenclature.

Remark 1. It is convenient in calculations to distinguish indices of matrices A from those of $\overline{A}$ of inequivalent representations Δ_+ and Δ_- of group $\mathsf{Spin}(p,q)$. To this end, one follows van der Waerden [20] and labels by dot indices in matrices $\overline{A}$ of the second representation Δ_-. Accordingly, indices of components of Weyl spinors are distinguished: for spinors $\xi \in \mathcal{V}_+$ and $\eta \in \mathcal{V}_-$, one denotes components as ξ_α and $\eta^{\dot\alpha}$ $(\alpha, \dot\alpha = 1, \ldots, 2^{\nu-1})$. The action of element $U \in \mathsf{Spin}(p,q)$, given by (6.2.37), in subspaces $\mathcal{V}_+ \subset \mathcal{F}$ and $\mathcal{V}_- \subset \mathcal{F}$ is written as follows:

$$\xi_\alpha \to \xi'_\alpha = A_\alpha{}^\beta \xi_\beta,\ \eta^{\dot\alpha} \to \eta'^{\dot\alpha} = \overline{A}^{\dot\alpha}{}_{\dot\beta}\eta^{\dot\beta}. \tag{6.2.39}$$

In Weyl representation (6.1.81), transformation (6.2.5) with U given by (6.2.37) reads

$$\gamma_m = \begin{pmatrix} \mathbf{0} & \mathbf{s}_m \\ \overline{\mathbf{s}}_m & \mathbf{0} \end{pmatrix} \to U\gamma_m U^{-1} = \begin{pmatrix} \mathbf{0} & A\,\mathbf{s}_m\,\overline{A}^{-1} \\ \overline{A}\,\overline{\mathbf{s}}_m\,A^{-1} & \mathbf{0} \end{pmatrix}. \tag{6.2.40}$$

This determines the assignment of indices in matrices $(\mathbf{s}_m)_{\alpha\dot\beta}$ and $(\overline{\mathbf{s}}_m)^{\dot\alpha\beta}$. With this assignment, the definition (6.2.36) of generators $\mathbf{s}_{mn}$, $\overline{\mathbf{s}}_{mn}$ of algebra $\mathsf{spin}(p,q)$ gives the index assignment for matrices $||(\mathbf{s}_{mn})_\alpha{}^\beta||$ and $||(\overline{\mathbf{s}}_{mn})^{\dot\alpha}{}_{\dot\beta}||$; this is consistent with index assignment for matrices A and $\overline{A}$, written in (6.2.39).

Remark 2. Let us now turn to odd case, $(p+1)+q = 2\nu+1$. Representation of group $\mathsf{Spin}(p+1,q)$ and its Lie algebra $\mathsf{spin}(p+1,q) = so(p+1,q)$ constructed in terms of representation (6.1.81) of even-dimensional Clifford algebra $\mathcal{C}\ell_{(p,q)}$ is irreducible. This follows from explicit form (6.1.86) of representation of the last generator, $\gamma_{2\nu+1}$, which shows that the matrices of elements $\rho(\Sigma_{mn})$ with $n = 2\nu+1$ (or $m = 2\nu+1$) are no longer block-diagonal. Instead, they read

$$\rho(\Sigma_{m,2\nu+1}) = \frac{1}{4}(\gamma_m\gamma_{2\nu+1} - \gamma_{2\nu+1}\gamma_m) = \frac{1}{2}\begin{pmatrix} \mathbf{0} & -\mathbf{s}_m \\ \overline{\mathbf{s}}_m & \mathbf{0} \end{pmatrix},\ m = 1,\ldots,2\nu. \tag{6.2.41}$$

Clearly, they act from space $\mathcal{V}_+$ to space $\mathcal{V}_-$ in $\mathcal{F}$, and vice versa. Irreducible complex representation (6.1.81), (6.2.41) of Lie algebra $\mathsf{spin}(p+1,q) = so(p+1,q)$ is equivalent (with $r = \nu$) to irreducible spinor representation $T_{(r)}$ of algebra $so(2r+1,\mathbb{C})$ with highest fundamental weight $\lambda_{(r)}$, announced in Section 3.4.2 (item 4) and constructed in Section 5.4.2.

Remark 3. Consider matrices (6.2.36) of $so(2\nu, \mathbb{C})$ generators L_{mn}, or, equivalently, Σ_{mn}, in the irreducible (and inequivalent) representations $\Delta_+(\Sigma_{mn}) = \mathbf{s}_{mn}$ and $\Delta_-(\Sigma_{mn}) = \bar{\mathbf{s}}_{mn}$. According to relations (6.1.83), we have equalities

$$\mathbf{s}_{1n} = -\bar{\mathbf{s}}_{1n}, \; \mathbf{s}_{mn} = \bar{\mathbf{s}}_{mn} \quad (n, m = 2, \ldots, 2\nu). \tag{6.2.42}$$

We now make use of the equivalence transformation, which changes numbering of generators: $\mathbf{s}_{1,j} \leftrightarrow \mathbf{s}_{2\nu,j}$ and $\bar{\mathbf{s}}_{1,j} \leftrightarrow \bar{\mathbf{s}}_{2\nu,j}$ (in the irreducible representation of Clifford algebra, the permutation of generators $\gamma_1 \leftrightarrow \gamma_{2\nu}$ is the equivalence transformation). Then the relations (6.2.42) are written as

$$\mathbf{s}_{2\nu,n} = -\bar{\mathbf{s}}_{2\nu,n}, \; \mathbf{s}_{mn} = \bar{\mathbf{s}}_{mn} \quad (n, m = 1, \ldots, 2\nu - 1). \tag{6.2.43}$$

It follows from these relations that algebra $so(2r, \mathbb{C})$ possesses the automorphism

$$L_{2\nu,n} \to -L_{2\nu,n}, \; L_{mn} \to L_{mn} \quad (n, m = 1, \ldots, 2\nu - 1), \tag{6.2.44}$$

which permutes the spinor representations $\mathbf{s}_{i,j} = \Delta_+(L_{ij})$ and $\bar{\mathbf{s}}_{ij} = \Delta_-(L_{ij})$. The automorphism (6.2.44) matches the one in Problem 87 of Section 3.4.2.

Problem 6.2.9. Show that in even case $p+q = 2\nu$, matrices $L_{mn} = \rho(\Sigma_{mn})$ given by (6.2.36), together with

$$P_m = \begin{pmatrix} \mathbf{0} & \mathbf{0} \\ \bar{\mathbf{s}}_m & \mathbf{0} \end{pmatrix}, \; K_m = \begin{pmatrix} \mathbf{0} & -\mathbf{s}_m \\ \mathbf{0} & \mathbf{0} \end{pmatrix}, \; D = \frac{1}{2}\begin{pmatrix} I_{2^{\nu-1}} & \mathbf{0} \\ \mathbf{0} & -I_{2^{\nu-1}} \end{pmatrix} = \frac{\gamma_{2\nu+1}}{2},$$

define matrix (spinor) representation of generators of conformal algebra conf$(\mathbb{R}^{p,q}) = so(p+1, q+1)$ (see Section I-3.4.1 in the first book) whose defining relations are

$$\begin{gathered} [D, P_m] = -P_m, \; [D, K_m] = K_m, \; [D, L_{jm}] = 0, \\ [K_j, K_m] = 0, \; [P_j, P_m] = 0, \; [P_k, K_m] = 2(\eta_{km} D - L_{km}), \\ [P_j, L_{km}] = \eta_{kj} P_m - \eta_{mj} P_k, \; [K_j, L_{km}] = \eta_{kj} K_m - \eta_{mj} K_k, \\ [L_{ij}, L_{k\ell}] = \eta_{jk} L_{i\ell} + \eta_{j\ell} L_{ki} + \eta_{ik} L_{\ell j} + \eta_{i\ell} L_{jk}. \end{gathered} \tag{6.2.45}$$

6.3. Conjugation Matrices

6.3.1. Conjugation matrices B, C, D for representations of algebra $\mathcal{C}\ell_{(p,q)}$, and their properties

Let ρ be 2^ν-dimensional irreducible representation of algebra $\mathcal{C}\ell_{2\nu}$: $\Gamma_m \to \rho(\Gamma_m)$ (even case). Consider new representations ρ', ρ'' obtained from ρ by complex conjugation and transposition

$$\Gamma_m \overset{\rho'}{\to} \eta_m \rho(\Gamma_m)^*, \; \Gamma_m \overset{\rho''}{\to} \eta'_m \rho(\Gamma_m)^\mathsf{T}, \tag{6.3.1}$$

where constants η_m, η'_m can take only two values ± 1, i.e., $\eta_m^2 = \eta_m'^2 = 1$ $(\forall m)$. Unlike in (6.2.12), complex conjugation in (6.3.1) is complex conjugation of all matrix elements. Mappings (6.3.1) preserve defining relations (6.1.9) and can be extended, as homomorphisms, to the whole algebra $\mathcal{C}\ell_{2\nu}$, so they indeed define representations. Due to Pauli theorem, Propositions 1.3.2 and 6.1.3, the initial representation ρ and two representations (6.3.1) are all equivalent.

Let us discuss this equivalence and take representation ρ given by (6.1.50), (6.1.51) as an example:

$$\Gamma_{2\alpha-1} = \sigma_3^{\otimes(\alpha-1)} \otimes \sigma_1 \otimes I_2^{\otimes(\nu-\alpha)}, \; \Gamma_{2\alpha} = \sigma_3^{\otimes(\alpha-1)} \otimes \sigma_2 \otimes I_2^{\otimes(\nu-\alpha)}, \tag{6.3.2}$$

where $\alpha = 1, \ldots, \nu$. In this representation, we have (see (6.1.53))

$$\Gamma_m^* = \Gamma_m^\mathsf{T} = (-1)^{m+1}\Gamma_m \quad (m = 1, \ldots, N = 2\nu), \tag{6.3.3}$$

while matrix of the longest element is

$$\Gamma_{N+1} = (-i)^\nu \Gamma_1 \cdots \Gamma_N = \sigma_3^{\otimes \nu} \;\Rightarrow\; \Gamma_{N+1}^2 = I_{2^\nu}, \; [\Gamma_{N+1}, \Gamma_m]_+ = 0. \tag{6.3.4}$$

Consider two particular cases of representations ρ', ρ'', with $\eta_m = \eta'_m = 1$ and $\eta_m = \eta'_m = -1$ $(\forall m)$. Our purpose is to show that $(2^\nu \times 2^\nu)$ matrices of equivalence of these representations and representation ρ are

$$C = \underbrace{\sigma_2 \otimes \sigma_1 \otimes \sigma_2 \otimes \sigma_1 \otimes \cdots}_{\nu}, \quad C^2 = I, \tag{6.3.5}$$

$$C' = \underbrace{\sigma_1 \otimes \sigma_2 \otimes \sigma_1 \otimes \sigma_2 \otimes \cdots}_{\nu}, \quad (C')^2 = I. \tag{6.3.6}$$

These matrices are related by

$$C' \cdot C = (-1)^\nu C \cdot C' = (-1)^{[\nu/2]}(i)^\nu \Gamma_{N+1}, \tag{6.3.7}$$

where $[\nu/2]$ is integer part of $\nu/2$. Besides (6.3.7), matrices C and C' obey

$$C^{\mathsf{T}} = C^* = (-1)^{[\frac{\nu+1}{2}]}C, \quad C'^{\mathsf{T}} = C'^* = (-1)^{[\frac{\nu}{2}]}C'. \tag{6.3.8}$$

They are related to gamma-matrices (6.3.2) as follows:

$$(-i)^k\Gamma_2\Gamma_4\cdots\Gamma_{2\nu} = \begin{cases} \sigma_1 \otimes \sigma_2 \otimes \sigma_1 \otimes \cdots \otimes \sigma_2 = C' & \text{for } \nu = 2k, \\ \sigma_2 \otimes \sigma_1 \otimes \sigma_2 \otimes \cdots \otimes \sigma_2 = C & \text{for } \nu = 2k+1, \end{cases} \tag{6.3.9}$$

$$i^k\Gamma_1\Gamma_3\cdots\Gamma_{2\nu-1} = \begin{cases} \sigma_2 \otimes \sigma_1 \otimes \sigma_2 \otimes \cdots \otimes \sigma_1 = C & \text{for } \nu = 2k, \\ \sigma_1 \otimes \sigma_2 \otimes \sigma_1 \otimes \cdots \otimes \sigma_1 = C' & \text{for } \nu = 2k+1. \end{cases} \tag{6.3.10}$$

Problem 6.3.1. Check formulas (6.3.9) and (6.3.10) by substituting (6.3.2) on their left-hand sides.

Proposition 6.3.1. *Generators* Γ_k *of algebra* $C\ell_{2\nu}$ *in representation* (6.3.2) *and conjugate generators* Γ_k^* *and* Γ_k^{T} *are related by equivalence matrices* (6.3.5), (6.3.6):

$$-C \cdot \Gamma_m \cdot C = C' \cdot \Gamma_m \cdot C' = \Gamma_m^* = \Gamma_m^{\mathsf{T}} \quad (m = 1, \ldots, N = 2\nu). \tag{6.3.11}$$

In other words, for $\eta_m = \eta'_m = +1$ *we have* $\rho'(\Gamma_m) = \rho''(\Gamma_m) = C' \cdot \Gamma_m \cdot C'$, *while for* $\eta_m = \eta'_m = -1$ *we obtain* $\rho'(\Gamma_m) = \rho''(\Gamma_m) = C \cdot \Gamma_m \cdot C$. *Also, the following relation holds*:

$$C \cdot \Gamma_{N+1} \cdot C = C' \cdot \Gamma_{N+1} \cdot C' = (-1)^{\nu}\Gamma_{N+1}. \tag{6.3.12}$$

Proof. We prove (6.3.11) for matrix C given by (6.3.5); then equality (6.3.11) for matrix C' follows from (6.3.7) and (6.3.4). When $\Gamma_{2\alpha-1}$ (in the representation (6.3.2)) commutes with C, we have two options: (1) factor σ_1 from $\Gamma_{2\alpha-1}$ meets factor σ_1 in C; then σ_1 in $\Gamma_{2\alpha-1}$ does not change sign when moving through C; (2) factor σ_1 from $\Gamma_{2\alpha-1}$ meets factor σ_2 in C; then σ_1 in $\Gamma_{2\alpha-1}$ changes sign when moving through C. In the former case, the number $(\alpha-1)$ and hence the number of factors σ_3 in $\Gamma_{2\alpha-1}$ is odd, while in the latter case this number is even. Since all factors σ_3 in $\Gamma_{2\alpha-1}$ change sign when moving through C, we have in both cases $\Gamma_{2\alpha-1}C = -C\Gamma_{2\alpha-1}$. Similar argument applied to $\Gamma_{2\alpha}$ gives $\Gamma_{2\alpha}C = C\Gamma_{2\alpha}$, which yields for all m

$$\Gamma_m C = (-1)^m C\Gamma_m, \tag{6.3.13}$$

and, in view of (6.3.3), proves equality (6.3.11) for matrix C. Relations (6.3.12) are consequences of matrix representations of C, C' and Γ_{N+1} given in (6.3.5), (6.3.6) and (6.3.4). □

It is instructive for what follows to study equivalence of irreducible representation ρ and conjugate representations ρ', ρ'' given by (6.3.1) of real Clifford algebra $\mathcal{C\ell}_{(p,q)}$ without specifying explicit forms of these representations. Let us consider even and odd cases of $(p+q) = N$ separately. We still omit notation ρ for "basic" representation, and write simply $\rho(\gamma_k) = \gamma_k$ wherever possible.

1. Conjugation matrices in even case $(p+q) = 2\nu$

Define irreducible representations ρ_i of Clifford algebra $\mathcal{C\ell}_{(p,q)}$ as mappings of its generators

$$\gamma_m \overset{\rho_1}{\to} \gamma_m^\dagger, \ \gamma_m \overset{\rho_2}{\to} -\gamma_m^\mathsf{T}, \ \gamma_m \overset{\rho_3}{\to} -\gamma_m^*, \tag{6.3.14}$$

that preserve defining relations (6.1.77). These mappings are extended to the entire algebra $\mathcal{C\ell}_{(p,q)}$ as homomorphism $\rho_i(a \cdot b) = \rho_i(a) \cdot \rho_i(b)$ (not as antihomomorphism $\rho_i(a \cdot b) = \rho_i(b) \cdot \rho_i(a)$, which would be the case for operations of transposition and Hermitian conjugation). In accordance with Pauli theorem, all irreducible representations ρ_i are equivalent to representation ρ, i.e., there exist matrices $\mathbf{D}, \mathbf{B}, \mathbf{C} \in \mathrm{Mat}_{2^\nu}(\mathbb{C})$ such that

$$\begin{gathered}\gamma_k^\dagger = \mathbf{D}^{-1} \cdot \gamma_k \cdot \mathbf{D}, \ \gamma_k^\mathsf{T} = -\mathbf{C}^{-1} \cdot \gamma_k \cdot \mathbf{C}, \ \gamma_k^* = -\mathbf{B}^{-1} \cdot \gamma_k \cdot \mathbf{B}, \\ k = 1, \ldots, 2\nu.\end{gathered} \tag{6.3.15}$$

Minus sign in the second and third mappings in (6.3.14), as well as in the second and third formulas in (6.3.15), is introduced for convenience; it can be removed by the redefinitions $\mathbf{B} \to \widetilde{\mathbf{B}} = \gamma_{2\nu+1}\mathbf{B}$, $\mathbf{C} \to \widetilde{\mathbf{C}} = \gamma_{2\nu+1}\mathbf{C}$, where we still use matrix $\gamma_{2\nu+1} = (-i)^{\nu+q}\gamma_1 \cdots \gamma_{2\nu}$ defined in (6.1.85). As we show below (see Section 6.4.2), this sign is relevant for defining Majorana representations of Clifford algebra $\mathcal{C\ell}_{(p,q)}$.

Relations (6.3.15) give for matrix $\gamma_{2\nu+1}$

$$\begin{aligned}\gamma_{2\nu+1}^\dagger &= (i)^{\nu+q}\mathbf{D}^{-1} \cdot \gamma_{2\nu} \cdots \gamma_1 \cdot \mathbf{D} = (-1)^q \mathbf{D}^{-1} \cdot \gamma_{2\nu+1} \cdot \mathbf{D}, \\ \gamma_{2\nu+1}^\mathsf{T} &= (-i)^{\nu+q}\mathbf{C}^{-1} \cdot \gamma_{2\nu} \cdots \gamma_1 \cdot \mathbf{C} = (-1)^\nu \mathbf{C}^{-1} \cdot \gamma_{2\nu+1} \cdot \mathbf{C}, \\ \gamma_{2\nu+1}^* &= (i)^{\nu+q}\mathbf{B}^{-1} \cdot \gamma_1 \cdots \gamma_{2\nu} \cdot \mathbf{B} = (-1)^{\nu+q} \mathbf{B}^{-1} \cdot \gamma_{2\nu+1} \cdot \mathbf{B}.\end{aligned} \tag{6.3.16}$$

Matrix $\mathbf{C}$ relating representations γ_k and $-\gamma_k^{\mathsf{T}}$ is called *charge conjugation* matrix. For particular representation (6.3.2) constructed in Section 6.1.2, matrix $\mathbf{C}$ coincides with matrix C given by (6.3.5), see Proposition 6.3.1.

Proposition 6.3.2. *Matrices* $\mathbf{D}, \mathbf{B}, \mathbf{C} \in \mathrm{Mat}_{2^\nu}(\mathbb{C})$ *are determined by* (6.3.15) *uniquely, modulo multiplication by numerical constants which can be chosen in such a way that*

$$\mathbf{D}^\dagger = \mathbf{D}, \ \mathbf{C}^{\mathsf{T}} = \epsilon' \mathbf{C}, \tag{6.3.17}$$

$$\mathbf{B}^* \cdot \mathbf{B} = \mathbf{B} \cdot \mathbf{B}^* = \epsilon I_{2^\nu}, \tag{6.3.18}$$

where numerical parameters $\epsilon' = \pm 1$ *and* $\epsilon = \pm 1$ *do not depend on the representation* ρ *of the algebra* $\mathcal{C}\ell_{(p,q)}$ *with matrices* γ_k.

Proof. Let there exist two matrices $\mathbf{B}$ and $\bar{\mathbf{B}} \in \mathrm{Mat}_{2^\nu}(\mathbb{C})$ that relate representation γ_m and representation $-\gamma_m^*$:

$$\gamma_k^* = -\mathbf{B}^{-1} \cdot \gamma_k \cdot \mathbf{B} = -\bar{\mathbf{B}}^{-1} \cdot \gamma_k \cdot \bar{\mathbf{B}} \quad (k = 1, \ldots, 2\nu). \tag{6.3.19}$$

These relations give

$$\gamma_m \bar{\mathbf{B}} \cdot \mathbf{B}^{-1} = \bar{\mathbf{B}} \cdot \mathbf{B}^{-1} \gamma_m.$$

Since complex representation ρ with matrices γ_k is irreducible, Schur's lemma ensures that $\bar{\mathbf{B}} \cdot \mathbf{B}^{-1} = \lambda I_{2^\nu}$, where λ is a numerical factor. So, $\bar{\mathbf{B}} = \lambda \mathbf{B}$ which proves the first part of proposition for matrix $\mathbf{B}$. The fact that $\mathbf{D}$ and $\mathbf{C}$ are also defined uniquely, modulo numerical factors, is proven in the same way.

Hermitian conjugation of the first of Eq. (6.3.15) and transposition of the second equality give $\mathbf{D}^\dagger \mathbf{D}^{-1} \gamma_k = \gamma_k \mathbf{D}^\dagger \mathbf{D}^{-1}$ and $\mathbf{C}^{\mathsf{T}} \mathbf{C}^{-1} \gamma_k = \gamma_k \mathbf{C}^{\mathsf{T}} \mathbf{C}^{-1}$. Again due to Schur's lemma, we have $\mathbf{D}^\dagger = \lambda \mathbf{D}$, $\mathbf{C}^{\mathsf{T}} = \epsilon' \mathbf{C}$, where $(\epsilon')^2 = 1$, while parameter $\lambda = \lambda^{*-1}$ (phase factor) can be set equal to 1 by redefinition $\mathbf{D} \to \lambda^{-1/2} \mathbf{D}$. The third of Eq. (6.3.15) and its complex conjugate yield $\gamma_m \mathbf{B} \cdot \mathbf{B}^* = \mathbf{B} \cdot \mathbf{B}^* \gamma_m$; then Eq. (6.3.18) with real ϵ again follows from Schur's lemma, and rescaling of $\mathbf{B}$ can be used to set $\epsilon = \pm 1$.

We now show that parameter ϵ in (6.3.18) is the same for all representations of algebra $\mathcal{C}\ell_{(p,q)}$ (the proof works also for parameter ϵ' in (6.3.17)). Two irreducible representations of algebra $\mathcal{C}\ell_{(p,q)}$ with matrices γ_k and γ_k' are related by equivalence transformation $\gamma_k' = T \gamma_k T^{-1}$. The third of

Eq. (6.3.15) for matrices γ'_k reads

$$(\gamma'_k)^* = T^*\gamma_k^*(T^{-1})^* = -(\bar{\mathbf{B}})^{-1}\gamma'_k\bar{\mathbf{B}}, \tag{6.3.20}$$

where

$$\bar{\mathbf{B}} = T\mathbf{B}(T^*)^{-1} \tag{6.3.21}$$

is the equivalence matrix relevant for representation γ'_k. The latter matrix obeys

$$(\bar{\mathbf{B}})^* \cdot \bar{\mathbf{B}} = (T^*\mathbf{B}^*T^{-1})(T\mathbf{B}(T^*)^{-1}) = T^*\mathbf{B}^*\mathbf{B}(T^*)^{-1} = \epsilon I_{2^\nu},$$

which proves the second part of proposition. □

Remark 1. We show below that parameters ϵ and ϵ' defined in Proposition 6.3.2 depend only on the values of p and q (on the metric signature for given $p+q=2\nu$).

To define yet another version of charge conjugation, and also define conjugation matrices in odd case, we make use of the following fact.

Proposition 6.3.3. *Let matrices* $\widetilde{\mathbf{D}}$, $\widetilde{\mathbf{C}}$ *and* $\widetilde{\mathbf{B}}$ *obey*

$$\gamma_k^\dagger = -\widetilde{\mathbf{D}}^{-1}\cdot\gamma_k\cdot\widetilde{\mathbf{D}},\ \gamma_k^\mathsf{T} = \widetilde{\mathbf{C}}^{-1}\cdot\gamma_k\cdot\widetilde{\mathbf{C}},\ \gamma_k^* = \widetilde{\mathbf{B}}^{-1}\cdot\gamma_k\cdot\widetilde{\mathbf{B}}, \\ \forall k \in (1,\ldots,2\nu), \tag{6.3.22}$$

while matrices $\mathbf{D}$, $\mathbf{C}$ *and* $\mathbf{B}$ *obey* (6.3.15), (6.3.17) *and* (6.3.18). *Then* $\widetilde{\mathbf{D}}$, $\widetilde{\mathbf{C}}$ *and* $\widetilde{\mathbf{B}}$ *are defined by* (6.3.22) *uniquely, modulo numerical factors, and can be chosen in such a way that*

$$\widetilde{\mathbf{D}} = \gamma_{2\nu+1}\mathbf{D},\ \widetilde{\mathbf{C}} = \gamma_{2\nu+1}\mathbf{C},\ \widetilde{\mathbf{B}} = \gamma_{2\nu+1}\mathbf{B}. \tag{6.3.23}$$

Matrices (6.3.23) *obey*

$$\widetilde{\mathbf{D}}^\dagger = (-1)^q\widetilde{\mathbf{D}},\ \widetilde{\mathbf{C}}^\mathsf{T} = \widetilde{\epsilon}'\,\widetilde{\mathbf{C}},\ \widetilde{\mathbf{B}}^*\,\widetilde{\mathbf{B}} = \widetilde{\mathbf{B}}\,\widetilde{\mathbf{B}}^* = \widetilde{\epsilon}I, \tag{6.3.24}$$

where parameters

$$\widetilde{\epsilon}' = \epsilon'(-1)^\nu, \quad \widetilde{\epsilon} = \epsilon(-1)^{\nu+q}, \tag{6.3.25}$$

are universal, i.e., they do not depend on the choice of representation of algebra $C\ell_{(p,q)}$.

Proof. The proof of uniqueness, modulo numerical factors, of matrices $\widetilde{\mathbf{D}}$, $\widetilde{\mathbf{C}}$, $\widetilde{\mathbf{B}}$ and of universality of parameters $\widetilde{\epsilon}'$ and $\widetilde{\epsilon}$ is the same as the proof of Proposition 6.3.2. Representation (6.3.23) of matrices $\widetilde{\mathbf{D}}$, $\widetilde{\mathbf{C}}$, $\widetilde{\mathbf{B}}$ as well as relations (6.3.24) are proven by direct calculation which relies on (6.3.15), (6.3.16),(6.3.17) and (6.3.18). □

Example. For particular representation (6.3.2) of algebra $\mathcal{C}\ell_{2\nu}$, matrix $\widetilde{\mathbf{C}}$ is proportional to matrix C' given by (6.3.6) (see Proposition 6.3.1). Recall that in this representation we have $\mathbf{C} = C$, so equalities (6.3.8) give

$$\epsilon' = (-1)^{[\frac{\nu+1}{2}]}, \quad \widetilde{\epsilon}' = (-1)^{[\frac{\nu}{2}]}. \tag{6.3.26}$$

These parameters are indeed related as claimed in Proposition 6.3.3: $\widetilde{\epsilon}' = \epsilon'(-1)^{\nu}$.

Remark 2. Transformations $\gamma_k \to i\gamma_k$ of a subset of γ-matrices correspond to the change of metric signature $s = p - q$. They do not modify the relations (6.3.15), (6.3.22) for matrices $\mathbf{C}$ and $\widetilde{\mathbf{C}}$. Hence, parameters ϵ' and $\widetilde{\epsilon}'$ in (6.3.26) do not depend on signature, and depend only on the dimension $N = 2\nu = p + q$. Note also that transformation $\gamma_k \to i\gamma_k$ made for all $k \in (1, \dots, N)$ relates generators of algebra $\mathcal{C}\ell_{(p,q)}$ to generators of algebra $\mathcal{C}\ell_{(q,p)}$. This transformation flips signs in (6.3.15), (6.3.22). So, matrices $\mathbf{D}, \mathbf{B}$ for algebra $\mathcal{C}\ell_{(p,q)}$ are matrices $\widetilde{\mathbf{D}}, \widetilde{\mathbf{B}}$ for algebra $\mathcal{C}\ell_{(q,p)}$ (with timelike and spacelike coordinates interchanged), and vice versa, and the parameters ϵ and $\widetilde{\epsilon}$ are interchanged, $\epsilon \leftrightarrow \widetilde{\epsilon}$.

Parameters ϵ, ϵ' and $\widetilde{\epsilon}$, $\widetilde{\epsilon}'$ play an important role in the theory. We show in Section 6.4.2 that ϵ and $\widetilde{\epsilon}$ determine whether or not one can define Majorana spinors — real representations of Clifford algebra $\mathcal{C}\ell_{(p,q)}$ and spinor group Spin(p, q). The value of ϵ' determines the type of group (symplectic or orthogonal) where group Spin(p, q) can be embedded.

Problem 6.3.2. Making use of (6.3.15) and (6.3.17), prove identities

$$\begin{aligned} \left(\mathbf{D}^{-1}\gamma_{m_1}\cdots\gamma_{m_r}\right)^{\dagger} &= (-1)^{r(r-1)/2}(\mathbf{D}^{-1}\gamma_{m_1}\cdots\gamma_{m_r}), \\ \left(\mathbf{C}^{-1}\gamma_{m_1}\cdots\gamma_{m_r}\right)^{\mathsf{T}} &= \epsilon'\,(-1)^{r(r+1)/2}\,(\mathbf{C}^{-1}\gamma_{m_1}\cdots\gamma_{m_r}), \end{aligned} \tag{6.3.27}$$

where $m_1 < m_2 < \cdots < m_r$. These identities ensure, in particular, that all matrices $(\mathbf{D}^{-1}\gamma_{m_1}\cdots\gamma_{m_r})$ are either Hermitian or anti-Hermitian, and that all matrices $(\mathbf{C}^{-1}\gamma_{m_1}\cdots\gamma_{m_r})$ are either symmetric or antisymmetric.

The identity $\gamma^\dagger = \gamma^{*\mathsf{T}}$ implies the relationship between matrices $\mathbf{B}$, $\mathbf{D}$ and $\mathbf{C}$.

Proposition 6.3.4. *Matrix* $\mathbf{B}$ *is expressed through matrices* $\mathbf{D}$ *and* $\mathbf{C}$ *in two ways:*

$$\mathbf{B} = \beta_1 \mathbf{C}(\mathbf{D}^{-1})^\mathsf{T}, \quad \mathbf{B} = \beta_2 \mathbf{D}(\mathbf{C}^{-1})^\dagger, \tag{6.3.28}$$

where $\beta_1, \beta_2 \in \mathbb{C}$ *are constants related by*

$$\beta_1^* \beta_2 = \epsilon\epsilon' \in \mathbb{R}. \tag{6.3.29}$$

Making use of the remaining freedom in the definitions of matrices $\mathbf{D}$, $\mathbf{C}$ *and* $\mathbf{B}$, *these constants can be chosen as follows:*

$$\beta_1 = \epsilon\epsilon', \quad \beta_2 = 1. \tag{6.3.30}$$

Proof. Matrices $\mathbf{B}$, $\mathbf{D}$ and $\mathbf{C}$ can be related in two ways, by writing $\gamma_k^* = (\gamma_k^\mathsf{T})^\dagger = (\gamma_k^\dagger)^\mathsf{T}$, which immediately gives (6.3.28). Equality (6.3.29) follows from (6.3.18) upon using (6.3.28) and (6.3.17):

$$\epsilon I_{2^\nu} = \mathbf{B}^*\mathbf{B} = \beta_1^*\beta_2 \mathbf{C}^* \left(\mathbf{D}^{-1}\right)^\dagger \mathbf{D} \left(\mathbf{C}^{-1}\right)^\dagger = \beta_1^*\beta_2 \epsilon' I_{2^\nu}. \tag{6.3.31}$$

Note that the definitions (6.3.17), (6.3.18) admit the following freedom:

$$\mathbf{B} \to e^{i\phi}\mathbf{B}, \ \mathbf{D} \to a\mathbf{D}, \ \mathbf{C} \to b\mathbf{C},$$

where $\phi, a \in \mathbb{R}$ and $b \in \mathbb{C}$. This freedom enables one to satisfy (6.3.30). □

Relations (6.3.28), (6.3.29) and (6.3.17) give also

$$\mathbf{B}^\dagger\mathbf{D}^{-1}\mathbf{B} = (\mathbf{D}^{-1}\mathbf{B})^\dagger\mathbf{B} = \beta_2^*\mathbf{C}^{-1}\mathbf{B} = \beta_2^*\beta_1(\mathbf{D}^{-1})^\mathsf{T} = \epsilon\epsilon'(\mathbf{D}^{-1})^\mathsf{T}; \tag{6.3.32}$$

this identity is used in what follows.

Problem 6.3.3. Show that matrices $\widetilde{\mathbf{D}}, \widetilde{\mathbf{B}}$ obey the following identities:

$$\widetilde{\mathbf{B}}\mathbf{C}^\dagger = \beta_2\widetilde{\mathbf{D}}, \ \widetilde{\mathbf{B}}\widetilde{\mathbf{D}}^\mathsf{T} = (-1)^\nu\beta_1\mathbf{C}.$$

Proposition 6.3.5. *Invariant parameters* ϵ' *and* ϵ *in* (6.3.17) *and* (6.3.18) *take the following values:*

$$\epsilon' = (-1)^k, \text{ where } k = \left[\frac{\nu+1}{2}\right] \Leftrightarrow \nu = 2k-1, 2k; \tag{6.3.33}$$

$$\epsilon = (-1)^{k+\beta}, \text{ where } \begin{cases} (1)\ k = \left[\frac{\nu+1}{2}\right] & \text{for } p = 2\beta - 1, \\ (2)\ k = \left[\frac{\nu}{2}\right] & \text{for } p = 2\beta; \end{cases} \tag{6.3.34}$$

here β *are integers, and* $[x]$ *is integer part of* x*. Let* $s = (p-q)$ *be metric signature, then formulas* (6.3.34) *for* ϵ *can be written as*

$$\epsilon = +1 \quad \text{for } s = -2, 0 \bmod(8); \quad \epsilon = -1 \quad \text{for } s = 2, 4 \bmod(8). \tag{6.3.35}$$

Proof. Since ϵ' and ϵ are the same in all irreducible representations of algebra $\mathcal{C}\ell_{(p,q)}$, $(p+q=2\nu)$, we calculate these parameters in a convenient representation. We choose representation (6.3.2) of complex algebra $\mathcal{C}\ell_{2\nu}$ and construct representation of algebra $\mathcal{C}\ell_{(p,q)}$ by making use of mapping (6.1.76). Matrices Γ_k are Hermitian in representation (6.3.2), so matrices γ_k (defined in (6.1.76)) obey

$$\gamma_1^\dagger = \gamma_1, \ldots, \gamma_p^\dagger = \gamma_p, \ \gamma_{p+1}^\dagger = -\gamma_{p+1}, \ldots, \ \gamma_{p+q}^\dagger = -\gamma_{p+q}. \tag{6.3.36}$$

Hence, the choice of matrix $\mathbf{D}$ in (6.3.15) is

$$\begin{aligned} &(1)\ \mathbf{D} = (-i)^\beta \Gamma_1 \cdots \Gamma_p \cdot \Gamma_{N+1} = I_2^{\otimes\beta} \otimes \sigma_3^{\otimes(\nu-\beta)} \quad \text{for } p = 2\beta, \\ &(2)\ \mathbf{D} = (-i)^{(\beta-1)} \Gamma_1 \cdots \Gamma_p = I_2^{\otimes(\beta-1)} \otimes \sigma_1 \otimes I_2^{\otimes(\nu-\beta)} \quad \text{for } p = 2\beta - 1. \end{aligned} \tag{6.3.37}$$

In both cases, $\mathbf{D}$ is normalized by Eq. (6.3.17). It follows from (6.3.11) that matrix $\mathbf{C}$ in (6.3.15) has the form (6.3.5) and obeys (6.3.8):

$$\begin{aligned} &(1)\ \mathbf{C} = (\sigma_2 \otimes \sigma_1)^{\otimes k}, \ \mathbf{C}^\mathsf{T} = \mathbf{C}^* = (-1)^{[\frac{\nu+1}{2}]}\mathbf{C} \quad \text{for } \nu = 2k; \\ &(2)\ \mathbf{C} = (\sigma_2 \otimes \sigma_1)^{\otimes k} \otimes \sigma_2, \ \mathbf{C}^\mathsf{T} = \mathbf{C}^* = (-1)^{[\frac{\nu+1}{2}]}\mathbf{C} \quad \text{for } \nu = 2k+1. \end{aligned} \tag{6.3.38}$$

This gives (6.3.33). Making use of (6.3.37) and (6.3.38), we see that in these cases matrices $\mathbf{D}$ and $\mathbf{C}$ obey

$$\mathbf{D}^2 = \mathbf{C}^2 = I_{2^\nu}, \ \mathbf{D}^\dagger = \mathbf{D}, \ \mathbf{D}^* = \mathbf{D}, \ \mathbf{C}^\dagger = \mathbf{C}. \tag{6.3.39}$$

The second of Eq. (6.3.28) with $\beta_2 = 1$ (the choice we made above) gives

$$\mathbf{B} = \mathbf{D} \cdot (\mathbf{C}^{-1})^\dagger = \mathbf{D} \cdot \mathbf{C},$$

and (6.3.18) and (6.3.38) yield

$$\begin{aligned} \epsilon\, I_{2^\nu} &= \mathbf{B} \cdot \mathbf{B}^* = \mathbf{D} \cdot \mathbf{C} \cdot \mathbf{D}^* \cdot \mathbf{C}^* = (-1)^{[\frac{\nu+1}{2}]} \mathbf{D} \cdot \mathbf{C} \cdot \mathbf{D} \cdot \mathbf{C} \\ &\Rightarrow \mathbf{D} \cdot \mathbf{C} = \epsilon (-1)^{[\frac{\nu+1}{2}]} \mathbf{C} \cdot \mathbf{D}. \end{aligned} \tag{6.3.40}$$

On the other hand, commutation rule $\mathbf{D} \cdot \mathbf{C} = \eta \mathbf{C} \cdot \mathbf{D}$ can be derived by making use of explicit representations (6.3.37) and (6.3.38) for $\mathbf{D}$ and $\mathbf{C}$. In this way we obtain the following: (1) for $p = 2\beta - 1$ with even (odd) β, we have $\eta = +1$ ($\eta = -1$), so $\eta = (-1)^\beta$; (2) for $p = 2\beta$ we have $\eta = (-1)^{\nu - \beta}$. Now, Eq. (6.3.40) gives $\eta = (-1)^{[\frac{\nu+1}{2}]} \epsilon$, and we arrive at (6.3.34).

Note that $[\frac{\nu+1}{2}]$ is even (odd) for $\nu = 3, 4 \bmod(4)$ ($\nu = 1, 2 \bmod(4)$), and $[\frac{\nu}{2}]$ is even (odd) for $\nu = 0, 1 \bmod(4)$ ($\nu = 2, 3 \bmod(4)$). So, in accordance with (6.3.34), parameter ϵ equals $+1$ in the following four cases:

$$\begin{aligned} &p = 1 \bmod(4), \ N = 2, 4 \bmod(8); \ p = 3 \bmod(4), \ N = 6, 8 \bmod(8); \\ &p = 0 \bmod(4), \ N = 0, 2 \bmod(8); \ p = 2 \bmod(4), \ N = 4, 6 \bmod(8). \end{aligned} \tag{6.3.41}$$

These cases can be written in unified way:

$$p = k \bmod(4), \ N = 2k, 2k + 2 \bmod(8) \quad (k = 0, 1, 2, 3). \tag{6.3.42}$$

We recall that the signature parameter is $s = p - q = 2p - N$, and write (6.3.42) as $s = 0, 6 \bmod(8)$; in other cases $s = 2, 4 \bmod(8)$ we have $\epsilon = -1$. This gives (6.3.35). □

Remark 3. The second formula in (6.3.25) for $\tilde{\epsilon}$ in view of (6.3.35) is written as

$$\tilde{\epsilon} = +1 \text{ for } s = 0, 2 \bmod(8); \quad \tilde{\epsilon} = -1 \text{ for } s = 4, 6 \bmod(8). \tag{6.3.43}$$

In other words, we have $\epsilon \leftrightarrow \tilde{\epsilon}$ under replacement $q \leftrightarrow p$.

Remark 4. The results (6.3.33) and (6.3.34) give for product $\epsilon \cdot \epsilon'$ entering (6.3.29)

$$(1)\ \epsilon \cdot \epsilon' = (-1)^{\beta} \text{ for } p = 2\beta - 1; \quad (2)\ \epsilon \cdot \epsilon' = (-1)^{\beta+\nu} \text{ for } p = 2\beta. \tag{6.3.44}$$

As we show below, equality $\epsilon = +1$ is the condition for the existence of Majorana spinors.[h]

Examples

(1) For any irreducible representation of algebra $\mathcal{C}\ell_{(p,q)}$ with even $p + q = 2\nu = N$, formulas (6.3.26) and (6.3.33) give

$$\epsilon' = -1 \text{ for } N = 2, 4 \bmod(8), \quad \epsilon' = +1 \text{ for } N = 6, 8 \bmod(8). \tag{6.3.45}$$

$$\widetilde{\epsilon}' = -1 \text{ for } N = 4, 6 \bmod(8), \quad \widetilde{\epsilon}' = +1 \text{ for } N = 0, 2 \bmod(8). \tag{6.3.46}$$

(2) In the Euclidean case $\mathbb{R}^N$, where $N = 2\nu$, all coordinates are timelike by Definition 6.2.4, and $p = 2\beta = 2\nu$, or $\beta = \nu$. So, parameters ϵ and $\widetilde{\epsilon}$ for any irreducible representation of algebra $\mathcal{C}\ell_N^{\mathbb{R}}$, as given by (6.3.34), (6.3.35) and (6.3.25), are

$$\epsilon = -1 \text{ for } N = 2, 4 \bmod(8), \quad \epsilon = +1 \text{ for } N = 6, 8 \bmod(8),$$

$$\widetilde{\epsilon} = -1 \text{ for } N = 4, 6 \bmod(8), \quad \widetilde{\epsilon} = +1 \text{ for } N = 2, 8 \bmod(8).$$

In that case, matrix $\mathbf{D}$ in (6.3.37) equals unit matrix, which agrees with the fact that representation (6.3.2) is Hermitian.

(3) For even-dimensional space $\mathbb{R}^{0,N}$ with all space-like coordinates ($p = 2\beta = 0$, $q = 2\nu$), we find from (6.3.34), (6.3.35) and (6.3.25) that[i]

$$\epsilon = +1 \text{ for } N = 2, 8 \bmod(8); \quad \epsilon = -1 \text{ for } N = 4, 6 \bmod(8),$$

$$\widetilde{\epsilon} = +1 \text{ for } N = 6, 8 \bmod(8); \quad \widetilde{\epsilon} = -1 \text{ for } N = 2, 4 \bmod(8).$$

[h] It is important here that coordinates $(x^1, \ldots, x^p)$ of space $\mathbb{R}^{p,q}$, that have positive signature in metric $\eta = I_{p,q}$, are interpreted as timelike (see Definition 6.2.4).

[i] These formulas coincide with those in Example 2 up to replacement $\epsilon \leftrightarrow \widetilde{\epsilon}$, which agrees with Remark 2 after Proposition 6.3.3.

In this case, representation of generators of algebra $\mathcal{C}\ell_{0,N}$ (obtained from representation (6.3.2) by replacement $\gamma_k = i\Gamma_k$) is anti-Hermitian: $\gamma_k^\dagger = -\gamma_k$, so we have from (6.3.37) that $\mathbf{D} = \Gamma_{N+1} = \sigma_3^{\otimes\nu}$.

(4) For even-dimensional Minkowski space $\mathbb{R}^{1,N-1}$ with $N = 2\nu$, we have $p = 2\beta - 1 = 1$, or $\beta = 1$. Therefore, parameters ϵ and $\widetilde{\epsilon}$ as given by (6.3.34), (6.3.35) and (6.3.25) are

$$\begin{aligned} &\epsilon = +1 \text{ for } N = 2,4\,\mathrm{mod}(8), \;\; \epsilon = -1 \text{ for } N = 6,8\,\mathrm{mod}(8), \\ &\widetilde{\epsilon} = +1 \text{ for } N = 2,8\,\mathrm{mod}(8), \;\; \widetilde{\epsilon} = -1 \text{ for } N = 4,6\,\mathrm{mod}(8), \end{aligned} \tag{6.3.47}$$

and in representation (6.3.37) we have $\mathbf{D} = \sigma_1 \otimes I_2^{\otimes(\nu-1)}$. As concerns the space $\mathbb{R}^{N-1,1}$ with $N = 2\nu$, in view of Remark 2 after Proposition 6.3.3 we have

$$\begin{aligned} &\widetilde{\epsilon} = +1 \text{ for } N = 2,4\,\mathrm{mod}(8), \;\; \widetilde{\epsilon} = -1 \text{ for } N = 6,8\,\mathrm{mod}(8), \\ &\epsilon = +1 \text{ for } N = 2,8\,\mathrm{mod}(8), \;\; \epsilon = -1 \text{ for } N = 4,6\,\mathrm{mod}(8). \end{aligned} \tag{6.3.48}$$

Problem 6.3.4. Show that for any irreducible representation of algebra $\mathcal{C}\ell_{2,N-2}$ associated with even-dimensional space $\mathbb{R}^{2,N-2}$, the following relations hold:

$$\begin{aligned} &\epsilon = -1 \text{ for } N = 2,8\,\mathrm{mod}(8), \;\; \epsilon = +1 \text{ for } N = 4,6\,\mathrm{mod}(8), \\ &\widetilde{\epsilon} = -1 \text{ for } N = 6,8\,\mathrm{mod}(8), \;\; \widetilde{\epsilon} = +1 \text{ for } N = 2,4\,\mathrm{mod}(8). \end{aligned}$$

Find the values of parameters ϵ, $\widetilde{\epsilon}$ for representation of algebra $\mathcal{C}\ell_{N-2,2}$.

2. Conjugation matrices in odd case $p + q = 2\nu + 1$

Let Γ_k be generators of algebra $\mathcal{C}\ell_{(p,q)}$, for odd $N = (p+q) = (2\nu+1)$, with defining relations (6.1.2) and matrices $(+\gamma_k) \in \mathrm{Mat}_{2^\nu}(\mathbb{C})$ or $(-\gamma_k) \in \mathrm{Mat}_{2^\nu}(\mathbb{C})$ be representations of generators $\Gamma_k \in \mathcal{C}\ell_{(p,q)}$:

$$\rho_+(\Gamma_k) = \gamma_k, \;\; \rho_-(\Gamma_k) = -\gamma_k, \quad k = 1, \ldots, 2\nu+1.$$

Representation ρ_- is not equivalent to representation ρ_+, since for central element given by (6.1.19), $\Gamma_{N+1} = e^{-i\frac{\pi}{2}(\nu+q)}\Gamma_1\Gamma_2\cdots\Gamma_{2\nu+1}$, we have $\rho_+(\Gamma_{N+1}) = -\rho_-(\Gamma_{N+1})$. As we have seen, in odd case there are only two inequivalent irreducible representations of algebra $\mathcal{C}\ell_{(p,q)}$ (see Proposition 6.1.3 and discussion after that proposition; these representations are explicitly constructed in (6.1.65), (6.1.66)). Therefore, eight representations with generators

$$\{\pm\gamma_k\}, \; \{\pm\gamma_k^\dagger\}, \; \{\pm\gamma_k^\top\}, \; \{\pm\gamma_k^*\},$$

are divided into two equivalence classes, one of which includes representation ρ_+, and another ρ_-. Equivalence class that includes representation ρ_+ is

$$\begin{aligned}
&(1)\ \text{even } \nu\ \big(N = 2\nu + 1 = 1\,(\text{mod } 4)\big),\ \text{ even } q\text{: } \{\gamma_k, \gamma_k^\dagger, \gamma_k^\mathsf{T}, \gamma_k^*\};\\
&(2)\ \text{even } \nu\ (N = 1\,(\text{mod } 4)),\ \text{ odd } q : \{\gamma_k, -\gamma_k^\dagger, \gamma_k^\mathsf{T}, -\gamma_k^*\};\\
&(3)\ \text{odd } \nu\ (N = 3\,(\text{mod } 4)),\ \text{ even } q : \{\gamma_k, \gamma_k^\dagger, -\gamma_k^\mathsf{T}, -\gamma_k^*\};\\
&(4)\ \text{odd } \nu\ (N = 3\,(\text{mod } 4)),\ \text{ odd } q : \{\gamma_k, -\gamma_k^\dagger, -\gamma_k^\mathsf{T}, \gamma_k^*\}.
\end{aligned} \tag{6.3.49}$$

This result follows directly from the relations

$$\gamma_1^\dagger\gamma_2^\dagger\cdots\gamma_{2\nu+1}^\dagger = (-1)^q\gamma_1\gamma_2\cdots\gamma_{2\nu+1},\ \gamma_1^\mathsf{T}\gamma_2^\mathsf{T}\cdots\gamma_{2\nu+1}^\mathsf{T} = (-1)^\nu\gamma_1\gamma_2\cdots\gamma_{2\nu+1},$$
$$\gamma_1^*\gamma_2^*\cdots\gamma_{2\nu+1}^* = (-1)^{\nu+q}\gamma_1\gamma_2\cdots\gamma_{2\nu+1}, \tag{6.3.50}$$

which follow from (6.3.15), (6.3.16) with $(\gamma_{2\nu+1})^2 = I$. These relations are valid also for $(\gamma_{2\nu+1})^2 = -I$, which can be seen by replacing $\gamma_{2\nu+1} \to (-i)\gamma_{2\nu+1}$, $q \to q+1$ in Eq. (6.3.16).

In even case, sign convention in (6.3.15) could be flipped by multiplying operators $\mathbf{D}$, $\mathbf{B}$ and $\mathbf{C}$ on the left by matrix $\gamma_{2\nu+1}$. This is no longer possible in odd case, because of inequivalence of representations $\{\gamma_k\}$ and $\{-\gamma_k\}$. In accordance with (6.3.49), the conjugation relations are now

$$\gamma_k^\dagger = (-1)^q\mathbf{D}'^{-1}\gamma_k\,\mathbf{D}',\ \ \gamma_k^\mathsf{T} = (-1)^\nu\mathbf{C}'^{-1}\gamma_k\,\mathbf{C}',\ \ \gamma_k^* = (-1)^{\nu+q}\mathbf{B}'^{-1}\gamma_k\,\mathbf{B}',$$
$$k = 1, \ldots, p+q = 2\nu+1. \tag{6.3.51}$$

Proposition 6.3.6. *Let operators* $\mathbf{D}$, $\mathbf{B}$ *and* $\mathbf{C}$ *be defined by* (6.3.15) *in even case* $N-1 = 2\nu$. *Let matrices* $\widetilde{\mathbf{D}}$, $\widetilde{\mathbf{C}}$, $\widetilde{\mathbf{B}}$ *be defined in* (6.3.23). *Then equivalence matrices* $\mathbf{D}'$, $\mathbf{B}'$ *and* $\mathbf{C}'$ *entering* (6.3.51) *in odd case* $N = p+q = 2\nu+1$ *are*

$$\begin{aligned}
&(1)\ \text{even } \nu\,(N = 1(\text{mod}\,4)),\ \text{ even } q : \mathbf{D}' = \mathbf{D},\ \mathbf{B}' = \tilde{\mathbf{B}},\ \mathbf{C}' = \tilde{\mathbf{C}};\\
&(2)\ \text{even } \nu\,(N = 1(\text{mod}\,4)),\ \text{ odd } q : \mathbf{D}' = \tilde{\mathbf{D}},\ \mathbf{B}' = \mathbf{B},\ \mathbf{C}' = \tilde{\mathbf{C}};\\
&(3)\ \text{odd } \nu\,(N = 3(\text{mod}\,4)),\ \text{ even } q : \mathbf{D}' = \mathbf{D},\ \mathbf{B}' = \mathbf{B},\ \mathbf{C}' = \mathbf{C};\\
&(4)\ \text{odd } \nu\,(N = 3(\text{mod}\,4)),\ \text{ odd } q : \mathbf{D}' = \tilde{\mathbf{D}},\ \mathbf{B}' = \tilde{\mathbf{B}},\ \mathbf{C}' = \mathbf{C}.
\end{aligned} \tag{6.3.52}$$

Proof. By making use of the relations (6.3.15) and (6.3.16) one shows that in odd case $N = (p+q) = 2\nu+1$, equivalence relations (1)–(4) in (6.3.49) involve matrices $\mathbf{D}$, $\mathbf{B}$, $\mathbf{C}$, $\widetilde{\mathbf{D}}$, $\widetilde{\mathbf{C}}$ and $\widetilde{\mathbf{B}}$ in the following way:

$$\begin{aligned}
&(1)\ \gamma_k^\dagger = \mathbf{D}^{-1}\gamma_k\mathbf{D},\ \gamma_k^\mathsf{T} = \widetilde{\mathbf{C}}^{-1}\gamma_k\widetilde{\mathbf{C}},\ \gamma_k^* = \widetilde{\mathbf{B}}^{-1}\gamma_k\widetilde{\mathbf{B}},\\
&(2)\ \gamma_k^\dagger = -\widetilde{\mathbf{D}}^{-1}\gamma_k\widetilde{\mathbf{D}},\ \gamma_k^\mathsf{T} = \widetilde{\mathbf{C}}^{-1}\gamma_k\widetilde{\mathbf{C}},\ \gamma_k^* = -\mathbf{B}^{-1}\gamma_k\mathbf{B},\\
&(3)\ \gamma_k^\dagger = \mathbf{D}^{-1}\gamma_k\mathbf{D},\ \gamma_k^\mathsf{T} = -\mathbf{C}^{-1}\gamma_k\mathbf{C},\ \gamma_k^* = -\mathbf{B}^{-1}\gamma_k\mathbf{B},\\
&(4)\ \gamma_k^\dagger = -\widetilde{\mathbf{D}}^{-1}\gamma_k\widetilde{\mathbf{D}},\ \gamma_k^\mathsf{T} = -\mathbf{C}^{-1}\gamma_k\mathbf{C},\ \gamma_k^* = \widetilde{\mathbf{B}}^{-1}\gamma_k\widetilde{\mathbf{B}}.
\end{aligned}$$

This immediately gives (6.3.52). □

In odd case $N = (p+q) = 2\nu+1$, the result similar to Proposition 6.3.2 is as follows.

Proposition 6.3.7. *Matrices* $\mathbf{D}', \mathbf{B}', \mathbf{C}' \in \mathrm{Mat}_{2^\nu}(\mathbb{C})$ *are determined by* (6.3.51) *uniquely, modulo multiplication by numerical constants, which can be chosen in such a way that*

$$\mathbf{D}'^\dagger = (-1)^q\mathbf{D}',\ \mathbf{C}'^\mathsf{T} = \epsilon'\mathbf{C}', \tag{6.3.53}$$

$$\mathbf{B}'^* \cdot \mathbf{B}' = \mathbf{B}' \cdot \mathbf{B}'^* = \epsilon I_{2^\nu}, \tag{6.3.54}$$

where parameters $\epsilon' = \pm 1$ *and* $\epsilon = \pm 1$ *take the same values as in even-dimensional case of one dimension smaller,* $(p-1)+q = 2\nu$ *(see* (6.3.17) *and* (6.3.18)*). So, the values of parameters* ϵ' *and* ϵ *do not depend on representation, but depend on* p *and* q*. Matrices* $\mathbf{D}', \mathbf{B}'$ *and* $\mathbf{C}'$ *obey the identities* (*cf.* (6.3.28), (6.3.29))

$$\mathbf{B}' = \beta_1'\mathbf{C}'(\mathbf{D}'^{-1})^\mathsf{T},\ \mathbf{B}' = \beta_2'\mathbf{D}'(\mathbf{C}'^{-1})^\dagger, \tag{6.3.55}$$

where parameters β_1' *and* β_2' *are related by* $(\beta_1')^*\,\beta_2' = \epsilon\,\epsilon'\,(-1)^q$.

Proof. The fact that matrices $\mathbf{D}', \mathbf{B}', \mathbf{C}'$ are determined by (6.3.51) uniquely, modulo multiplication by numerical constants, is proven in the same way as similar fact in even case, see Proposition 6.3.2. Relations (6.3.53), (6.3.54) follow from (6.3.17), (6.3.18), (6.3.24) and definitions (6.3.52). Let us prove, as an example, relation (6.3.54). For odd $(\nu+q)$, we see from (6.3.52) that matrix $\mathbf{B}'$ in (6.3.51) coincides with matrix $\mathbf{B}$ from (6.3.15) with even number of dimensions $(p-1)+q = 2\nu$, so relation (6.3.54) is simply the identity (6.3.18). For even $(\nu+q)$, we get from (6.3.52) that matrix $\mathbf{B}'$ in (6.3.51) coincides with $\widetilde{\mathbf{B}}$. In view of (6.3.24), we again have

$$\mathbf{B}'^* \cdot \mathbf{B}' = \widetilde{\mathbf{B}}^*\widetilde{\mathbf{B}} = (-1)^{\nu+q}\epsilon I_{2^\nu} = \epsilon\, I_{2^\nu}.$$

So, parameter ϵ in odd case $N = p + q = 2\nu + 1$ is the same as parameter ϵ in even case $N - 1 = (p-1) + q = 2\nu$.

Finally, identities (6.3.55) are straightforwardly established by making use of (6.3.51), while the relation $\beta_1'^* \beta_2' = \epsilon\epsilon'(-1)^q$ is obtained by chain of equalities analogous to (6.3.31). □

Examples

1. In accordance with Proposition 6.3.7 and relations (6.3.42), we obtain that $\epsilon = +1$ in odd case $N = p + q = 2\nu + 1$, provided that

$$\begin{gathered} p - 1 = k \bmod(4),\ N - 1 = 2k, (2k+2) \bmod(8)\ (k = 0,1,2,3) \\ \Rightarrow\ s = (p-q) = 2p - N = -1, 1 \bmod(8). \end{gathered} \tag{6.3.56}$$

Since $(\nu + q)$ is even (odd) for $(p - q) = 1 \bmod(4)$ $((p-q) = -1 \bmod(4))$, respectively, we find that

$$\begin{gathered} \epsilon = +1,\ \gamma_k^* = \mathbf{B}'^{-1}\gamma_k\mathbf{B}'\ \text{ for }\ (p-q) = 1 \bmod(8), \\ \epsilon = +1,\ \gamma_k^* = -\mathbf{B}'^{-1}\gamma_k\mathbf{B}'\ \text{ for }\ (p-q) = -1 \bmod(8). \end{gathered} \tag{6.3.57}$$

2. For odd-dimensional Minkowski space $\mathbb{R}^{1,N-1}$, we recall the result in Example 3 after Proposition 6.3.5 and obtain, in agreement with (6.3.56),

$$\epsilon = +1\ \text{ for }\ N = 1,3 \bmod(8); \quad \epsilon = -1\ \text{ for }\ N = 5,7 \bmod(8). \tag{6.3.58}$$

6.3.2. Conjugation matrices B, C, D and structure of groups Spin(p,q). Group Spin(8)

Proposition 6.3.8. *Elements U of group $CL(p,q)$ (see Definition 6.2.1) in representation generated by irreducible representation ρ of algebra $C\ell_{(p,q)}$ can be normalized in such a way that U obey*

$$U^\dagger\, \mathbf{D}^{-1}\, U = \lambda(U)\, \mathbf{D}^{-1},\ U^\mathsf{T}\, \mathbf{C}^{-1}\, U = \mathbf{C}^{-1}, \tag{6.3.59}$$

where $\lambda(U) = \pm 1$ and matrices $\mathbf{D}$ and $\mathbf{C}$ are defined by (6.3.15) in even case and are obtained by replacement $\mathbf{D} \to \mathbf{D}'$ and $\mathbf{C} \to \mathbf{C}'$ in odd case, with matrices $\mathbf{D}'$ and $\mathbf{C}'$ defined by (6.3.51).

Proof. We first prove the second equality in (6.3.59) in even case; the proof is analogous in odd case. We write transformations (6.2.5) in terms of generators γ_m^T in two ways. On the one hand, we have

$$\gamma_m^\mathsf{T} \to (U^{-1})^\mathsf{T}\gamma_m^\mathsf{T}U^\mathsf{T} = -(U^{-1})^\mathsf{T}\mathbf{C}^{-1}\gamma_m\mathbf{C}U^\mathsf{T}, \tag{6.3.60}$$

and on the other

$$\gamma_m^\mathsf{T} \to (\gamma_k\Lambda^k_m)^\mathsf{T} = -\mathbf{C}^{-1}\gamma_k\mathbf{C}\Lambda^k_m = -\mathbf{C}^{-1}U\gamma_mU^{-1}\mathbf{C}. \tag{6.3.61}$$

By comparing the two expressions, we have $[\mathbf{C}U^\mathsf{T}\mathbf{C}^{-1}U, \gamma_m] = 0$, and Schur's lemma gives

$$\mathbf{C}U^\mathsf{T}\mathbf{C}^{-1}U = \rho_U I, \tag{6.3.62}$$

where $\rho_U \equiv \rho(U)$ is some scalar function on group $CL(p,q)$. We redefine $U \to \rho_U^{1/2}U$ and arrive at the second equality in (6.3.59).

Likewise, the transformation (6.2.5) in terms of generators $\gamma_m^\dagger$, written in two ways, gives $[\mathbf{D}U^\dagger\mathbf{D}^{-1}U, \gamma_m] = 0$ and using Schur's lemma we have

$$\mathbf{D}U^\dagger\mathbf{D}^{-1}U = \lambda_U I, \tag{6.3.63}$$

where $\lambda_U \equiv \lambda(U)$ is some function on group $CL(p,q)$ such that $\lambda_U\lambda_{U'} = \lambda_{U\cdot U'}$. We make Hermitian conjugate of (6.3.63) and recall that $\mathbf{D}^\dagger = \mathbf{D}$ (cf. (6.3.17)) to obtain $\lambda_U^* = \lambda_U$. So, λ_U is real one-dimensional representation of group $CL(p,q)$. We calculate determinants of the two sides in (6.3.62) (with $\rho_U = 1$) and (6.3.63) and obtain the relations $\det(U)^2 = 1$ and $\det(U)^*\det(U) = (\lambda_U)^d$, where $d = 2^\nu$ is dimension of representation ρ. These equalities show that $(\lambda_U)^d = 1$, and since λ_U is real, we have $\lambda_U = \pm 1$. This gives the first relation in (6.3.59). □

Remark. The value $\lambda_U = -1$ in the first of Eq. (6.3.59) occurs, e.g., when element $U \in CL(p,q)$ in (6.2.5) describes a reflection. In particular, the longest element $U = \gamma_{N+1} \in CL(p,q)$ generates, in even case $N = (p+q) = 2\nu$ and $p = 1, q = N-1$, matrix $\Lambda = -I$ of reflection of all coordinates, and since $\mathbf{D} = \gamma_0$ in this case (see Section 6.4.3), we have $\lambda(U) = -1$ in the first formula in (6.3.59).

Elements of subgroup $\mathsf{Spin}(p,q) \subset CL(p,q)$ are, by definition, normalized by (6.3.59) with $\lambda(U) = -1$:

$$U^\dagger\mathbf{D}^{-1}U = \mathbf{D}^{-1}, \quad U^\mathsf{T}\mathbf{C}^{-1}U = \mathbf{C}^{-1}. \tag{6.3.64}$$

Indeed, in view of relations (6.3.15) in even case, and relations (6.3.51) in odd case, we obtain the following properties of generators $\rho(\Sigma_{mn})$ of Lie algebra $\mathsf{spin}(p,q)$:

$$\rho(\Sigma_{mn})^{\dagger} = -\mathbf{D}^{-1}\,\rho(\Sigma_{mn})\,\mathbf{D}, \;\; \rho(\Sigma_{mn})^{\mathsf{T}} = -\mathbf{C}^{-1}\,\rho(\Sigma_{mn})\,\mathbf{C}. \qquad (6.3.65)$$

So, elements (6.2.30) of group $\mathsf{Spin}(p,q)$ indeed obey (6.3.64). Let us also give, for future references, complex conjugation rules for matrices $\rho(\Sigma_{mn}) \in \mathsf{spin}(p,q)$ and $U \in \mathsf{Spin}(p,q)$, which follow from the third equalities in (6.3.15) and (6.3.51):

$$\rho(\Sigma_{mn})^{*} = \mathbf{B}^{-1}\,\rho(\Sigma_{mn})\,\mathbf{B} \;\Rightarrow\; \mathbf{B}U^{*} = U\,\mathbf{B}, \;(\mathbf{B} \leftrightarrow \mathbf{B}'). \qquad (6.3.66)$$

The third equality in (6.3.22) gives

$$\rho(\Sigma_{mn})^{*} = \widetilde{\mathbf{B}}^{-1}\,\rho(\Sigma_{mn})\,\widetilde{\mathbf{B}} \;\Rightarrow\; \widetilde{\mathbf{B}}U^{*} = U\,\widetilde{\mathbf{B}}. \qquad (6.3.67)$$

Now, in view of (6.3.17), (6.3.38) and (6.3.45) we have

$$\mathbf{C}^{\mathsf{T}} = -\mathbf{C} \text{ for } N = 2,4\,\mathrm{mod}(8), \;\; \mathbf{C}^{\mathsf{T}} = \mathbf{C} \text{ for } N = 6,8\,\mathrm{mod}(8),$$

and the second equality in Eq. (6.3.64) shows that

$$U \in Sp'(2^{\nu},\mathbb{C}) \text{ for } N = 2,4\,\mathrm{mod}(8), \;\; U \in SO'(2^{\nu},\mathbb{C}) \text{ for } N = 6,8\,\mathrm{mod}(8), \qquad (6.3.68)$$

where $SO'(N,\mathbb{C})$ and $Sp'(2r,\mathbb{C})$ are groups which are isomorphic to orthogonal and symplectic groups $SO(N,\mathbb{C})$ and $Sp(2r,\mathbb{C})$, respectively, defined in (5.1.13) and (5.3.1). In what follows, we omit primes in similar cases to simplify notations. Explicit form (6.3.37) of Hermitian matrix $\mathbf{D}^{\dagger} = \mathbf{D}$ shows that for $p \neq 2\nu$ it has $2^{\nu-1}$ eigenvalues $+1$ and the same number of eigenvalues -1, so that $U \in SU(2^{\nu-1},2^{\nu-1})$, i.e., group $\mathsf{Spin}(p,q)$ is a subgroup of pseudo-unitary group $SU(2^{\nu-1},2^{\nu-1})$. In the case $p = 2\nu$, there is Hermitian representation (6.3.2), where $\mathbf{D} = I_{2^{\nu}}$, and therefore $\mathsf{Spin}(2\nu) \subset SU(2^{\nu})$, i.e., group $\mathsf{Spin}(2\nu)$ is compact. To summarize, for $p \neq 2\nu = N$ we have

$$\begin{array}{l} \mathsf{Spin}(p,q) \subset Sp(2^{\nu},\mathbb{C}) \cap SU(2^{\nu-1},2^{\nu-1}) \;\text{ for } N = 2,4\,\mathrm{mod}(8), \\ \mathsf{Spin}(p,q) \subset SO(2^{\nu},\mathbb{C}) \cap SU(2^{\nu-1},2^{\nu-1}) \;\text{ for } N = 6,8\,\mathrm{mod}(8), \end{array} \qquad (6.3.69)$$

while for $p = 2\nu = N$ the embedding property is

$$\begin{array}{l} \mathsf{Spin}(2\nu) \subset Sp(2^{\nu},\mathbb{C}) \cap SU(2^{\nu}) = USp(2^{\nu}) \;\text{ for } N = 2,4\,\mathrm{mod}(8), \\ \mathsf{Spin}(2\nu) \subset SO(2^{\nu},\mathbb{C}) \cap SU(2^{\nu}) = SO(2^{\nu}) \;\text{ for } N = 6,8\,\mathrm{mod}(8) \end{array} \qquad (6.3.70)$$

(recall that the groups on the right-hand sides here are isomorphic, but not identical to $USp(2^\nu)$ and $SO(2^\nu)$).

Problem 6.3.5. $\star$ Let matrices $U \in \mathrm{Mat}_M(\mathbb{C})$ satisfy the condition $\det(U) = 1$ and relations (6.3.64) for $\mathbf{D} = I_M$:

$$U^\dagger U = I_M, \;\; U^\mathsf{T}\, \mathbf{C}^{-1} U = \mathbf{C}^{-1},$$

where $\mathbf{C}$ is real matrix and $\mathbf{C}^\mathsf{T} = \mathbf{C}$. The set of all such U's generates the group $SU(M) \cap SO(M,\mathbb{C})$. Prove that $SO(M,\mathbb{C}) \cap SU(M) = SO(M)$. Hint: decompose matrix $A \in SO(M,\mathbb{C}) \cap SU(M)$ into real and imaginary parts.

Consider, in even case $p+q = 2\nu$, Weyl representation of Clifford algebra $\mathcal{C}\ell_{(p,q)}$ with generators given by (6.1.81), (6.1.86), i.e.,

$$\gamma_1 = \begin{pmatrix} \mathbf{0} & \mathbf{1} \\ \mathbf{1} & \mathbf{0} \end{pmatrix}, \; \gamma_m = \begin{pmatrix} \mathbf{0} & \mathbf{s}_m \\ \bar{\mathbf{s}}_m & \mathbf{0} \end{pmatrix} \;\; (m = 2, \dots, 2\nu) \;\Rightarrow\; \gamma_{2\nu+1} = \begin{pmatrix} \mathbf{1} & \mathbf{0} \\ \mathbf{0} & -\mathbf{1} \end{pmatrix}. \tag{6.3.71}$$

Here $\mathbf{1} \equiv I_{2^{\nu-1}}$, and $2^{\nu-1}$-dimensional blocks $\mathbf{s}_m$ and $\bar{\mathbf{s}}_m$ are components of vectors (6.1.84):

$$\begin{aligned} \mathbf{s} &= (\mathbf{s}_1, \dots, \mathbf{s}_{2\nu}) = (I_{2^{\nu-1}}, -i\Gamma'_1, \dots, -i\Gamma'_{p-1}, \Gamma'_p \dots, \Gamma'_{2\nu-1}), \\ \bar{\mathbf{s}} &= (\bar{\mathbf{s}}_1, \dots, \bar{\mathbf{s}}_{2\nu}) = (I_{2^{\nu-1}}, i\Gamma'_1, \dots, i\Gamma'_{p-1}, -\Gamma'_p \dots, -\Gamma'_{2\nu-1}). \end{aligned} \tag{6.3.72}$$

They obey (6.1.82), (6.1.83). Here matrices $\{\Gamma'_1, \dots, \Gamma'_{2\nu-1}\}$ generate Hermitian representation of generators of algebra $\mathcal{C}\ell_{2\nu-1}$, so blocks $\mathbf{s}_m$ and $\bar{\mathbf{s}}_m$ obey $\mathbf{s}^\dagger_m = \eta_{mm}\bar{\mathbf{s}}_m$, where η_{mm} are diagonal elements of metric $\eta = I_{p,q}$. The latter property gives $\gamma^\dagger_m = \eta_{mm}\gamma_m$ $(m = 1, \dots, 2\nu)$. Making use of (6.3.72), representation (6.3.71) can be written in compact form (see (6.1.79),(6.1.80))

$$\begin{gathered} \gamma_1 = \sigma_1 \otimes \mathbf{1}, \; \gamma_{m+1} = \sigma_2 \otimes \gamma'_m = \sigma_2 \otimes \Gamma'_m \;\; (m = 1, \dots, p-1), \\ \gamma_{m+1} = \sigma_2 \otimes \gamma'_m = i\,\sigma_2 \otimes \Gamma'_m \; (m = p, \dots, 2\nu - 1) \;\Rightarrow\; \gamma_{2\nu+1} = \sigma_3 \otimes \mathbf{1}, \end{gathered} \tag{6.3.73}$$

where matrices γ'_k $(k = 1, \dots, 2\nu - 1)$ are representations of generators of algebra $\mathcal{C}\ell_{(p-1,q)}$, and matrix $\gamma'_{2\nu-1}$ is constructed in terms of generators $\{\gamma'_1, \dots, \gamma'_{2\nu-2}\}$ in accordance with (6.1.80).

Proposition 6.3.9. *In even case, and in Weyl representation, operators* $\mathbf{D}$, $\mathbf{C}$ *and* $\mathbf{B}$, *which obey*

$$\gamma_k^\dagger = \mathbf{D}^{-1} \cdot \gamma_k \cdot \mathbf{D}, \ \ \gamma_k^\mathsf{T} = -\mathbf{C}^{-1} \cdot \gamma_k \cdot \mathbf{C}, \ \ \gamma_k^* = -\mathbf{B}^{-1} \cdot \gamma_k \cdot \mathbf{B}, \tag{6.3.74}$$

$$\begin{gathered}\gamma_{2\nu+1}^\dagger = (-1)^q \mathbf{D}^{-1} \cdot \gamma_{2\nu+1} \cdot \mathbf{D}, \ \ \gamma_{2\nu+1}^\mathsf{T} = (-1)^\nu \mathbf{C}^{-1} \cdot \gamma_{2\nu+1} \cdot \mathbf{C}, \\ \gamma_{2\nu+1}^* = (-1)^{\nu+q} \mathbf{B}^{-1} \cdot \gamma_{2\nu+1} \cdot \mathbf{B}, \end{gathered} \tag{6.3.75}$$

are given by

$$(1)\ q, p \in 2\mathbb{Z}_{\geq 0} \ \Rightarrow \ \mathbf{D} = I_2 \otimes \mathbf{d}; \ \ q, p \in 2\mathbb{Z}_{\geq 0} + 1 \ \Rightarrow \ \mathbf{D} = \sigma_1 \otimes \mathbf{g}; \tag{6.3.76}$$

$$(2)\ \nu \in 2\mathbb{Z}_{\geq 0} \ \Rightarrow \ \mathbf{C} = \sigma_3 \otimes \mathbf{c}; \ \ \nu \in 2\mathbb{Z}_{\geq 0} + 1 \ \Rightarrow \ \mathbf{C} = i\sigma_2 \otimes \mathbf{h}; \tag{6.3.77}$$

$$(3)\ \nu + q \in 2\mathbb{Z}_{\geq 0} \ \Rightarrow \ \mathbf{B} = \sigma_3 \otimes \mathbf{b}; \ \ \nu + q \in 2\mathbb{Z}_{\geq 0} + 1 \ \Rightarrow \ \mathbf{B} = i\sigma_2 \otimes \mathbf{f}, \tag{6.3.78}$$

where $2\mathbb{Z}_{\geq 0}$ *and* $2\mathbb{Z}_{\geq 0} + 1$ *are sets of even and odd nonnegative numbers, respectively;* $\mathbf{d}, \mathbf{g}, \mathbf{c}, \mathbf{h}, \mathbf{b}, \mathbf{f} \in \mathrm{Mat}_{2^{\nu-1}}(\mathbb{C})$ *and*

$$\mathbf{d}^\dagger = \mathbf{d}, \ \ \mathbf{g}^\dagger = \mathbf{g}, \ \ \mathbf{c}^T = \epsilon' \mathbf{c}, \ \ \mathbf{h}^T = -\epsilon' \mathbf{h}, \ \ \mathbf{b}^* \mathbf{b} = \epsilon I_{2^{\nu-1}} = -\mathbf{f}^* \mathbf{f}, \tag{6.3.79}$$

$$\begin{gathered}(1)\ q, \nu \in 2\mathbb{Z}_{\geq 0} \ \Rightarrow \ \mathbf{b} = \mathbf{d}\mathbf{c}^{\dagger -1}; \ \ q, \nu \in 2\mathbb{Z}_{\geq 0} + 1 \Rightarrow \mathbf{b} = -\mathbf{g}\mathbf{h}^{\dagger -1}; \\ (2)\ q, (\nu+1) \in 2\mathbb{Z}_{\geq 0} \ \Rightarrow \ \mathbf{f} = \mathbf{d}\mathbf{h}^{\dagger -1}; \ \ (q+1), \nu \in 2\mathbb{Z}_{\geq 0} \Rightarrow \mathbf{f} = -\mathbf{g}\mathbf{c}^{\dagger -1}. \end{gathered} \tag{6.3.80}$$

Proof. Before coming to the proof, we notice that properties $(\nu+q) \in 2\mathbb{Z}_{\geq 0}$ and $(\nu + q) \in 2\mathbb{Z}_{\geq 0} + 1$ are equivalent to $s/2 \in 2\mathbb{Z}$ and $s/2 \in 2\mathbb{Z} + 1$, respectively, where $s = (p-q)$ is signature parameter. We prove Eq. (6.3.76); the proof of (6.3.77) and (6.3.78) is similar. We write matrix $\mathbf{D}$ in block form, $\mathbf{D} = \begin{pmatrix} \mathbf{d} & \mathbf{g} \\ \mathbf{g}' & \mathbf{d}' \end{pmatrix}$, where $\mathbf{d}, \mathbf{g}, \mathbf{d}', \mathbf{g}' \in \mathrm{Mat}_{2^{\nu-1}}(\mathbb{C})$. Then the first relation in (6.3.74) with $k = 1$ gives, in view of (6.3.71), that $\mathbf{d}' = \mathbf{d}$ and $\mathbf{g}' = \mathbf{g}$. Then the first relation in (6.3.75) and explicit formula (6.3.71) for $\gamma_{2\nu+1}$ yield (6.3.76). Upon inserting matrices (6.3.76), (6.3.77) and (6.3.78) in (6.3.17), (6.3.18) we obtain (6.3.79). Relations (6.3.80) follow from the second formula in (6.3.28). □

Corollary 6.3.10. *We use block forms* (6.3.76)–(6.3.78) *of* $\mathbf{D}, \mathbf{C}$ *and* $\mathbf{B}$ *and recall the expressions* (6.3.73) *for gamma-matrices in Weyl representation. Then we obtain the following form of* (6.3.74):

$$\begin{aligned}
&(1)\ q,p \in 2\mathbb{Z}_{\geq 0} \Rightarrow \gamma_k'^{\dagger} = \mathbf{d}^{-1}\gamma_k'\mathbf{d};\ \ q,p \in 2\mathbb{Z}_{\geq 0}+1 \Rightarrow \gamma_k'^{\dagger} = -\mathbf{g}^{-1}\gamma_k'\mathbf{g};\\
&(2)\ \nu \in 2\mathbb{Z}_{\geq 0} \Rightarrow \gamma_k'^{\mathsf{T}} = -\mathbf{c}^{-1}\gamma_k'\mathbf{c};\ \ \nu \in 2\mathbb{Z}_{\geq 0}+1 \Rightarrow \gamma_k'^{\mathsf{T}} = \mathbf{h}^{-1}\gamma_k'\mathbf{h};\\
&(3)\ \nu+q \in 2\mathbb{Z}_{\geq 0} \Rightarrow \gamma_k'^{*} = -\mathbf{b}^{-1}\gamma_k'\mathbf{b};\ \ \nu+q \in 2\mathbb{Z}_{\geq 0}+1 \Rightarrow \gamma_k'^{*} = \mathbf{f}^{-1}\gamma_k'\mathbf{f},
\end{aligned} \tag{6.3.81}$$

where γ_k' $(k = 1,\ldots,2\nu-2)$ are generators of algebra $\mathcal{C}\ell_{(p-1,q-1)}$. So, matrices $\mathbf{d}$, $\mathbf{c}$ and $\mathbf{b}$ (matrices $\mathbf{g}$, $\mathbf{h}$ and $\mathbf{f}$ multiplied on the left by $\gamma_{2\nu-1}'$) are analogs of matrices $\mathbf{D}, \mathbf{C}$ and $\mathbf{B}$ but now for algebra $\mathcal{C}\ell_{(p-1,q-1)}$. Thus, relations (6.3.76)–(6.3.78) can be treated as recurrence relations that enable one to construct matrices $\mathbf{D}, \mathbf{C}$ and $\mathbf{B}$ for algebra $\mathcal{C}\ell_{(p,q)}$ in terms of similar matrices $\mathbf{d}$, $\mathbf{c}$ and $\mathbf{b}$ ($\mathbf{g}$, $\mathbf{h}$ and $\mathbf{f}$) for algebra $\mathcal{C}\ell_{(p-1,q-1)}$.

Problem 6.3.6. * Let matrices Γ_k' $(k = 1,\ldots,2\nu-2)$ in (6.3.72) be taken in representation (6.3.2) (with replacement $\nu \to \nu-1$). Prove that

$$\begin{aligned}
&p = 2\beta : \mathbf{d} = I_2^{\otimes(\beta-1)} \otimes \sigma_1 \otimes I_2^{\otimes(\nu-\beta-1)}\ (\beta = 1,\ldots,\nu-1),\\
&\mathbf{d} = I_{2^{\nu-1}}\ (\beta = \nu),\\
&p = 2\beta+1 : \mathbf{g} = \sigma_3^{\otimes\beta} \otimes I_2^{\otimes(\nu-\beta-1)}\ (\beta = 1,\ldots,\nu-1),
\end{aligned} \tag{6.3.82}$$

$$\begin{aligned}
\mathbf{c} &= (\sigma_2 \otimes \sigma_1)^{\otimes(k-1)} \otimes \sigma_2\ \ (\nu = 2k),\\
\mathbf{h} &= (\sigma_1 \otimes \sigma_2)^{\otimes k}\ \ (\nu = 2k+1).
\end{aligned} \tag{6.3.83}$$

Problem 6.3.7. Check the relations (6.3.79), with parameters ϵ, ϵ' given by (6.3.33), (6.3.34), for representations (6.3.82), (6.3.83).

Formulas (6.3.76), (6.3.77) and (6.3.78), together with (6.3.64), (6.3.66), give the following properties of blocks A and $\overline{A}$ of irreducible inequivalent representations ρ' and ρ'' of group $\mathsf{Spin}(p,q)$ entering (6.2.37) (see Proposition 6.2.7):

$$(1)\ q,p \in 2\mathbb{Z}_{\geq 0} \Rightarrow A^{\dagger}\mathbf{d}A = \mathbf{d},\ \overline{A}^{\dagger}\mathbf{d}\overline{A} = \mathbf{d};\ \ q,p \in 2\mathbb{Z}_{\geq 0}+1 \Rightarrow A^{\dagger}\mathbf{g}\overline{A} = \mathbf{g}; \tag{6.3.84}$$

$$(2)\ \nu \in 2\mathbb{Z}_{\geq 0} \Rightarrow A^{\mathsf{T}}\mathbf{c}A = \mathbf{c},\ \overline{A}^{\mathsf{T}}\mathbf{c}\overline{A} = \mathbf{c};\ \ \nu \in 2\mathbb{Z}_{\geq 0}+1 \Rightarrow A^{\mathsf{T}}\mathbf{h}\overline{A} = \mathbf{h}; \tag{6.3.85}$$

$$\begin{aligned}&(3)\ \nu+q\in 2\mathbb{Z}_{\geq 0}\Rightarrow \mathbf{b}A^*=A\mathbf{b},\ \mathbf{b}\overline{A}^*=\overline{A}\mathbf{b};\\ &\nu+q\in 2\mathbb{Z}_{\geq 0}+1\Rightarrow \mathbf{f}\overline{A}^*=A\mathbf{f},\ \mathbf{f}A^*=\overline{A}\mathbf{f}.\end{aligned} \tag{6.3.86}$$

Consider $2^{\nu-1}$-dimensional representations Δ_+ and Δ_- of group $\mathsf{Spin}(p,q)$ defined in Proposition 6.2.7. We denote the representation contragredient to $\Delta_\pm$ by $\widetilde{\Delta}_\pm$, and representations complex conjugate to $\Delta_\pm$ and $\widetilde{\Delta}_\pm$ by $\Delta^*_\pm$ and $\widetilde{\Delta}^*_\pm$. So, if A is an element of $U\in\mathsf{Spin}(p,q)$ in representation Δ_+, then $\Delta^*_+(U)=A^*$, $\widetilde{\Delta}_+(U)=(A^\mathsf{T})^{-1}$ and $\widetilde{\Delta}^*_+(U)=(A^\dagger)^{-1}$. Then relations (6.3.84)–(6.3.86) imply the following property.

Proposition 6.3.11. *There are the following equivalence relations for irreducible representations of group* $\mathsf{Spin}(p,q)$ *with* $(p+q)=2\nu$:

(1) $\Delta_+=\widetilde{\Delta}^*_+$, $\Delta_-=\widetilde{\Delta}^*_-$ *for even p,q and* $\Delta_-=\widetilde{\Delta}^*_+$ *for odd p,q;*
(2) $\Delta_+=\widetilde{\Delta}_+$, $\Delta_-=\widetilde{\Delta}_-$ *for even* ν *and* $\Delta_-=\widetilde{\Delta}_+$ *for odd* ν;
(3) $\Delta_+=\Delta^*_+$, $\Delta_-=\Delta^*_-$ *for even* $(\nu+q)$ *(even* $(p-q)/2$*) and* $\Delta_-=\Delta^*_+$ *for odd* $(\nu+q)$ *(odd* $(p-q)/2$*).*

Note that the third property here is a direct consequence of the first two.

Problem 6.3.8. Making use of explicit forms of matrices $\mathbf{d}$, Eq. (6.3.82), and $\mathbf{c}$, Eq. (6.3.83), as well as relations (6.3.84), (6.3.85) and (6.3.33), show that

$$A,\overline{A}\in SU(2^{\nu-2},2^{\nu-2})\ (p=2\beta,\ 1\leq\beta\leq\nu-1),$$
$$A,\overline{A}\in SU(2^{\nu-1})\ (p=2\nu),$$
$$A,\overline{A}\in Sp(2^{\nu-1},\mathbb{C})\ \nu=2\,\mathrm{mod}(4),\ A,\overline{A}\in SO(2^{\nu-1},\mathbb{C})\ \nu=4\,\mathrm{mod}(4).$$

Examples

1. Let us discuss the structure of the group $\mathsf{Spin}(1,3)$. Consider Weyl representation (6.1.3) of algebra $\mathcal{C}\ell_{(1,3)}$:

$$\gamma_0=\sigma_1\otimes I_2,\ \gamma_1=i\sigma_2\otimes\sigma_1,\ \gamma_2=i\sigma_2\otimes\sigma_2,\ \gamma_3=i\,\sigma_2\otimes\sigma_3. \tag{6.3.87}$$

We have $\gamma^\dagger_m=(-1)^{\delta_{m,0}+1}\gamma_m$, $\gamma^\mathsf{T}_m=(-1)^m\gamma_m$, $\gamma^*_m=\gamma^{\mathsf{T}\dagger}_m=(-1)^{\delta_{m,2}}\gamma_m$. So, using formulas (6.3.74), and (6.3.76)–(6.3.78), with $p=1,q=3,\nu=2$, we obtain the following expressions for matrices $\mathbf{D}$, $\mathbf{C}$ and $\mathbf{B}$:

$$\begin{aligned}&\mathbf{D}=\sigma_1\otimes\mathbf{g}=\sigma_1\otimes I_2=\gamma_0,\ \mathbf{C}=\sigma_3\otimes\mathbf{c}=-\sigma_3\otimes\sigma_2=\gamma_0\gamma_2,\\ &\mathbf{B}=i\sigma_2\otimes\mathbf{f}=i\sigma_2\otimes\sigma_2=\gamma_2.\end{aligned} \tag{6.3.88}$$

Problem 6.3.9. Show that matrices (6.3.88) obey identities (6.3.28) and (6.3.17), (6.3.18) with $\epsilon = -\epsilon' = +1$.

In view of (6.3.84)–(6.3.86), matrices A and $\overline{A}$, that define element $U \in \mathsf{Spin}(1,3)$ (see (6.2.37)), obey

$$\overline{A} = (A^\dagger)^{-1}, \quad A^\mathsf{T}\sigma_2 A = \sigma_2 = \overline{A}^\mathsf{T}\,\sigma_2\,\overline{A}, \quad \overline{A} = \sigma_2\,A^*\,\sigma_2. \tag{6.3.89}$$

These reduce to two conditions $\overline{A} = (A^\dagger)^{-1}$ and $\det(A) = 1$, i.e., $A \in SL(2,\mathbb{C}) = Sp(2,\mathbb{C})$. The transformations (6.2.3), defining group $\mathsf{Spin}(1,3)$ read (6.2.40):

$$\begin{aligned}\gamma_m x^m &= \begin{pmatrix} \mathbf{0} & \sigma_m x^m \\ \tilde{\sigma}_m x^m & \mathbf{0}\end{pmatrix} \to U(\gamma_m x^m)\,U^{-1} \\ &= \begin{pmatrix} \mathbf{0} & A\,\sigma_m x^m\,A^\dagger \\ (A^\dagger)^{-1}\,\tilde{\sigma}_m x^m\,A^{-1} & \mathbf{0}\end{pmatrix},\end{aligned}$$

where matrices σ_m and $\tilde{\sigma}_m$ are defined in (6.1.4). Thus, group $\mathsf{Spin}(1,3)$ is isomorphic to group $SL(2,\mathbb{C})$ (which has to be treated as six-dimensional real Lie group).

2. To discuss the structure of the group $\mathsf{Spin}(4)$ we take Weyl representation (6.1.75) of algebra $\mathcal{C}\ell_{(4,0)}$,

$$\gamma_0 = \sigma_1 \otimes I_2, \quad \gamma_1 = \sigma_2 \otimes \sigma_1, \quad \gamma_2 = \sigma_2 \otimes \sigma_2, \quad \gamma_3 = \sigma_2 \otimes \sigma_3, \tag{6.3.90}$$

obeys $\gamma_m^\dagger = \gamma_m$ and $\gamma_m^\mathsf{T} = \gamma_m^* = (-1)^m\,\gamma_m$. So, in accordance with (6.3.76)–(6.3.78) with $p = 4, q = 0, \nu = 2$, we obtain

$$\mathbf{D} = I_2 \otimes \mathbf{d} = I_2 \otimes I_2, \quad \mathbf{C} = \sigma_3 \otimes \mathbf{c} = \sigma_3 \otimes \sigma_2 = \sigma_3 \otimes \mathbf{b} = \mathbf{B}. \tag{6.3.91}$$

Using these in (6.3.84)–(6.3.86), we derive the properties of blocks A and $\overline{A}$ determining an element $U \in \mathsf{Spin}(4)$:

$$A^\dagger\,A = I = \overline{A}^\dagger\,\overline{A}, \; A^\mathsf{T}\,\sigma_2\,A = \sigma_2 = \overline{A}^\mathsf{T}\,\sigma_2\,\overline{A}, \; A^* = \sigma_2\,A\,\sigma_2, \; \overline{A}^* = \sigma_2\,\overline{A}\,\sigma_2. \tag{6.3.92}$$

These properties show that $A, \overline{A} \in SU(2)$, so we have $\mathsf{Spin}(4) = SU(2) \times SU(2)$.

3. Consider group Spin(8) which has $N = p = 8$ and $q = 0$. Generators of algebra $\mathcal{C}\ell_{(8,0)}$ in Weyl representation (6.1.69) are

$$\Gamma_1 = \sigma_1 \otimes I_2^{\otimes 3},\ \Gamma_2 = \sigma_2 \otimes \sigma_1 \otimes I_2^{\otimes 2},\ \Gamma_3 = \sigma_2^{\otimes 2} \otimes \sigma_1 \otimes I_2,\ \Gamma_4 = \sigma_2^{\otimes 3} \otimes \sigma_1,$$
$$\Gamma_5 = \sigma_2^{\otimes 4},\ \Gamma_6 = \sigma_2^{\otimes 3} \otimes \sigma_3,\ \Gamma_7 = \sigma_2^{\otimes 2} \otimes \sigma_3 \otimes I_2,\ \Gamma_8 = \sigma_2 \otimes \sigma_3 \otimes I_2^{\otimes 2}. \tag{6.3.93}$$

They obey the relations $\Gamma_m^{\mathsf{T}} = \Gamma_m^* = (-1)^{m+1}\Gamma_m$, $\Gamma_m^\dagger = \Gamma_m$. Therefore, in accordance with definitions (6.3.74) of matrices $\mathbf{D}$, $\mathbf{C}$ and $\mathbf{B}$ and formulas (6.3.76)–(6.3.78) with $p = 8, q = 0$ and $\nu = 4$, we obtain the following expressions:

$$\mathbf{C} = \mathbf{B} = \Gamma_1\Gamma_3\Gamma_5\Gamma_7 = -\sigma_3 \otimes \sigma_2 \otimes I \otimes \sigma_2,\ \ \mathbf{c} = \mathbf{b} = -\sigma_2 \otimes I \otimes \sigma_2,$$
$$\mathbf{D} = I^{\otimes 4},\ \mathbf{d} = I^{\otimes 3},\ \ I \equiv I_2. \tag{6.3.94}$$

For these matrices, we have $\epsilon = +1$ and $\epsilon' = +1$, which agrees with the general analysis (see Proposition 6.3.5). In view of the relations (6.3.84)–(6.3.86), diagonal blocks A and $\overline{A}$ of element $U \in \mathsf{Spin}(8)$ (see (6.2.37)) obey the relations

$$\det(A) = 1,\ A^\dagger A = I_8,\ \ \overline{A}^\dagger \overline{A} = I_8, \tag{6.3.95}$$
$$A^{\mathsf{T}}\,\mathbf{c}\,A = \mathbf{c} = \overline{A}^{\mathsf{T}}\,\mathbf{c}\,\overline{A},\ \mathbf{c}\,A^* = A\,\mathbf{c},\ \mathbf{c}\,\overline{A}^* = \overline{A}\,\mathbf{c}, \tag{6.3.96}$$

where, according to (6.3.94), matrix $\mathbf{c}$ is real and symmetric, $\mathbf{c}^{\mathsf{T}} = \mathbf{c}$. Note that blocks $A \in \Delta_+(\mathsf{Spin}(8))$ and $\overline{A} \in \Delta_-(\mathsf{Spin}(8))$ have dimension 8×8. Indeed, in accordance with the definition of irreducible spinor representations Δ_+ and Δ_- of group $\mathsf{Spin}(2\nu)$ (see Proposition 6.2.7), their dimension is $2^{\nu-1}$. So, the dimension of spinor representation of group $\mathsf{Spin}(8)$ equals $2^3 = 8$. This is the same dimension as the dimension of defining representation of group $SO(8)$. The latter fact distinguishes the group $\mathsf{Spin}(8)$ among all groups $\mathsf{Spin}(N)$.

It follows from the first of relations (6.3.96) that $A, \bar{A} \in SO(8, \mathbb{C})$. Moreover, in view of (6.3.95) we have $A, \bar{A} \in SU(8)$. Therefore, $A, \bar{A} \in SU(8) \cap SO(8, \mathbb{C})$, and due to the result of Problem 6.3.5 (see also its solution), the two sets of matrices A and $\bar{A}$, obeying (6.3.95) and (6.3.96), form two groups, each isomorphic to $SO(8, \mathbb{R}) \equiv SO(8)$. In other words, each of the spinor representations $\Delta_+(\mathsf{Spin}(8))$ and $\Delta_-(\mathsf{Spin}(8))$ is mapped, one-to-one, to defining representation of group $SO(8)$.

In Weyl representation (6.3.93), transformation (6.2.5) determined by element $U \in \mathsf{Spin}(8)$ in (6.2.37) is explicitly written as

$$\Gamma_m = \begin{pmatrix} \mathbf{0} & \mathbf{s}_m \\ \overline{\mathbf{s}}_m & \mathbf{0} \end{pmatrix} \to U\Gamma_m U^{-1} = \begin{pmatrix} \mathbf{0} & A\mathbf{s}_m\overline{A}^{-1} \\ \overline{A}\overline{\mathbf{s}}_m A^{-1} & \mathbf{0} \end{pmatrix} = \Gamma_k \Lambda^k{}_m, \tag{6.3.97}$$

where $||\Lambda^k_m|| \in SO(8)$, and explicit form of matrices $\mathbf{s}_m$ and $\overline{\mathbf{s}}_m$ is obtained from (6.3.93):

$$\begin{aligned} \mathbf{s}_1 &= \overline{\mathbf{s}}_1 = I^{\otimes 3}, \\ (\overline{\mathbf{s}}_2, \ldots, \overline{\mathbf{s}}_8) &= -(\mathbf{s}_2, \ldots, \mathbf{s}_8) \\ &= i((\sigma_1 \otimes I^{\otimes 2}), (\sigma_2 \otimes \sigma_1 \otimes I), (\sigma_2^{\otimes 2} \otimes \sigma_1), (\sigma_2^{\otimes 3}), (\sigma_2^{\otimes 2} \otimes \sigma_3), \\ &\quad (\sigma_2 \otimes \sigma_3 \otimes I), (\sigma_3 \otimes I^{\otimes 2})). \end{aligned}$$

The last relation in (6.3.97) together with the first relation in (6.3.96) give

$$A\mathbf{s}_m\overline{A}^{-1} = \mathbf{s}_k\Lambda^k_m, \quad \overline{A}\overline{\mathbf{s}}_m A^{-1} = \overline{\mathbf{s}}_k\Lambda^k_m, \tag{6.3.98}$$

where 8×8-matrices $A(\Lambda) \in \Delta_+(\mathsf{Spin}(8))$, $\overline{A}(\Lambda) \in \Delta_+(\mathsf{Spin}(8))$ and $\Lambda \in SO(8)$ belong to spinor and defining representations of algebra $so(8,\mathbb{C})$.

Root system and system of weights of algebra $so(8,\mathbb{C})$ are determined by simple roots and fundamental weights see (3.1.27) and (3.4.47))

$$\begin{aligned} \alpha^{(1)} &= e^{(1)} - e^{(2)}, \quad \alpha^{(2)} = e^{(2)} - e^{(3)}, \\ \alpha^{(3)} &= e^{(3)} - e^{(4)}, \quad \alpha^{(4)} = e^{(3)} + e^{(4)}, \\ \lambda_{(1)} &= (1,0,0,0), \quad \lambda_{(2)} = (1,1,0,0), \\ \lambda_{(3)} &= \left(\frac{1}{2},\frac{1}{2},\frac{1}{2},-\frac{1}{2}\right), \quad \lambda_{(4)} = \left(\frac{1}{2},\frac{1}{2},\frac{1}{2},\frac{1}{2}\right). \end{aligned} \tag{6.3.99}$$

These systems are invariant under an arbitrary permutation of simple roots $\alpha^{(i)}$ $(i = 1,3,4)$, which does not touch the root $\alpha^{(2)}$, and under arbitrary permutation of fundamental weights $\lambda_{(i)}$ $(i = 1,3,4)$ (weight $\lambda_{(2)}$ stays intact). This invariance corresponds to the symmetry of Dynkin diagram of algebra $so(8,\mathbb{C})$:

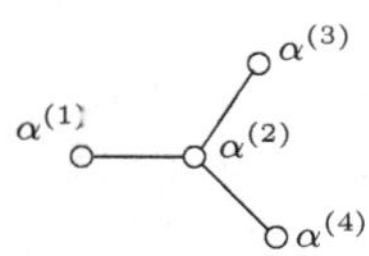

Clearly, the transformations we consider leave intact the whole weight lattice of algebra $so(8,\mathbb{C})$. A particular case of this symmetry is permutation of two roots $\alpha^{(3)}$ and $\alpha^{(4)}$, while $\alpha^{(1)}$ and $\alpha^{(2)}$ staying fixed. This symmetry property is valid for all algebras $so(2r,\mathbb{C})$, as we discuss in the end of item 3 of Section 3.4.2 (see Problem 3.4.8). This case gives the automorphism of algebra $so(8,\mathbb{C})$ which leads to the interchange of two spinor representations $T_{(3)}$ and $T_{(4)}$ whose weights are $\lambda_{(3)}$ and $\lambda_{(4)}$ (see Remark 3 in the end of Section 6.2.2).

Consider now the symmetry which corresponds to transposition of roots $\alpha^{(1)} \leftrightarrow \alpha^{(3)}$, $\alpha^{(2)} \to \alpha^{(2)}$, $\alpha^{(4)} \to \alpha^{(4)}$ and transposition of fundamental weights $\lambda_{(1)} \leftrightarrow \lambda_{(3)}$, $\lambda_{(2)} \to \lambda_{(2)}$, $\lambda_{(4)} \to \lambda_{(4)}$. This transposition is obtained by the action, in root space, $\mathfrak{V}_4$, of the operator

$$A = \begin{pmatrix} 1 & 1 & 1 & -1 \\ 1 & 1 & -1 & 1 \\ 1 & -1 & 1 & 1 \\ -1 & 1 & 1 & 1 \end{pmatrix}.$$

Under this action, weight (and root) lattice of algebra $so(8,\mathbb{C})$ transforms into itself. The latter invariance implies that there is automorphism of algebra $so(8,\mathbb{C})$ which makes a linear transformation of its generators (including Cartan elements) in such a way that the eight-dimensional spinor representation $T_{(3)}$ and eight-dimensional defining representation $T_{(1)}$ are interchanged. This automorphism is explicitly constructed in [7, Section 25.2].

Problem 6.3.10. * Prove isomorphism Spin$(2,4) = SU(2,2)$. Hint: make use of Weyl representation of algebra $\mathcal{C}\ell_{(2,4)}$, find matrices $\mathbf{d}$, $\mathbf{h}$ and $\mathbf{f}$ in this representation and consider relations (6.3.84)–(6.3.86) for blocks $A, \overline{A}$.

We now turn to *odd case* $(p+q) = 2\nu + 1$. We find from (6.3.51) that generators $\Sigma_{mn} = \frac{1}{4}(\gamma_m\gamma_n - \gamma_n\gamma_m)$ of Lie algebra spin(p,q) obey

$$\mathbf{C}'\,\Sigma_{mn}^{\mathsf{T}} + \Sigma_{mn}\,\mathbf{C}' = 0, \;\; \mathbf{D}'\,\Sigma_{mn}^{\dagger} + \Sigma_{mn}\,\mathbf{D}' = 0. \tag{6.3.100}$$

This gives for matrices $U \in$ Spin(p,q) defined in (6.2.30)

$$U^{\mathsf{T}}\,\mathbf{C}'^{-1}\,U = \mathbf{C}'^{-1}, \;\; U^{\dagger}\,\mathbf{D}'^{-1}\,U = \mathbf{D}'^{-1}. \tag{6.3.101}$$

This implies that in odd case we have $(U^{\mathsf{T}})^{-1} = \mathbf{C}'^{-1}\,U\,\mathbf{C}'$ and $(U^{\dagger})^{-1} = \mathbf{D}'^{-1}U\mathbf{D}'$. That is, there is equivalence $\rho = \tilde{\rho} = \tilde{\rho}^*$ between the defining representation ρ of group $\mathsf{Spin}(p,q)$, representation $\tilde{\rho}$ contragredient to ρ and representation $\tilde{\rho}^*$ complex conjugated to $\tilde{\rho}$:

$$\rho(U) = U, \ \ \tilde{\rho}(U) = (U^{\mathsf{T}})^{-1}, \ \ \tilde{\rho}^*(U) = (U^{\dagger})^{-1}.$$

In view of (6.3.53), matrices $\mathbf{D}'$ and $i\mathbf{D}'$ can be viewed as (pseudo-) unitary metric for even and odd q, respectively, and matrix $\mathbf{C}'$ as metric of orthogonal and symplectic group for $\epsilon' = +1$ and $\epsilon' = -1$, respectively. Then we have "odd" analogs of embeddings (6.3.69), (6.3.70)

$$\begin{gathered}
\mathsf{Spin}(p,q) \subset Sp(2^\nu,\mathbb{C}) \cap SU(2^{\nu-1},2^{\nu-1}) \quad \text{for } N = 3,5 \operatorname{mod}(8),\ q \neq 0,\\
\mathsf{Spin}(p,q) \subset SO(2^\nu,\mathbb{C}) \cap SU(2^{\nu-1},2^{\nu-1}) \quad \text{for } N = 1,7 \operatorname{mod}(8),\ q \neq 0,\\
\mathsf{Spin}(2\nu+1) \subset Sp(2^\nu,\mathbb{C}) \cap SU(2^\nu) = USp(2^\nu) \quad \text{for } N = 3,5 \operatorname{mod}(8),\\
\mathsf{Spin}(2\nu+1) \subset SO(2^\nu,\mathbb{C}) \cap SU(2^\nu) \quad \text{for } N = 1,7 \operatorname{mod}(8).
\end{gathered} \tag{6.3.102}$$

Problem 6.3.11. Derive (6.3.102) by making use of (6.3.101), (6.3.53) and values of ϵ' given by (6.3.33) in even case $(p-1)+q = 2\nu$. Hint: evaluate spectrum of matrix $\mathbf{D}'$ for even q and matrix $i\mathbf{D}'$ for odd q, using explicit forms (6.3.37) of matrix $\mathbf{D}$ and (6.3.4) of matrix $\gamma_{2\nu+1}$ in representation (6.3.2), (6.1.76).

Equivalence matrices $\mathbf{B}$, $\mathbf{C}$ and $\mathbf{D}$ play important role in defining Dirac conjugation of spinors, as well as operations of discrete symmetries: charge conjugation C, spatial reflection P and inversion of time T. Also, matrices $\mathbf{B}$ and $\mathbf{C}$ are used in construction of Majorana (real) representations of algebra $\mathcal{C}\ell_{(p,q)}$. These aspects are studied in the following section.

6.4. Dirac, Weyl and Majorana Spinors in Spaces $\mathbb{R}^{p,q}$

This section presents the theory of spinors in N-dimensional spaces $\mathbb{R}^{p,N-p}$. This material is based on classic works [19, 56]; see also [57, Appendix; 52–54].

6.4.1. Spinors in spaces $\mathbb{R}^{p,q}$ and their tensor products

Consider irreducible representation ρ of Clifford algebra $\mathcal{C}\ell_{(p,q)}$ with matrices of generators γ_k, obeying (see (6.1.77)):

$$\gamma_m\gamma_n + \gamma_n\gamma_m = 2\eta_{mn}I_{2^\nu}, \quad ||\eta_{mn}|| = I_{p,q}. \tag{6.4.1}$$

Vectors in space $\mathcal{F}$ of this representation, by Definition 5.4.1, are *Dirac spinors*. In accordance with (6.2.27) and (6.2.30), this representation of algebra $\mathcal{C}\ell_{(p,q)}$ determines representation ρ of group Spin(p, q):

$$\psi \to \psi' = U\,\psi, \tag{6.4.2}$$

where $U \in \mathsf{Spin}(p, q)$. Consider matrix $\mathbf{D}$ obeying the first of Eq. (6.3.15) (for even case) or matrix $\mathbf{D}'$ obeying the first of Eq. (6.3.51) (for odd case). We use in what follows a common notation $\mathbf{D}$ for matrices $\mathbf{D}$ and $\mathbf{D}'$, unless the opposite is specified. We make use of matrix $\mathbf{D}^{-1}$ to define Dirac conjugate spinors,

$$\bar{\psi} = \psi^\dagger \mathbf{D}^{-1}. \tag{6.4.3}$$

In view of the first formula in (6.3.64) which is valid in both even and odd cases, these spinors transform, under transformations (6.4.2), as contragredient objects:

$$\bar{\psi} \to \bar{\psi} U^{-1}. \tag{6.4.4}$$

Spinors $\bar{\psi}$ form space $\tilde{\mathcal{F}}$ dual to $\mathcal{F}$, which is the space of contragredient representation $\tilde{\rho}$ of group Spin(p, q). We then define Spin(p, q)-invariant scalar product in space $\mathcal{F}$,

$$\bar{\psi}\phi = \psi^\dagger \mathbf{D}^{-1} \phi, \tag{6.4.5}$$

where $\phi \in \mathcal{F}$ is Dirac spinor, i.e., it transforms in the same way as spinor ψ in (6.4.2).

It follows from (6.4.2), (6.4.4) and (6.2.5) that objects

$$t_{k_1 \dots k_\ell} = \overline{\psi}(\gamma_{k_1} \dots \gamma_{k_\ell})\phi = \psi^\dagger\, \mathbf{D}^{-1}\, (\gamma_{k_1} \dots \gamma_{k_\ell})\, \phi \tag{6.4.6}$$

transform under $SO(p, q)$ as rank-ℓ tensors (cf. (5.1.24)):

$$\begin{aligned} t_{k_1 \dots k_\ell} \to t'_{k_1 \dots k_\ell} &= \overline{\psi}'(\gamma_{k_1} \dots \gamma_{k_\ell})\phi' = \overline{\psi} U^{-1}(\gamma_{k_1} \dots \gamma_{k_\ell}) U\phi \\ &= \overline{\psi}(\gamma_{i_1} \dots \gamma_{i_\ell})\phi \cdot (\Lambda^{-1})^{i_1}_{k_1} \cdots (\Lambda^{-1})^{i_\ell}_{k_\ell} \\ &= t_{i_1 \dots i_\ell} \cdot (\Lambda^{-1})^{i_1}{}_{k_1} \cdots (\Lambda^{-1})^{i_\ell}{}_{k_\ell}. \end{aligned} \tag{6.4.7}$$

Here we use the identities $\gamma'_k = U^{-1}\gamma_k U = \gamma_i(\Lambda^{-1})^i{}_k$ that follow from (6.2.5).

Problem 6.4.1. Prove the identities

$$\overline{\psi}(\gamma_{k_1}\dots\gamma_{k_\ell})\phi = (-1)^{q+\ell}\overline{\psi}'(\gamma_{k_1}\dots\gamma_{k_\ell})\,\phi', \;\; N = 2\nu, \tag{6.4.8}$$

$$\overline{\psi}(\gamma_{k_1}\dots\gamma_{k_\ell})\phi = \overline{\psi}'(\gamma_{k_1}\dots\gamma_{k_\ell})\,\phi', \;\; N = 2\nu+1, \tag{6.4.9}$$

where $\psi' = \gamma_{N+1}\psi$, $\phi' = \gamma_{N+1}\phi$ and $\gamma_{N+1} = (-i)^{q+\nu}\gamma_1\cdots\gamma_N$. Hint: make use of (6.3.16) and (6.3.51).

Note that Eq. (6.4.1) guarantees that tensors (6.4.6) with components $t_{k_1\dots k_\ell}$ are always decomposed in linear combinations of totally antisymmetric tensors with components $t_{[k_1\dots k_n]} = \overline{\psi}\gamma_{[k_1}\cdots\gamma_{k_n]}\phi$, where we introduce antisymmetrized products of gamma-matrices:

$$\gamma_{[k_1}\cdots\gamma_{k_n]} = \frac{1}{n!}\sum_{\sigma\in S_n}(-1)^{p(\sigma)}\gamma_{k_{\sigma(1)}}\cdots\gamma_{k_{\sigma(n)}}. \tag{6.4.10}$$

Here $p(\sigma)$ is parity of permutation $\sigma\in S_n$. So, without loss of generality, tensors of the type (6.4.6) can be understood as totally antisymmetric in their indices $\{k_1,\cdots,k_\ell\}$. The latter tensors form space of irreducible representations $T^{([1^\ell])}$ of group $SO(p,q)$ associated with Young diagrams $[1^\ell]$; this class includes defining representation that acts in space of tensors with components $t_k = \overline{\psi}\gamma_k\phi$. Representations $T^{([1^\ell])}$ of group $SO^\uparrow(p,q)$ are at the same time representations of group $\mathsf{Spin}(p,q)$, embedded in representation $\rho\otimes\tilde{\rho}$, and since unit matrix I_{2^ν} and antisymmetric products of generators γ_k form basis in algebra $\mathrm{Mat}_{2^\nu}(\mathbb{C})$, we have the following decompositions:

$$\begin{aligned}\rho\otimes\tilde{\rho} &= T^{([\emptyset])} + T^{([1])} + \cdots + T^{([1^N])} = 2T^{([\emptyset])} + \cdots + 2T^{([1^{r-1}])}\\ &\quad + T_+^{([1^r])} + T_-^{([1^r])}, \;\; N = 2r,\end{aligned} \tag{6.4.11}$$

$$\begin{aligned}\rho\otimes\tilde{\rho} &= T^{([\emptyset])} + T^{([1^2])} + \cdots + T^{([1^{2r}])} = T^{([\emptyset])} + T^{([1])}\\ &\quad + \cdots + T^{([1^r])}, \;\; N = 2r+1,\end{aligned} \tag{6.4.12}$$

where $r \equiv \nu = [N/2]$ is rank of group $SO(p, N-p)$, and we make use of equivalence of representations of group $SO(p, N-p)$:

$$T^{([1^m])} = T^{([1^{N-m}])}. \tag{6.4.13}$$

The latter equivalence is discussed in Section 5.1.3 (see (5.1.36)). To write (6.4.12), we use the fact that in odd case, basis in algebra $\mathrm{Mat}_{2^\nu}(\mathbb{C})$ is made of either even or odd[j] antisymmetrized products of $\gamma_{[k_1} \cdots \gamma_{k_i]}$. For $N = 2r$ we also make use of the fact that representation $T^{([1^r])}$ is reducible, and breaks up into a sum of selfdual and anti-selfdual representations $T_+^{([1^r])}$ and $T_-^{([1^r])}$. Expansion (6.4.11) is written in explicit form below, in Section 6.4.4, see (6.4.78).

Problem 6.4.2. Show that multiplicity of representations $T^{([1^m])}$ in $(\rho \otimes \tilde{\rho})$ is correct in (6.4.11) and (6.4.12). Hint: compare dimensions of representations on the left- and right-hand sides of (6.4.11) and (6.4.12); use identities $\dim(\rho \otimes \tilde{\rho}) = 2^{2r}$ and $\dim(T^{([1^m])}) = \frac{N!}{m!(N-m)!}$.

Problem 6.4.3. * Prove the identity for $\mathcal{C}\ell_{(p,q)}$:

$$\gamma_{[k_1} \cdots \gamma_{k_m]} \cdot \gamma_{N+1} = \alpha_m \varepsilon_{k_1 \ldots k_m k_{m+1} \ldots k_N} \gamma^{k_{m+1}} \cdots \gamma^{k_N},$$
$$\alpha_m = \frac{(-i)^{\nu+q}(-1)^{[\frac{m}{2}]+q}}{(N-m)!}, \tag{6.4.14}$$

where $N = (p+q) = 2\nu = 2r$, $\gamma_{N+1} = (-i)^{\nu+q}\gamma_0 \cdots \gamma_{N-1}$, $[\frac{m}{2}]$ is integer part of $\frac{m}{2}$, and $\varepsilon_{01 \ldots N-1} = 1$.

We note that equivalence (6.4.13) does not hold for the whole group $O(p, N-p)$ with $N = 2r$. There is equivalence (5.1.38) instead: $T^{([1^k])} = T^{(0)} \cdot T^{([1^{N-k}])}$, where $T^{(0)}(\Lambda) = \det(\Lambda) = \pm 1$ is one-dimensional determinant representation of group $O(p, N-p)$. In our case, space of representation $T^{[\emptyset]}$ consists of scalars $\overline{\psi}\phi$, while space of representation $T^{[1^N]}$ consists of pseudo-scalars $\overline{\psi}\gamma_{N+1}\phi$; space of representation $T^{([1])}$ is space of vectors $\overline{\psi}\gamma_k\phi$, while space of representation $T^{([1^{N-1}])}$ is space of pseudo-vectors $\overline{\psi}\gamma_k\gamma_{N+1}\phi$, etc. In general, formula (6.4.14) tells that space of representation $T^{([1^{N-m}])}$ with $m < [N/2]$ consists of pseudo-tensors with components $\overline{\psi}(\gamma_{[k_1} \cdots \gamma_{k_m]}\gamma_{N+1})\phi$:

$$\overline{\psi}(\gamma^{[k_{m+1}} \cdots \gamma^{k_N]})\phi = (-1)^m \, \alpha_{N-m} \, \varepsilon^{k_1 \ldots k_m k_{m+1} \ldots k_N} \, \overline{\psi}(\gamma_{[k_1} \cdots \gamma_{k_m]} \cdot \gamma_{N+1})\phi. \tag{6.4.15}$$

[j]In odd case $N = (p+q) = (2r+1)$, there is one-to-one correspondence $b = \gamma_{1 \ldots N} \cdot a$ between even elements a and odd elements b of algebra $\mathcal{C}\ell_{(p,q)}$, where central element $\gamma_{1 \ldots N} \equiv \gamma_1 \cdots \gamma_N$ in irreducible representations $\rho_\pm$ equals ± 1 or $\pm i$, see Remark 2 in Section 6.2.1.

This fact is relevant for physics applications. So, expansion (6.4.11) in the case of the whole group $O(p, N-p)$, $N = 2r$, reads

$$\rho \otimes \tilde{\rho} = (T^{([\emptyset])} + T^{(0)} \cdot T^{([\emptyset])}) + \cdots + (T^{([1^{r-1}])} + T^{(0)} \cdot T^{([1^{r-1}])}) + T^{([1^r])}, \tag{6.4.16}$$

where representation $T^{([1^r])}$ no longer splits into two irreducible representations $T_{\pm}^{([1^r])}$.

In odd case $N = 2r+1$ and group $SO(p,q)$, representation $T^{([1^m])}$ on tensors $\overline{\psi}\gamma_{[k_1\ldots k_m]}\phi$ and representation $T^{([1^{N-m}])}$ on tensors $\overline{\psi}\gamma_{[k_1\ldots k_m]}\gamma_{N+1}\phi$, where $\gamma_{[k_1\ldots k_m]} \equiv \gamma_{[k_1}\cdots\gamma_{k_m]}$, are also equivalent, since the element γ_{N+1} is central and takes values $\rho_{\pm}(\gamma_{N+1}) = \pm I_{2^r}$ in irreducible representations $\rho = \rho_{\pm}$, so that $\overline{\psi}\gamma_{[k_1\ldots k_m]}\gamma_{N+1}\phi = \pm\overline{\psi}\gamma_{[k_1\ldots k_m]}\phi$. Under improper transformations (6.2.9) from group $O(p,q)$, central element changes sign, $\gamma_{N+1} \to -\gamma_{N+1}$, so expansion (6.4.12) for the full orthogonal group with $N = 2r+1$ is written as

$$\begin{aligned} \rho \otimes \tilde{\rho} &= T^{([\emptyset])} + T^{([1^2])} + \cdots + T^{([1^{2r}])} \\ &= T^{([\emptyset])} + T^{(0)} \cdot T^{([1])} + T^{([1^2])} + T^{(0)} \cdot T^{([1^3])} + \cdots, \end{aligned} \tag{6.4.17}$$

where the summation on the right-hand side ends on $T^{([1^r])}$ for even r and on $T^{(0)} \cdot T^{([1^r])}$ for odd r.

In even case $N = (p+q) = 2r$, irreducible representation ρ of Clifford algebra $\mathcal{C}\ell_{(p,q)}$ determines reducible representation of group $\mathsf{Spin}(p,q)$ (see Proposition 6.2.7 in Section 6.2.2): $\rho = \Delta_+ \oplus \Delta_-$, where $\Delta_{\pm}$ are inequivalent irreducible Weyl (semispinor) representations of group $\mathsf{Spin}(p,q)$. Recall that in even case $N = (p+q) = 2r$, irreducible representations $T^{([1^m])}$ ($m = 1, \ldots, r-2$) of group $SO(p,q)$ in spaces of tensors $t_{[k_1\ldots k_m]} = \overline{\psi}\gamma_{[k_1}\cdots\gamma_{k_m]}\phi$ are equivalent to representations $T_{(m)}$ of algebra $so(2r,\mathbb{C})$ with highest fundamental weights $\lambda_{(m)}$ given by (3.4.47). Representations $T^{([1^{r-1}])}$ and $T^{([1^r])}$ in spaces of tensors $t_{[k_1\ldots k_{r-1}]}$ and $t_{[k_1\ldots k_r]}$ correspond to irreducible representations $T'_{(r-1)}$ and $T'_{(r)}$ of algebra $so(2r,\mathbb{C})$ with composite highest weights (3.4.50):

$$\lambda'_{(r-1)} = \lambda_{(r-1)} + \lambda_{(r)} = (1,1,\ldots,1,0), \quad \lambda'_{(r)} = 2\,\lambda_{(r)} = (1,1,\ldots,1).$$

These are highest weights of representations extracted from tensor products $(\Delta_+ \otimes \Delta_-)$ and $(\Delta_+ \otimes \Delta_+)$, respectively: highest weights of representations $\Delta_+ = T_{(r-1)}$ and $\Delta_- = T_{(r)}$ are $\lambda_{(r-1)}$ and $\lambda_{(r)}$. Note that highest weight vectors of representations $T'_{(r-1)}$ and $T'_{(r)}$ are tensor products of highest weight vectors of representations Δ_+ and Δ_-. Representations

Δ_+ and Δ_- act in $2^{(r-1)}$-dimensional spaces $\mathcal{V}_+$ and $\mathcal{V}_-$ of Weyl spinors (see Section 6.2.2). Weyl spinors $\psi \in \mathcal{V}_+$ and $\phi \in \mathcal{V}_-$ obey

$$P_-\psi = 0, \;\; P_+\phi = 0 \;\Rightarrow\; \psi = P_+\psi, \; \phi = P_-\phi,$$

where projection operators $P_\pm = \frac{1}{2}(1 \pm \gamma_{N+1})$ are introduced in (6.2.38), and we treat Weyl spinors ψ and ϕ as embedded in the space of Dirac spinors $\mathcal{V}_+ \oplus \mathcal{V}_-$.

Contragredient representation $\tilde{\rho}$ of group Spin(p,q) is also reducible in even case $(p+q) = 2r$, and splits into sum $\tilde{\rho} = \widetilde{\Delta}_+ \oplus \widetilde{\Delta}_-$, where $\widetilde{\Delta}_+$ and $\widetilde{\Delta}_-$ are two irreducible contragredient Weyl representations on spinors $\overline{\xi}$ and $\overline{\eta}$ which obey

$$\overline{\xi}\,\frac{1}{2}(1-\gamma_{N+1}) = \overline{\xi}\,P_- = 0, \;\; \overline{\eta}\,\frac{1}{2}(1+\gamma_{N+1}) = \overline{\eta}\,P_+ = 0,$$

i.e., $\overline{\xi} = \overline{\xi}P_+$ and $\overline{\eta} = \overline{\eta}P_-$. Due to the properties $P_\pm\gamma_k = \gamma_k P_\mp$ and $P_\pm P_\mp = 0$, we have (upper and lower signs are correlated)

$$\overline{\psi}_\pm \gamma_{[k_1} \cdots \gamma_{k_{2i}]}\phi_\mp = 0, \quad \overline{\psi}_\pm \gamma_{[k_1} \cdots \gamma_{k_{2i+1}]}\phi_\pm = 0, \tag{6.4.18}$$

where $\overline{\psi}_\pm = \overline{\psi}P_\pm$ and $\phi_\pm = P_\pm\phi$. Also, formula (6.4.14) with $m = r$ gives

$$\frac{(-i)^{r+q}}{r!}\varepsilon_{i_1\ldots i_r k_1 \ldots k_r}(\overline{\psi}\gamma^{[k_1}\cdots\gamma^{k_r]}\phi_\pm) = \pm(-1)^{[\frac{r}{2}]+q}\,(\overline{\psi}\gamma_{[i_1}\cdots\gamma_{i_r]}\phi_\pm), \tag{6.4.19}$$

i.e., components $\mathsf{t}_{(+)i_1\ldots i_r} = \overline{\psi}\,\gamma_{[i_1}\cdots\gamma_{i_r]}\,\phi_+$ make selfdual (anti-selfdual) tensors for even (odd) $([\frac{r}{2}]+q)$, while tensors $\mathsf{t}_{(-)i_1\ldots i_r} = \overline{\psi}\,\gamma_{[i_1}\cdots\gamma_{i_r]}\,\phi_-$ are selfdual (anti-selfdual) for odd (even) $([\frac{r}{2}]+q)$.

Due to relations (6.4.18) and (6.4.19), some terms cancel out in the expansion of the tensor products $(\Delta_\pm \otimes \widetilde{\Delta}_\pm)$ and $(\Delta_\pm \otimes \widetilde{\Delta}_\mp)$ that are obtained from (6.4.11) upon substitution $\rho = (\Delta_+ + \Delta_-)$ and $\tilde{\rho} = (\tilde{\Delta}_+ + \tilde{\Delta}_-)$. This leads to the following result.

Proposition 6.4.1. *Let $r \equiv \nu = N/2$ be even. Then*

$$\begin{aligned}
\Delta_\pm \otimes \widetilde{\Delta}_\mp &= T^{[1]} + T^{[1^3]} + \cdots + T^{[1^{r-1}]}, \\
\Delta_\pm \otimes \widetilde{\Delta}_\pm &= T^{[\emptyset]} + T^{[1^2]} + \cdots + T^{[1^{r-2}]} \\
&\quad + \begin{cases} T_\pm^{[1^r]} & \text{for } \left(\frac{r}{2}+q\right) \in 2\mathbb{Z}_{\geq 0}, \\ T_\mp^{[1^r]} & \text{for } \left(\frac{r}{2}+q\right) \in 2\mathbb{Z}_{\geq 0}+1. \end{cases}
\end{aligned} \tag{6.4.20}$$

For odd $r \equiv \nu = N/2$, we have

$$\Delta_\pm \otimes \widetilde{\Delta}_\mp = T^{[1]} + \cdots + T^{[1^{r-2}]} + \begin{cases} T_\pm^{[1^r]} & \text{for } \left(\left[\frac{r}{2}\right] + q\right) \in 2\mathbb{Z}_{\geq 0}, \\ T_\mp^{[1^r]} & \text{for } \left(\left[\frac{r}{2}\right] + q\right) \in 2\mathbb{Z}_{\geq 0} + 1, \end{cases}$$

$$\Delta_\pm \otimes \widetilde{\Delta}_\pm = T^{[\emptyset]} + T^{[1^2]} + \cdots + T^{[1^{r-1}]}. \tag{6.4.21}$$

Upper and lower signs in (6.4.20) and (6.4.21) are correlated, and we removed parenthesis to simplify notation in $T^{([1^k])}$.

Explicit formulas for expansions (6.4.20) and (6.4.21) are obtained below in Section 6.4.4, Proposition 6.4.4.

Problem 6.4.4. Check that multiplicities of representations $T^{[1^m]}$ in $\Delta_\pm \otimes \widetilde{\Delta}_\mp$ and $\Delta_\pm \otimes \widetilde{\Delta}_\pm$ are correct in (6.4.20) and (6.4.21). Hint: compare dimension $2^{r-1} \cdot 2^{r-1} = 2^{2r-2}$ of representations on the left-hand sides of (6.4.20) and (6.4.21) with dimensions of representations on the right-hand sides. Make use of identities $\dim(T^{[1^m]}) = \frac{N!}{m!(N-m)!}$ and $\dim(T_\pm^{[1^r]}) = \frac{N!}{2(r!)^2}$.

Problem 6.4.5. Expand tensor products $(\Delta_+ \otimes \Delta_-)$ and $(\Delta_\pm \otimes \Delta_\pm)$ in irreducible representations, by using (6.4.20), (6.4.21) and equivalences in Proposition 6.3.11, item 2. Show that defining representation $T^{[1]}$ of $SO(p,q)$ is contained in $(\Delta_+ \otimes \Delta_-)$ for even $r = \nu$ and in $(\Delta_\pm \otimes \Delta_\pm)$ for odd $r = \nu \geq 3$.

Problem 6.4.6. Show that complex conjugation rules for tensors (6.4.6) in even case are

$$(\overline{\psi}\,(\gamma_{k_1} \cdots \gamma_{k_\ell})\,\phi)^* = (-1)^{\ell(\ell-1)/2+g}\,\overline{\phi}\,(\gamma_{k_1} \cdots \gamma_{k_\ell})\psi, \tag{6.4.22}$$

where factor $(-1)^g$ is determined from the relation $\phi\psi = (-1)^g\psi\phi$, i.e., $(-1)^g$ equals -1 $(+1)$, if spinors ψ and ϕ anticommute (commute). The anticommuting case corresponds to Grassmann variables (see Section 1.3.3) and has to do with Fermi–Dirac statistics. Hint: make use of the first formula in (6.3.27).

We now turn to odd case $(p+q) = 2r+1$. Irreducible representations $T^{[1^m]}$ $(m = 1, \ldots, r-1)$ of group $SO(p,q)$ correspond to representations $T_{(m)}$ of algebra $so(2r+1, \mathbb{C})$ with highest weights $\lambda_{(m)}$ given by (3.4.51). Representation $T^{[1^r]}$ of group $SO(p,q)$ corresponds to irreducible representation $T'_{(r)}$ of algebra $so(2r+1, \mathbb{C})$ with composite highest weight (3.4.53). The latter is contained in tensor product $(\rho \otimes \rho)$ of two irreducible spinor representations[k] $\rho = T_{(r)}$, each with fundamental highest weight $\lambda_{(r)}$ given by (3.4.51) (see Remark 2 in Section 6.2.2). Recall that for group $\mathsf{Spin}(p, q)$ $(p+q = 2r+1)$ the defining spinor representation $\rho = T_{(r)}$ and the contragredient representation $\tilde{\rho}$ are equivalent (see (6.3.101)). Therefore, expansion (6.4.11) of tensor product $(\rho \otimes \tilde{\rho})$ is valid also for $(\rho \otimes \rho)$. So, defining representation $T^{([1])}$ of group $SO(p, q)$ and algebra $so(p, q)$ is contained in $(\rho \otimes \rho)$. This proves the statement (see item 4 in Section 3.4.2) that all finite-dimensional irreducible representations of Lie algebra $so(p, q)$ can be constructed as subrepresentations in tensor products $T_{(r)}^{\otimes M}$ of spinor representations $\rho = T_{(r)}$.

To end this section, we point out that formulas (6.4.11), (6.4.12), (6.4.16), (6.4.17), (6.4.20) and (6.4.21) are, in fact, formulas for Clebsch–Gordan expansion of tensor products of spinor and Weyl (semispinor) representations of algebra $so(p, q) = \mathsf{spin}(p, q)$ (group $\mathsf{Spin}(p, q)$).

6.4.2. Charge conjugation of spinors in spaces $\mathbb{R}^{p,q}$

Making use of matrices $\mathbf{C}$ or $\mathbf{C}'$ defined in (6.3.15) or (6.3.51), we can define yet another conjugation operation for spinors, in addition to (6.4.3) (in what follows, we again use unified notation $\mathbf{C}$ for both even and odd dimensions):

$$\psi \to \widetilde{\psi} = \psi^{\mathsf{T}} \mathbf{C}^{-1}, \tag{6.4.23}$$

where $\widetilde{\psi}$, like $\bar{\psi}$ turns out to be spinor, contragredient to ψ. Indeed, in view of the second formula in (6.3.64), spinor $\widetilde{\psi}$ transforms according to (6.4.4) under the action (6.4.2) of group $\mathsf{Spin}(p, q)$ on spinor ψ. So, objects

$$\widetilde{t}_{k_1 \ldots k_\ell} = \widetilde{\psi}(\gamma_{k_1} \cdots \gamma_{k_\ell})\phi = \psi^{\mathsf{T}} \mathbf{C}^{-1}(\gamma_{k_1} \cdots \gamma_{k_\ell})\phi, \tag{6.4.24}$$

are again components of rank-ℓ $SO(p, q)$-tensor. In particular, scalar product $\psi^{\mathsf{T}} \mathbf{C}^{-1} \phi$ is invariant under $\mathsf{Spin}(p, q)$-transformations (6.4.2), i.e., it is scalar under $SO(p, q)$.

---

[k]Recall that for odd $N = 2r+1$, two inequivalent representations $\rho_\pm$ of algebra $\mathcal{C}\ell_{(p,q)}$ determine two equivalent representations (i.e., one and the same representation $\rho = T_{(r)}$) of algebra $so(p, q)$ and group $\mathsf{Spin}(p, q)$.

Problem 6.4.7. Show that in even case, components (6.4.24) transform, under transformation of spinors (6.4.2), in accordance with (6.4.7) and obey the identities

$$(\widetilde{\psi}\gamma_{k_1}\dots\gamma_{k_\ell}\phi) = \epsilon'(-1)^{\ell(\ell+1)/2+g}\,(\widetilde{\phi}\gamma_{k_1}\dots\gamma_{k_\ell}\psi), \tag{6.4.25}$$

where factor $(-1)^g$ is the same as in Problem 6.4.6. Hint: make use of the second formula in (6.3.27).

We consider in what follows even case $N = p + q = 2\nu$, unless the opposite is stated explicitly. Making use of matrix $\mathbf{B}$ defined in (6.3.15), we construct a new spinor

$$\psi_C = \eta'\mathbf{B}\psi^*, \tag{6.4.26}$$

where η' is a phase factor, $|\eta'| = 1$. According to the rules (6.3.66), spinor ψ_C transforms under $\mathsf{Spin}(p,q)$-action (6.4.2) in the same way as ψ : $\psi_C \to U\,\psi_C$. We point out without proof that for $p = 1$ and with our Definition 6.2.4 of timelike and spacelike coordinates in $\mathbb{R}^{p,q}$, spinor (6.4.26) describes spinor antiparticle with respect to particle described by spinor[1] ψ; for this reason, spinor (6.4.26) is called charge-conjugate to ψ.

Charge conjugate spinor (6.4.26) can also be defined in terms of matrix $\mathbf{C}$. To this end, we write charge conjugation operation (6.4.26) as follows:

$$\begin{aligned} \psi \overset{C}{\to} \psi_C &= \eta'\,\mathbf{B}\,\psi^* = \eta\,\mathbf{C}(\mathbf{D}^{-1})^{\mathsf{T}}\,\psi^* = \eta\,\mathbf{C}\,\bar{\psi}^{\mathsf{T}}, \\ \bar{\psi} \overset{C}{\to} \bar{\psi}_C &= \psi_C^\dagger\,\mathbf{D}^{-1} = (\eta')^*\,\psi^{\mathsf{T}}\,\mathbf{B}^\dagger\,\mathbf{D}^{-1} = (\eta')^*\,\psi^{\mathsf{T}}\,\mathbf{C}^{-1}, \end{aligned} \tag{6.4.27}$$

where η, $\eta' = \epsilon\epsilon'\eta$ are phase factors, and we use (6.3.28), (6.3.30) in deriving (6.4.27). In view of (6.4.27), matrix $\mathbf{C}$ is called *charge conjugation matrix.* Note that the conjugation operation C applied twice gives

$$(\psi_C)_C = \eta'\mathbf{B}\psi_C^* = \mathbf{B}\mathbf{B}^*\psi = \epsilon\psi,$$

so the identity $C^2 = I$ is possible for spinors only for $\epsilon = +1$. Since the charge conjugate spinor ψ_C transforms under $\mathsf{Spin}(p,q)$-transformations in the same way as ψ : $\psi_C \to U\,\psi_C$ one can introduce an analog of reality condition in complex space $\mathcal{F}$ of defining representation of group

[1]Strictly speaking, we should talk here about spinor fields $\psi(x)$ and $\psi_C(x)$, where $x \in \mathbb{R}^{p,q}$, and introduce Dirac equation.

Spin(p,q): $\psi_C = \eta'' \psi$. Here, the phase factor η'' can be set equal to 1 without loss of generality, by redefinition of factor η' in (6.4.26).

Definition 6.4.1. Spinor ψ equal to its charge conjugate spinor ψ_C, i.e., obeying reality condition

$$\psi = \psi_C \;\Leftrightarrow\; \psi = \eta' \mathbf{B}\, \psi^*, \tag{6.4.28}$$

is called *Majorana spinor*. Representation of group Spin(p,q) in the space of Majorana spinors is *Majorana representation.*

Majorana spinors describe particles which are identical to their antiparticles. Such particles have zero electric charges (in general, antiparticle charge has opposite sign to particle charge, while absolute values of these charges are equal).

The second equation in (6.4.28) and its complex conjugate together with (6.3.18) give $\psi = \mathbf{B}\mathbf{B}^*\psi = \epsilon\psi$, which means that Majorana spinors can be introduced only for $\epsilon = +1$. Dimensions and metric signatures of spaces $\mathbb{R}^{p,q}$, where this is the case, are listed in (6.3.35), (6.3.42). In particular, Majorana spinors exist in N-dimensional Minkowski space $\mathbb{R}^{(1,N-1)}$ (with even N) for

$$N = 2, 4 \bmod(8).$$

Note that formulas (6.3.28) enable one to cast Majorana condition (6.4.28) in the form

$$\psi^{\mathsf{T}}\mathbf{C}^{-1} = \zeta\, \psi^\dagger \mathbf{D}^{-1} \;\Leftrightarrow\; \widetilde{\psi} = \zeta\, \bar{\psi}, \tag{6.4.29}$$

where $\zeta = \eta'\epsilon'\beta_1 = \eta'\epsilon$ is the phase factor. In other words, Dirac conjugation (6.4.3) of Majorana spinors coincides, modulo factor ζ, with conjugation operation (6.4.23). Thus, in the Majorana case, tensors with components (6.4.6) are equal, modulo phase factor ζ, to tensors with components (6.4.25). Generally speaking, this is not the case for Dirac spinors.

‖ **Problem 6.4.8.** Derive the condition (6.4.29) from (6.4.28).

In Minkowski space, scalars $(\bar{\psi}\psi)$ and $(\widetilde{\psi}\psi)$, given by (6.4.6) and (6.4.25), respectively, with $\ell = 0$, are often dubbed as *Dirac and Majorana mass terms.* In view of (6.4.25), Majorana mass term $(\widetilde{\psi}\psi)$ does not vanish only if $\epsilon' = (-1)^g$. In the fermionic case $(-1)^g = -1$, we have $\epsilon' = -1$, which is possible, in accordance with Proposition 6.3.5 (see (6.3.33) and (6.3.45)), for $N = 2, 4 \bmod(8)$.

Remark 1. If there are two spinors ψ_i $(i = 1, 2)$, then, instead of reality condition (6.4.28) for each of them, one can impose mixed reality condition ($SU(2)$-reality [58])

$$\psi_i = \eta' \varepsilon_{ij} \mathbf{B}\, \psi_j^* \; (\varepsilon_{ij} = -\varepsilon_{ji},\; \varepsilon_{12} = 1), \tag{6.4.30}$$

which is consistent only for $\epsilon = -1$ (one considers Eq. (6.4.30) and its complex conjugate). So, condition (6.4.30) may replace the standard Majorana condition, which requires $\epsilon = +1$.

Remark 2. Making use of matrix $\widetilde{\mathbf{B}}$ defined in (6.3.22), i.e., $\gamma_k^* = \widetilde{\mathbf{B}}^{-1} \cdot \gamma_k \cdot \widetilde{\mathbf{B}}$, $\forall k \in (1, \ldots, 2\nu)$, we define yet another charge conjugation (cf. (6.4.26))

$$\psi_{\widetilde{C}} = \widetilde{\eta}' \, \widetilde{\mathbf{B}} \, \psi^*, \tag{6.4.31}$$

where $\widetilde{\eta}'$ is a phase factor, $|\widetilde{\eta}'| = 1$. Spinor (6.4.31) defines spinor antiparticle, if one defines timelike and spacelike coordinates in a way opposite to ours, namely, spacelike (timelike) coordinates x^k correspond to positive (negative) metric components[m]. We still denote the number of positive (now spacelike) and negative (now timelike) components of metric by p and q, so the alternative assignment of coordinates corresponds to the same space $\mathbb{R}^{p,q}$ with metric

$$\eta = (\underbrace{+1, \ldots, +1}_{p}, \underbrace{-1, \cdots -1}_{q}).$$

In view of (6.3.67), charge conjugate spinor (6.4.31) transforms under $\mathsf{Spin}(q, p)$ as $\psi_{\widetilde{C}} \to U\psi_{\widetilde{C}}$, i.e., in the same way as spinor ψ. This enables one to introduce another analog of reality (Majorana) condition, cf. (6.4.28):

$$\psi = \psi_{\widetilde{C}} \;\; \Leftrightarrow \;\; \psi = \widetilde{\eta}' \, \widetilde{\mathbf{B}} \, \psi^*. \tag{6.4.32}$$

Spinors obeying this condition are sometimes called *pseudo-Majorana*. Complex conjugation of the second equality here gives $\psi = |\widetilde{\eta}'|^2 \, \widetilde{\mathbf{B}}\widetilde{\mathbf{B}}^* \, \psi = \widetilde{\epsilon}\, \psi$, where we make use of (6.3.24). So, the new Majorana condition (6.4.32) is consistent only for $\widetilde{\epsilon} = +1$. Under the replacement $p \leftrightarrow q$ we have

[m]Explaining why charge conjugation (6.4.26) must be replaced by (6.4.31) is beyond the scope of this book. We point out only that flipping metric signature yields the replacement of relativistic mass shell condition $p_k p^k - \mathsf{m}^2 = 0$ with $p_k p^k + \mathsf{m}^2 = 0$, where p_k are momentum components and m is mass. This dictates modification of the form of Dirac equation.

$\epsilon \leftrightarrow \widetilde{\epsilon}$ (see Remark 3 after Proposition 6.3.5) so the condition $\widetilde{\epsilon} = \pm 1$ for representation of $\mathcal{C}\ell_{(p,q)}$ is obtained from (6.3.35) and reads (see (6.3.43))

$$\widetilde{\epsilon} = +1 \text{ for } q - p = -2, 0 \,\mathrm{mod}(8); \;\; \widetilde{\epsilon} = -1, \text{ for } q - p = 2, 4 \,\mathrm{mod}(8). \tag{6.4.33}$$

In particular, for $\mathcal{C}\ell_{(N-1,1)}$ we have $\widetilde{\epsilon} = +1$ for $1 - (N-1) = -2, 0$ mod(8), i.e., for $N = 2, 4$ mod(8) (cf. (6.3.47) valid for $\mathcal{C}\ell_{(1,N-1)}$). Thus, the conditions for existence of Majorana spinors are the same irrespectively of whether Minkowski metric is chosen as $\mathrm{diag}(+1, -1, \dots, -1)$ or as $\mathrm{diag}(-1, +1, \dots, +1)$ (this is true also for odd N; see Proposition 6.4.2).

Remark 3. In odd case $N = p + q = 2\nu + 1$ charge conjugate spinor (6.4.26) and charge conjugation operation (6.4.27) are defined in terms of matrices $\mathbf{C}'$ and $\mathbf{B}'$:

$$\psi \overset{C}{\to} \psi_C = \eta\, \mathbf{C}'\, \bar{\psi}^{\mathsf{T}} = \eta\, \mathbf{C}'(\mathbf{D}'^{-1})^{\mathsf{T}}\, \psi^* = \eta'\, \mathbf{B}'\, \psi^*, \tag{6.4.34}$$

where η, η' are phase factors, and we use (6.3.55). Majorana condition now reads

$$\psi = \psi_C \;\Leftrightarrow\; \psi = \eta' \mathbf{B}' \psi^*. \tag{6.4.35}$$

Like in even case, it is consistent, in view of (6.3.54), only for $\epsilon = +1$, i.e., when (6.3.56), (6.3.57) are satisfied. As an example, for odd-dimensional Minkowski space $\mathbb{R}^{1,N-1}$, the condition $\epsilon = +1$ holds only for $N = 1$, $3 \,\mathrm{mod}(8)$ (see (6.3.58)).

To summarize, we list all cases when there exist Majorana spinors, see also [58].

Proposition 6.4.2. *With metric $\eta = I_{p,q}$, the Majorana conditions are as follows.*

(1) *Even dimensions* $(p+q) = 2\nu = N$:

$$\begin{gathered} \psi = \eta_1 \mathbf{B}\, \psi^*, \;\; \gamma_k^* = -\mathbf{B}^{-1}\, \gamma_k\, \mathbf{B}, \;\; \epsilon = +1 \;\Rightarrow\; (p - q) = -2, 0 \,\mathrm{mod}(8), \\ \psi = \eta_2\, \widetilde{\mathbf{B}}\, \psi^*, \;\; \gamma_k^* = \widetilde{\mathbf{B}}^{-1}\, \gamma_k\, \widetilde{\mathbf{B}}, \;\; \widetilde{\epsilon} = +1 \;\Rightarrow\; (p - q) = 0, 2 \,\mathrm{mod}(8). \end{gathered} \tag{6.4.36}$$

(2) *Odd dimensions* $(p+q) = 2\nu + 1 = N$:

$$\begin{gathered} \psi = \eta_3\, \mathbf{B}'\, \psi^*, \;\; \gamma_k^* = -\mathbf{B}'^{-1}\, \gamma_k\, \mathbf{B}', \;\; \epsilon = +1 \;\Rightarrow\; (p - q) = -1 \,\mathrm{mod}(8), \\ \psi = \eta_4\, \mathbf{B}'\, \psi^*, \;\; \gamma_k^* = \mathbf{B}'^{-1}\, \gamma_k\, \mathbf{B}', \;\; \epsilon = +1 \;\Rightarrow\; (p - q) = 1 \,\mathrm{mod}(8), \end{gathered} \tag{6.4.37}$$

where parameters ϵ, $\widetilde{\epsilon}$ *are the same as in* (6.3.18), (6.3.24) *and* (6.3.54), *while* η_k *are phase factors,* $\eta_k \, \eta_k^* = 1$.

Let us come back to even case $(p+q) = 2\nu$. We use Weyl representation (6.3.71) and define two projectors (6.2.38):

$$P_L = \frac{1}{2}(I_{2^\nu} + \gamma_{2\nu+1}) = \begin{pmatrix} I_{2^{\nu-1}} & \mathbf{0} \\ \mathbf{0} & \mathbf{0} \end{pmatrix}, \; P_R = \frac{1}{2}(I_{2^\nu} - \gamma_{2\nu+1}) = \begin{pmatrix} \mathbf{0} & \mathbf{0} \\ \mathbf{0} & I_{2^{\nu-1}} \end{pmatrix}. \tag{6.4.38}$$

These projectors divide 2^ν-dimensional Dirac spinors ψ (associated with space $\mathbb{R}^{p,q}$) into two $2^{\nu-1}$-dimensional Weyl spinors ξ and η:

$$\psi = \begin{pmatrix} \xi \\ \eta \end{pmatrix} \equiv \begin{pmatrix} \xi_\alpha \\ \eta^{\dot\alpha} \end{pmatrix} \Rightarrow \psi_L = P_L \psi = \begin{pmatrix} \xi_\alpha \\ 0 \end{pmatrix}, \; \psi_R = P_R \psi = \begin{pmatrix} 0 \\ \eta^{\dot\alpha} \end{pmatrix}. \tag{6.4.39}$$

where $\alpha, \dot\alpha = 1, \ldots, 2^{\nu-1}$. Indices with and without dot are introduced in Remark 1 in Section 6.2.2. Spinors ψ_L and ψ_R are left and right chiral spinors in physics nomenclature (this nomenclature is superficial in our context).

It follows from definition (6.4.39) and properties of projectors $P_{R,L}$ that

$$\frac{1}{2}(I_{2^\nu} + \gamma_{2\nu+1})\, \psi_R = 0, \;\; \frac{1}{2}(I_{2^\nu} - \gamma_{2\nu+1})\psi_L = 0. \tag{6.4.40}$$

Let chiral spinor ψ_R (or ψ_L) be at the same time Majorana spinor, i.e., obey reality condition (6.4.28). Then it follows from (6.4.40) and the last formula in (6.3.16) that

$$\begin{aligned} 0 &= (I_{2^\nu} \pm \gamma^*_{2\nu+1})\psi^*_{R(L)} = (I_{2^\nu} \pm (-1)^{\nu+q} \mathbf{B}^{-1} \gamma_{2\nu+1} \mathbf{B})\, \psi^*_{R(L)} \\ &= \eta'^* \mathbf{B}^{-1} (I_{2^\nu} \pm (-1)^{\nu+q} \gamma_{2\nu+1}) \psi_{R(L)}, \end{aligned} \tag{6.4.41}$$

where we use (6.4.28) in the last equality. So, Weyl conditions (6.4.40) are consistent with Majorana conditions (6.4.28) only for even $(\nu + q)$, or $(p - q)/2 = 2k$. In other words,

$$N - 2p = 0 \,\mathrm{mod}(4) \;\Leftrightarrow\; s \equiv p - q = 0 \,\mathrm{mod}(4). \tag{6.4.42}$$

We recall (6.4.36) and conclude that Majorana–Weyl spinors exist if and only if

$$p = k \,\mathrm{mod}(4), \; N = 2k \,\mathrm{mod}(8) \;\Leftrightarrow\; s \equiv p - q = 0 \,\mathrm{mod}(8), \tag{6.4.43}$$

where $k = 0, 1, 2, 3$. In particular, Majorana–Weyl spinors exist in $\mathbb{R}^{1,1}$ and $\mathbb{R}^{1,9}$. Likewise, pseudo-Majorana spinors (6.4.32) can obey Weyl conditions (6.4.40) and Majorana condition $\psi = \tilde{\eta}' \widetilde{\mathbf{B}} \psi^*$ also only in the case (6.4.42). So, in view of the second line in (6.4.36), pseudo-Majorana–Weyl spinors exist in the same cases (6.4.43). We note that the number of independent real components of Majorana–Weyl spinor is $2^{\nu-1}$, which is four times smaller than the number of independent real components of Dirac spinor.

Examples

1. Consider the case $p = 1$ (Minkowski space $\mathbb{R}^{1,N-1}$). Majorana spinors exist only for $N = 2, 4 \,\mathrm{mod}(8)$ (see (6.3.42)), while consistency condition (6.4.42) reads $N = 2 \,\mathrm{mod}(4)$. So, Majorana–Weyl spinors exist only for $N = 2 \,\mathrm{mod}(8) = 2, 10, 18, 26, \ldots$, which agrees with (6.4.43).
2. Consider the case $p = N$, $q = 0$ (Euclidean space $\mathbb{R}^N$). It follows from (6.4.43) that Majorana–Weyl spinors exist only for $N = 0 \,\mathrm{mod}(8) = 8, 16, \ldots$. The same result is obtained in the case $p = 0$, $q = N$.

Problem 6.4.9. Describe explicitly Majorana–Weyl spinors in two-dimensional Minkowski space $\mathbb{R}^{1,1}$.

6.4.3. Algebra $\mathcal{C}\ell_{(1,N-1)}$ and spinor group Spin$(1, N-1)$. Spinors in Minkowski space $\mathbb{R}^{1,N-1}$

In this section, we discuss separately real algebra $\mathcal{C}\ell_{(1,N-1)}$ associated with Minkowski space $\mathbb{R}^{1,N-1}$ with pseudo-Euclidean metric $||\eta_{mn}|| = \mathrm{diag}(+1, -1, \ldots, -1)$. This case deserves special consideration, since it is of great interest for physics applications. We begin with the even case $N = 2\nu$.

In the first place, we recall (6.1.8) and write the generators of $\mathcal{C}\ell_{(1,N-1)}$ as follows:

$$\gamma_0 = \Gamma_1, \ \gamma_k = i\Gamma_{k+1} \ (k = 1, \ldots, N-1), \tag{6.4.44}$$

where elements Γ_k obey defining relations (6.1.9) of algebra $\mathcal{C}\ell_{2\nu}$ with Euclidean metric. We also modify element (6.1.85) by a factor $(-i)^N = (-1)^\nu$:

$$\gamma_{N+1} = (-i)^{(\nu-1)} \gamma_0 \gamma_1 \cdots \gamma_{N-1} \ \Rightarrow \ \gamma_{N+1}^2 = I_{2^\nu}. \tag{6.4.45}$$

Generators of Clifford algebra $\mathcal{C}\ell_{1,N-1}$ obey (see (6.4.1))

$$\gamma_m \gamma_n + \gamma_n \gamma_m = 2\eta_{mn} \, I_{2^\nu}, \quad ||\eta_{mn}|| = I_{1,N-1}. \tag{6.4.46}$$

We choose irreducible 2^ν-dimensional representation (6.3.2) of elements Γ_k, which according to (6.4.44) leads to the representation of γ_k. Then (6.1.49), (6.1.54), (6.3.3), (6.3.11) and (6.3.13) give

$$\gamma_{N+1} = (-1)^\nu \, \sigma_3^{\otimes \nu},$$

$$\gamma_m^* = (-1)^m \, \eta_{mm} \, \gamma_m = (-1)^m \, \gamma_0 \gamma_m \gamma_0, \;\; \gamma_m^\mathsf{T} = (-1)^m \gamma_m, \tag{6.4.47}$$

$$\gamma_m^\dagger = \eta_{mm} \gamma_m = \gamma_0 \gamma_m \gamma_0, \tag{6.4.48}$$

$$-C\gamma_m \, C = (-1)^m \, \gamma_m, \;\; \gamma_m^\mathsf{T} = -C\gamma_m C, \;\; C^2 = I_{2^\nu}, \tag{6.4.49}$$

where $m = 0, 1, \ldots, N-1$, and matrix C is introduced in (6.3.5). The second equality in (6.4.49) follows from the first one and the last identity in (6.4.47). In accordance with (6.3.9), (6.3.10), matrix C is chosen as follows:

$$\begin{aligned} C &= (-i)^{k-1} \gamma_0 \gamma_2 \cdots \gamma_{N-2} \quad (\nu = 2k), \\ C &= -(i)^k \gamma_1 \gamma_3 \cdots \gamma_{N-1} \quad (\nu = 2k-1), \end{aligned} \tag{6.4.50}$$

where $k = 1, 2, 3, \ldots$.

Problem 6.4.10. Prove the first and third of Eq. (6.4.49) for matrices (6.4.50), by employing only defining relations (6.4.46).

We compare (6.4.47), (6.4.48) and (6.4.49) with (6.3.15) and find explicit expressions for matrices $\mathbf{D}$, $\mathbf{C}$ and $\mathbf{B}$:

$$\mathbf{D} = \gamma_0, \;\; \mathbf{C} = C, \;\; \mathbf{B} = \gamma_0 C, \tag{6.4.51}$$

which are valid for the particular representation ρ of algebra $C\ell_{(1,N-1)}$. Recall that matrix $\mathbf{B}$ obeys (6.3.18) where parameter ϵ is the same for all representations, and in our case is given by (6.3.47):

$$\epsilon = (-1)^{k+1} \text{ for } \nu = 2k-1, 2k. \tag{6.4.52}$$

This result can be obtained from (6.3.8) and follows from general formulas (6.3.34) proven in Proposition 6.3.5. Note that in the particular representation we consider, we have, in accordance with (6.3.8) and (6.3.44), the following relations:

$$\epsilon' = -\epsilon = (-1)^k \text{ for } \nu = 2k-1, 2k \tag{6.4.53}$$

(cf. (6.4.52) and (6.3.45)).

Problem 6.4.11. Calculate parameters ϵ, ϵ' explicitly. Hint: make use of the relations (6.4.47), (6.4.44), (6.3.8) and formula (6.4.51) for $\mathbf{B}$. In view of (6.3.5) and (6.4.51), write $\mathbf{B}$ in explicit form

$$\mathbf{B} = \Gamma_1 C = i\,\underbrace{\sigma_3 \otimes \sigma_1 \otimes \sigma_2 \otimes \sigma_1 \otimes \sigma_2 \otimes \dots}_{\nu}. \tag{6.4.54}$$

Prove that matrices (6.4.51) obey (6.3.28) with $\beta_2 = -\beta_1 = 1$ and $\epsilon\epsilon' = \beta_1\beta_2 = -1$ (cf. (6.4.53)).

Remark 1. Recall that Majorana spinors in N-dimensional Minkowski space with even N exist only for $N = 2, 4\,\mathrm{mod}(8)$, and Majorana–Weyl spinors exist for $N = 2\,\mathrm{mod}(8)$ (see Example 1 in Section 6.4.2).

For odd-dimensional Minkowski space $\mathbb{R}^{1,N-1}$ with $N = 2\nu + 1$ and metric η_{1,N_1}, analogs $\mathbf{B}'$, $\mathbf{C}'$ and $\mathbf{D}'$ of matrices $\mathbf{B}$, $\mathbf{C}$ and $\mathbf{D}$ are defined by (6.3.51):

$$\gamma_k^\dagger = \mathbf{D}'^{-1}\,\gamma_k\,\mathbf{D}',\ \ \gamma_k^\top = (-1)^\nu \mathbf{C}'^{-1}\,\gamma_k\,\mathbf{C}',\ \ \gamma_k^* = (-1)^\nu\,\mathbf{B}'^{-1}\,\gamma_k\,\mathbf{B}', \tag{6.4.55}$$

where we set $q = 2\nu$, since $p = 1$. Here matrices $\gamma_k (k = 0, \dots, 2\nu - 1)$ are taken in representation (6.4.44), (6.3.2), while the last generator is

$$\gamma_{2\nu} = (-i)^\nu\,\gamma_0\gamma_1\cdots\gamma_{2\nu-1},\ \ \gamma_{2\nu}^\dagger = -\gamma_{2\nu} = \gamma_0\,\gamma_{2\nu}\,\gamma_0. \tag{6.4.56}$$

The latter differs from (6.4.45) by an overall factor that ensures $\gamma_{2\nu}^2 = -I_{2^\nu}$ corresponding to metric component $\eta_{2\nu 2,\nu} = -1$. Making use of (6.4.51) for matrices $\mathbf{B}$, $\mathbf{C}$ and $\mathbf{D}$ in even case $N = 2\nu$ and following rules (1) and (3) in (6.3.52), we obtain explicit forms of matrices $\mathbf{B}'$, $\mathbf{C}'$ and $\mathbf{D}'$:

$$\mathbf{D}' = \gamma_0,\ \begin{cases} \mathbf{C}' = C,\ \mathbf{B}' = \gamma_0\,C, & \text{odd } \nu,\ N = 3\,\mathrm{mod}(4), \\ \mathbf{C}' = \gamma_{2\nu} C,\ \mathbf{B}' = \gamma_{2\nu}\,\gamma_0\,C, & \text{even } \nu,\ N = 1\,\mathrm{mod}(4), \end{cases} \tag{6.4.57}$$

where matrix $\gamma_{2\nu}$ is defined in (6.4.56).

Let us again consider the case of both even and odd dimension N. Without repeating the general analysis, we only recall that elements $U \in$ Spin$(1, N-1)$ are written, in a neighborhood of unity, in the exponential form (6.2.30):

$$U = \exp\left(\frac{1}{2}\omega^{mn}\Sigma_{mn}\right),\quad \Sigma_{mn} = \frac{1}{4}(\gamma_m\gamma_n - \gamma_n\gamma_m). \tag{6.4.58}$$

Here $\omega^{mn} = -\omega^{mn}$ are real parameters which can be viewed as parameters of the Lorentz group $SO(1, N-1)$, and generators Σ_{mn} of Lie algebra $\mathsf{spin}(1, N-1)$ obey (6.2.28) and (6.2.29). Note that due to the properties (6.4.48), (6.4.56) of γ-matrices under Hermitian conjugation, we have

$$\Sigma^\dagger_{mn} = -\gamma_0\, \Sigma_{mn}\, \gamma_0. \tag{6.4.59}$$

Therefore, pseudo-unitarity property (6.3.64) for $U \in \mathsf{Spin}(1, N-1)$ reads

$$U^\dagger\, \gamma_0\, U = U\, \gamma_0\, U^\dagger = \gamma_0. \tag{6.4.60}$$

Problem 6.4.12. Prove that $2^{[N/2]}$-dimensional matrices U obeying (6.4.60) make a subgroup in noncompact pseudo-unitary group $SU(n, n)$, where $n = 2^{[N/2]-1}$. Hint: make use of representation $\gamma_0 = \Gamma_1 = \sigma_1 \otimes I_2^{\otimes(\nu-1)}$ (see (6.3.2) or (6.1.67)), and diagonalize γ_0.

Like in the general case (see Section 6.2.1), groups $SO^\uparrow(1, N-1)$ and $\mathsf{Spin}(1, N-1)$ are locally isomorphic, and group $\mathsf{Spin}(1, N-1)$ covers group $SO^\uparrow(1, N-1)$ twice. Moreover, group $\mathsf{Spin}(1, N-1)$ is universal covering of $SO^\uparrow(1, N-1)$.

Consider now Dirac spinors ψ in space $\mathbb{R}^{1,N-1}$. Making use of (6.4.3), (6.4.51) and (6.4.57), one defines Dirac conjugation

$$\overline{\psi} = \psi^\dagger\, \gamma_0, \tag{6.4.61}$$

Then, in view of (6.4.60), group $\mathsf{Spin}(1, N-1)$ acts on spinors ψ and $\bar{\psi}$ as follows:

$$\psi \to \psi' = U\, \psi, \; \overline{\psi} \to \overline{\psi}' = \overline{\psi}\, U^{-1}, \;\; U \in \mathsf{Spin}(1, N-1). \tag{6.4.62}$$

It follows from the general analysis of Section 6.4.1 (see (6.4.7)) that objects $t_{k_1 \ldots k_r} = \overline{\psi}(\gamma_{k_1} \ldots \gamma_{k_r})\psi$ are transformed as rank-r tensors.

Remark 2. Consider even case $N = 2\nu$. Matrix $\mathbf{B}$, defined by (6.4.51), obeys

$$\mathbf{B}^\mathsf{T} = \epsilon \mathbf{B}, \; \mathbf{B} \cdot \mathbf{B}^\dagger = I_{2^\nu}. \tag{6.4.63}$$

The second identity is equivalent to the first one due to (6.3.18). Let us prove the first identity. We use (6.3.27) and write

$$(C\gamma_{m_1}\cdots\gamma_{m_r})^{\mathsf{T}} = \epsilon'(-1)^{\frac{r(r+1)}{2}}\,C\,\gamma_{m_1}\cdots\gamma_{m_r}$$
$$= \epsilon\,(-1)^{\frac{(r-1)(r-2)}{2}}\,C\,\gamma_{m_1}\cdots\gamma_{m_r}, \qquad (6.4.64)$$

where $m_1 < m_2 < \cdots < m_r$ and we recall that $C^{-1} = C$. For $r = 1$ we get $(C\gamma_m)^{\mathsf{T}} = \epsilon C\gamma_m$, which, in accordance with (6.4.49), can be written as $(\gamma_m C)^{\mathsf{T}} = \epsilon\gamma_m C$. Using the definition (6.4.51) of matrix $\mathbf{B}$, we obtain the first identity in (6.4.63).

Remark 3. An equivalence transformation $\gamma_m \to T\gamma_m T^{-1}$ changes matrix $\mathbf{B}$ as follows (see (6.3.21)):

$$\mathbf{B} \to T\,\mathbf{B}\,(T^*)^{-1}. \qquad (6.4.65)$$

If matrix $\mathbf{B}$ can be transformed in this way to unit matrix, then, in view of (6.3.15), (6.3.74) matrices γ_m are pure imaginary $\gamma_m^* = -\gamma_m$, and generators $\Sigma_{mn} \in \mathsf{spin}(1, N-1)$ are real matrices. Also, in accordance with definition (6.4.28), the case $\mathbf{B} = I$ and $\eta' = 1$ corresponds to real Majorana spinors, $\psi^* = \mathbf{B}\psi = \psi$. In view of the relation (see (6.3.18))

$$\mathbf{B}^*\cdot\mathbf{B} = \mathbf{B}\cdot\mathbf{B}^* = \epsilon I_{2^\nu}, \qquad (6.4.66)$$

necessary condition for equivalence of $\mathbf{B}$ to unit matrix is $\epsilon = +1$, which, in the case of Minkowski space metric $p = 1$, $q = (N-1)$, is valid (see (6.3.41)) for $N = 2, 4\,\mathrm{mod}(8)$, when there exist Majorana representations. Let us show that this condition is, in fact, sufficient. We write matrix $\mathbf{B}$ in terms of its real and imaginary parts, $\mathbf{B} = \mathbf{B}_1 + i\mathbf{B}_2$, where $\mathbf{B}_{1,2}$ are real and in view of (6.4.63) with $\epsilon = +1$ they are symmetric. Then (6.4.66) gives $[\mathbf{B}_1, \mathbf{B}_2] = 0$. So, matrices $\mathbf{B}_{1,2}$ can be diagonalized simultaneously by real orthogonal transformation $O : \mathbf{B}_i \to O\,\mathbf{B}_i\,O^{-1}$, and so does complex matrix $\mathbf{B}$. Since $\mathbf{B}$ is unitary (see (6.4.63)), we get $\mathbf{B} \to O\,\mathbf{B}\,O^{-1} = \mathrm{diag}(e^{i\beta_1}, e^{i\beta_2}, \ldots)$. We then choose $T' = \mathrm{diag}(e^{-i\beta_1/2}, e^{-i\beta_2/2}, \ldots)$, and find that $\mathbf{B}$ is transformed to unit matrix by transformation (6.4.65) with $T = T'\,O$. This completes the proof: matrix $\mathbf{B}$ is a unit matrix in appropriate representation, if and only if $N = 2, 4\ \mathrm{mod}(8)$; in that case Majorana representation is simply real: $\psi^* = \psi$.

Problem 6.4.13. * Find explicit form of matrix T in (6.4.65) that transforms matrix $\mathbf{B}$ given in (6.4.54) to unit matrix in the cases $N = 2, 4\,\mathrm{mod}(8)$. Construct pure imaginary representation for generators γ_k of algebras $\mathcal{C}\ell_{(1,1)}$ and $\mathcal{C}\ell_{(1,3)}$. Hint: transform to unit matrix the matrices $B_1 = i\,\sigma_3$, $B_2 = \sigma_1$, $B_3 = \sigma_2 \otimes \sigma_1 \otimes \sigma_2$ whose tensor product gives matrix $\mathbf{B}$ in (6.4.54) for the cases $N = 2, 4\ \mathrm{mod}(8)$.

6.4.4. Fierz identities for multi-dimensional spinors

Fierz identities are instrumental in physics applications. From group theory viewpoint, Fierz identities can be considered as explicit formulas for the expansions of tensor products of representations of spinor groups $\mathsf{Spin}(p,q)$ in sums of representations of $\mathsf{Spin}(p,q)$ (similar to Clebsch–Gordan expansion). Throughout this section, we consider Fierz identities only in *even-dimensional case* $N = 2\nu$.

Let matrices $\{\gamma_0, \ldots \gamma_{N-1}\}$ form an irreducible representation of generators of Clifford algebra $\mathcal{C}\ell_{(p,q)}$, $p+q = N = 2\nu$. Matrix γ_{N+1} is defined in (6.1.85).

Proposition 6.4.3. *Traces of products of even number of γ-matrices obey recurrence relation*

$$\begin{aligned}\mathrm{Tr}(\gamma_1 \cdots \gamma_{2k}) = \eta_{12}\mathrm{Tr}(\gamma_3 \cdots \gamma_{2k}) - \eta_{13}\mathrm{Tr}(\gamma_2\gamma_4 \cdots \gamma_{2k}) \\ +\eta_{14}\mathrm{Tr}(\gamma_2\gamma_3\gamma_5 \cdots \gamma_{2k}) - \cdots + \eta_{1,2k}\mathrm{Tr}(\gamma_2 \cdots \gamma_{2k-1}),\end{aligned} \tag{6.4.67}$$

where we denote indices m_ℓ by ℓ for brevity, i.e., $\eta_{m_1m_2} \to \eta_{12}$, $\gamma_{m_3} \to \gamma_3$, etc. Products of odd number of γ-matrices are traceless[n]:

$$\mathrm{Tr}(\gamma_{m_1} \cdots \gamma_{m_{2k+1}}) = 0. \tag{6.4.68}$$

Proof. Formula (6.4.68) follows from the chain of equalities:

$$\begin{aligned}\mathrm{Tr}(\gamma_{m_1} \cdots \gamma_{m_{2k+1}}) &= \mathrm{Tr}(\gamma_{N+1}\gamma_{N+1}\gamma_{m_1} \cdots \gamma_{m_{2k+1}}) \\ &= \mathrm{Tr}(\gamma_{N+1}\gamma_{m_1} \cdots \gamma_{m_{2k+1}}\gamma_{N+1}) \\ &= -\mathrm{Tr}(\gamma_{N+1}\gamma_{N+1}\gamma_{m_1} \cdots \gamma_{m_{2k+1}}) \\ &= -\mathrm{Tr}(\gamma_{m_1} \cdots \gamma_{m_{2k+1}}) = 0,\end{aligned} \tag{6.4.69}$$

[n] In odd-dimensional case $N = 2\nu + 1$, formula (6.4.68) is valid for all $k < \nu$ and is not valid for $k = \nu$ if all m_i are different, since in this case the product $\gamma_{m_1} \cdots \gamma_{m_{2\nu+1}}$ is central element in $\mathcal{C}\ell_{2\nu+1}$.

where we use cyclic property of trace in the second equality and anticommutation property $[\gamma_{N+1}, \gamma_m]_+ = 0$ ($\forall m = 0, \ldots, N-1$) in the third equality.

Formula (6.4.67) follows from the identity

$$\mathrm{Tr}(\gamma_1 \cdots \gamma_{2k}) = \frac{1}{2}(\mathrm{Tr}(\gamma_1\, \gamma_2 \cdots \gamma_{2k}) + \mathrm{Tr}(\gamma_2 \cdots \gamma_{2k}\, \gamma_1)), \tag{6.4.70}$$

which is again written in shorthand notations and is due to cyclic property of trace. Making use of defining relations for γ-matrices, one moves matrix γ_1 in the second term on the right-hand side of (6.4.70) to the left,

$$\begin{aligned} \mathrm{Tr}(\gamma_2 \cdots \gamma_{2k-1}\gamma_{2k}\, \gamma_1) &= -\mathrm{Tr}(\gamma_2 \cdots \gamma_{2k-1}\gamma_1\gamma_{2k}) + 2\eta_{1,2k}\mathrm{Tr}(\gamma_2 \cdots \gamma_{2k-1}) \\ &= \cdots = -\mathrm{Tr}(\gamma_1 \cdots \gamma_{2k}) + 2\eta_{12}\mathrm{Tr}(\gamma_3 \cdots \gamma_{2k}) \\ &\quad - \cdots + 2\eta_{1,2k}\mathrm{Tr}(\gamma_2 \cdots \gamma_{2k-1}), \end{aligned}$$

then the first term in (6.4.70) cancels out, and one obtains (6.4.67). □

Trace of any product of even number of γ-matrices is straightforwardly calculated by making use of (6.4.67). One begins with obvious formula $\mathrm{Tr}(\gamma_m\gamma_n) = 2^\nu \eta_{mn}$, and then writes

$$\begin{aligned} \mathrm{Tr}(\gamma_m\gamma_n\gamma_k\gamma_l) &= 2^\nu(\eta_{mn}\eta_{kl} - \eta_{mk}\eta_{nl} + \eta_{ml}\eta_{nk}), \\ \mathrm{Tr}(\gamma_1\gamma_2\gamma_3\gamma_4\gamma_5\gamma_6) &= 2^\nu[\eta_{12}(\eta_{34}\eta_{56} - \eta_{35}\eta_{46} + \eta_{36}\eta_{45}) \\ &\quad -\eta_{13}(\eta_{24}\eta_{56} - \eta_{25}\eta_{46} + \eta_{26}\eta_{45}) \\ &\quad +\eta_{14}(\eta_{23}\eta_{56} - \eta_{25}\eta_{36} + \eta_{26}\eta_{35}) - \cdots], \end{aligned}$$

where indices $1, 2, 3, \ldots$ still mean $m_1, m_2, m_3, \ldots$.

‖ **Problem 6.4.14.** Complete the last formula for $\mathrm{Tr}(\gamma_1 \cdots \gamma_6)$.

Recall that in view of the isomorphism $\mathcal{C}\ell^{\mathbb{C}}_{(p,q)} = \mathrm{Mat}_{2^\nu}(\mathbb{C})$ (see Remark 4 in Section 6.1.3) the following set of 2^N matrices (6.1.78) forms the basis°

°In odd case $N = (2\nu + 1)$, complete system of basis elements in $\mathrm{Mat}_{2^\nu}(\mathbb{C})$ is formed by products of even number (or odd number) of matrices $\gamma_k \in \mathcal{C}\ell_{(p,q)}$.

in the linear space of complex matrices $\mathrm{Mat}_{2^\nu}(\mathbb{C})$:

$$\Gamma_A = \{I,\, \gamma_{m_1},\, \gamma_{m_1}\gamma_{m_2},\, \gamma_{m_1}\gamma_{m_2}\gamma_{m_3},\, \dots\}, \quad 0 \le m_1 < m_2 < m_3 < \dots \le N-1, \tag{6.4.71}$$

where multi-index $A = \{m_1, m_2, \dots\}$ labels basis elements $\gamma_{m_1}\gamma_{m_2}\cdots$. We introduce a set of dual matrices $\Gamma^{\overline{A}}$ (with upper index):

$$\Gamma^{\overline{A}} = \{I,\, \gamma^{m_1},\, \gamma^{m_2}\gamma^{m_1},\, \gamma^{m_3}\gamma^{m_2}\gamma^{m_1},\, \gamma^{m_4}\gamma^{m_3}\gamma^{m_2}\gamma^{m_1}, \dots\}, \quad 0 \le m_1 < m_2 < m_3 < \dots \le N-1. \tag{6.4.72}$$

Let $A_k = \{m_1, \dots, m_k\}$, where $m_1 < \dots < m_k$, be multi-index which numerate basis monomials $(\gamma_{m_1}\cdots\gamma_{m_k})$ of order k. In view of Proposition 6.4.3, the set (6.4.72) is normalized by

$$\frac{1}{2^\nu}\mathrm{Tr}(\Gamma^{\overline{B}_\ell}\,\Gamma_{A_k}) = \delta^{B_\ell}_{A_k}. \tag{6.4.73}$$

Here

$$\Gamma^{\overline{B}_k} = (\gamma^{n_k}\gamma^{n_{k-1}}\dots\gamma^{n_1}),\ \Gamma_{A_k} = (\gamma_{m_1}\gamma_{m_2}\cdots\gamma_{m_k}), \tag{6.4.74}$$

$\delta^{B_k}_{A_k} = \delta^{n_1}_{m_1}\delta^{n_2}_{m_2}\cdots\delta^{n_k}_{m_k}$ and $\delta^{B_k}_{A_\ell} = 0$ for monomials of different orders $k \neq \ell$. We emphasize that bar over multi-index $A_k = \{m_1, \dots, m_k\}$ shows that indices in $\overline{A}_k$ are ordered in the opposite way, $\overline{A}_k = \{m_k, \dots, m_1\}$, cf. (6.4.71) and (6.4.72). Now, every matrix $\Gamma \in \mathrm{Mat}_{2^\nu}(\mathbb{C})$ can be expanded in basis matrices (6.4.71):

$$\begin{aligned}\Gamma &= \sum_A C^A\Gamma_A \equiv \sum_k\sum_{A_k} C^{A_k}\Gamma_{A_k}\\ &\equiv \sum_k \sum_{m_1<m_2\cdots<m_k} C^{m_1\dots m_k}\gamma_{m_1}\gamma_{m_2}\cdots\gamma_{m_k},\end{aligned} \tag{6.4.75}$$

and in view of (6.4.73), coefficients C^A are given by

$$C^A = \frac{1}{2^\nu}\mathrm{Tr}(\Gamma^{\overline{A}}\,\Gamma) \;\Rightarrow\; \Gamma = \frac{1}{2^\nu}\sum_A \mathrm{Tr}(\Gamma^{\overline{A}}\,\Gamma)\Gamma_A.$$

Since matrix Γ is arbitrary, the latter relation implies completeness property of the sets $\{\Gamma_A\}$ and $\{\Gamma^{\overline{A}}\}$:

$$\delta_\kappa{}^\beta\,\delta_\alpha{}^\rho = \frac{1}{2^\nu}\sum_A (\Gamma^{\overline{A}})_\kappa{}^\rho\,(\Gamma_A)_\alpha{}^\beta \;\Leftrightarrow\; P_{12} = \frac{1}{2^\nu}\sum_A (\Gamma^{\overline{A}})_1(\Gamma_A)_2$$
$$\Leftrightarrow\; P = \frac{1}{2^\nu}\sum_A \Gamma^{\overline{A}}\otimes\Gamma_A \tag{6.4.76}$$

(here α,β,ρ,κ are spinor indices). These relations are understood as equalities between operators acting in tensor product $\mathcal{F}\otimes\mathcal{F}$ of spaces of Dirac spinors, then P_{12} is the matrix $(P_{12})^{\rho\,\beta}_{\kappa\alpha} = \delta^\beta_\kappa\,\delta^\rho_\alpha$ of the permutation operator P (see (4.2.12)), and indices $1,2$ label vector spaces $\mathcal{F}$ in the tensor product $\mathcal{F}\otimes\mathcal{F}$ where matrices $(\Gamma^{\overline{A}})_1 = \Gamma^{\overline{A}}\otimes I_{2^\nu}$ and $(\Gamma_A)_2 = I_{2^\nu}\otimes\Gamma_A$ act nontrivially. As we show below relations (6.4.76) contain all information about Fierz identities.

Problem 6.4.15. Check the normalization coefficient $2^{-\nu}$ in (6.4.76). Hint: set $\rho=\alpha$, perform summation over α and make use of equality

$$\sum_{A_k}\Gamma^{\overline{A_k}}\,\Gamma_{A_k} = \frac{1}{k!}\sum_{\ell_1,\cdots,\ell_k}(\gamma^{[\ell_k}\cdots\gamma^{\ell_1]}\cdot\gamma_{[\ell_1}\cdots\gamma_{\ell_k]}) = \frac{N!}{k!(N-k)!}\,I_{2^\nu}, \tag{6.4.77}$$

where summation on the left-hand side runs over all ordered sets $A_k = (0\le m_1<m_2<\cdots<m_k\le N-1)$, antisymmetrized combinations $\gamma_{[\ell_1}\cdots\gamma_{\ell_k]}$ are defined in (6.4.10), and summation in the second expression runs over all ℓ_i such that $\ell_i\neq\ell_j$ for $i\neq j$ (the ordering $0\le\ell_1<\cdots<\ell_k\le N-1$ is not imposed here).

Proposition 6.4.4. *Expansions* (6.4.11), (6.4.20), (6.4.21) *of representations* $\rho\otimes\tilde\rho$, $\Delta_\pm\otimes\tilde\Delta_\pm$ *and* $\Delta_\mp\otimes\tilde\Delta_\pm$ *are given by the following explicit formulas:*

(i) *for* $\rho\otimes\tilde\rho$:

$$\bar\psi^\rho\phi_\kappa = \frac{1}{2^\nu}\sum_{k=0}^{2\nu}\frac{1}{k!}\,(\gamma^{[m_k}\cdots\gamma^{m_1]})_\kappa{}^\rho\,(\bar\psi\gamma_{[m_1}\cdots\gamma_{m_k]}\phi), \tag{6.4.78}$$

where ϕ *and* $\bar\psi$ *are spinors in the spaces of representations* ρ *and* $\tilde\rho$, *respectively;*

(ii) *for* $(\Delta_\pm \otimes \tilde{\Delta}_\pm)$:

$$\text{odd } \nu: \quad (\bar{\psi}_\pm)^\rho \, (\phi_\pm)_\kappa = \sum_{k=0}^{\frac{\nu-1}{2}} (C_\pm^{[m_{2k}\dots m_1]})_\kappa{}^\rho (\bar{\psi}_\pm \, \gamma_{[m_1\dots m_{2k}]}\phi_\pm), \tag{6.4.79a}$$

$$\begin{aligned} \text{even } \nu: \quad (\bar{\psi}_\pm)^\rho \, (\phi_\pm)_\kappa &= \sum_{k=0}^{\frac{\nu}{2}-1} (C_\pm^{[m_{2k}\dots m_1]})_\kappa{}^\rho \, (\bar{\psi}_\pm \, \gamma_{[m_1\dots m_{2k}]}\phi_\pm) \\ &\quad + (C_\pm^{[m_\nu\dots m_1]})_\kappa{}^\rho \, (\bar{\psi}_\pm \, \gamma_{[m_1\dots m_\nu]}\phi_\pm); \end{aligned} \tag{6.4.79b}$$

(iii) *for* $(\Delta_\mp \otimes \tilde{\Delta}_\pm)$:

$$\text{even } \nu: \quad (\bar{\psi}_\pm)^\rho \, (\phi_\mp)_\kappa = \sum_{k=1}^{\frac{\nu}{2}} (C_\pm^{[m_{2k-1}\dots m_1]})_\kappa{}^\rho \, (\bar{\psi}_\pm \, \gamma_{[m_1\dots m_{2k-1}]}\phi_\mp),$$

$$\begin{aligned} \text{odd } \nu: \quad (\bar{\psi}_\pm)^\rho \, (\phi_\mp)_\kappa &= \sum_{k=1}^{\frac{\nu-1}{2}} (C_\pm^{[m_{2k-1}\dots m_1]})_\kappa{}^\rho \, (\bar{\psi}_\pm \, \gamma_{[m_1\dots m_{2k-1}]}\phi_\mp) \\ &\quad + (C_\pm^{[m_\nu\dots m_1]})_\kappa{}^\rho (\bar{\psi}_\pm \, \gamma_{[m_1\dots m_\nu]}\phi_\mp), \end{aligned} \tag{6.4.80}$$

where $\bar{\psi}_\pm = \bar{\psi}\, P_\pm$, $\phi_\pm = P_\pm \, \phi$ *are spinors in the representation spaces* $\tilde{\Delta}_\pm$ *and* $\Delta_\mp$, *respectively, projectors* $P_\pm = \frac{1}{2}(I \pm \gamma_{N+1})$ *are introduced in (6.2.38)* , $\gamma_{[m_1\dots m_\ell]} \equiv \gamma_{[m_1} \cdots \gamma_{m_\ell]}$ *and*

$$\begin{aligned} &C_\pm^{[m_k\dots m_1]} \\ &\quad \equiv \tfrac{1}{2^\nu} \left(\frac{1}{(k)!} \gamma^{[m_k\dots m_1]} \pm \frac{\alpha_{2\nu-k}}{(2\nu-k)!} \gamma_{[n_{2\nu-k}\dots n_1]} \varepsilon^{n_1\dots n_{2\nu-k} m_1\dots m_k} \right), \end{aligned} \tag{6.4.81}$$

coefficients α_m *are defined in* (6.4.14).

Proof. We make a contraction of the first identity in (6.4.76) with Dirac spinors $\bar{\psi}^\alpha$ and ϕ_β, and, in view of the first equality in (6.4.77), obtain expansion (6.4.78). We now come to (6.4.79); formulas (6.4.80) are proven in a similar way. We make a contraction of the first identity in (6.4.76) with chiral spinors $\bar{\psi}_\pm^\alpha$ and $(\phi_\pm)_\beta$ and make use of the fact that

$(\bar{\psi}_\pm \Gamma_{A_{2k+1}} \phi_\pm) = 0$. Using the first identity in (6.4.77), we get

$$(\bar{\psi}_\pm)^\rho (\phi_\pm)_\kappa = \frac{1}{2^\nu} \sum_{k=0}^{\nu} \sum_{A_{2k}} (\Gamma^{\bar{A}_{2k}} P_\pm)_\kappa{}^\rho (\bar{\psi}_\pm \Gamma_{A_{2k}} \phi_\pm)$$

$$= \frac{1}{2^\nu} \sum_{k=0}^{\nu} \frac{1}{(2k)!} (\gamma^{[m_{2k} \dots m_1]} P_\pm)_\kappa{}^\rho (\bar{\psi}_\pm \gamma_{[m_1 \dots m_{2k}]} \phi_\pm). \quad (6.4.82)$$

In accordance with (6.4.14), we have an identity

$$\begin{aligned}(\bar{\psi}\, \gamma_{[m_1 \dots m_k]} \phi_\pm) &= \pm\, (\bar{\psi}\, \gamma_{[m_1 \dots m_k]} \gamma_{N+1} \phi_\pm) \\ &= \pm \alpha_k\, \varepsilon_{m_1 \dots m_k i_1 \dots i_{N-k}} (\bar{\psi}\, \gamma^{i_1 \dots i_{N-k}} \phi_\pm). \qquad (6.4.83)\end{aligned}$$

It enables us to write (6.4.82) in the form (6.4.79). Formula (6.4.79b) agrees with expansion in the second line in (6.4.21). Note that the last term on the right-hand side of (6.4.79b), in view of (6.4.81), can be written as

$$\begin{aligned}C_\pm^{[m_\nu \dots m_1]} (\bar{\psi}\, \gamma_{[m_1 \dots m_\nu]} \phi_\pm) = \frac{(-1)^{[\frac{\nu}{2}]}}{2^\nu \nu!} &(\gamma^{[m_1 \dots m_\nu]} \pm \alpha_\nu \gamma_{[n_1 \dots n_\nu]} \varepsilon^{n_1 \dots n_\nu m_1 \dots m_\nu}) \\ &\times (\bar{\psi}\, \gamma_{[m_1 \dots m_\nu]} \phi_\pm).\end{aligned}$$

We recall (6.4.19) and observe that it defines (anti-)selfdual projections of tensors with components $\bar{\psi}\, \gamma_{[m_1 \dots m_\nu]} \phi_\pm$. So, formula (6.4.79b) agrees with the second line in (6.4.20). □

Identity (6.4.76) not only yields expansions (6.4.78), (6.4.79) and (6.4.80), but also enables one to obtain numerous identities involving spinors, known as *Fierz identities.* As an example, consider an operator in space $\mathcal{F} \otimes \mathcal{F}$, which can always be written as a linear combination of tensor products of two monomials:

$$X_{A_\ell}^{\overline{B}_k} (\Gamma^{\overline{A}_\ell} \otimes \Gamma_{B_k}) \equiv X_{A_\ell}^{\overline{B}_k} (\Gamma^{\overline{A}_\ell})_1 (\Gamma_{B_k})_2 = (\widetilde{\Gamma}^{\overline{B}_k})_1 (\Gamma_{B_k})_2, \qquad (6.4.84)$$

where $X_{A_\ell}^{\overline{B}_k}$ are numerical coefficients, $\widetilde{\Gamma}^{\overline{B}_k} \equiv X_{A_\ell}^{\overline{B}_k} \Gamma^{\overline{A}_\ell}$ and, like in (6.4.77), contraction of multi-index $A_k = \{m_1, \dots, m_k\}$ with $\bar{A}_k = \{m_k, \dots, m_1\}$, and also B_k with $\bar{B}_k$, is understood. We multiply the operator (6.4.84) on the right by permutation operator P_{12}. Using $O_1 P_{12} = P_{12} O_2$ and

$O_2P_{12} = P_{12}O_1$, we find

$$(\widetilde{\Gamma}^{\overline{B}_k})_1(\Gamma_{B_k})_2\, P_{12} = (\Gamma_{B_k})_2\, P_{12}\, (\widetilde{\Gamma}^{\overline{B}_k})_2 = (\widetilde{\Gamma}^{\overline{B}_k})_1\, P_{12}\, (\Gamma_{B_k})_1. \qquad (6.4.85)$$

We now insert formula (6.4.76) for P_{12} into each of the terms in (6.4.85) and get the identities

$$\begin{aligned}(\widetilde{\Gamma}^{\overline{B}_k})_1\, (\Gamma_{B_k})_2\, P_{12} &= \frac{1}{2^\nu}\sum_A (\widetilde{\Gamma}^{\overline{B}_k}\, \Gamma^{\overline{A}})_1(\Gamma_{B_k}\, \Gamma_A)_2 \\ = \frac{1}{2^\nu}\sum_A (\Gamma^{\overline{A}})_1(\Gamma_{B_k}\, \Gamma_A\, \widetilde{\Gamma}^{\overline{B}_k})_2 &= \frac{1}{2^\nu}\sum_A (\widetilde{\Gamma}^{\overline{B}_k}\, \Gamma^{\overline{A}}\, \Gamma_{B_k})_1(\Gamma_A)_2.\end{aligned} \qquad (6.4.86)$$

In component form, identities (6.4.86) can be written as

$$\begin{aligned}(\widetilde{\Gamma}^{\overline{B}_k})_\kappa{}^\beta\, (\Gamma_{B_k})_\alpha{}^\lambda &= \frac{1}{2^\nu}\sum_A (\widetilde{\Gamma}^{\overline{B}_k}\Gamma^{\overline{A}})_\kappa{}^\lambda\, (\Gamma_{B_k}\Gamma_A)_\alpha{}^\beta \\ = \frac{1}{2^\nu}\sum_A (\Gamma^{\overline{A}})_\kappa{}^\lambda\, (\Gamma_{B_k}\Gamma_A\widetilde{\Gamma}^{\overline{B}_k})_\alpha{}^\beta &= \frac{1}{2^\nu}\sum_A (\widetilde{\Gamma}^{\overline{B}_k}\Gamma^{\overline{A}}\Gamma_{B_k})_\kappa{}^\lambda(\Gamma_A)_\alpha{}^\beta.\end{aligned} \qquad (6.4.87)$$

As an example, the component form of the first identity is obtained by using (6.4.76):

$$(\widetilde{\Gamma}^{\overline{B}_k})_\kappa{}^\beta(\Gamma_{B_k})_\alpha{}^\lambda = (\widetilde{\Gamma}^{\overline{B}_k})_\kappa{}^\rho\, (\Gamma_{B_k})_\alpha{}^\sigma \delta_\rho^\beta\, \delta_\sigma^\lambda$$

$$= (\widetilde{\Gamma}^{\overline{B}_k})_\kappa{}^\rho\, (\Gamma_{B_k})_\alpha{}^\sigma \cdot \left[\frac{1}{2^\nu}\sum_A (\Gamma^{\overline{A}})_\rho{}^\lambda(\Gamma_A)_\sigma{}^\beta\right] = \frac{1}{2^\nu}\sum_A (\widetilde{\Gamma}^{\overline{B}_k}\Gamma^{\overline{A}})_\kappa{}^\lambda\, (\Gamma_{B_k}\Gamma_A)_\alpha{}^\beta.$$

By contracting the identities (6.4.87) with components of spinors $(\overline{\Psi}_1)^\kappa, (\Psi_2)_\beta, (\overline{\Psi}_3)^\alpha$ and $(\Psi_4)_\lambda$ we obtain Fierz identities, valid in any even-dimensional space $\mathbb{R}^{p,q}$:

$$\begin{aligned}(\overline{\Psi}_1\widetilde{\Gamma}^{\overline{B}_k}\Psi_2)(\overline{\Psi}_3\Gamma_{B_k}\Psi_4) &= \frac{(-1)^g}{2^\nu}\sum_A (\overline{\Psi}_1\widetilde{\Gamma}^{\overline{B}_k}\Gamma^{\overline{A}}\Psi_4)(\overline{\Psi}_3\Gamma_{B_k}\Gamma_A\Psi_2) \\ &= \frac{(-1)^g}{2^\nu}\sum_A (\overline{\Psi}_1\Gamma^{\overline{A}}\Psi_4)(\overline{\Psi}_3\Gamma_{B_k}\Gamma_A\widetilde{\Gamma}^{\overline{B}_k}\Psi_2) \\ &= \frac{(-1)^g}{2^\nu}\sum_A (\overline{\Psi}_1\widetilde{\Gamma}^{\overline{B}_k}\Gamma^{\overline{A}}\Gamma_{B_k}\Psi_4)(\overline{\Psi}_3\Gamma_A\Psi_2),\end{aligned} \qquad (6.4.88)$$

where, as before, $g = 0$ for commuting spinors, and $g = 1$ for anticommuting spinors. In particular cases $\widetilde{\Gamma}^{\overline{B}_k} \otimes \Gamma_{B_k} = I \otimes I$ and $\widetilde{\Gamma}^{\overline{B}_k} \otimes \Gamma_{B_k} = \gamma^m \otimes \gamma_m$,

Fierz identities read

$$(\overline{\Psi}_1\Psi_2)(\overline{\Psi}_3\Psi_4) = \frac{(-1)^g}{2^\nu}\sum_A(\overline{\Psi}_1\Gamma^{\overline{A}}\Psi_4)(\overline{\Psi}_3\Gamma_A\Psi_2), \qquad (6.4.89)$$

$$(\overline{\Psi}_1\gamma^m\Psi_2)(\overline{\Psi}_3\gamma_m\Psi_4) = \frac{(-1)^g}{2^\nu}\sum_A(\overline{\Psi}_1\Gamma^{\overline{A}}\Psi_4)(\overline{\Psi}_3\gamma_m\Gamma_A\gamma^m\Psi_2). \qquad (6.4.90)$$

Identity (6.4.90) can be simplified by using the following result.

Proposition 6.4.5. *There is an identity*

$$\sum_{m=0}^{N-1}\gamma_m\Gamma_{A_k}\gamma^m = (-1)^k\,(N-2k)\,\Gamma_{A_k}, \qquad (6.4.91)$$

where monomials Γ_{A_k} *are defined in* (6.4.74).

Proof. We recall that $0 \le m_1 < m_2 < \cdots < m_k \le N-1$ for multi-index A_k and write

$$\begin{aligned}\sum_{m=0}^{N-1}\gamma_m\,\Gamma_{A_k}\,\gamma^m = &\sum_{m\notin\{m_1,\ldots,m_k\}}\gamma_m(\gamma_{m_1}\gamma_{m_2}\cdots\gamma_{m_k})\gamma^m\\ &+\sum_{m\in\{m_1,\ldots,m_k\}}\gamma_m(\gamma_{m_1}\gamma_{m_2}\cdots\gamma_{m_k})\gamma^m\\ =&\ (-1)^k\,(N-k)\,\Gamma_{A_k}+\gamma_{m_1}(\gamma_{m_1}\gamma_{m_2}\cdots\gamma_{m_k})\gamma^{m_1}\\ &+\cdots+\gamma_{m_k}(\gamma_{m_1}\gamma_{m_2}\cdots\gamma_{m_k})\gamma^{m_k}.\end{aligned}$$

In each of these terms, except for the first one, we move γ^{m_r} from right to left until it hits γ_{m_r}, and use the identity $\gamma_{m_r}\gamma^{m_r} = I$ (no summation over m_r). The result is

$$\begin{aligned}\sum_{m=0}^{N-1}\gamma_m\,\Gamma_{A_k}\,\gamma^m &= (-1)^k\,(N-k)\,\Gamma_{A_k}+(-1)^{k-1}\gamma_{m_1}\cdots\gamma_{m_k}\\ &\quad+(-1)^{k-2}\gamma_{m_2}(\gamma_{m_1}\gamma_{m_3}\cdots\gamma_{m_k})\\ &\quad+\cdots+(-1)\,\gamma_{m_{k-1}}(\gamma_{m_1}\gamma_{m_2}\cdots\gamma_{m_{k-2}}\gamma_{m_k})\\ &\quad+\gamma_{m_k}(\gamma_{m_1}\gamma_{m_2}\cdots\gamma_{m_{k-1}})\\ &= (-1)^k(N-k)\,\Gamma_{A_k}+k\,(-1)^{k-1}\,\gamma_{m_1}\cdots\gamma_{m_k}\\ &= (-1)^k(N-2k)\Gamma_{A_k},\end{aligned}$$

which coincides with (6.4.91). □

By inserting (6.4.91) in identity (6.4.90), we get the simplified Fierz identity

$$(\overline{\Psi}_1\gamma^m\Psi_2)(\overline{\Psi}_3\gamma_m\Psi_4) = \frac{(-1)^g}{2^\nu}\sum_{k=0}^{N}(-1)^k(N-2k) \times \sum_{A_k}(\overline{\Psi}_1\Gamma^{\overline{A}_k}\Psi_4)(\overline{\Psi}_3\,\Gamma_{A_k}\Psi_2). \quad (6.4.92)$$

Problem 6.4.16. Show that in *four-dimensional case* $N = 4$, Fierz identities (6.4.89) and (6.4.92) read

$$(\overline{\Psi}_1\,\Psi_2)(\overline{\Psi}_3\Psi_4) = \frac{(-1)^g}{4}\,((\overline{\Psi}_1\Psi_4)(\overline{\Psi}_3\Psi_2) + (\overline{\Psi}_1\gamma^m\Psi_4)(\overline{\Psi}_3\gamma_m\Psi_2) - \frac{1}{2}(\overline{\Psi}_1\gamma^{m_1m_2}\Psi_4)(\overline{\Psi}_3\,\gamma_{m_1m_2}\Psi_2) - (\overline{\Psi}_1\gamma^m\gamma_5\Psi_4)(\overline{\Psi}_3\gamma_m\gamma_5\Psi_2) + (\overline{\Psi}_1\gamma_5\Psi_4)(\overline{\Psi}_3\gamma_5\Psi_2))\,, \quad (6.4.93)$$

$$(\overline{\Psi}_1\gamma^m\Psi_2)(\overline{\Psi}_3\gamma_m\Psi_4) = (-1)^g((\overline{\Psi}_1\,\Psi_4)(\overline{\Psi}_3\Psi_2) - \frac{1}{2}(\overline{\Psi}_1\gamma^m\Psi_4)(\overline{\Psi}_3\,\gamma_m\,\Psi_2) - \frac{1}{2}(\overline{\Psi}_1\gamma^m\gamma_5,\Psi_4)(\overline{\Psi}_3\,\gamma_m\,\gamma_5\Psi_2) - (\overline{\Psi}_1\gamma_5\Psi_4)(\overline{\Psi}_3\,\gamma_5\,\Psi_2)), \quad (6.4.94)$$

where $\gamma_{m_1m_2} = \frac{1}{2}(\gamma_{m_1}\gamma_{m_2} - \gamma_{m_2}\gamma_{m_1})$ and $\gamma_5 = (-i)\gamma_0\,\gamma_1\,\gamma_2\,\gamma_3$ (see (6.4.45) with $N = 4$). Hint: derive and make use of identities

$$\sum_{A_4}(\overline{\Psi}_1\Gamma^{\overline{A}_4}\Psi_4)(\overline{\Psi}_3\,\Gamma_{A_4}\,\Psi_2) = (\overline{\Psi}_1\gamma_5\Psi_4)(\overline{\Psi}_3\,\gamma_5\,\Psi_2),$$

$$\sum_{A_3}(\overline{\Psi}_1\Gamma^{\overline{A}_3}\Psi_4)(\overline{\Psi}_3\,\Gamma_{A_3}\,\Psi_2) = -(\overline{\Psi}_1\gamma^m\,\gamma_5\Psi_4)(\overline{\Psi}_3\,\gamma_m\,\gamma_5\,\Psi_2),$$

$$\sum_{A_2}(\overline{\Psi}_1\Gamma^{\overline{A}_2}\Psi_4)(\overline{\Psi}_3\,\Gamma_{A_2}\,\Psi_2) = -\frac{1}{2}(\overline{\Psi}_1\gamma^{m_1m_2}\Psi_4)(\overline{\Psi}_3\,\gamma_{m_1m_2}\,\Psi_2).$$

For Weyl and Majorana spinors, Fierz identities get simplified. In the rest of this section, we consider Minkowski case $p = 1$ and $q = N - 1 = 2\nu - 1$.

1. Chiral Weyl spinors

Let Ψ_K ($K = 1, \ldots, 4$) in (6.4.89), (6.4.92) be chiral Weyl spinors defined in (6.4.39) and (6.4.40). If Ψ_K obeys $(1 \mp \gamma_{N+1})\Psi_K = 0$, then we have for conjugate spinor

$$\Psi_K^\dagger(1 \mp \gamma_{N+1}^\dagger)\mathbf{D}^{-1} = \Psi_K^\dagger\mathbf{D}^{-1}(1 \mp (-1)^q\,\gamma_{N+1}) = \overline{\Psi}_K(1 \pm \gamma_{N+1}) = 0.$$

Therefore, there is an identity

$$\overline{\Psi}_K\,\gamma_{m_1}\gamma_{m_2}\cdots\gamma_{m_r}\Psi_{K'} = 0 \qquad (m_i = 0,\ldots,N-1),$$

for all even r, provided that spinors Ψ_K and $\Psi_{K'}$ have the same chirality, and for all odd r with spinors Ψ_K and $\Psi_{K'}$ having opposite chiralities. Then some of the terms on the right-hand sides of (6.4.89), (6.4.92) vanish.

Problem 6.4.17. Starting from (6.4.94), obtain the following identities valid for $N = 4$:

$$\begin{aligned}&(\overline{\Psi}_1\gamma^m(1+\gamma_5)\Psi_2)(\overline{\Psi}_3\gamma_m(1+\gamma_5)\Psi_4)\\ &\quad= (-1)^{g+1}\,(\overline{\Psi}_1\gamma^m(1+\gamma_5)\Psi_4)(\overline{\Psi}_3\,\gamma_m\,(1+\gamma_5)\Psi_2),\\ &(\overline{\Psi}_1\gamma^m(1+\gamma_5)\Psi_2)(\overline{\Psi}_3\gamma_m(1-\gamma_5)\Psi_4)\\ &\quad= 2(-1)^g(\overline{\Psi}_1\,(1-\gamma_5)\Psi_4)(\overline{\Psi}_3\,(1+\gamma_5)\Psi_2).\end{aligned}$$

Hint: make replacement $\Psi_K \to (1 \pm \gamma_5)\Psi_K$, $\overline{\Psi}_K \to \overline{\Psi}_K(1 \pm \gamma_5)$ in (6.4.94).

2. Majorana spinors

These spinors exist for $N = 2, 4\,\mathrm{mod}(8)$, when $\epsilon = +1$. Making use of (6.4.29) and (6.4.64), we get

$$\begin{aligned}\overline{\Psi}_K\,\gamma_{m_1}\cdots\gamma_{m_r}\Psi_{K'} &= \zeta\Psi_K^{\mathsf{T}}C\,\gamma_{m_1}\cdots\gamma_{m_r}\,\Psi_{K'}\\ &= (-1)^g\,\zeta\,\Psi_{K'}^{\mathsf{T}}\,(C\gamma_{m_1}\cdots\gamma_{m_r})^{\mathsf{T}}\Psi_K\\ &= (-1)^{g+\frac{(r-1)(r-2)}{2}}\overline{\Psi}_{K'}\,\gamma_{m_1}\cdots\gamma_{m_r}\,\Psi_K,\end{aligned} \qquad (6.4.95)$$

where ζ is a phase factor, and $g = 1$ ($g = 0$) for anticommuting (commuting) spinors. The relation (6.4.95) is conveniently written as two identities

$$\overline{\Psi}_K\,\gamma_{m_1}\cdots\gamma_{m_r}\,\Psi_{K'} = (-1)^g\overline{\Psi}_{K'}\,\gamma_{m_1}\cdots\gamma_{m_r}\,\Psi_K\ (r = 4k+1, 4k+2), \qquad (6.4.96)$$

$$\overline{\Psi}_K\,\gamma_{m_1}\cdots\gamma_{m_r}\,\Psi_{K'} = (-1)^{g+1}\overline{\Psi}_{K'}\,\gamma_{m_1}\cdots\gamma_{m_r}\,\Psi_K\ (r = 4k, 4k+3). \qquad (6.4.97)$$

Therefore, one can make a transposition of the Majorana spinors in Fierz identities (6.4.89), (6.4.90). Moreover, if $\Psi_K = \Psi_{K'}$, then the expressions in (6.4.96) vanish for $g = 1$ (anticommuting spinors), while expressions in (6.4.97) vanish for $g = 0$ (commuting spinors), so some terms on the right-hand side of Fierz identities vanish.

Problem 6.4.18. Write explicitly the Fierz identities (6.4.89) and (6.4.92) in the case of Majorana–Weyl spinors (see Remark 1 in Section 6.4.3).

Chapter 7

Solutions to Selected Problems

7.1. Problem 1.3.6 of Section 1.3.3

Show that the integral on the right-hand side of

$$\langle\Phi|\Psi\rangle = \int \left(\prod_{\alpha} d\theta_\alpha d\bar{\theta}_\alpha\right) e^{-\theta_\alpha\bar{\theta}_\alpha} \langle\Phi|\bar{\theta}\rangle \langle\theta|\Psi\rangle$$
$$= \int \left(\prod_{\alpha} d\theta_\alpha d\bar{\theta}_\alpha\right) e^{-\theta_\alpha\bar{\theta}_\alpha} \bar{\Phi}(\bar{\theta})\, \Psi(\theta) \tag{7.1.1}$$

(*see* (1.3.53)) *gives*

$$\langle\Phi|\Psi\rangle = \sum_{k=0}^{\nu} \sum_{\alpha_1<\cdots<\alpha_k} \phi^*_{\alpha_1,\ldots,\alpha_k} \psi_{\alpha_1,\ldots,\alpha_k} \tag{7.1.2}$$

(*see* (1.3.31)).

Consider integral

$$\int d\theta d\bar{\theta}\, e^{-\theta_\alpha\bar{\theta}_\alpha}\, \bar{\theta}_{\beta_\ell}\cdots\bar{\theta}_{\beta_1}\,\theta_{\alpha_1}\cdots\theta_{\alpha_k}, \tag{7.1.3}$$

where $d\theta d\bar{\theta} \equiv \prod_\alpha d\theta_\alpha d\bar{\theta}_\alpha$, $\beta_1 < \cdots < \beta_\ell$ and $\alpha_1 < \cdots < \alpha_k$. In accordance with the definition (1.3.54), integral (7.1.3) does not vanish only if $\ell = k$ and $\beta_i = \alpha_i$ for all $i = 1, \ldots, k$. Therefore, this integral can be written as

(no summation over indices $\alpha_1, \ldots, \alpha_k$, and $1 \le \alpha_1 < \cdots < \alpha_k \le d$)

$$\delta_{\ell,k}\, \delta_{\alpha_1\beta_1} \cdots \delta_{\alpha_k\beta_k} \int d\theta d\bar{\theta}\, e^{\sum\limits_{\alpha} \bar{\theta}_\alpha \theta_\alpha}\, \bar{\theta}_{\alpha_1}\theta_{\alpha_1} \cdots \bar{\theta}_{\alpha_k}\theta_{\alpha_k}$$

$$= \delta_{\ell,k}\, \delta_{\alpha_1\beta_1} \cdots \delta_{\alpha_k\beta_k} \int d\theta d\bar{\theta}\, \frac{1}{(d-k)!} \left(\sum_{\alpha} \bar{\theta}_\alpha \theta_\alpha\right)^{d-k} \bar{\theta}_{\alpha_1}\theta_{\alpha_1} \cdots \bar{\theta}_{\alpha_k}\theta_{\alpha_k}$$

$$= \delta_{\ell,k}\, \delta_{\alpha_1\beta_1} \cdots \delta_{\alpha_k\beta_k}. \tag{7.1.4}$$

Using this formula, we obtain for arbitrary functions (1.3.49) and (1.3.50) that

$$\int d\theta d\bar{\theta}\, e^{-\theta_\alpha \bar{\theta}_\alpha}\, \bar{\Phi}(\bar{\theta})\, \Psi(\theta) = \sum_{k=0}^{d} \sum_{\alpha_1 < \cdots < \alpha_k} \phi^*_{\alpha_1,\ldots,\alpha_k} \psi_{\alpha_1,\ldots,\alpha_k}, \tag{7.1.5}$$

as desired.

7.2. Problem 1.3.7 of Section 1.3.3

Making use of integral representations

$$\det(A) = \int d\theta d\bar{\theta}\ \exp(\bar{\theta}_\alpha A_{\alpha\beta}\theta_\beta) = \frac{1}{d!} \varepsilon_{\alpha_1\cdots\alpha_d}\, A_{\alpha_1\beta_1} \cdots A_{\alpha_d\beta_d}\, \varepsilon_{\beta_1\cdots\beta_d}, \tag{7.2.1}$$

where $d\theta d\bar{\theta} \equiv \prod_\alpha d\theta_\alpha d\bar{\theta}_\alpha$ *and*

$$\int d\theta d\bar{\theta}\ \Psi(\theta, \bar{\theta}) = \int d\theta_d \int d\bar{\theta}_d \cdots \int d\theta_1 \int d\bar{\theta}_1\ \Psi(\theta, \bar{\theta}), \tag{7.2.2}$$

see (1.3.57) *and* (1.3.55), *prove identity* $\det(A) = (\mathrm{Pf}(A))^2$ *for antisymmetric* $(2n \times 2n)$ *matrix* A.

We begin with identity

$$\bar{\theta}_\alpha A_{\alpha\beta}\theta_\beta = \frac{1}{2}\theta^+_\alpha A_{\alpha\beta}\theta^+_\beta - \frac{1}{2}\theta^-_\alpha A_{\alpha\beta}\theta^-_\beta, \tag{7.2.3}$$

where

$$\theta^+_\alpha = \frac{1}{\sqrt{2}}(\bar{\theta}_\alpha + \theta_\alpha), \quad \theta^-_\alpha = \frac{1}{\sqrt{2}}(\bar{\theta}_\alpha - \theta_\alpha). \tag{7.2.4}$$

Under the change of variables (7.2.4), we have (no summation over α)

$$\bar{\theta}_\alpha \theta_\alpha = \theta^-_\alpha\, \theta^+_\alpha \Rightarrow \frac{\partial}{\partial \theta_\alpha} \frac{\partial}{\partial \bar{\theta}_\alpha} = \frac{\partial}{\partial \theta^+_\alpha} \frac{\partial}{\partial \theta^-_\alpha} \Rightarrow d\theta_\alpha\, d\bar{\theta}_\alpha = d\theta^+_\alpha\, d\theta^-_\alpha,$$

where the last step follows from the fact that integral over Grassmann variable θ is equivalent to derivative $\frac{\partial}{\partial\theta}$. So, measure $d\theta d\bar{\theta}$ in integral (7.2.2) reads

$$d\theta_d d\bar{\theta}_d \cdots d\theta_1 d\bar{\theta}_1 = d\theta_d^+ d\theta_d^- \cdots d\theta_1^+ d\theta_1^- = (-1)^n \, d\theta_d^+ \cdots d\theta_1^+ \, d\theta_d^- \cdots d\theta_1^- , \tag{7.2.5}$$

where $d = 2n$. We insert (7.2.3) and (7.2.5) in the expression (7.2.1) for determinant and get

$$\begin{aligned}
\det(A) &= \int d\theta d\bar{\theta} \, \exp(\bar{\theta}_\alpha A_{\alpha\beta}\theta_\beta) \\
&= \int d\theta_d^+ \cdots d\theta_1^+ \int (-1)^n \, d\theta_d^- \cdots d\theta_1^- e^{(\frac{1}{2}\theta_\alpha^+ A_{\alpha\beta}\theta_\beta^+ - \frac{1}{2}\theta_\alpha^- A_{\alpha\beta}\theta_\beta^-)} \\
&= \int d\theta_d^+ \cdots d\theta_1^+ e^{\frac{1}{2}\theta_\alpha^+ A_{\alpha\beta}\theta_\beta^+} \int (-1)^n \, d\theta_d^- \cdots d\theta_1^- e^{-\frac{1}{2}\theta_\alpha^- A_{\alpha\beta}\theta_\beta^-} \\
&= (\mathrm{Pf}(A))^2.
\end{aligned}$$

This is the desired result.

7.3. Problem 2.2.13 of Section 2.2.3

Derive the identities (see (2.2.72)),

$$e^{\lambda e_+} \cdot \Psi(z) = \Psi(e^{\lambda e_+} \, z \, e^{-\lambda e_+}) \, e^{\lambda e_+} \cdot 1, \tag{7.3.1}$$

and (see (2.2.73)),

$$e^{\lambda e_+} \, z \, e^{-\lambda e_+} = \frac{z}{(1+\lambda z)}, \quad e^{\lambda e_+} \cdot 1 = (1+\lambda z)^{2j}. \tag{7.3.2}$$

Making use these identities, obtain equation (see (2.2.71)):

$$e^{a_+ e_+} \cdot \Psi(z) = e^{a_+(-z^2\,\partial_z + 2\,j\,z)} \cdot \Psi(z) = (1 + a_+ z)^{2j} \, \Psi\left(\frac{z}{1+a_+ z}\right). \tag{7.3.3}$$

Meromorphic function $\Psi(z)$ in a vicinity of any point $z = a$ can be expanded in Laurent series $\Psi(z) = \sum_{k=-\infty}^{\infty} c_k \, (z-a)^k$. We have for every term in this series

$$e^{\lambda e_+} \cdot (z-a)^k = e^{\lambda e_+} \, (z-a)^k \, e^{-\lambda e_+} \, e^{\lambda e_+} \cdot 1 = (e^{\lambda e_+} \, z \, e^{-\lambda e_+} - a)^k \, e^{\lambda e_+} \cdot 1,$$

which immediately gives (7.3.1). We remind the reader that dot denotes the action of an operator on a function, e.g., $\partial_z \cdot 1 = 0$, $\partial_z \cdot z^n = n\, z^{n-1}$, etc.

If a linear differential operator has the form $\hat{O} = v(z)\partial_z$, then the action of operator $e^{\hat{O}}$ on product of functions $f(z)$ and $h(z)$ reads

$$e^{\hat{O}}(f \cdot h) = (e^{\hat{O}} f) \cdot (e^{\hat{O}} h). \tag{7.3.4}$$

Indeed, one expands the left-hand side in $\hat{O}$ and obtains

$$\begin{aligned} e^{\hat{O}}(f \cdot h) &= \sum_{n=0}^{\infty} \frac{1}{n!} \hat{O}^n (f \cdot h) \\ &= \sum_{n=0}^{\infty} \frac{1}{n!} \sum_{k=0}^{n} C_n^k (\hat{O}^k f) \cdot (\hat{O}^{n-k} h) \\ &= \sum_{n=0}^{\infty} \sum_{k=0}^{n} \frac{1}{k!} \frac{1}{(n-k)!} (\hat{O}^k f) \cdot (\hat{O}^{n-k} h), \end{aligned}$$

which coincides with the right-hand side of (7.3.4) upon the change of summation variables.

We now turn to identities (7.3.2). We write the differential operator e_+, given by (2.2.64), as follows: $e_+ = -z^2\,\partial_z + 2\,j\,z = z^{2j}\,(-z^2\,\partial_z)\,z^{-2j}$. Then we find

$$\begin{aligned} & e^{\lambda e_+} = z^{2j}\, e^{-\lambda\, z^2\, \partial_z}\, z^{-2j} \\ \Rightarrow\ & e^{\lambda e_+}\, z\, e^{-\lambda e_+} = z^{2j} \big(e^{-\lambda z^2\, \partial_z}\, z\, e^{\lambda z^2\, \partial_z}\big) z^{-2j} = g(\lambda, z), \end{aligned} \tag{7.3.5}$$

where the function $g(\lambda, z)$ is

$$g(\lambda, z) = e^{-\lambda z^2\, \partial_z}\, z\, e^{\lambda z^2\, \partial_z} = e^{-\lambda z^2\, \partial_z} \cdot z = z - \lambda z^2 + \lambda^2 z^3 - \cdots . \tag{7.3.6}$$

The series on the right-hand side can be straightforwardly summed up. However, we make use of more general approach for calculating $g(\lambda, z)$, as this approach is useful in various other circumstances. Consider operator $e^{\lambda v(z)\partial_z}$, where $v(z)$ is a polynomial in z. This operator is an element of diffeomorphism group of complex plane Diff($\mathbb{C}$) and for $v(z) = -z^2$ it coincides with the operator $e^{-\lambda z^2\, \partial_z}$ in (7.3.6). For arbitrary function $\Psi(z)$, we have

$$\exp(\lambda v(z)\partial_z) \cdot \Psi(z) = \Psi\big(e^{\lambda v(z)\partial_z}\, z\, e^{-\lambda v(z)\partial_z}\big)\, e^{\lambda v(z)\partial_z} \cdot 1 = \Psi\big(g(\lambda, z)\big), \tag{7.3.7}$$

where function $g(\lambda, z)$ is defined by (cf. (7.3.6))

$$g(\lambda, z) = e^{\lambda v(z)\partial_z}\, z\, e^{-\lambda v(z)\partial_z} = e^{\lambda v(z)\partial_z} \cdot z. \tag{7.3.8}$$

This definition leads to equations that relate $v(z)$ and $g(\lambda, z)$. Indeed, upon differentiating both sides of (7.3.8) over λ and making use of equality

$$v(z)\,\partial_z \cdot g(\lambda, z) = e^{\lambda v(z)\partial_z}\,(v(z)\,\partial_z)\cdot z = e^{\lambda v(z)\partial_z}\cdot v(z) = v\big(g(\lambda, z)\big),$$

we get

$$\partial_\lambda\, g(\lambda, z) = v(z)\,\partial_z \cdot g(\lambda, z) = v\big(g(\lambda, z)\big), \quad g(0, z) = z.$$

This shows that functions $v(z)$ and $g(\lambda, z)$ are related by integral equation

$$\int_z^{g(\lambda,z)} \frac{dz'}{v(z')} = \lambda. \tag{7.3.9}$$

This equation is known in quantum field theory (in renormalization group context) as *Gell-Mann–Low equation*; it determines "invariant charge" $g(\lambda, z)$ through "beta-function" $v(z)$. Thus, Eq. (7.3.9) demonstrates the relationship between renormalization group and diffeomorphism group Diff($\mathbb{C}$).

We now insert $v(z) = -z^2$ into Eq. (7.3.9) and obtain

$$\frac{1}{g(\lambda, z)} - \frac{1}{z} = \lambda \Rightarrow g(\lambda, z) = \frac{z}{1 + \lambda\, z}. \tag{7.3.10}$$

This proves the first relation of Eq. (7.3.2). The second relation in Eq. (7.3.2) follows from (7.3.7) and (7.3.10):

$$e^{\lambda e_+}\cdot 1 = z^{2j}\, e^{-\lambda\, z^2\,\partial_z}\cdot z^{-2j} = z^{2j}\,(g(\lambda, z))^{-2j} = (1 + \lambda\, z)^{2j}.$$

Finally, by making use of (7.3.1) and (7.3.2), we obtain (7.3.3):

$$e^{a_+ e_+}\,\Psi(z) = \Psi(e^{a_+ e_+}\, z\, e^{-a_+ e_+})\, e^{a_+ e_+}\cdot 1 = \Psi\left(\frac{z}{1 + a_+ z}\right)\,(1 + a_+ z)^{2j}.$$

7.4. Problem 2.3.4 of Section 2.3.2

Show that elements $\mathcal{D}^{(j)}_{m'm}(\alpha, \beta)$ *in* (2.3.50), *namely,*

$$\mathcal{D}^{(j)}_{m'm}(\alpha, \beta) = \sum_k (-1)^k \frac{\sqrt{(j+m)!(j-m)!(j+m')!(j-m')!}}{k!(j-k-m')!(j+m-k)!(k-m+m')!} \times \alpha^{j+m-k}\beta^{k-m+m'}(\beta^*)^k(\alpha^*)^{j-k-m'}, \tag{7.4.1}$$

can be written in compact form (see (2.3.51)),

$$\mathcal{D}^{(j)}_{m'm}(\alpha,\beta) = A^{(j)}_{m'm}(\alpha,\beta)\,(\partial_y)^{j+m'}(y-1)^{j+m}\,(y+1)^{j-m}\Big|_{y=2\alpha\alpha^*-1},$$
$$A^{(j)}_{m'm}(\alpha,\beta) \equiv \frac{2^{m'-j}\sqrt{(j-m')!}}{\sqrt{(j+m')!(j-m)!(j+m)!}}\cdot\alpha^{m'+m}\beta^{m'-m}, \tag{7.4.2}$$

where $\partial_y = \frac{\partial}{\partial y}$ *and* $|\alpha|^2+|\beta|^2=1$.

It follows from (2.3.49) and (2.2.26) that elements $\mathcal{D}^{(j)}_{mm'}(\alpha,\beta)$, given by (7.4.1), can be written as

$$\mathcal{D}^{(j)}_{mm'}(\alpha,\beta) = \frac{\sqrt{(j-m')!}}{\sqrt{(j+m')!}}\left(\frac{\partial}{\partial s}\right)^{j+m'} T^j_m(s',t')\Big|_{s=0,t=1}, \tag{7.4.3}$$

where s' and t' are defined in (2.3.48). We introduce homogeneous variable $z=s/t$ and write the latter formula in the following form:

$$\mathcal{D}^{(j)}_{mm'}(\alpha,\beta) = \frac{\sqrt{(j-m')!}}{\sqrt{(j+m')!(j+m)!(j-m)!}}\left(\frac{\partial}{\partial z}\right)^{j+m'}$$
$$\times(\alpha z-\beta^*)^{j+m}(\beta z+\alpha^*)^{j-m}\Big|_{z=0}. \tag{7.4.4}$$

Let us change variables $z\to y$:

$$y+1 = 2\alpha(\beta z+\alpha^*) \Rightarrow y-1 = 2\beta(\alpha z-\beta^*), \tag{7.4.5}$$

where the second formula is obtained from the first one by using $\alpha\alpha^*+\beta\beta^*=1$. By inserting (7.4.5) in (7.4.4), we obtain (7.4.2).

7.5. Problem 2.3.6 of Section 2.3.3

Making use of formulas (2.3.58), (2.3.66), (2.3.72), (2.3.73) *and* (2.3.74), *prove the addition theorem for spherical harmonics (see* (2.3.77)),

$$\sum_m Y^{(\ell)}_{-m}(-\phi_1,\theta_1)Y^{(\ell)}_m(\phi_2,\theta_2) = \sum_m e^{im(\phi_1+\phi_2)}P^{(\ell)}_{-m}(\cos\theta_1)P^{(\ell)}_m(\cos\theta_2)$$
$$= P^{(\ell)}(\cos\omega_{12}), \tag{7.5.1}$$

where $\cos\omega_{12} = \cos\theta_1\cos\theta_2 - \sin\theta_1\sin\theta_2\cos(\phi_1+\phi_2)$.

Making use of formulas (2.3.58), (2.3.66), (2.3.72) and (2.3.73), we write the following chain of equalities:

$$\begin{aligned}
&\sum_m Y^{(\ell)}_{-m}(-\phi_1,\theta_1)Y^{(\ell)}_m(\phi_2,\theta_2)\\
&\quad=\sum_m \mathcal{D}^{(\ell)}_{-m0}(-\phi_1,\theta_1,0)\mathcal{D}^{(\ell)}_{m0}(\phi_2,\theta_2,0)\\
&\quad=\sum_m \mathcal{D}^{(\ell)}_{0m}(0,\theta_1,\phi_1)\mathcal{D}^{(\ell)}_{m0}(\phi_2,\theta_2,0)\\
&\quad=\sum_m \langle \ell,0|U_{\vec{e}_2}(\theta_1)\cdot U_{\vec{e}_3}(\phi_1)|\ell,\,m\rangle\langle \ell,m|U_{\vec{e}_3}(\phi_2)\cdot U_{\vec{e}_2}(\theta_2)|\ell,\,0\rangle\\
&\quad=\langle \ell,0|U_{\vec{e}_2}(\theta_1)\cdot U_{\vec{e}_3}(\phi_1+\phi_2)\cdot U_{\vec{e}_2}(\theta_2)|\ell,\,0\rangle. \qquad (7.5.2)
\end{aligned}$$

Every element of group $SU(2)$ is uniquely parameterized by Euler angles, so element written in the last expression in (7.5.2) has the following representation:

$$U_{\vec{e}_2}(\theta_1)\cdot U_{\vec{e}_3}(\phi_1+\phi_2)\cdot U_{\vec{e}_2}(\theta_2)=U_{\vec{e}_3}(\alpha_{12})\cdot U_{\vec{e}_2}(\omega_{12})\cdot U_{\vec{e}_3}(\beta_{12}), \qquad (7.5.3)$$

where, as before, matrices $U_{\vec{e}_2}(\theta)$ and $U_{\vec{e}_3}(\phi)$ are defined in (2.3.17), while angles ω_{12},α_{12} and β_{12} are determined by

$$\begin{aligned}
e^{i\frac{\alpha_{12}+\beta_{12}}{2}}\cos\frac{\omega_{12}}{2}&=e^{i\frac{\phi_1+\phi_2}{2}}\cos\frac{\theta_1}{2}\cos\frac{\theta_2}{2}-e^{-i\frac{\phi_1+\phi_2}{2}}\sin\frac{\theta_1}{2}\sin\frac{\theta_2}{2},\\
e^{i\frac{\alpha_{12}-\beta_{12}}{2}}\sin\frac{\omega_{12}}{2}&=e^{i\frac{\phi_1+\phi_2}{2}}\cos\frac{\theta_1}{2}\sin\frac{\theta_2}{2}+e^{-i\frac{\phi_1+\phi_2}{2}}\sin\frac{\theta_1}{2}\cos\frac{\theta_2}{2}.
\end{aligned}$$

The latter relations give

$$\cos\omega_{12}=\cos\theta_1\cos\theta_2-\sin\theta_1\sin\theta_2\cos(\phi_1+\phi_2).$$

We insert the identity (7.5.3) on the right-hand side of (7.5.2) and make use of (2.3.70). The result is

$$\begin{aligned}
\sum_m Y^{(\ell)}_{-m}(-\phi_1,\theta_1)Y^{(\ell)}_m(\phi_2,\theta_2)&=\langle \ell,0|U_{\vec{e}_2}(\omega_{12})|\ell,\,0\rangle\\
&=Y^{(\ell)}_0(\omega_{12})=P^{(\ell)}(\cos\omega_{12}),
\end{aligned}$$

which proves (7.5.1).

We note that identities similar to (7.5.3) play important role in theory of quantum integrable systems and are called “star–triangle equations”.

7.6. Problem 2.3.7 of Section 2.3.3

Let an element $U = \exp(\frac{i}{2}\, t^k\, \sigma_k) = \exp(-t^k\, \tau_k)$ *of group* $SU(2)$ *be close to unit element, i.e., parameters* t^i *be small. Show that functions* ϕ', θ', ψ' *in the relation (see* (2.3.78)*)*

$$U \cdot U_{\vec{e}_3}(\phi)\, U_{\vec{e}_2}(\theta)\, U_{\vec{e}_3}(\psi) = U_{\vec{e}_3}(\phi')\, U_{\vec{e}_2}(\theta')\, U_{\vec{e}_3}(\psi'), \tag{7.6.1}$$

are equal, to the first order in t^i, *to (see* (2.3.80)*)*

$$\begin{aligned} \phi' &= \phi + (t^1\, \cos(\phi) - t^2\, \sin(\phi))\, \cot(\theta) + t^3 + \cdots\,, \\ \theta' &= \theta + t^1\, \sin(\phi) + t^2\, \cos(\phi) + \cdots\,, \\ \psi' &= \psi + (-t^1\, \cos(\phi) + t^2\, \sin(\phi))\, \sin^{-1}(\theta) + \cdots\,, \end{aligned} \tag{7.6.2}$$

where dots denote terms of second and higher orders in variables t^1, t^2, t^3.

We differentiate both sides of (7.6.1) in t^k ($k = 1, 2, 3$) and then set $t^k = 0$ to obtain

$$\begin{aligned} \sigma_k \cdot U(\phi, \theta, \psi) = {}& \partial_k\phi'\, \sigma_3 \cdot U(\phi, \theta, \psi) \\ &+ \partial_k\theta'\, U_{\vec{e}_3}(\phi)\sigma_2 U_{\vec{e}_2}(\theta)\, U_{\vec{e}_3}(\psi) \\ &+ \partial_k\psi'\, U_{\vec{e}_3}(\phi)\, U_{\vec{e}_2}(\theta)\, \sigma_3\, U_{\vec{e}_3}(\psi), \end{aligned} \tag{7.6.3}$$

where $U(\phi, \theta, \psi) = U_{\vec{e}_3}(\phi)\, U_{\vec{e}_2}(\theta)\, U_{\vec{e}_3}(\psi)$, and we use the notations

$$\partial_k\phi' = \left.\frac{\partial\phi'}{\partial t^k}\right|_{t^k=0}, \quad \partial_k\theta' = \left.\frac{\partial\theta'}{\partial t^k}\right|_{t^k=0}, \quad \partial_k\psi' = \left.\frac{\partial\psi'}{\partial t^k}\right|_{t^k=0}.$$

Using formulas (2.3.17), one straightforwardly finds

$$\begin{aligned} U_{\vec{e}_3}(\phi)\, \sigma_2\, U_{\vec{e}_3}(-\phi) &= \cos\phi\, \sigma_2 + \sin\phi\, \sigma_1, \\ U_{\vec{e}_2}(\theta)\, \sigma_3\, U_{\vec{e}_2}(-\theta) &= \cos\theta\, \sigma_3 - \sin\theta\, \sigma_1, \\ U_{\vec{e}_3}(\phi)\, \sigma_1\, U_{\vec{e}_3}(-\phi) &= \cos\phi\, \sigma_1 - \sin\phi\, \sigma_2. \end{aligned} \tag{7.6.4}$$

These enable one to write Eq. (7.6.3) as

$$\begin{aligned} \sigma_k \cdot U(\phi, \theta, \psi) = {}& [\partial_k\phi'\, \sigma_3 + \partial_k\theta'(\cos\phi\, \sigma_2 + \sin\phi\, \sigma_1) \\ &+ \partial_k\psi'(\cos\theta\, \sigma_3 - \sin\theta\, \cos\phi\, \sigma_1 + \sin\theta\, \sin\phi\, \sigma_2)] \cdot U(\phi, \theta, \psi). \end{aligned} \tag{7.6.5}$$

We multiply both sides of this equality by $U(\phi,\theta,\psi)^{-1}$ on the right, and set $k = 1, 2, 3$. Since operators σ_i are linear independent, the resulting system of equations takes the form

$$\begin{pmatrix} \partial_1\phi' & \partial_1\theta' & \partial_1\psi' \\ \partial_2\phi' & \partial_2\theta' & \partial_2\psi' \\ \partial_3\phi' & \partial_3\theta' & \partial_3\psi' \end{pmatrix} \begin{pmatrix} 0 & 0 & 1 \\ \sin\phi & \cos\phi & 0 \\ -\sin\theta\cos\phi & \sin\theta\sin\phi & \cos\theta \end{pmatrix} = \begin{pmatrix} 1 & 0 & 0 \\ 0 & 1 & 0 \\ 0 & 0 & 1 \end{pmatrix} \tag{7.6.6}$$

(Eqs. (7.6.5) are equivalent to (7.6.6), since they are obtained by multiplying (7.6.6) on column $(\sigma_1, \sigma_2, \sigma_3)$). The solution to Eq. (7.6.6) is

$$\begin{pmatrix} \partial_1\phi' & \partial_1\theta' & \partial_1\psi' \\ \partial_2\phi' & \partial_2\theta' & \partial_2\psi' \\ \partial_3\phi' & \partial_3\theta' & \partial_3\psi' \end{pmatrix} = \begin{pmatrix} \cos\phi\cot\theta & \sin\phi & -\cos\phi/\sin\theta \\ -\sin\phi\cot\theta & \cos\phi & \sin\phi/\sin\theta \\ 1 & 0 & 0 \end{pmatrix}, \tag{7.6.7}$$

which yields formulas (7.6.2).

7.7. Problem 2.3.11 of Section 2.3.3

Show that Laplacian (see (2.3.86)) *on sphere* S^3 *(group* $SU(2)$) *in Euler angle parameterization* (2.3.23), *given by*

$$\Delta_{SU(2)} = \left(\frac{1}{\sin^2\theta}\partial_\phi^2 + \cot\theta\partial_\theta + \partial_\theta^2\right) + \frac{1}{\sin^2\theta}\left(\partial_\psi - 2\cos\theta\,\partial_\phi\right)\partial_\psi, \tag{7.7.1}$$

is equal, modulo factor 4, to the standard Laplacian on sphere S^3, *written in spherical angle parameterization* (2.3.13):

$$\begin{aligned} \Delta'_{SU(2)} &= \frac{1}{\sin^2\chi}\partial_\chi(\sin^2\chi\cdot\partial_\chi) + \frac{1}{\sin^2\chi}\frac{1}{\sin\vartheta}\partial_\vartheta(\sin\vartheta\cdot\partial_\vartheta) \\ &\quad + \frac{1}{\sin^2\chi}\frac{1}{\sin^2\vartheta}\partial_\varphi^2. \end{aligned} \tag{7.7.2}$$

The precise relation is $4\cdot\Delta_{SU(2)} = \Delta'_{SU(2)}$.

We write an arbitrary vector $\vec{x} = (x_1, \ldots, x_n) \in \mathbb{R}^n$ of length r as $\vec{x} = r\vec{\mathsf{x}}$, where $\vec{\mathsf{x}} = (\mathsf{x}_1, \ldots, \mathsf{x}_n)$ is unit vector in $\mathbb{R}^n$ that defines a point on unit sphere S^{n-1}. Components of vector $\vec{\mathsf{x}}$ are usually parameterized by hyperspherical angles (as we discuss in Section I-7.2 of the first book for any number of dimensions). The following relations hold:

$$\hat{\partial}_i = (\hat{\partial}_i x_k)\,\partial_k = r\,\partial_i, \qquad \partial_r = (\partial_r x_i)\,\partial_i = \mathsf{x}_i\,\partial_i \Rightarrow r\,\partial_r = x_i\,\partial_i, \tag{7.7.3}$$

where we introduce shorthand notations $\partial_r = \frac{\partial}{\partial r}$, $\hat{\partial}_i = \frac{\partial}{\partial \mathsf{x}_i}$ and $\partial_i = \frac{\partial}{\partial x_i}$. Consider now standard differential realization for generators of algebra $so(n)$:

$$L_{ij} = x_i\partial_j - x_j\partial_i = \mathsf{x}_i\hat{\partial}_j - \mathsf{x}_j\hat{\partial}_i. \tag{7.7.4}$$

Note that operators L_{ij} depend only on variables x_i that parameterize sphere S^{n-1} (and do not depend on radial variable r) and define tangent vector fields to unit sphere S^{n-1} at a point $\vec{\mathsf{x}} \in S^{n-1}$. Let us calculate quadratic Casimir operator of algebra $so(n)$ in differential parameterization (7.7.4):

$$\begin{aligned} L_{ij}^2 &= (x_i\partial_j - x_j\partial_i)^2 = 2(x_i^2\partial_j^2 + x_i\partial_i - n(x_i\partial_i) - x_ix_j\partial_j\partial_i) \\ &= 2(x_i^2\partial_j^2 + (2-n)\,(x_i\partial_i) - (x_i\partial_i)^2) \\ &= 2(r^2\partial_j^2 + (2-n)\,(r\partial_r) - (r\partial_r)^2), \end{aligned} \tag{7.7.5}$$

where we use the second equality in (7.7.3). The latter relation gives the formula for Laplace operator in space $\mathbb{R}^n$ in spherical coordinates r and x_i:

$$\partial_j^2 = \frac{1}{r^2}\,(r\partial_r)^2 + (n-2)\frac{1}{r}\,\partial_r + \frac{1}{2r^2}L_{ij}^2 = \partial_r^2 + (n-1)\frac{1}{r}\,\partial_r + \frac{1}{2r^2}L_{ij}^2. \tag{7.7.6}$$

Note that the dependence on "angular" variables x_i in (7.7.6) is contained in the Casimir operator L_{ij}^2, which is naturally called Laplace operator[a] on sphere S^{n-1}.

Let us first derive formula (7.7.2) for Laplace operator on sphere S^3, using parameterization of S^3 in terms of usual spherical angles (2.3.13)

[a]This operator is constructed as quadratic combination of tangent vectors L_{ij} on sphere S^{n-1} and it commutes with all these tangent vectors.

in $\mathbb{R}^4$:

$$\mathsf{x}_0 = \cos\chi, \quad \mathsf{x}_1 = \sin\chi\,\sin\vartheta\,\cos\varphi, \\ \mathsf{x}_2 = \sin\chi\,\sin\vartheta\,\sin\varphi, \quad \mathsf{x}_3 = \sin\chi\,\cos\vartheta. \tag{7.7.7}$$

The relations between derivatives with respect to angles and $\hat{\partial}_k = \partial/\partial\mathsf{x}_k$ are

$$\begin{aligned} \partial_\chi &= -s_\chi\,\hat{\partial}_0 + c_\chi\,c_\vartheta\,\hat{\partial}_3 + c_\chi\,s_\vartheta\,c_\varphi\,\hat{\partial}_1 + c_\chi\,s_\vartheta\,s_\varphi\,\hat{\partial}_2, \\ \partial_\vartheta &= -s_\chi\,s_\vartheta\,\hat{\partial}_3 + s_\chi\,c_\vartheta\,c_\varphi\,\hat{\partial}_1 + s_\chi\,c_\vartheta\,s_\varphi\,\hat{\partial}_2, \\ \partial_\varphi &= -s_\chi\,s_\vartheta\,s_\varphi\,\hat{\partial}_1 + s_\chi\,s_\vartheta\,c_\varphi\,\hat{\partial}_2, \end{aligned} \tag{7.7.8}$$

where, as usual, we use shorthand notations $s_\vartheta := \sin\vartheta$, $c_\varphi := \cos\varphi$, etc. Making use of (7.7.4), (7.7.7) and (7.7.8), we obtain the representation of generators of algebra $so(4)$ in terms of spherical angles:

$$\begin{gathered} (x_1\partial_2 - x_2\partial_1) = \partial_\varphi, \quad (x_3\partial_1 - x_1\partial_3) = c_\varphi\,\partial_\vartheta - \frac{c_\vartheta s_\varphi}{s_\vartheta}\partial_\varphi, \\ (x_3\partial_2 - x_2\partial_3) = s_\varphi\,\partial_\vartheta + \frac{c_\vartheta c_\varphi}{s_\vartheta}\partial_\varphi, \quad (x_0\partial_3 - x_3\partial_0) = c_\vartheta\,\partial_\chi - \frac{c_\chi s_\vartheta}{s_\chi}\partial_\vartheta, \\ (x_0\partial_1 - x_1\partial_0) = s_\vartheta c_\varphi\partial_\chi + \frac{c_\chi\,c_\varphi\,c_\vartheta}{s_\chi}\partial_\vartheta - \frac{c_\chi\,s_\varphi}{s_\chi\,s_\vartheta}\partial_\varphi, \\ (x_0\partial_2 - x_2\partial_0) = s_\varphi\,s_\vartheta\,\partial_\chi + \frac{c_\chi\,c_\vartheta\,s_\varphi}{s_\chi}\partial_\vartheta + \frac{c_\chi c_\varphi}{s_\chi s_\vartheta}\partial_\varphi. \end{gathered}$$

Finally, Laplace operator on sphere S^3, in terms of spherical angles, reads

$$\begin{aligned} L_{ij}^2 &= \partial_\varphi^2 + \left(c_\varphi\,\partial_\vartheta - \frac{c_\vartheta\,s_\varphi}{s_\vartheta}\,\partial_\varphi\right)^2 + \left(s_\varphi\,\partial_\vartheta + \frac{c_\vartheta\,c_\varphi}{s_\vartheta}\,\partial_\varphi\right)^2 \\ &\quad + (c_\vartheta\,\partial_\chi - \frac{c_\chi s_\vartheta}{s_\chi}\partial_\vartheta)^2 + \left(s_\vartheta c_\varphi\partial_\chi + \frac{c_\chi\,c_\varphi\,c_\vartheta}{s_\chi}\partial_\vartheta - \frac{c_\chi\,s_\varphi}{s_\chi\,s_\vartheta}\partial_\varphi\right)^2 \\ &\quad + \left(s_\varphi\,s_\vartheta\,\partial_\chi + \frac{c_\chi\,c_\vartheta\,s_\varphi}{s_\chi}\partial_\vartheta + \frac{c_\chi c_\varphi}{s_\chi s_\vartheta}\partial_\varphi\right)^2 \\ &= \partial_\varphi^2 + \partial_\vartheta^2 + \cot\vartheta\,\partial_\vartheta + \cot^2\vartheta\,\partial_\varphi^2 + \partial_\chi^2 + 2\frac{c_\chi}{s_\chi}\partial_\chi + \frac{c_\chi^2 s_\vartheta^2}{s_\chi^2}\partial_\vartheta^2 \\ &\quad + \left(\frac{c_\chi\,c_\varphi\,c_\vartheta}{s_\chi}\partial_\vartheta - \frac{c_\chi\,s_\varphi}{s_\chi\,s_\vartheta}\partial_\varphi\right)^2 + \left(\frac{c_\chi\,c_\vartheta\,s_\varphi}{s_\chi}\partial_\vartheta + \frac{c_\chi c_\varphi}{s_\chi s_\vartheta}\partial_\varphi\right)^2 \\ &= \partial_\vartheta^2 + \cot\vartheta\,\partial_\vartheta + \frac{1}{s_\vartheta^2}\,\partial_\varphi^2 + \partial_\chi^2 + 2\frac{c_\chi}{s_\chi}\partial_\chi + \frac{c_\chi^2}{s_\chi^2}\partial_\vartheta^2 + \frac{c_\chi^2\,c_\vartheta}{s_\chi^2\,s_\vartheta}\partial_\vartheta + \frac{c_\chi^2}{s_\chi^2 s_\vartheta^2}\partial_\varphi^2, \end{aligned} \tag{7.7.9}$$

which agrees with $\Delta'_{SU(2)}$ in (7.7.2).

Let us now find Laplace operator on sphere S^3 in Euler angle parameterization (2.3.23):

$$\begin{aligned} \mathrm{x}_0 &= \cos\frac{\theta}{2}\,\cos\frac{\phi+\psi}{2}, \quad \mathrm{x}_1 = \sin\frac{\theta}{2}\,\sin\frac{\phi-\psi}{2}, \\ \mathrm{x}_2 &= \sin\frac{\theta}{2}\,\cos\frac{\phi-\psi}{2}, \quad \mathrm{x}_3 = \cos\frac{\theta}{2}\,\sin\frac{\varphi+\psi}{2}. \end{aligned} \tag{7.7.10}$$

We set $\tau = \frac{\theta}{2}$ and $\phi_\pm = \frac{\phi\pm\psi}{2}$ and write

$$\partial_\tau = r(-s_\tau\, c_{\phi_+}\,\partial_0 + c_\tau\, s_{\phi_-}\,\partial_1 + c_\tau\, c_{\phi_-}\,\partial_2 - s_\tau\, s_{\phi_+}\,\partial_3), \tag{7.7.11}$$

$$\begin{aligned} \partial_{\phi_+} &= r(-c_\tau\, s_{\phi_+}\,\partial_0 + c_\tau\, c_{\phi_+}\,\partial_3) = -x_3\,\partial_0 + x_0\,\partial_3, \\ \partial_{\phi_-} &= r(s_\tau\, c_{\phi_-}\,\partial_1 - s_\tau\, s_{\phi_-}\,\partial_2) = x_2\,\partial_1 - x_1\,\partial_2. \end{aligned} \tag{7.7.12}$$

We combine the relations (7.7.11) and (7.7.12), and make use of (7.7.10) to obtain

$$\begin{aligned} \frac{s_\tau\, s_{\phi_-}}{c_\tau}\,\partial_{\phi_+} &= s_{\phi_+}\,(x_0\partial_1 - x_1\partial_0) + c_{\phi_+}\,(x_1\partial_3 - x_3\partial_1), \\ s_{\phi_-}\,\partial_\tau + \frac{c_\tau\, c_{\phi_-}}{s_\tau}\,\partial_{\phi_-} &= c_{\phi_+}(x_0\,\partial_1 - x_1\,\partial_0) - s_{\phi_+}(x_1\,\partial_3 - x_3\,\partial_1), \\ \frac{s_\tau\, c_{\phi_-}}{c_\tau}\,\partial_{\phi_+} &= s_{\phi_+}\,(x_0\,\partial_2 - x_2\,\partial_0) + c_{\phi_+}\,(x_2\,\partial_3 - x_3\,\partial_2), \\ c_{\phi_-}\,\partial_\tau - \frac{c_\tau\, s_{\phi_-}}{s_\tau}\,\partial_{\phi_-} &= c_{\phi_+}\,(x_0\,\partial_2 - x_2\,\partial_0) - s_{\phi_+}\,(x_2\,\partial_3 - x_3\,\partial_2), \end{aligned} \tag{7.7.13}$$

or, in matrix form,

$$\begin{pmatrix} \dfrac{s_\tau\, s_{\phi_-}}{c_\tau}\,\partial_{\phi_+} \\ s_{\phi_-}\,\partial_\tau + \dfrac{c_\tau\, c_{\phi_-}}{s_\tau}\,\partial_{\phi_-} \end{pmatrix} = \begin{pmatrix} c_{\phi_+} & s_{\phi_+} \\ -s_{\phi_+} & c_{\phi_+} \end{pmatrix} \begin{pmatrix} L_{13} \\ L_{01} \end{pmatrix}, \tag{7.7.14}$$

$$\begin{pmatrix} \dfrac{s_\tau\, c_{\phi_-}}{c_\tau}\,\partial_{\phi_+} \\ c_{\phi_-}\,\partial_\tau - \dfrac{c_\tau\, s_{\phi_-}}{s_\tau}\,\partial_{\phi_-} \end{pmatrix} = \begin{pmatrix} c_{\phi_+} & s_{\phi_+} \\ -s_{\phi_+} & c_{\phi_+} \end{pmatrix} \begin{pmatrix} L_{23} \\ L_{02} \end{pmatrix}. \tag{7.7.15}$$

Since the right-hand sides of (7.7.14) and (7.7.15) contain matrices of two-dimensional rotations, we immediately write

$$
\begin{aligned}
L_{13}^2 + L_{01}^2 &= \left(\frac{s_\tau\, s_{\phi_-}}{c_\tau}\right)^2 \partial_{\phi_+}^2 + \left(s_{\phi_-}\,\partial_\tau + \frac{c_\tau\, c_{\phi_-}}{s_\tau}\,\partial_{\phi_-}\right)^2 \\
&\quad - \frac{s_\tau\, s_{\phi_-}}{c_\tau}\left(s_{\phi_-}\,\partial_\tau + \frac{c_\tau\, c_{\phi_-}}{s_\tau}\,\partial_{\phi_-}\right), \\
L_{23}^2 + L_{02}^2 &= \left(\frac{s_\tau\, c_{\phi_-}}{c_\tau}\right)^2 \partial_{\phi_+}^2 + \left(c_{\phi_-}\,\partial_\tau - \frac{c_\tau\, s_{\phi_-}}{s_\tau}\,\partial_{\phi_-}\right)^2 \\
&\quad - \frac{s_\tau\, c_{\phi_-}}{c_\tau}\left(c_{\phi_-}\,\partial_\tau - \frac{c_\tau\, s_{\phi_-}}{s_\tau}\,\partial_{\phi_-}\right).
\end{aligned}
\tag{7.7.16}
$$

The last terms on the right-hand sides of (7.7.16) are due to the fact that matrices of two-dimensional rotations in (7.7.14) and (7.7.15) do not commute with two-dimensional vectors involving derivatives ∂_{ϕ_+}. Finally, we use (7.7.12) and (7.7.16) to write Laplace operator on S^3 in Euler angle parameterization:

$$
\begin{aligned}
L_{ij}^2 &= \partial_{\phi_+}^2 + \partial_{\phi_-}^2 + \left(\frac{s_\tau\, s_{\phi_-}}{c_\tau}\right)^2 (\partial_{\phi_+})^2 + \left(s_{\phi_-}\,\partial_\tau + \frac{c_\tau\, c_{\phi_-}}{s_\tau}\,\partial_{\phi_-}\right)^2 \\
&\quad + \left(\frac{s_\tau\, c_{\phi_-}}{c_\tau}\right)^2 (\partial_{\phi_+})^2 + \left(c_{\phi_-}\,\partial_\tau - \frac{c_\tau\, s_{\phi_-}}{s_\tau}\,\partial_{\phi_-}\right)^2 \\
&\quad - \frac{s_\tau\, c_{\phi_-}}{c_\tau}\left(c_{\phi_-}\,\partial_\tau - \frac{c_\tau\, s_{\phi_-}}{s_\tau}\,\partial_{\phi_-}\right) - \frac{s_\tau\, s_{\phi_-}}{c_\tau}\left(s_{\phi_-}\,\partial_\tau + \frac{c_\tau\, c_{\phi_-}}{s_\tau}\,\partial_{\phi_-}\right) \\
&= \frac{1}{c_\tau^2}\partial_{\phi_+}^2 + \left(1 + \frac{c_\tau^2}{s_\tau^2}\right)\partial_{\phi_-}^2 + \partial_\tau^2 + \left(\frac{c_\tau}{s_\tau} - \frac{s_\tau}{c_\tau}\right)\partial_\tau \\
&= 4\left(\frac{1}{\sin^2\theta}\partial_\phi^2 + \cot\theta\,\partial_\theta + \partial_\theta^2\right) + \frac{4}{\sin^2\theta}\left(\partial_\psi - 2\cos\theta\,\partial_\phi\right)\partial_\psi = 4\Delta_{SU(2)},
\end{aligned}
\tag{7.7.17}
$$

where we change variables

$$
\phi_\pm = \frac{1}{2}(\phi \pm \psi), \quad \tau = \frac{\theta}{2} \Rightarrow \partial_{\phi_\pm} = \partial_\phi \pm \partial_\psi, \quad \partial_\tau = 2\partial_\theta,
$$

and operator $\Delta_{SU(2)}$ is defined in (7.7.1). Since the left-hand sides of (7.7.9) and (7.7.17) coincide, their right-hand sides prove the identity $\Delta_{SU(2)} = \Delta'_{SU(2)}/4$.

7.8. Problem 2.5.4 of Section 2.5.2

6-j symbol (2.5.20) is represented by tetrahedron in Fig. 2.5.6, which we reproduce here:

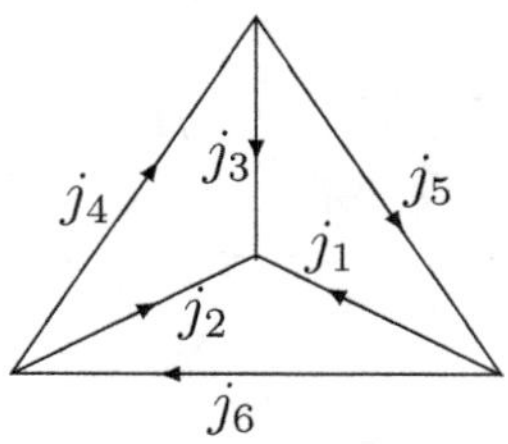

$$= \begin{Bmatrix} j_1 & j_2 & j_3 \\ j_4 & j_5 & j_6 \end{Bmatrix}$$

Fig. 7.8.1. Diagrammatic representation of 6-j symbol.

Using this representation, show that 6-j symbol is symmetric under certain permutations of spins j_k:

(1) *6-j symbol is invariant under permutations of columns:*

$$\begin{Bmatrix} j_1 & j_2 & j_3 \\ j_4 & j_5 & j_6 \end{Bmatrix} = \begin{Bmatrix} j_2 & j_1 & j_3 \\ j_5 & j_4 & j_6 \end{Bmatrix} = \begin{Bmatrix} j_1 & j_3 & j_2 \\ j_4 & j_6 & j_5 \end{Bmatrix}; \tag{7.8.1}$$

(2) *6-j symbol is invariant under permutation of the first and second rows in the first two columns:*

$$\begin{Bmatrix} j_1 & j_2 & j_3 \\ j_4 & j_5 & j_6 \end{Bmatrix} = \begin{Bmatrix} j_4 & j_5 & j_3 \\ j_1 & j_2 & j_6 \end{Bmatrix}. \tag{7.8.2}$$

Show that all compositions of transformations (7.8.1) and (7.8.2) make a group of symmetries of tetrahedron which is isomorphic to symmetric group S_4.

Symmetry of 6-j symbol under cyclic permutation of columns,

$$\begin{Bmatrix} j_1 & j_2 & j_3 \\ j_4 & j_5 & j_6 \end{Bmatrix} = \begin{Bmatrix} j_3 & j_1 & j_2 \\ j_6 & j_4 & j_5 \end{Bmatrix} = \begin{Bmatrix} j_2 & j_3 & j_1 \\ j_5 & j_6 & j_4 \end{Bmatrix}, \tag{7.8.3}$$

follows from invariance of diagram in Fig. 7.8.1 under its rotations by angles $2\pi/3$ and $4\pi/3$ in a plane around the vertex (j_1, j_2, j_3). The invariant contraction of 3-j symbols does not change under this rotation, while spins j_k are permuted in a way given by (7.8.3). Let us represent the diagram for 6-j symbol shown in Fig. 7.8.1 as a tetrahedron with vertices (A, B, C, D) as

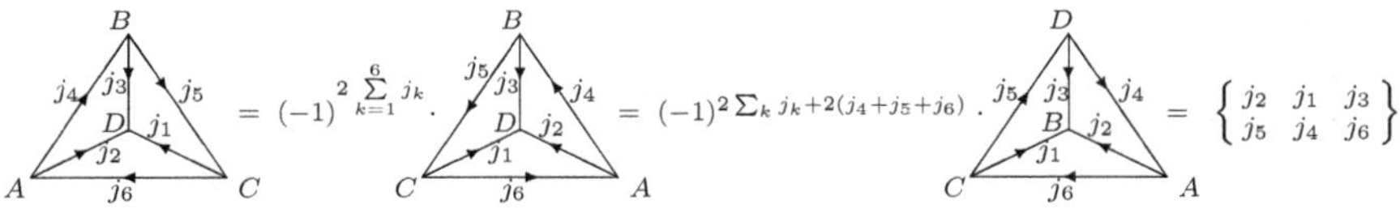

Fig. 7.8.2. Derivation of symmetry (7.8.1) of 6-j symbol.

shown on the left part of Fig. 7.8.2. Then rotations of diagram in Fig. 7.8.1 by angles $2\pi/3$ and $4\pi/3$ are equivalent to rotations of tetrahedron on the left part of Fig. 7.8.2 by the same angles around the axis passing through vertex D and center of the opposite side (A, B, C). Now, we make mirror reflection of the diagram for 6-j symbol on the left part of Fig. 7.8.2 against the vertical axis, which is equivalent to reflection of the tetrahedron against the plane passing through the edge (B, D) and center of the opposite edge (A, C). As a result, we obtain the invariant convolution of 3-j symbols shown in the second diagram in Fig. 7.8.2. The latter convolution is equal to the original 6-j symbol, modulo sign factor which accounts for the change of orientation of all vertices under reflection (see Fig. 2.5.1).

We reverse the directions of edges denoted by j_4, j_5 and j_6 in the second diagram of Fig. 7.8.2 (in accordance with the rules shown in Fig. 2.5.7), and make use of identity $2\sum_k j_k + 2(j_4 + j_5 + j_6) = 0 \bmod(2)$. In this way we obtain, on the right part of Fig. 7.8.2, 6-j symbol with permuted columns. This proves the first equality in (7.8.1). The second equality in (7.8.1) follows from the first one and cyclic symmetry (7.8.3).

Relation (7.8.2) is obtained by moving one of the edges of the triangle (A, B, C) in the first diagram in Fig. 7.8.3, (say, edge (A, C)) over the opposite vertex B. In this way, we obtain the second diagram in Fig. 7.8.3. Comparison of these two diagrams shows that the pertinent invariant contractions of 3-j symbols coincide. The third diagram in Fig. 7.8.3 is obtained from the second one by reversing the directions of edges j_2, j_3, j_5 and j_6. The sign factor there is equal to 1 due to identity $2(j_2+j_3+j_5+j_6) \bmod(2) = 4j_1 \bmod(2) = 0 \bmod(2)$. Now, tetrahedra shown in the first and third diagrams in Fig. 7.8.3 are related to each other by rotation by angle π around the symmetry axis passing through the centers of edges (A, C) and (B, D). Compositions of tetrahedron transformations shown in Figs. 7.8.2 and 7.8.3 plus rotations about the axis that passes through the vertex D and center of the opposite edge (A, B, C) enable one to interchange the vertices (A, B, C, D) in an arbitrary way. Therefore, these transformations, and the pertinent 6-j symbol transformations (7.8.1),

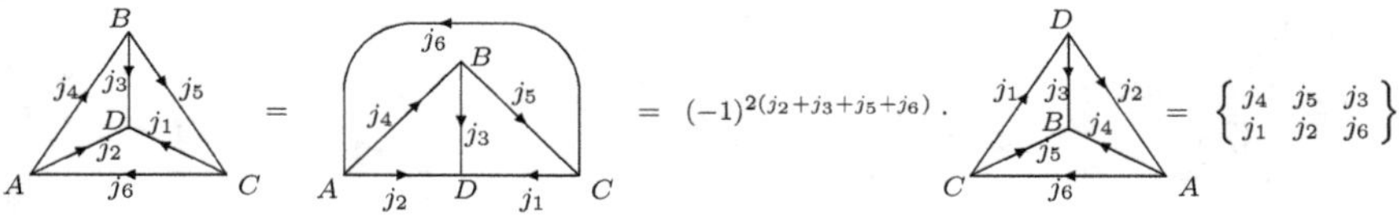

Fig. 7.8.3. Derivation of symmetry (7.8.2) of 6-j symbol.

(7.8.2) and (7.8.3) make tetrahedron symmetry group which is isomorphic to S_4.

7.9. Problem 2.5.15 of Section 2.5.4

Derive the expression (see (2.5.76))

$$\det(I_d + w \cdot z) = \left[\mathrm{Pf}\begin{pmatrix} z & -I_d \\ I_d & w \end{pmatrix}\right]^2 \tag{7.9.1}$$

for determinant $\det(I_d + w \cdot z)$. *Obtain the equality (see* (2.5.78))

$$\mathrm{Pf}\begin{pmatrix} z & -I_d \\ I_d & w \end{pmatrix} = (-1)^{d(d+1)/2}\left(1 + \sum_{k=1}^{[\frac{d}{2}]} \frac{(-1)^k}{(d-2k)!}\, \widetilde{w}_{i_1\cdots i_{d-2k}} \cdot \widetilde{z}_{i_1\cdots i_{d-2k}}\right), \tag{7.9.2}$$

where $[\frac{d}{2}]$ *is integer part of* $\frac{d}{2}$ *and*

$$\widetilde{z}_{i_1\cdots i_{d-2k}} = \frac{1}{2^k\, k!} \sum_{j_1,\ldots,j_{2k}} \varepsilon_{i_1\cdots i_{d-2k}\, j_1 j_2 j_3 j_4 \cdots j_{2k}} \left(z_{j_1 j_2} z_{j_3 j_4} \cdots z_{j_{2k-1} j_{2k}}\right).$$

Let A, B, C and D be four $d \times d$ matrices, where matrix A is invertible. Then the following identity holds:

$$\det\begin{pmatrix} A & B \\ C & D \end{pmatrix} = \det(A)\, \det(D - C \cdot A^{-1} \cdot B). \tag{7.9.3}$$

It is a consequence of decomposition of a block $2d \times 2d$ matrix

$$\begin{pmatrix} A & B \\ C & D \end{pmatrix} = \begin{pmatrix} A & 0 \\ C & I_d \end{pmatrix} \begin{pmatrix} I_d & A^{-1} \cdot B \\ 0 & D - C \cdot A^{-1} \cdot B \end{pmatrix},$$

where 0 is zero $d \times d$ block. We set $A = -I_d$, $B = z$, $C = w$ and $D = I_d$, then (7.9.3) gives

$$\det\begin{pmatrix} -I_d & z \\ w & I_d \end{pmatrix} = \det(-I_d)\, \det(I_d + w \cdot z) = (-1)^d\, \det(I_d + w \cdot z).$$

Permuting the columns on the left-hand side, we obtain identity (7.9.1):

$$\det(I_d + w \cdot z) = \det \begin{pmatrix} z & -I_d \\ I_d & w \end{pmatrix} = \left[\mathrm{Pf} \begin{pmatrix} z & -I_d \\ I_d & w \end{pmatrix} \right]^2 . \tag{7.9.4}$$

Here we make use of the fact that $Z(z,w) = \left(\begin{smallmatrix} z & -I_d \\ I_d & w \end{smallmatrix} \right)$ is even-dimensional antisymmetric matrix, for which one can define Pfaffian (1.3.58), (2.5.77) (see Section 1.3.3 and Section I-1.2.2 of the first book), such that $\det(Z) = [\mathrm{Pf}(Z)]^2$. To obtain identity (7.9.2) one makes use of the definition (1.3.58) of Pfaffian. We leave the derivation of (7.9.2) to the reader.

7.10. Problem 3.3.1 of Section 3.3

Prove that expansion of simple roots $\alpha^{(i)}$ in fundamental weights $\lambda_{(k)}$ reads

$$\alpha^{(i)} = \sum_{k=1}^{r} K_{ik}\, \lambda_{(k)}, \tag{7.10.1}$$

where

$$K_{ik} = \frac{2(\alpha^{(i)},\, \alpha^{(k)})}{(\alpha^{(k)})^2} \tag{7.10.2}$$

are elements of Cartan matrix (see Section I-6.2.2 *of the first book). Show that order of group Λ/Λ_r is equal to* $\det ||K_{ik}||$, *where Λ and Λ_r are weight and root lattices, respectively.*

Making use of the formula (7.10.2) and definition of fundamental weights (3.3.2), we obtain for any simple root $\alpha^{(j)}$

$$\left(\alpha^{(i)} - \sum_{k=1}^{r} K_{ik}\, \lambda_{(k)},\ \alpha^{(j)} \right) = 0. \tag{7.10.3}$$

Since simple roots $\{\alpha^{(j)}\}$ make complete system of vectors in r-dimensional root space $\mathfrak{V}_r$, and metric g_{ij} in space $\mathfrak{V}_r$ is Euclidean, the relation (7.10.3) yields (7.10.1). It follows from the latter equality that fundamental weights $\{\lambda_{(k)}\}$ also form complete system of vectors in r-dimensional root space $\mathfrak{V}_r$.

The proof of the relation $\mathrm{ord}(\Lambda/\Lambda_r) = \det(K)$, given below, is valid for any lattice Λ and its sublattice Λ_r defined by an arbitrary integer-valued matrix $||K_{ik}||$ with $\det(K) > 0$. We associate groups of shifts to lattices Λ and Λ_r in a standard way. Elements of group Λ are shifts in $\mathfrak{V}_r$

along vectors

$$\lambda_{(\vec{\mathsf{n}})} = \sum_{k=1}^{r} \mathsf{n}_k \lambda_{(k)}, \quad \mathsf{n}_k \in \mathbb{Z},$$

while elements of subgroup $\Lambda_r \subset \Lambda$ are shifts along vectors

$$\alpha_{(\vec{\mathsf{m}})} = \sum_{i=1}^{r} \mathsf{m}_i \alpha^{(i)} = \sum_{i,k=1}^{r} \mathsf{m}_i K_{ik}\, \lambda_{(k)}, \quad \mathsf{m}_k \in \mathbb{Z}.$$

In what follows, we say for brevity that vectors $\lambda_{(\vec{\mathsf{n}})}$ are elements of group Λ, while vectors $\alpha_{(\vec{\mathsf{m}})}$ are elements of subgroup Λ_r. Coset of subgroup Λ_r with respect to element $\lambda_{(\vec{\mathsf{n}})} \in \Lambda$ has elements

$$\lambda_{(\vec{\mathsf{n}}')} = \lambda_{(\vec{\mathsf{n}})} + \alpha_{(\vec{\mathsf{m}})} = \sum_{k=1}^{r} \mathsf{n}_k \lambda_{(k)} + \sum_{i,k=1}^{r} \mathsf{m}_i K_{ik}\, \lambda_{(k)}, \quad \forall \mathsf{m}_i \in \mathbb{Z}. \tag{7.10.4}$$

Since fundamental weights make basis in $\mathfrak{V}_r$, it follows from (7.10.4) that elements $\lambda_{(\vec{\mathsf{n}})}$ and $\lambda_{(\vec{\mathsf{n}}')}$ belong to the same coset if and only if

$$\mathsf{n}'_k = \mathsf{n}_k + \sum_{i=1}^{r} \mathsf{m}_i K_{ik}, \quad \mathsf{m}_i \in \mathbb{Z}. \tag{7.10.5}$$

Order of group Λ/Λ_r equals the number of *different* cosets. We write (7.10.5) as

$$\vec{e}_k \mathsf{n}'_k = \vec{e}_k \mathsf{n}_k + \mathsf{m}_i K_{ik} \vec{e}_k, \tag{7.10.6}$$

where vectors $\vec{e}_i$ $(i = 1, \ldots, r)$ make basis in $\mathbb{R}^r$ and summation over repeated indices is assumed. Consider lattice Λ' in $\mathbb{R}^r$ whose sites are determined by vectors $\vec{e}_k \mathsf{n}_k$, $\forall \mathsf{n}_k \in \mathbb{Z}$. Then sites corresponding to vectors $\vec{\epsilon}_k \mathsf{m}_k$, $\forall \mathsf{m}_k \in \mathbb{Z}$, with

$$\vec{\epsilon}_k = \sum_{i=1}^{r} \vec{e}_i K_{ik}, \tag{7.10.7}$$

form sublattice $\Lambda'_r \subset \Lambda'$. By construction, groups Λ/Λ_r and Λ'/Λ'_r are isomorphic, so $\mathrm{ord}(\Lambda/\Lambda_r) = \mathrm{ord}(\Lambda'/\Lambda'_r)$. Consider now r-dimensional parallelepiped which is defined as the set of points

$$\vec{x} = x_k\, \vec{\epsilon}_k \in \mathbb{R}^r, \quad x_k \in [0,1] \ \forall k. \tag{7.10.8}$$

This parallelepiped is called *fundamental region* of lattice Λ'_r. Clearly, $\mathrm{ord}(\Lambda'/\Lambda'_r)$ is equal to the number N of sites of lattice Λ' inside this

$$||K_{ij}|| = \begin{pmatrix} 6 & 1 \\ 2 & 4 \end{pmatrix},$$

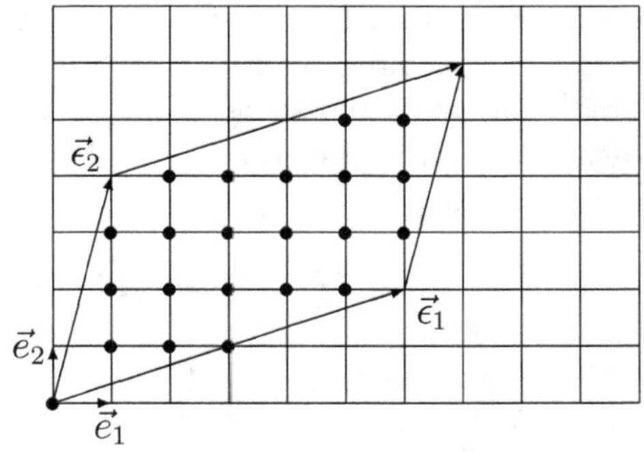

Fig. 7.10.1. Fundamental parallelepiped (parallelogram) of lattice Λ'_r defined by vectors $\vec{\epsilon}_k$ in (7.10.7). Number N of sites of lattice Λ' with basis $\vec{e}_k$ inside this parallelogram, which define cosets belonging to Λ'/Λ'_r, is equal to the area $\det(K)$ of the fundamental region of Λ'_r.

r-dimensional parallelepiped. This number includes the site at origin and also part of sites that sit on faces of parallelepiped (7.10.8). As an example, site $\vec{e}_k \mathsf{n}_k \in \Lambda'$ that sits on some face must have one of coordinates, say x_i, equal to either zero or one. If the site $\vec{e}_k \mathsf{n}_k \in \Lambda'$ sits on face and has $x_i = 0$, then there exists site $\vec{e}_k \mathsf{n}_k + \vec{\epsilon}_i \in \Lambda'$ which also sits on face of fundamental region, but now with $x_i = 1$. The site $\vec{e}_k \mathsf{n}_k + \vec{\epsilon}_i \in \Lambda'$ should not be counted in N, since it belongs to the same coset as the element $\vec{e}_k \mathsf{n}_k \in \Lambda'$; see Fig. 7.10.1.

We choose basis $\vec{e}_k$ and its normalization in such a way that the volume of the parallelepiped composed of these vectors (volume of fundamental region of lattice Λ') is equal to 1. Let us show that in this case, the number of sites N equals the volume of fundamental region of lattice Λ'_r, i.e., the volume of parallelepiped (7.10.8), which, as usual, equals the absolute value of determinant of matrix whose columns are equal to coordinates of vectors $\vec{\epsilon}_k$ in (7.10.7); see Fig. 7.10.1. Thus, we would like to show that $\text{ord}(\Lambda/\Lambda_r) = N = |\det ||K_{ik}|||$.

We first consider the case when the integer-valued matrix $||K_{ik}||$ is diagonal, $K_{ik} = n_i \delta_{ij}$, where $n_i \in \mathbb{Z}$. Then it is clear that $\Lambda'/\Lambda'_r = \mathbb{Z}_{n_1} \times \cdots \times \mathbb{Z}_{n_r}$ and we have

$$\text{ord}(\Lambda'/\Lambda'_r) = N = |n_1 \cdots n_r| = |\det ||K_{ik}|||. \tag{7.10.9}$$

Now, let the integer-valued matrix $||K_{ik}||$ be arbitrary. Under the change of basis of lattice Λ',

$$\vec{e}_i \to \vec{e}_i^{\,\prime} = \vec{e}_k \, C_{ki} \Rightarrow \vec{e}_i = \vec{e}_k^{\,\prime} \, C_{ki}^{-1}, \tag{7.10.10}$$

with integer-valued matrix C such that $\det(C) = \pm 1$ (matrix C^{-1} is therefore also integer-valued), lattice Λ' is transformed into itself. Indeed, sites of lattices with bases $\vec{e}_i{}'$ and $\vec{e}_i$ coincide, since these bases are related to each other via integer-valued matrices. Likewise, the basis transformation of lattice Λ'_r,

$$\vec{\epsilon}_i \to \vec{\epsilon}_i{}' = \vec{\epsilon}_k\, C'_{ki} \Rightarrow \vec{\epsilon}_i = \vec{\epsilon}_k{}'\, C'^{-1}_{ki}, \tag{7.10.11}$$

with integer-valued matrix C' such that $\det(C') = \pm 1$, also transforms lattice Λ'_r into itself. Under basis transformations (7.10.10) and (7.10.11), basis vectors $\vec{e}_i{}'$ and $\vec{\epsilon}_k{}'$ of lattices Λ' and Λ'_r are related by $\vec{\epsilon}_k{}' = \vec{e}_i{}' K'_{ik}$, where

$$K'_{ik} = C^{-1}_{i\ell}\, K_{\ell m}\, C'_{mk}. \tag{7.10.12}$$

Now, there exist integer-valued matrices C and C' with $\det(C) = \pm 1$ and $\det(C') = \pm 1$. such that the transformation (7.10.12) leads to diagonalization of matrix K: $K'_{ij} = n_i \delta_{ij}$. Let us show that this is indeed the case (see also Ref. [29]). The action on the right in (7.10.12) by integer-valued matrix C' with $\det(C') = \pm 1$ on matrix K is reduced to three elementary operations:

(1) addition, to any column of matrix K, of another column multiplied by an integer;
(2) permutation of any two columns of matrix K;
(3) multiplication of any column of matrix K by (-1).

Likewise, the action in (7.10.12) of integer-valued matrix C^{-1}, with $\det(C^{-1}) = \pm 1$ on the left on matrix K is reduced to the same elementary operations, but now for rows. Consider now the first row in integer-valued matrix K, where K_{1j} is the maximal (in absolute value) element. If this row has an element K_{1k} such that $K_{1j} = q \cdot K_{1k}$, with $q \in \mathbb{Z}$, then we subtract kth column multiplied by q from jth column and obtain $K'_{1j} = 0$. If the element K_{1k} divides K_{1j} with remainder, i.e., $K_{1j} = q \cdot K_{1k} + r$, then subtracting kth column multiplied by q from jth column, we obtain $K'_{1j} = r < |K_{1k}|$, so the element K_{1j} is no longer the largest in the absolute value. Continuing with this procedure, and permuting columns, we obtain in the end that all elements in the first row vanish, except for one element K'_{11} (otherwise matrix K is degenerate). Applying the same procedure to the first column of matrix K, we make all but one elements of the first column vanish, so $r \times r$ matrix K takes the form

$$\left(\begin{array}{c|ccc} K'_{11} & 0 & \cdots & 0 \\ \hline 0 & & & \\ \vdots & & ||K'_{ab}|| & \\ 0 & & & \end{array}\right),$$

where $||K'_{ab}||$ is nondegenerate integer-valued $(r-1) \times (r-1)$-matrix. Continuing this procedure further, we make the whole matrix K' diagonal, $K'_{ij} = n_i \delta_{ij}$. We recall formula (7.10.9) for diagonal matrix K'_{ij}. Since transformations (7.10.12) yield $\det(K) = \pm \det(K')$, the formula (7.10.9) is valid for nondiagonal matrix K_{ij} as well. For Cartan matrix K, we have $\det(K) > 0$, and formula (7.10.9) gives the desired result $\mathrm{ord}(\Lambda/\Lambda_r) = \det(K)$.

7.11. Problem 3.4.5 of Section 3.4.2

Prove that vector (*see* (3.4.33))

$$|\lambda_{(f)}\rangle = \frac{1}{f!} \sum_{\sigma \in S_f} (-1)^{p(\sigma)} \epsilon_{(\sigma_1)} \otimes \epsilon_{(\sigma_2)} \otimes \cdots \otimes \epsilon_{(\sigma_f)} = \epsilon_{(1)} \wedge \epsilon_{(2)} \wedge \cdots \wedge \epsilon_{(f)}, \tag{7.11.1}$$

obeys (*see* (3.4.34)):

$$T^{\otimes f}_{(1)}\left(E_{\alpha^{(i)}}\right) |\lambda_{(f)}\rangle = 0 \tag{7.11.2}$$

for all simple roots $\alpha^{(i)}$ *of algebra* $s\ell(r+1, \mathbb{C})$.

Making use of antisymmetrizer (3.4.28), one writes vector (7.11.1) as follows:

$$\begin{aligned} |\lambda_{(f)}\rangle &= \frac{1}{f!} \sum_{\sigma \in S_f} (-1)^{p(\sigma)} \epsilon_{(\sigma_1)} \otimes \epsilon_{(\sigma_2)} \otimes \cdots \otimes \epsilon_{(\sigma_f)} \\ &= A \cdot (\epsilon_{(1)} \otimes \epsilon_{(2)} \otimes \cdots \otimes \epsilon_{(f)}). \end{aligned} \tag{7.11.3}$$

Note that if any two of the vectors $\vec{v}_1, \vec{v}_2, \ldots, \vec{v}_f \in \mathcal{V}_{r+1}$ coincide, $\vec{v}_k = \vec{v}_j$ for $k \neq j$, then the definition (3.4.28) of antisymmetrizer A ensures that

$$A \cdot (\vec{v}_1 \otimes \vec{v}_2 \otimes \cdots \otimes \vec{v}_f)|_{\vec{v}_k = \vec{v}_j} = 0. \tag{7.11.4}$$

We make use of the formula (3.4.24) for operator $T^{\otimes f}_{(1)}(E_{\alpha^{(i)}})$, formula (3.1.7) for defining representation of generator $E_{\alpha^{(i)}}$ and equalities (3.4.29) and (7.11.3) to rewrite the left-hand side of (7.11.2):

$$\begin{aligned}T_{(1)}^{\otimes f}(E_{\alpha^{(i)}})\,|\lambda_{(f)}\rangle &= T_{(1)}^{\otimes f}(E_{\alpha^{(i)}})\cdot A\cdot(\epsilon_{(1)}\otimes\cdots\otimes\epsilon_{(f)})\\ &= A\cdot\left(\sum_{k=1}^{f}(I_{r+1})^{\otimes(k-1)}\otimes e_{i,i+1}\otimes(I_{r+1})^{\otimes(f-k)}\right)\\ &\quad\times(\epsilon_{(1)}\otimes\cdots\otimes\epsilon_{(f)}). \qquad (7.11.5)\end{aligned}$$

The action of matrix unit e_{ik} on basis vector $\epsilon_{(m)}\in\mathcal{V}_{r+1}$ is given by

$$e_{ik}\cdot\epsilon_{(m)}=\delta_{km}\,\epsilon_{(i)}. \qquad (7.11.6)$$

Then, if $i+1>f$, the right-hand side of (7.11.5) vanishes, and we obtain the desired result (7.11.2). If $i<f$, then we use (7.11.6) on the right-hand side of Eq. (7.11.5) and find

$$\begin{aligned}A\cdot&\left(\sum_{k=1}^{f}I_{r+1}^{\otimes(k-1)}\otimes e_{i,i+1}\otimes I_{r+1}^{\otimes(f-k)}\right)\\ &\times(\epsilon_{(1)}\otimes\cdots\otimes\epsilon_{(i)}\otimes\epsilon_{(i+1)}\otimes\cdots\otimes\epsilon_{(f)})\\ &=A\cdot(\epsilon_{(1)}\otimes\cdots\otimes\epsilon_{(i)}\otimes\epsilon_{(i)}\otimes\cdots\otimes\epsilon_{(f)})=0,\end{aligned}$$

where we use the relation (7.11.4) to obtain the last equality.

7.12. Problems 4.3.2 and 4.3.3 of Section 4.3.1

Prove the following identity for total symmetrizer

$$\begin{aligned}E_{P[r]}&=\frac{1}{r!}(1+\sigma_{r-1}+\sigma_{r-2}\sigma_{r-1}+\sigma_{r-3}\sigma_{r-2}\sigma_{r-1}+\cdots+\sigma_1\cdots\sigma_{r-1})\\ &\quad\times(1+\sigma_{r-2}+\sigma_{r-3}\sigma_{r-2}+\cdots+\sigma_1\cdots\sigma_{r-2})\cdots(1+\sigma_1)\\ &=\frac{1}{r!}\,Y(\mathsf{T}_{[r]}). \qquad (7.12.1)\end{aligned}$$

Prove that total antisymmetrizer $E_{Q[1^r]}$ is obtained from total symmetrizer $E_{P[r]}$ by replacement $\sigma_i\to-\sigma_i$. Check that $Y(\mathsf{T}_{[r]})$ in (7.12.1) is a sum of all $r!$ elements of symmetric group S_r.

For proving that element $E_{P[r]}$ given by (7.12.1) is total symmetrizer in $\mathbb{C}[S_r]$, we have to show that the following relations hold:

$$\sigma_k\cdot E_{P[r]}=E_{P[r]}\cdot\sigma_k=E_{P[r]},\quad\forall k=1,\ldots,r-1. \qquad (7.12.2)$$

Normalization factor $1/r!$ in (7.12.1) is chosen in such a way that once (7.12.2) is satisfied, then $E_{P[r]}^2=E_{P[r]}$. Note that Eq. (7.12.1) can be written as recurrence relation

$$E_{P[r]} = \frac{1}{r}(1 + \sigma_{r-1} + \sigma_{r-2}\sigma_{r-1} + \sigma_{r-3}\sigma_{r-2}\sigma_{r-1} + \cdots + \sigma_1 \cdots \sigma_{r-1}) \\ \times E_{P[r-1]}. \tag{7.12.3}$$

Hereafter, we write 1 instead of unity e. Consider now elements

$$J_{m\to r} = \sigma_m \cdots \sigma_{r-2} \cdot \sigma_{r-1} \cdot \sigma_{r-2} \cdots \sigma_m = \sigma_m \cdot J_{m+1\to r} \cdot \sigma_m \\ (m = 1, \ldots, r-2), \\ J_{r-1\to r} = \sigma_{r-1}, \quad J_{r\to r} = 1.$$

The sum

$$Z_r = \sum_{m=1}^{r} J_{m\to r} = 1 + \sigma_{r-1} + \sigma_{r-2}\sigma_{r-1}\sigma_{r-2} + \cdots + \sigma_1 \cdot \sigma_{r-2}\sigma_{r-1}\sigma_{r-2} \cdot \sigma_1$$

belongs to center of subalgebra $\mathbb{C}[S_{r-1}] \subset \mathbb{C}[S_r]$, i.e., $[Z_r, \sigma_k] = 0$ ($\forall k = 1, \ldots, r-2$), but $[Z_r, \sigma_{r-1}] \neq 0$. This fact follows from equalities

$$\sigma_k \cdot J_{m\to r} = J_{m\to r} \cdot \sigma_k, \quad \forall k \neq m, m-1, r-1, \\ (J_{k\to r} + J_{k+1\to r}) \cdot \sigma_k = \sigma_k \cdot (J_{k\to r} + J_{k+1\to r}), \tag{7.12.4}$$

which in their turn follow from defining relations (4.3.13). As an example, let us check the first equality in (7.12.4) for $k = m+1$ (for other $k = m+2, \ldots, r-2$ the check is similar):

$$\sigma_{m+1} \cdot J_{m\to r} = \underline{\sigma_{m+1} \cdot (\sigma_m \cdot \sigma_{m+1}} \cdot \sigma_{m+2} \cdots \sigma_{r-2}\sigma_{r-1}\sigma_{r-2} \cdots \sigma_m) \\ = \sigma_m \cdot \sigma_{m+1} \cdot \sigma_m \cdot \sigma_{m+2} \cdots \sigma_{r-1} \cdots \sigma_m \\ = \sigma_m \cdots \sigma_{r-1} \cdots \underline{\sigma_m \cdot \sigma_{m+1} \cdot \sigma_m} = J_{m\to r} \cdot \sigma_{m+1},$$

where we underline combinations which are transformed by making use of braid relations (see (4.3.13)):

$$\sigma_{m+1} \cdot \sigma_m \cdot \sigma_{m+1} = \sigma_m \cdot \sigma_{m+1} \cdot \sigma_m. \tag{7.12.5}$$

We now prove (7.12.2) by induction. For $r = 2$, we have

$$E_{P[2]} = \frac{1}{2}(1 + \sigma_1) \Rightarrow \sigma_1 \cdot E_{P[2]} = E_{P[2]} \cdot \sigma_1 = \frac{1}{2}(\sigma_1 + \sigma_1^2) = E_{P[2]}.$$

Let relations (7.12.2) hold for $E_{P[r-1]}$:

$$\sigma_k \cdot E_{P[r-1]} = E_{P[r-1]} \cdot \sigma_k = E_{P[r-1]}, \quad \forall k = 1, \ldots, r-2. \tag{7.12.6}$$

We would like to show that relations (7.12.2) are valid for $E_{P[r]}$. By making use of the definition of element Z_r and formula (7.12.6), we write formula (7.12.3) as follows:

$$E_{P[r]} = \frac{1}{r}\, Z_r \;\cdot E_{P[r-1]} = \frac{1}{r}\, E_{P[r-1]} \;\cdot Z_r, \tag{7.12.7}$$

where the second equality is a consequence of the fact that Z_r is central element in $\mathbb{C}[S_{r-1}]$ and $E_{P[r-1]} \in \mathbb{C}[S_{r-1}]$. Note that the second equality in (7.12.7) can be written in the form similar to (7.12.3):

$$\begin{aligned} E_{P[r]} &= \frac{1}{r}\, E_{P[r-1]} \cdot (1 + \sigma_{r-1} + \sigma_{r-1}\sigma_{r-2} + \sigma_{r-1}\sigma_{r-2}\sigma_{r-3} \\ &\quad + \cdots + \sigma_{r-1} \cdots \sigma_1). \end{aligned} \tag{7.12.8}$$

Since $E_{P[r-1]}$ obeys (7.12.6), we obtain, in accordance with (7.12.7), the relation (7.12.2) for $k = 1, \ldots, r-2$:

$$\sigma_k {\cdot} E_{P[r]} = \frac{1}{r}\, \sigma_k {\cdot} E_{P[r-1]} {\cdot} Z_r = E_{P[r]}, \quad E_{P[r]} {\cdot} \sigma_k = \frac{1}{r}\, Z_r {\cdot} E_{P[r-1]} {\cdot} \sigma_k = E_{P[r]}.$$

Finally, for $k = (r-1)$, we employ representation (7.12.3), braid relations (7.12.5) and equality $\sigma_{r-2} \cdot E_{P[r-1]} = E_{P[r-1]}$ to obtain

$$\begin{aligned} \sigma_{r-1} \cdot E_{P[r]} &= \frac{1}{r}\,\sigma_{r-1} \cdot \big(1 + \sigma_{r-1} + (1 + \sigma_{r-3} + \sigma_{r-4}\sigma_{r-3} \\ &\quad + \cdots + \sigma_1 \cdots \sigma_{r-3})\sigma_{r-2}\sigma_{r-1}\big) \times E_{P[r-1]} \\ &= \frac{1}{r}\,\big(\sigma_{r-1} + 1 + (1 + \sigma_{r-3} + \cdots + \sigma_1 \cdots \sigma_{r-3})\sigma_{r-2}\sigma_{r-1}\sigma_{r-2}\big) \\ &\quad \times E_{P[r-1]} = E_{P[r]}. \end{aligned}$$

The relation $E_{P[r]} \cdot \sigma_{r-1} = E_{P[r]}$ is proven in a similar way; in that case one makes use of representation (7.12.8).

Let us substitute $\sigma_i \to -\sigma_i$ in identity (7.12.2). Since this change is automorphism of algebra $\mathbb{C}(S_r)$ defined by (4.3.13), identity (7.12.2) transforms into an identity, which now reads

$$\sigma_k \cdot E_{Q[r]} = E_{Q[r]} \cdot \sigma_k = -E_{Q[r]}, \quad \forall k = 1, \ldots, r-1, \tag{7.12.9}$$

where element $E_{Q[r]}$ is obtained from element $E_{P[r]}$ by substitution $\sigma_i \to -\sigma_i$. In view of (7.12.9), element $E_{Q[r]}$ replaces any neighboring transposition σ_k by (-1), so it replaces any odd permutation by (-1) and even permutation by $(+1)$. Thus, element $E_{Q[r]}$ is total antisymmetrizer.

Finally, we show that $Y(\mathsf{T}_{[r]})$ in (7.12.1) is a sum of all $r!$ elements of group S_r (sum of all permutations of r numbers $(1, 2, \ldots, r)$). To this end, we write (7.12.3) as follows:

$$\begin{aligned} Y(\mathsf{T}_{[r]}) = (1 &+ \sigma_{r-1} + \sigma_{r-2}\sigma_{r-1} + \sigma_{r-3}\sigma_{r-2}\sigma_{r-1} \\ &+ \cdots + \sigma_1 \cdots \sigma_{r-1}) \cdot Y(\mathsf{T}_{[r-1]}), \end{aligned} \tag{7.12.10}$$

and again use induction. Let element $Y(\mathsf{T}_{[r-1]})$ be a sum of all $(r-1)!$ permutations of $(r-1)$ numbers $(1, 2, 3, \ldots, r-1)$. The idea then is that all elements in factor $(1 + \sigma_{r-1} + \sigma_{r-2}\sigma_{r-1} + \cdots + \sigma_1 \cdots \sigma_{r-1})$ in product (7.12.10), except for unity 1, transpose the last number r with one of the first $(r-1)$ numbers $(1, 2, \ldots, r-1)$. We leave it to the reader to complete the proof.

7.13. Problem 4.3.10 of Section 4.3.2

Prove the identity (*see* (4.3.62))

$$F(h_1, \ldots, h_m) = \sum_{i=1}^{m} F(h_1, \ldots, h_i - 1, \ldots, h_m), \tag{7.13.1}$$

for functions (*see* (4.3.57))

$$F(h_1, \ldots, h_m) = \frac{r!}{(h_1! \cdots h_m!)} \prod_{i<j} (h_i - h_j). \tag{7.13.2}$$

We insert explicit expression (7.13.2) for function $F(h_1, \ldots, h_m)$ in (7.13.1) and observe that we have to prove that

$$r\, \Delta(h_1, \ldots, h_m) = \sum_{i=1}^{m} h_i\, \Delta(h_1, \ldots, h_i - 1, \ldots, h_m), \tag{7.13.3}$$

where the total number of boxes r is expressed in terms of hook lengths h_i (for boxes from the first column of diagram λ) via $r = \sum_i h_i - \frac{m(m-1)}{2}$, and we introduce notation

$$\Delta(h_1, \ldots, h_m) = \prod_{i<j} (h_i - h_j),$$

for Vandermonde determinant (3.5.38). The identity (7.13.3) indeed holds, as it follows from more general identity

$$\left(\sum_{i=1}^{m} h_i + t\, C_m^2 \right) \Delta(h_1, \ldots, h_m) = \sum_{i=1}^{m} h_i\, \Delta(h_1, \ldots, h_i + t, \ldots, h_m), \tag{7.13.4}$$

where t is a parameter, and $C_m^2 \equiv \frac{m(m-1)}{2}$. To prove (7.13.4) we note that its left-hand side is first-order polynomial in t, and the first two terms of the expansion of the right-hand side in t coincide with the left-hand side:

$$\sum_{i=1}^{m} h_i\,\Delta(h_1,\dots,h_i+t,\dots,h_m)|_{t=0} = \left(\sum_{i=1}^{m} h_i\right)\Delta(h_1,\dots,h_m),$$

$$\partial_t \sum_{i=1}^{m} h_i\,\Delta(h_1,\dots,h_i+t,\dots,h_m)|_{t=0}$$

$$= \left(\sum_{i=1}^{m} h_i\,\partial_i\right)\Delta(h_1,\dots,h_m) = \frac{(m-1)m}{2}\,\Delta(h_1,\dots,h_m).$$

In the third line here we use the fact that $\Delta(h_1,\dots,h_m)$ is a homogeneous polynomial in h_i of order $\frac{(m-1)m}{2} \equiv C_m^2$, so the action on it of the operator $\sum_i h_i\,\partial_i$ is equivalent to multiplication of $\Delta(h_1,\dots,h_m)$ by C_m^2. Thus, it remains to be shown that the terms proportional to t^n with $n = 2, 3, \dots$ in the expansion of the right-hand side of (7.13.4), vanish. We show this by writing the right-hand side of (7.13.4) as contour integral

$$\sum_{i=1}^{m} h_i\,\Delta(h_1,\dots,h_i+t,\dots,h_m)$$

$$= \left(\sum_{i=1}^{m} h_i \prod_{m\neq i} \frac{(h_i+t-h_m)}{(h_i-h_m)}\right)\Delta(h_1,\dots,h_m)$$

$$= \left(\oint_{\mathcal{C}} \frac{dz}{2\pi i}\,\frac{z}{t} \prod_{k=1}^{m} \frac{z+t-h_k}{z-h_k}\right)\Delta(h_1,\dots,h_m), \qquad (7.13.5)$$

where contour $\mathcal{C}$ surrounds all points $z = h_k$ $(k = 1,\dots,m)$. Now, we expand in t the product on the right-hand side of (7.13.5):

$$\prod_{k=1}^{m} \frac{z+t-h_k}{z-h_k} = 1 + t\sum_{k=1}^{m}\frac{1}{z-h_k} + \frac{t^2}{2}\sum_{k\neq j}\frac{1}{(z-h_k)(z-h_j)}$$

$$+ \frac{t^3}{3!}\sum_{k\neq j\neq \ell}\frac{1}{(z-h_k)(z-h_j)(z-h_\ell)} + \cdots$$

$$= \sum_{n=0}^{\infty}\frac{t^n}{n!}\sum_{k_1,\dots,k_n}{}' \frac{1}{(z-h_{k_1})(z-h_{k_2})\cdots(z-h_{k_n})}. \qquad (7.13.6)$$

Here prime in the notation $\sum'$ means that the summation runs over all $k_1, \ldots, k_n$ such that $k_i \neq k_j$ for $i \neq j$. We insert (7.13.6) in (7.13.5) and obtain

$$\sum_{i=1}^{m} h_i \, \Delta(\ldots, h_i + t, \ldots) = \left(a_{-1}\frac{1}{t} + a_0 + a_1 t + a_2 t^2 + \cdots\right) \Delta(h_1, \ldots, h_m), \tag{7.13.7}$$

where the coefficients a_i are evaluated as contour integrals:

$$\begin{aligned} a_{-1} &= \oint_C z \, \frac{dz}{2\pi i} = 0, \quad a_0 = \oint_C \left(\sum_{k=1}^{m} \frac{z}{z - h_k}\right) \frac{dz}{2\pi i} = \sum_{k=1}^{m} h_k, \\ a_1 &= \frac{1}{2} \oint_C \left(\sum_{k \neq j} \frac{z}{(z - h_k)(z - h_j)}\right) \frac{dz}{2\pi i} = \frac{1}{2} \sum_{k \neq j} 1 = \frac{(m-1)m}{2}, \\ a_2 &= \frac{1}{3!} \oint_C \left(\sum_{k \neq j \neq \ell} \frac{z}{(z - h_k)(z - h_j)(z - h_\ell)}\right) \frac{dz}{2\pi i} = 0, \quad a_n = 0 \quad \forall n > 2. \end{aligned} \tag{7.13.8}$$

The coefficient a_2 as well as other coefficients a_n, $\forall n > 2$, vanish, since the integration contour can be moved to infinity $z = \infty$ where integrand decay as $1/z^n$. We use (7.13.8) in (7.13.7) and obtain (7.13.4).

7.14. Problem 4.3.11 of Section 4.3.2

Let $p(r)$ be the number of different Young diagrams $\lambda \vdash r$. Prove that the generating function for numbers $p(r)$ is given by Euler formula (see, e.g., [41])

$$\sum_{r=0}^{\infty} p(r) \, x^r = \prod_{k=1}^{\infty} \frac{1}{1 - x^k}. \tag{7.14.1}$$

We expand every factor, $\frac{1}{1-x^k} = (1 + x^k + x^{2k} + x^{3k} + \cdots)$, and write the right-hand side of (7.14.1) as product

$$\begin{aligned} &(1 + x^1 + x^{1\cdot 2} + x^{1\cdot 3} + x^{1\cdot 4} + \cdots)(1 + x^2 + x^{2\cdot 2} + \cdots) \\ &\quad \times (1 + x^3 + x^{3\cdot 2} + \cdots)(1 + x^4 + \cdots) \cdots . \end{aligned} \tag{7.14.2}$$

To begin with, let us show that coefficient of x^4 in the expansion of the product (7.14.2) equals $p(4) = 5$, which is the number of all Young diagrams with four boxes, [4], [3, 1], [2^2], [$2, 1^2$], [1^4]. Monomial x^4 comes only from the first four factors $\frac{1}{(1-x^k)}$ with $k = 1, 2, 3, 4$, which are explicitly written in (7.14.2). Diagram [4] with one row corresponds to choice of term x^4 in the fourth factor in (7.14.2) with trivial terms (equal to 1) in the first three factors; diagram [3, 1] corresponds to the product of x^3 from the third factor in (7.14.2) and $x^{1\cdot 1}$ from the first factor; diagram [2^2] corresponds to the choice of term $x^{2\cdot 2}$ in the second factor in (7.14.2), finally diagrams [$2, 1^2$] and [1^4] correspond to the choice of terms $x^{1\cdot 2} \cdot x^2$ and $x^{1\cdot 4}$, respectively.

This example shows that every Young diagram $\lambda = [\lambda_1^{n_1}, \lambda_2^{n_2}, \ldots, \lambda_m^{n_m}] \vdash r$, with $\lambda_1 > \lambda_2 > \cdots > \lambda_m > 0$ can be associated with the monomial x^r on the right-hand side of (7.14.1) that emerges as a product of factor $x^{n_1 \cdot \lambda_1}$ from expansion of $\frac{1}{(1-x^{\lambda_1})}$, factor $x^{n_2 \cdot \lambda_2}$ from expansion of $\frac{1}{(1-x^{\lambda_2})}$, etc., up to factor $x^{n_m \cdot \lambda_m}$ from expansion of $\frac{1}{(1-x^{\lambda_m})}$.

On the other hand, every monomial x^r in the expansion of the right-hand side of (7.14.1) in x is obtained as a product of terms $x^{n_k k}$ coming from the expansion of factors $\frac{1}{1-x^k}$ $(k = 1, \ldots, r)$, such that

$$r = n_1 \cdot 1 + n_2 \cdot 2 + \cdots + n_r \cdot r = n_{\lambda_m} \lambda_m + \cdots + n_{\lambda_2} \lambda_2 + n_{\lambda_1} \lambda_1, \tag{7.14.3}$$

where $n_i \in \mathbb{Z}_{\geq 0}$ and the right-hand side of (7.14.3) involves finite set of nonzero multiplicities $n_{\lambda_1}, n_{\lambda_2}, \ldots, n_{\lambda_m}$, which are ordered as $\lambda_1 > \lambda_2 > \cdots > \lambda_m$ and $m \leq r$. Then this monomial is associated with Young diagram $\lambda = [\lambda_1^{n_{\lambda_1}}, \lambda_2^{n_{\lambda_2}}, \ldots, \lambda_m^{n_{\lambda_m}}] \vdash r$, where λ_k are lengths of rows of diagram λ. So, every monomial x^r in the expansion of the right-hand side of (7.14.1) in x is indeed associated with a certain Young diagram. We see that there is one-to-one correspondence between monomials x^r (with given r) on the right-hand side of (7.14.1) and set $\mathcal{M}_r$ of Young diagrams $\lambda \vdash r$. Therefore, the coefficient of x^r in the expansion of the right-hand side of (7.14.1) in x equals the number $p(r)$ of elements in the set $\mathcal{M}_r$. This proves Euler formula (7.14.1).

7.15. Problem 4.5.7 of Section 4.5.1

Prove that the left ideal $\mathcal{L}_{[2,1]_2}$ is two dimensional, and (for $\varepsilon = 0$) its basis elements can be chosen as in (4.5.45), *i.e.,*

$$\epsilon_3 = \frac{1}{3}(1 - \sigma_1)(1 + \sigma_2), \quad \epsilon_4 = \frac{1}{3}(1 - \sigma_2)\,\sigma_1\,(1 + \sigma_2).$$

Show that these elements obey (see (4.5.46), (4.5.47) *and* (4.5.48)):

$$\sigma_1 \cdot \epsilon_3 = -\epsilon_3, \quad \sigma_1 \cdot \epsilon_4 = \epsilon_3 + \epsilon_4, \quad \sigma_2 \cdot \epsilon_3 = \epsilon_3 + \epsilon_4, \quad \sigma_2 \cdot \epsilon_4 = -\epsilon_4, \tag{7.15.1}$$

$$\epsilon_3 \cdot \epsilon_3 = \epsilon_3, \quad \epsilon_4 \cdot \epsilon_4 = 0, \quad \epsilon_3 \cdot \epsilon_4 = 0, \quad \epsilon_4 \cdot \epsilon_3 = \epsilon_4, \tag{7.15.2}$$

$$\epsilon_1 \cdot \epsilon_3 = \epsilon_3 \cdot \epsilon_1 = 0, \quad \epsilon_2 \cdot \epsilon_3 = \epsilon_4 \cdot \epsilon_1 = 0, \quad \epsilon_3 \cdot \epsilon_2 = 3\,\epsilon_2, \quad \epsilon_1 \cdot \epsilon_4 = 3\,\epsilon_4. \tag{7.15.3}$$

Prove equivalence of representations determined by left ideals $\mathcal{L}_{[2,1]_1}$ *and* $\mathcal{L}_{[2,1]_2}$ *for* $\epsilon = 0$.

There is an automorphism in algebra $\mathbb{C}[S_3]$ given by

$$\sigma_1 \to \sigma_2 = S \cdot \sigma_1 \cdot S^{-1}, \quad \sigma_2 \to \sigma_1 = S \cdot \sigma_2 \cdot S^{-1}, \quad S \equiv \sigma_1 \cdot \sigma_2 \cdot \sigma_1. \tag{7.15.4}$$

Under this automorphism, one has

$$\epsilon_1 \to \epsilon_3 = (1-\sigma_1)(1+\sigma_2), \quad \epsilon_2 \to \epsilon_4 = (1-\sigma_2)\,\sigma_1\,(1+\sigma_2), \tag{7.15.5}$$

whereas (4.5.43) and (4.5.42) become (7.15.1) and (7.15.2). Making use of (4.5.43) and (7.15.1), we find that the left ideal $\mathcal{L}_{[2,1]_2}$ generated by idempotent (see (4.5.37))

$$e_{[2,1]_2} = \frac{1}{3}(\epsilon_3 - \varepsilon\,\epsilon_2) \tag{7.15.6}$$

is also two dimensional; basis in $\mathcal{L}_{[2,1]_2}$ has two elements $(\epsilon_3 - \varepsilon\,\epsilon_2)$ and $(\epsilon_4 - \varepsilon\,\epsilon_1)$. We set $\varepsilon = 0$ for simplicity. Then, in accordance with (7.15.1), every element $x' \in \mathcal{L}_{[2,1]_2}$ can be written as

$$x' = x_3\,\epsilon_3 + x_4\,\epsilon_4, \quad x_3, x_4 \in \mathbb{C}.$$

Automorphism (7.15.4), (7.15.5) (with $\varepsilon = 0$) moves idempotents $e_{[2,1]_1}$ and $e_{[2,1]_2}$ into each other and establishes equivalence of representations corresponding to left ideals $\mathcal{L}_{[2,1]_1}$ and $\mathcal{L}_{[2,1]_2}$. Relations (7.15.2) are proven by direct computation.

7.16. Problem 4.6.9 of Section 4.6.5

Prove the second equalities in (4.6.93), (4.6.94) *and* (4.6.96) *for* $n = 2$, *i.e.*,

$$\begin{aligned} \frac{1}{2}\,\mathbf{Tr}(g_1\, g_2\, (1+P_1)) &= \frac{1}{2}\,(\mathrm{Tr}\, g)^2 + \frac{1}{2}\,\mathrm{Tr}(g^2), \\ \frac{1}{2}\,\mathbf{Tr}(g_1\, g_2\, (1-P_1)) &= \frac{1}{2}\,(\mathrm{Tr}\, g)^2 - \frac{1}{2}\,\mathrm{Tr}(g^2); \end{aligned} \tag{7.16.1}$$

for $n = 3$:

$$\begin{aligned} \frac{1}{6}\,\mathbf{Tr}(g_1\, g_2\, g_3\, (1+P_2+P_1P_2)(1+P_1)) &= \frac{1}{6}\,p_1^3 + \frac{1}{2}\,p_1\,p_2 + \frac{1}{3}\,p_3, \\ \frac{1}{6}\,\mathbf{Tr}(g_1\, g_2\, g_3\, (1-P_2+P_1P_2)(1-P_1)) &= \frac{1}{6}\,p_1^3 - \frac{1}{2}\,p_1\,p_2 + \frac{1}{3}\,p_3, \end{aligned} \tag{7.16.2}$$

$$\begin{aligned} \frac{1}{3}\mathbf{Tr}(g_1\, g_2\, g_3(1-P_2+\varepsilon(1-P_1)\,P_2)(1+P_1)) &= \frac{1}{3}p_1^3 - \frac{1}{3}\,p_3, \\ \frac{1}{3}\mathbf{Tr}(g_1\, g_2\, g_3(1-P_1)(1+P_2-\varepsilon\,P_2\,(1+P_1))) &= \frac{1}{3}p_1^3 - \frac{1}{3}\,p_3, \end{aligned} \tag{7.16.3}$$

where $g_k = I^{\otimes(k-1)} \otimes g \otimes I^{\otimes(n-k)}$ *and* $p_k = \mathrm{Tr}(g^k)$.

Let us prove the first equality in (7.16.2); the proofs of other equalities in (7.16.1), (7.16.2) and (7.16.3) are completely analogous. We make use of cyclic property of trace (4.6.86) and the following relations:

$$\begin{gathered} P_k\, g_1\, g_2 \cdots g_n = g_1\, g_2 \cdots g_n\, P_k \quad (k = 1, \ldots, n-1), \\ \mathrm{Tr}_{(k)} P_k = \mathrm{Tr}_{(k+1)} P_k = I^{\otimes(n-1)}, \\ P_k\, g_k = g_{k+1}\, P_k, \quad P_k\, g_{k+1} = g_k\, P_k, \quad P_k^2 = I^{\otimes n}, \end{gathered}$$

where trace $\mathrm{Tr}_{(m)}$ is in space $\mathcal{V}$ which occupies mth place in the product $\mathcal{V}^{\otimes n}$. The latter partial trace is defined in (4.6.85). We have, in particular,

$$\begin{aligned} \mathbf{Tr}(g_1 \cdots g_n\, P_k\, P_{k+1}) &= \mathbf{Tr}(P_{k+1}\, g_1 \cdots g_n\, P_k) = \mathbf{Tr}(g_1 \cdots g_n\, P_{k+1}\, P_k) \\ &= (\mathrm{Tr}(g))^{n-3} \cdot \mathrm{Tr}_{k,k+1,k+2}(g_k g_{k+1} g_{k+2}\, P_{k+1} P_k) \\ &= (\mathrm{Tr}(g))^{n-3} \cdot \mathrm{Tr}_{k,k+1,k+2}(g_k g_{k+1}\, P_{k+1} g_{k+1}\, P_k) \\ &= (\mathrm{Tr}(g))^{n-3} \cdot \mathrm{Tr}_{k,k+1}(g_k g_{k+1}\, \mathrm{Tr}_{(k+2)}(P_{k+1})\, g_{k+1}\, P_k) \\ &= (\mathrm{Tr}(g))^{n-3} \cdot \mathrm{Tr}_{k,k+1}(g_k\, P_k\, g_k\, g_k) \\ &= (\mathrm{Tr}(g))^{n-3}\, \mathrm{Tr}(g^3). \end{aligned}$$

Here $\mathrm{Tr}_{k_1,k_2,\dots,k_r}$ is the trace in the space $\mathcal{V}_{k_1,k_2,\dots,k_r} \equiv \mathcal{V}_{k_1} \otimes \mathcal{V}_{k_2} \otimes \cdots \otimes \mathcal{V}_{k_r}$, which is tensor product of spaces $\mathcal{V}$ with numbers $\{k_1, k_2, \dots, k_r\}$ in the product $\mathcal{V}^{\otimes n}$. We also use the convention that operators P_{k_s}, $g_{k_s}, \dots$ entering the traces $\mathrm{Tr}_{k_1,k_2,\dots,k_r}$ act in the subspace $\mathcal{V}_{k_1,k_2,\dots,k_r}$ only. With all these rules, we obtain the following chain of equalities for the left-hand side of the first expression in (7.16.2):

$$\begin{aligned}
&\mathbf{Tr}(g_1\, g_2\, g_3\, (1 + P_2 + P_1 P_2)(1 + P_1)) \\
&\quad = \mathbf{Tr}(g_1\, g_2\, g_3\, (1 + P_2)(1 + P_1)) + \mathbf{Tr}(g_1\, g_2\, g_3\, P_2(1 + P_1)\, P_1) \\
&\quad = \mathbf{Tr}(g_1\, g_2\, g_3\, (1 + 2\, P_2)(1 + P_1)) = \mathbf{Tr}(g_1\, g_2\, g_3\, (1 + 2\, P_2 + P_1 + 2\, P_2\, P_1)) \\
&\quad = \mathbf{Tr}(g_1\, g_2\, g_3) + 2\, \mathbf{Tr}(g_1\, g_2\, P_2\, g_2) + \mathbf{Tr}(g_1\, P_1\, g_1\, g_3) + 2\, \mathbf{Tr}(g_1\, g_2\, g_3\, P_2\, P_1) \\
&\quad = (\mathrm{Tr}(g))^3 + 3\, \mathrm{Tr}(g)\, \mathrm{Tr}(g^2) + 2\, \mathrm{Tr}(g^3).
\end{aligned}$$

Upon multiplying this by a factor 1/6, we obtain the right-hand side in the first line in (7.16.2).

7.17. Problem 4.6.11 of Section 4.6.5

Without assuming that a matrix g is diagonalizable, show that the generating function $E(z)$ for characters $\chi_{[1^k]}(g)$ is given by the characteristic polynomial obtained by expanding in z the determinant

$$\det(I_N + z\, g) \equiv E(z) = \sum_{k=0}^{N} \chi_{[1^k]}(g)\, z^k. \tag{7.17.1}$$

In the case of diagonalizable $(N \times N)$ matrix g, prove the characteristic identity (Hamilton–Cayley theorem):

$$\sum_{r=0}^{N} (-g)^r\, \chi_{[1^{(N-r)}]}(g) = 0. \tag{7.17.2}$$

Proof of identity (7.17.1) that does not assume that matrix g is diagonalizable, is useful for various applications. Let s be representation $\mathbb{C}[S_N] \to \mathcal{V}^{\otimes N}$, defined in (4.2.6). Consider antisymmetrizers $e([1^k]) \in \mathbb{C}[S_N]$ (for $k \leq N$) defined in (4.6.84), and projectors $P_{[1^k]} = \mathsf{s}(e([1^k]))$, acting in $\mathcal{V}^{\otimes N}$. We note in the first place that, for the largest possible $k = N$, the projector $P_{[1^N]} = \mathsf{s}(e([1^N]))$ that acts in space $\mathcal{V}^{\otimes N}$ has rank 1.

Therefore, we have

$$\left(P_{[1^N]}\right)^{i_1\cdots i_N}_{k_1\cdots k_N} = \frac{1}{N!}\varepsilon^{i_1\cdots i_N}\, \varepsilon_{k_1\cdots k_N}, \tag{7.17.3}$$

where $\varepsilon_{k_1\cdots k_N}$ are components of rank-N totally antisymmetric tensor. Let us calculate, one by one, the traces of projector (7.17.3) in spaces with numbers $N, N-1, \ldots, r+1$. In accordance with the definition (4.6.84) of antisymmetrizers $e([1^k])$, we obtain at each step projectors $P_{[1^k]}$, and as a result we get

$$\begin{aligned} \mathrm{Tr}_{N,N-1,\ldots,r+1}(P_{[1^N]}) &= \frac{(N-r)!r!}{N!} P_{[1^r]} \\ \Rightarrow (P_{[1^r]})^{i_1\cdots i_r}_{k_1\cdots k_r} &= \frac{1}{(N-r)!r!}\,\varepsilon^{i_1\cdots i_r j_{r+1}\cdots j_N}\,\varepsilon_{k_1\cdots k_r j_{r+1}\cdots j_N}. \end{aligned} \tag{7.17.4}$$

Making use of representation (7.17.3), formula (7.17.4) and definition of determinant

$$\det(A) = \frac{1}{N!}\varepsilon_{k_1\cdots k_N}\, A^{k_1}_{\ i_1}\cdots A^{k_N}_{\ i_N}\,\varepsilon^{i_1\cdots i_N},$$

we write the left-hand side of (7.17.1) as follows:

$$\begin{aligned} \det(I_N + z\,g) &= \mathbf{Tr}\big(P_{[1^N]}\,(I_N + z\,g)^{\otimes N}\big) \\ &= \sum_{r=0}^{N} C_N^r\, z^r\, \mathbf{Tr}\big(P_{[1^N]}\, g^{\otimes r}\otimes I_N^{\otimes (N-r)}\big) \\ &= \sum_{r=0}^{N} z^r\, \mathrm{Tr}_{1,\ldots,r}\big(P_{[1^r]}\, g^{\otimes r}\big) = \sum_{r=0}^{N} z^r\, \chi_{[1^r]}(g). \end{aligned} \tag{7.17.5}$$

This is precisely (7.17.1). Note that the coefficients of the largest power z^N and of the linear term z on the right-hand side of (7.17.5) are equal to $\chi_{[1^N]}(g) = \det(g)$ and $\chi_{[1]}(g) = \mathrm{Tr}(g)$, respectively.

Let us now prove the identity (7.17.2) in a simple case of diagonalizable matrix g. We write the identity

$$\prod_{i=1}^{N}(z + x_i) = \sum_{r=0}^{N} z^{N-r}\,\chi_{[1^r]}(g), \tag{7.17.6}$$

where, as before,

$$\chi_{[1^r]}(g) = \sum_{i_1<\cdots<i_r} x_{i_1}\cdots x_{i_r},$$

and x_i are eigenvalues of matrix g. The identity (7.17.6) is obtained by expanding the left-hand side in z. It is valid for any formal parameter z, which can be a matrix as well. We set $z = -g$ in identity (7.17.6) and make use of the characteristic identity $\prod_{i=1}^{N}(g - x_i) = 0$. Upon changing the summation index on the right-hand side of (7.17.6), $r \to (N-r)$, we obtain the Hamilton–Cayley identity (7.17.2).

7.18. Problem 4.7.1 of Section 4.7

Construct Gelfand–Tsetlin basis in spaces $\mathcal{V}^{([3])}$ and $\mathcal{V}^{([2,1])}$ of representations of algebra $u(3)$ labeled by Young diagrams [3] *and* [2,1]. *Find the correspondence between Gelfand–Tsetlin schemes (basis vectors of spaces $\mathcal{V}^{([3])}$ and $\mathcal{V}^{([2,1])}$) and members of baryon decuplet and octet (see formulas* (4.4.81)–(4.4.84) *and* (4.4.86) *in Section* 4.4.4).

Elements of spaces $\mathcal{V}^{([2,1])}$ and $\mathcal{V}^{([3])}$ of the irreducible representations of algebra $u(3)$ are tensors with components $B^{i_1 i_2 i_3}$ and $D^{i_1 i_2 i_3}$, associated with Young tableaux

$$B^{i_1 i_2 i_3} = \begin{array}{|c|c|}\hline i_1 & i_2 \\ \hline i_3 & \multicolumn{1}{c}{} \\ \cline{1-1}\end{array}\,, \qquad D^{i_1 i_2 i_3} = \begin{array}{|c|c|c|}\hline i_1 & i_2 & i_3 \\ \hline\end{array}\,. \tag{7.18.1}$$

When subalgebra $u(2)$ is embedded in $u(3)$ in accordance with (4.7.3), the action of $u(2)$ on tensors (7.18.1) does not affect those indices of components $\psi^{i_1 i_2 i_3}$ that take value 3. Likewise, when subalgebra $u(1)$ is embedded in $u(2) \subset u(3)$ the action of $u(1)$ on tensors (7.18.1) does not touch tensor indices which take values 2 and 3. Making use of these observations, and applying the rules of construction of Gelfand–Tsetlin schemes described in Section 4.7, we arrive at the correspondence between Gelfand–Tsetlin schemes and components of tensors symmetrized in accordance with (7.18.1). It is worth recalling that irreducible representations of subalgebra $u(2) \subset u(3)$ are characterized by isospin I and are associated with Young diagrams in the second line of Gelfand–Tsetlin scheme.

Baryon octet (4.4.86), associated with the first tensor in (7.18.1) In this case, we have the following branching of representations for embeddings $u(1) \subset u(2) \subset u(3)$:

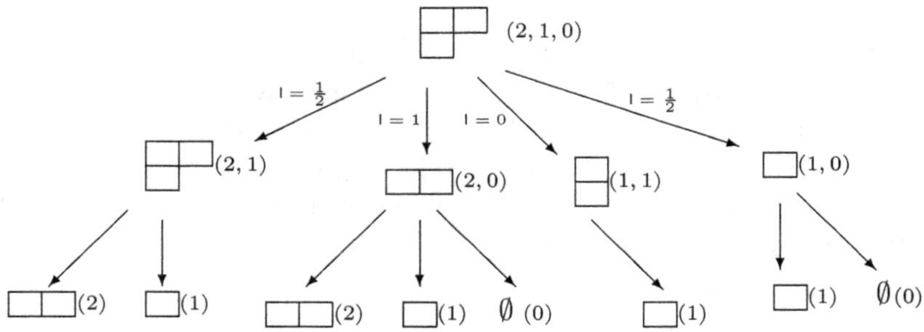

(we show in parentheses near diagrams the partitions that enter the rows of Gelfand–Tsetlin schemes). In accordance with this branching, we establish the correspondence of Gelfand–Tsetlin schemes and components $B^{i_1 i_2 i_3}$ (we omit spin and color indices which are irrelevant here).

(1) Representation with zero strangeness $s = 0$ (no s-quark) and isospin $\mathsf{I} = 1/2$, with all indices i_1, i_2, i_3 in $B^{i_1 i_2 i_3}$ taking values 1, 2:

$$\mathsf{I} = \tfrac{1}{2}: \quad p = B^{112} = \begin{smallmatrix} 2 & & 1 & & 0 \\ & 2 & & 1 & \\ & & 2 & & \end{smallmatrix}, \quad n = B^{122} = \begin{smallmatrix} 2 & & 1 & & 0 \\ & 2 & & 1 & \\ & & 1 & & \end{smallmatrix}.$$

(2) Representations with strangeness $s = -1$ (one s-quark), with one of the indices i_1, i_2, i_3 in $B^{i_1 i_2 i_3}$ taking value 3, and isospins $\mathsf{I} = 1$ and $\mathsf{I} = 0$:

$$\mathsf{I} = 1: \quad \Sigma^+ = B^{113} = \begin{smallmatrix} 2 & & 1 & & 0 \\ & 2 & & 0 & \\ & & 2 & & \end{smallmatrix}, \quad \Sigma^- = B^{223} = \begin{smallmatrix} 2 & & 1 & & 0 \\ & 2 & & 0 & \\ & & 0 & & \end{smallmatrix},$$

$$\Sigma^0 = 2 \cdot B^{(12)3} = \begin{smallmatrix} 2 & & 1 & & 0 \\ & 2 & & 0 & \\ & & 1 & & \end{smallmatrix},$$

$$\mathsf{I} = 0: \quad \Lambda^0 = B^{132} = 2 \cdot B^{[12]3} = \begin{smallmatrix} 2 & & 1 & & 0 \\ & 1 & & 1 & \\ & & 1 & & \end{smallmatrix}.$$

(3) Representation with strangeness $s = -2$ (two s-quarks), with two indices taking value 3, and isospin $\mathsf{I} = 1/2$:

$$\mathsf{I} = \tfrac{1}{2}: \quad \Xi^0 = B^{133} = \begin{smallmatrix} 2 & & 1 & & 0 \\ & 1 & & 0 & \\ & & 1 & & \end{smallmatrix}, \quad \Xi^- = B^{233} = \begin{smallmatrix} 2 & & 1 & & 0 \\ & 1 & & 0 & \\ & & 0 & & \end{smallmatrix}.$$

Baryon decuplet (4.4.81)–(4.4.84), associated with the second tensor in (7.18.1)

We invite the reader to draw representation branching graph for embeddings $u(1) \subset u(2) \subset u(3)$. The correspondence between Gelfand–Tsetlin schemes and components $D^{i_1 i_2 i_3}$ is as follows.

(1) Representation with zero strangeness $s = 0$ (no s-quark) and isospin $\mathsf{I} = 3/2$ with all indices i_1, i_2, i_3 in $D^{i_1 i_2 i_3}$ taking values $1, 2$:

$$\Delta^{++} = D^{(111)} = \begin{smallmatrix} 3 & & 0 & & 0 \\ & 3 & & 0 & \\ & & 3 & & \end{smallmatrix}, \quad \Delta^{-} = D^{(222)} = \begin{smallmatrix} 3 & & 0 & & 0 \\ & 3 & & 0 & \\ & & 0 & & \end{smallmatrix},$$

$$\Delta^{+} = D^{(112)} = \begin{smallmatrix} 3 & & 0 & & 0 \\ & 3 & & 0 & \\ & & 2 & & \end{smallmatrix}, \quad \Delta^{0} = D^{(122)} = \begin{smallmatrix} 3 & & 0 & & 0 \\ & 3 & & 0 & \\ & & 1 & & \end{smallmatrix}.$$

(2) Representation with strangeness $s = -1$ (one s-quark) and isospin $\mathsf{I} = 1$; one index among i_1, i_2, i_3 equals 3 and the other two take values $1, 2$:

$$\Sigma^{*+} = D^{(113)} = \begin{smallmatrix} 3 & & 0 & & 0 \\ & 2 & & 0 & \\ & & 2 & & \end{smallmatrix}, \quad \Sigma^{*-} = D^{(223)} = \begin{smallmatrix} 3 & & 0 & & 0 \\ & 2 & & 0 & \\ & & 0 & & \end{smallmatrix},$$

$$\Sigma^{*0} = D^{(123)} = \begin{smallmatrix} 3 & & 0 & & 0 \\ & 2 & & 0 & \\ & & 1 & & \end{smallmatrix}.$$

(3) Representation with strangeness $s = -2$ (two s-quarks) and isospin $\mathsf{I} = 1/2$; two indices take value 3:

$$\Xi^{*0} = D^{(133)} = \begin{smallmatrix} 3 & & 0 & & 0 \\ & 1 & & 0 & \\ & & 1 & & \end{smallmatrix}, \quad \Xi^{*-} = D^{(233)} = \begin{smallmatrix} 3 & & 0 & & 0 \\ & 1 & & 0 & \\ & & 0 & & \end{smallmatrix}.$$

(4) Representation with strangeness $s = -3$ (three s-quarks) and isospin $\mathsf{I} = 0$:

$$\Omega^{-} = D^{333} = \begin{smallmatrix} 3 & & 0 & & 0 \\ & 0 & & 0 & \\ & & 0 & & \end{smallmatrix}.$$

7.19. Problem 5.1.3 of Section 5.1.3

Show that irreducible finite-dimensional representations of group $O(N, \mathbb{C})$ (and its real subgroups $O(p, q)$ with $p + q = N$) are associated with Young diagrams whose sums of numbers of boxes in the first two rows do not exceed N. Hint: derive the formula for dimension of representation $T^{(\lambda')}$ of $O(N, \mathbb{C})$ associated with diagram λ' with two columns of heights a and b

$(a \geq b)$:

$$\dim(T^{(\lambda')}) = \frac{(a-b+1)}{(a+1)!b!} \frac{N!(N+2)!\,(N-a-b+1)}{(N-a+1)!(N-b+2)!}. \tag{7.19.1}$$

In the course of solving this problem, we need the following fact.

Lemma 7.19.1. *Let the diagram $\lambda \vdash r$ has n columns, and the numbers in standard tableau T_λ are ordered in columns (see discussion after Definition 4.5.2). Consider standard tableau $\mathsf{T}_{\lambda'}$ which is obtained from T_λ by removing all $(k+1)$th, $(k+2)$th, ..., nth columns on the right. So, the remaining tableau $\mathsf{T}_{\lambda'}$ contains only $k < n$ left columns of tableau T_λ. Hence, tableau $\mathsf{T}_{\lambda'}$ is embedded in tableau T_λ (diagram $\lambda' \vdash r'$ is embedded in diagram $\lambda \vdash r$; clearly $r' < r$). Let $Y(\mathsf{T}_\lambda) \in \mathbb{C}[S_r]$ and $Y(\mathsf{T}_{\lambda'}) \in \mathbb{C}[S_{r'}]$ be Young symmetrizers associated with tableaux T_λ and $\mathsf{T}_{\lambda'}$, where subgroup $S_{r'}$ is embedded in S_r in an obvious way. Then, the following identity holds:*

$$Y(\mathsf{T}_\lambda) = (\alpha_{\lambda'})^{-1}\, Y(\mathsf{T}_{\lambda'}) \cdot Y(\mathsf{T}_\lambda). \tag{7.19.2}$$

Proof. In accordance with definition (4.3.11), we have

$$Y(\mathsf{T}_{\lambda'}) = Q' \cdot P', \quad Y(\mathsf{T}_\lambda) = Q \cdot P.$$

Here horizontal symmetrizers P and P' of tableaux T_λ and $\mathsf{T}_{\lambda'}$ obey $P = P' \cdot P$, while vertical antisymmetrizers Q and Q' are related by $Q = Q' \cdot Q_{k+1} \cdots Q_n$, where Q_i antisymmetrizes numbers in ith column of tableau T_λ. Since numbers from the right columns numbered by $\ell > k$ in tableau T_λ do not exist in tableau $\mathsf{T}_{\lambda'}$, we have $P' \cdot Q_\ell = Q_\ell \cdot P'$. Therefore,

$$\begin{aligned} Y(\mathsf{T}_\lambda) &= Q \cdot P = Q' \cdot Q_{k+1} \cdots Q_n \cdot P' \cdot P = Q' \cdot P' \cdot Q_{k+1} \cdots Q_n \cdot P \\ &= Y(\mathsf{T}_{\lambda'}) \cdot Q_{k+1} \cdots Q_n \cdot P = (\alpha_{\lambda'})^{-1}\, Y(\mathsf{T}_{\lambda'}) \cdot Y(\mathsf{T}_{\lambda'}) \\ &\quad \times Q_{k+1} \cdots Q_n \cdot P \\ &= (\alpha_{\lambda'})^{-1}\, Y(\mathsf{T}_{\lambda'}) \cdot Y(\mathsf{T}_\lambda), \end{aligned}$$

which proves (7.19.2). □

Now we start to solve the problem. For representation of $O(N, \mathbb{C})$ acting in space of antisymmetric tensors and associated with Young diagram $[1^k]$ with just one column, the result of the problem follows from the fact that any rank-k antisymmetric tensor vanishes for $k > N$. We turn to Young diagrams with two or more columns and begin with diagram

$\lambda' = [2^b, 1^{a-b}]$ with two columns of heights a and b ($a \geq b$); see (7.19.3) below. Consider space $\bar{\mathcal{V}}^{(\lambda')}$ of tensors symmetrized with Young projector $\mathsf{s}(Y(\mathsf{T}_{\lambda'}))$ (symmetrization with projector $P(\mathsf{T}_{\lambda'})$ is considered in the same way). In accordance with the rules given in Section 4.4.1, we represent components of tensor $\mathsf{t} \in \bar{\mathcal{V}}^{(\lambda')}$ graphically in terms of Young tableau of shape λ' (see (4.4.10)):

$$\mathsf{t}^{[k_1,\dots,k_a]\,[i_1,\dots,i_b]} = \begin{array}{|c|c|} \hline k_1 & i_1 \\ \hline \vdots & \vdots \\ \hline k_b & i_b \\ \hline \vdots & \multicolumn{1}{c}{} \\ \cline{1-1} k_a & \multicolumn{1}{c}{} \\ \cline{1-1} \end{array} . \tag{7.19.3}$$

We indicate explicitly on the left-hand side of (7.19.3) that components of tensor t are antisymmetric in indices $(k_1, \dots, k_a)$ and separately $(i_1, \dots, i_b)$, i.e., they are antisymmetrized in columns of the Young diagram. Let us calculate the number of tensors symmetrized with Young projector $\mathsf{s}(Y(\mathsf{T}_{\lambda'}))$, or, in other words, dimension of space $\bar{\mathcal{V}}^{(\lambda')}$ of representation of group $SL(N, \mathbb{C})$ (not $SO(N, \mathbb{C})$ yet!). Making use of hook formula (4.4.21) we get

$$\dim(\bar{\mathcal{V}}^{(\lambda')}) = \frac{(a-b+1)}{(a+1)!b!} \frac{N!(N+1)!}{(N-a)!(N-b+1)!}. \tag{7.19.4}$$

To find the dimension of subspace $\mathcal{V}^{(\lambda')} \subset \bar{\mathcal{V}}^{(\lambda')}$ of representation of $SO(N, \mathbb{C})$, we have to count traceless tensors in $\bar{\mathcal{V}}^{(\lambda')}$. To this end, we subtract from dimension (7.19.4) the number of "tracelessness" conditions (5.1.32) imposed on tensors $\mathsf{t} \in \bar{\mathcal{V}}^{(\lambda')}$. Contraction of tensor t in two indices in the same column of diagram (7.19.3) vanishes identically due to antisymmetry in these indices, so we get no conditions in this way. Again due to antisymmetry in columns, convolutions in two indices from different columns of diagram (7.19.3) give equivalent "tracelessness" conditions, irrespectively of the choice of an index from the set $(k_1, \dots, k_a)$ and an index from the set $(i_1, \dots, i_b)$. Thus, the only set of nontrivial "tracelessness" conditions is

$$\eta_{k_1 i_1} \mathsf{t}^{[k_1,\dots,k_a]\,[i_1,\dots,i_b]} = 0. \tag{7.19.5}$$

Tensors t' on the left-hand side of (7.19.5) have rank $r = a + b - 2$ and components $\mathsf{t}'^{\,[k_2,\dots,k_a]\,[i_2,\dots,i_b]}$. These tensors have symmetry associated with Young tableau obtained from tableau (7.19.3) by removing the first row with indices k_1 and i_1. The space $\mathcal{V}^{(\lambda'')}$ of these tensors t' is nothing but the space of representation of $SL(N, \mathbb{C})$ associated with Young diagram $\lambda'' = [2^{b-1}, 1^{a-b}]$. The number of independent relations (7.19.5) is equal to dimension of space $\mathcal{V}^{(\lambda'')}$ and is given by the formula (7.19.4) with substitution $a \to (a-1)$ and $b \to (b-1)$:

$$\dim(\bar{\mathcal{V}}^{(\lambda'')}) = \frac{(a-b+1)}{a!(b-1)!} \frac{N!(N+1)!}{(N-a+1)!(N-b+2)!}. \tag{7.19.6}$$

Dimension of representation $T^{(\lambda')}$ of group $SO(N, \mathbb{C})$ is given by $\dim(T^{(\lambda')}) \equiv \dim(\mathcal{V}^{(\lambda')}) = \dim(\bar{\mathcal{V}}^{(\lambda')}) - \dim(\bar{\mathcal{V}}^{(\lambda'')})$ which gives (7.19.1). Formula (7.19.1) shows that for $(a+b) > N$ we have $\dim(T^{(\lambda')}) \le 0$, i.e., representation $T^{(\lambda')}$ does not exist in that case. This proves the result of the problem for representations associated with Young diagrams with two columns.

Finally, consider diagram $\lambda \vdash r$ with three or more columns. Let its first two columns form Young diagram $\lambda' = [2^b, 1^{a-b}]$ shown in (7.19.3). Let T_λ be standard Young tableau whose numbers in boxes are ordered in columns (see discussion after Definition 4.5.2) and $\mathsf{T}_{\lambda'}$ be its standard subtableau made of the two first columns. According to Lemma 7.19.1 we have for Young symmetrizers the identity (7.19.2) which gives

$$\mathsf{s}(Y(\mathsf{T}_\lambda)) = (\alpha_{\lambda'})^{-1}\, \mathsf{s}(Y(\mathsf{T}_{\lambda'})) \cdot \mathsf{s}((Y(\mathsf{T}_\lambda)). \tag{7.19.7}$$

This means, in turn, that components of rank-r tensor t symmetrized with projector $\mathsf{s}(Y(\mathsf{T}_\lambda))$, can be written as

$$\mathsf{t}^{[k_1,\ldots,k_a]\,[i_1,\ldots,i_b]\,j_1,\ldots,j_{r-a-b}}, \tag{7.19.8}$$

where first $(a+b)$ indices $(k_1, \ldots, k_a, i_1, \ldots, i_b)$ are symmetrized with operator $\mathsf{s}(Y(\mathsf{T}_{\lambda'}))$ that corresponds to tableau (7.19.3). For any fixed indices $(j_1, \ldots, j_{r-a-b})$ in (7.19.8), we have tensors of the type (7.19.3) whose number of independent components is again given by (7.19.1). Such tensors do not exist for $a+b > N$, so for the tableau T_λ considered, the result of the problem is valid. Now, all representations associated with one and the same diagram shape λ are equivalent, which completes the proof of the result of the problem.

7.20. Problem 5.1.5 of Section 5.1.3

Let $\mathsf{t}^{k_1\cdots k_r;\ell_1\cdots\ell_r}$ be components of traceless rank-$2r$ tensor $\mathsf{t} \in \mathcal{V}^{\otimes 2r}$, where $\mathcal{V}$ is the space of defining representation of group $SO(2r-q, q)$ (or group

$SO(2r,\mathbb{C})$). Let these components be symmetrized in accordance with the tableau

$$\begin{array}{|c|c|}\hline k_1 & \ell_1 \\ \hline \vdots & \vdots \\ \hline k_r & \ell_r \\ \hline \end{array}\,.$$

Prove the identity

$$(P_+)^{i_1\cdots i_r}{}_{k_1\cdots k_r}\,(P_-)^{j_1\cdots j_r}{}_{\ell_1\cdots \ell_r}\;\mathsf{t}^{k_1\cdots k_r;\ell_1\cdots\ell_r}=0, \tag{7.20.1}$$

where (anti-) selfdual projectors $P_\pm$ are (see (5.1.41))

$$(P_\pm)^{i_1\cdots i_r}{}_{k_1\cdots k_r}\equiv A^{i_1\cdots i_r}{}_{k_1\cdots k_r}\pm\frac{\alpha}{r!}\varepsilon^{i_1\cdots i_r}{}_{k_1\cdots k_r}.$$

Here $A^{i_1\cdots i_r}{}_{k_1\cdots k_r}$ is rank-r antisymmetrizer and $\alpha=e^{-i\pi(r+q)/2}$.

We represent components of tensor $\mathsf{t}\in\mathcal{V}^{\otimes 2r}$ symmetrized in accordance with Young diagram $[2^r]$ as follows:

$$\mathsf{t}^{1,3,\dots,2r-1;\,2,4,\dots,2r}=\begin{array}{|c|c|}\hline 1 & 2 \\ \hline 3 & 4 \\ \hline \vdots & \vdots \\ \hline \cdots & 2r \\ \hline \end{array}=A_{1,3,\dots,2r-1}\cdot A_{2,4,\dots,2r}$$

$$\times\,P^{(+)}_1P^{(+)}_3\cdots P^{(+)}_{2r-1}\,\psi^{1,2,\dots,2r}, \tag{7.20.2}$$

where ψ is an arbitrary tensor in $\mathcal{V}^{\otimes 2r}$; instead of tensor indices $i_1,i_2,\dots$ we write the numbers $1,2,\dots$ of pertinent spaces $\mathcal{V}$ in tensor product $\mathcal{V}^{\otimes 2r}$; $A_{a_1,\dots,a_r}$ are rank-r antisymmetrizers in spaces $\mathcal{V}$ with numbers $a_1,\dots,a_r$; operators $P^{(+)}_k=\frac{1}{2}(I+P_k)$ symmetrize indices i_k,i_{k+1} in tensor components $\psi^{i_1\cdots i_k i_{k+1}\cdots i_{2r}}$, i.e., symmetrize spaces $\mathcal{V}$ with numbers k and $(k+1)$ in $\mathcal{V}^{\otimes 2r}$ (permutation P_k is defined in (4.6.92)). Tracelessness conditions for tensors (7.20.2) reduce to the conditions

$$\eta_{k_i\ell_j}\mathsf{t}^{k_1\cdots k_r;\ell_1\cdots\ell_r}=0\Leftrightarrow\eta_{k_1\ell_1}\mathsf{t}^{k_1\cdots k_r;\ell_1\cdots\ell_r}=\mathsf{t}_{\ell_1}{}^{k_2\cdots k_r;\ell_1,\ell_2\cdots\ell_r}=0. \tag{7.20.3}$$

Tensors (7.20.2) are symmetric under simultaneous transposition of all indices in columns of Young tableau:

$$\begin{array}{|c|c|}\hline k_1 & \ell_1 \\ \hline \vdots & \vdots \\ \hline k_r & \ell_r \\ \hline \end{array}=\mathsf{t}^{k_1\cdots k_r;\ell_1\cdots\ell_r}=\mathsf{t}^{\ell_1\cdots\ell_r;k_1\cdots k_r}=\begin{array}{|c|c|}\hline \ell_1 & k_1 \\ \hline \vdots & \vdots \\ \hline \ell_r & k_r \\ \hline \end{array}. \tag{7.20.4}$$

Indeed, projectors $P_k^{(+)}$ obey $P_k^{(+)} = P_k \, P_k^{(+)}$, so for tensor (7.20.2) we have

$$\begin{aligned}
&\mathsf{t}^{1,\dots,2r-1;\,2,\dots,2r} \\
&\quad = A_{1,\dots,2r-1} \cdot A_{2,\dots,2r} \cdot (P_1 \cdots P_{2r-1}) \cdot P_1^{(+)} \cdots P_{2r-1}^{(+)} \, \psi^{1,2,\dots,2r} \\
&\quad = (P_1 \cdots P_{2r-1}) \cdot A_{2,\dots,2r} \cdot A_{1,\dots,2r-1} \cdot P_1^{(+)} \cdots P_{2r-1}^{(+)} \, \psi^{1,2,\dots,2r} \\
&\quad = \mathsf{t}^{2,\dots,2r;\,1,\dots,2r-1},
\end{aligned} \tag{7.20.5}$$

where we take into account that $[A_{2,\dots,2r},\ A_{1,\dots,2r-1}] = 0$. Making use of symmetry (7.20.4), one can write (7.20.1) as

$$(P_-)^{i_1\cdots i_r}{}_{k_1\cdots k_r} \, (P_+)^{j_1\cdots j_r}{}_{\ell_1\cdots\ell_r} \, \mathsf{t}^{k_1\cdots k_r;\ell_1\cdots\ell_r} = 0. \tag{7.20.6}$$

Given the explicit form of (anti-)selfdual projectors $P_\pm$, equalities (7.20.1) and (7.20.6) are equivalent to

$$\mathsf{t}^{i_1\cdots i_r;j_1\cdots j_r} = \frac{(-1)^{r+q}}{(r!)^2} \, \varepsilon^{i_1\cdots i_r}{}_{k_1\cdots k_r} \, \varepsilon^{j_1\cdots j_r}{}_{\ell_1\cdots\ell_r} \, \mathsf{t}^{k_1\cdots k_r;\ell_1\cdots\ell_r}, \tag{7.20.7}$$

$$\varepsilon^{i_1\cdots i_r}{}_{k_1\cdots k_r} \mathsf{t}^{k_1\cdots k_r;j_1\cdots j_r} = \varepsilon^{j_1\cdots j_r}{}_{\ell_1\cdots\ell_r} \, \mathsf{t}^{i_1\cdots i_r;\ell_1\cdots\ell_r}. \tag{7.20.8}$$

Note that relations (7.20.7) and (7.20.8) follow from each other. So, we have to prove one of these relations, say, (7.20.7).

We prove (7.20.7) for group $SO(2r)$, i.e., for $q = 0$ and Euclidean metric $\eta_{ik} = \delta_{ik}$ (the general case $SO(2r-q, q)$ with $q \neq 0$ is considered in the same way). We make use of the identity (see Section I-1.2.2 in the first book)

$$\varepsilon^{i_1\dots i_r}{}_{k_1\dots k_r} \cdot \varepsilon^{j_1\dots j_r}{}_{\ell_1\dots\ell_r} = \det \left(\begin{array}{ccc|ccc}
\delta^{i_1 j_1} & \cdots & \delta^{i_1 j_r} & \delta^{i_1}_{\ell_1} & \cdots & \delta^{i_1}_{\ell_r} \\
\vdots & \vdots & \vdots & \vdots & \vdots & \vdots \\
\delta^{i_r j_1} & \cdots & \delta^{i_r j_r} & \delta^{i_r}_{\ell_1} & \cdots & \delta^{i_r}_{\ell_r} \\
\hline
\delta^{j_1}_{k_1} & \cdots & \delta^{j_r}_{k_1} & \delta_{k_1\ell_1} & \cdots & \delta_{k_1\ell_r} \\
\vdots & \vdots & \vdots & \vdots & \vdots & \vdots \\
\delta^{j_1}_{k_r} & \cdots & \delta^{j_r}_{k_r} & \delta_{k_r\ell_1} & \cdots & \delta_{k_r\ell_r}
\end{array} \right)$$

and write the right-hand side of (7.20.7) as

$$\frac{(-1)^r}{(r!)^2} \det \begin{pmatrix} B^{i_1\cdots i_r}_{j_1\cdots j_r} & B^{i_1\cdots i_r}_{\ell_1\cdots\ell_r} \\ B^{j_1\cdots j_r}_{k_1\cdots k_r} & \mathbf{0} \end{pmatrix} \mathsf{t}^{k_1\cdots k_r;\ell_1\cdots\ell_r}, \tag{7.20.9}$$

where we introduce $r \times r$ matrices

$$B^{i_1 \cdots i_r}_{j_1 \cdots j_r} \equiv \begin{pmatrix} \delta^{i_1}_{j_1} & \cdots & \delta^{i_1}_{j_r} \\ \vdots & \cdots & \vdots \\ \delta^{i_r}_{j_1} & \cdots & \delta^{i_r}_{j_r} \end{pmatrix},$$

and zero $r \times r$ matrix $\mathbf{0}$ emerges due to the tracelessness property (7.20.3) of tensor t. By permutation of rows, the determinant in (7.20.9) can be written in the form

$$\begin{aligned} \det \begin{pmatrix} B^{i_1 \cdots i_r}_{j_1 \cdots j_r} & B^{i_1 \cdots i_r}_{\ell_1 \cdots \ell_r} \\ B^{j_1 \cdots j_r}_{k_1 \cdots k_r} & \mathbf{0} \end{pmatrix} &= (-1)^r \det \begin{pmatrix} B^{j_1 \cdots j_r}_{k_1 \cdots k_r} & \mathbf{0} \\ B^{i_1 \cdots i_r}_{j_1 \cdots j_r} & B^{i_1 \cdots i_r}_{\ell_1 \cdots \ell_r} \end{pmatrix} \\ &= (-1)^r \det \left(B^{j_1 \cdots j_r}_{k_1 \cdots k_r} \right) \cdot \det \left(B^{i_1 \cdots i_r}_{\ell_1 \cdots \ell_r} \right) \\ &= (-1)^r (r!)^2 A^{j_1 \cdots j_r}_{k_1 \cdots k_r} A^{i_1 \cdots i_r}_{\ell_1 \cdots \ell_r}, \end{aligned} \tag{7.20.10}$$

where $A^{j_1 \cdots j_r}_{k_1 \cdots k_r} \equiv \frac{1}{r!} \det \left(B^{j_1 \cdots j_r}_{k_1 \cdots k_r} \right)$ is rank-r antisymmetrizer (see (3.4.28)). We insert (7.20.10) into (7.20.9) and find for the right-hand side of (7.20.7) that

$$A^{j_1 \cdots j_r}_{k_1 \cdots k_r} A^{i_1 \cdots i_r}_{\ell_1 \cdots \ell_r} \, \mathsf{t}^{k_1 \cdots k_r; \ell_1 \cdots \ell_r} = \mathsf{t}^{j_1 \cdots j_r; i_1 \cdots i_r},$$

which is equal to the left-hand side of (7.20.7) due to symmetry (7.20.4) Thus, we have proven the relation (7.20.7) and hence (7.20.8). By combining these relations, we arrive at (7.20.1).

7.21. Problem 5.3.4 of Section 5.3

Prove that components $t^{k_1 k_2 \cdots k_r}$ of antisymmetric and traceless tensor t coincide with components of dual tensor $\tilde{t}$

$$\tilde{t}^{k_1 \cdots k_r} = (-1)^{[\frac{r}{2}]} \frac{\mathrm{Pf}(J)}{r!} J^{k_1 \ell_1} \ldots J^{k_r \ell_r} \varepsilon_{\ell_1 \cdots \ell_r i_1 \cdots i_r} t^{i_1 \cdots i_r}, \tag{7.21.1}$$

where $||J^{ki}||$ is inverse matrix to the symplectic metric $||J_{ki}||$, and $[\frac{r}{2}]$ is the integer part of $\frac{r}{2}$.

According to definition (1.3.58) of Pfaffian (see also Section I-1.2.2 in [1]), we have

$$\begin{aligned} \varepsilon_{i_1 i_2 \cdots i_{2r}} J_{i_1 i_2} \cdots J_{i_{2r-1} i_{2r}} &= r! \, 2^r \, \mathrm{Pf}(||J_{ik}||), \\ \varepsilon_{i_1 i_2 \cdots i_{2r}} J^{i_1 i_2} \ldots J^{i_{2r-1} i_{2r}} &= r! \, 2^r \, \mathrm{Pf}(||J^{ik}||), \end{aligned} \tag{7.21.2}$$

where $\varepsilon_{i_1 i_2 \cdots i_{2r}}$ are components of completely antisymmetric tensor, which is fixed by condition $\varepsilon_{12\ldots 2r} = +1$. Define ε-tensor with upper indices in a standard way:

$$\varepsilon^{i_1 i_2 \cdots i_{2r}} := J^{i_1 k_1} \ldots J^{i_{2r} k_{2r}} \varepsilon_{k_1 k_2 \cdots k_{2r}} = \det ||J^{ik}|| \cdot \varepsilon_{i_1 i_2 \cdots i_{2r}}. \tag{7.21.3}$$

Then the contraction of ε-tensors is

$$\varepsilon_{i_1 i_2 \cdots i_{2r}} \varepsilon^{i_1 i_2 \cdots i_{2r}} = \varepsilon_{i_1 i_2 \cdots i_{2r}} J^{i_1 k_1} \ldots J^{i_{2r} k_{2r}} \varepsilon_{k_1 k_2 \cdots k_{2r}} = (2r)! \cdot \det ||J^{ik}||,$$

and complete antisymmetrizer is given by formula

$$A^{i_1 i_2 \cdots i_{2r}}_{k_1 k_2 \cdots k_{2r}} = \frac{\det ||J_{ik}||}{(2r)!} \, \varepsilon^{i_1 i_2 \cdots i_{2r}} \, \varepsilon_{k_1 k_2 \cdots k_{2r}} \Rightarrow A^2 = A. \tag{7.21.4}$$

From relations (7.21.3) and (7.21.2) one obtains the following identity:

$$\begin{aligned} \varepsilon^{i_1 i_2 \cdots i_{2r}} J_{i_1 i_2} \cdots J_{i_{2r-1} i_{2r}} &= \det ||J^{ik}|| \cdot \varepsilon_{i_1 i_2 \cdots i_{2r}} J_{i_1 i_2} \cdots J_{i_{2r-1} i_{2r}} \\ &= r! \, 2^r \det ||J^{ik}|| \cdot \mathrm{Pf}(||J_{ik}||), \end{aligned} \tag{7.21.5}$$

which we use below. Consider the antisymmetric combination

$$\mathcal{A}(J_{i_1 i_2} \cdots J_{i_{2r-1} i_{2r}}) = \sum_{\sigma \in S_{2r}} (-1)^{p(\sigma)} J_{i_{\sigma(1)} i_{\sigma(2)}} \cdots J_{i_{\sigma(2r-1)} i_{\sigma(2r)}}, \tag{7.21.6}$$

where $p(\sigma)$ is parity of permutation σ. For example, for $r = 2$ we find: $\mathcal{A}(J_{i_1 i_2} J_{i_3 i_4}) = 2\big(J_{i_1 i_2} J_{i_3 i_4} + cycle(i_2, i_3, i_4)\big)$. For combination (7.21.6), we have identity

$$\mathcal{A}(J_{i_1 i_2} \cdots J_{i_{2r-1} i_{2r}}) = r! \, 2^r \, \mathrm{Pf}(||J_{ik}||) \, \varepsilon_{i_1 i_2 \cdots i_{2r}}. \tag{7.21.7}$$

Indeed, one obtains this identity as follows:

$$\begin{aligned} &\sum_{\sigma \in S_{2r}} (-1)^{p(\sigma)} J_{i_{\sigma(1)} i_{\sigma(2)}} \cdots J_{i_{\sigma(2r-1)} i_{\sigma(2r)}} \\ &\quad = (2r)! A^{k_1 k_2 \cdots k_{2r}}_{i_1 i_2 \cdots i_{2r}} J_{k_1 k_2} \cdots J_{k_{2r-1} k_{2r}} \\ &\quad = \det ||J_{ik}|| \, \varepsilon_{i_1 i_2 \cdots i_{2r}} \, \varepsilon^{k_1 k_2 \cdots k_{2r}} J_{k_1 k_2} \cdots J_{k_{2r-1} k_{2r}} \\ &\quad = \varepsilon_{i_1 i_2 \cdots i_{2r}} \, r! \, 2^r \cdot \mathrm{Pf}(||J_{ik}||), \end{aligned}$$

where in the second equality we make use the expression for antisymmetrizer (7.21.4), while in the third equality we take into account (7.21.5).

Now, consider contraction $\varepsilon_{\ell_1\cdots\ell_r i_1\cdots i_r} t^{i_1\cdots i_r}$ entering the right-hand side of (7.21.1) and substitute it in the expression (7.21.7) for $\varepsilon_{\ell_1\cdots\ell_r i_1\cdots i_r}$:

$$\begin{aligned}
&\varepsilon_{\ell_1\cdots\ell_r i_1\cdots i_r} t^{i_1\cdots i_r}\\
&= \frac{1}{r!2^r \mathrm{Pf}(J)} \sum_{\sigma\in S_{2r}} (-1)^{p(\sigma)} J_{\ell_{\sigma(1)}\ell_{\sigma(2)}} \cdots J_{\ell_{\sigma(2r-1)}\ell_{\sigma(2r)}} t^{\ell_{r+1}\cdots\ell_{2r}}\\
&= \frac{1}{r!2^r \mathrm{Pf}(J)} 2^r \sum_{\sigma,\sigma'\in S_r} (-1)^{p(\sigma)+p(\sigma')+r(r-1)/2} J_{\ell_{\sigma(1)} i_{\sigma'(1)}} \cdots J_{\ell_{\sigma(r)} i_{\sigma'(r)}} t^{i_1\cdots i_r}\\
&= \frac{(-1)^{[r/2]}}{\mathrm{Pf}(J)} \sum_{\sigma\in S_r} (-1)^{p(\sigma)} J_{\ell_{\sigma(1)} i_1} \cdots J_{\ell_{\sigma(r)} i_r} t^{i_1\cdots i_r}\\
&= (-1)^{[\frac{r}{2}]} \frac{r!}{\mathrm{Pf}(J)} J_{\ell_1 i_1} \cdots J_{\ell_r i_r} t^{i_1\cdots i_r},
\end{aligned} \tag{7.21.8}$$

where in the second equality we make use of the fact that indices of type i_k can occupy in components $J_{\ell_{\sigma(k)}\ell_{\sigma(k+1)}}$ only one place out of two, otherwise expression (7.21.8) is equal to zero in view of tracelessness of $t^{i_1\cdots i_r}$, and also we take into account all permutations $J_{\ell i} \to J_{i\ell}$ for r factors $J_{\ell_{\sigma(k)} i_{\sigma'(k)}}$, that gives factor 2^r (factor $(-1)^{r(r-1)/2}$ appears in view of the additional permutation of indices $(\ell_1, \ell_2, \ell_3, \ldots, \ell_{2r}) \to (\ell_1, \ell_{r+1}, \ell_2, \ell_{r+2}, \ldots, \ell_r, \ell_{2r})$); in the third equality we make use of antisymmetry of $t^{i_1\cdots i_r}$ and sum up over all σ', that gives factor $r!$; finally, in the last equality we again make use of antisymmetry of $t^{i_1\cdots i_r}$, that gives us possibility to sum up over σ and yields another factor $r!$. Now, we substitute (7.21.8) into the right-hand side of (7.21.1) and obtain the desired result

$$\begin{aligned}
&J^{k_1\ell_1} \ldots J^{k_r\ell_r} \varepsilon_{\ell_1\cdots\ell_r i_1\cdots i_r} t^{i_1\cdots i_r}\\
&\quad = \frac{(-1)^{[\frac{r}{2}]} r!}{\mathrm{Pf}(J)} J^{k_1\ell_1} \ldots J^{k_r\ell_r} J_{\ell_1 i_1} \cdots J_{\ell_r i_r} t^{i_1\cdots i_r} = \frac{(-1)^{[\frac{r}{2}]} r!}{\mathrm{Pf}(J)} t^{k_1\cdots k_r}.
\end{aligned}$$

7.22. Problem 6.2.2 of Section 6.2.1

Prove that in odd case, every element $U \in CL(p,q)$ can be written as $U = z\cdot a$, where z is central element of algebra $\mathcal{C}\ell_{(p,q)}$ and $a \in CL^{(0)}(p,q)$.

Let us consider element $U = a + b \in CL(p,q)$ where a and b are even and odd elements of algebra $\mathcal{C}\ell_{(p,q)}$, respectively. For $a = 0$ or $b = 0$, the statement of proposition is trivial. Let $a \neq 0$, $b \neq 0$, then transformation

(6.2.8), like in even case, gives two relations (cf. (6.2.15))

$$a \cdot \gamma_m = \gamma_k \, \Lambda^k{}_m \cdot a, \quad b \cdot \gamma_m = \gamma_k \, \Lambda^k{}_m \cdot b. \tag{7.22.1}$$

Odd element b can be written as $b = \gamma_{1 \cdots N} \cdot a'$, where a' is even element, and $a' \cdot \gamma_m = \gamma_k \, \Lambda^k{}_m \cdot a'$. If $a' = c \cdot a$ where c is a central element, then $U = a + b = (1 + c \cdot \gamma_{1 \cdots N})a$, and since element U is invertible, then element a is also invertible, so that $a \in CL^{(0)}(p,q)$. Let $a' \neq c \cdot a$, then element a in (7.22.1) is not uniquely defined, and it can be made invertible by a shift $a \to a + \alpha a' \equiv \tilde{a}$, where α is an appropriate parameter.[b] So, we have $\tilde{a} \in CL^{(0)}(p,q)$. Now, it follows from (7.22.1) that element $\tilde{a}^{-1} \cdot b$ is central, and since element $\tilde{a}^{-1} \cdot b$ is odd, we have $b = \beta \gamma_{1 \cdots N} \cdot \tilde{a}$. Since U is invertible, we observe that $\beta^2 \neq (-1)^{\nu + q}$. So, we obtain $U = (1 + \beta \gamma_{1 \cdots N}) \cdot \tilde{a}$ where $\tilde{a} \in CL^{(0)}(p,q)$, as required.

7.23. Problem 6.2.4 of Section 6.2.1

Show that if $U \in \mathsf{Spin}(p,q)$, then $(-U)$ also belongs to group $\mathsf{Spin}(p,q)$. Show that any element $\gamma_{i_1} \cdots \gamma_{i_{2k}} \in \mathsf{Pin}^{(0)}(2\nu)$ of even order can be represented in the exponential form (6.2.30).

We have to show that operator $(-I)$ belongs to group $\mathsf{Spin}(p,q)$, i.e., there exists smooth curve in $\mathsf{Spin}(p,q)$ connecting it to unit operator $I \in \mathsf{Spin}(p,q)$. We leave the special case $p = 1$ and $q = 1$ for a reader to consider. Let us take $p > 1$ (the case $q > 1$ is treated in the same way). Then in representation given by (6.1.76),

$$\gamma_1 = \Gamma_1, \ldots, \quad \gamma_p = \Gamma_p, \quad \gamma_{p+1} = i \, \Gamma_{p+1}, \ldots, \quad \gamma_{p+q} = i \, \Gamma_{p+q}, \tag{7.23.1}$$

with matrices Γ_m given by (6.1.50), (6.1.51), we have

$$\Sigma_{1,2} = \frac{1}{4} (\Gamma_1 \Gamma_2 - \Gamma_2 \Gamma_1) = \frac{i}{2} (\sigma_3 \otimes I_2^{\otimes(\nu-1)}). \tag{7.23.2}$$

Clearly, smooth curve

$$g(t) = \exp(t \, \Sigma_{1,2}) = \exp\left(\frac{i}{2} \, t \, \sigma_3 \otimes I_2^{\otimes(\nu-1)}\right) = \exp\left(\frac{i}{2} t \sigma_3\right) \otimes I_2^{\otimes(\nu-1)} \tag{7.23.3}$$

---

[b]Since $U = (a + \gamma_{1 \cdots N} \cdot a')$ is invertible, operators a and a' cannot have common zero modes (eigenvectors with zero eigenvalues). Therefore, even if each of the operators a and a' is not invertible, there always exists linear combination $(a + \alpha a')$ which does not have zero modes, and hence is invertible.

belongs to group $\mathsf{Spin}(p,q)$ and connects elements $g(0) = I$ and $g(2\pi) = -I$, i.e., $-I \in \mathsf{Spin}(p,q)$. Therefore, element $-U = g(2\pi)\cdot U$ is connected with element U by a smooth curve in group $\mathsf{Spin}(p,q)$, and $-U \in \mathsf{Spin}(p,q)$.

We leave it to the reader to prove the general proposition that *any* even monomial $\gamma_{i_1}\cdots\gamma_{i_{2k}} \in \mathsf{Pin}^{(0)}(2\nu)$ can be represented in the exponential form (6.2.30), and give the proof for a second-order monomial $(\gamma_1\cdot\gamma_2)$. Consider generators $\gamma_m \in \mathcal{C}\ell_{(2\nu,0)}$ in representation $\gamma_m = \Gamma_m$, where matrices Γ_m are again defined in (6.1.50), (6.1.51). Then, in accordance with (7.23.2), we have $\gamma_1\cdot\gamma_2 = \Gamma_1\cdot\Gamma_2 = i\,(\sigma_3\otimes I_2^{\otimes(\nu-1)})$. The required representation is obtained by setting $t=\pi$ in (7.23.3):

$$g(\pi) = \exp\left(i\frac{\pi}{2}\sigma_3\right)\otimes I_2^{\otimes(\nu-1)} = \begin{pmatrix} e^{i\pi/2} & 0 \\ 0 & e^{-i\pi/2}\end{pmatrix}\otimes I_2^{\otimes(\nu-1)}$$

$$= i\,\sigma_3\otimes I_2^{\otimes(\nu-1)} = \gamma_1\cdot\gamma_2.$$

7.24. Problem 6.3.5 of Section 6.3.2

Let matrices $U\in \mathrm{Mat}_M(\mathbb{C})$ *satisfy the condition* $\det(U)=1$ *and relations* (6.3.64) *for* $\mathbf{D}=I_M$:

$$U^\dagger\, U = I_M, \quad U^\top\,\mathbf{C}^{-1}\,U = \mathbf{C}^{-1}, \tag{7.24.1}$$

where $\mathbf{C}$ *is real matrix and* $\mathbf{C}^\top = \mathbf{C}$. *The set of all such* U*'s generates the group* $SU(M)\cap SO(M,\mathbb{C})$. *Prove that* $SO(M,\mathbb{C})\cap SU(M) = SO(M)$. *Hint: decompose matrix* $A\in SO(M,\mathbb{C})\cap SU(M)$ *into real and imaginary parts.*

We multiply the second relation in (7.24.1) by U^{-1} on the right and then use $U^{-1}=U^\dagger$:

$$U^\top\mathbf{C}^{-1} = \mathbf{C}^{-1}U^\dagger \overset{\mathsf{T}}{\Rightarrow} U\,\mathbf{C} = \mathbf{C}\,U^*. \tag{7.24.2}$$

The second relation in (7.24.2) and its complex conjugate together with the property that $\mathbf{C}$ is real give $[U,\mathbf{C}^2]=0$, i.e., $\mathbf{C}^2 = \alpha I_M$, where $\alpha\in\mathbb{R}$. Since $\mathbf{C}^\top = \mathbf{C}$, we have $\mathbf{C}\mathbf{C}^\top = \alpha I_M$, so $\alpha>0$, and one can normalize the matrix $\mathbf{C}$ in (7.24.1) in such a way that $\mathbf{C}^2 = I_M$ (cf. matrix (6.3.5) in representation (6.3.2) for $N=6,8\ \mathrm{mod}(8)$).

Consider first the simplest case $\mathbf{C}=I_M$. We set $U = X+iY$, where $X,Y\in\mathrm{Mat}_M(\mathbb{R})$, then the relations (7.24.1) and (7.24.2) give $X^\top X = I_M$ and $Y=0$. Therefore, $U = X\in SO(M)$, i.e., we indeed arrive at $SO(M,\mathbb{C})\cap SU(M) = SO(M)$.

Now, consider the general case $\mathbf{C} \neq I_M$. We again set $U = X + iY$ with $X, Y \in \mathrm{Mat}_M(\mathbb{R})$. Then (7.24.2) gives

$$XC = CX, \quad YC = -CY, \tag{7.24.3}$$

while equalities (7.24.1) are equivalent to the following relations:

$$X^\top X + Y^\top Y = I_M, \quad Y^\top X - X^\top Y = 0. \tag{7.24.4}$$

We now define matrix $\tilde{U}$ related to U by equivalence transformation

$$\tilde{U} = A\,U\,A^{-1}, \tag{7.24.5}$$

where $A = P_+ + iP_-$, $A^{-1} = P_+ - iP_-$, and we introduce projectors $P_\pm \equiv \frac{1}{2}(I_M \pm \mathbf{C})$. The latter obey the identities $P_\pm^2 = P_\pm$, $P_\pm P_\mp = 0$ and $P_+ + P_- = I_M$. Relations (7.24.1) for U, in view of the properties of matrix $\mathbf{C}$, are written in terms of $\tilde{U}$ as follows:

$$\tilde{U}^\dagger\,\tilde{U} = I_M, \quad \tilde{U}^\top\,\tilde{U} = I_M, \tag{7.24.6}$$

while the property $\det(U) = +1$ gives $\det(\tilde{U}) = +1$. Thus, the matrix $\tilde{U}$ has the same properties that matrix U in our previous case $\mathbf{C} = I_M$. Therefore, by making use of the decomposition $\tilde{U} = \tilde{X} + i\tilde{Y}$, with $\tilde{X}, \tilde{Y} \in \mathrm{Mat}_M(\mathbb{R})$, we find $\tilde{Y} = 0$, while $\tilde{U} = \tilde{X}$ is a real orthogonal matrix with unit determinant, i.e., $\tilde{U} \in SO(M)$.

We now find the explicit expression for matrix $\tilde{X}$ in terms of matrix U. We insert the decomposition $U = X + iY$ on the right-hand side of (7.24.5), and make use of (7.24.3) and the projector properties of $P_\pm$ to write

$$\begin{aligned}
\tilde{U} &= (P_+ + iP_-)\,(X + iY)\,(P_+ - iP_-) \\
&= (P_+XP_+ + P_-XP_- + P_+YP_- - P_-YP_+) \\
&\quad + i(P_+YP_+ + P_-YP_- - P_+XP_- + P_-XP_+) \\
&= (P_+ + P_-)X + (P_+ - P_-)Y = (X + \mathbf{C}Y) \in \mathrm{Mat}_M(\mathbb{R}).
\end{aligned}$$

This shows that $\tilde{U} = \tilde{X} = (X + \mathbf{C}Y) \in SO(M)$. Relation (7.24.5) defines homomorphism π: $SO(M, \mathbb{C}) \cap SU(M) \to SO(M)$ such that $\pi(U) = A\,U\,A^{-1} = X + \mathbf{C}Y$. This homomorphism is invertible. Indeed, any matrix $\tilde{X} \in SO(M)$ can be written as $\tilde{X} = X + \mathbf{C}Y$, where matrices

$$X = P_+\tilde{X}P_+ + P_-\tilde{X}P_-, \quad Y = P_+\tilde{X}P_- - P_-\tilde{X}P_+,$$

obey (7.24.3), (7.24.4) and uniquely determine matrix $U = X + iY$ obeying (7.24.1). Thus, π is isomorphism, and we have $SO(M,\mathbb{C}) \cap SU(M) = SO(M)$.

7.25. Problem 6.3.6 of Section 6.3.2

Let matrices Γ'_k $(k = 1,\ldots,2\nu-2)$ in (6.3.72) be taken in representation (6.3.2) (with replacement $\nu \to \nu - 1$). Prove that

$$\begin{aligned} &p = 2\beta\colon \mathbf{d} = I_2^{\otimes(\beta-1)} \otimes \sigma_1 \otimes I_2^{\otimes(\nu-\beta-1)} \quad (\beta = 1,\ldots,\nu-1),\\ &\mathbf{d} = I_{2^{\nu-1}} \quad (\beta = \nu), \qquad\qquad (7.25.1)\\ &p = 2\beta+1\colon \mathbf{g} = \sigma_3^{\otimes\beta} \otimes I_2^{\otimes(\nu-\beta-1)} \quad (\beta = 1,\ldots,\nu-1),\end{aligned}$$

$$\left.\begin{aligned} &\mathbf{c} = (\sigma_2 \otimes \sigma_1)^{\otimes(k-1)} \otimes \sigma_2 \quad (\nu = 2k),\\ &\mathbf{h} = (\sigma_1 \otimes \sigma_2)^{\otimes k} \quad (\nu = 2k+1).\end{aligned}\right. \qquad (7.25.2)$$

We insert gamma-matrices (6.3.71) and matrices **D** (6.3.76) and **C** (6.3.77) in (6.3.74) and (6.3.75) and recall the properties (6.3.3) of Hermitian matrices Γ'_k in representation (6.3.2). In this way, we obtain

$$\begin{aligned} &(1)\ \text{even } q\ (p = 2\beta) \Rightarrow \mathbf{d}\cdot\Gamma'_k = \zeta(k)\Gamma'_k\cdot\mathbf{d},\\ &(2)\ \text{odd } q\ (p = 2\beta+1) \Rightarrow \mathbf{g}\cdot\Gamma'_k = -\zeta(k)\Gamma'_k\cdot\mathbf{g},\end{aligned}$$

where $\zeta(k) = +1 \quad (k = 1,\ldots,p-1), \quad \zeta(k) = -1 \quad (k = p,\ldots,2\nu-1);$

$$\begin{aligned} &(3)\ \text{even } \nu \Rightarrow \mathbf{c}\cdot\Gamma'_k = (-1)^k\Gamma'_k\cdot\mathbf{c} \quad (k = 1,\ldots,2\nu-1),\\ &(4)\ \text{odd } \nu \Rightarrow \mathbf{h}\cdot\Gamma'_k = (-1)^{k+1}\Gamma'_k\cdot\mathbf{h} \quad (k = 1,\ldots,2\nu-1).\end{aligned} \qquad (7.25.3)$$

These relations determine matrices **d**, **g**, **c** and **h**. Making use of explicit form of matrices Γ'_k $(k = 1,\ldots,2\nu-2)$ given by (6.3.2) (with $\nu \to \nu-1$) and matrix $\Gamma'_{2\nu-1}$ in (6.1.68), one makes sure that blocks **d** and **g** written in (7.25.1) obey relations (1) and (2) in (7.25.3). The explicit form (7.25.2) of matrices **c** and **h** is obtained by comparing relations (3) and (4) in (7.25.3) with formulas (6.3.11), and is dictated by the form of matrices C, C' given by (6.3.5), (6.3.6).

7.26. Problem 6.3.10 of Section 6.3.2

Prove isomorphism $\mathsf{Spin}(2,4) = SU(2,2)$. *Hint: make use of Weyl representation of algebra $C\ell_{(2,4)}$, find matrices* **d**, **h** *and* **f** *in this representation and consider relations* (6.3.84)–(6.3.86) *for blocks $A, \overline{A}$.*

Making use of (6.1.69), we construct Weyl representation of algebra $\mathcal{C}\ell_{(2,4)}$:

$$\begin{gathered}\gamma_0 = \sigma_1 \otimes I_2^{\otimes 2}, \quad \gamma_1 = \sigma_2 \otimes \sigma_1 \otimes I_2, \quad \gamma_2 = i\sigma_2^{\otimes 2} \otimes \sigma_1, \\ \gamma_3 = i\sigma_2^{\otimes 3}, \quad \gamma_4 = i\sigma_2^{\otimes 2} \otimes \sigma_3, \quad \gamma_5 = i\sigma_2 \otimes \sigma_3 \otimes I_2.\end{gathered} \tag{7.26.1}$$

In this representation, we have $\gamma_m^\dagger = \gamma_m$ $(m = 0, 1)$, $\gamma_m^\dagger = -\gamma_m$ $(m = 2, \ldots, 5)$, $\gamma_m^\mathsf{T} = (-1)^m \gamma_m$ and $\gamma_m^* = \gamma_m$ $(m = 0, 3, 5)$, $\gamma_m^* = -\gamma_m$ $(m = 1, 2, 4)$. Therefore, in view of (6.3.74) and (6.3.76)–(6.3.78), we obtain, by setting $p = 2, q = 4, \nu = 3$, that $\mathbf{D} \sim \gamma_2\gamma_3\gamma_4\gamma_5$, $\mathbf{C} \sim \gamma_1\gamma_3\gamma_5$, $\mathbf{B} \sim \gamma_1\gamma_2\gamma_4$. Therefore, with appropriate normalization, the conjugation matrices are

$$\begin{gathered}\mathbf{D} = I_2 \otimes \mathbf{d} = I_2 \otimes \sigma_1 \otimes I_2, \quad \mathbf{C} = i\sigma_2 \otimes \mathbf{h} = i\sigma_2 \otimes I_2 \otimes \sigma_2, \\ \mathbf{B} = i\sigma_2 \otimes \mathbf{f} = i\sigma_2 \otimes \sigma_1 \otimes \sigma_2.\end{gathered} \tag{7.26.2}$$

We insert matrices $\mathbf{d}, \mathbf{h}$ and $\mathbf{f}$ defined in (7.26.2) into (6.3.84)–(6.3.86), and find that diagonal blocks $A, \overline{A} \in \mathrm{Mat}_4(\mathbb{C})$ of element $U \in \mathsf{Spin}(2,4)$ obey

$$A^\dagger\,(\sigma_1 \otimes I_2)\,A = I_2^{\otimes 2} = \overline{A}^\dagger\,(\sigma_1 \otimes I_2)\,\overline{A}, \tag{7.26.3}$$

$$\overline{A} = (I_2 \otimes \sigma_2)\,(A^\mathsf{T})^{-1}\,(I_2 \otimes \sigma_2), \quad A^* = (\sigma_1 \otimes \sigma_2)\,\overline{A}\,(\sigma_1 \otimes \sigma_2). \tag{7.26.4}$$

The property (7.26.3) shows that $A, \overline{A} \in U(2,2)$, while the formula (7.26.4) gives one-to-one relations between matrices A and $\overline{A}$, i.e., element $U \in \mathsf{Spin}(2,4)$ is determined by one of these matrices, say, $A \in U(2,2)$. Note that if matrices A are treated as elements of group $U(2,2)$ in defining representation, then, in accordance with the first formula in (7.26.4), matrices $\overline{A}$ belong to representation equivalent to contragredient representation. Upon imposing normalization $\det(A) = 1$, and recalling that groups $SU(2,2)$ and $\mathsf{Spin}(2,4)$ have the same dimension (equal to 15) and both are connected manifolds, we conclude that $\mathsf{Spin}(2,4) = SU(2,2)$.

7.27. Problem 6.4.3 of Section 6.4.1

Prove the identity for $\mathcal{C}\ell_{(p,q)}$:

$$\gamma_{[k_1} \cdots \gamma_{k_m]} \cdot \gamma_{N+1} = \alpha_m\, \varepsilon_{k_1 \cdots k_m k_{m+1} \cdots k_N}\, \gamma^{k_{m+1}} \cdots \gamma^{k_N}, \tag{7.27.1}$$

$$\alpha_m = \frac{1}{(N-m)!}(-i)^\nu\, i^q\, (-1)^{[\frac{m}{2}]}, \tag{7.27.2}$$

where $N = (p+q) = 2\nu = 2r$, $\gamma_{N+1} = (-i)^{\nu+q}\gamma_0 \cdots \gamma_{N-1}$, $[\frac{m}{2}]$ *is integer part of* $\frac{m}{2}$, *and* $\varepsilon_{01\cdots N-1} = 1$.

Since γ_{N+1} is a product of all generators γ_k, we recall the defining relations (6.4.1) and observe that the left-hand side of (7.27.1) is proportional to the product of all those generators γ_k whose indices k do not belong to the set $(k_1 \cdots k_m)$, i.e., it is proportional to the right-hand side of (7.27.1). It remains to calculate numerical coefficient (7.27.2). To this end, we multiply both sides of (7.27.1) on the left by $\gamma^{[k_m} \cdots \gamma^{k_1]}$ and sum up over indices $(k_1, \ldots, k_m)$. In this way, we get in right-hand side

$$\begin{aligned} &\alpha_m\, \gamma^{[k_m} \cdots \gamma^{k_1]}\, \varepsilon_{k_1 \cdots k_m k_{m+1} \cdots k_N}\, \gamma^{k_{m+1}} \cdots \gamma^{k_N} \\ &\quad = (-1)^{\frac{m(m-1)}{2}} \alpha_m\, \varepsilon_{k_1 \cdots k_N}\, \gamma^{k_1} \cdots \gamma^{k_N} = (-1)^{\frac{m(m-1)}{2}}\, i^\nu (-i)^q\, N!\, \alpha_m\, \gamma_{N+1}, \end{aligned} \tag{7.27.3}$$

where we make use of (6.1.85). The left-hand side is

$$\gamma^{[k_m} \cdots \gamma^{k_1]}\, \gamma_{[k_1} \cdots \gamma_{k_m]} \cdot \gamma_{N+1} = \frac{N!}{(N-m)!}\, \gamma_{N+1}. \tag{7.27.4}$$

Here, we employ definition (6.4.10) and identity (cf. (6.4.77))

$$\sum_{k_1,\ldots,k_m} \gamma^{[k_m} \cdots \gamma^{k_1]} \cdot \gamma_{[k_1} \cdots \gamma_{k_m]} = \frac{N!}{(N-m)!}\, I_{2^\nu}. \tag{7.27.5}$$

The latter identity is due to the fact that the sum on the left-hand side (over all k_ℓ such that $k_i \neq k_j$ for $i \neq j$) contains $\frac{N!}{(N-m)!}$ terms, each equal to I_{2^ν}. By comparing (7.27.3) and (7.27.4), we obtain (7.27.2).

7.28. Problem 6.4.13 of Section 6.4.3

Find explicit form of matrix T *in* (6.4.65):

$$\mathbf{B} \to T\, \mathbf{B}\, (T^*)^{-1} \tag{7.28.1}$$

that transforms matrix $\mathbf{B}$ *given in* (6.4.54) *to unit matrix in the cases* $N = 2, 4 \bmod(8)$. *Construct pure imaginary representation for generators* γ_k *of algebras* $C\ell_{(1,1)}$ *and* $C\ell_{(1,3)}$. *Hint: transform to unit matrix the matrices* $B_1 = i\,\sigma_3$, $B_2 = \sigma_1$, $B_3 = \sigma_2 \otimes \sigma_1 \otimes \sigma_2$ *whose tensor product gives matrix* $\mathbf{B}$ *in* (6.4.54) *for the cases* $N = 2, 4 \bmod(8)$.

Matrix $\mathbf{B}$ given by (6.4.54) has the following form in dimensions $d = 2, 4 \bmod(8)$:

$$\mathbf{B} = \begin{pmatrix} i & 0 \\ 0 & -i \end{pmatrix} \otimes \begin{pmatrix} 0 & 1 \\ 1 & 0 \end{pmatrix} \otimes B' \otimes \begin{pmatrix} 0 & 1 \\ 1 & 0 \end{pmatrix} \otimes B' \otimes \cdots , \tag{7.28.2}$$

where blocks B' are

$$B' = \begin{pmatrix} 0 & -1 \\ 1 & 0 \end{pmatrix} \otimes \begin{pmatrix} 0 & 1 \\ 1 & 0 \end{pmatrix} \otimes \begin{pmatrix} 0 & 1 \\ -1 & 0 \end{pmatrix}. \tag{7.28.3}$$

There exist matrices T_1 and T_2 that, being used in (7.28.1), transform to unit matrix the matrices $B_1 = \left(\begin{smallmatrix} i & 0 \\ 0 & -i \end{smallmatrix}\right)$ and $B_2 = \left(\begin{smallmatrix} 0 & 1 \\ 1 & 0 \end{smallmatrix}\right)$. The explicit forms are

$$T_1 = r \cdot \begin{pmatrix} e^{-i\pi/4} & 0 \\ 0 & e^{i\pi/4} \end{pmatrix}, \quad T_2 = \begin{pmatrix} a & a^* \\ d^* & d \end{pmatrix}.$$

Here $a, d \in \mathbb{C}$ and $(ad - a^*d^*) \neq 0$; $r \in \mathbb{R}$ and $r \neq 0$. Since matrix B_2 can be transformed to unit matrix, matrix B' can be transformed to unit matrix, provided there exists matrix T_3 such that $T_3\, B_3\, T_3^{*-1} = I_4$, where

$$B_3 = \begin{pmatrix} 0 & -1 \\ 1 & 0 \end{pmatrix} \otimes \begin{pmatrix} 0 & 1 \\ -1 & 0 \end{pmatrix} = \begin{pmatrix} 0 & 0 & 0 & -1 \\ 0 & 0 & 1 & 0 \\ 0 & 1 & 0 & 0 \\ -1 & 0 & 0 & 0 \end{pmatrix}.$$

Matrix T_3 does exist, and has block structure

$$T_3 = \begin{pmatrix} A & B \\ C & D \end{pmatrix}, \quad B = A^* \cdot \sigma, \quad D = C^* \cdot \sigma, \quad \sigma \equiv \begin{pmatrix} 0 & -1 \\ 1 & 0 \end{pmatrix},$$

where the only constraint on matrices $A, C \in \mathrm{Mat}_2(\mathbb{C})$ is $\det(T_3) \neq 0$.

In two and four dimensions, $N = 2$ and $N = 4$, the expressions (6.4.44), (6.3.2) for generators of $\mathcal{C}\ell_{(1,1)}$ and $\mathcal{C}\ell_{(1,3)}$ are $\gamma_0 = \sigma_1$, $\gamma_1 = i\sigma_2$ and

$$\gamma_0 = \sigma_1 \otimes I_2, \quad \gamma_1 = i\sigma_2 \otimes I_2, \quad \gamma_2 = i\sigma_3 \otimes \sigma_1, \quad \gamma_3 = i\sigma_3 \otimes \sigma_2,$$

respectively. In the case $N = 2$, transformation (7.28.1) with matrix $T = T_1$ gives pure imaginary representation of generators of algebra $\mathcal{C}\ell_{(1,1)}$,

$$\gamma_0' = T_1\sigma_1 T_1^{-1} = \sigma_2, \quad \gamma_1' = i\, T_1\sigma_2 T_1^{-1} = i\sigma_1.$$

For $N = 4$, transformation (7.28.1) with matrix $T = T_1 \otimes T_2$, where we choose $a = i$ and $d = 1$ for definiteness, gives pure imaginary representation for generators of algebra $\mathcal{C}\ell_{(1,3)}$, namely,

$$\gamma_0' = T(\sigma_1 \otimes I_2)T^{-1} = \sigma_2 \otimes I_2, \quad \gamma_1' = i\,T(\sigma_2 \otimes I_2)T^{-1} = i\sigma_1 \otimes I_2,$$
$$\gamma_2' = i\,T(\sigma_3 \otimes \sigma_1)T^{-1} = -i\sigma_3 \otimes \sigma_3, \quad \gamma_3' = i\,T(\sigma_3 \otimes \sigma_2)T^{-1} = i\sigma_3 \otimes \sigma_1.$$

The construction in the general case $N = 2, 4 \bmod(8)$ is then straightforward.

Selected Bibliography

Here we give far from being a complete list of monographs and reviews, which addresses the issues raised in our book.

- P.I. Golod and A.U. Klimyk, *Mathematical Foundations of Symmetry Theory*, Regular and Chaotic Dynamics, Moscow (2001), in Russian.
- M.M. Postnikov, *Lie Groups and Lie Algebras. Lectures in Geometry, Semester V*, Editorial URSS (1994).
- I.L. Buchbinder and S.M. Kuzenko, *Ideas and Methods of Supersymmetry and Supergravity, Or a Walk Through Superspace*, IOP, Bristol (1995).
- J.-P. Serre, *Linear Representations of Finite Groups* (Graduate Texts in Mathematics, Volume 42), Springer-Verlag (1977).
- I.M. Gelfand, M.I. Graev and N.Ya. Vilenkin, *Generalized Functions, Volume 5: Integral Geometry and Representation Theory* (AMS Chelsea Publishing, Volume 381), American Mathematical Society, Providence, RI (1966).
- M.A. Naimark, Unitary representations of noncompact groups, in *Proc. Inter. School on Theoretical Physics* (Yalta, 1966), Fizika Vysokikh Energii i Teoriya Elementarnykh Chastits, ed. V.P. Shelest, Kiev, Naukova Dumka (1967), pp. 183–215, in Russian.
- I.R. Shafarevich, *Basic Notions of Algebra*, Springer-Verlag, Berlin (2005).
- B.A. Dubrovin, A.T. Fomenko and S.P. Novikov, *Modern Geometry — Methods and Applications*, Springer-Verlag (1992).

- S.P. Novikov and I.A. Taimanov, *Modern Geometric Structures and Fields* (Graduate Studies in Mathematics, Volume 71), American Mathematical Society, Providence, RI (2006).
- F. Gürsey, Introduction to theory of groups, in *Relativity, Groups and Topology*, eds. C. De Witt and B. De Witt, Gordon & Breach publishers, New York (1964).
- L.C. Biedenharn and J.D. Louck, *Angular Momentum in Quantum Physics: Theory and Application*, Cambridge University Press, Cambridge (1984).
- J.P. Elliott and P.G. Dawber, *Symmetry in Physics*, Volume 1 and 2, Macmillan Press, London (1979).
- L.B. Okun, *Leptons and Quarks*, Elsevier Science Publishers (2013).
- N.N. Bogolubov, A.A. Logunov, A.I. Oksak and I. Todorov, *General Principles of Quantum Field Theory*, Volume 10, Springer Science and Business Media (2012).
- N.N. Bogolubov and D.V. Shirkov, *Introduction to the Theory of Quantized Fields*, (Interscience Monographs in Physics and Astronomy, Volume 3), Wiley, New York (1980).
- Yu.B. Rumer and A.I. Fet, *Group Theory and Quantized Fields*, Nauka, Moscow (1977), in Russian.
- Yu.V. Novozhilov, *Introduction to Elementary Particle Theory* (International Series of Monographs in Natural Philosophy, Volume 78), Elsevier (2013).
- S. Schweber, *An Introduction to Relativistic Quantum Field Theory* (Dover Books on Physics), Dover Publications (2005).
- M.A. Naimark and A.I. Stern, *Theory of Group Representations* (Comprehensive Studies in Mathematics, Volume 246), Springer (1982).
- E. Wigner, *Group Theory and its Application to the Quantum Mechanics of Atomic Spectra*, Academic Press (1959).
- R. Penrose and W. Rindler, *Spinors and Space-Time. Volume 1, Two-Spinor Calculus and Relativistic Fields; Volume 2, Spinor and Twistor Methods in Space-Time Geometry*, Cambridge University Press (1986).
- R. Goodman and N.R. Wallach, *Symmetry, Representations, and Invariants*, Graduate Texts in Mathematics, Vol. 255, Springer (2009).

References

[1] A.P. Isaev and V.A. Rubakov, *Theory of Groups and Symmetries: Finite Groups, Lie groups, and Lie algebras.* World Scientific (2018).

[2] L.S. Pontryagin, *Topological Groups*, 3rd Edition, Gordon & Breach Science Publishers (1986).

[3] J.E. Hamphreys, *Introduction to Lie Algebras and Representation Theory* (Graduate Texts in Mathematics, Volume 9), Springer-Verlag (1994).

[4] N. Bourbaki, *Lie Groups and Lie Algebras: Chapters 4–6.* Vol. 3, Springer (2008).

[5] G. Racah, *Group Theory and Spectroscopy.* Lectures delivered at the Institute for Advanced Study, Princeton, 1951. Preprint CERN 61-8 (1961).

[6] H. Weyl, *Representation theory of continuous semi-simple groups through linear transformations*, Uspekhi Mat. Nauk, no. 4 (1938), pp. 201–257, in Russian.

[7] H. Georgi, *Lie Algebras in Particle Physics*, Westview Press, Advanced Book Program (1999).

[8] D. Olive, *Gauge Theories and Lie Algebras with some Applications to Spontaneous Symmetry Breaking and Integrable Dynamical Systems.* Lectures given at the University of Virginia, 1982. Preprint (1983).

[9] J.-P. Serre, *Lie Algebras and Lie Groups*, 1964, Lectures given at Harvard University. Springer (2009).

[10] M. Hamermesh, *Group Theory and Its Application to Physical Problems*, Addison-Wesley Publishing Company (1962).

[11] N.N. Bogoliubov, Theory of symmetry of elementary particles, in *Proc. Int. School on Theoretical Physics* (Yalta, 1966), Fizika Vysokikh Energii i Teoriya Elementarnykh Chastits, ed. V.P. Shelest, Kiev, Naukova Dumka (1967), pp. 5–112; N.N. Bogoliubov, *Collection of Scientific Works*, Quantum Theory, Volume 12, Moscow, Nauka (2009), pp. 11–140, in Russian.

[12] S. Ogawa, S. Sawada and M. Nakagawa, *Composite Models of Elementary Particles*, Iwanami Shoten, Tokyo (1980).

[13] J. Beringer, *et al.*, *Review of Particle Physics*, Phys. Rev. D **86**(1) (2012) 010001.

[14] A. Barut and R. Raczka, *Theory of Group Representations and Applications*, 2nd Revised Edition, World Scientific (1986).

[15] L.D. Faddeev and A.A. Slavnov, *Gauge Fields, Introduction to Quantum Theory* (Frontiers in Physics, Volume 50), Addison-Wesley, Reading, MA (1980).

[16] F.A. Berezin, *The Method of Second Quantization*, Pure Appl. Phys. **24** (1966) 1–228.

[17] L. Takhtajan, *Quantum Mechanics for Mathematicians* (Graduate Studies in Mathematics, Volume 95), American Mathematical Society Providence, RI (2008).

[18] C. Itzykson and J.-B. Zuber, *Quantum Field Theory* (International Series in Pure and Applied Physics) McGraw-Hill (1985).

[19] R. Brauer and H. Weyl, *Spinors in n Dimensions*, Amer. J. Math. **57**(2) (1935) 425–449.

[20] B.L. Van Der Waerden, *Group Theory and Quantum Mechanics*, Volume 214. Springer Science and Business Media (2012).

[21] L.D. Landau and E.M. Lifshitz, *The Quantum Mechanics: Nonrelativistic Theory* (*Volume 3 of A Course of Theoretical Physics*), Elsevier (2013).

[22] S. Weinberg, *The Quantum Theory of Fields, Volume I. Foundations*, Cambridge University Press (1995).

[23] V.A. Rubakov, *Classical Theory of Gauge Fields*, Princeton University Press, Princeton (2002).

[24] A. Jucys, I.B. Levinson and V.V. Vanagas, *Mathematical Apparatus of the Theory of Angular Momentum*. Israel Program for Scientific Translations (1962).

[25] A.N. Kirillov and N.Yu. Reshetikhin, Representations of the algebra $U_q(s\ell(2))$, q-orthogonal polynomials and invariants of links, in *Infinite-Dimensional Lie Algebras and Groups* (Luminy-Marseille, 1988), pp. 285–339, World Scientific Publishing, Teaneck, NJ (1989).

[26] H. Weyl, *The Theory of Groups and Quantum Mechanics*, Dover Publications (1950).

[27] H. Weyl, *The Classical Groups: Their Invariants and Representations*, Princeton University Press (1997).

[28] W. Fulton and J. Harris, *Representation theory. A First Course* (Graduate Texts in Mathematics), Springer (1991).

[29] E.B. Vinberg, *A Course in Algebra*, No.56, American Mathematical Society, Providence, RI (2003).

[30] P.A.M. Dirac, *The Principles of Quantum Mechanics.* No. 27, Oxford University Press (1981).

[31] D.P. Zhelobenko, *Lectures in Lie Group Theory*, JINR Publishing Department, Dubna (1965), in Russian.

[32] D.P. Zhelobenko, *Compact Lie Groups and Their Representations*, Volume 40, American Mathematical Society, Providence, RI (1973).

[33] M. Gaudin, *The Bethe Wavefunction*, Cambridge University Press (2014).

[34] H. Bethe, *Zur Theorie der Metalle. I. Eigenwerte und Eigenfunktionen der Linearen Atomkette*, Zeit. Phys. **71**, (3–4) (1931) 205–206.

[35] G. Ponzano and T. Regge, Semi-classical limit of Racah coefficients, in *Spectroscopic and Group Theoretical Methods in Physics*, edited by F. Block, (North-Holland, Amsterdam, 1968); T. Regge and R.M. Williams, *Discrete Structures in Gravity*, J. Math. Phys. **41**(6) (2000) 3964–3984.

[36] V.G. Turaev and O.Y. Viro, *State Sum Invariants of 3-Manifolds and Quantum 6j-Symbols*, Topology **31**(4) (1992) 865–902.

[37] J. Schwinger, *On Angular Momentum*, U.S. Atomic Energy Commission, NYO-3071 (1952); in *Quantum Theory of Angular Momentum*, eds. L.C. Biedenharn and H. Van Dam (Academic Press, 1969).

[38] V. Bargmann, *On the Representations of the Rotation Group*, Rev. Mod. Phys. **34**(4) (1962) 829 – 845.

[39] Ya.A. Smorodinsky and L.A. Shelepin, *Clebsch–Gordan Coefficients, Viewed from Different Sides*, Soviet Physics Uspekhi **15**(1) (1972) 1.

[40] T. Regge, Nuovo Cimento **10** (1958) 544; Nuovo Cimento **11** (1959) 116.

[41] B.E. Sagan, *The Symmetric Group: Representations, Combinatorial Algorithms, and Symmetric Functions.* Wadsworth & Brooks/Cole (1991).

[42] J.S. Frame, G. de B. Robinson and R.M. Thrall, *The Hook Graphs of the Symmetric Group*, Canadian J. Math. **6** (1954) 316–325.

[43] S. Weinberg, *The Quantum Theory of Fields. Volume 2. Modern Applications*, Cambridge University Press (2000).

[44] O.V. Ogievetsky and P.N. Pyatov, Lecture on Hecke algebras, in *Proc. Int. School "Symmetries and Integrable Systems"*, Dubna (June 8–11, 1999), JINR, D2, 5-2000-218; preprint MPIM (Bonn), MPI 2001–40, http://www.mpim-bonn.mpg.de/.

[45] A.P. Isaev and O.V. Ogievetsky, Representations of A-type Hecke algebras, in *Proc. Int. Workshop "Supersymmetries and Quantum Symmetries"* (Dubna, July 27–31, 2005) 2006, pp. 132–141; arXiv:0912.3701[math.QA].

[46] W. Fulton, *Young Tableaux: With Applications to Representation Theory and Geometry*, Volume 35, Cambridge University Press (1997).

[47] A. Okounkov and A. Vershik, *A New Approach to Representation Theory of Symmetric Groups*, Selecta Mathematica New Ser. **2**(4) (1996) 581–605.

[48] M. Nazarov, *Young's Orthogonal Form for Brauer's Centralizer Algebra*, J. Algebra **182** (1996) 664–693.

[49] A.P. Isaev and O.V. Ogievetsky, *Jucys–Murphy Elements for Birman–Murakami–Wenzl Algebras*, Phys. Particles Nucl. Lett. **8**(3) (2011) 234.

[50] A.P. Isaev and A.I. Molev, *Fusion Procedure for the Brauer Algebra*, St. Petersburg Math. J. **22** (2011) 437–446.

[51] A.P. Isaev, A.I. Molev and O.V. Ogievetsky, *A New Fusion Procedure for the Brauer Algebra and Evaluation Homomorphisms*, Int. Math. Res. Not. (2012), 2571–2606.

[52] D.S. Shirokov, *Lectures on Clifford Algebras and Spinors*, (Lecture Courses in Scientific University Center, Volume 19), MIRAN, Moscow (2012) pp. 3–179.

[53] P. Lounesto, *Clifford Algebras and Spinors*, Cambridge University Press (2001).

[54] V.S. Varadarajan, *Supersymmetry for Mathematicians: An Introduction*, American Mathematical Society, Providence, RI (2004).

[55] A.L. Onishchik and E.B. Vinberg, *Seminar on Lie Groups and Algebraic Groups*, Springer-Verlag, 1990.
[56] E. Cartan, *The Theory of Spinors*, Hermann, Paris (1966).
[57] R. Penrose and W. Rindler, *Spinors and Space-Time, Volume 2: Spinor and Twistor Methods in Space-Time Geometry*, Cambridge University Press (1986).
[58] T. Kugo and P. Townsend, *Supersymmetry and the Division Algebras*, Nucl. Phys. B **221** (1983) 357–380.
[59] A.S. Schwarz, *Quantum Field Theory and Topology*, Vol. 307, Springer Science and Business Media (2013).

Index

www.ingramcontent.com/pod-product-compliance
Lightning Source LLC
LaVergne TN
LVHW010339220926
848484LV00004BB/930
* 9 7 8 9 8 1 1 2 1 7 4 0 1 *